DIVERSITY-ORIENTED SYNTHESIS

DIVERSITY-ORIENTED SYNTHESIS

Basics and Applications in Organic Synthesis, Drug Discovery, and Chemical Biology

Edited by

ANDREA TRABOCCHI
University of Florence
Sesto Fiorentino, Florence, Italy

WILEY

Published by John Wiley & Sons, Inc., Hoboken, New Jersey.
Published simultaneously in Canada.

Library of Congress Cataloging-in-Publication Data:

Diversity-oriented synthesis : basics and applications in organic synthesis, drug discovery, and chemical biology / edited by Andrea Trabocchi, University of Florence, Sesto Fiorentino, Florence, Italy.

pages cm

Includes bibliographical references and index.

ISBN 978-1-118-14565-4 (hardback)

1. Organic compounds–Synthesis. 2. Drug development. 3. Biosynthesis. I. Trabocchi, Andrea.

QD262.D58 2013

547′.2–dc23

2012048231

10 9 8 7 6 5 4 3 2 1

CONTENTS

PART I CHEMICAL METHODOLOGY IN DIVERSITY-ORIENTED SYNTHESIS

PART IV APPLICATIONS IN CHEMICAL BIOLOGY AND DRUG DISCOVERY

CONTRIBUTORS

Jürgen Bajorath, Rheinische Friedrich-Wilhelms-Universität, Department of Life Science Informatics, B-IT, LIMES Program Unit Chemical Biology and Medicinal Chemistry, Dahlmannstrasse 2, D-53113, Bonn, Germany

Naděžda Cankařová, Palacký University, Department of Organic Chemistry, Faculty of Science, 17. Listopadu 12, 771 46, Olomouc, Czech Republic

Duccio Cavalieri, Fondazione Edmund Mach, Research and Innovation Centre, Via E. Mach 1, 38010 S. Michele all'Adige, Trento, Italy

Eamon Comer, Broad Institute of Harvard and MIT, Chemical Biology Platform, 7 Cambridge Center, Cambridge, MA 02142

Sivaraman Dandapani, Broad Institute of Harvard and MIT, Chemical Biology Platform, 7 Cambridge Center, Cambridge, MA 02142

Carlotta De Filippo, Fondazione Edmund Mach, Research and Innovation Centre, Via E. Mach 1, 38010 S. Michele all'Adige, Trento, Italy

Mark Dow, University of Leeds, School of Chemistry and Astbury Centre for Structural Molecular Biology, Leeds, LS2 9JT, United Kingdom

Lingyan Du, Duke University, Department of Chemistry, 124 Science Drive, Box 90346, 2102 French Family Science Center, Durham, NC 27708

Jeremy R. Duvall, Broad Institute of Harvard and MIT, Chemical Biology Platform, 7 Cambridge Center, Cambridge, MA 02142

Warren R. J. D. Galloway, University of Cambridge Department of Chemistry, Lensfield Road, Cambridge, CB2 1EW, United Kingdom

John R. Goodell, University of Pittsburgh, Center for Chemical Methodologies and Library Development, Department of Chemistry, Parkman Avenue, Pittsburgh, PA 15260

Susanne Heynen-Genel, Sanford-Burnham Medical Research Institute, Conrad Prebys Center for Chemical Genomics, 10901 North Torrey Pines Road, La Jolla, CA 92037

Nicholas Hill, Duke University, Department of Chemistry, 124 Science Drive, Box 90346, 2102 French Family Science Center, Durham, NC 27708

John M. Knapp, University of California, Department of Chemistry, One Shields Avenue, Davis, CA 95616

Viktor Krchňák, University of Notre Dame, Department of Chemistry and Biochemistry, 251 Nieuwland Science Center, Notre Dame, IN 46556

Mark J. Kurth, University of California, Department of Chemistry, One Shields Avenue, Davis, CA 95616

Ohyun Kwon, University of California–Los Angeles, Department of Chemistry and Biochemistry, 607 Charles E. Young Drive East, Los Angeles, CA 90095-1569

Matthew G. LaPorte, University of Pittsburgh, Center for Chemical Methodologies and Library Development, Department of Chemistry, Parkman Avenue CSC 658, Pittsburgh, PA 15260

Luca Mannocci, Philochem AG, Libernstrasse 3, CH-8112, Otelfingen, Switzerland

Francesco Marchetti, University of Leeds, School of Chemistry and Astbury Centre for Structural Molecular Biology, Leeds, LS2 9JT, United Kingdom

Eric Marsault, Institut de Pharmacologie de Sherbrooke, Université de Sherbrooke, 3001 12e av nord, Sherbrooke, QC J1H 5N4, Canada

José Luis Medina-Franco, Torrey Pines Institute for Molecular Studies, Port St. Lucie, FL 34987

Giovanni Muncipinto, Broad Institute of Harvard and MIT, Chemical Biology Platform, 7 Cambridge Center, Cambridge, MA 02142

Adam Nelson, University of Leeds, School of Chemistry and Astbury Centre for Structural Molecular Biology, Leeds, LS2 9JT, United Kingdom

Kieron M. G. O'Connell, University of Cambridge, Department of Chemistry, Lensfield Road, Cambridge, CB2 1EW, United Kingdom

Stuart L. Schreiber, Howard Hughes Medical Institute, Broad Institute, Cambridge, MA 02142

Eduard A. Sergienko, Sanford-Burnham Medical Research Institute, Conrad Prebys Center for Chemical Genomics, 10901 North Torrey Pines Road, La Jolla, CA 92037

Jared T. Shaw, University of California, Department of Chemistry, One Shields Avenue, Davis, CA 95616

David R. Spring, University of Cambridge, Department of Chemistry, Lensfield Road, Cambridge, CB2 1EW, United Kingdom

Irene Stefanini, Fondazione Edmund Mach, Research and Innovation Centre, Via E. Mach 1, 38010 S. Michele all'Adige, Trento, Italy

Hongyan Sun, Department of Biology and Chemistry, City University of Hong Kong, 83 Tat Chee Avenue, Kowloon, Hong Kong, P. R. China

Andrea Trabocchi, University of Florence, Department of Chemistry "Ugo Schiff", Via della Lastruccia 13, I-50019 Sesto Fiorentino, Florence, Italy

Sammi Tsegay, University of Pittsburgh, Center for Chemical Methodologies and Library Development, Department of Chemistry, Parkman Avenue, Pittsburgh, PA 15260

Qiu Wang, Duke University, Department of Chemistry, 124 Science Drive, Box 90346, 2102 French Family Science Center, Durham, NC 27708

Zhiming Wang, School of Petrochemical Engineering, Changzhou University, No. 1 Gehu Road, Changzhou, Jiangsu, 213164, P. R. China

Peter Wipf, University of Pittsburgh, Center for Chemical Methodologies and Library Development, Department of Chemistry, Parkman Avenue, Pittsburgh, PA 15260

Ashkaan Younai, University of California, Department of Chemistry, One Shields Avenue, Davis, CA 95616

FOREWORD

The gap between insights into human disease and therapeutics that arise from these insights is closing, but it cannot close fast enough. Society has patiently invested in science. But its expectation of scientists—that we mitigate suffering from disease—must be met if we expect to receive its support in the future.

Advances in human biology are revealing novel insights into the cause of disease and requirements for the maintenance of disease. But the therapeutic targets that are arising, such as transcription factors and RNA molecules, do not fit conveniently into what we believe is currently achievable in drug discovery. Overcoming this belief is the twenty-first century challenge for organic chemistry, organic synthesis, and chemical biology. If we can do so, drug discovery and human health will be transformed.

The insights provided in *Diversity-Oriented Synthesis: Basics and Applications in Organic Synthesis, Drug Discovery, and Chemical Biology* leave me feeling optimistic. I can sense the fearlessness and audacity of the authors as they undertake the impossible. The three-dimensional world of biological macromolecules is now interfaced with the three-dimensional world of small molecules to a far greater degree. Therapeutic targets are now seeing the full force of modern organic chemistry. Using simple concepts exploited by nature's synthesis of naturally occurring small molecules, small molecules with the physical properties required of drugs yet with the topographic properties needed for achieving the impossible are now accessible. Bravo!

Stuart L. Schreiber

Broad Institute and Harvard University
Cambridge, Massachusetts
October 2012

PREFACE

Since the early reports by Stuart L. Schreiber, diversity-oriented synthesis (DOS) has become a new paradigm for developing large collections of structurally diverse small molecules as probes to investigate biological pathways and to provide a larger array of the chemical space in drug discovery issues. The principles of DOS have evolved from the concept of generating structurally diverse compounds from a divergent approach consisting of a complexity-generating reaction followed by cyclization steps and appendage diversity, to the development of different cyclic structures through the build/couple/pair approach. The concept of expanding the molecular complexity to explore the chemical space more thoroughly produced new advances in generating chemical libraries. Moreover, technology advances followed the need of automation in this field, thus producing high-tech instrumentation for library development and compound management, as well as improving high-throughput screening facilities. The possibility of creating new highly diverse and complex molecular platforms and the achievement of hundreds to thousands to millions of compounds is producing significant advances in chemical biology and drug discovery. This is due primarily to improvement in the quality of chemical libraries, which are more stereochemically rich and structurally complex. Moreover, advances in bioinformatics and systems biology are enabling an interdisciplinary setting between chemistry and biology in advancing the knowledge about the functions of biological systems and the correlation between genes and function. Finally, drug discovery is also taking advantage of DOS concepts in several medicinal chemistry programs, which in the near future will produce advances in both target and ligand discovery.

The book has been conceived in four parts, encompassing synthetic methods to achieve small-molecule collections according to DOS principles, strategies to develop

DOS libraries, screening methods for ligand identification, and selected significant applications of small molecules in drug discovery and chemical biology.

The first chapter deals with the basics of diversity-oriented synthesis, including definitions of molecular diversity and chemical space, discussing how DOS relates to classic combinatorial chemistry and showing significant approaches that have been developed for expanding the chemical diversity, including the well-known build/couple/pair concept introduced by Schreiber.

Part I encompasses key chemical methods addressing the generation of small molecules according to DOS principles and also important classes of molecules generated through DOS approaches, including peptidomimetics and macrocycles. Accordingly, important topics for accessing complexity and diversity have been taken into account. Chapter 2 reports the application of multicomponent reactions as a powerful tool to introduce chemical diversity and multifunctional building blocks in a DOS approach. Chapter 3 covers the use of cycloaddition reactions in the fields of DOS as a key approach to provide cyclic and heterocyclic compounds with a high degree of structural complexity and skeletal diversity. Phosphine organocatalysis is described in Chapter 4 as a valid approach encompassing catalytic methods in the DOS area, and stimulating examples with a wide array of building blocks are reported, together with some applications in chemical biology. Chapter 5 introduces the role of domino reactions in DOS as a concept devoted to the generation of small molecules in few synthetic steps, taking advantage of pericyclic, anionic, radical, or transition metal–mediated domino processes. Finally, solid-phase methods are reported in Chapter 7 to present the use of this important technique in generating large collections of small molecules according to DOS principles. The application of DOS to achieve specific classes of compounds is exemplified in Chapters 6 and 8, where the generation of peptidomimetics and macrocyclic structures, respectively, are reported.

In Part II the concept of diversity-oriented synthesis is expanded to describe chemical libraries and how these two elements are related. Chapter 9 presents a synthesis of chemical libraries inspired by natural products as a key platform in addressing both chemical diversity and molecular complexity. Chapter 10 deals with chemoinformatic methods of analyzing the chemical space, and several methods for representing small-molecule libraries are outlined. Chapter 11 reports the approach of DNA-encoded chemical libraries as an innovative technology addressing the need of huge libraries for drug discovery issues and the requirement of a fast deconvolution method.

Part III is dedicated to modern approaches for screening DOS libraries, including the basics of high-throughput and high-content screening (Chapter 12), small-molecule microarrays (Chapter 13), and the use of yeast as a model in smart screening assays encompassing chemical genetics and chemical genomics (Chapter 14). *In silico* methods are described in Chapters 15 and 16, which are connected to the chemoinformatic concepts reported in Chapter 10, and they present, respectively, the virtual screening of chemical libraries and the concepts of activity landscapes and activity cliffs as powerful methods for the analysis of structure–activity relationship data.

Finally, Part IV presents significant applications of DOS libraries and small molecules in the fields of drug discovery (Chapter 17) and chemical biology (Chapter 18), reporting selected key studies in these research areas, and giving a picture of the prominent role of diversity-oriented synthesis in present and future biomedical research.

I express my thanks to the authors who contributed the careful and detailed reviews presented in this book. These presentations should interest not only those readers who currently work in the field of diversity-oriented synthesis, but also those who are considering this approach in the fields of drug discovery and chemical biology. I hope that these chapters will stimulate further advances in this rapidly developing field.

Also, I would like to thank my mentor, professor Antonio Guarna, for kind support during the development of this book, and throughout my career in research.

ANDREA TRABOCCHI

Florence, Italy
October 2012

ABBREVIATIONS

μw	Microwave irradiation
1,3-DNB	1,3-Dinitrobenzene
3CR	Three-component reaction
3D	Three-dimensional
4CR	Four-component reaction
AcOH	Acetic acid
$AcONH_4$	Ammonium acetate
AD	Activating domain
AD-mix	Asymmetric dihydroxylation-mix
ADME	Absorption, distribution, metabolism, and elimination
AIBN	2,2′-Azobis(isobutyronitrile)
AIDS	Acquired immunodeficiency syndrome
AIV	Avian influenza virus
All	Allyl
ALPHA	Amplified luminescent proximity homogeneous assay
ATP	Adenosin triphosphate
B/C/P	Build/couple/pair
BCL-2	β-Cell lymphoma 2
BD	Binding domain
BEMP	2-*t*-Butylimino-2-diethylamino-1,3-dimethylperhydro-1,3,2-diazaphosphorine
BIOS	Biology-oriented synthesis
BMMSG	Bipartite matching molecular series graph
Bn	Benzyl
Boc	*t*-Butoxycarbonyl

Bredereck's reagent	*t*-Butoxybis(dimethyamino)methane
BRET	Bioluminescence resonance energy transfer
BRo5	"Beyond the rule of 5"
Bs	Brosyl
BTPP	*t*-Butyliminotri(pyrrolidino)phosphorane
Bts	Benzothiazole-2-sulfonyl
Bz	Benzoyl
cAMP	Cyclic adenosine monophosphate
CAP	Complementary ambiphile pairing
CAS	Chemical Abstracts Service
CC	Commercial compounds
Cdc42	Cell division cycle 42
CDK1	Cyclin-dependent kinase 1
CHO	Chinese hamster ovary cells
CM	Cross metathesis
CMLD-BU	Center for Chemical Methodology for Library Development at Boston University
CNG	Chemical neighborhood graph
CNS	Central nervous system
COD	Cyclooctadiene
COX-1	Cyclooxigenase-1
Cp	Cyclopentadienyl
CPCCG	Conrad Prebys Center for Chemical Genomics
CuAAC	Copper-catalyzed azide–alkyne cycloaddition
CXCR4	CXC chemokine receptor 4
Da	Dalton
DABCO	1,4-Diazabicyclo[2.2.2]octane
DAD	Dual activity difference
DAmP	Decreased abundance by *m*RNA perturbation
DBU	1,8-Diazabicyclo[5.4.0]undec-7-ene
DCC	*N,N'*-Dicyclohexylcarbodiimide
DCE	1,2-Dichloroethane
DCM	Dichloromethane
DDQ	2,3-Dichloro-5,6-dicyano-1,4-benzoquinone
Ddz	α,α-Dimethyl-3,5-dimethoxybenzyloxycarbonyl
DEAD	Diethyl azodicarboxylate
Dess–Martin periodinane	1,1,1-Triacetyloxy-1,1-dihydro-1,2-benziodoxol-3(1*H*)-one
DH-PH	Dbl homology/pleckstrin homology
DHFR	Dihydrofolate reductase
(DHQD)PHAL	Hydroquinidine 1,4-phthalazinediyl diether
DIAD	Diisopropyl azodicarboxylate
DIC	*N,N'*-Diisopropylcarbodiimide
DIPEA	*N,N*-Diisopropylethylamine
DKP	Diketopiperazine

DMAD	Dimethylacetylenedicarboxylate
DMAP	*N,N*-Dimethylaminopyridine
DME	Dimethoxyethane
DMEDA	*N,N*-dimethylethylenediamine
DMF	*N,N*-Dimethylformamide
DMS	dimethylsulfide
DMSO	Dimethyl sulfoxide
DMT	4,4′-Dimethoxytrityl
DNA	Deoxyribonucleic acid
DNMT	DNA methyltransferase
DOPA	3,4-Dihydroxyphenylalanine
DOS	Diversity-oriented synthesis
DPC	DNA-programmed chemistry platform
DPPA	Diphenylphosphoryl azide
DPPP	Diphenylphosphinopropane
dr	Diastereomeric ratio
DRCS	Delimited reference chemical spaces
DSC	Differential scanning calorimetry
DTPA	Diethylenetriamine pentaacetic acid
DTS	DNA-templated synthesis
DTT	Dithiothreitol
EC	Endothelial cell
ECL3	Extracellular loop
EDCI	1-Ethyl-3-(3-dimethylaminopropyl)carbodiimide
EGFP	Enhanced green fluorescent protein
ELSD	Evaporative light scattering detection
ER	Endoplasmic reticulum
ERK	Extracellular signal-regulated kinase
ESAC	Encoded self-assembling chemical libraries
ESR	Electron spin resonance
F-SPE	Fluorous solid-phase extraction
FACS	Fluorescence-activated sorting instrument
FBDD	Fragment-based drug discovery
FGI	Functional group interconversion
FI	Fluorescence intensity
FKBP	FK506-binding protein
Fmoc	Fluorenylmethyloxycarbonyl
FOS	Function-oriented synthesis
FP	Fluorescence polarization
FRET	Fluorescence resonance energy transfer
FTase	Farnesyltransferase
GBP	Glycan-binding protein
GEF	Guanine nucleotide exchange factor
GFP	Green fluorescent protein
GGTase	Geranylgeranyltransferase

Gli	Glial transcription factor
Glu	Glutamic acid
GLUT	Glucose transporters
GNF	Genomics Institute of the Novartis Research Foundation
GPCRs	G-protein-coupled receptors
Grb2	Growth factor receptor-bound protein 2
Grubbs I	Benzylidene–bis(tricyclohexylphosphine) dichlororuthenium; bis(tricyclohexylphosphine) benzylidine ruthenium(IV) dichloride
Grubbs II	[1,3-Bis(2,4,6-trimethylphenyl)-2-imidazolidinylidene] dichloro(phenylmethylene)(tricyclohexylphosphine) ruthenium
GSIS	Glucose-stimulated insulin secretion
GSK	GlaxoSmithKline
GST	Glutathione *S*-transferase
GST-PBD	GST fusion protein of the p21-binding domain of PAK1
GTPase	Guanine triphosphatase
HA	Hemagglutinin
HaM	Heck-aza-Michael
HATU	*N,N,N′,N′*-Tetramethyl-*O*-(7-azabenzotriazol-1-yl)uronium hexafluorophosphate
HBA	Hydrogen-bond acceptors
HBD	Hydrogen-bond donors
HCS	High-content screening
HCV	Hepatitis C virus
HDAC	Histone deacetylase
Hh	Hedgehog
HIP	Haploinsufficiency profiling
HIV	Human immunodeficiency virus
HMDS	hexamethyldisilazide
HMG-CoA	3-Hydroxy-3-methylglutaryl coenzyme A
HMPT	Hexamethylphosphorus triamide
hMSC	Human mesenchymal stem cell
HOP	Homozygous profiling
Hoveyda–Grubbs II	[1,3-Bis(2,4,6-trimethylphenyl)-2-imidazolidinylidene]dichloro(*o*-isopropoxyphenylmethylene)ruthenium
HPLC	High-performance liquid chromatography
HPNCC	Hereditary nonpolyposis colorectal cancer
Hsc	Heat shock cognate protein
Hsp	Heat shock protein
HTS	High-throughput screening
ICCB	Harvard Medical School's Institute for Chemistry and Cell Biology
IDPCR	Interaction-dependent PCR

IEDDA	Inverse electron-demand aza-Diels–Alder
IFNγ	Interferon-gamma
IL	Interleukin
IMAC	Immobilized metal ion affinity chromatography
IMDA	Intramolecular Diels–Alder
INC	Intramolecular nitrone cycloaddition
INOC	Intramolecular nitrile-oxide cycloaddition
ISAC	Identification of structure-based activity cliffs
ITC	Isothermal titration calorimetry
KEGG	Kyoto encyclopedia of genes and genomes
LA	Lewis acid
LBVS	Ligand-based virtual screening
LC-MS	Liquid chromatography–mass spectrometry
LDA	Lithium diisopropylamide
LF	Lethal factor
LiRIf	Ligand–receptor interaction fingerprint
LOD	Limit of detection
LOLS	Libraries from libraries
LUMO	Lowest unoccupied molecular orbital
MACCS	Molecular access system
MALDI	Matrix-assisted laser desorption/ionization
MAPK	Mitogen-activated protein kinase
MBP	Maltose-binding protein
MCAP	Multicomponent assembly process
*m*CPBA	3-Chloroperoxybenzoic acid
MCR	Multicomponent reaction
MCSL	Multiple-core structure library
MDR	Multidrug resistance
MeCN	Acetonitrile
MEND	Modular enhancement of nature's diversity
MEQI	Molecular equivalent indices
MGD	Mouse genome database
mHTS	Mechanistic HTS
MiB	Multicomponent macrocyclization
MIC_{50}	Minimum inhibitory concentration to inhibit the growth of 50% of the organisms
MIP1α	Macrophage inflammatory protein 1-alpha
MLPCN	Molecular Libraries Production Center Network
MLSMR	Molecular libraries small-molecule repository
MMP	Matrix metalloprotease
MMS	Matching molecular series
MOA	Mechanism of action
MQNs	Molecular quantum numbers
*m*RNA	Messenger ribonucleic acid
MRSA	Methicillin-resistant *Staphylococcus aureus*

MS	Molecular sieves
MSP	Multicopy suppression profiling
MSSA	Methicillin-susceptible *Staphylococcus aureus*
MTAD	4-Methyl-1,2,4-triazoline-3,5-dione
MW	Molecular weight
N-WASP	Neuronal Wiskott–Aldrich syndrome protein
NADH	Nicotinamide adenine dinucleotide
NADPH	Nicotinamide adenine dinucleotide phosphate
NBS	*N*-Bromosuccinimide
NCI	National Cancer Institute
NCS	*N*-Chlorosuccinimide
NFAT	Nuclear factor of activated T-cells
NF-κB	Nuclear factor κB
NGF	Nerve growth factor
NGS	Next-generation sequencing
NMDAR	*N*-methyl-D-aspartate glutamate receptor
NMO	4-Methylmorpholine-*N*-oxide
NMP	*N*-Methyl-2-pyrrolidone
NMR	Nuclear magnetic resonance
NOe	Nuclear Overhauser effect
NP	Natural products
Ns-Cl	Nitrobenzenesulfonyl chloride
NSG	Network-like similarity graph
OAc	Acetate
O.D.	Optical density
on	Overnight
ORF	Open reading frame
OTf	Trifuoromethanesulfonate or triflate
OTIPS	Triisopropylsilyloxy
OXYPHOS	Oxidative phosphorylation
PAGE	Polyacrylamide gel electrophoresis
PAK1	P21-activated kinase 1
PCA	Principal components analysis
PCP	Planar cell polarity
PCR	Polymerase chain reaction
PDAC	Pancreatic ductal adenocarcinoma
PG	Protecting group
PIP2	Phosphatidylinositol 4,5-bisphosphate
PK	Pharmacokinetics
PLGA	Poly(lactic acid)–poly(glycolic acid) copolymer
PMB	*p*-Methoxybenzyl
PMI	Principal moment of inertia
PNA	Peptide nucleic acid
PPIs	Protein–protein interactions
PPTS	Pyridinium *p*-toluenesulfonate

PS	Polystyrene
Ptc1	Patched receptor protein 1
Py	Pyridine
PyBOP	(Benzotriazol-1-yloxy)tripyrrolidinophosphonium hexafluorophosphate
PyBrOP	Bromotripyrrolidinophosphonium hexafluorophosphate
PZQ	Praziquantel core structure
QC	Quality control
qHTS	Quantitative high-throughput screening
QM	Quinone methide
QSAR	Quantitative structure–activity relationship
QSPR	Quantitative structure–property relationship
QUINAP	1-(2-Diphenylphosphino-1-naphthyl)isoquinoline
R&D	Research and development
RabGGTase	Rab geranylgeranyltransferase
RANK	Receptor activator of NF-κB
RANKL	Receptor of NF-κB ligand
Ras	Rat sarcoma
RB	Number of rotatable bonds
RCM	Ring-closing metathesis
Rho	Ras-homologous
RNA	Ribonucleic acid
RNA*i*	Ribonucleic acid interference
ROM	Ring-opening metathesis
rt	Room temperature
RuAAC	Ruthenium-catalyzed azide–alkyne cycloaddition
RWG	Resonant waveguide grating
S/B	Signal-to-background ratio
S/N	Signal-to-noise ratio
SAG	Smo agonist
SALI	Structure–activity landscape index
SAM	Self-assembled monolayer
SAR	Structure–activity relationship
SAS	Structure–activity similarity
SBMRI	Sanford–Burnham Medical Research Institute
SBVS	Structure-based virtual screening
scFv	Single-chain variable fragment
SE	Shannon entropy
SGD	*Saccharomyces* genome database
SH2	Sarcoma homology 2 (protein domain)
Shh	Sonic hedgehog protein
S log *P*	Octanol/water partition coefficient
SmALI	Structure multiple-activity landscape index
SmAS	Structure multiple-activity similarity maps
SMM	Small-molecule microarray

Smo	Smoothened receptor
SMR	Small-molecule repository
S_NAr	Aromatic nucleophilic substitution
SOMs	Self-organizing maps
SPE	Solid-phase extraction
SPR	Surface plasmon resonance (*also* Structure–property relationship)
SPS	Split–pool (or split-and-pool) synthesis
SPT	Similarity-potency tree
Src	Sarcoma
SSAR	Stereostructure–activity relationships
SSE	Scaled Shannon entropy
STRE	Stress response element
Su	Succinimide
SV40	Simian virus 40
SVM	Support vector machine
TA	Tail-anchored protein
TACO-1	Tryptophan-aspartate containing coat protein 1
TAD	Triple activity difference
TAg	Large T antigen
TBAF	Tetrabutylammonium fluoride
TBDMS	*t*-Butyldimethylsilyl
TBPB	1-(1-2-methylbenzyl)-1,4-bipiperidin-4-yl)-1*H* benzo[*d*]imidazol-2(3*H*)-one
TBS	*t*-Butylsilyl
TCM	Traditional Chinese medicine
TES	Triethylsilane
TFA	Trifluoroacetic acid
TFMSA	Trifluoromethanesulfonic acid
TGD	Typed graph distance
TGN	trans-Golgi network
THF	Tetrahydrofuran
TIPS	Triisopropylsilyl
TIPSOTf	Triisopropylsilyl trifluoromethanesulfonate
TLC	Thin-layer chromatography
TMOF	Trimethyl orthoformate
TMS	Trimethylsilyl
TMSOK	Potassium trimethylsilanolate
TMSOTf	Trimethylsilyl trifluoromethanesulfonate
TNFα	Tumor necrosis factor-alpha
TOS	Target-oriented synthesis
TPSA	Topological polar surface area
TR-FRET	Time-resolved fluorescence resonance energy transfer
TRAP	Tartrate-resistant acid phosphatase
TRF	Time-resolved fluorescence

TRH	Thyrotropin-releasing hormone
TrkA	Tyrosine receptor kinase A
TrkC	Tyrosine receptor kinase C
Ts	*p*-Toluenesulfonyl
TS	Transition state
TsOH	*p*-Toluenesulfonic acid
UDC	Ugi/de-Boc/cyclize
UMAM	Ugi/Michael/aza-Michael
UPCMLD	University of Pittsburgh Center for Chemical Methodologies and Library Development
VCA domain	Verprolin homology cofilin homology acidic domain
VEGFR2	Vascular endothelial growth factor receptor-2
VS	Virtual screening
VSG	Variant surface glycoprotein
VSVG	Vesicular stomatitis virus glycoprotein
Y2H	Yeast two-hybrid
Y3H	Yeast three-hybrid
yR	YoctoReactor technology

1

THE BASICS OF DIVERSITY-ORIENTED SYNTHESIS

Kieron M. G. O'Connell, Warren R. J. D. Galloway, and David R. Spring

1.1 INTRODUCTION

In this chapter, the underlying ideas behind diversity-oriented synthesis are introduced. The relationship between diversity-oriented synthesis and combinatorial chemistry is discussed, and the rationale behind the use of diversity-oriented synthesis as a tool for the discovery of biologically active molecules is explained. Common synthetic strategies for the efficient generation of structurally diverse compound collections are then introduced. In the second part of the chapter we discuss recent examples of diversity-oriented syntheses, with examples taken from our own research and from the wider community. These examples seek to illustrate the imaginative ways in which the various synthetic strategies have been implemented and to represent the current state of the art in diversity-oriented synthesis.

1.2 WHAT IS DIVERSITY-ORIENTED SYNTHESIS?

The term *diversity-oriented synthesis* (DOS) first appeared in the chemical literature in the year 2000 in an article written by Stuart Schreiber [1]. In this article, which was written with a particular focus on drug discovery, the term was used to distinguish between compound libraries (or single compounds) synthesized with the intention of interacting with preselected protein targets [called *target-oriented synthesis* (TOS)] and those libraries used in "efforts to identify simultaneously therapeutic protein

Diversity-Oriented Synthesis: Basics and Applications in Organic Synthesis, Drug Discovery, and Chemical Biology, First Edition. Edited by Andrea Trabocchi.

targets and their small molecule regulators" [1]. For TOS compounds, knowledge of the preselected targets can lead to some degree of rational design being implemented; however, according to Schreiber, the second class of library should benefit from high levels of structural diversity within the compound collection, so the deliberate synthesis of such libraries can be considered to be DOS. An alternative definition was offered later by our group, where it was suggested that "diversity-oriented synthesis involves the deliberate, simultaneous and efficient synthesis of more than one target compound in a diversity-driven approach to answer a complex problem" [2].

This statement leaves some room for interpretation; however, as the very nature of diversity, in a chemical sense, is to some degree subjective, it provides a useful general definition that can be applied across the majority of the examples of DOS that are published today. The "complex problem" mentioned in this definition usually refers to the discovery of novel biologically relevant compounds, and this is the context in which DOS is usually discussed. However, as mentioned, this does not have to be the case, as the DOS approach could potentially be applied to other problems, such as the discovery of a novel ligand or catalyst for a reaction [2].

1.3 SMALL MOLECULES AND BIOLOGY

The term *small molecules* has no strict definition; however, it usually refers to potentially orally bioavailable compounds that have a molecular weight of less than 1500 Da [3] and that are distinct from naturally occurring biological macromolecules: DNA, RNA, and proteins [4]. The label "small molecule" can therefore be applied across the vast majority of synthetic drugs and naturally occurring secondary metabolites.

The ability of small molecules to interact with biological macromolecules, in particular, proteins, and consequentially to exert specific effects, often in a selective and dose-dependent manner, has led to them being regarded as powerful tools for the study and manipulation of biological systems [5–7]. Indeed, the use of small molecules in this way, to modulate biological function deliberately and selectively, underpins the fields of medicinal chemistry (where molecules are used to treat disease states) and chemical genetics (where molecules are used as probes to study biological systems) [3,8,9].

The discovery of novel molecular entities or structural classes capable of these specific interactions represents a significant challenge. In cases where the biological target is well defined and understood, the rational design of ligands is sometimes possible, especially when the structure of a native ligand or its single protein target is known [6]. However, for other, less well understood disease states, or if a novel mode of binding or biological target is sought, this is not possible. In these cases high-throughput screening (HTS) of small-molecule libraries can provide an effective solution [10]. Clearly, the composition of these libraries, in terms of the chemical structures included within them, is an extremely important consideration [2]. As the biological activity of a given molecule is intrinsically related to its chemical structure, the greater the degree of structural variation between compounds within a library, the higher the likelihood of achieving broad-ranging and distinct biological activity across that library [11–13]. The presence of multiple structural classes within a library being tested against a single target also increases the likelihood of discovering a molecule capable of binding to that target in a novel manner [14].

The molecules that comprise these libraries, or indeed any compound collection, may be obtained from either natural (natural products) or nonnatural (chemical synthesis) sources. Nature has produced huge numbers of biologically relevant secondary metabolites that have evolved to have specific and exquisite biological activity. These compounds have been used medicinally for millennia, still provide many lead compounds and drugs today [15,16], and almost certainly will continue to do so long into the future [17]. Also, there can be no argument that natural products do not represent a truly structurally diverse collection of compounds. Taking these factors into account, the screening of natural products for biological activity is clearly an extremely valuable thing to do. However, it is not realistic to suggest the production of large libraries based solely on natural products, due predominantly to difficulties in sourcing, isolating, and identifying the bioactive components, as well as in purifying and chemically modifying these often extremely complex structures [18]. Therefore, in terms of producing large numbers of compounds for screening, deliberate chemical synthesis is generally considered to be the most efficient approach [13,19].

With the advent of combinatorial chemistry in the 1990s, it became possible for chemists to produce very large numbers of compounds in an efficient manner. The use of split-and-pool techniques and advances in automation made it possible to synthesize literally millions of compounds in a short period of time [20]. These libraries were, however, generally made up of broadly similar structures, resulting in a fairly limited biological profile across the library. The molecules were usually synthesized by combining a number of building blocks in different ways using the same synthetic methods to generate distinct structures, so achieving a multiplicative increase in the final number of compounds synthesized with an additive increase in the number of building blocks used. This approach usually resulted in the variation of substituents (R-groups) around a common scaffold. Libraries of this sort have had limited success in the discovery of novel biologically active agents, a fact that is attributed primarily to the relative lack of structural diversity within the libraries [21]. It is therefore believed that the quality of the compounds that make up these screening collections, in terms of structural complexity and diversity, is as, if not more important than the total number of compounds present [2].

There are many commercially available or proprietary compound collections that can be seen to represent a source of small molecules without the need for de novo synthesis [12]. The compounds comprising these libraries are generally synthesized in a combinatorial fashion and so suffer from the limitations described above in terms of structural and hence functional diversity. These collections also suffer from limitations that can be attributed to the desire to produce very large numbers of compounds quickly: They are generally "flat" (based around aromatic core structuers containing few stereocentres), structurally simple, and similar [22].

There is also a belief that these compound collections are too heavily biased toward traits that are traditionally perceived to be desirable in drug-like molecules, such as the Lipinski rules for oral bioavailability [12,14,23]. These collections are therefore heavily weighted toward known bioactive chemical space (the region of chemical space spanned by known drug molecules and bioactive natural products). By definition, the exploration of this region has been a fruitful endeavor for the discovery of biologically active and medicinally useful compounds; however, to restrict all screening campaigns to molecules with a relatively narrow range of properties risks the omission of many potentially biologically active molecules that reside in underrepresented and underexplored regions of chemical space [5]. Expanding the region of chemical space explored by screening collections may help to discover small-molecule modulators for classically undruggable targets and so serve to expand the "druggable" genome [24]: a key challenge in chemical biology [9]. In fact, a number of DOS campaigns have already discovered small molecules capable of modulating nontraditional drug targets such as protein–DNA interactions [25,26] and protein–protein interactions [27].

While both natural products and proprietary compound collections obviously have their place in the discovery of novel biologically active compounds, the deliberate synthesis of libraries of high-quality compounds (in terms of structural complexity and variety) represents a third, distinct option that can prove superior for some applications. This is where DOS comes in. As noted above, the term *diversity-oriented synthesis* did not appear in the literature until 2000 [1,28]; however, it is fair to say that many of the ideas behind DOS had existed for some time before then.

A review article from 1997 by Spaller et al. suggested that combinatorial libraries may be suggested to fall into two categories: *focused libraries*, where a number of closely related compounds based on a privileged structure are synthesized with a known target in mind, and *prospecting libraries*, where an entirely new lead compound is sought, so the objective is to screen a large number of structurally varied compounds in the hope of finding a lead with a novel mode of action [29]. The focused libraries can be considered to be the result of a "classical" combinatorial chemistry approach, and further reference made here to combinatorial chemistry refers to the synthesis of this type of library. On the other hand, it can be argued that the syntheses of these prospecting libraries could probably be considered to be early examples of diversity-oriented syntheses, as they were produced with aims similar to those of contemporary DOS libraries: to achieve high levels of structural variety and bioactive chemical space coverage. However, it was when the term *DOS* was coined in 2000 that the ideas and strategies underpinning modern diversity-driven synthesis began to become more formalized.

1.4 COMPARING DOS, TOS, AND COMBINATORIAL CHEMISTRY: FOCUSED LIBRARY SYNTHESIS

The aim of efficiently synthesizing large numbers of structurally diverse compounds capable of effectively interrogating useful areas of chemical space is not easy to realize, for a number of reasons. Principal among them is the fact that the synthetic challenge of producing a biologically relevant DOS library has to be approached from the direction opposite to that of more traditional chemical synthesis [1,2]. In both TOS (of natural and unnatural products) and focused library synthesis, a target structure (or structures) are in mind at the beginning of the synthetic campaign. These structures are then broken down rationally into simpler starting materials and building

There also exists something of a hierarchy within these types of diversity that is based on both synthetic ease and the relative perceived value of each type. Appendage diversity is viewed as the easiest to achieve but is the least important when it comes to producing functionally (biologically) diverse compounds, and it is widely accepted that scaffold diversity is by far the most important and most difficult to achieve [2,32,33]. For this reason there are many published examples of DOS that focus almost entirely on producing diverse molecular skeletons [31,34].

Scaffold diversity is considered the most important diversity element because biomacromolecules are (on a molecular scale) large three-dimensional environments with more or less defined binding regions, pockets, and surfaces; as such, they will interact only with molecules that have complementary three-dimensional structure [13,35]. Therefore, it is the overall shape of a molecule that is the most important factor in terms of determining its biological effects, and this is linked intrinsically to the molecular scaffold or skeleton that the molecule possesses [36]. Libraries that contain large numbers of distinct molecular scaffolds should then cover the widest range of potential binding partners.

To provide a conceptually simple and easily interpretable comparison of the relative molecular diversity incorporated into different compound collections, Spandl et al. suggested the consideration of molecular diversity as a spectrum [19]. At one extreme of the spectrum is a single compound occupying a single point in chemical space, and at the fartherest extreme are all possible compounds giving the maximum chemical space coverage possible (Figure 1.2).

In this context, DOS aims to produce small-molecule libraries that occupy a position toward the right-hand side of the spectrum. This qualitative representation of molecular diversity on a sliding scale shows clearly the idea that DOS libraries should be considerably more diverse than their traditional combinatorial counterparts; however, it is not possible to use this spectrum to compare the relative diversity of compound collections in any meaningful way. More quantitative assessment of the relative diversity of compound collections can be achieved by looking at their comparative molecular descriptors and using them computationally to generate a visual representation of their positions in chemical space.

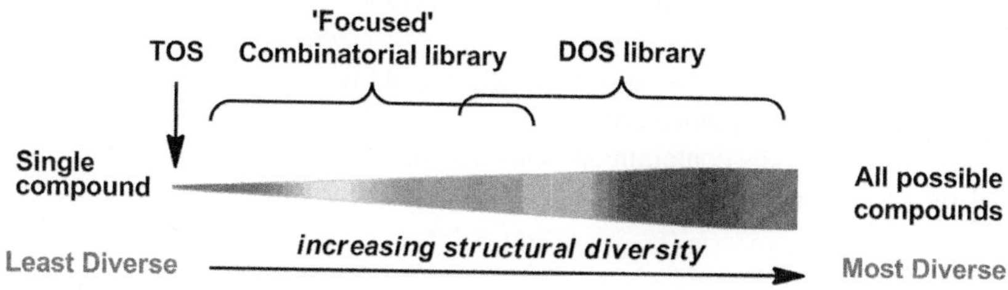

FIGURE 1.2 Molecular diversity spectrum: a representation of the relative degrees of molecular diversity achieved using TOS, focused library synthesis, and DOS. (From [19], with permission of The Royal Society of Chemistry.) (*See insert for color representation of the figure.*)

1.6 MOLECULAR DIVERSITY AND CHEMICAL SPACE

Chemical space, or more properly, *multidimensional descriptor space*, encompasses all theoretically possible compounds and is therefore essentially infinite, limited only by the imagination of chemists and current synthetic methodologies [37,38]. Molecules occupy discrete points within chemical space with "similar" molecules grouped together and "dissimilar" molecules farther apart.† Molecules' positions in chemical space are determined by their comparable physical properties, such as molecular weight, log P, and polarizability as well as their topological features [37,39].

An algorithm based on a large number of these descriptors can be used to create a representation of chemical space based on the descriptors used and the limits placed on them. A molecule's position within this particular multidimensional descriptor space can then be calculated. To give a visually accessible representation of multidimensional descriptor space, it is necessary to use principal components analysis (PCA) [40] to condense the information into two- or three-dimensional scatter plots. These plots then provide a means to easily compare the relative coverage of multidimensional descriptor space achieved by compound collections. It is, however, worth mentioning that these plots are not absolute assessments of diversity or chemical space coverage, as there is the potential for a molecule's relative and absolute position to move depending on the molecular descriptors chosen and any weighting scheme applied to the analysis [41]. Because of this potential, scatter plots are often produced with two or more compound collections superimposed on each other so that their relative diversity can be compared. Figure 1.3 shows an example of chemical space analysis produced using chemical descriptors and PCA.

1.7 SYNTHETIC STRATEGIES FOR CREATING MOLECULAR DIVERSITY

As noted earlier, the challenge of creating molecular diversity efficiently is a considerable one, requiring strategies that differ from the majority of traditional chemical syntheses. Since the beginnings of DOS in the early 2000s, two distinct strategies for the generation of molecular diversity (in particular, skeletal diversity) have been identified in the literature [5]. They are: (1) a *reagent-based approach*, where subjecting a given molecule to a range of reaction conditions allows the synthesis of a number of distinct compounds; and (2) a *substrate-based approach*, where a number of starting materials containing preencoded skeletal information are transformed under the same conditions into a range of molecular structures (Figure 1.4). These

†The words *similar* and *dissimilar* are used with caution, as these terms require a point of reference against which to compare. As such, the same set of molecules could be considered similar or dissimilar, depending on how they are compared (the descriptors used). However, within a given analysis more "similar" molecules should group closer together than those with traits that are more different.

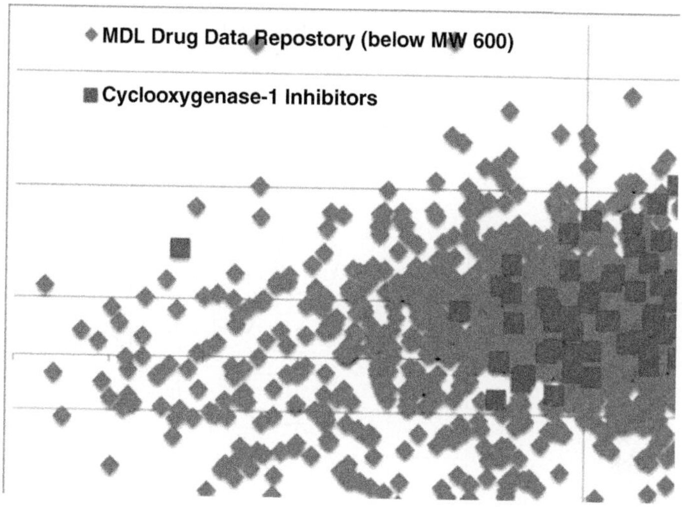

FIGURE 1.3 Chemical space analysis plot of cyclooxygenase-1 (COX-1) inhibitors (red squares) and MDDR compounds (blue diamonds), created using chemical descriptors and principal components analysis. The plot shows that COX-1 inhibitors occupy a wide range of chemical space. (*See insert for color representation of the figure.*)

strategies are not orthogonal to each other, and many DOS campaigns will contain aspects of both. Reagent-based diversification (also known as a *branching reaction pathway*) can be used at any stage of a DOS; it can be used in the early stages to create diverse functionality or in later stages to transform prefunctionalized molecules into distinct molecular scaffolds. Generally, there are considered to be two approaches to reagent-based diversification: the use of "pluripotent" functionality, where a single functional group can be transformed under a range of reaction conditions to give distinct functionality or molecular scaffolds; and the use of densely functionalized molecules, where different functional groups can be transformed orthogonally. The latter approach is generally used to pair functional groups and so create diverse molecular skeletons [19,30]. Substrate-based diversification is generally used in the later stages of a DOS to react strategically placed functional groups intramolecularly and so *fold* compounds into distinct molecular structures. For this reason, it is often referred to as a *folding reaction pathway*.

Some of these ideas were further refined by Schreiber when he identified the use of a *build/couple/pair strategy* as a common feature in the production of small-molecule collections for biological screening [42]. In the *build* stage, the required, ideally chiral starting materials are synthesized or obtained from the chiral pool. These starting materials are then *coupled* together to produce densely functionalized molecules; multicomponent reactions are often used at this stage to couple three or more building blocks together. The *pair* stage then involves intramolecular reactions of the attached functional groups to generate distinct molecular

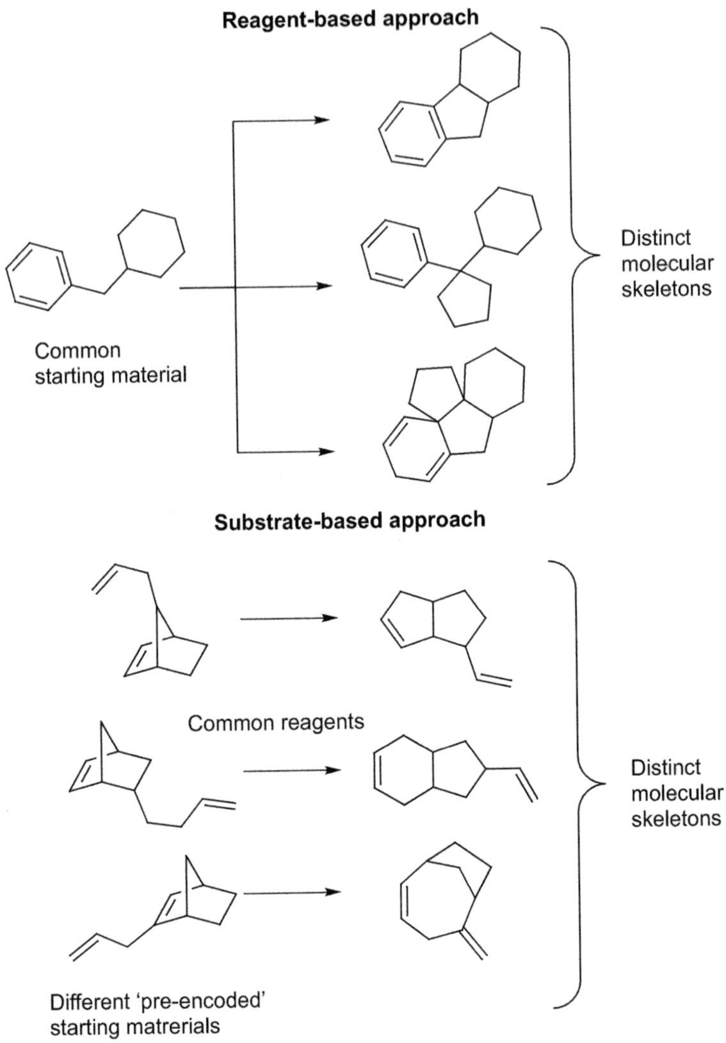

FIGURE 1.4 Representation of the two general strategies for the creation of chemically and skeletally diverse molecules.

scaffolds. The pair stage can generate diversity either by reagent-based pathways, where different functional groups can be paired under orthogonal reaction conditions, or by substrate-based pathways, where the same functional groups are paired under common reaction conditions (Figure 1.5). In the latter case, the diversity generated in the pair stage is due to the relative positions of the functional groups that are paired together. The build/couple/pair approach has subsequently been widely adopted in the literature [43–45].

Important examples of DOS using these various strategies are discussed next.

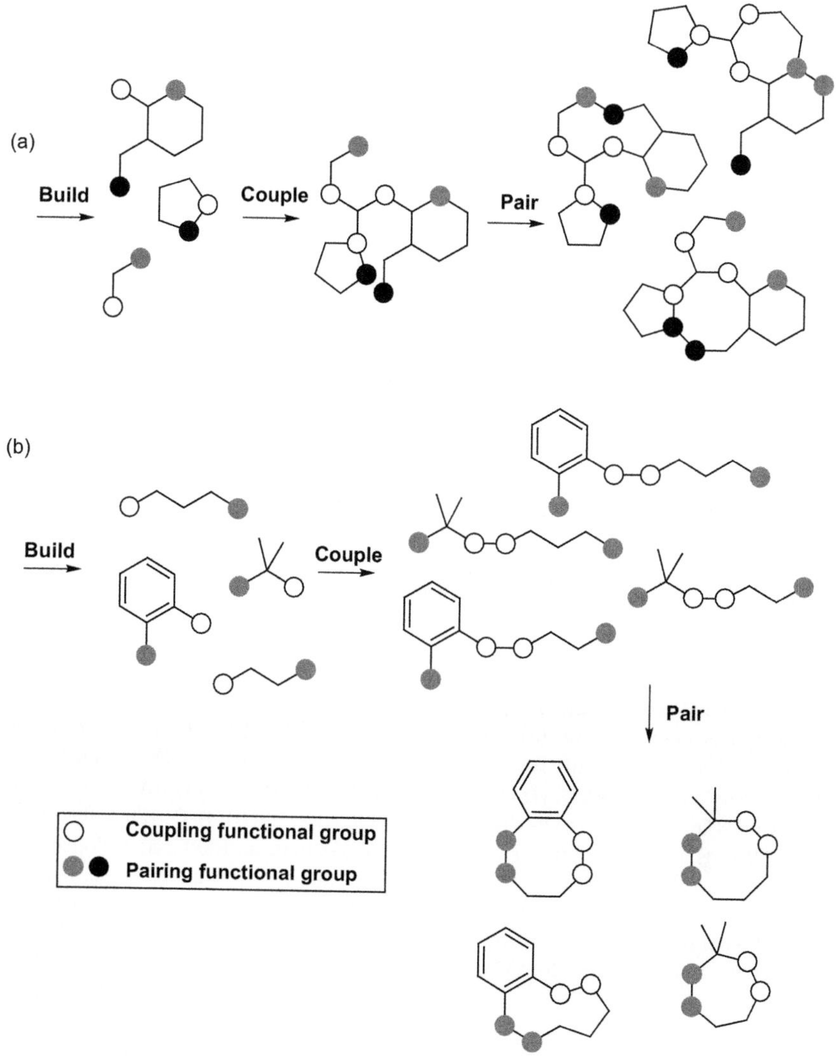

FIGURE 1.5 Build/couple/pair strategy utilizing (a) reagent-based diversification and (b) substrate-based diversification.

1.8 REAGENT-BASED APPROACHES TO DIVERSITY GENERATION

1.8.1 Use of Pluripotent Functional Groups

The use of pluripotent functionality in DOS can be neatly illustrated by two libraries. These libraries were produced from small simple starting materials, which were then subjected to a number of complexity-generating reactions to give a range of diverse and structurally complex compounds in a small number of synthetic steps.

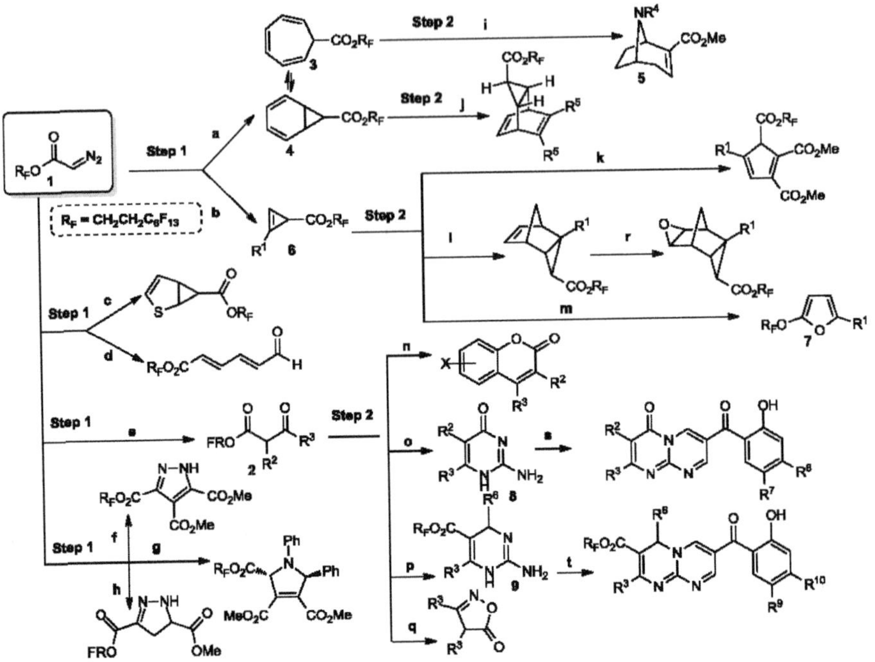

SCHEME 1.1 DOS of a library of small molecules from a simple diazoacetate starting material **1**. Step 1 refers to the first step of the DOS, Step 2 refers to the second step of the DOS. Reagents and conditions: (a) C_6H_6, $Rh_2(OCOCF_3)_4$; (b) R^1CCH, $Rh_2(OAc)_4$, CH_2Cl_2; (c) thiophene, $Rh_2(OAc)_4$; (d) furan, $Rh_2(OAc)_4$ then I_2; (e) LDA $-78°C$ then R^2COR^3, THF then $Rh_2(OAc)_4$, CH_2Cl_2; (f) DMAD; (g) PhCHO, $PhNH_2$ then DMAD, $Rh_2(OAc)_4$ or PhMe $[Cu(OTf)]_2$, CH_2Cl_2; (h) methyl acrylate; (i) R^4NH_2, NaOH, H_2O, 180°C then MeOH, H_2SO_4, 60°C; (j) dienophile, toluene, reflux; (k) DMAD, toluene, 100°C; (l) cyclopentadiene, CH_2Cl_2, 0°C to rt; (m) Grubbs's second-generation catalyst, toluene, ethylene, reflux; (n) phenol derivative, conc. H_2SO_4; (o) guanidine, EtOH, reflux; (p) guanidine, R^6CHO, DMF, 75°C; (q) NH_2OH, THF, reflux; (r) *m*CPBA, CH_2Cl_2, rt; (s) substituted 3-formyl chromone, EtOH, reflux; (t) substituted 3-formyl chromone, EtOH, reflux.

The first library, synthesized in 2006 by Wyatt et al., used a fluorous-tagged diazoacetate species (**1**) as a two-carbon starting unit (Scheme 1.1) [46]. This compound can be considered to be pluripotent, as under a range of conditions it is able to act as both a nucleophilic and an electrophilic species. In total, a library of 223 small molecules was synthesized, based around 30 distinct molecular skeletons. This synthesis was achieved in two to four synthetic steps from the diazoacetate species, clearly exemplifying the powerful nature of this type of approach to molecular diversity generation.

An initial three-way branching strategy was employed, involving three-membered ring formation by addition of rhodium carbenoids to alkene species; α-deprotonation and subsequent quenching with nucleophiles, followed by carbene formation and extrusion of nitrogen to give compounds with general structure **2**; and 1,3-dipolar cycloaddition with a range of dipolarophiles. The products of these reactions were then subjected to further complexity-generating reactions to complete the library synthesis. These further transformations included the trapping of cyclohexatriene **3** (generated by electrocyclic ring opening of fused cyclopropane **4**) with primary amines to give ecgonine-type scaffolds **5**; an unusual Grubbs II–mediated rearrangement of cyclopropene **6** to give furan **7**; and Biginelli-type three-component reactions to give dihydropyrimidines **8** and **9**.

The second library to use pluripotent functionality was published in 2008 by Thomas et al. This library used an *E*-selective Horner–Wadsworth–Emmons reaction to generate the solid-supported enone substrate (**10**) that they used as the pluripotent functional group in their initial branching pathway [47]. This group was then transformed using three catalytic enantioselective processes: a Sharpless asymmetric dihydroxylation to give **11**, a [3 + 2] cycloaddition with an imino ester to give substituted pyrrolidine **12**, and a [4 + 2] cycloaddition with cyclopentadiene to give bridged bicycle **13** (Scheme 1.2). These initial compounds (and variations of them) were eventually transformed into a library of 242 compounds based on 18 distinct molecular skeletons, including a novel *cis-trans*-fused 7-5-7 tricycle (**14**) generated by ring opening–ring closing metathesis of a decorated norbornene.

The compounds produced in these libraries were screened for their effects against three strains of UK epidemic *Staphylococcus aureus*: methicillin-susceptible *S. aureus* (MSSA), and two strains of methicillin-resistant *S. aureus* (EMRSA-15 and EMRSA-16). Of the 223 compounds screened from the library of Wyatt et al., 64 were found to modulate the growth of EMRSA-15 and EMRSA-16 at concentrations between 10 and 100 μM [48]. Of these active species, the vast majority were based around four nitrogen heterocycle frameworks. Inspection of these compounds led to the identification of a number of structural features generally associated with higher levels of antibacterial activity, so an additional focused library of 35 compounds was synthesized. The screening of these compounds against the same strains of bacteria led to the discovery of a number of more potent compounds, the most potent of which was named *emmacin* [48]. Mode-of-action studies suggested that emmacin acts as a prokaryote-selective dihydrofolate reductase (DHFR) inhibitor. The nitrogen heterocycle core of emmacin is reminiscent of that of other reported DHFR inhibitors [49,50]; however, the exact heterocycle, a dihydropyrimidine, is believed to represent a new structural subclass.

The library of Thomas et al. produced a lower hit rate; however, three compounds that reproducibly inhibited the growth of the strains were discovered. The most active compound, *gemmacin*, showed comparable activity against both strains of MRSA to the widely used antibiotics erythromycin and oxacillin. In the original DOS, gemmacin was made racemically, but the enantiomerically pure compounds were subsequently synthesized and showed comparable activity, with (–)-gemmacin being slightly more potent. Structure–activity relationship (SAR) studies were then carried out on gemmacin, resulting in the discovery of the analog gemmacin B, which

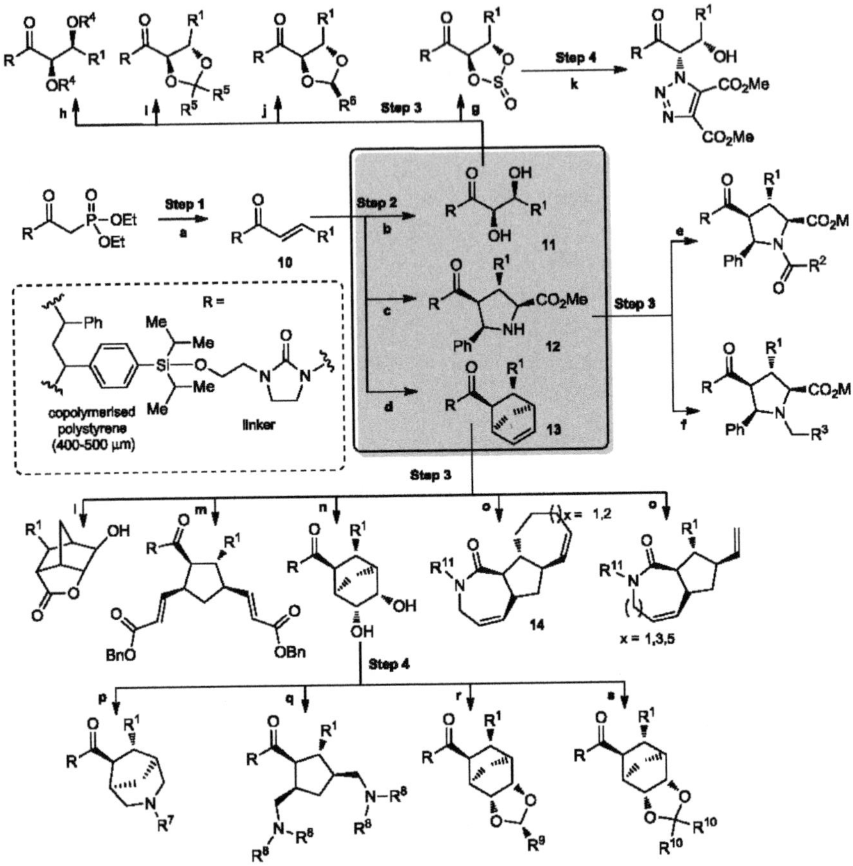

SCHEME 1.2 Diversity-oriented synthesis of 242 compounds based around 18 discrete molecular frameworks by Thomas et al. Conditions: (a) LiBr, 1,8-diazabicyclo[5.4.0]undec-7-ene, R^1CHO, MeCN; (b); AD-mix, (DHQD)PHAL, THF/H_2O (1:1); c) (*R*)-QUINAP, AgOAc, *i*-Pr_2NEt, THF, −78°C → 25°C; (d) chiral bis(oxazoline), $Cu(OTf)_2$, 3 Å MS, CH_2Cl_2, C_5H_6; (e) R^2COCl, DMAP, pyridine, CH_2Cl_2; (f) R^3CHO, BH_3·pyridine, MeOH; (g) $SOCl_2$, pyridine, CH_2Cl_2, 40°C; (h) R^4Br, Ag_2O, CH_2Cl_2, 40°C; (i) $R^5C(O)R^5$, TsOH, DMF, 65°C; (j) R^6CHO, TsOH, DMF, 65°C; (k) NaN_3, DMF, 100°C then DMAD, PhMe, 65°C; (l) *m*CPBA, CH_2Cl_2 then MeOH, 65°C; (m) $CH_2{=}CHCO_2Bn$, PhMe, 120°C, Grubbs II, $CH_2{=}CH_2$; (n) OsO_4, NMO, $CH_3C(O)CH_3/H_2O$ (10 : 1); (o) RNH_2, Me_2AlCl, PhMe, 120°C, then NaH, R^{11}X, DMF, THF then PhMe, 120°C, Grubbs II, $CH_2{=}CH_2$; (p) $NaIO_4$, THF/H_2O (1 : 1) then R^7NH_2, $NaBH(OAc)_3$, CH_2Cl_2; (q) $NaIO_4$, THF/H_2O (1:1) then R^8NHR^8, $NaBH(OAc)_3$, CH_2Cl_2; (r) R^9CHO, DMF, TsOH, 60°C; (s) $R^{10}C(O)R^{10}$, DMF, TsOH, 60°C. (From [47], with permission of John Wiley & Sons; copyright (© 2008 John Wiley & Sons.)

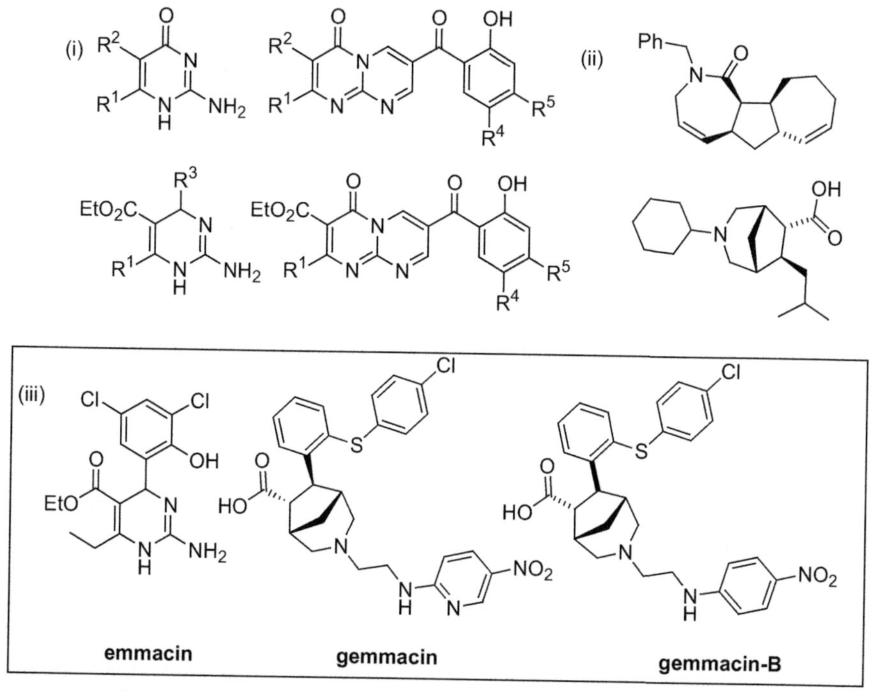

FIGURE 1.6 Examples of compounds and scaffolds that exhibited anti-MRSA activity, including (i) the four nitrogen heterocycle scaffolds from the library of Wyatt et al. [46]; (ii) active compounds from the library of Thomas et al. [47]; (iii) emmacin, (−)-gemmacin, and (±)-gemmacin B.

showed increased efficacy in restricting bacterial growth [51]. Assays for common antibacterial modes of action were performed (such as DHFR reductase inhibition, protein synthesis, and ATP synthesis decoupling), but gemmacin proved inactive in all of these assays. However, gemmacin did show activity in an assay to test for the generation of reactive oxygen species, which suggests that gemmacin (and gemmacin B) may act as bacterial cell-membrane disruptors [51]. The discovery of these two antibacterial compounds, both of which represent a novel structural class (or subclass), illustrates the power of the DOS approach for the discovery of novel bioactive species. Figure 1.6 shows an overview of the structures of the antibacterial compounds produced by the two DOS libraries, and the MIC_{50} values of the most active compounds against MSSA and two strains of MRSA are reported in Table 1.1.

1.8.2 Use of Densely Functionalized Molecules

A recent example of the use of a reagent-based pathway to generate diversity from densely functionalized molecules can be found in the work of Schreiber's

TABLE 1.1 The Comparable Effects of Emmacin, the Enantiomers of Gemmacin, Gemmacin B, Erythromycin, and Oxacillin on Three Strains of *Staphylococcus aureus*

	MIC_{50} (μg/mL)		
	MSSA	EMRSA-15	EMRSA-16
Emmacin	2	9	9
(±)-Gemmacin B	Not determined	8	8
(±)-Gemmacin	2	16	32
Erythromycin	0.5	>64	>64
Oxacillin	0.5	>32	>32

group [52]. Their synthesis could also be considered to be an example of the build/couple/pair approach. In what could be identified as the build and couple stages, they synthesized the four possible diastereomers of *N*-allylpropargylic amino alcohol **15a–d** in six steps from (*R*)- and (*S*)-phenylalanine using standard methods [53]. They were then able to transform these compounds into a range of interesting scaffolds, by pairing functional groups using transition metal–catalyzed enyne cyclizations. When **15a** was used, they saw unusual endoselectivity in an enyne metathesis cyclization to give seven-membered ring product **16**. Then, protecting the free hydroxyl group in **15a** as the corresponding acetate **17** gave complete selectivity for the more usual six-membered exoproduct **18**, and also allowed bicyclic cyclopentenone **19** to be synthesized via a cobalt-catalyzed Pauson–Khand reaction. Treating **15a** with $InCl_3$ under microwave conditions gave **20**, and treatment with TBAF resulted in a Smiles rearrangement to give **21**. Interesting bridged bi- and tricyclic structures **23** and **24** were synthesized by initial S_NAr cyclization to give **22**, followed by subsequent enyne metathesis and Pauson–Khand cyclizations (Scheme 1.3).

Similar cyclizations were carried out on other isomers of **6** to give a small library of structurally complex single stereoisomer small molecules. This DOS pathway provides a useful illustration of the versatility of enyne functionality; this versatility makes the use of enyne functionality very popular in DOS campaigns [54,55].

1.8.3 Twelve-fold Branching Strategy

The strategies described above are intended to highlight the general themes of pluripotent reactivity and dense functionality commonly utilized in DOS. However, these concepts are not exhaustive, and there are many examples of DOS campaigns that resist such generalization, as they do not fall neatly within either category. One such example is the 12-fold branching strategy reported in 2011 by Robbins et al. (Scheme 1.4) [56]. In this work, the authors combined two-directional synthesis and tandem reactions to transform a symmetrical linear ketone **25** into 12 distinct scaffolds. Their strategy involved the initial transformation of the central ketone group

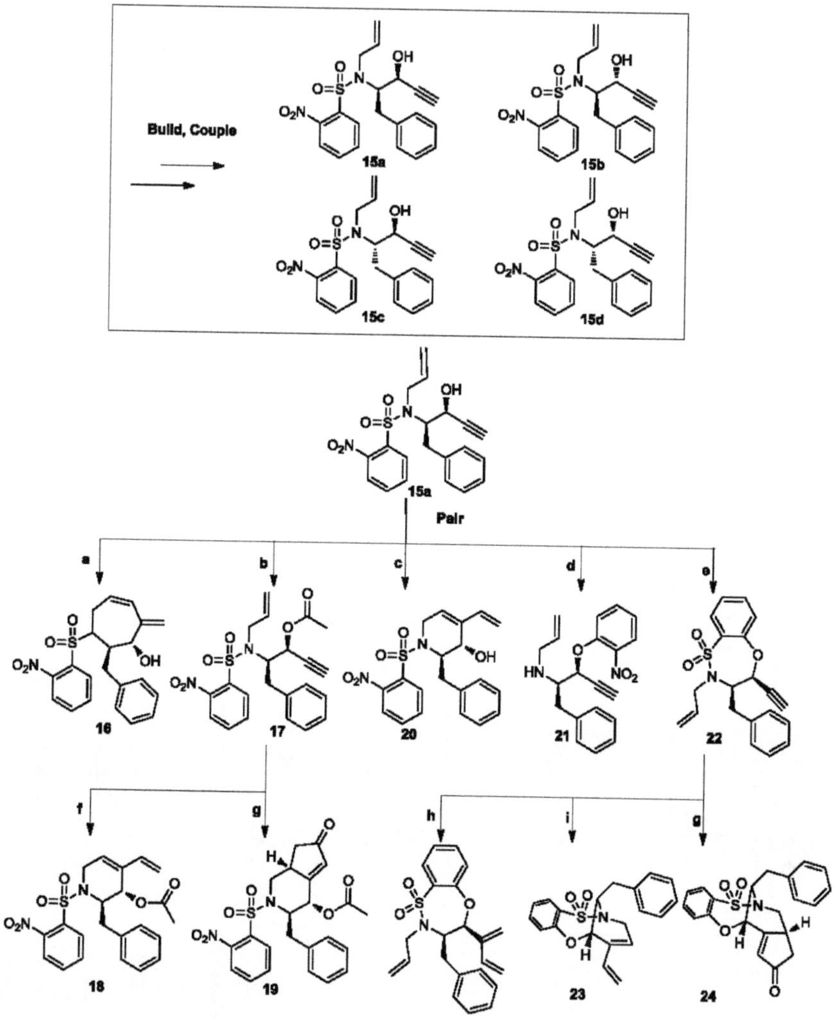

SCHEME 1.3 Reagent-based diversification using a densely functionalized molecule. (a) Hoveyda–Grubbs II, ethylene, toluene, rt; (b) Ac_2O, Et_3N, DMAP, CH_2Cl_2 0°C; (c) $InCl_3$, 1,2-DCE, 90°C μw; (d) TBAF, THF, 0°C; (e) NaH, THF, −10°C; (f) Hoveyda–Grubbs II, ethylene, CH_2Cl_2, 45°C; (g) $Co_2(CO)_8$, NMO, THF, rt; (h) Hoveyda–Grubbs II, ethylene, benzene, rt; (i) Grubbs I, CH_2Cl_2, rt, then $Pb(OAc)_4$. (Adapted from [52], with permission; copyright © 2010 American Chemical Society.)

into various nucleophilic functionalities able to react with the α,β-unsaturated esters at the chain termini. In some cases this led to essentially symmetrical products such as **26** being produced, and in other cases, such as the oxime formation followed by tandem aza-Michael reaction and [3 + 2] cycloaddition to give tricycle **27**, desymmetrization was achieved efficiently.

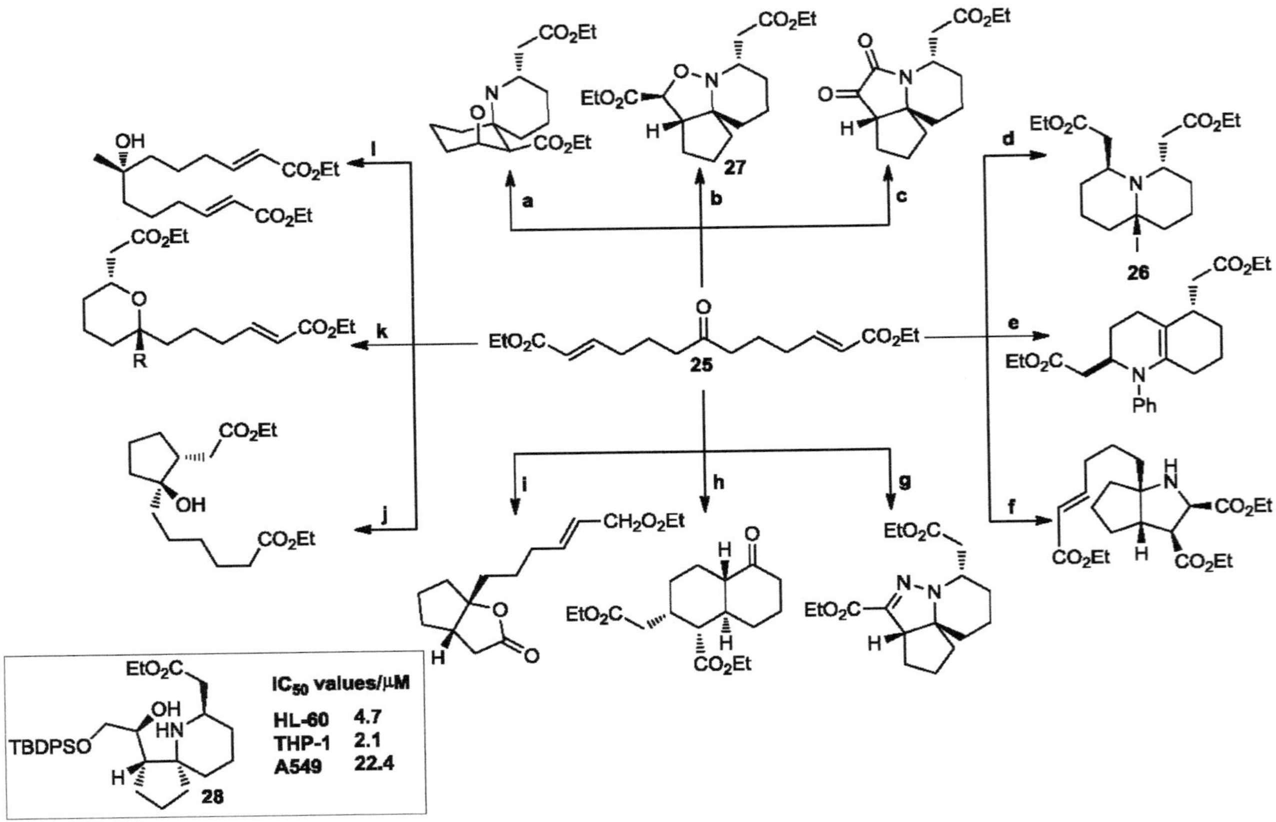

SCHEME 1.4 Twelve-fold branching strategy employed by Robbins et al. (a) $NH_2OH\cdots HCl$, NaOAc, MeCN then toluene 140°C, μw; (b) $NH_2OH\cdots HCl$, NaOAc, MeCN, 60°C; (c) $NH_2OH\cdots HCl$, NaOEt, EtOH; (d) $NaBH_4$, NH_3, EtOH, $Ti(OEt)_4$ then AcOH; (e) $PhNH_2$, $TiCl_4$, CH_2Cl_2, rt; (f) DIPEA, $H_2NCH_2CO_2Et$; (g) NH_2NHTs, toluene, reflux; (h) NaH, THF; (i) SmI_2 (2 equiv), THF, MeOH, −78°C; (j) SmI_2 (5 equiv), THF, MeOH, −78°C; (k) superhydride, THF; (l) MeMgBr, THF. (From [56], with permission of The Royal Society of Chemistry.)

The authors suggest the concept of considering the linear ketone as a "molecular rope" which they were then able to "tie into knots" using tandem reactions, and so produce a range of three-dimensional scaffolds. In some ways this approach could be considered to be a combination of the use of pluripotent functionality (provided by the ketone and α,β-unsaturated ester groups) and densely functionalized molecules (generated in situ), followed by tandem pairing reactions. However, such a classification is largely irrelevant, and regardless of nomenclature, this 12-fold pathway represents an ingenious use of reagent-based diversification. Overall the work generated a range of mono-, bi-, and tricyclic scaffolds possessing fused bridged and cyclic architecture from a single linear substrate. They then synthesized a set of analogs of compound **27** for testing against three cancer cell lines (HL-60, THP-1, A549), which resulted in the discovery of a number of compounds showing low micromolar activity against all three cell lines, including their best compound **28**, which showed sub-10-μM activity against two of the cell lines.

1.9 SUBSTRATE-BASED APPROACH TO SKELETAL DIVERSITY GENERATION

An impressive example of a folding pathway in the generation of skeletal diversity was devised by Morton et al., who used metathesis cascade processes to produce a library consisting of over 80 distinct scaffolds [57]. This was achieved by attaching two of a number of building blocks ("called a propagating" building block in the center of the molecule and a "capping" building block on the end) to a fluorous tagged linker to give densely functionalized linear intermediates containing several unsaturated groups able to take part in the metathesis cascade (Scheme 1.5). The fluorous tagged linker group was also designed to take part in the metathesis cascade, with the final cyclization intended to cleave the linker and allow easy generic purification of the desired products from uncyclized material.

In total, 86 linear substrates were prepared, and from these 96 final products were obtained, based around 84 molecular scaffolds. The molecules also contain a high degree of stereochemical diversity and structural complexity and as such can be considered to be "natural product–like." This work represents the largest number of scaffolds present to date in a deliberately synthesized DOS library.

As for the work of Schreiber's group, this DOS library could be considered to follow the build/couple/pair synthetic plan. First, the fluorous tag and the propagating and capping building blocks are built, they are then coupled together to give the linear precursors, and finally, the unsaturated groups are paired in the metathesis cascade.

1.10 OTHER BUILD/COUPLE/PAIR EXAMPLES

As indicated by some of the examples discussed above, the build/couple/pair paradigm has become very widespread in DOS. Two additional examples published in 2011 will now be discussed; these are intended to illustrate the range of structural

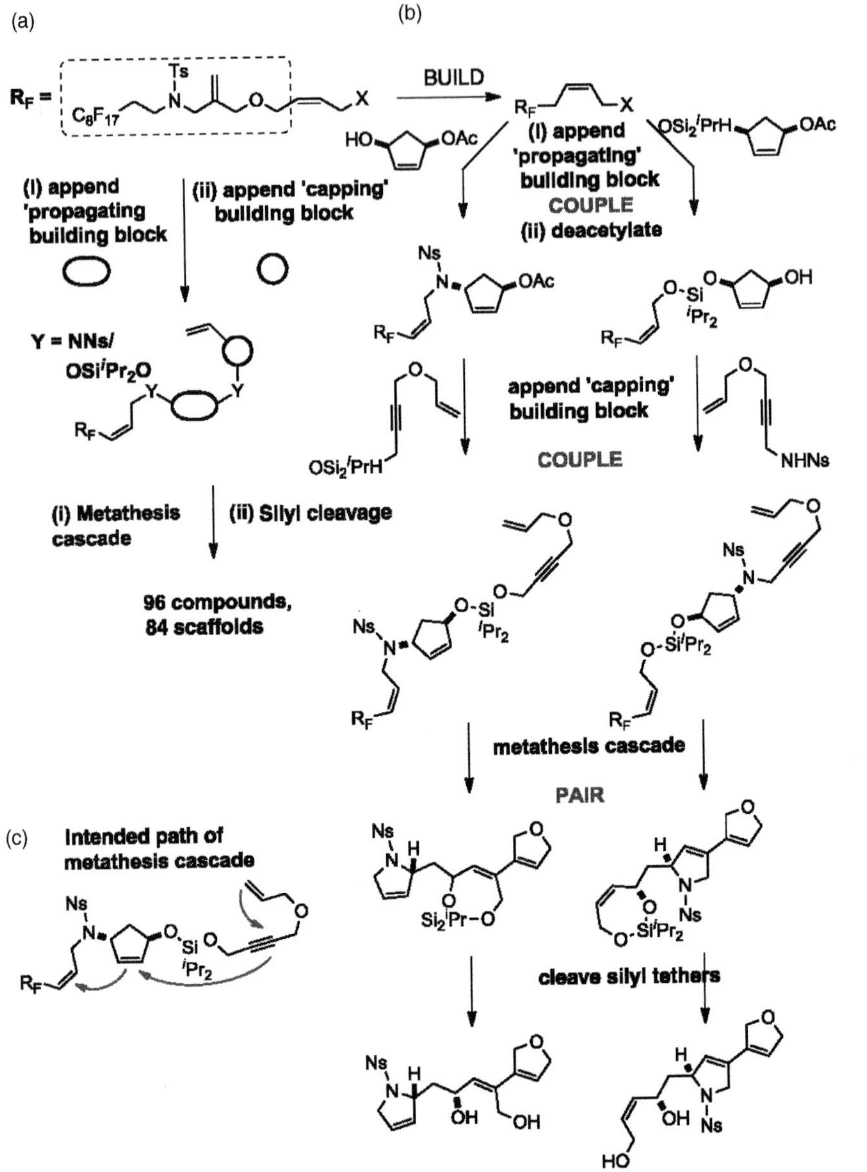

SCHEME 1.5 Folding pathway for skeletal diversity generation employed by Morton et al. [57]: (a) general scheme for the synthesis; (b) synthesis of 2 out of 96 library members with the build/couple/pair stages highlighted; (c) the intended path for the metathesis cascade, culminating in cleavage of the fluorous tag.

classes to which the build/couple/pair approach can be applied. The first example is focused toward the synthesis of three-dimensional fragments for potential use in fragment-based drug discovery. The work highlighted was carried out by Hung et al. and is part of a larger synthetic effort aimed at producing a range of chiral bicyclic corestructures that could be used to enrich existing fragment collections that are generally biased toward "flat"sp^2-rich compounds [58]. The authors employed the build/couple/pair approach to efficiently provide a number of fused bicyclic and spirocyclic compounds that were compliant with the fragment rule of 3, the number of physicochemical parameters that can be applied to fragment-based drug discovery (Scheme 1.6) [59].

The build stage of the synthesis involved the synthesis of **29** from proline employing Seebach et al.'s concept of self-reproduction of chirality [60] or the purchase of the commercially available derivative **30**. The couple stage then involved the addition of other latently reactive appendages, by either functionalization of the proline nitrogen or by peptide coupling of amine groups to the free carboxylic acid of **30**. These groups were then paired to generate the bicyclic architectures desired. In the majority of cases, this was achieved by the ring-closing metathesis of alkene groups but two other approaches: hydantoin formation and oxy-Michael addition to a vinyl sulfone. In a slight addition to the standard build/couple/pair protocol, the authors suggested a post-pairing stage, where the functional group diversity of the compound collection is increased. Two post-pairing modifications—methyl ester hydrolysis and reduction of the alkene groups to give the saturated species—were implemented, which altered the electronic and conformational properties of the fragments but did not significantly change the molecular weight. Then the authors used computational methods to compare their compounds to an existing fragment collection and found that although, as expected, the shapes of the compounds in the various collections were mutually exclusive, their physical properties remained comparable and thus within the desired range for fragments [58].

The second example was targeted toward the synthesis of a small library of macrocyclic compounds. In this work, Isidro-Llobet et al. produced a small library of macrocyclic peptidomimetics in an efficient manner employing the build/couple/pair approach (Scheme 1.7) [61]. In the build stage they produced a number of alkyne-acid and azido-amine building blocks using standard methods. These building blocks were then coupled to give the required linear azido-alkyne precursor (**31**), and macrocyclization was achieved by the pairing of these functional groups in two variations of the azide-alkyne cycloaddition reaction to produce a triazole. The copper-catalyzed azide-alkyne cycloaddition (CuAAC) provided the 1,4-isomer of the triazole (**32**), and the ruthenium-catalyzed variant (RuAAC) gave the 1,5-isomer (**33**). This use of different catalysts for essentially the same process to produce molecular diversity has been dubbed "catalyst control" [62].

Further diversity was then introduced into their compound set when the attached ester and and amine functionalities were paired to give diketopiperazine (DKP) moieties (**34** and **35**). In total, they were able to produce a small proof-of-concept library of 14 macrocyclic compounds.

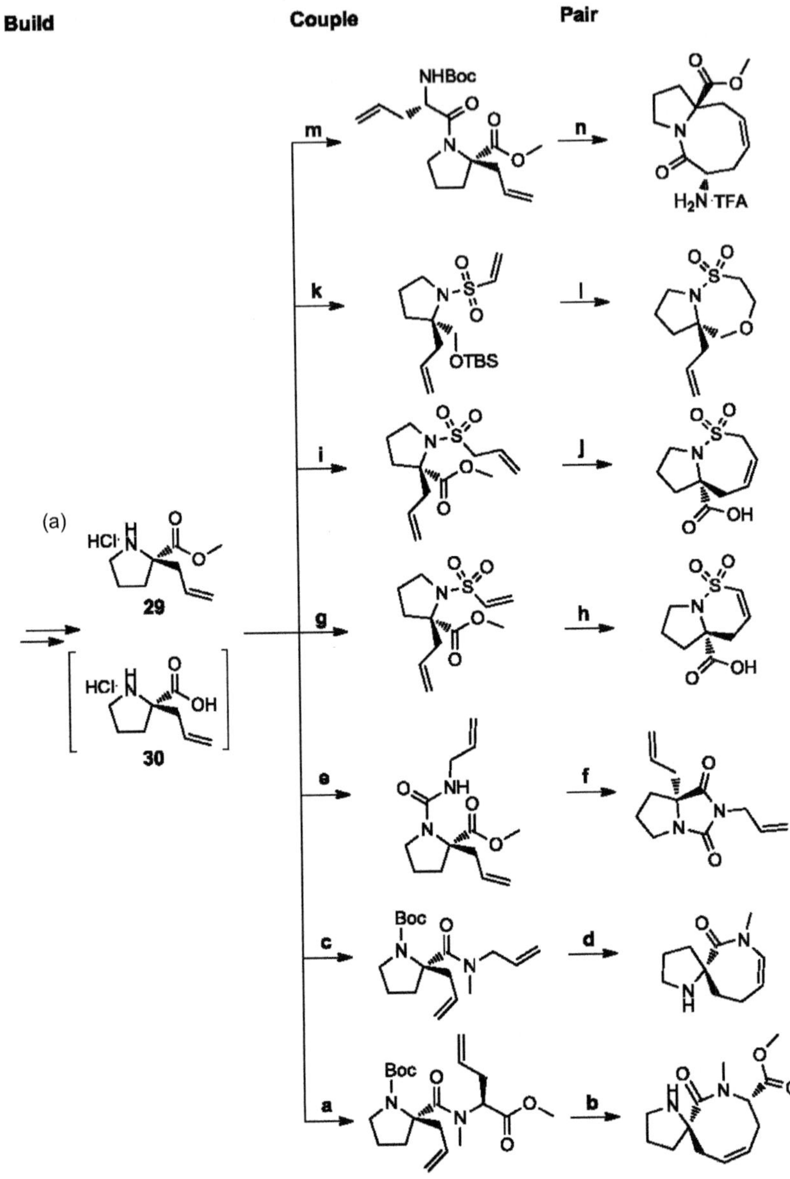

SCHEME 1.6 Synthesis of structurally diverse bicyclic "fragments" by Hung et al. [58] using the build/couple/pair approach. (a) (*S*)-Allylglycine methyl ester, 1-ethyl-3-(3-dimethylaminopropane), ethyl(hydroxyimino)cyanoacetate, Et_3N, CH_2Cl_2; NaH, MeI, DMF; (b) Grubbs II, CH_2Cl_2 reflux; TFA; (c) allylamine, 1-ethyl-3-(3-dimethylaminopropane), ethyl(hydroxyimino)cyanoacetate Et_3N, CH_2Cl_2; NaH, MeI, DMF; (d) Grubbs II, toluene, TFA; (e) allyl isocyanate, CH_2Cl_2; (f) NaH, DMF; (g) 2-chlorosulfonyl chloride, Et_3N, CH_2Cl_2; (h) Grubbs II, CH_2Cl_2, reflux; (i) prop-2-ene-1-sulfonyl chloride, Et_3N, CH_2Cl_2; (j) Grubbs II, CH_2Cl_2, reflux; (k) $LiAlH_4$, THF, *t*-butyldimethylsilylchloride, Et_3N, CH_2Cl_2; 2-chlorosulfonyl chloride, Et_3N, CH_2Cl_2; (l) tetrabutylammonium fluoride, THF; (m) (*S*)-*N*-Boc-allylglycine, 1-ethyl-3-(3-dimethylaminopropane), ethyl(hydroxyimino)cyanoacetate Et_3N, CH_2Cl_2; (n) Grubbs II, CH_2Cl_2, TFA.

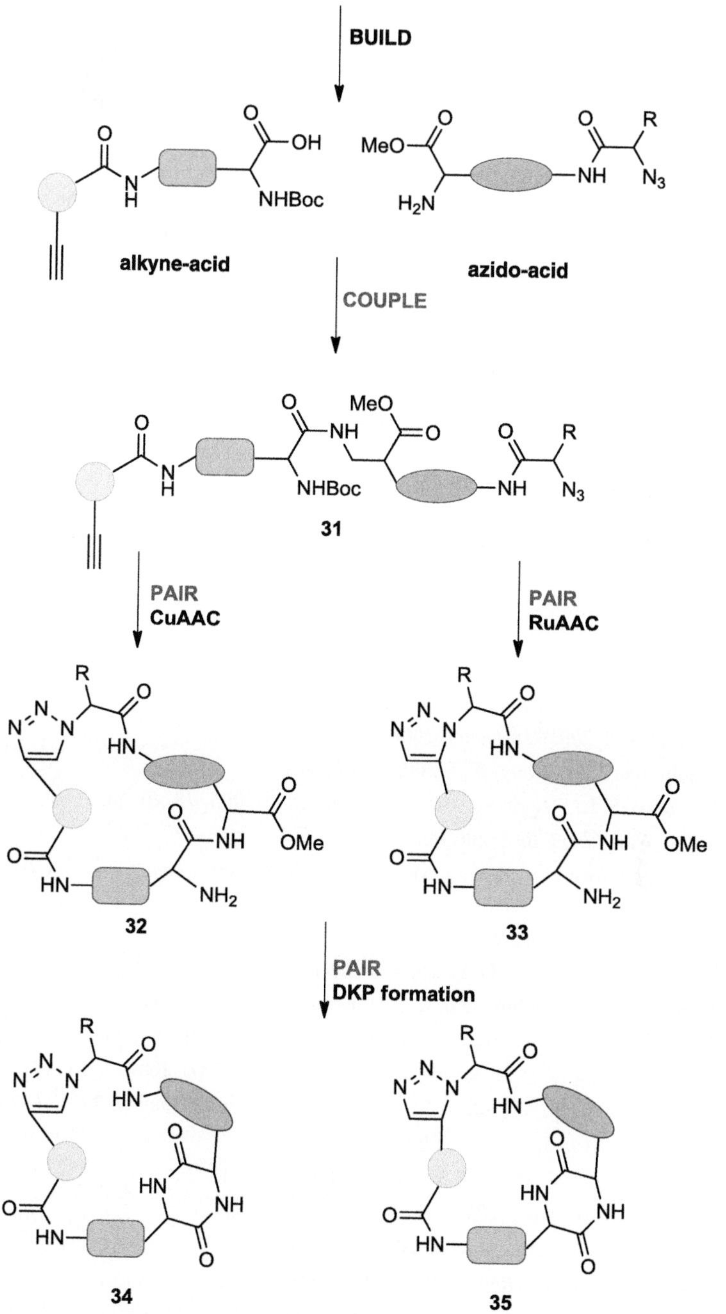

SCHEME 1.7 Overview of Isidro-Llobet et al.'s synthesis of a library of macrocyclic peptidomimetic compounds. (From [61]; copyright © 2011 National Academy of Sciences, U.S.A.)

1.11 CONCLUDING REMARKS

In the decade since the DOS concept was introduced, the field has matured considerably, with numerous sophisticated approaches to the generation of molecular diversity being developed. Emphasis on the generation of scaffold diversity has become integral to the field, as there is an overriding, and justified, opinion that the incorporation of this type of diversity into a compound collection is most important in terms of producing a functionally (biologically) diverse library. As such, many modern examples focus on the production of compound collections containing a high degree of three-dimensional scaffold and shape diversity. As the examples highlighted in this chapter show, the DOS approach has been applied toward the synthesis of a range of compound collections with varying molecular parameters. In the majority of cases, the libraries produced are intended to span known drug-like chemical space; however, as the later examples in this chapter show, the ideas can also be applied effectively to the synthesis of other structural classes, such as low-molecular-weight fragments or larger macrocyclic compounds.

REFERENCES

1. S. L. Schreiber, *Science* **2000**, *287*, 1964–1969.
2. D. R. Spring, *Org. Biomol. Chem.* **2003**, *1*, 3867–3870.
3. B. R. Stockwell, *Nat. Rev. Genet.* **2000**, *1*, 116–125.
4. S. L. Schreiber, *Nat. Chem. Biol.* **2005**, *1*, 64–66.
5. M. D. Burke, S. L. Schreiber, *Angew. Chem. Int. Ed.* **2004**, *43*, 46–58.
6. D. S. Tan, *Nat. Chem. Biol.* **2005**, *1*, 74–84.
7. W. R. J. D. Galloway, A. Bender, M. Welch, D. R. Spring, *Chem. Commun.* **2009**, 2446–2462.
8. D. R. Spring, *Chem. Soc. Rev.* **2005**, *34*, 472–482.
9. C. J. O'Connor, L. Laraia, D. R. Spring, *Chem. Soc. Rev.* **2011**, *40*, 4332–4345.
10. W. P. Walters, M. Namchuk, *Nat. Rev. Drug Discov.* **2003**, *2*, 259–266.
11. W. H. Sauer, M. K. Schwarz, *J. Chem. Inf. Comput. Sci.* **2003**, *43*, 987–1003.
12. W. R. J. D. Galloway, D. R. Spring, *Exp. Opin. Drug Discov.* **2009**, *4*, 467–472.
13. W. R. J. D. Galloway, A. Isidro-Llobet, D. R. Spring, *Nat. Commun.* **2011**, *1*, 80.
14. C. Lipinski, A. Hopkins, *Nature* **2004**, *432*, 855–861.
15. D. J. Newman, G. M. Cragg, *J. Nat. Prod.* **2007**, *70*, 461–477.
16. A. L. Harvey, *Curr. Opin. Chem. Biol.* **2007**, *11*, 480–484.
17. J. W.-H. Li, J. C. Vederas, *Science* **2009**, *325*, 161–165.
18. M. S. Butler, *Nat. Prod. Rep.* **2005**, *22*, 162–195.
19. R. J. Spandl, A. Bender, D. R. Spring, *Org. Biomol. Chem.* **2008**, *6*, 1149–1158.
20. D. S. Tan, M. A. Foley, M. D. Shair, S. L. Schreiber, *J. Am. Chem. Soc.* **1998**, *120*, 8565–8566.
21. S. Borman, *Chem. Eng. News* **2004**, *82*, 32–40.

22. S. L. Schreiber, *Nature* **2009**, *457*, 153–154.
23. C. A. Lipinski, F. Lombardo, B. W. Dominy, P. J. Feeney, *Adv. Drug Deliv. Rev.* **1997**, *23*, 3–25.
24. K.-H. Altmann, J. Buchner, H. Kessler, F. Diederich, B. Kräutler, S. Lippard, R. Liskamp, K. Müller, E. M. Nolan, B. Samorì, G. Schneider, S. L. Schreiber, H. Schwalbe, C. Toniolo, C. A. A. van Boeckel, H. Waldmann, C. T. Walsh, *ChemBioChem* **2009**, *10*, 16–29.
25. P. Y. Ng, Y. Tang, W. M. Knosp, H. S. Stadler, J. T. Shaw, *Angew. Chem. Int. Ed.* **2007**, *46*, 5352–5355.
26. A. N. Koehler, A. F. Shamji, S. L. Schreiber, *J. Am. Chem. Soc.* **2003**, *125*, 8420–8421.
27. F. G. Kuruvilla, A. F. Shamji, S. M. Sternson, P. J. Hergenrother, S. L. Schreiber, *Nature* **2002**, *416*, 653–657.
28. D. Lee, J. K. Sello, S. L. Schreiber, *Org. Lett.* **2000**, *2*, 709–712.
29. M. R. Spaller, M. T. Burger, M. Fardis, P. A. Bartlett, *Curr. Opin. Chem. Biol.* **1997**, *1*, 47–53.
30. R. J. Spandl, M. Díaz-Gavilán, K. M. O'Connell, G. L. Thomas, D. R. Spring, *Chem. Rec.* **2008**, *8*, 129–142.
31. R. J. Spandl, H. Rudyk, D. R. Spring, *Chem. Commun.* **2008**, 3001–3003.
32. J. P. Kennedy, L. Williams, T. M. Bridges, R. N. Daniels, D. Weaver, C. W. Lindsley, *J. Comb. Chem.* **2008**, *10*, 345–354.
33. A. A. Shelat, R. K. Guy, *Nat. Chem. Biol.* **2007**, *3*, 442–446.
34. M. Díaz-Gavilán, W. R. J. D. Galloway, K. M. O'Connell, J. T. Hodkingson, D. R. Spring, *Chem. Commun.* **2010**, *46*, 776–778.
35. M. D. Burke, E. M. Berger, S. L. Schreiber, *Science* **2003**, *302*, 613–618.
36. J. A. Haigh, B. T. Pickup, J. A. Grant, A. Nicholls, *J. Chem. Inf. Model.* **2005**, *45*, 673–684.
37. C. M. Dobson, *Nature* **2004**, *432*, 824–828.
38. T. I. Oprea, J. Gottfries, *J. Comb. Chem.* **2001**, *3*, 157–166.
39. S. Fergus, A. Bender, D. R. Spring, *Curr. Opin. Chem. Biol.* **2005**, *9*, 304–309.
40. S. J. Haggarty, *Curr. Opin. Chem. Biol.* **2005**, *9*, 296–303.
41. T. I. Oprea, *Curr. Opin. Chem. Biol.* **2002**, *6*, 384–389.
42. T. E. Nielsen, S. L. Schreiber, *Angew. Chem. Int. Ed.* **2008**, *47*, 48–56.
43. T. Luo, S. L. Schreiber, *J. Am. Chem. Soc.* **2009**, *131*, 5667–5674.
44. T. Uchida, M. Rodriquez, S. L. Schreiber, *Org. Lett.* **2009**, *11*, 1559–1562.
45. A. Zhou, D. Rayabarapu, P. R. Hanson, *Org. Lett.* **2009**, *11*, 531–534.
46. E. E. Wyatt, S. Fergus, W. R. J. D. Galloway, A. Bender, D. J. Fox, A. T. Plowright, A. S. Jessiman, M. Welch, D. R. Spring, *Chem. Commun.* **2006**, 3296–3298.
47. G. L. Thomas, R. J. Spandl, F. G. Glansdorp, M. Welch, A. Bender, J. Cockfield, J. A. Lindsay, C. Bryant, D. F. Brown, O. Loiseleur, H. Rudyk, M. Ladlow, D. R. Spring, *Angew. Chem. Int. Ed.* **2008**, *47*, 2808–2812.
48. E. E. Wyatt, W. R. J. D. Galloway, G. L. Thomas, M. Welch, O. Loiseleur, A. T. Plowright, D. R. Spring, *Chem. Commun.* **2008**, 4962–4964.
49. P. C. Wyss, P. Gerber, P. G. Hartman, C. Hubschwerlen, H. Locher, H.-P. Marty, M. Stahl, *J. Med. Chem.* **2003**, *46*, 2304–2312.
50. P. Schneider, S. Hawser, K. Islam, *Bioorg. Med. Chem. Lett.* **2003**, *13*, 4217–4221.

51. A. Robinson, G. L. Thomas, R. J. Spandl, M. Welch, D. R. Spring, *Org. Biomol. Chem.* **2008**, *6*, 2978–2981.
52. D. Pizzirani, T. Kaya, P. A. Clemons, S. L. Schreiber, *Org. Lett.* **2010**, *12*, 2822–2825.
53. F. D'Aniell, A. Mann, M. Taddei, *J. Org. Chem.* **1996**, *61*, 4870–4871.
54. E. Comer, E. Rohan, L. Deng, J. A. Porco, Jr., *Org. Lett.* **2007**, *9*, 2123–2126.
55. N. Kumagai, G. Muncipinto, S. L. Schreiber, *Angew. Chem. Int. Ed.* **2006**, *45*, 3635–3638.
56. D. Robbins, A. F. Newton, C. Gignoux, J.-C. Legeay, A. Sinclair, M. Rejzek, C. A. Laxon, S. K. Yalamanchili, W. Lewis, M. A. O'Connell, R. A. Stockman, *Chem. Sci.* **2011**, *2*, 2232–2235.
57. D. Morton, S. Leach, C. Cordier, S. Warriner, A. Nelson, *Angew. Chem. Int. Ed.* **2009**, *48*, 104–109.
58. A. W. Hung, A. Ramek, Y. Wang, T. Kaya, J. A. Wilson, P. A. Clemons, D. W. Young, *Proc. Natl. Acad. Sci. U.S.A.* **2011**, *108*, 6799–6804.
59. M. Congreve, R. Carr, C. Murray, H. Jhoti, *Drug Discov. Today* **2003**, *8*, 876–877.
60. D. Seebach, M. Boes, R. Naef, W. B. Schweizer, *J. Am. Chem. Soc.* **1983**, *105*, 5390–5398.
61. A. Isidro-Llobet, T. Murillo, P. Bello, A. Cilibrizzi, J. T. Hodgkinson, W. R. J. D. Galloway, A. Bender, M. Welch, D. R. Spring, *Proc. Natl. Acad. Sci. U.S.A.* **2011**, *108*, 6793–6798.
62. A. R. Kelly, J. Wei, S. Kesavan, J.-C. Marié, N. Windmon, D. W. Young, L. A. Marcaurelle, *Org. Lett.* **2009**, *11*, 2257–2260.

PART I

CHEMICAL METHODOLOGY IN DIVERSITY-ORIENTED SYNTHESIS

2

STRATEGIC APPLICATIONS OF MULTICOMPONENT REACTIONS IN DIVERSITY-ORIENTED SYNTHESIS

JOHN M. KNAPP, MARK J. KURTH, JARED T. SHAW, AND ASHKAAN YOUNAI

2.1 INTRODUCTION

Multicomponent reactions (MCRs), chemical processes in which three or more reagents react selectively to form a single product, are extremely efficient transformations in organic synthesis. Although this concise definition will be adopted for the purposes of this chapter, the reader is advised that the term *MCR* is not subject to strict (e.g., mechanistic) delineation in the manner of reactions that are classified as catalytic or concerted. In the strictest sense, an MCR involves the admixture of at least three chemicals without regard to precise control of the order of addition and with the further restriction that each component be variable. As such, a three-component reaction (3CR) should involve three reagents, each of which can be structurally varied. Examples of commonly occurring variable components include primary and secondary amines, ketones, aldehydes, carboxylic acids, and isonitriles. In the most liberal definition, any series of reactions involving at least three reagents in one reaction vessel that contributes to the final product structure might be referred to as an MCR. This would include many tandem, "one-pot," and cascade processes, some of which amount to two separate reactions performed without isolation of an intermediate. In this chapter we discuss MCRs that meet the strictest definition and

Diversity-Oriented Synthesis: Basics and Applications in Organic Synthesis, Drug Discovery, and Chemical Biology, First Edition. Edited by Andrea Trabocchi.

several related processes that are included in the more liberal definition and still exemplify the high level of efficiency that is sought by the general strategy of combining multiple reactive chemicals to produce a single complex product.

We also provide instructive examples of how MCRs have affected diversity-oriented synthesis (DOS). The ensuing description of selected reactions and sequences is not intended to be an exhaustive discussion of any individual MCR or an exhaustive list of every instance in which an MCR has been employed in DOS, library synthesis, hit-to-lead studies, assessment of structure–activity relationships (SAR), and all of the activities that accompany the use of libraries in high-throughput screening. Instead, two general areas are explored: (1) MCR products as direct library inputs, and (2) MCR products as starting points for subsequent reactions leading to structurally diverse products. The goal is to delineate recurrent themes in the use of MCRs in DOS with a specific emphasis on synthetic strategy. This chapter should set the stage for the design of new strategies, tactics, pathways, and perhaps even new MCRs that will continue to enable this rich sector of organic synthesis to be a vital part of drug discovery and chemical biology.

Unlike DOS or its progenitor combinatorial chemistry, MCRs have been a part of organic chemistry for over a century [1]. Prototypical examples include Hantsch (1882), Biginelli (1893), and Passerini (1921) reactions. Unfortunately, these reactions, and other MCRs discovered though the early 1990s, would remain relatively obscure and their discoverers would never know of the huge impact of their discoveries. The seminal publications of Houghten et al. [2] and Furka et al. [3] on solid-phase split-and-pool synthesis (SPS) paired with a nearly simultaneous and symbiotic rise in high-throughput screening (HTS) ushered in a wave of effort directed at library synthesis. SPS, of peptides initially and of myriad small molecules later, offered an opportunity for the efficient preparation of thousands, and in some cases millions (hypothetically), of compounds for screening by multistep synthesis. Ivar Ugi, who discovered the four-component reaction (4CR) that bears his name, as well as many mechanistically related reactions, recognized that MCRs offered access to small-molecule libraries by a complementary strategy [4]. SPS relies on a *multistep* sequence of reactions that usually introduce a single point of diversity in each step, assuming no activation, deprotection, or redox transformations and neglecting steps for loading of starting materials and cleavage of products. An MCR, on the other hand, could create a high level of diversity by introducing three or more points of diversity in a *single step!* As such, the broad interest in MCRs can be traced to the early 1990s, shortly after the discovery of SPS and the ensuing groundswell of effort directed at library synthesis. Interest in SPS has since leveled off and declined, due to the many challenges associated with this strategy, including the necessarily small scale of SPS reactions, tracking the identity of individual compounds, the need for highly specialized equipment, adaptation of solution-phase reactions to solid-phase substrates, difficulty in assessing the progress or yield of individual reaction steps, and highly variable final yields and purities of SPS products. Conversely, interest in MCRs has continued to grow, offering new vistas for target-directed synthesis, mechanistic studies, catalysis, and the ever-increasing demands for the preparation of libraries for HTS in both academia and industry.

2.2 MCR PRODUCTS FOR HTS

2.2.1 MCRs and HTS: The Real and the Virtual

The most direct application of MCRs in DOS is to use MCR products directly for HTS. This strategy carries the obvious advantage of executing a one-step synthesis that is amenable to automation with simple liquid handling and subsequent purification. Other immediate advantages include short time frames for scale-up and SAR studies. Liabilities include structural diversity that is limited to variation in the appendages of a single core structure as well as products that are usually racemic. MCRs are, on the whole, difficult to render asymmetric by either auxiliary or catalyst control. Racemic molecules are often viewed as an advantage for HTS in that they double the chances for discovering a high-affinity ligand or otherwise biologically active small molecule in an HTS experiment. That said, the discovery of activity immediately requires the identification of the active enantiomer, which has required the use of preparative chiral chromatography.

Many important discoveries in HTS are MCR products that are used directly or with minimal modification in HTS. Although it is difficult to quantify, early screening libraries from commercial vendors were based largely on the availability of small molecules, to which single-step synthetic compounds, including MCR products, contributed greatly. One prominent example is found in monastrol (**1**), a dihydropyrimidinone that is formed by the Biginelli reaction of urea, *m*-hydroxybenzaldehyde, and ethyl acetoacetate. This compound was found to induce a "monoastral" phenotype in a microscopy-based phenotypic screen at Harvard Medical School's Institute for Chemistry and Cell Biology (ICCB) using a library of compounds acquired from commercial sources [5]. After this initial discovery, the active enantiomer was discerned [6], a co-crystal structure with the protein target (Eg5/MSK) revealed that monsatrol bound in an unfavorable conformation [7], and a cyclic product that mimicked this state was revealed to have superior activity [8]. More recently, Priaxon, an MCR-focused company engaged in drug discovery, identified compound **5** as one of several inhibitors of the proteins HDM2 and p53 [9]. Compound **5** is formed by a Shaw 4CR, followed directly by amide coupling. Priaxon has a stated focus specifically on MCRs for the advantages that are conferred (i.e., ease of library synthesis as well as subsequent follow-up studies) on the assumption that discovery time lines will necessarily be shortened. Furthermore, their current repertoire of MCRs enables access to (theoretically) 1×10^9 compounds, assuming reasonable reaction efficiency for all available components [10]. Clearly, this number will grow with the discovery of each new MCR. In addition to these two examples, several drugs (approved or in development) have been synthesized by MCRs (Figure 2.1), including the antihelminthic drug praziquantel [11] (**11**), GlaxoSmithKline's (GSK's) oxytocin antagonist [12] (**12**), the antiplatelet agent clopidogrel [13] (**13**), and the antiandrogen bicalutamide [14] (**14**).

The basic idea of accelerating discovery by focusing on MCRs in HTS has recently been applied prospectively by linking docking or "virtual" screening to synthesis. The use of computation to guide discovery ligands for protein targets *in silico* has

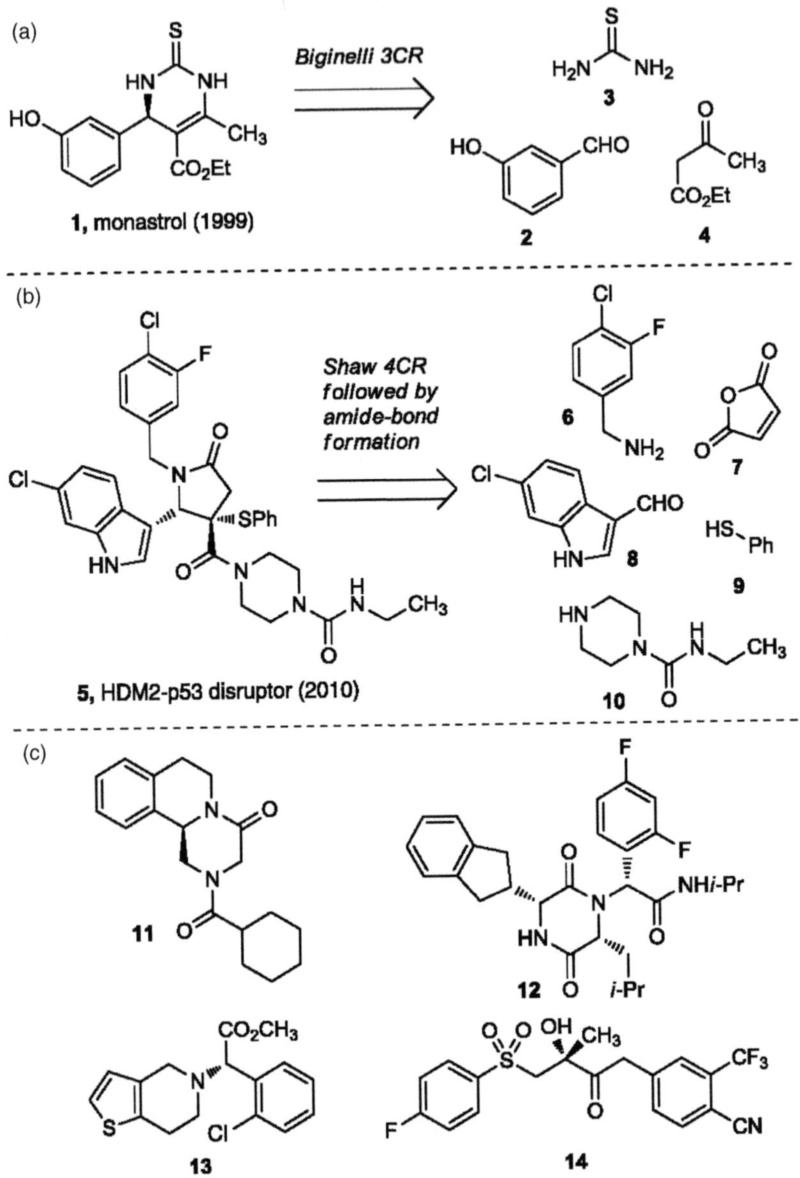

FIGURE 2.1 (a) Mitotic spindle kinesin (Eg5) inhibitor monastrol, made by the Biginelli 3CR. (b) Priaxon's MDM2-p53 inhibitor, made by the Shaw 4CR and amide coupling. (c) Praziquantel (**11**, Ugi 4CR), GSK's oxytocin antagonist (**12**, Ugi 4CR), clopidogrel (**13**, Ugi 3CR and Petasis 3CR), and bicalutamide (**14**, Passerini 3CR).

gradually been gaining importance in HTS and, as a consequence, is influenced by DOS strategies. The ZINC database is a freely available collection of over 1 million commercially available small molecules that can be applied to docking experiments [15]. A more focused use of MCRs is found in the ANCHOR database [16], which was created by Alexander Dömling, who received doctoral training from Ugi. ANCHOR is a database that is populated with MCR products that are "enumerated," or synthesized *in silico* from commercially available components. The majority of the compounds have never been synthesized but all can, in theory, be prepared in a one-step synthesis from one of 22 different MCRs (Figure 2.2). Although the collection can probably be used in many different docking experiments, it is constructed to use key peptide side chains (e.g., the phenyl, indole, or isopropyl groups of Phe, Trp, or Ile/Val, respectively) in protein–protein reactions as "anchors" or fixed appendages that are used to identify small molecules that mimic one side of a protein–protein interface. The software ranks MCR products in the database based on their ability to properly position the anchored appendage. Users can then synthesize individual compounds or small collections of small molecules using one or more MCRs for evaluation in their assay of choice. Although similar functionality is possible with any chemical reaction, MCRs provide ANCHOR with the unique opportunity to populate the database with an enormous quantity of compounds that are, conceivably, extremely easy to synthesize.

2.2.2 Expanding Accessible Diversity: New MCRs

One of the most versatile compounds seen in a wide array of MCRs is the isonitrile, with two of the best-known MCRs, the Passerini and Ugi, featuring these components [17]. The unique ability of isocyanides to serve as C-nucleophiles toward imines and aldehydes, resulting in nitrilium ion electrophiles, enables the Ugi 4CR and Passerini 3CR, respectively. This topic has been reviewed extensively, and a great number of recently disclosed MCRs incorporate this unique reactivity to access a diverse array of heterocyclic structures [18]. Recent examples have expanded the core diversity available from MCRs and pushed the limits of complexity that are attainable in a single transformation. The work of Santra and Andreana in the area of novel isocyanide-based MCRs is noteworthy, due to their lab's ability to access diverse scaffolds from the same components. This is achieved through the use of bifunctional components that participate in a Michael-type cyclization step directly following a Ugi reaction [19]. The Ugi cascades are also unique due to the use of microwave irradiation, the absence of any additives, and the use of water as the solvent (Scheme 2.1). More recently, Santra and Andreana have expanded this work to the preparation of natural product–like [20] fused azaspiro-tricycles by a Ugi/Michael/aza-Michael (UMAM) cascade reaction [21]. This particular cascade process generates a quaternary center, four stereogenic centers, and six contiguous bonds, and provides moderate to high yields, excellent regioselectivities, and significant diastereoselectivity (Scheme 2.2).

In addition to the isocyanide-based MCRs of Andreana's lab, several others have been reported in recent years. In one such example, functionalized spirobenzofuranones were prepared from 3-cyanochromones, acetylenecarboxylates, and

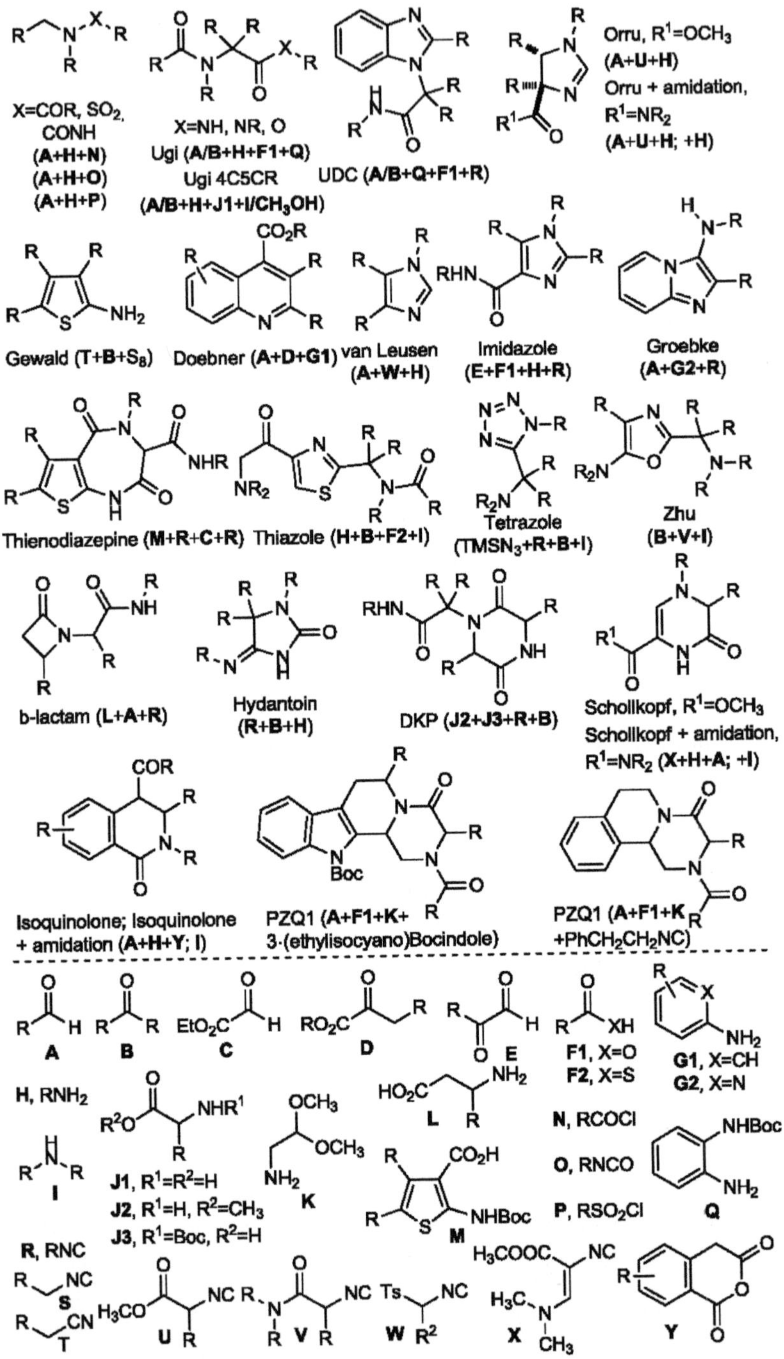

FIGURE 2.2 MCR products used to populate the ANCHOR database and the building blocks from which they are composed.

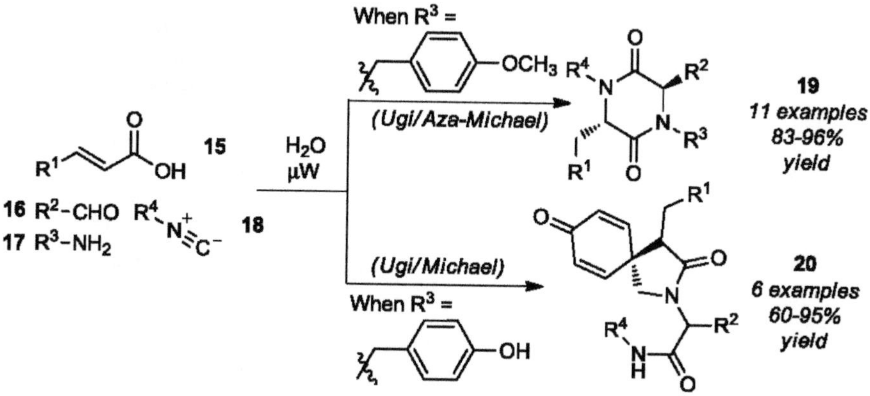

SCHEME 2.1 Andreana's use of bifunctional substrates in an Ugi reaction leading to a subsequent Michael-type cyclization.

isocyanides (Scheme 2.3) [22]. It is worth mentioning that spirocyclic systems are found in a wide variety of natural products; thus, a number of synthetic methods have been devised for the synthesis of this scaffold [23]. Spirobenzofurans in particular have garnered attention due to their biological properties, a notable example being the antifungal drug griseofulvin (**28**), which has been reported to have antiproliferative effects in cancer cells [24].

A Mumm rearrangement is the last step in the Ugi reaction, and Basso et al. [25] have been able to interrupt (i.e., precluding the Mumm rearrangement) the Ugi reaction by restricting the amine and acid components within a single molecule. As a result, the Mumm rearrangement cannot proceed because of strain in the potential intermediate. The opportunity to use the interrupted Ugi product as an acylating agent is therefore accessed in this reaction (Scheme 2.4). These reactions are run in methanol; thus, the acyl product formed is a methyl ester.

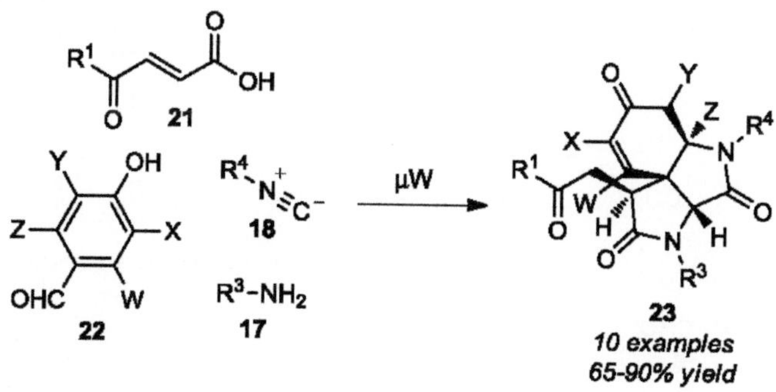

SCHEME 2.2 Andreana's use of bifunctional substrates in a Ugi/Michael/aza-Michael cascade reaction leading to the formation of a complex fused azaspiro-tricycle.

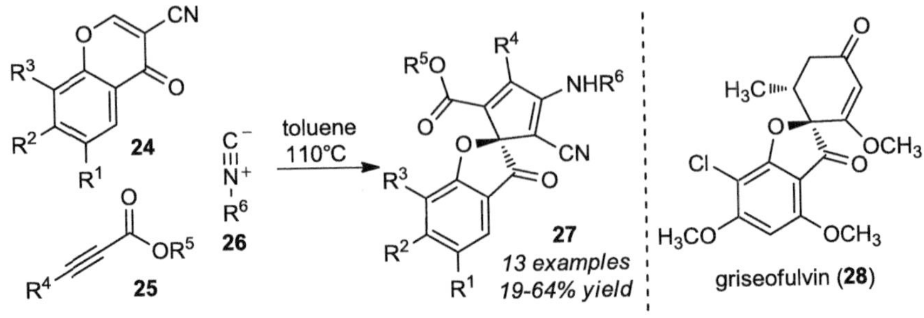

SCHEME 2.3 Formation of spirobenzofuranones via an isocyanide-based 3CR, a scaffold seen in the antifungal drug griseofulvin (**28**).

One of the more interesting electrophiles seen in MCRs with isonitriles are arynes, which, when allowed to react, generate a highly reactive 1,3-zwitterionic intermediate that can lead to a wide variety of structures, depending on the third component present in the reaction [26]. Stoltz's group discovered a novel 3CR of arynes, isocyanides, and phenyl esters to provide phenoxy-iminoisobenzofurans in good yield [27]. The arynes were generated in situ from 2-(trimethylsilyl)aryl triflates [28] using a fluoride source, and the formation of the unusual heterocyclic scaffold took place under mild conditions. Interestingly, when the ester component of this MCR was replaced by an electron-deficient alkyne (internal or terminal), the reaction provided carbocyclic iminoindenones in good yield. Worthy of note is the selective addition of isocyanides to arynes in the presence of electrophilic alkynes (Scheme 2.5).

Another interesting example of an MCR involving arynes and isocyanides was reported by Sha and Huang [29], wherein polysubstituted pyridines and isoquinolines were obtained with excellent selectivity using terminal alkynes as the third component (Scheme 2.6). The selectivity observed between these two heterocycles was dependent on the reaction conditions, as pyridines were formed with an excess of terminal alkynes, and isoquinolines were formed with an excess of arynes. This MCR serves as a mild preparation of these heterocycles, which often involve harsh reaction conditions or complicated procedures [30].

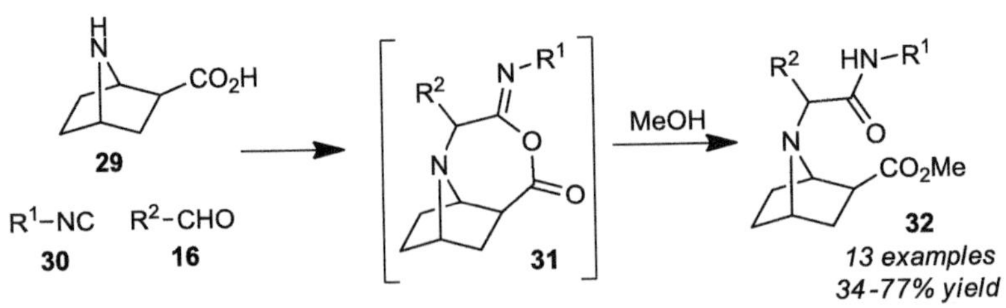

SCHEME 2.4 Interrupted Ugi-acylation reaction.

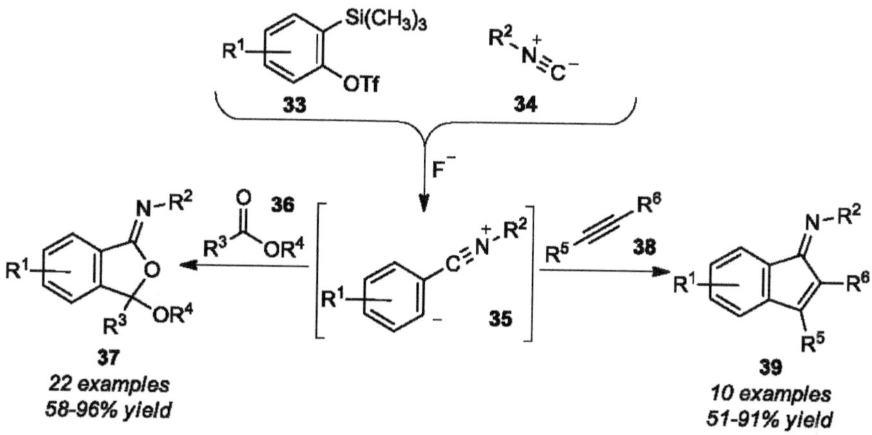

SCHEME 2.5 MCR reported by Stoltz's group [27] involving arynes and isocyanides leading to the formation of phenoxyiminoisobenzofurans or iminoindenones.

Although many of the recently disclosed MCRs have been inspired by the Ugi and Passerini reactions, the isonitrile components of these MCRs are often difficult to work with, due to their highly unpleasant odor. Several novel isonitrile-free MCRs have been developed in recent years that use other highly reactive components, such as amines and highly enolizable carbonyl compounds. A recent report by Wei and Shaw of a 4CR between aldehydes, amines, thiols, and maleic anhydrides to produce γ-lactams is one example [31]. Both tetra- and penta-substituted lactams were obtained in high yield and diastereoselectivity, with water as the only by-product (Scheme 2.7).

In another recent example, benzaldehydes, malononitrile (or ethyl cyanoacetate), 6-methyl-2,3,4,9-tetrahydro-1*H*-carbazol-1-one, and ammonium acetate react in the presence of L-proline to form pyridocarbazoles [32]. The carbazole heterocycle is a scaffold found in many biologically active compounds of diverse origins, and chemists have been interested in these structures because of their potential applications as pharmaceutical agents [33]. Although the benzaldehyde is the only component that

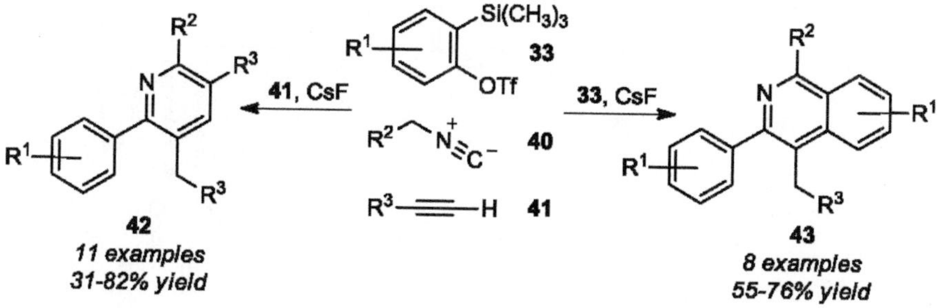

SCHEME 2.6 MCR reported by Sha and Huang [29] involving arynes and isocyanides, leading to the formation of polysubstituted pyridines and isoquinolines.

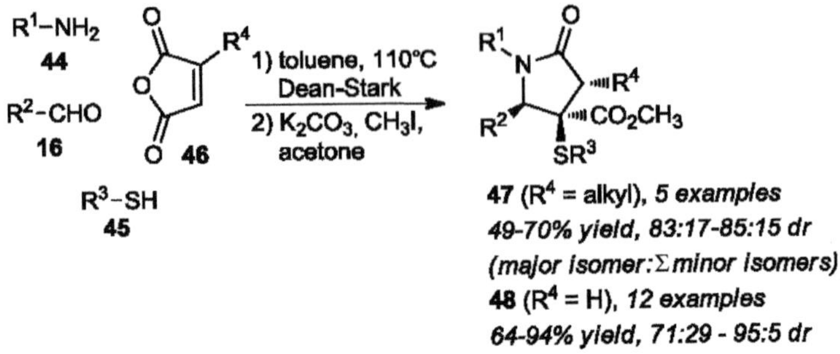

SCHEME 2.7 4CR reported by Wei and Shaw [31], leading to the formation of γ-lactams.

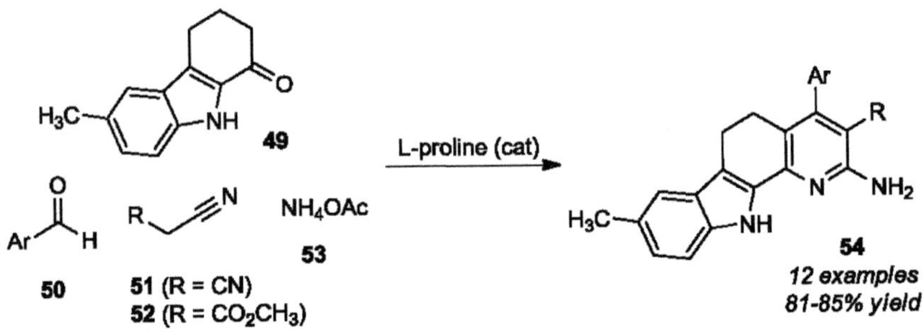

SCHEME 2.8 4CR catalyzed by L-proline for the formation of pyridocarbazoles.

is highly variable in this particular reaction, all the carbazole products were prepared in high yield (Scheme 2.8), and a structure–activity relationship was established for the antimicrobial, antioxidant, and cytotoxic properties of this scaffold.

Knapp and Kurth have also developed a one-pot 2·3CR that effectively assembles in one pot isoxazolines from four components (Scheme 2.9) [34]. Here a two-component reaction between an acyl ketene and amine forms a β-ketoamide. In this

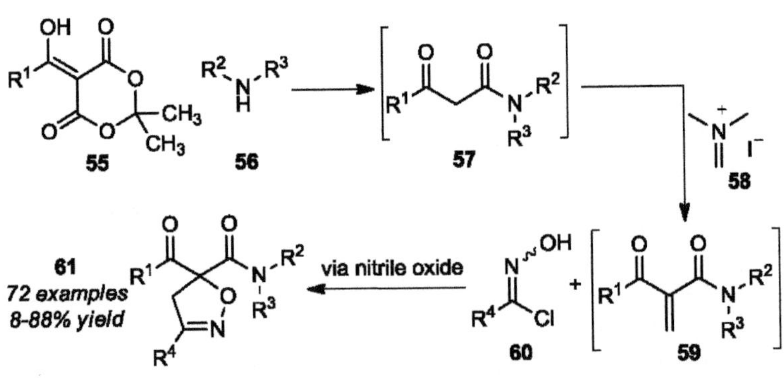

SCHEME 2.9 2·3 Component synthesis of β-ketoamide isoxazolines.

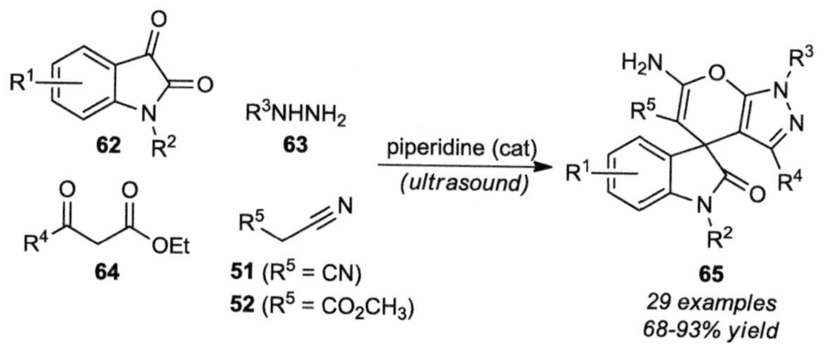

SCHEME 2.10 4CR catalyzed by piperidine for the synthesis of complex spirooxindole-fused heterocycles.

first process, an acyl-Meldrum's acid (**55**) cycloreverts to an acyl ketene upon heating with loss of acetone and CO_2, which is trapped by the amine. Subsequent addition of Eschenmoser's salt and a chlorooxime to the flask initiates a three-component reaction yielding a four-component assembly product, a β-ketoamide isoxazoline. In the 3CR reaction, Eshenmoser's salt undergoes a Mannich reaction, and the elimination of dimethyl ammonium iodide gives a α,β-unsaturated β-ketoamide. This dipolarophile then participates in a nitrile oxide (formed in situ from chlorooxime **60**) 1,3-dipolar cycloaddition.

Another recent example of an "isonitrile-free" MCR involved reaction of hydrazine, β-keto ester, isatin, and malononitrile (or ethyl cyanoacetate) under ultrasound irradiation to form spirooxindole-fused heterocycles [35]. The spirooxindole system is at the core of many pharmacological agents and natural products; thus, a number of methods have been reported for the preparation of spirooxindole-fused heterocycles [36–38]. This particular MCR is noteworthy due to the ease of purification, as simply washing the crude solid products with ethanol avoids the need for chromatography or recrystallization (Scheme 2.10).

Recently, an efficient enantioselective inverse electron-demand aza-Diels–Alder (IEDDA) reaction with isoeugenol derivatives as the dienophile catalyzed by chiral phosphoric acid (**68**) has been developed by He et al. [39]. This cycloaddition proceeds with a wide range of aldehydes and anilines, providing a highly diastereo- and enantioselective method to 2,3,4-trisubstituted 4-aryl-tetrahydroquinolines (Scheme 2.11).

2.3 MCRs AS STARTING POINTS FOR DOS

MCRs can serve as key transformations in versatile approaches to DOS. One common strategy involves the application of inputs having functional groups that are orthogonal to the MCR but will participate in subsequent bond-forming events. In early examples, these processes were often spontaneous under MCR conditions or induced thermally by warming the reaction mixture. In later DOS approaches, functional groups that

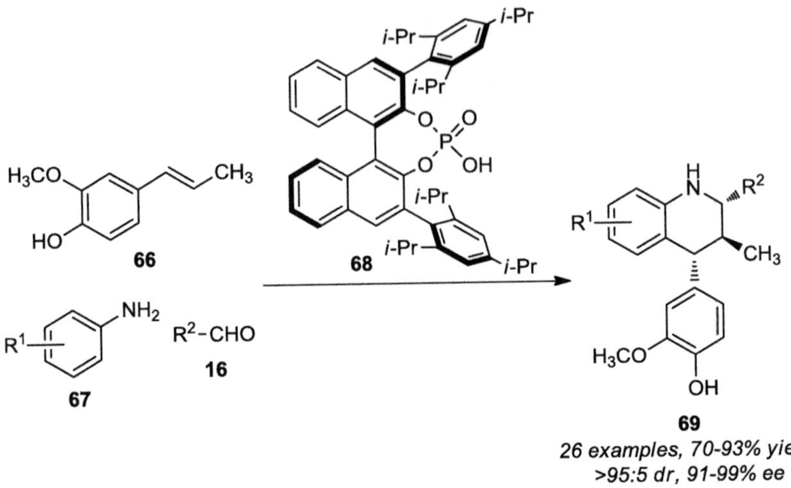

SCHEME 2.11 3CR catalyzed by chiral phosphoric acid leading to the formation of trisubstituted tetrahydroquinolines.

would undergo a variety of metal-mediated reactions were installed for a subsequent metal-catalyzed reaction. The latter strategy was termed *build/couple/pair*: *build* a core structure, use *coupling* reactions to attach reactive appendages, and finally, *pair* the various functional groups with a catalyst.

Bararjanian et al. [40] have also made oxindoles by a Ugi/Heck/Sonogashira/nucleophilic addition protocol. In this reaction, the amine and halide are positioned ortho to each other on the aromatic ring. A Ugi reaction forms the intramolecular Heck precursor, and the palladium catalytic cycle terminates with a Sonogashira reaction. Conjugate addition to the resulting oxindole by a secondary amine yields a single stereoisomer (Scheme 2.12).

Bonnaterre et al. [41] have developed a tandem Ugi 4CR followed by Buchwald–Hartwig amidation in their synthesis of oxindoles. *o*-Iodobenzaldehydes were used as the aldehyde portion of the Ugi 4CR. The resulting amide nitrogen atom contained in the isonitrile is strategically positioned to form a five-membered ring after Ugi 4CR by an intramolecular Buchwald–Hartwig reaction (Scheme 2.13). Attempts at copper-catalyzed cyclization failed or gave low yields compared to the palladium-catalyzed variant. Thirteen examples were reported in this communication.

The earliest applications of MCRs as starting points in DOS involved Diels–Alder reactions that result from components containing dienes and dienophiles. Paulvannan first demonstrated this approach by combining fumaric acid monoethyl ester **82** and heterocyclic diene-containing aldehydes with various amines and aldehydes (Scheme 2.14a). Under the reaction conditions, the Ugi product was not isolated and underwent a spontaneous Diels–Alder reaction to provide oxabicyclic products **86** in good yield and with high diastereoselectivities. When the dienophile is an alkyne, the expected cyclohexadiene products oxidize under the reaction conditions,

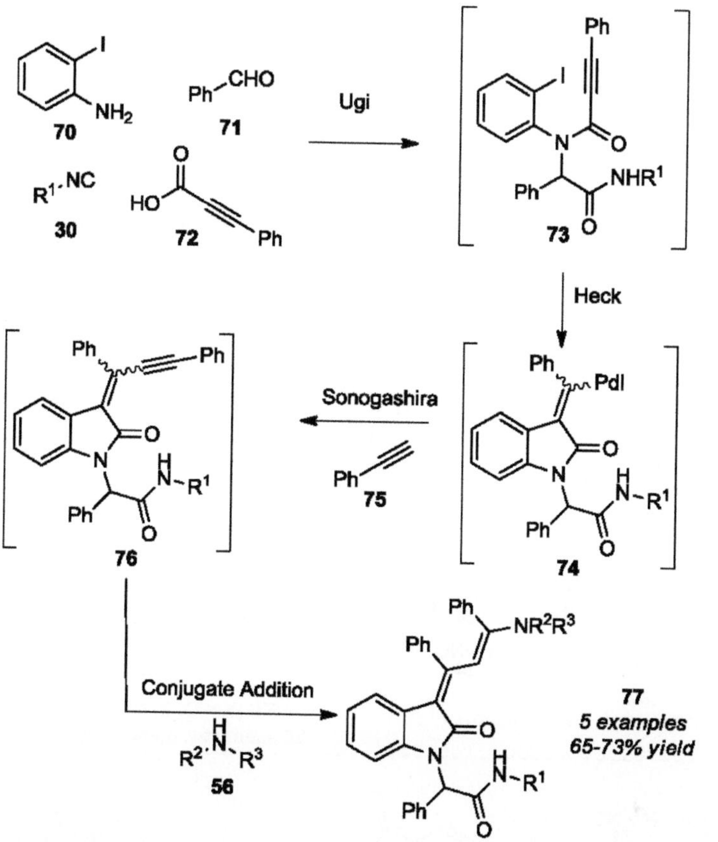

SCHEME 2.12 Ugi/Heck/Sonagashira/conjugate addition sequence.

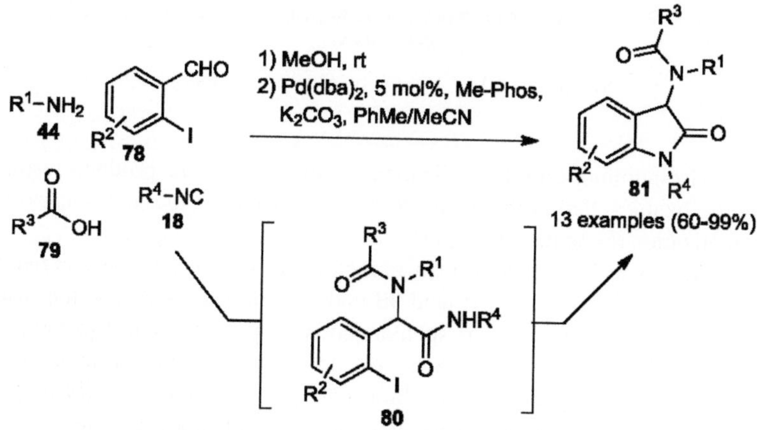

SCHEME 2.13 Ugi organometallic amide arylation.

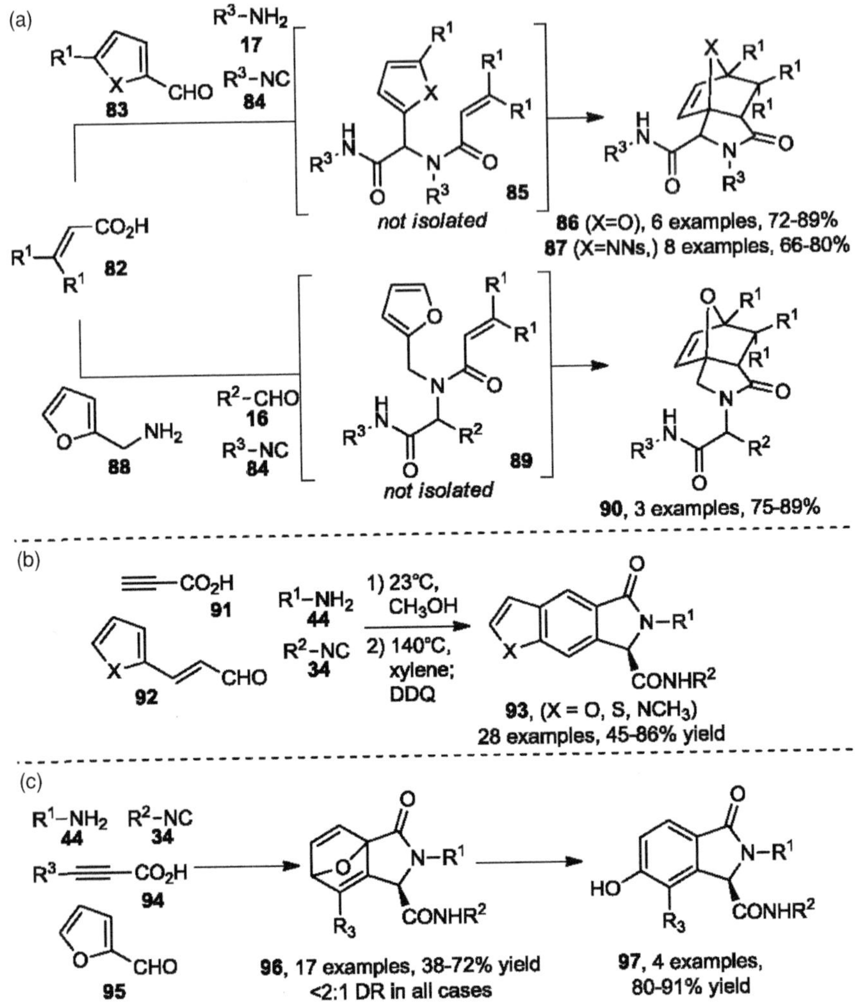

SCHEME 2.14 Ugi Diels–Alder methodology.

resulting in isoindolinone products (Scheme 2.14b) [42]. Similar products result from treating the oxabicyclic intermediates **96** with a Lewis acid to induce ring opening and aromatization (Scheme 2.14c) [43].

Heterocycles may be formed from inverse electron-demand Diels–Alder reactions. Propargyl amine and heterocyclic acid **98** undergo the Ugi 4CR, and the resulting bicyclic intermediate (not shown) spontaneously loses molecular nitrogen to produce substituted pyridine products (Scheme 2.15a) [44]. Sun et al. observed a related process after a 3CR in which the nitrilium ion intermediate of the Ugi reaction is trapped to form oxazole **106** [45]. This intermediate could be acylated with an unsaturated acid chloride to form substrates for an intramolecular Diels–Alder reaction that

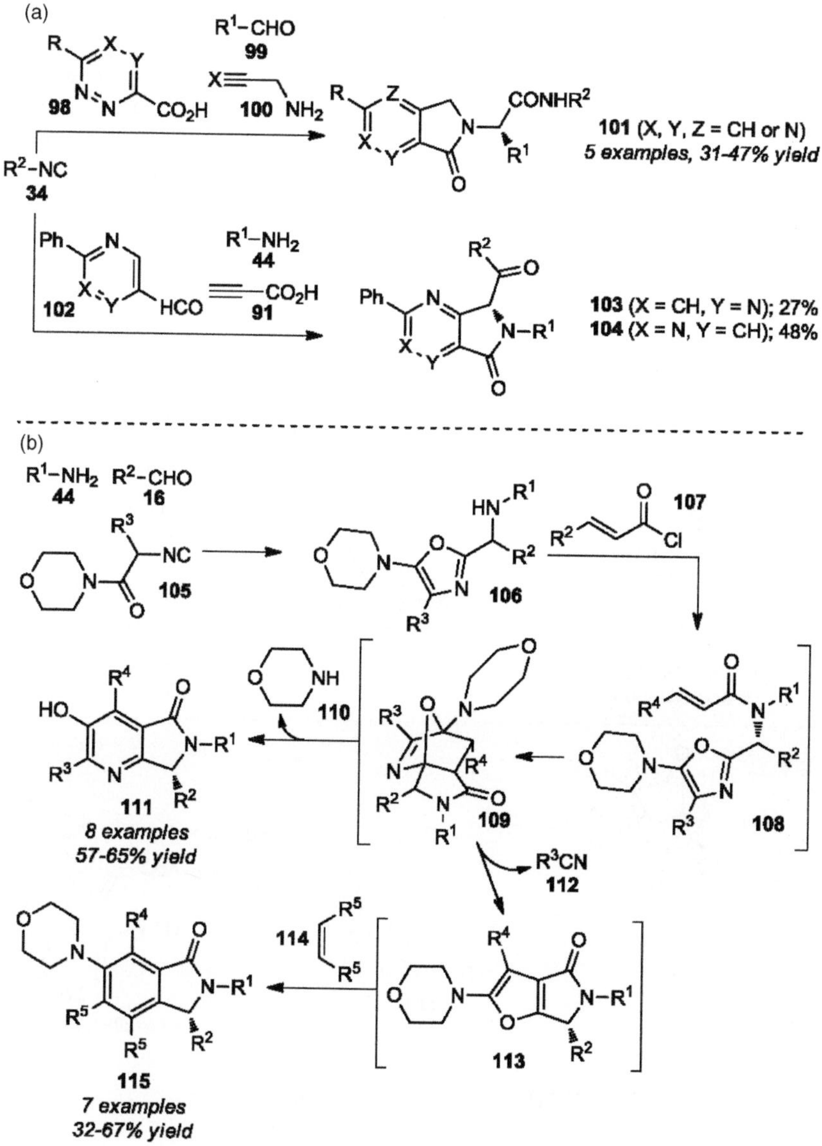

SCHEME 2.15 Ugi Diels–Alder rearomatization.

eventually eliminates morpholine to give densely substituted pyridine products (**111**, Scheme 2.15b) [44]. A subsequent report demonstrated that oxazole **106** could be treated with an acid chloride in the presence of a symmetric dienophile to form hexasubstituted benzene products after heating. This one-pot process proceeds via furan **113**, which is generated from the elimination of a nitrile to form bicyclic intermediate **115** [46].

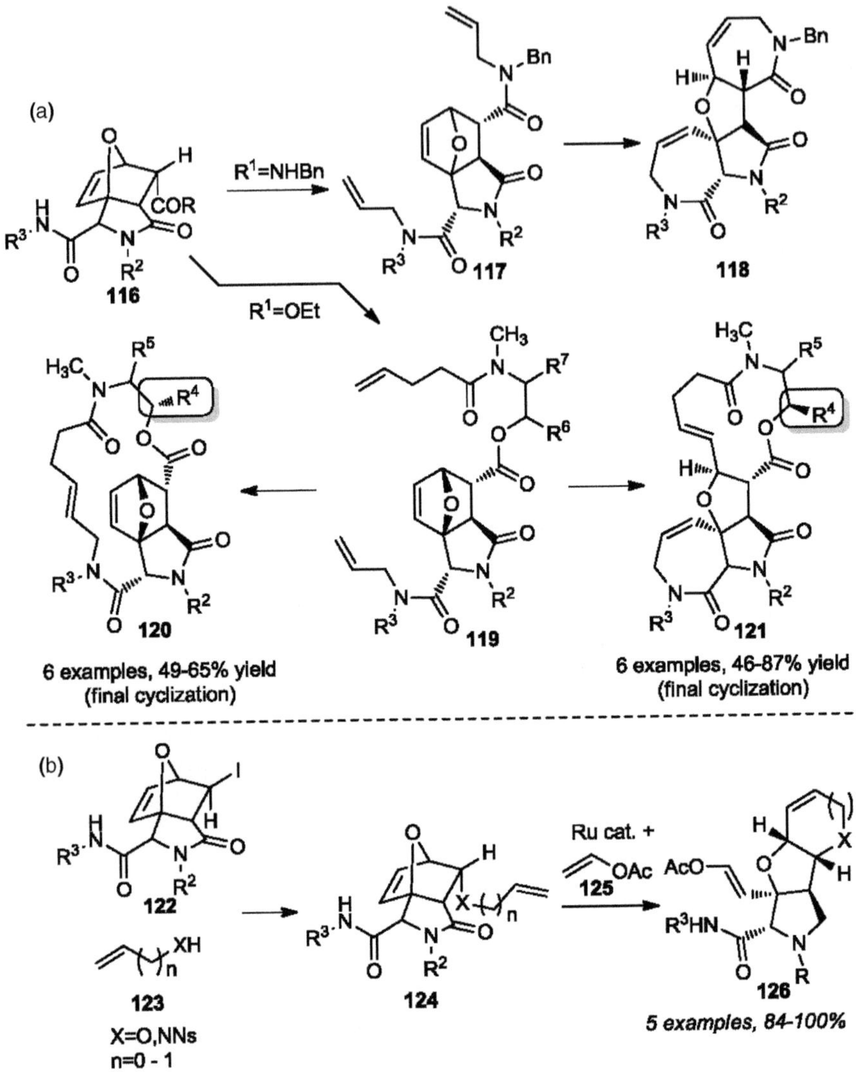

SCHEME 2.16 Ugi metathesis methodology.

Lee et al. have demonstrated that the products of the Ugi–Diels–Alder process are excellent staring points for the preparation of products that rival the complexity of natural products. Intermediates from the process developed by Paulvannan (i.e., when the acid is fumaric acid monobenzylamide) can be double-allylated with base and allyl bromide (Scheme 2.16). Treatment of **117** with Grubbs's metathesis catalyst results in a ring-closing/ring-opening/ring-closing metathesis cascade that forms two new seven-membered rings [47]. In a subsequent example, exquisite control of metathesis regiochemistry could be exerted, based on the diastereomeric configuration of the starting material. Specifically, the dienophile ester could be cleaved and appended with an acylated amino alcohol to intermediate **119**, depending on the stereochemistry of the amino alcohol [48]. In most cases, the racemic core structure was coupled with an enantiomerically pure amino alcohol, and the resulting diastereomers were separated. The stereochemistry at R^4 completely dictated whether the metathesis reaction proceeded similarly to **121** (i.e., by formation of two new rings or following an alternative macrocyclization pathway to give **120**). In related studies, Ikoma et al. showed that the two amide termini could be allylated selectively by tuning the N-substituents (i.e., by replacing R^3 and Bn of **116** with various substituents) [49]. Later studies used (*Z*)-2-iodoacrylic acid in the initial Ugi reaction to produce **122**, which was used to alkylate allyl and butenyl amides and alcohols. These intermediates underwent high-yielding ring-opening-, ring-closing-, and cross metathesis when treated with Grubbs's catalyst and vinyl acetate [50,51]. Although the final products are formed in short sequences from the initial 4CR, final products **120**, **121**, and **126** barely resemble the Ugi products from which they emanate.

Propargyl/allyl amines have found applications in Ugi reactions [52]. By running a Ugi reaction with nitrocarboxylic acid **128**, the product is situated conveniently to form either a six- or a seven-membered ring by an intramolecular nitrile oxide cyclization (INOC) reaction (Scheme 2.17). Treatment of the alkyl nitro group with phosphorus oxychloride forms the dipole, which proceeds to do a cycloaddition on the dipolarophile from the amine-containing component **127**.

The Ugi reaction has been combined with the click reaction to easily form a variety of diverse triazoles. Compounds **131–135** were prepared by an initial Ugi reaction in methanol at room temperature for 48 h (Figure 2.3). The Ugi product containing an azide and alkyne can then form the triazole by a copper-free intramolecular click reaction. The cycloaddition was performed in benzene at reflux. Seven examples were presented in this paper by Akritopoulu-Zanze et al. [53].

The combination of multiple multicomponent reactions can exponentially increase product diversity. Orru's group [54,55] has demonstrated one-pot procedures of these tandem MCRs, as shown in Scheme 2.18. Imidazolines **139** can be formed by a Van Leusen type reaction [56]. An acetate salt can be carried unaffected through this procedure and can then participate in the following three-component Ugi reaction. A bis-isonitrile (**140**) was utilized in this four-component reaction, and the second isonitrile participated in a series ending Ugi reaction. This unique transformation can

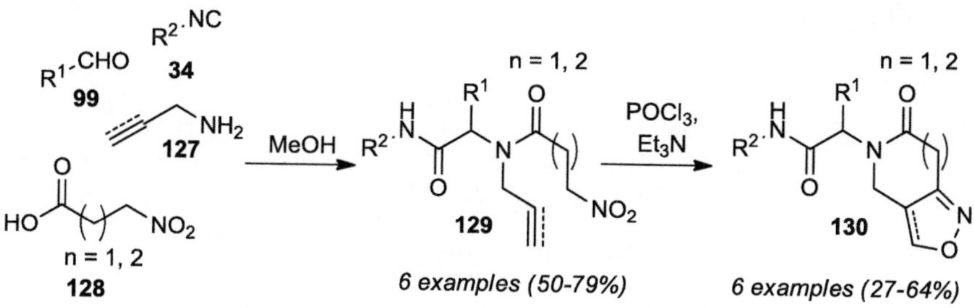

SCHEME 2.17 Ugi-dipolar cycloaddition.

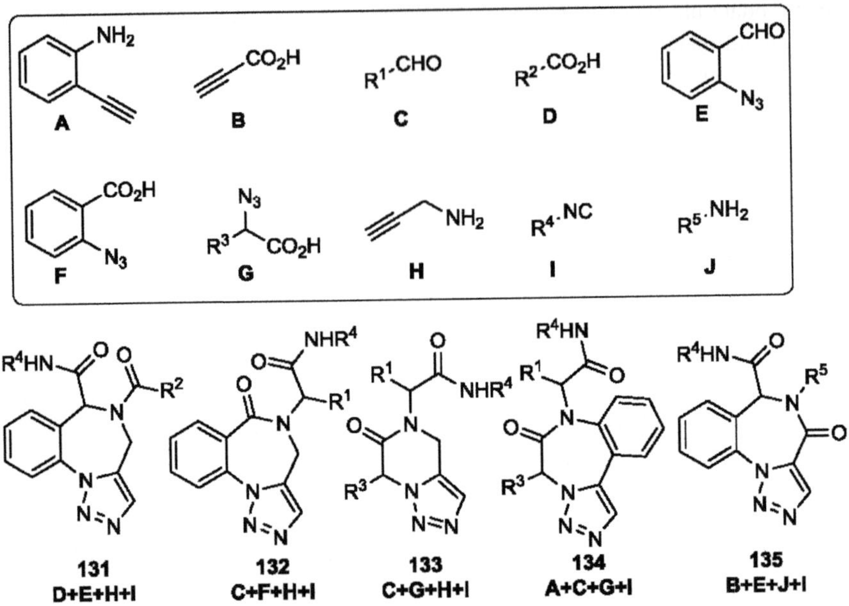

7 examples, (33-81%) Ugi Reaction, (86-98%) Dipolar Cycloaddition

FIGURE 2.3 Ugi-click reaction.

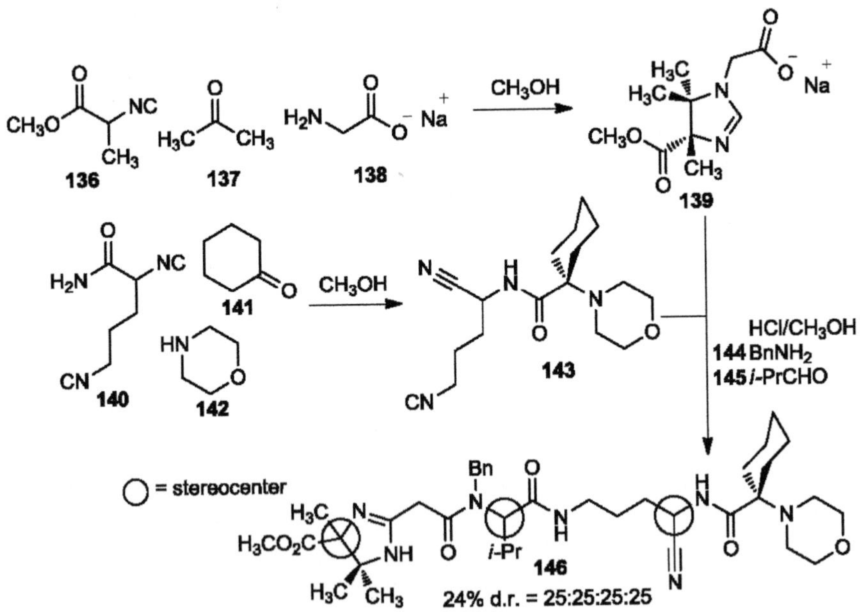

SCHEME 2.18 Orru one-pot eight-component reaction.

accommodate up to eight components in a single flask. The use of racemic starting materials ultimately gives four diasteromeric products **146**. However, this method can easily be used to build up structural diversity.

In addition to the direct use of MCRs to access diverse and complex small-molecule scaffolds, the build/couple/pair (B/C/P) strategy has gained some attention [57]. This strategy, outlined by Nielsen and Schreiber, begins with the asymmetric synthesis of building blocks incorporating defined stereogenic units and tailored reactive functionalities (build phase). These building blocks are then combined (couple phase) through intermolecular bond-forming processes to yield a complete matrix of stereoisomers of the main carbon framework. Finally, intramolecular joining of strategically positioned functional groups (pair phase) takes place. When this strategy is utilized with a multicomponent couple phase, one can obtain a diverse array of compounds for high-throughput screening.

One of the more prominent MCRs in combinatorial chemistry and drug discovery, especially as part of a B/C/P strategy, is the Petasis reaction. Here, an amine, aldehyde, and vinyl- or arylboronic acid come together to form substituted amines. First reported in 1993 as a practical method for the synthesis of the geometrically pure antifungal agent naftifine [58], the Petasis reaction can be ascribed as a variation of the Mannich reaction. However, rather than generating an enolate to form the substituted amine product, the boronic acid serves as the nucleophile. In comparison to other methods of generating substituted amines, the Petasis reaction tolerates a multifunctional scaffold with a variety of amines and organoboronic acids as potential starting materials. Additionally, the reaction does not require anhydrous or inert conditions, which makes it a mild and selective synthesis.

In the first B/C/P library synthesis, the build phase consisted of the stereoselective synthesis of α-hydroxyaldehyde **147** (protected as the lactol) and phenylalanine methyl ester **148** [59]. A Petasis 3CR between these two components and (*E*)-2-cyclopropylvinylboronic acid **149**, followed by propargylation of the resulting amine, was used in the couple phase. Subsequent reagent-controlled rearrangement reactions (pair phase) between the polar and nonpolar functionalities present provided a wide range of structurally complex small molecules (Scheme 2.19).

A more recent example of library synthesis utilizing a B/C/P strategy with a Petasis 3CR involved a Ru-alkylidene-catalyzed ring-closing metathesis reaction for the pair phase [60]. By varying the location of the terminal alkenes, diverse arrays of five- and seven-membered heterocycles were obtained. In addition, changing the catalyst for the seven-membered ring substrate led to the formation of bicyclic aminals via a tandem isomerization/*N*-alkyliminium ion cyclization sequence, providing even more diversity from the Petasis amino alcohol products (Scheme 2.20).

Although the strictest classification of an MCR requires a mixture of at least three chemicals without regard to precise control of the order of addition, there are many processes that incorporate multiple components in the same product in which the order of addition is crucial. These components can still be highly variable and lead to a diverse array of products. The specific order of addition distinguishes this process from that of a typical MCR. This strategy, pioneered by Martin et al. in 1991 and referred to as a *multicomponent assembly process* (MCAP) [61], has been used in

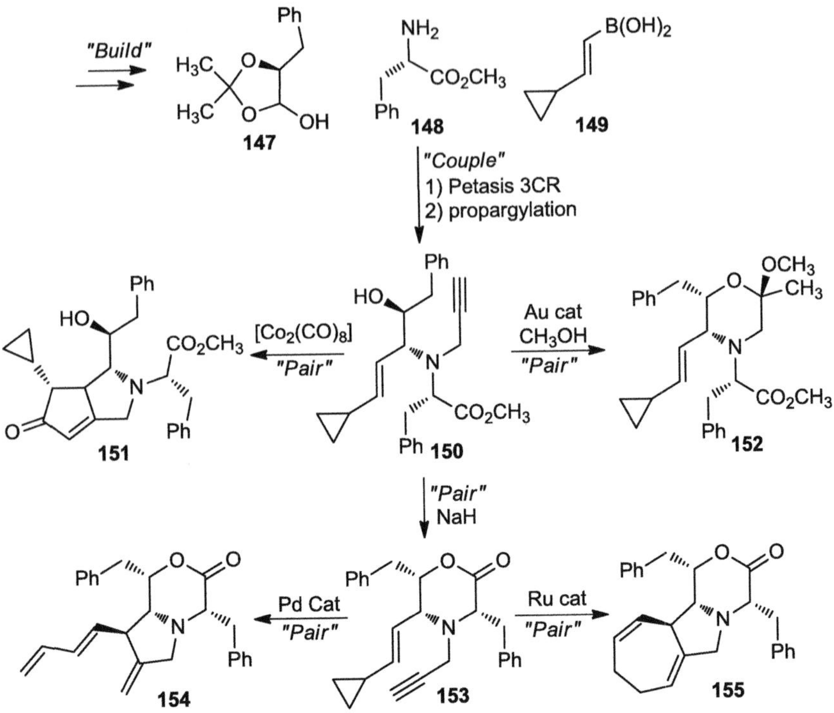

SCHEME 2.19 First reported example of Schreiber's build/couple/pair library synthesis.

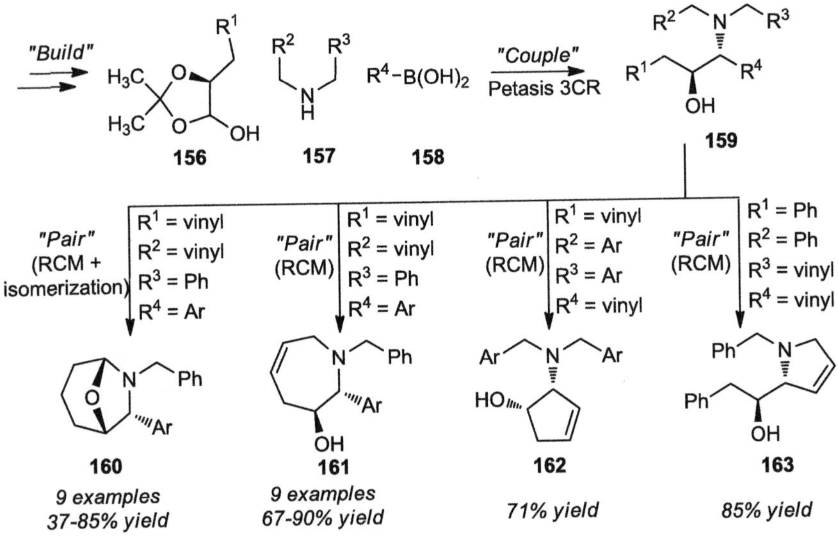

SCHEME 2.20 B/C/P library synthesis using ring-closing metathesis in the pair phase.

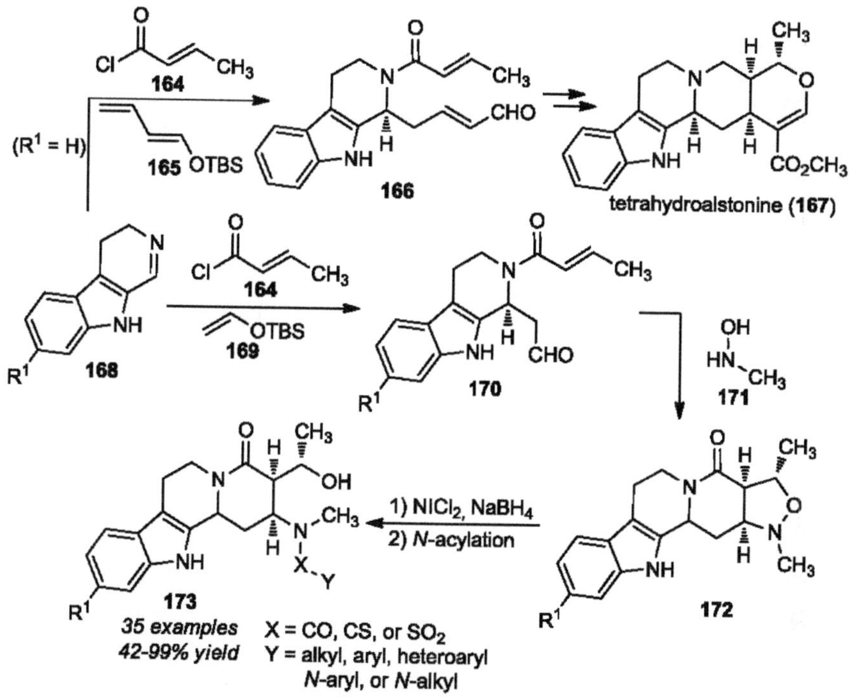

SCHEME 2.21 MCAP reported by Martin et al. [61] as the key step in the total synthesis of the indole alkaloid tetrahydroalstonine and as a means of generating a diverse library of compounds based on the natural product scaffold.

both target- and diversity-oriented synthesis [62,63]. Recently, an MCAP was used as the key step in the total synthesis of the indole alkaloid tetrahydroalstonine and has also been used to generate a diverse library of compounds based on the natural product scaffold (Scheme 2.21) [64].

Another MCAP library synthesis recently reported by Martin's group utilizes a reductive amination and azide-alkyne dipolar (Huisgen) cycloaddition sequence to access a 1,2,3-triazolo-1,4-benzodiazepine core [65]. The MCR products were further diversified using N-functionalization and α-aminonitrile chemistry. Compounds containing the 1,4-benzodiazepine ring system bind to many different targets, including G protein–coupled receptors (GPCRs), ligand-gated ion channels, and enzymes [66]. This ring system is also a common structural subunit in numerous pharmaceutical agents, biological probes, and bioactive natural products. The use of a MCAP in this case led to the rapid assembly of a library of over 100 compounds for high-throughput screening (Scheme 2.22).

Although the term *MCAP* has only been applied to research reported by Martin's group, other efforts in diversity-oriented synthesis follow a very similar strategy. For instance, Shaw's group has recently reported a formal cycloaddition between cyanosuccinic anhydride and imines to form γ-lactams with multiple stereocenters

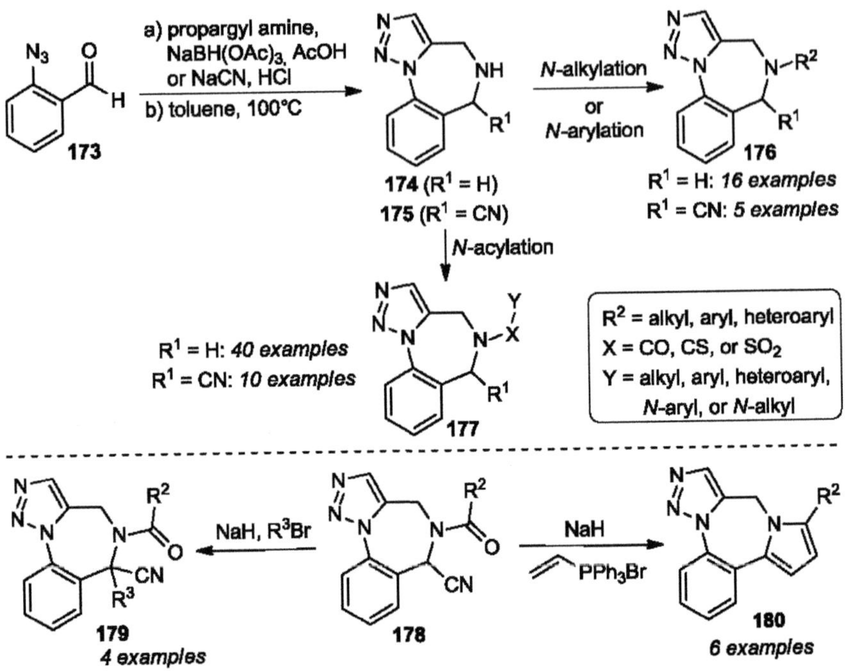

SCHEME 2.22 MCAP library synthesis reported by Martin's group [65] which utilizes a reductive amination and azide–alkyne dipolar (Huisgen) cycloaddition sequence to access 1,2,3-triazolo-1,4-benzodiazepines which are functionalized further.

(Scheme 2.23) [67]. The aldehyde and amine components that form the imine prior to reaction with the anhydride are highly variable, which is the starting point for the synthesis of a diverse library of small molecules.

Imines have been utilized quite often in library synthesis as highly variable electrophiles. The corresponding nucleophiles can possess additional functionality that

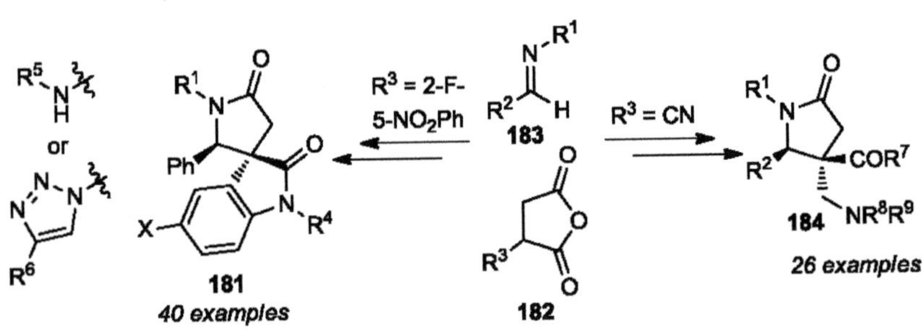

SCHEME 2.23 MCAP type of library synthesis reported by Shaw's group [67] that utilizes a formal cycloaddition between cyanosuccinic anhydride and imines to form γ-lactams with multiple stereocenters.

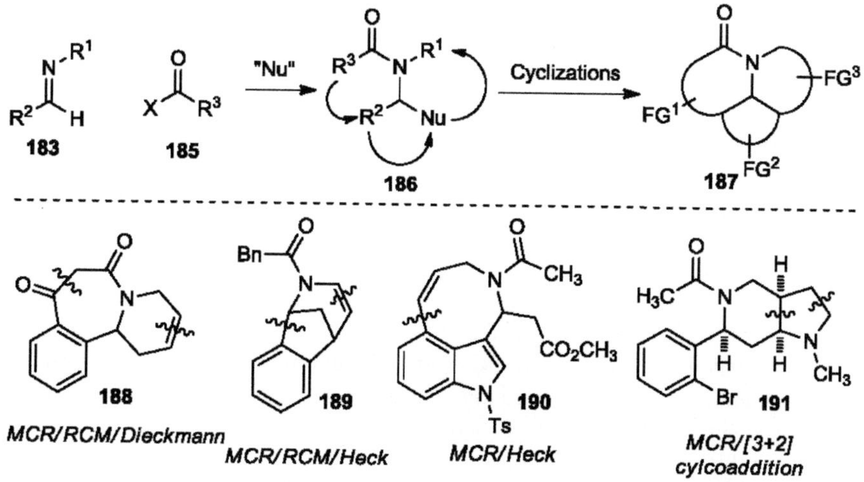

SCHEME 2.24 Sequential MCR/cyclization strategy for DOS reported by Sunderhaus et al. [68].

can be exploited for further diversification, taking advantage of the pair phase of the B/C/P strategy. In a 2007 report by Sunderhaus et al. sequential reactions of imines, acid chlorides, and nucleophiles were developed to prepare multifunctional substrates that were employed in subsequent ring-forming reactions [68]. These subsequent reactions included RCM, Dieckmann and Heck reactions, and Diels–Alder and dipolar cycloadditions to assemble a varied collection of functionalized heterocyclic scaffolds (Scheme 2.24).

The biological importance of indoles has led to a large number of strategies to form the indole ring system [69,70]. One branch of these strategies is the employment of transition metal catalysts to mediate formation of the indole ring system; indeed, multicomponent variants of these reactions have emerged. Kaspar and Ackermann have developed a three-component reaction starting with *o*-dihaloarenes involving a Sonagashira coupling followed by Buchwald–Hartwig coupling and ending with an intramolecular hydroamination (Scheme 2.25) [71].

In a related reaction, Knapp and Kurth utilized *o*-dihaloarenes with amines and aldehydes or ketones in a three-component reaction [72]. This transformation first

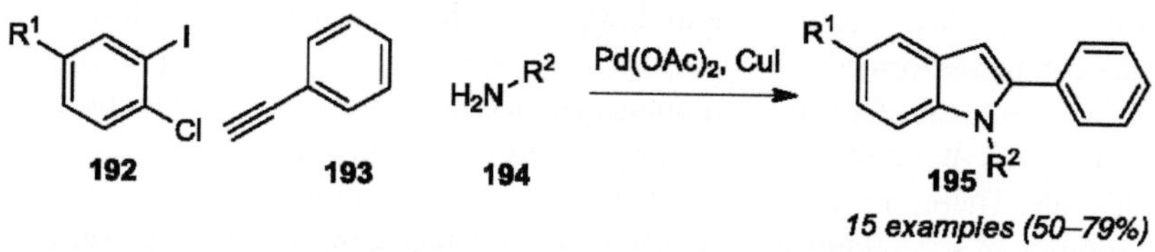

SCHEME 2.25 3CR Sonagashira/Buchwald–Hartwig/hydroamination indole synthesis.

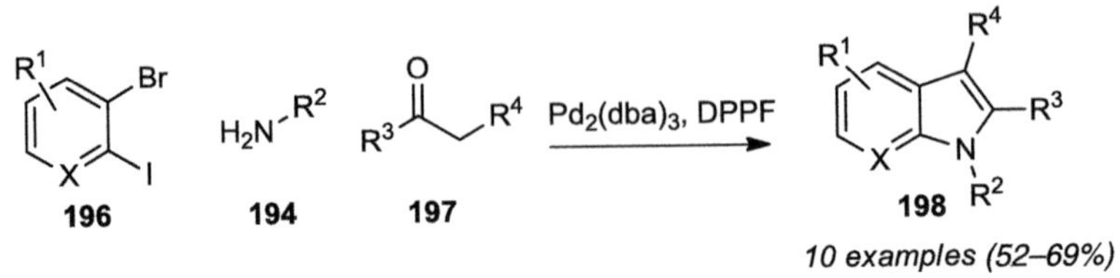

SCHEME 2.26 3CR Buchwald–Hartwig/condensation/arene–alkene coupling indole synthesis.

forms an *o*-haloaniline via Buchwald–Hartwig coupling. Subsequent condensation of the aryl amine and carbonyl group yields an enamine poised to ring-close to the indole by an arene-alkene coupling (Scheme 2.26). A proposed mechanism was also modeled in this work using quantum chemical calculations.

Several methods to produce a diverse array of macrocycles by MCRs have been developed in the laboratories of Zhu and Wessjohann. The Ugi reaction and other isonitrile-based MCRs have played a central role in these macrocycle-forming MCRs. A general strategy has been to use long linear reactants with bifunctionality on opposite ends. As in Scheme 2.27A, one reactant contains a terminal alkyne and a terminal isonitrile. One oxazole-forming MCR occurs and the macrocycle then closes with an azide dipolar [3 + 2] cycloaddition [73]. Scheme 2.27B demonstrates the same oxazole-forming MCR reaction performed twice to make symmetrical macrocycles (i.e., via one intermolecular MCR followed by one intramolecular MCR) [74]. Wessjohann's group has taken advantage of the Ugi reaction in Scheme 2.27C to make tri-macrocylic cage compounds from four Ugi MCRs [75].

In a particularly interesting MCR, Barluenga et al. [76] used a palladium catalyst to synthesize spirofused ketals. Here, an alkynyl alcohol, in the presence of a transition metal catalyst, forms a vinyl ether, and a condensation between an aniline and aromatic aldehyde forms an imine. This simultaneous in situ formation of a nucleophile and an electrophile leads to a reaction producing the spiro system in a 1 : 1 mixture of diastereomers. Epimerization provides access to a single diastereomer **217** by treatment with dry perchloric acid with an overall yield of 87% (Scheme 2.28).

Ugi products have found use as precursors to benzoxazoles. Salcedo et al. [77] subjected *o*-iodobenzoic acids and *o*-iodophenylisonitrile **218** to Ugi conditions, giving Ugi products with bifunctional aryl iodides **220** in good yields (Scheme 2.29). In the presence of a copper catalyst, the benzoxazole can be formed as the sole product with no evidence of the seven-membered benzamide ring. The acidity of the α-proton to the benzoxazole is thus lowered, such that a palladium-catalyzed ring closure can be performed.

A 3CR isonitrile-free multicomponent reaction is presented by Preciado et al. (Scheme 2.30). What initially starts as a Povarov reaction is ultimately prevented by Bredt's rule. Therefore, an opportunity to trap the oxonium intermediate is available. The addition of several nitriles reacts with intermediate **228**, followed by cyclization. This Mannich–Ritter reaction leads to a single diastereomer in moderate yields [78].

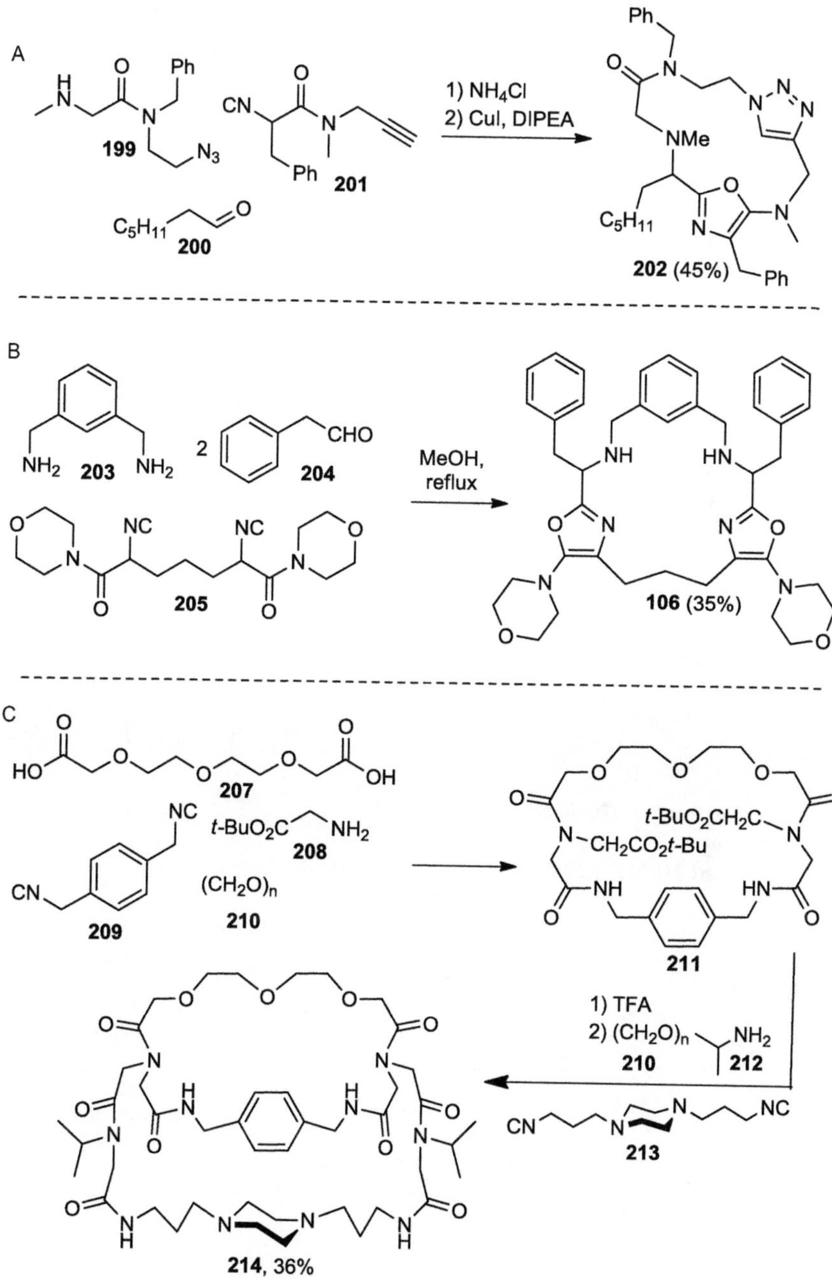

SCHEME 2.27 Multicomponent reactions forming macrocycles.

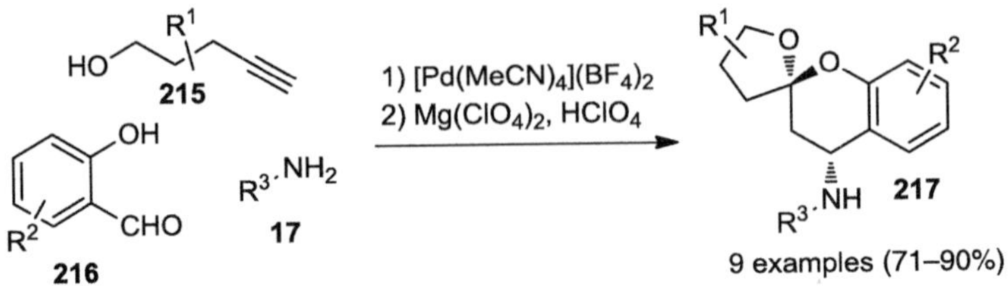

SCHEME 2.28 3CR organometallic synthesis of spirocyclic acetals.

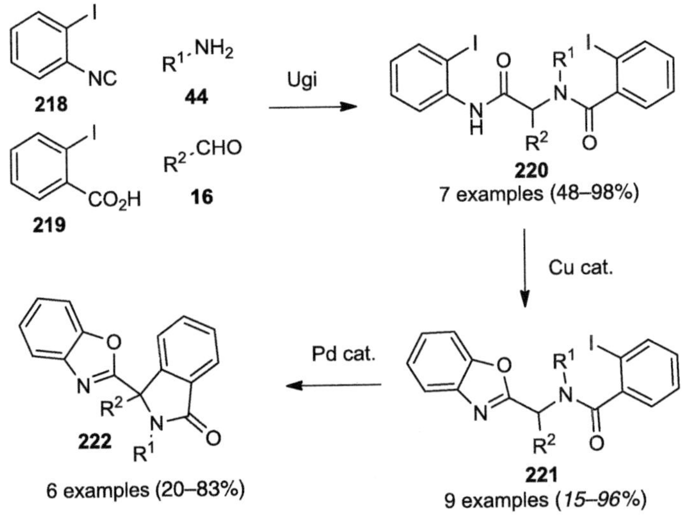

SCHEME 2.29 Ugi/benzooxazole–isoindolinone synthesis.

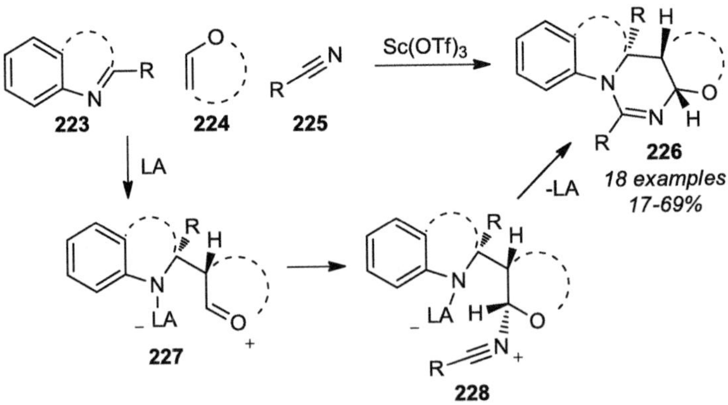

SCHEME 2.30 Mannich–Ritter MCR; utilization of Bredt's rule to interrupt Povarov reaction.

2.4 CONCLUSIONS

Multicomponent reactions have heavily influenced DOS and will continue to do so in the future. MCRs can produce diverse compounds for screening in a single step or serve as a nucleating event for bringing multiple functional groups into proximity for subsequent reactions. The recent advent of new and mechanistically distinct MCRs, including those that avoid moving beyond isonitriles and/or involve organic or transition metal catalysts, will increase exponentially the diversity available from these processes. Although clever technology for liquid handling and robotics can be applied to streamline production of diverse molecules for screening and other applications, the fascinating chemistry of MCRs will continue to offer access to unexplored regions of chemical space.

REFERENCES

1. J. Zhu, H. Bienaymé, Eds., *Multicomponent Reactions*, Wiley-VCH, Weinheim, Germany, **2005**.
2. (a) R. A. Houghten, C. Pinilla, S. E. Blondelle, J. R. Appel, C. T. Dooley, J. H. Cuervo, *Nature* **1991**, *354*, 84–86; (b) R. A. Houghten, *Proc. Natl. Acad. Sci. U.S.A.* **1985**, *82*, 5131–5135.
3. A. Furka, F. Sebestyen, M. Asgedom, G. Dibo, *Int. J. Pept. Protein Res.* **1991**, *37*, 487–493.
4. A. Dömling, I. Ugi, *Angew. Chem. Int. Ed.* **2000**, *39*, 3168–3210.
5. T. U. Mayer, T. M. Kapoor, S. J. Haggarty, R. W. King, S. L. Schreiber, T. J. Mitchison, *Science* **1999**, *286*, 971–974.
6. Z. Maliga, T. M. Kapoor, T. J. Mitchison, *Chem. Biol.* **2002**, *9*, 989–996.
7. Y. Yan, V. Sardana, B. Xu, C. Homnick, W. Halczenko, C. A. Buser, M. Schaber, G. D. Hartman, H. E. Huber, L. C. Kuo, *J. Mol. Biol.* **2004**, *335*, 547–554.
8. M. Gartner, N. Sunder-Plassmann, J. Seiler, M. Utz, I. Vernos, T. Surrey, A. Giannis, *ChemBioChem* **2005**, *6*, 1173–1177.
9. C. Burdack, C. Kalinski, G. Ross, L. Weber, V. Khazak (Ed.: EPO), Priaxon AG, Germany. Application: WO, **2010**, 171 pp.
10. http://www.priaxon.com/content/technology/molecular-space/index.php?; accessed June 16, 2012.
11. M. Woelfle, J.-P. Seerden, J. de Gooijer, K. Pouwer, P. Olliaro, M. H. Todd, *PLoS Negl. Trop. Dis.* **2011**, *5*, e1260.
12. Y. Huang, K. Khoury, A. Dömling, *Top. Heterocycl. Chem.* **2010**, *23*, 85–127.
13. J.-M. Pereillo, M. Maftouh, A. Andrieu, M.-F. Uzabiaga, O. Fedeli, P. Savi, M. Pascal, J.-M. Herbert, J.-P. Maffrand, C. Picard, *Drug Metab. Dispos.* **2002**, *30*, 1288–1295.
14. I. D. Cockshott, *Clin. Pharmacokinet.* **2004**, *43*, 855–878.
15. (a) J. J. Irwin, B. K. Shoichet, *J. Chem. Inf. Model.* **2005**, *45*, 177–182; (b) J. J. Irwin, T. Sterling, M. M. Mysinger, E. S. Bolstad, R. G. Coleman, *J. Chem. Inf. Model.* **2012**, *52*, 1757–1768.
16. http://anchorquery.csb.pitt.edu/; accessed June 12, 2012.

17. J. E. Biggs-Houck, A. Younai, J. T. Shaw, *Curr. Opin. Chem. Biol.* **2010**, *14*, 371–382.
18. I. Akritopoulou-Zanze, *Curr. Opin. Chem. Biol.* **2008**, *12*, 324–331.
19. S. Santra, P. R. Andreana, *Org. Lett.* **2007**, *9*, 5035–5038.
20. H. E. Pelish, J. R. Peterson, S. B. Salvarezza, E. Rodriguez-Boulan, J.-L. Chen, M. Stamnes, E. Macia, Y. Feng, M. D. Shair, T. Kirchhausen, *Nat. Chem. Biol.* **2006**, *2*, 39–46.
21. S. Santra, P. R. Andreana, *Angew. Chem. Int. Ed.* **2011**, *50*, 9418–9422.
22. T. Zarganes-Tzitzikas, M. A. Terzidis, J. Stephanidou-Stephanatou, C. A. Tsoleridis, G. E. Kostakis, *J. Org. Chem.* **2011**, *76*, 9008–9014.
23. S. Kotha, A. C. Deb, K. Lahiri, E. Manivannan, *Synthesis* **2009**, 165–193.
24. B. Rebacz, T. O. Larsen, M. H. Clausen, M. H. Rønnest, H. Löffler, A. D. Ho, A. Krämer, *Cancer Res.* **2007**, *67*, 6342–6350.
25. A. Basso, L. Banfi, G. Guanti, R. Riva, *Org. Biomol. Chem.* **2009**, *7*, 253–258.
26. S. S. Bhojgude, A. T. Biju, *Angew. Chem. Int. Ed.* **2012**, *51*, 1520–1522.
27. K. M. Allan, C. D. Gilmore, B. M. Stoltz, *Angew. Chem. Int. Ed.* **2011**, *50*, 4488–4491.
28. D. Peña, A. Cobas, D. Pérez, E. Guitián, *Synthesis* **2002**, 1454–1458.
29. F. Sha, X. Huang, *Angew. Chem. Int. Ed.* **2009**, *48*, 3458–3461.
30. Q. Huang, R. C. Larock, *J. Org. Chem.* **2002**, *68*, 980–988, and references therein.
31. J. Wei, J. T. Shaw, *Org. Lett.* **2007**, *9*, 4077–4080.
32. T. Indumathi, V. S. Jamal Ahamed, S.-S. Moon, F. R. Fronczek, K. J. Rajendra Prasad, *Eur. J. Med. Chem.* **2011**, *46*, 5580–5590.
33. H.-J. Knölker, K. R. Reddy, *Chem. Rev.* **2002**, *102*, 4303–4427.
34. J. M. Knapp, J. S. Zhu, A. B. Wood, M. J. Kurth, *ACS Comb. Sci.* **2012**, *14*, 85–88.
35. Y. Zou, Y. Hu, H. Liu, D. Shi, *ACS Comb. Sci.* **2012**, *14*, 38–43.
36. J. J. Badillo, G. E. Arevalo, J. C. Fettinger, A. K. Franz, *Org. Lett.* **2011**, *13*, 418–421.
37. A. K. Franz, P. D. Dreyfuss, S. L. Schreiber, *J. Am. Chem. Soc.* **2007**, *129*, 1020–1021.
38. J. J. Badillo, A. Silva-Garcia, B. H. Schupe, J. C. Fettinger, A. K. Franz, *Tetrahedron Lett.* **2011**, *52*, 5550–5553.
39. L. He, M. Bekkaye, P. Retailleau, G. Masson, *Org. Lett.* **2012**, *14*, 3158–3161.
40. M. Bararjanian, S. Balalaie, F. Rominger, B. Movassagh, H. R. Bijanzadeh, *J. Org. Chem.* **2010**, *75*, 2806–2812.
41. F. Bonnaterre, M. Bois-Choussy, J. Zhu, *Org. Lett.* **2006**, *8*, 4351–4354.
42. K. Lu, T. Luo, Z. Xiang, Z. You, R. Fathi, J. Chen, Z. Yang, *J. Comb. Chem.* **2005**, *7*, 958–967.
43. D. L. Wright, C. V. Robotham, K. Aboud, *Tetrahedron Lett.* **2002**, *43*, 943–946.
44. I. Akritopoulou-Zanze, Y. Wang, H. Zhao, S. W. Djuric, *Tetrahedron Lett.* **2009**, *50*, 5773–5776.
45. X. Sun, P. Janvier, G. Zhao, H. Bienaymé, J. Zhu, *Org. Lett.* **2001**, *3*, 877–880.
46. P. Janvier, H. Bienaymé, J. Zhu, *Angew. Chem. Int. Ed.* **2002**, *41*, 4291–4294.
47. D. Lee, J. K. Sello, S. L. Schreiber, *Org. Lett.* **2000**, *2*, 709–712.
48. J. K. Sello, P. R. Andreana, D. Lee, S. L. Schreiber, *Org. Lett.* **2003**, *5*, 4125–4127.
49. M. Ikoma, M. Oikawa, M. Sasaki, *Eur. J. Org. Chem.* **2009**, 72–84.

50. M. Oikawa, M. Ikoma, M. Sasaki, M. B. Gill, G. T. Swanson, K. Shimamoto, R. Sakai, *Eur. J. Org. Chem.* **2009**, 5531–5548.
51. M. Oikawa, M. Ikoma, M. Sasaki, M. B. Gill, G. T. Swanson, K. Shimamoto, R. Sakai, *Bioorg. Med. Chem.* **2010**, *18*, 3795–3804.
52. I. Akritopoulou-Zanze, V. Gracias, J. D. Moore, S. W. Djuric, *Tetrahedron Lett.* **2004**, *45*, 3421–3423.
53. I. Akritopoulu-Zanze, V. Gracias, S. W. Djuric, *Tetrahedron Lett.* **2004**, *45*, 8439–8441.
54. N. Elders, D. van der Born, L. J. D. Hendrickx, B. J. J. Timmer, A. Krause, E. Janssen, F. J. J. de Kanter, E. Ruijter, R. V. A. Orru, *Angew. Chem. Int. Ed.* **2009**, *48*, 5856–5859.
55. M. Paravidino, R. Scheffelaar, R. F. Schmitz, F. J. J. de Kanter, M. B. Groen, E. Ruijter, R. V. A. Orru, *J. Org. Chem.* **2007**, *72*, 10239–10242.
56. A. M. van Leusen, J. Wildeman, O. H. Oldenziel, *J. Org. Chem.* **1977**, *42*, 1153–1159.
57. T. E. Nielsen, S. L. Schreiber, *Angew. Chem. Int. Ed.* **2008**, *47*, 48–56.
58. N. A. Petasis, I. Akritopoulou, *Tetrahedron Lett.* **1993**, *34*, 583–586.
59. N. Kumagai, G. Muncipinto, S. L. Schreiber, *Angew. Chem. Int. Ed.* **2006**, *45*, 3635–3638.
60. E. Ascic, S. T. Le Quement, M. Ishoey, M. Daugaard, T. E. Nielsen, *ACS Comb. Sci.* **2012**, *14*, 253–257.
61. S. F. Martin, J. E. Hunter, B. Benage, L. S. Geraci, M. Mortimore, *J. Am. Chem. Soc.* **1991**, *113*, 6161–6171.
62. B. Cheng, J. D. Sunderhaus, S. F. Martin, *Org. Lett.* **2010**, *12*, 3622–3625.
63. J. D. Sunderhaus, C. Dockendorff, S. F. Martin, *Tetrahedron* **2009**, *65*, 6454–6469.
64. B. A. Granger, K. Kaneda, S. F. Martin, *ACS Comb. Sci.* **2011**, *14*, 75–79.
65. J. R. Donald, R. R. Wood, S. F. Martin, *ACS Comb. Sci.* **2012**, *14*, 135–143.
66. D. A. Horton, G. T. Bourne, M. L. Smythe, *Chem. Rev.* **2003**, *103*, 893–930.
67. D. Q. Tan, A. L. Atherton, A. J. Smith, C. Soldi, K. A. Hurley, J. C. Fettinger, J. T. Shaw, *ACS Comb. Sci.* **2012**, *14*, 218–223.
68. J. D. Sunderhaus, C. Dockendorff, S. F. Martin, *Org. Lett.* **2007**, *9*, 4223–4226.
69. D. A. Horton, G. T. Bourne, M. L. Smythe, *Chem. Rev.* **2003**, *103*, 893–930.
70. H.-J. Knölker, K. R. Reddy, *Chem. Rev.* **2002**, *102*, 4303–4428.
71. L. T. Kaspar, L. Ackermann, *Tetrahedron* **2005**, *61*, 11311–11316.
72. J. M. Knapp, J. S. Zhu, D. J. Tantillo, M. J. Kurth, *Angew. Chem. Int. Ed.* **2012**, *51*, 10588–10591.
73. T. Pirali, G. C. Tron, J. Zhu, *Org. Lett.* **2006**, *8*, 4145–4148.
74. P. Janvier, M. Bois-Choussy, H. Bienaymé, J. Zhu, *Angew. Chem. Int. Ed.* **2003**, *42*, 811–814.
75. D. G. Rivera, L. A. Wessjohann, *J. Am. Chem. Soc.* **2009**, *131*, 3721–3732.
76. J. Barluenga, A. Mendoza, F. Rodríguez, F. J. Fañanás, *Angew. Chem. Int. Ed.* **2009**, *48*, 1644–1647.
77. A. Salcedo, L. Neuville, J. Zhu, *J. Org. Chem.* **2008**, *73*, 3600–3603.
78. S. Preciado, E. Vicente-Garcia, S. Llabres, F. J. Luque, R. Lavilla, *Angew. Chem. Int. Ed.* **2012**, *51*, 6874–6877.

3

CYCLOADDITION REACTIONS IN DIVERSITY-ORIENTED SYNTHESIS

GIOVANNI MUNCIPINTO

3.1 INTRODUCTION

Cycloaddition reactions are fundamental processes in organic chemistry, as they are among the most powerful methodologies that can be used to provide rapid access to natural and unnatural compounds of medicinal interest. Their ability to provide cyclic and heterocyclic compounds with a high degree of structural complexity and skeletal and stereochemical diversity have made cycloaddition reactions some of the most useful and widely utilized reactions in diversity-oriented synthesis (DOS) [1]. Large and unexploited regions of chemical space need to be addressed to increase the chances of succeeding in drug discovery and to take advantage of the extraordinary insights made by genomics in understanding the biological causes underlying human diseases [2]. Recently, there has been a growing consensus that the crucial factor for drug discovery success is not library size but a library's structural diversity [3]. DOS aims to generate such structural diversity in an efficient manner. It is therefore not surprising that most papers in the early days of DOS reported cycloaddition reactions as tools to access structurally complex and diverse polyheterocyclic systems, frameworks that are also present in biologically relevant natural products. Diels–Alder and 1,3-dipolar cycloaddition reactions represent the ideal reactions for this purpose. In fact, a large number of DOS publications report them as pairing reactions of strategically placed appendages in building blocks. In the build/couple/pair strategy [4], cycloaddition reactions have been used both as diversity-generating reactions (same → diverse) and

Diversity-Oriented Synthesis: Basics and Applications in Organic Synthesis, Drug Discovery, and Chemical Biology, First Edition. Edited by Andrea Trabocchi.

as complexity-generating reactions (simple → complex), contributing to the achievement of structural diversity in the first case and of structural complexity in the second. In addition, with their ability to generate new stereocenters, cycloaddition reactions also contribute to the development of stereo structure–activity relationships during screening campaigns. Following a chronological order, in this review we present an overview of Diels–Alder and inverse electron-demand Diels–Alder reactions, and 1,3-dipolar cycloaddition reactions, including Huisgen [3 + 2] cycloadditions, nitrone and nitrile oxide cycloadditions, and azomethine ylide cycloadditions. In the final section, miscellaneous cyloaddition reactions, such as [2 + 2 + 2], [2 + 2], formal [4 + 3], and [3 + 3], are presented.

3.2 [4 + 2] CYCLOADDITION REACTIONS

[4 + 2] Cycloadditions represent one of the most important reactions in the area, as is evident from the large number of papers, reviews, and book chapters in the field. Due to the fact that [4 + 2] cycloaddition reactions usually furnish a high degree of chemo-, regio-, and stereoselectivities, and generate up to four stereocenters in a single step, these reactions have long been adopted for DOS.

3.2.1 Diels–Alder Reaction

One of the first examples of a Diels–Alder reaction in DOS was reported by Schreiber's group in 2000 (Scheme 3.1) [5]. Aiming to generate structural complexity, the authors reported an example of the pairwise use of complexity-generating reactions involving the Ugi four-component, intramolecular Diels–Alder, and ring-opening/ring-closing olefin metathesis reactions. Fundamental to the success of this pathway is the introduction of a diene and dienophile in two of the four components in the Ugi reaction [6]. The intramolecular Diels–Alder product **1** was synthesized as a single stereoisomer with good yield.

Using a one bead–one stock solution technology platform [7], Schreiber et al. reported an efficient synthesis of 29,400 polycyclic compounds [8] based on the methodology developed by Fallis's group for the use of consecutive Diels–Alder reactions (Scheme 3.2) [9]. Fallis-type trienes were loaded on macrobeads, followed by the Diels–Alder reaction. Double cycloaddition occurred when disubstituted cyclic dienophiles were utilized (**3**), whereas monocycloaddition was found with tri- or tetrasubstituted cyclic dienophiles (**2**). Monocyclic dienes **2** were then reacted with a second dienophile to yield tetracycles, generated from two different dienophiles (**4**). Single and double cycloaddition both proceeded stereoselectively, and bi- and tetracyclic compounds with up to six stereocenters were synthesized.

The discovery by Winterfeldt's group [10] of a Diels–Alder and transannular retro-Diels–Alder reaction sequence of a steroidal diene enabled Kumar et al. to develop a new DOS pathway generating fused bicyclic compounds containing a five-membered ring and a 14-membered ring paracyclophane (Scheme 3.3) [11]. The Diels–Alder reaction between the stereoidal dienes **5** and ynones is catalyzed by the Lewis acid Et_2AlCl, which accelerates the Diels–Alder reaction ahead of the retro-Diels–Alder reaction. This allowed for the formation and isolation of intermediate Diels–Alder

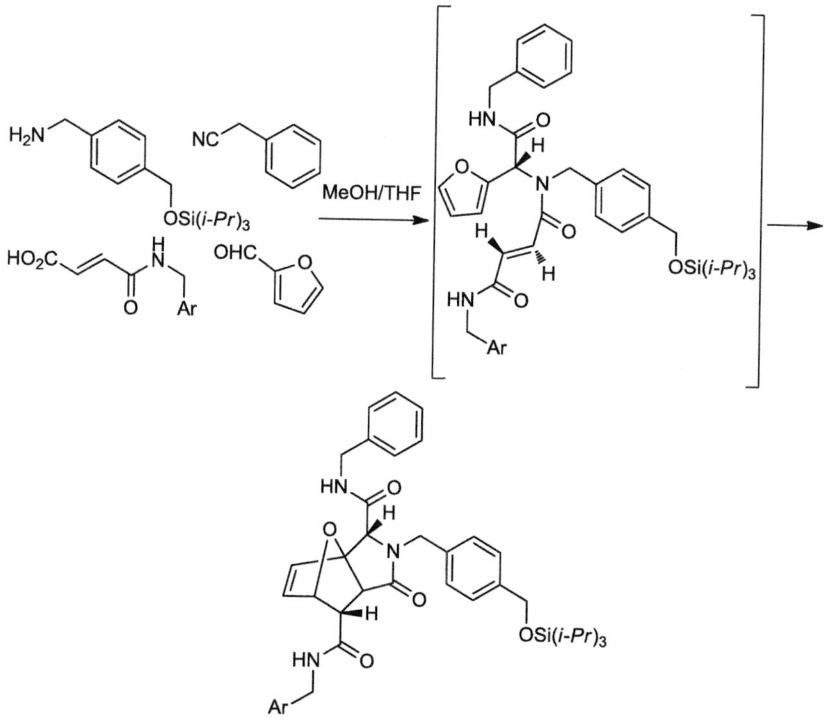

1

SCHEME 3.1 Pairwise use of complexity-generating reactions involving Ugi four-component and intramolecular Diels–Alder reactions.

adducts **6** as single regio- and diastereoisomers, which were efficiently converted to paracyclophanes **7** upon heating. More than 4000 skeletally diverse small molecules were synthesized through this strategy.

Aubé et al. reported tandem reaction sequences in which they combined a Diels–Alder reaction with an intramolecular Schmidt reaction [12] to access *Stemona* [13] alkaloid-like scaffolds (Scheme 3.4) [14]. When the Diels–Alder reaction was followed by an imide acylation reaction, octahydroisoquiol-1-one-8-carboxylic acids were afforded (Scheme 3.5) [15]. With regard to the azido-Schmidt Diels–Alder reaction sequence, the authors reported two ways to combine the individual reactions. One strategy used the Diels–Alder cycloaddition reaction to join together the azide and ketone present on the diene and dienophile, respectively, followed by the intramolecular Schmidt reaction (Scheme 3.4a). In the second strategy, the azide and ketone, the latter in the form of enone, were both present on the dienophile, and the Schmidt reaction occurred only after elimination of the conjugation during the Diels–Alder reaction (Scheme 3.4b). Both strategies, catalyzed by Lewis acids, proceeded through an endo-selective Diels–Alder cycloaddition followed by stereoselective Schmidt ring expansion, generating polycyclic compounds (**8** and **9**) with good yields and high diastereoselectivity.

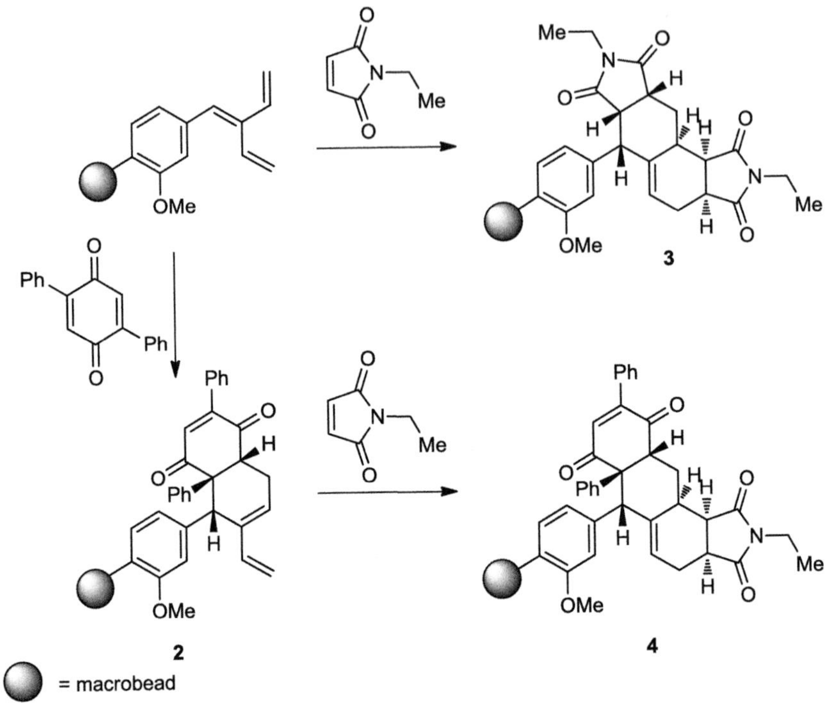

SCHEME 3.2 Use of consecutive Diels–Alder reactions.

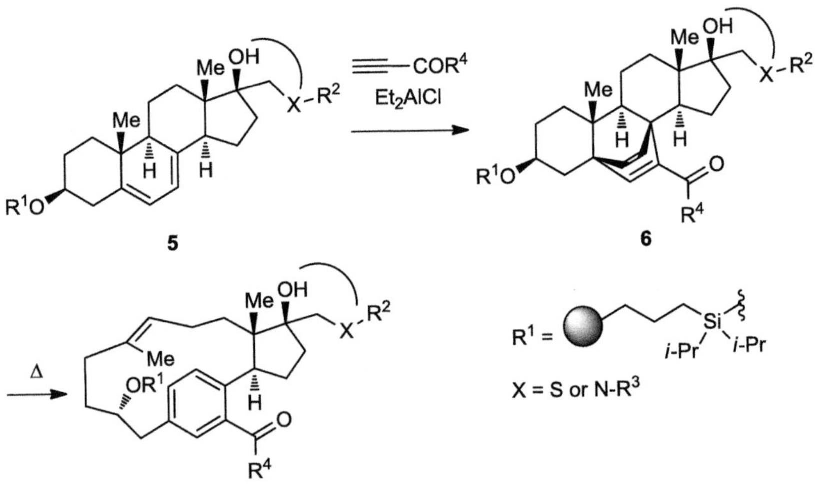

SCHEME 3.3 Diels–Alder and transannular retro-Diels–Alder reaction sequence.

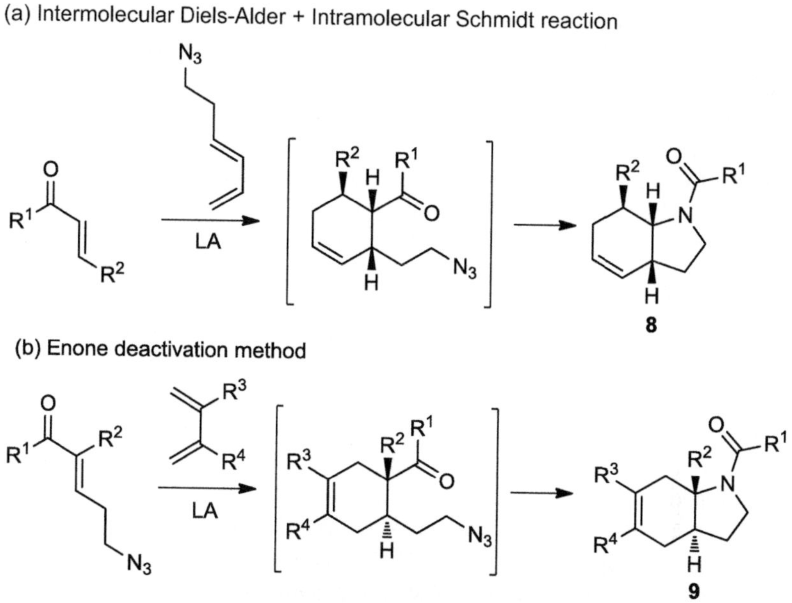

SCHEME 3.4 Tandem Diels–Alder/Schmidt sequence. (Adapted from [14b], with permission; copyright © 2004 American Chemical Society.)

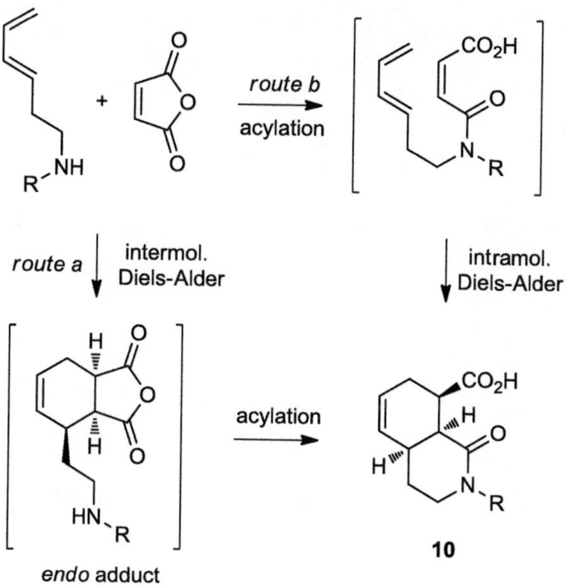

SCHEME 3.5 Tandem Diels–Alder/acylation sequence.

In the tandem Diels–Alder/acylation sequence approach to building small focused libraries (Scheme 3.5), Aubé et al. began with the hypothesis that the Diels–Alder cycloaddition of highly reactive dienophiles with amine-containing dienes would give only the endo product, with the amine side chain in a cis orientation with respect to the reactive carbonyl group, thus promoting the acylation step. Following the optimization of reaction conditions, a one-step synthesis of an isoquinolone (**10**) containing a carboxylic acid for further diversification was achieved. In contrast with the authors' original hypothesis, they did not find any evidence that the reaction proceeds by the initial Diels–Alder cycloaddition reaction followed by the intramolecular acylation.

Recent work by Cui et al. demonstrated the power of combining several reactions to achieve both skeletal diversity and structural complexity (Scheme 3.6) [16]. The Diels–Alder cycloaddition reaction was used along with 1,6-enyne cycloisomerization, Os-catalyzed dihydroxylation, and Sn-catalyzed carbamoylation in a unique approach that generated 191 skeletally and structurally diverse compounds with high diastereoselectivity and employing only 16 building blocks. Principal moment of inertia (PMI) computational analysis showed that the library covers a broad chemical space, suggesting shape diversity [17]. The 1,3-dienes (**11–14**) were synthesized through transition metal–catalyzed cycloisomerizations of 1,6-enynes. Four dienes were reacted with five selected dienophiles (**15–19**) with high efficiency and

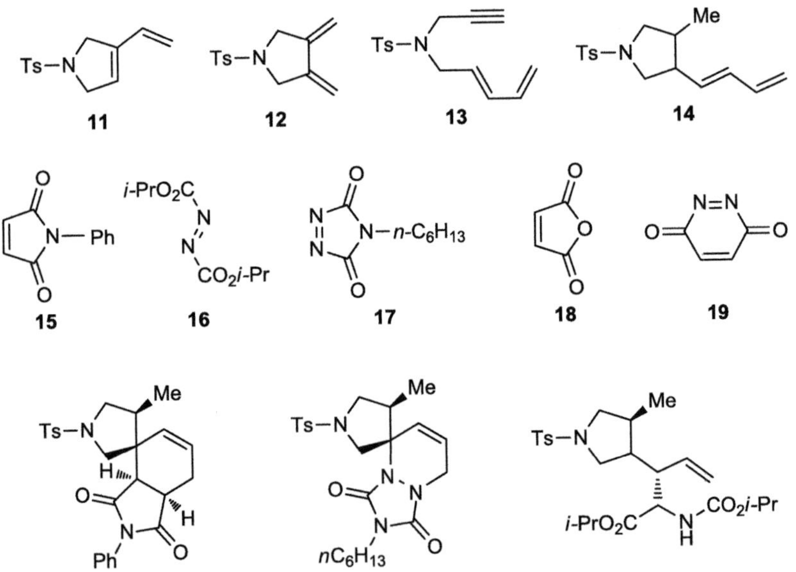

SCHEME 3.6 Diels–Alder cycloaddition reactions of dienes generated through transition metal–catalyzed cycloisomerizations of 1,6-enynes.

diastereoselectivity; the only exception was the exocyclic diene **14**, presumably due to the difficulty of achieving the required s-cis conformation. Indeed, in two cases (**20** and **21**), activation with lithium bis-trifluoromethanesulfonimide in acetone was required [18]. This activation did not succeed when the azodicarboxylate **16** was used as a dienophile, and resulted in the formation of the ene-reaction product **22**. The 19 new scaffolds were then used as substrates for subsequent diversity-generating reactions, such as dihydroxylation and carbamoylation. A small molecule that suppresses glycolytic production of ATP and lactate in a CHO-K1 cell line [19] was discovered.

Recently, Kwon's group reported a branching reaction pathway in which they combined phosphine-catalyzed annulations, Tebbe reactions, and Diels–Alder cycloaddition reactions to generate skeletal and stereochemical diversity (Scheme 3.7) [20]. Ninety-one heterocyclic compounds, including 16 different scaffolds, were synthesized. Alkoxy dienes **23** and **24** were generated by the Tebbe reaction. They reacted diastereoselectively with several dienophiles (specifically, benzoquinone, imines, maleimides, triazolinediones, and tetracyanoethylenes), and produced polycyclic compounds (e.g., **25** and **26**) that possessed high chemical diversity in terms of both scaffold and stereochemical diversity. Several hits targeting breast cancer cells were found within the library, demonstrating how diversity and complexity might be more important than library size for success.

In the context of developing new synthetic strategies that may allow for a more systematic exploration of the chemical space, Nelson's group reported the use of intra- and intermolecular Diels–Alder cycloaddition reactions in combination with metathesis cascade chemistry (Scheme 3.8) [21]. The 1,3-dienes (**27** to **29**), generated through the metathesis cascades, were reacted using three different approaches.

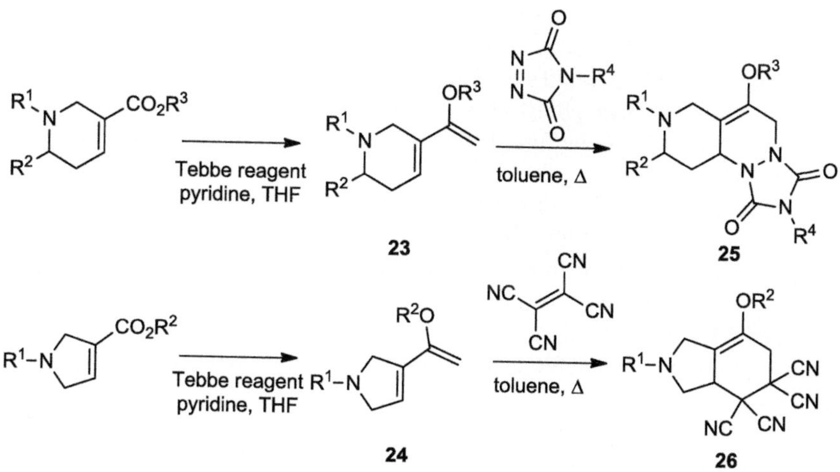

SCHEME 3.7 Branch reaction pathway: combination of Tebbe reaction with Diels–Alder cycloaddition reaction to generate polycyclic compounds.

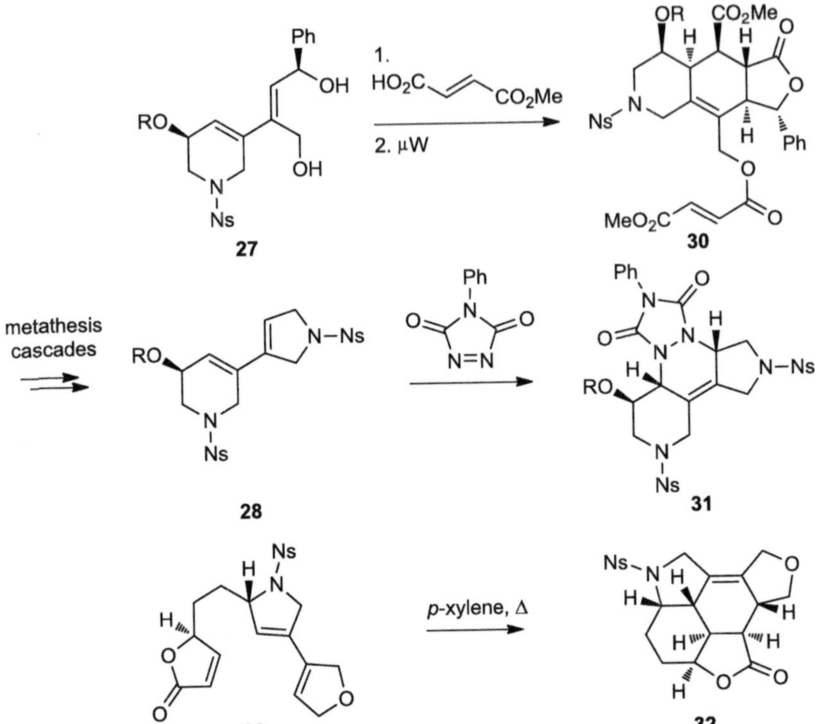

SCHEME 3.8 Use of intra- and intermolecular Diels–Alder cycloaddition reactions in combination with metathesis cascade chemistry to produce polycyclic compounds.

First, a dienophile was attached to the substrate followed by an intramolecular Diels–Alder cycloaddition (**30**). Second, intermolecular Diels–Alder reactions took place with different dienophiles (**31**). Third, intramolecular Diels–Alder reactions occurred between the diene and dienophile, which were both formed in cascade reactions (**32**).

The Diels–Alder cycloaddition reaction is often used as the last pairing step in build/couple/pair pathways. Kumagai et al. reported several examples where Diels–Alder cycloadditions followed enyne ring-closing metathesis reactions. Diene **33** reacted with 4-methyl-1,2,4-triazolin-3,5-dione (MTAD) to afford the tricyclic compound **34** as a single diastereomer (Scheme 3.9) [22]. The diastereoselectivity originated from the dienophile approaching the less-hindered side of the pyrroline ring **33**.

The second build/couple/pair pathway was generated by coupling three simple monomers, each with electrophilic and nucleophilic functionalities (Scheme 3.10) [23]. The pair phase, which focused on joining the nonpolar alkene and alkyne groups using Ru-catalyzed metathesis reactions, generated four final dienes (e.g., **36**) that underwent a Diels–Alder reaction with MTAD (e.g., **37**).

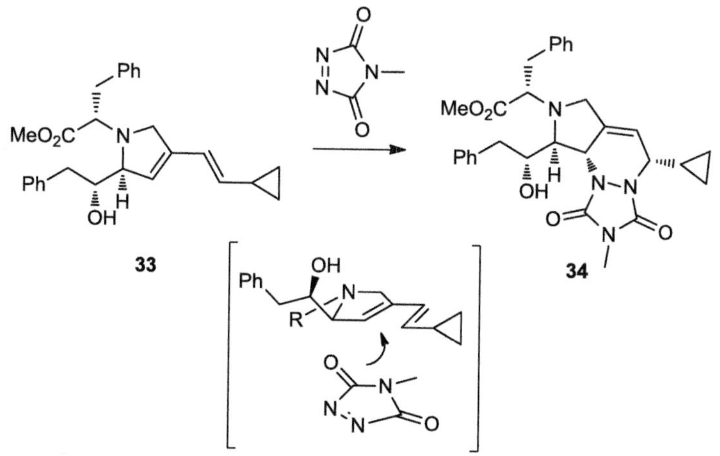

SCHEME 3.9 Diastereoselective Diels–Alder cycloaddition reaction of a diene generated through enyne ring-closing metathesis reactions.

3.2.2 Inverse Electron-Demand Diels–Alder Reaction

Lindsley et al. reported the synthesis of carpanone-like molecules using an endo-selective inverse electron-demand Diels–Alder cycloaddition reaction (Scheme 3.11) [24]. The Diels–Alder precursor was synthesized through diastereoselective oxidative

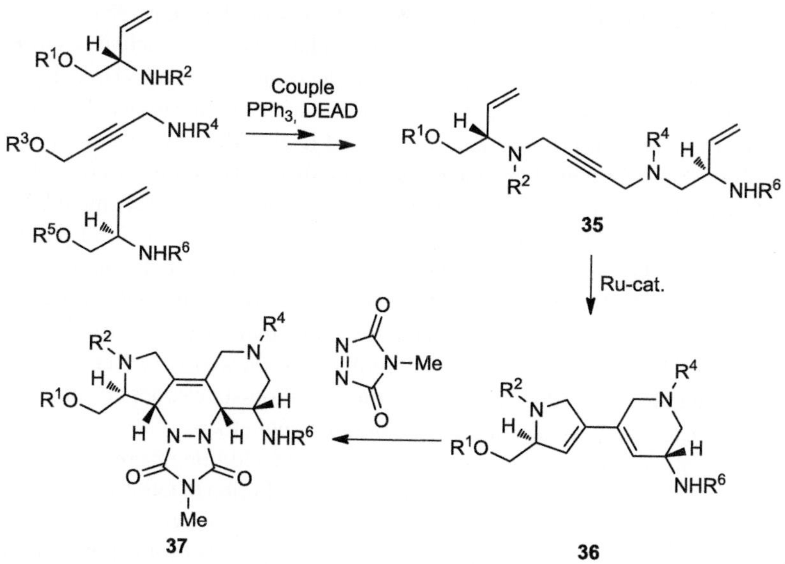

SCHEME 3.10 Oligomer-based build/couple/pair pathway. Diels–Alder reaction of dienes synthesized by Ru-catalyzed metathesis reactions.

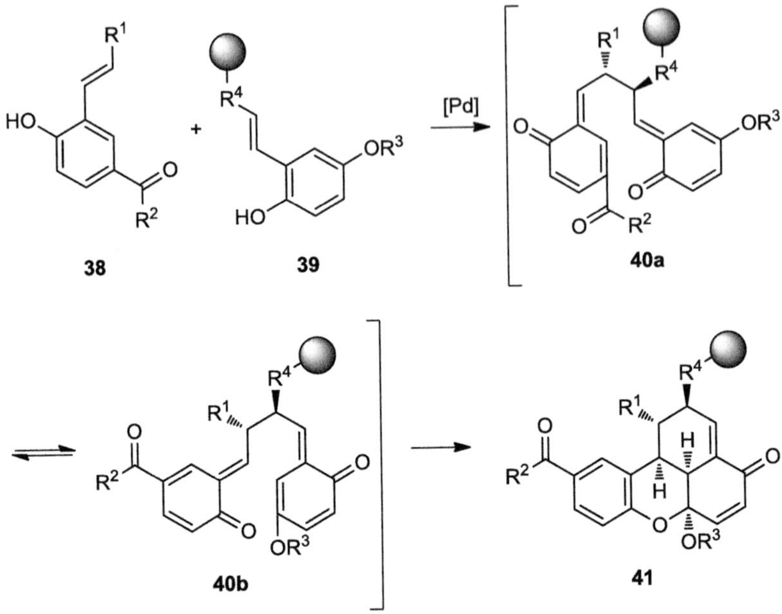

SCHEME 3.11 Synthesis of carpanone-like molecules using an endo-selective inverse electron-demand Diels–Alder cycloaddition reaction. (Adapted from [24], with permission; copyright © 2000 American Chemical Society.)

heterocoupling between electron-rich (**39**) and electron-deficient (**38**) phenols. To reduce its tendency toward homocoupling, the oxidatively more reactive electron-rich phenol **39** was immobilized on solid phase. The electronically matched transition state **40b** yielded tetracycle **41** as a single stereoisomer with four positions of diversity.

Another example of inverse electron-demand Diels–Alder cycloaddition applied to DOS was reported by Stavenger and Schreiber, who synthesized more than 4000 dihydropyrancarboxamides in an enantioselective fashion (Scheme 3.12) [25]. Asymmetric cycloaddition between vinyl ethers, immobilized on solid phase, and β,γ-unsaturated ketoesters was performed in the presence of (*t*-BuBOX)Cu(OTf_2) complex [(*S*) or (*R*)] to provide cycloadducts **42** to **45** with high diastereo- and enantioselectivities. Interestingly, unlike the *E*-enol ethers that provided a high level of diastereoselectivity, the *Z*-configured enol ethers showed only low-to-moderate diastereoselection. The endo–exo switch in the transition-state structure could explain the lower diastereoselectivity for the *Z*-configured enol ethers.

Sarkar et al. developed a hetero-Diels–Alder cycloaddition reaction of *N*-alkenyl iminium ions with alkenes, followed by nucleophilic addition of tetrahydropyridinium ion cycloadduct intermediates (Scheme 3.13) [26]. Highly functionalized piperidines were generated efficiently through three-component [4 + 2] cycloaddition/iminium ion addition reactions. Cycloaddition with electron-rich dienophiles was highly endo selective but lacked facial selectivity during nucleophilic addition (**46**). Steric factors

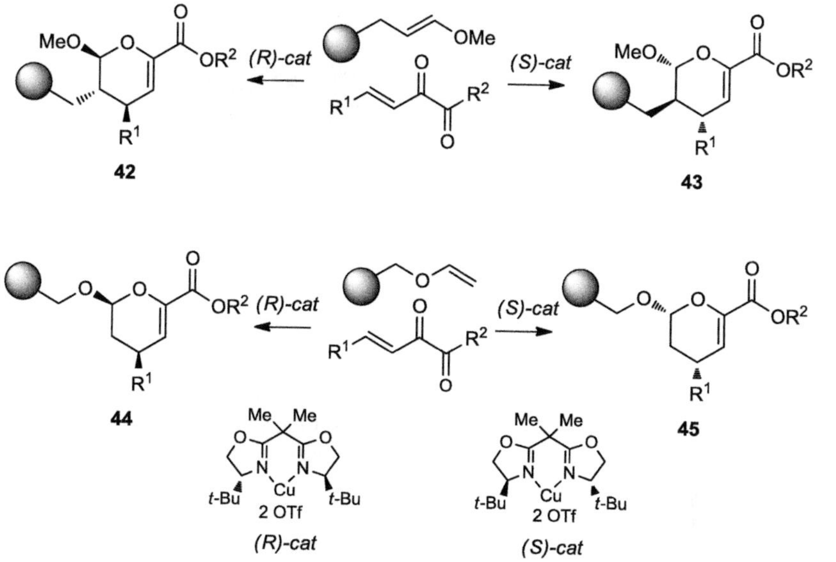

SCHEME 3.12 Asymmetric cycloaddition reaction of vinyl ethers and β,γ-unsaturated ketoesters to generate dihydropyrancarboxamides.

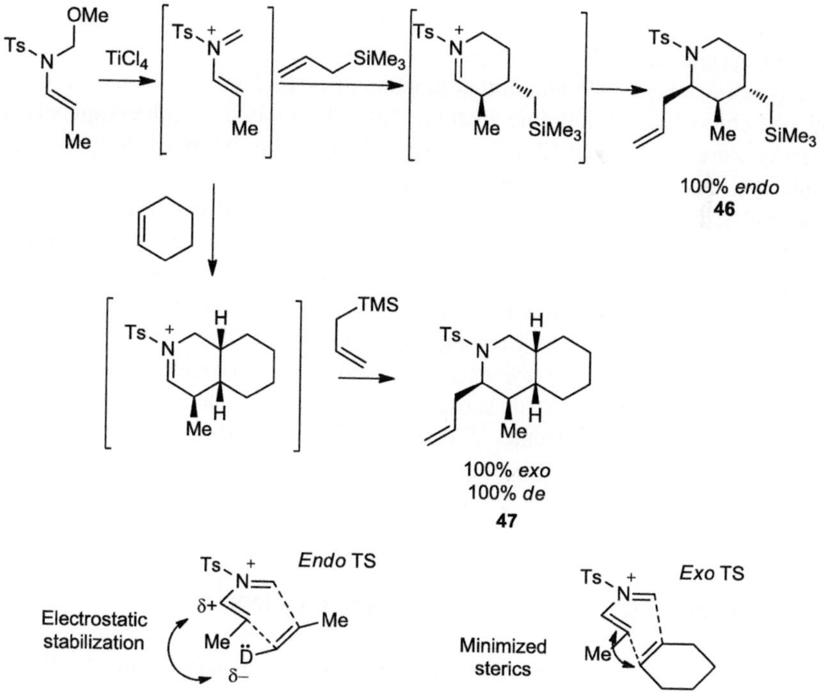

SCHEME 3.13 Hetero-Diels–Alder cycloaddition reaction of *N*-alkenyl iminium ions with alkenes followed by nucleophilic addition of tetrahydropyridinium ion cycloadduct intermediates.

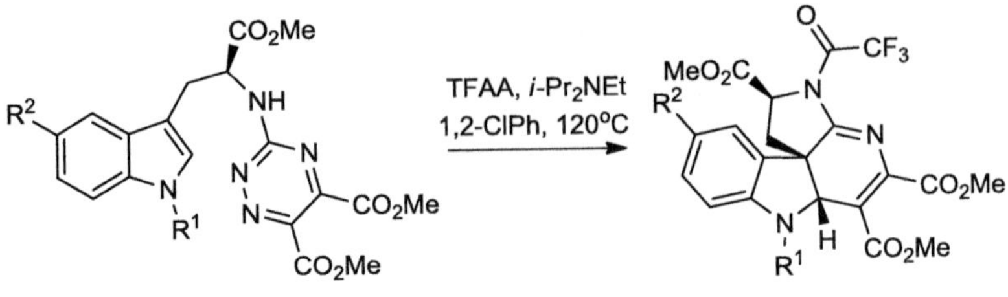

48 **49**

SCHEME 3.14 Intramolecular inverse electron-demand Diels–Alder reaction of tryptophan derivative triazines.

appeared to be critical in the transition state, explaining the high *exo* selectivity when cyclohexene, norbornene, and cyclooctene were used as dienophiles. In contrast to electron-rich dienophiles, facial selectivity was observed with unactivated olefins, and a single diastereomer was obtained (**47**).

Snyder's group reported a synthetic pathway in which tryptophan derivatives were joined to triazines before an intramolecular inverse electron-demand Diels–Alder reaction occurred (Scheme 3.14) [27]. The acylation of the inverse electron-demand Diels–Alder precursor **48** was required to lower the lowest unoccupied molecular orbital (LUMO) of the triazines, allowing for the cycloaddition reaction to occur upon heating. Using this general synthetic approach, three main scaffolds were synthesized (e.g., **49**) and further functionalized on *N*14 with the use of sulfonyl and acyl chlorides. 199 Indoline alkaloid scaffolds possessing high complexity and diversity were synthesized [28]. Several hits that weakly inhibited the growth of a number of *Plasmodium falciparum* lines were found.

Recently, Dow, Murrison, and co-workers demonstrated that when the synthesis and folding of substrates took place in a single reaction, the efficiency could increase. Accordingly, they reported the synthesis of several alkaloid-like scaffolds using a three-component reaction of secondary amines, carbonyl compounds, and triazines (Scheme 3.15) [21b,29]. Inverse electron-demand Diels–Alder cycloaddition reactions of triazines with previously generated enamines yielded 2-azadienes **50**, which were ideal substrates for subsequent folding reactions. When a dienophile is present within the molecule, 2-azadienes **50** can undergo an intramolecular Diels–Alder reaction, thus giving highly complex and diverse alkaloid-like scaffolds (**51** to **53**).

3.3 1,3-DIPOLAR CYCLOADDITION REACTIONS

The 1,3-dipolar cycloaddition reaction is an important transformation for the construction of polyheterocyclic molecules. The concurrent formation of two carbon–carbon bonds, yielding cyclic or even bicyclic heterocycles, makes this reaction a good choice for DOS.

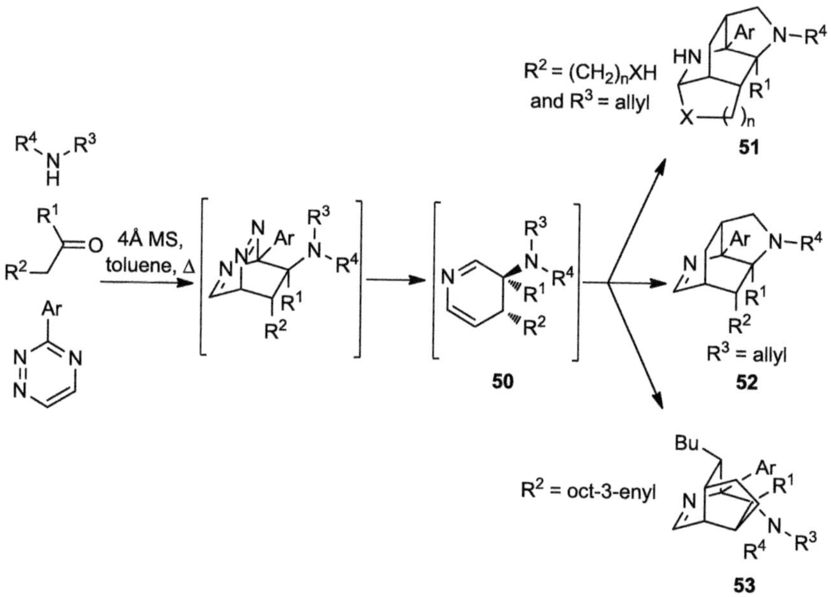

SCHEME 3.15 Diels–Alder cycloaddition reactions to yield alkaloid-like scaffolds.

In 1998, during the early days of DOS, Tan et al. reported a tandem acylation/1,3-dipolar cycloaddition with nitrones to yield tetracyclic scaffolds **55** (Scheme 3.16) [30]. Over 2 million small molecules were synthesized through further functionalization of chemical handles (i.e., iodobenzyl, γ-butyrolactone, and epoxide) within the tetracycles. Epoxycyclohexenol **54** was loaded on beads and then reacted with several nitrone carboxylic acids to generate tetracyclic compounds. There was no evidence that the reaction first proceeded by acylation, followed by 1,3-dipolar cycloaddition; nevertheless, the tandem reaction occurred with complete regio- and stereoselectivity.

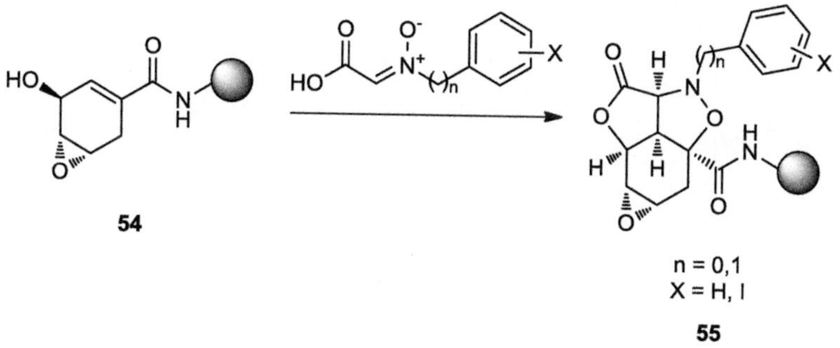

SCHEME 3.16 Tandem acylation/1,3-dipolar cycloaddition sequence of epoxycyclohexenol with nitrones to yield tetracyclic scaffolds.

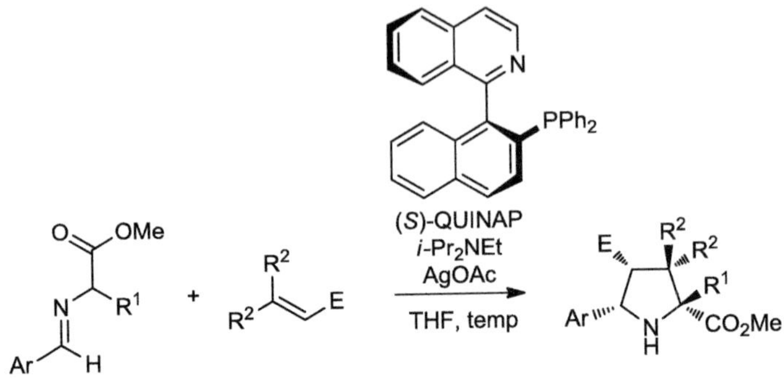

56

SCHEME 3.17 Synthesis of substituted pyrrolidines through silver(I)-catalyzed azomethine cycloaddition.

Pyrrolidines are common scaffolds present both in natural products and in small-molecule pharmaceuticals. For this reason, continuous efforts have been directed toward the development of new synthetic protocols, which could generate pyrrolidines efficiently and with complete control over the stereochemistry. In this context, Chen et al. described the synthesis of pyrrolidine-based scaffolds **56** through 1,3-dipolar cycloaddition of unactivated olefins with azometine ylides. The reaction proceeded enantioselectively, and up to four stereocenters were generated (Scheme 3.17) [31]. Based on the Ag(I)-catalyzed enentioselective cycloaddition developed by Longmire et al. [32], the authors found Ag(I) acetate/QUINAP to be the best catalyst system when a series of different α-imino esters, derived from a large variety of aromatic aldehydes, were utilized with various dipolarophiles. Good enantioselectivity was also observed when glycinate was replaced by various amino esters. This was the first example of a quaternary center at the 2-position of pyrrolidine generated by a catalytic asymmetric [3 + 2] cycloaddition reaction.

A few years later, Garner et al. reported an Ag(I)-catalyzed asymmetric coupling synthesis of highly functionalized pyrrolidine structures **57** (Scheme 3.18) [33]. The

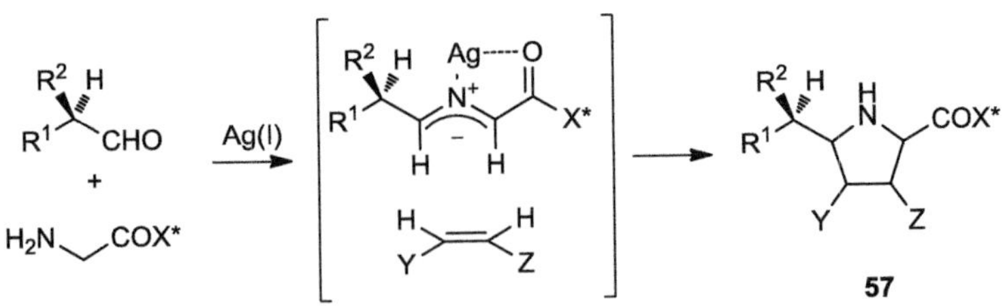

57

X* = Oppolzer's D- or L-camphorsultam

SCHEME 3.18 Silver(I)-catalyzed cycloaddition reaction of in situ–generated azomethine ylides with electron-deficient alkenes.

key step in the three-component reaction is the 1,3-dipolar cycloaddition reaction of an electron-deficient alkene with the azomethine ylide generated in situ upon treatment of the imine with Ag(I). Oppolzer's chiral glicyl sultam, used as the amine component, played an important role in this reaction, as this chiral auxiliary controlled the absolute stereochemistry and facilitated the reaction cascade by reducing the nucleophilicity of the amine, thereby preventing Michael addition, and by increasing the α-acidity of the intermediate imine, thus facilitating the formation of azomethine ylides. In addition, it was useful as a chemical handle for further synthetic transformations. The reaction exhibited high endo selectivity, tolerated several aldehydes, including sterically hindered, heteroalkyl-substituted, and chiral aldehydes, and showed high regioselectivity when monoactivated alkenes were used.

The presence of tetrahydro-1,2-oxazines in biologically relevant natural products and the lack of this heterocycle in common chemical libraries make them particularly interesting scaffolds. Young and Kerr reported a facile and versatile synthesis of highly functionalized tetrahydro-1,2-oxazines **58** through intermolecular 1,3-dipolar cycloaddition reaction of freshly prepared nitrones and cyclopropane diesters in the presence of $Yb(OTf)_3$ (Scheme 3.19) [34]. The cis product was the sole diastereomer produced in most cases. Among the new 26 isoxazines, skeletal congeners of FR900482 [35], an antitumor and antibiotic natural product, were synthesized.

Using rhodium(II)-catalyzed consecutive cyclization–cycloaddition reactions developed by Padwa's group [36], Oguri and Schreiber developed a folding pathway to generate indole alkaloid-like skeletons [37]. Rh(II)-catalyzed cyclization of rhodium carbenoid, generated from α-diazoketocarbonyl, with adjacent carbonyl oxygen yielded a carbonyl ylide, which underwent a 1,3-dipolar cycloaddition with the double bond at the 2- and 3-positions of the neighboring indole. Using the same reaction conditions, six different modes of intramolecular reaction were possible, based on all combinations of indole and α-diazo ketocarbonyl groups (Figure 3.1). Three different modes were reported, and in all cases the tandem cyclization–cycloaddition proceeded with excellent yield and complete diastereoselectivity (**59** to **61**) (Scheme 3.20).

Several studies propose that macrocycles, with their restricted conformation, seem to perform uniquely in biological assays [38]. To test this hypothesis, Looper et al. developed an elegant macrocycloaddition strategy that combines a 1,3-dipolar

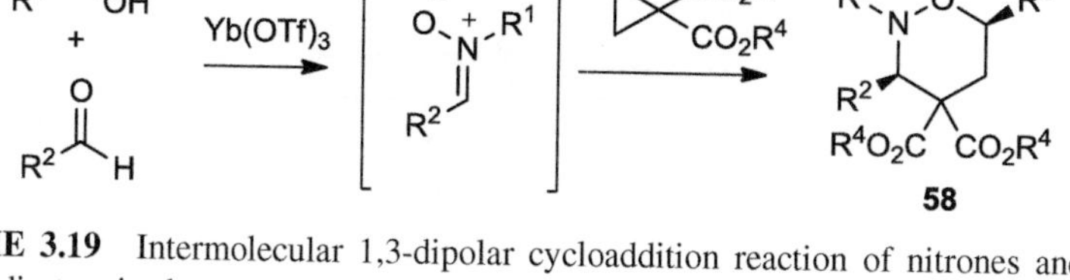

58

SCHEME 3.19 Intermolecular 1,3-dipolar cycloaddition reaction of nitrones and cyclopropane diesters in the presence of $Yb(OTf)_3$ to yield highly functionalized tetrahydro-1,2-oxazines.

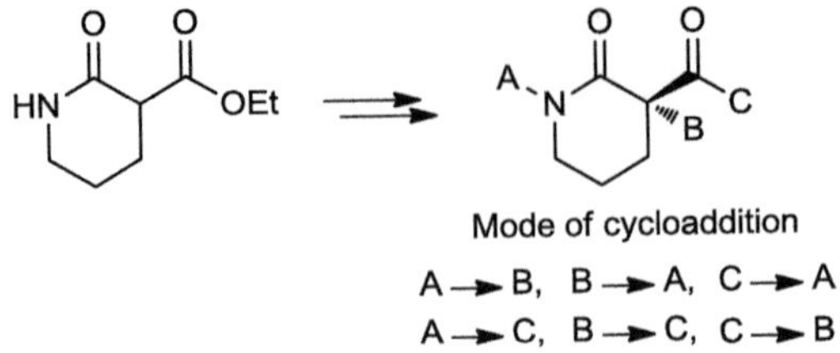

FIGURE 3.1 Multiple modes of cycloaddition.

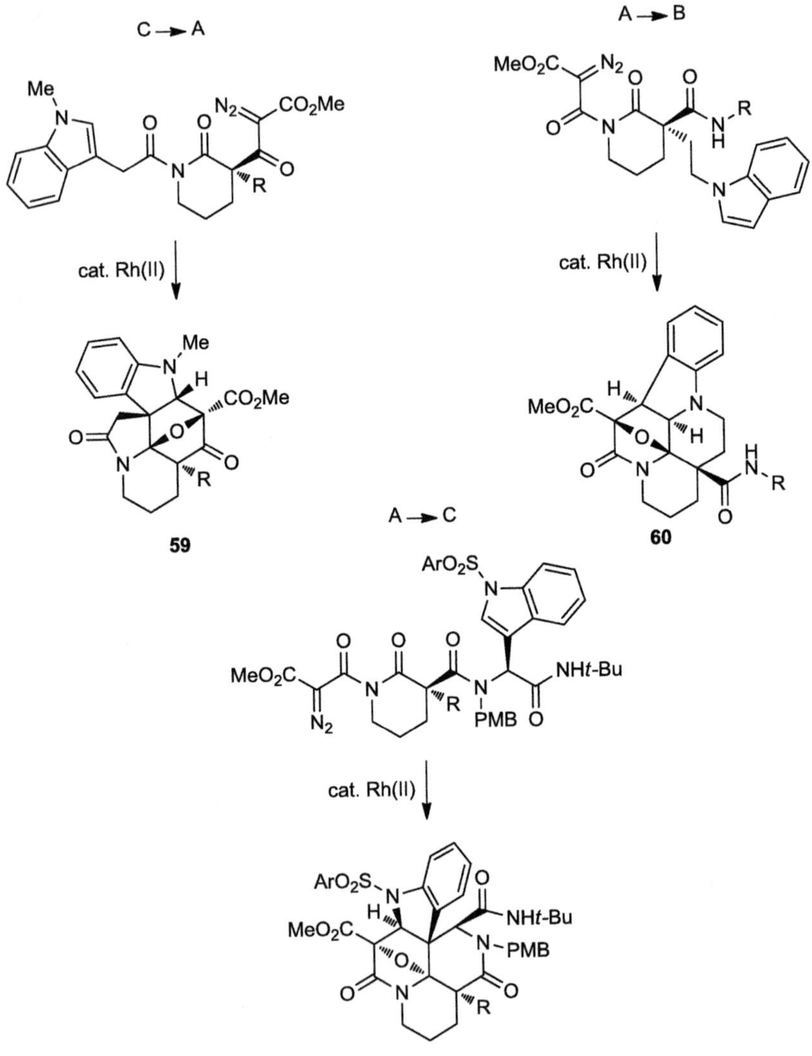

SCHEME 3.20 Folding pathway to generate indole alkaloid-like skeletons.

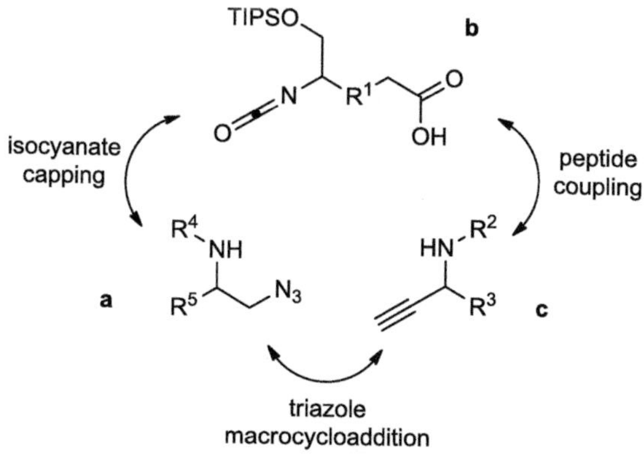

FIGURE 3.2 Synthetic strategy for assembling macrocyclic triazoles.

cycloaddition reaction of intermediate ylides with alkynes followed by retro-[4 + 2], with Cu(I)-catalyzed [3 + 2] cycloaddition of alkynes with azides, to provide an efficient pathway for the synthesis of macrocyclic triazoles [39]. A three-subunit system was used to study this macrocycloaddition and its tolerance to several substitutents and stereochemistry (Figure 3.2).

Two different options of cyclization were reported, depending on subunits A and C, which can be interchangeable. Subunit B, on the other hand, contains two functional groups, a carboxylic acid and an isocyanate, which can react with the amines in A and C to generate the cyclization precursor. A cycloaddition–cycloreversion strategy was developed to construct linear subunit B. First, a [3 + 2] cycloaddition reaction occurred between several alkynes and the carbonyl ylide; the latter was obtained by treating a diazoimido ester with Rh catalyst. Then the intermediate underwent a [4 + 2] cycloreversion reaction, generating an isocyanate that was trapped by several amines (**62** and **63**) (Scheme 3.21). The third subunit was installed by a simple coupling reaction of amines and carboxylic acids. Therefore, by using and interchanging several amino-azide and amino-alkyne subunits, a large number of precursors ready to cyclize were synthesized. The Cu(I)-catalyzed macrocycloaddition of alkynes with azides was successful regardless of stereochemistry, substituents, and ring size. Surprisingly, a significant amount of cyclodimers was obtained only when the macrocycloaddition produced a 21-membered monomer. No cyclodimers and good yield were obtained with 17-membered rings **64** and **65**.

Using functional group pairing reactions of enantioenriched Michael adducts, Comer et al. developed a build/couple/pair pathway to access five- to 10-membered fused and bridged rings (Figure 3.3) [40]. β-Nitrostyrenes, with bromine, allyl, and alkynyl substituents in the *ortho* position, were used in a stereoselective Michael addition using *Cinchona* alkaloid derivatives as the catalyst [41]. The conjugate addition of nucleophiles to β-nitrostyrenes proceeded with excellent enantioselectivity, thus providing highly functionalized scaffolds (**66**) suitable for selective functional group pairing reactions, such as ring-closing and enyne metathesis,

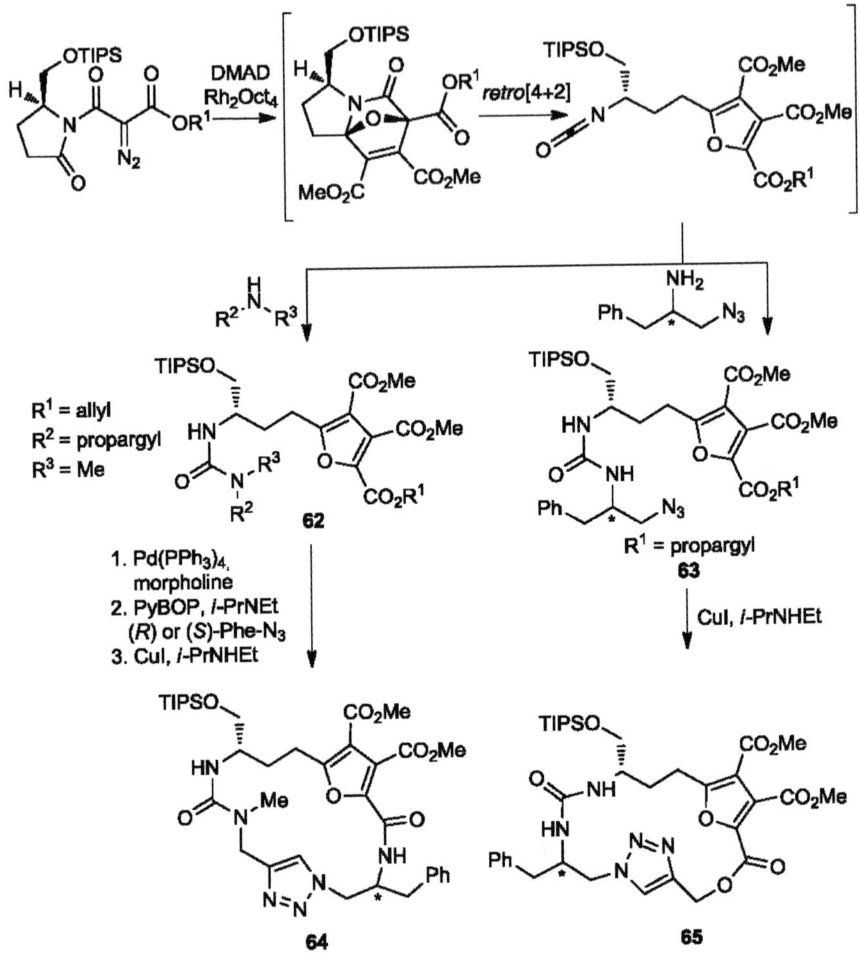

SCHEME 3.21 Synthetic route to 17-membered macrocyclic triazoles.

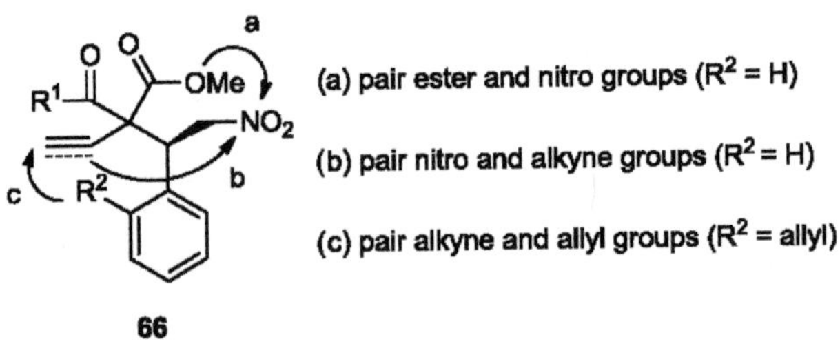

FIGURE 3.3 Functional group-pairing reactions of enantioenriched Michael adducts. (Adapted from [40], with permission; copyright © 2007 American Chemical Society.)

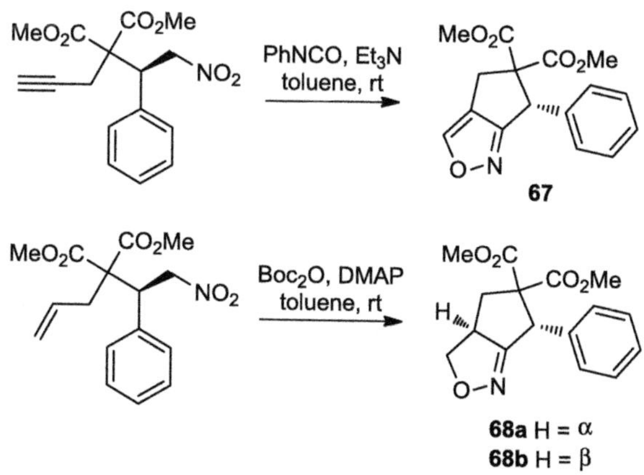

SCHEME 3.22 Intramolecular nitrile oxide cycloaddition reactions.

Pauson–Khand reaction, and 1,3-dipolar cycloaddition. Isoxazole **67** and isoxazolines **68a** and **68b** were synthesized through 1,3-dipolar cycloaddition of the derived nitrile oxides with the alkynyl and alkenyl substituents on the dicarbonyl compounds (Scheme 3.22).

Marcaurelle et al. reported a stereocontrolled [3 + 2] azomethine ylide-alkene cycloaddition that generated highly functionalized pyrrolidine templates **69**. Subsequent pyridone cyclization provided tricyclic and bridged bicyclic scaffolds (**70** and **71**) that, once loaded on lanterns, generated a library of 15,000 small molecules (Scheme 3.23) [42]. The reaction of glycine-derived azomethine ylides with α,β-unsaturated esters was promoted by AgOAc/DBU and provided cycloadducts in high yield and with excellent diastereoselectivity.

Brawn et al. developed a three-component reaction of enantioenriched allenylsilanes **72**, aldehydes, and silyl ethers, focusing on enantioenriched allenylsilanes as chiral carbon nucleophiles, which generated homopropargylic alcohols **75** and **76**. When an azide was embedded within the aldehyde components (**73**), a [3 + 2] cycloaddition of the newly generated alkyne with the pendant azide occurred, yielding tricyclic compounds with 1,2,3-triazoles (Scheme 3.24) [43]. The dipolar cyclization reaction resulted in the formation of triazole **77** as a single regioisomer. Electron-withdrawing and electron-donating groups on the aromatic ring were well tolerated. In addition, aliphatic acetals with embedded azides (**74**) proved to be successful in this reaction sequence, thus demonstrating the versatility of this methodology (**78**).

In their continued efforts at the Broad Institute to design and build a small-molecule screening collection, Marcaurelle et al. reported a synthetic pathway to access medium-sized and large ring systems [44]. Macrocycles are of interest in drug discovery both because their rigid structure often leads to high affinity and selectivity for protein targets and because they are present extensively in biologically relevant natural products. A build/couple/pair strategy was used successfully and a variety of

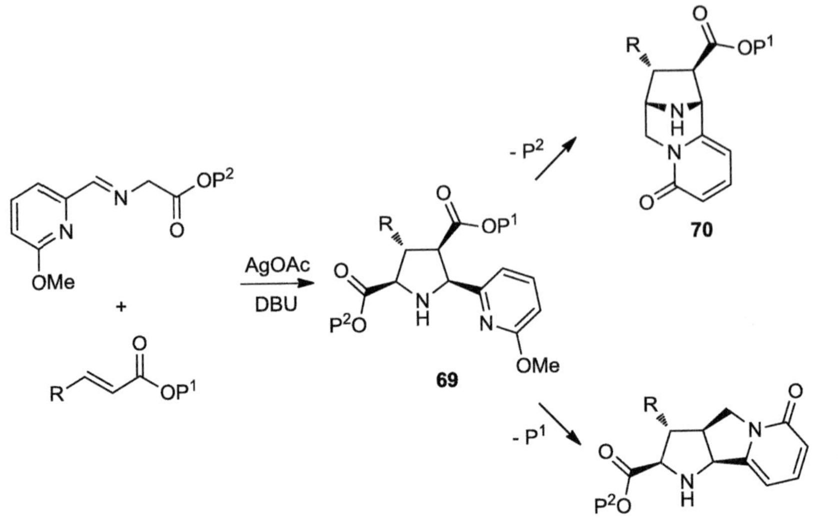

SCHEME 3.23 Silver (I)-promoted [3 + 2] cycloaddition reaction of azomethine ylides with unactivated alkenes to generate highly reactive pyrrolidine intermediates.

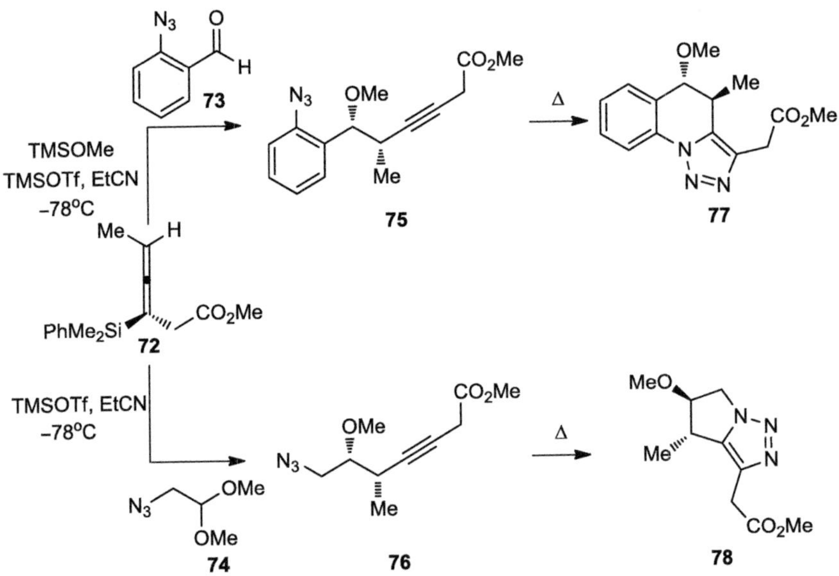

SCHEME 3.24 Homopropargylic alcohols are generated through a three-component reaction of enantioenriched allenylsilanes, aldehydes, and silyl ethers. When a pendant azide is present, an intramolecular [3 + 2] cycloaddition reaction occurs to generate bi- and tricyclic compounds.

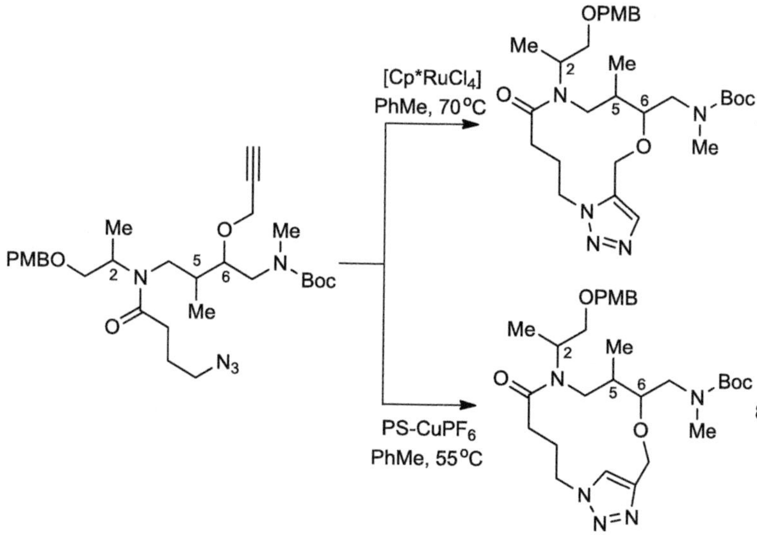

SCHEME 3.25 Regioselective Huisgen [3 + 2] cycloaddition reactions to yield 12- and 13-membered rings.

eight- to 14-membered rings were synthesized. In the pair phase three different reactions were utilized successfully in ring formation: nucleophilic aromatic substitution, ring-closing metathesis, and Huisgen [3 + 2] cycloaddition. In the latter, the authors expanded their previous study [45] to extend regioselective macrocyclization under catalyst control to a more structurally and stereochemically complex system (Scheme 3.25). 1,4- and 1,5-Macrocyclic triazoles were synthesized successfully with appreciable differences in shape based on a molecular shape-diversity analysis derived from a normalized PMI [17]. RuAAC afforded 12-membered rings. Using $[Cp^{*}RuCl]_4$ as the catalyst, all four diastereoisomers underwent macrocyclization regioselectively (**79a** to **d**). *syn*-Aldol-derived substrates were much better substrates than *anti*-aldol-derived substrates, which yielded triazoles in only modest yield because of dimer formation. In contrast with the 12-membered-rings, 13-membered macrocyclic triazoles were generated by intramolecular CuAAC. To achieve pseudodilution conditions and overcome dimer formation, a polystyrene-bound copper catalyst, PS-$CuPF_6$, was selected. Macrocyclization on all four diastereomers of azido-alkyne was successful with an opposite reactivity with respect to stereochemistry (**80a** to **d**). Unlike the *anti*-aldol-derived substrates, which easily underwent intramolecular CuAAC to yield triazoles without any dimerization, the *syn*-aldol-derived substrates afforded the desired triazoles in moderate yield with dimer formation.

Muncipinto et al., using the build/couple/pair strategy, recently reported an efficient and modular synthetic pathway in which readily synthesized and highly functionalized amino alcohols underwent allylic alcohol rearrangements and intramolecular 1,3-dipolar cycloadditions to yield densely functionalized isoxazoles, isoxazolines,

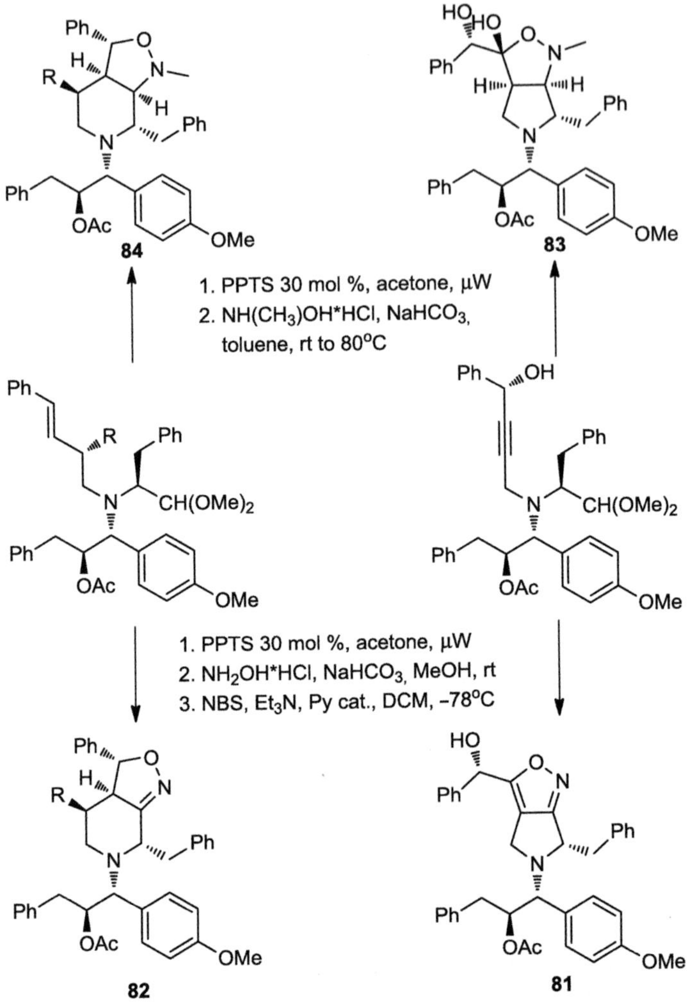

SCHEME 3.26 Intramolecular nitrone and nitrile oxide cycloaddition reactions.

and isoxazolidines **81** to **84** (Scheme 3.26) [46]. Intramolecular nitrile oxide (INOC) and nitrone (INC) cycloaddition reactions were explored. With few exceptions, all 1,3-dipolar cycloadditions proceeded in good yield and with excellent diastereoselectivity.

In addition, a suggested mechanism for the intramolecular nitrone cycloaddition reaction is shown in Figure 3.4. As illustrated in the transition states proposed, a major steric interaction exists between the oxygen and the benzyl group when the allylic group approaches the nitrone from the *si* side. In contrast, the minor steric interaction between the nitrone oxygen and the hydrogen at position 2 makes transition state (a) more favorable. The cis orientation at positions 3, 7, and 8 and the

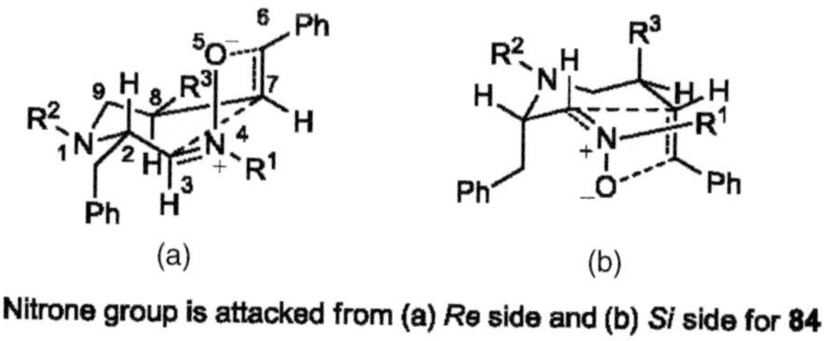

FIGURE 3.4 Suggested mechanism for intramolecular nitrone cycloaddition reaction. (Adapted from [46], with permission; copyright © 2010 American Chemical Society.)

trans orientation at positions 2 and 3 were assigned for **84** from coupling constants and NOE measurements. Additional evidence of this conformation is the fact that the substituent at positions 2 and 8 are both quasi-equatorial, whereas in the *si* side attack they are quasi-axial.

Several recent studies have focused on synthesizing collections of small molecules containing privileged scaffolds [47]. Privileged scaffolds are particular frameworks with drug-like and versatile binding properties, which lead to high hit rates and more drug-like libraries. In these efforts, Martin et al. recently reported the design and synthesis of a DOS pathway that combines a Mannich-type multicomponent assembly process (MCAP) [48] with several pairing cyclization reactions to afford highly functionalized heterocyclic ring systems. Benzodiazepine-, tetrahydropyridine-, aryl piperidine-, and tetrahydroisoquinoline-based scaffolds were synthesized. A thermally induced intramolecular Huisgen [3 + 2] cycloaddition reaction generated the 1,2,3-triazole–1,4-benzodiazepine scaffold **85**, which was further functionalized through *N*-diversification [49]. Using silyl-protected enols in MCAP, a free aldehyde was generated that, upon heating with *N*-methylhydroxylamine, underwent an intramolecular dipolar cycloaddition reaction, which generated the isoxazolidine **86** as a single diastereomer. Condensation of the aldehyde with sarcosine, followed by thermolysis of the oxazolidinone intermediate, afforded the pyrrolidine-fused scaffold **87** (Scheme 3.27) [50].

When a tetrahydroisoquinoline core was introduced in a MCAP, a series of complex heterocyclic scaffolds were synthesized in which tetrahydroisoquinolines are fused to quinazolone, dihydropyrimidine-2,4-dione, 1,4-diazepine-2,5-dione, and 1,5-benzodiazepin-2,5-dione rings (Scheme 3.28) [51]. 7-Bromodihydroisoquinoline **88** was used for these MCAPs, because it allowed for further functionalization via cross-coupling reactions. The three-component reaction of **88**, *trans*-crotonyl chloride, and silyl enol ether catalyzed by TMSOTf provided the aldehyde **89**. Then the aldehyde underwent intramolecular nitrone cycloaddition when *N*-methylhydroxylamine was added, affording isoxazolidine **90**. Compounds with this tricylic ring system are potent α_2-adrenoceptor antagonists [52]. Moreover, when **88** reacted with either zinc phenylacetylide or ethynylmagnesium bromide, the intermediate generated was trapped with *o*-azidobenzylchloride to yield an amide. Upon warming to room temperature, the amide easily underwent a [3 + 2] cycloaddition

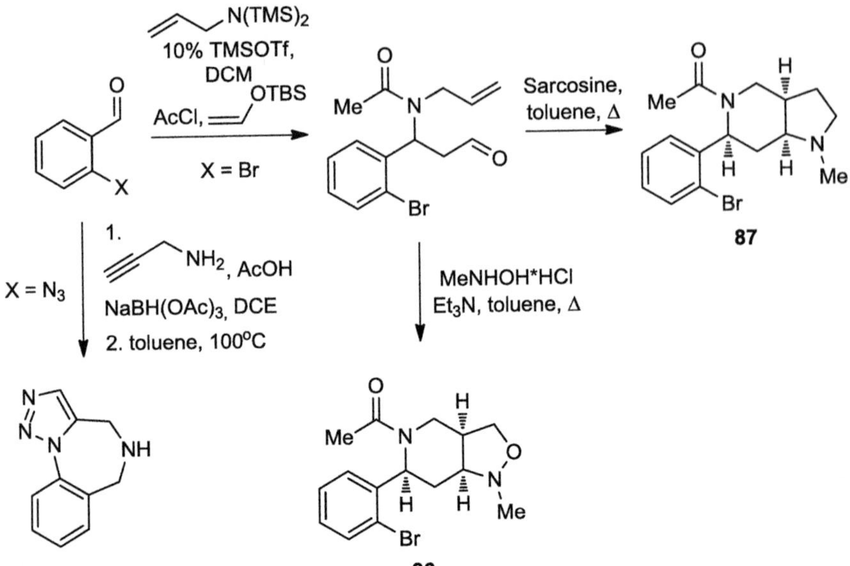

SCHEME 3.27 Tandem MCR/intramolecular [3 + 2] dipolar cycloaddition sequence.

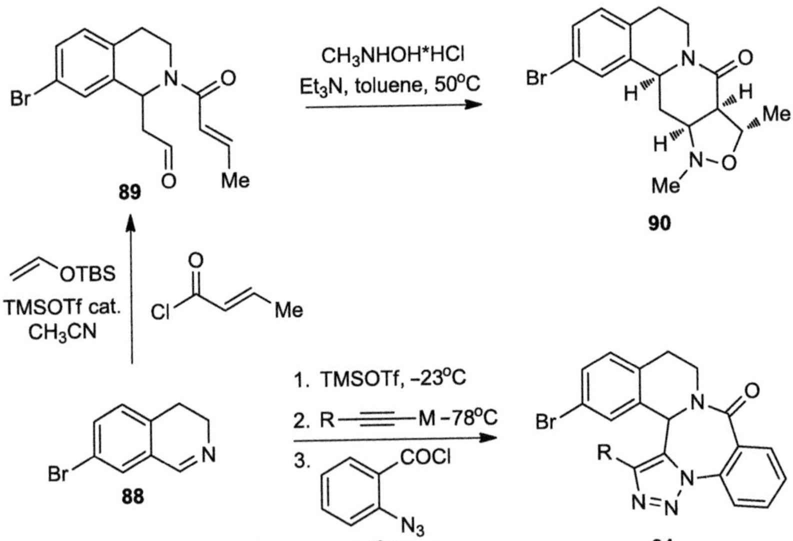

SCHEME 3.28 Tandem MCR/intramolecular nitrone cycloaddition and MCR/Huisgen [3 + 2] cycloaddition sequences.

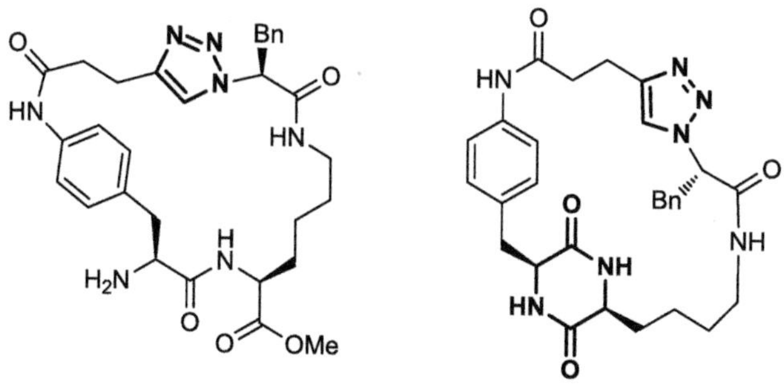

FIGURE 3.5 Synthesis of macrocyclic peptidomimetic compounds; triazole and monoketopiperazine moieties are introduced as amide bioisosteres.

reaction to afford triazolo-1,5-benzodiazepine-2-ones **91**. This scaffold is present in compounds that inhibit serine proteases and bind to the benzodiazepine receptor [53].

Inspired by the large number of biologically active molecules containing peptide motifs, Isidro-Llobet and co-workers designed and developed a DOS strategy for the generation of macrocyclic peptidomimetic compounds, which are still underrepresented in current small-molecule libraries [54]. A triazole ring and a ketopiperazine moiety were introduced as bioisosteres of the amide to improve the instability of the peptides [55]. Besides conferring rigidity, triazoles also mimic either the cis- or trans-like configuration of the amide bond (Figure 3.5). In a proof-of-concept study, they reported the synthesis of 12 compounds with four different scaffolds: *cis*-DKPs, *trans*-DKPs, and 1,4- and 1,5-triazoles.

A click-type 1,3-dipolar cycloaddition reaction was used to build the macrocyclic triazole-based peptidomimetic. Cu(I) and [Cp*$RuCl_4$] were used as catalysts to synthesize 1,4- and 1,5-disubstituted triazoles, respectively. The 1,3-dipolar cycloaddition reactions were fully regioselective, and no epimerization of the stereogenic center adjacent to the azide occurred (Scheme 3.29).

3.4 MISCELLANEOUS CYCLOADDITIONS

Several other cycloaddition reactions have been used in DOS for their ability to introduce structural and skeletal diversity. In line with their interest in the reactivity of siloxy alkynes, Sweis et al. reported the discovery of the first silver-catalyzed [2 + 2] cycloaddition of siloxy alkynes with esters, nitriles, and ketones [56]. This transformation is an efficient and versatile method to access highly functionalized siloxy cyclobutenes (Scheme 3.30). $AgNTf_2$ was found to best effect the cycloaddition of α,β-unsaturated ketones, nitriles, and esters with siloxy alkynes to yield the corresponding siloxy cyclobutenes **92**. In addition, [4.2.0]bicyclic silyl enol ether was synthesized when using cyclohexenone. Aryl-substituted siloxy alkynes were also well tolerated. (*E*)- and (*Z*)-crotonates gave the same trans-substituted siloxy cyclobutene, suggesting that the reaction proceeds via a stepwise mechanism. Silver

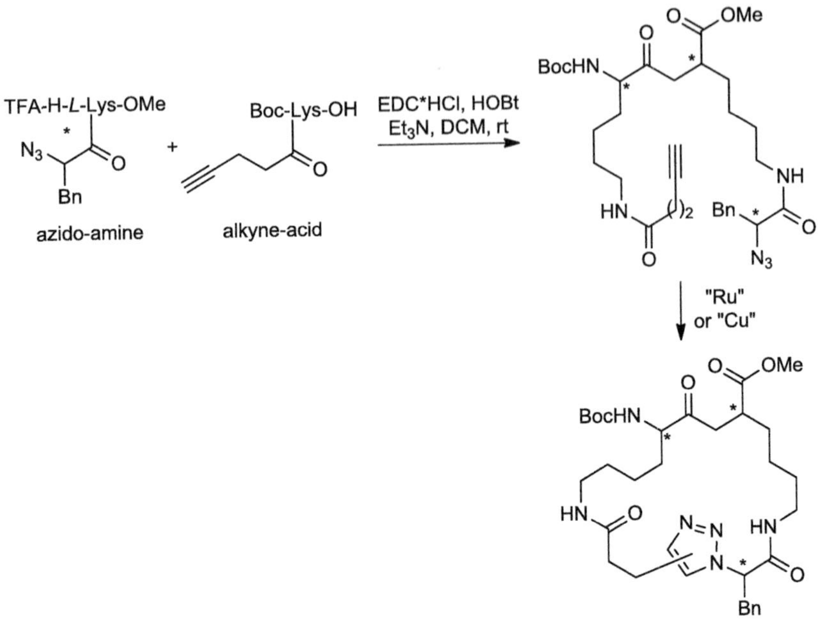

SCHEME 3.29 Regioselective 1,3-dipolar cycloaddition reaction to generate 1,4- and 1,5-disubstituted triazoles.

activates the siloxy alkyne, followed by 1,4-addition and trapping of the ketenium ion. Interestingly, the resulting siloxy cyclobutenes can be further functionalized through their ester, nitrile, and ketone functionalities, allowing for the synthesis of small libraries.

Using intramolecular pairing cyclization reactions of nonpolar–nonpolar and polar–nonpolar groups strategically placed on highly functionalized amino alcohols, Kumagai et al. generated a short modular synthetic pathway to access 15 different scaffolds in only five steps [22]. A ruthenium-catalyzed [5 + 2] cycloaddition reaction between the nonpolar alkene and alkyne groups afforded the cyclic

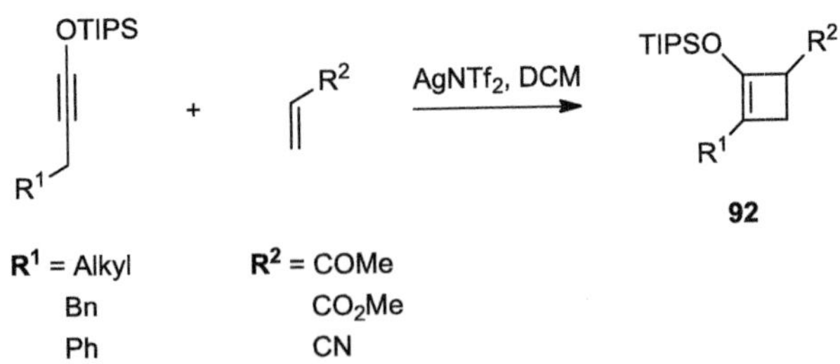

SCHEME 3.30 Silver-catalyzed [2 + 2] cycloaddition of siloxy alkynes with esters, nitriles, and ketones.

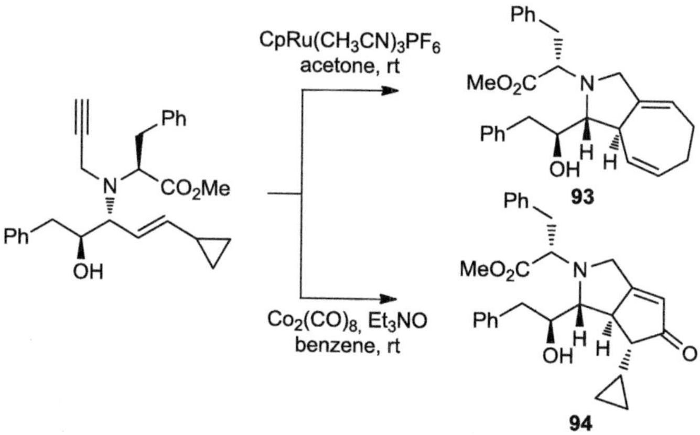

SCHEME 3.31 Ruthenium-catalyzed [5 + 2] cycloaddition reaction and Pauson–Khand reaction to generate azabicyclic compounds.

diene **93** through a cyclopropyl ring-opening/reductive elimination sequence (Scheme 3.31). The reaction proceeded diastereolectively to afford a single diastereomer. A Pauson–Khand reaction, a [2 + 2 + 1] cycloaddition, was also reported. In the presence of trimethylamine *N*-oxide, the azabicyclo-[3.3.0] **94** was synthesized with high diastereoselectivity (>10 : 1 dr).

Zhou et al. developed a cobalt-catalyzed intramolecular [2 + 2 + 2] cyclization reaction of dialkynylnitriles to provide tetrahydronaphthyridines scaffolds, which, compared to fully aromatized naphthyridine, have not received particular attention in drug discovery (Scheme 3.32) [57]. $CpCo(CO)_2$ was found to be the best catalyst, and by changing the length of the aminonitriles, the authors were able to access 5,6,7,8-tetrahydro-1,6-naphthyridines ($n = 2$), 6,7-dihydro-5*H*-pyrrolo[3,4-*b*]pyridines ($n = 1$), and 6,7,8,9-tetrahydro-5*H*-pyrido[2,3-*d*]azepines ($n = 3$). Tetrahydropyranonaphtyridine ($n = 2$) **95** was chosen for library production. With this strategy, several substituents at *C*6 and *C*8 were introduced and the secondary amine at *N*2 was subsequently used as the diversification point for the formation of ureas, amides, and sulfonamides.

Gray et al. described a build/couple/pair strategy to synthesize highly substituted pyridines **96** through a transition metal–mediated [2 + 2 + 2] cycloaddition reaction

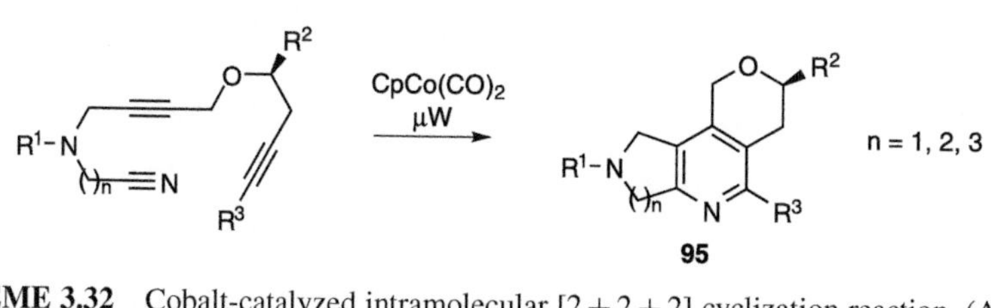

SCHEME 3.32 Cobalt-catalyzed intramolecular [2 + 2 + 2] cyclization reaction. (Adapted from [57], with permission; copyright © 2008 American Chemical Society.)

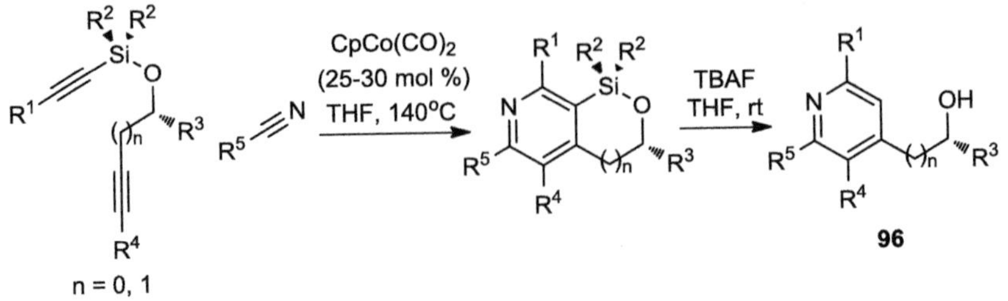

SCHEME 3.33 Cobalt-catalyzed [2 + 2 + 2] cycloaddition reaction of silyl ether–tethered diynes with nitriles.

of silyl ether–tethered diynes with nitriles (Scheme 3.33) [58]. The tether plays a key role by controlling the regioselectivity and allowing for transformation of the bicyclic intermediates into final highly functionalized monocyclic pyridines. $CpCo(CO)_2$ was found to be the best catalyst, and the use of THF as a solvent, inducing catalyst turnover, allowed an irradiation-free reaction. A large variety of nitriles was well tolerated for this reaction, with the exception of strongly deactivated and pyrazine-containing nitriles, and nitriles with free hydroxyl and amine groups. For example, aromatic, heteroaromatic, benzylic, and primary and secondary aliphatic nitriles easily underwent cycloaddition reaction, giving the desired products. Large substituents on the silylated alkyne were also well tolerated, and since the distal alkyne did not need to be terminal, its substituents allowed access to fully substituted pyridines. One of the desilylated pyridines was discovered to inhibit the neuregulin/Erb4 pathway, which appeared likely to be involved in breast cancer and schizophrenia [59].

Guo et al. reported an elegant phosphine-promoted [3 + 3] annulation of aziridines and allenoates to generate highly substituted tetrahydropyridines **99** (Scheme 3.34) [60]. The authors started with the idea that the phosphonium enolate zwitterionic intermediate may interact with the aziridines and give the desired product. PPh_3 was found to be the best phosphine source when *N*-nosylaziridine **97** and diethyl 2-vinylidenesuccinate **98** were used. The reaction proceeded with excellent diastereoselctivity (trans/cis 10 : 1). Interestingly, the presence of the α, β, and β′ carbon atoms of the starting allenoate in the final core, rather than α, β, and γ as with

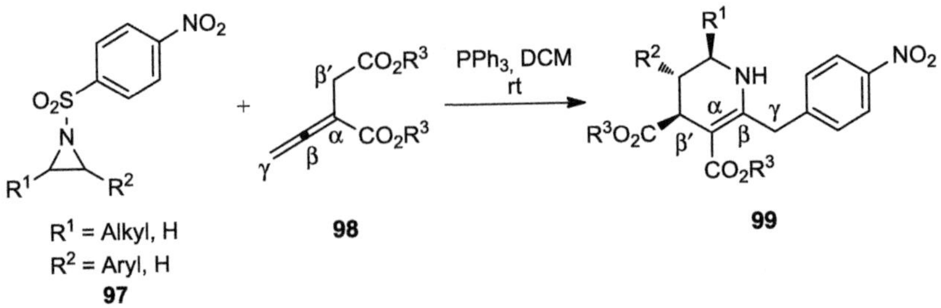

SCHEME 3.34 Phosphine-promoted [3 + 3] annulation of aziridines and allenoates to generate highly substituted tetrahydropyridines.

the phosphine-mediated [3 + 2] annulations, suggested a new reaction mechanism. In addition, the loss of SO_2 was observed with the *p*-nitrobenzene attached to the γ-carbon. Aryl-substituted aziridines worked very well with good-to-excellent yield and good 1,2-trans-diastereoselectivity. Electron-withdrawing or electron-donating groups in the ortho, meta, and para positions were all well tolerated. Loss of diastereoselectivity occurred when an alkyl-substituted aziridine was used, whereas the unsubstituted aziridine gave poor yield. It is noteworthy that these tetrahydropyridines can be functionalized further through the NH and both ester groups.

Bauer et al. developed a unique approach to yield 10 different classes of polycyclic scaffolds using transition metal–mediated cycloaddition and cyclization reactions of enynes and diynes (Scheme 3.35) [61]. A *t*-butylsulfinamide [62] group (**100**, **101**) was used as a unique lynchpin, due to the fact that it gives asymmetric induction during substrate synthesis, can be deprotonated and *N*-alkylated, and can readily be deprotected or oxidized. The authors performed a reactivity study in several transition metal–mediated reactions of the enynes and diynes and found that these substrates were suitable for Ru-, Co-, Rh-, and Ni-catalyzed reactions. Among over 25 reactions investigated, eight appeared to be selective and efficient for library production. All enynes readily underwent Pauson–Khand reaction, yielding [5,5]-bicyclic

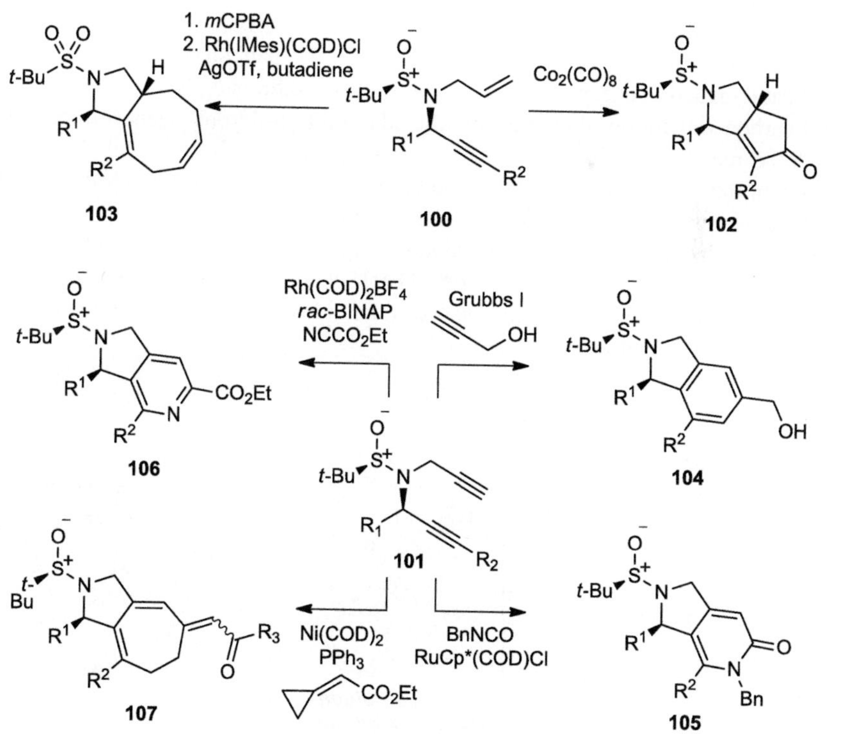

SCHEME 3.35 Transition metal–mediated cycloaddition and cyclization reactions of enynes and diynes.

cyclopentapyrrolidinone scaffolds as single diastereomers (**102**). The Rh-catalyzed butadiene [4 + 2 + 2] cycloaddition reaction developed by Evans [63], after oxidation to the corresponding *t*-butylsulfonamides, afforded [5,8]-bicyclic cyclooctapyrrolidine scaffolds (**103**) diastereoselectivily, but with moderate yield. Several [2 + 2 + 2] cyclotrimerization were also identified for the diynes. Reaction of diyne **101** with propargyl alcohol in the presence of Grubbs's first-generation catalyst provided [5,6]-bicyclic isoindoline scaffolds (**104**) with complete regioselectivity [64]. Treatment of the diyne with benzyl isocyanate and Yamamoto's Ru(II) catalyst afforded [5,6]-bicyclic pyrrolopyridone scaffolds (**105**) [65]. The diyne also cyclotrimerized with ethyl cyanoformate efficiently using Tanaka's Rh(I) catalyst [66], providing [5,6]-bicyclic pyrrolopyridine scaffolds (**106**) regioselectively. Finally, [5,7]-bicyclic cycloheptapyrrolidine scaffolds (**107**) were synthesized through [3 + 2 + 2] Ni(0)-catalyzed cyclotrimerization [67] of the diyne with ethyl cyclopropylideneacetate. The reaction provided single regioisomers, but inseparable *E*/*Z* mixtures.

In an effort to develop new chemical methodologies for the production of diverse sultam-based libraries, Samarakoon et al. recently reported a one-pot, complementary ambiphile-pairing (CAP) reaction to produce 5,2,1-dibenzooxathiazocine-2,2-dioxides (**108**) in a formal [4 + 4] cyclization pathway (Scheme 3.36) [68]. In this CAP strategy, *o*-quinone methides (*o*-QMs) and *o*-fluorobenzensulfonamides were utilized as ambiphilic synthons. *o*-QMs are powerful intermediates both as Michael acceptors and as dienes in cycloaddition reactions, but they have never been reported as ambiphiles in a hetero [4 + 4] cyclization. In the pair phase, a formal [4 + 4] cyclization reaction occurred in which aza-Michael addition at the *exo*-methylene *o*-QM carbon was followed by nucleophilic addition to *o*-fluorobenzenesulfonamides via S_NAr reaction.

When the *o*-QM was replaced by an epoxide, a formal [4 + 3] epoxide cascade reaction occurred with epoxide ring opening followed by either an S_NAr or oxa-Michael cyclization to provide benzothiaoxazepine-1,1′-dioxide and oxathiazepine-1,1′-dioxide scaffolds, respectively (**109**, **110**) (Scheme 3.37) [69]. In both cases, the reaction tolerated several epoxides and sulfonamides, demonstrating the versatility and extent of this pathway.

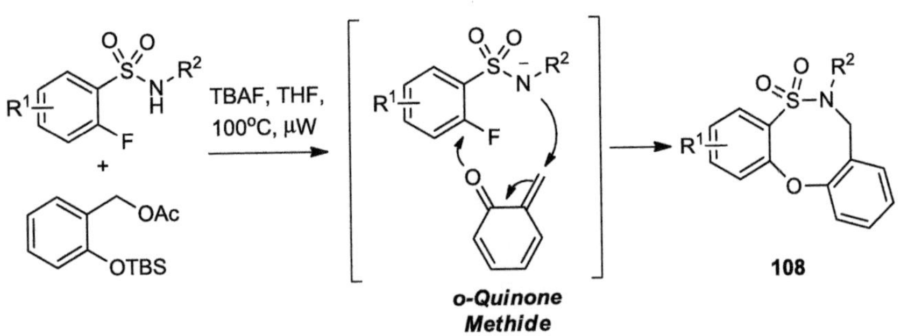

SCHEME 3.36 Formal [4 + 4] cyclization of *o*-quinone methides and *o*-fluorobenzensulfonamides.

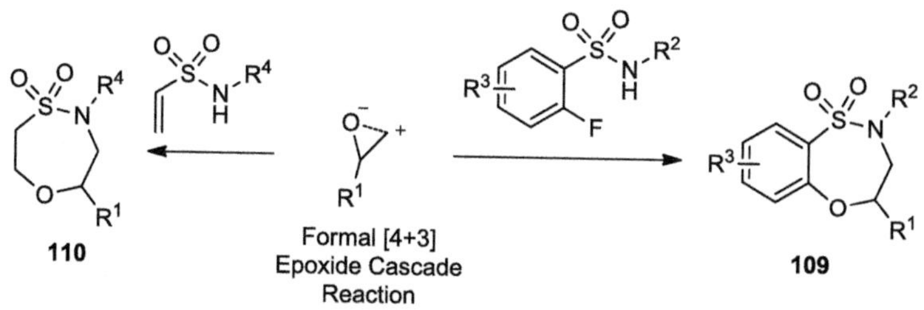

SCHEME 3.37 Formal [4 + 3] epoxide cascade followed by either an S_NAr or oxa-Michael cyclization.

Zhang and Panek developed a Lewis acid–catalyzed [5 + 2] annulation of substituted *N*-methylisatin dimethyl ketals and enantioenriched crotyl sylanes [70] to access spirooxindoles with excellent stereocontrol (Scheme 3.38) [71]. The cis isomer **111** was the main product when lower temperature and a shorter reaction time were employed. A longer reaction time or more polar solvents favored the trans isomer **112**, suggesting that the cis isomer might epimerize through a spiro-ring-opening mechanism. Interestingly, the cis product **111** could easily be converted to the trans isomer **112** in the presence of $BF_3{\cdot}OEt_2$ under microwave irradiation. This made it possible to access all possible stereoisomers for elucidating stereostructure–activity relationships during screening campaigns.

A proposed mechanism is illustrated in Figure 3.6. Because of the absence of peri-like interaction, the transition state **113** is favored over the (*E*)-oxonium intermediate **114**, and generates the cis as the main diastereomer under kinetic conditions. Finally, the complexity of the spirooxindoles can be enhanced by employing different combinations of functionalized silyl alcohols and functionalized isatines.

Panek's group later extended this work by combining the use of enantioenriched crotyl sylanes with rhodium(II)-catalyzed asymmetric cyclopropanation to generate cyclopentene compounds as building blocks for the synthesis of small-molecule libraries (Scheme 3.39) [72,73]. Cyclopropanation reaction, followed by Lewis

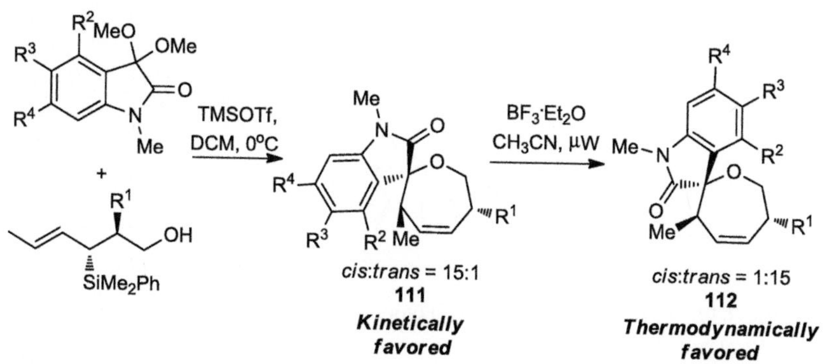

SCHEME 3.38 [5 + 2] Annulation of substituted *N*-methylisatin dimethyl ketals and enantioenriched crotyl sylanes.

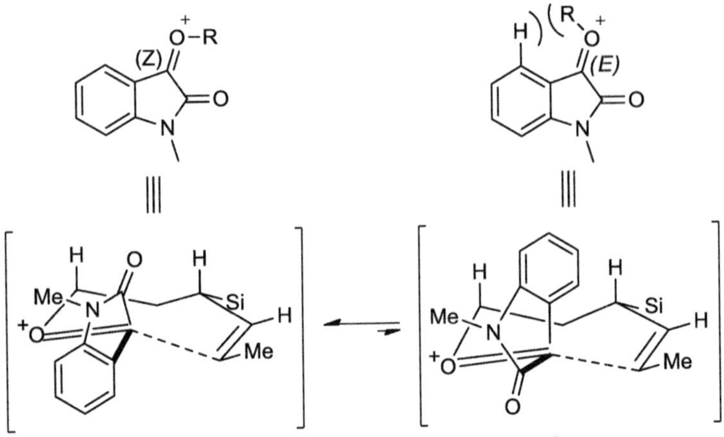

cis
113

trans
114

FIGURE 3.6 Proposed mechanism of the [5 + 2] annulation. (Adapted from [71], with permission; copyright © 2009 American Chemical Society.)

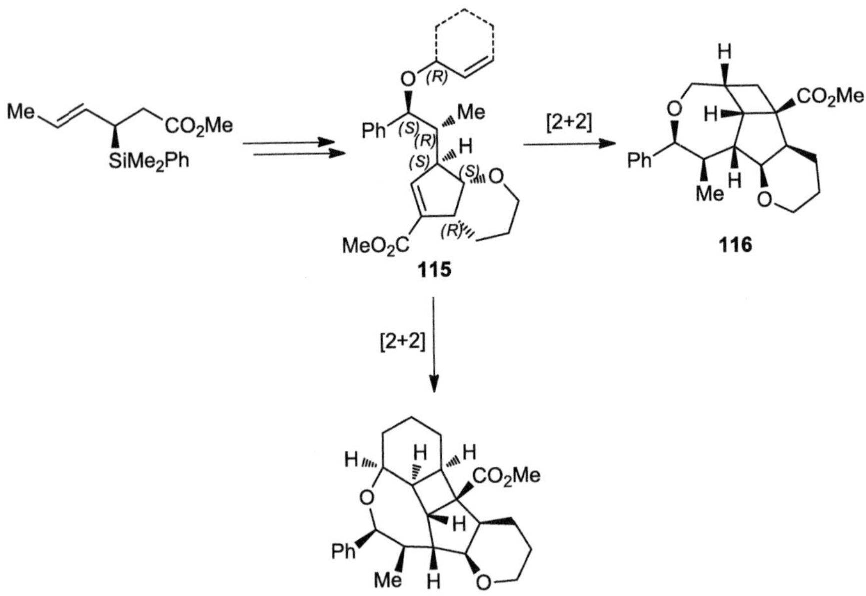

115

116

117

SCHEME 3.39 Intramolecular [2 + 2] photocycloaddition reactions of α,β-unsaturated esters and isolated olefins.

acid–catalyzed rearrangement, provided intermediates such as **115**. Bromoaryl and allyl ether functionalities were introduced for further ring construction. [2 + 2] Photocycloaddition reactions between α,β-unsaturated esters and isolated olefins occurred, producing complex and condensed polycyclic systems not accessible by other procedures. The [2 + 2] intramolecular cycloaddition reaction of the cyclopentene with the allyl group proceeded easily to give the tetracyclic **116** with a cis–syn–cis structure. Photocycloaddition with the cyclohexene resulted in the fused-ring product **117** as a single diastereomer, but with poor yield. Surprisingly, only the *S*-allylic ether underwent the photocycloaddition.

3.5 CONCLUSIONS

In recent years, due primarily to the decreasing cost of DNA sequencing [74], there have been extraordinary advances in genomics that have allowed the identification and validation of new biological targets underlying human diseases [2]. Unfortunately, many of these new therapeutic targets are considered "undruggable" since there are no compounds in screening desks able to modulate them [75]. This scenario demonstrates a clear need for novel molecular scaffolds able to interact specifically with these targets. Chemical libraries with high structural complexity and skeletal and stereochemical diversity allow for the exploration of larger regions of chemical space, thereby increasing the chances of success. Certainly, DOS, with its ability to generate small-molecule libraries with a high degree of structural and functional diversity, represents one of the best strategies for achieving this goal. Within DOS, cycloaddition reactions represent a powerful tool to achieve the structural diversity needed in compound collections. Moreover, the rigidity and restricted conformation of cyclic and heterocyclic compounds, generated through cycloaddition reactions, also seem to be responsible for specific and unique interactions with biological targets, conferring a higher probability of success in screening campaigns. In addition, as illustrated in this chapter, many DOS compounds address targets that have traditionally been considered too difficult to hit, thus demonstrating how this approach might be the best path to follow to take advantage of recent advances in genomics and to take a step toward meeting society's expectations for scientific research [76].

REFERENCES

1. (a) S. L. Schreiber, *Science* **2000**, *287*, 1964–1969; (b) M. D. Burke, S. L. Schreiber, *Angew. Chem. Int. Ed.* **2004**, *43*, 46–58.
2. (a) R. I. Tepper, R. Roubenoff, in: *Genomics and Personalized Medicine*, H. F. Willard, G. S. Ginsburg, Eds., Elsevier, New York, **2009**, pp. 335–342; (b) A. T. Bansal, M. R. Barnes, *Curr. Opin. Drug Discov. Dev.* **2008**, *11*, 303–310.
3. (a) W. R. J. D. Galloway, A. Isidro-Llobet, D. R. Spring, *Nat. Commun.* **2000**, *1*, 80; (b) C. Lipinski, A. Hopkins, *Nature* **2004**, *432*, 855–861; (c) S. J. Haggarty, *Curr. Opin.*

Chem. Biol. **2005**, *9*, 296–303; (d) F. Lovering, J. Bikker, C. Humblet, *J. Med. Chem.* **2009**, *52*, 6752–6756.

4. (a) T. E. Nielsen, S. L. Schreiber, *Angew. Chem. Int. Ed.* **2008**, *47*, 48–56; (b) S. L. Schreiber, *Nature*, **2009**, *457*, 153–154.
5. D. Lee, J. K. Sello, S. L. Schreiber, *Org. Lett.* **2000**, *2*, 709–712.
6. K. Paulvannan, *Tetrahedron Lett.* **1999**, *40*, 1851–1854.
7. H. E. Blackwell, L. Perez, R. A. Stavenger, J. A. Tallarico, E. C. Eatough, M. A. Foley, S. L. Schreiber, *Chem. Biol.* **2001**, *8*, 1167–1182.
8. O. Kwon, S. B. Park, S. L. Schreiber, *J. Am. Chem. Soc.* **2002**, *124*, 13402–13404.
9. S. Woo, N. Squires, A. G. Fallis, *Org. Lett.* **1999**, *1*, 573–575.
10. (a) D. Schomburg, M. Thielmann, E. Winterfeldt, *Tetrahedron Lett.* **1985**, *26*, 1705–1706; (b) A. Prelle, E. Winterfeldt, *Heterocycles* **1989**, *28*, 333–346.
11. N. Kumar, M. Kiuchi, J. A. Tallarico, S. L. Schreiber, *Org. Lett.* **2005**, *7*, 2535–2538.
12. (a) J. Aubé, G. L. Milligan, C. J. Mossman, *J. Org. Chem.* **1992**, *57*, 1635–1637; (b) P. Desai, K. Schildknegt, K. A. Agrios, C. Mossman, G. L. Milligan, J. Aubé, *J. Am. Chem. Soc.* **2000**, *122*, 7226–7232.
13. (a) H. Greger, *Planta Med.* **2006**, *72*, 99–113; (b) K. J. Frankowski, J. E. Golden, Y. Zeng, Y. Lei, J. Aubé, *J. Am. Chem. Soc.* **2008**, *130*, 6018–6024.
14. (a) J. E. Golden, J. Aubé, *Angew. Chem. Int. Ed.* **2002**, *41*, 4316–4318; (b) Y. Zeng, D. S. Reddy, E. E. Hirt, J. Aubé, *Org. Lett.* **2004**, *6*, 4993–4995; (c) K. J. Frankowski, V. Setola, J. M. Evans, B. Neuenswander, B. L. Roth, J. Aubé, *Proc. Natl. Acad. Sci. U.S.A.* **2011**, *108*, 6727–6732.
15. K. J. Frankowski, E. E. Hirt, Y. Zeng, B. Neuenswander, D. Fowler, F. Schoenen, J. Aubé, *J. Comb. Chem.* **2007**, *9*, 1188–1192.
16. J. Cui, J. Hao, O. A. Ulanovskaya, J. Dundas, J. Liang, S. A. Kozmin, *Proc. Natl. Acad. Sci. U.S.A.* **2011**, *108*, 6763–6768.
17. W. H. Sauer, M. K. Schwarz, *J. Chem. Inf. Comput. Sci.* **2003**, *43*, 987–1003.
18. S. T. Handy, P. A. Grieco, C. Mineur, L. Ghosez, *Synlett* **1995**, 565–567.
19. M. G. Vander Heiden, L. C. Cantley, C. B. Thompson, *Science* **2009**, *324*, 1029–1033.
20. (a) Z. Wang, S. Castellano, S. S. Kinderman, C. E. Argueta, A. B. Beshir, G. Fenteany, O. Kwon, *Chem. Eur. J.* **2011**, *17*, 649–654; (b) D. Cruz, Z. Wang, J. Kibbie, R. Modlin, O. Kwon, *Proc. Natl. Acad. Sci. U.S.A.* **2011**, *108*, 6769–6774.
21. (a) C. O'Leary-Steele, P. J. Pedersen, T. James, T. Lanyon-Hogg, S. Leach, J. Hayes, A. Nelson, *Chem. Eur. J.* **2010**, *16*, 9563–9571; (b) M. Dow, M. Fisher, T. James, F. Marchetti, A. Nelson, *Org. Biomol. Chem.* **2012**, *10*, 17–28.
22. N. Kumagai, G. Muncipinto, S. L. Schreiber, *Angew. Chem. Int. Ed.* **2006**, *45*, 3635–3638.
23. D. A. Spiegel, F. C. Schroeder, J. R. Duvall, S. L. Schreiber, *J. Am. Chem. Soc.* **2006**, *128*, 14766–14767.
24. C. W. Lindsley, L. K. Chan, B. C. Goess, R. Joseph, M. D. Shair, *J. Am. Chem. Soc.* **2000**, *122*, 422–423.
25. R. A. Stavenger, S. L. Schreiber, *Angew. Chem. Int. Ed.* **2001**, *40*, 3417–3421.
26. N. Sarkar, A. Banerjee, S. G. Nelson, *J. Am. Chem. Soc.* **2008**, *130*, 9222–9223.
27. (a) S. C. Benson, L. Lee, L. Yang, J. K. Snyder, *Tetrahedron* **2000**, *56*, 1165–1180; (b) L. E. Brown, K. C.-C. Cheng, W.-G. Wei, P. Yuan, P. Dai, R. Trilles, F. Ni, J. Yuan,

R. MacArthur, R. Guha, R. L. Johnson, X.-Z. Su, M. M. Dominguez, J. K. Snyder, A. B. Beeler, S. E. Schaus, J. Inglese, J. A. Porco, Jr., *Proc. Natl. Acad. Sci. U.S.A.* **2011**, *108*, 6775–6780.

28. W. Wei, C. Cai, S. Kota, V. Takahashi, F. Ni, A. D. Strosberg, J. K. Snyder, *Bioorg. Med. Chem. Lett.* **2009**, *19*, 6926–6930.
29. S. Murrison, S. K. Maurya, C. Einzinger, B. McKeever-Abbas, S. Warriner, A. Nelson, *Eur. J. Org. Chem.* **2011**, 2354–2359.
30. (a) D. S. Tan, M. A. Foley, M. D. Shair, S. L. Schreiber, *J. Am. Chem. Soc.* **1998**, *120*, 8565–8566; (b) D. S. Tan, M. A. Foley, B. R. Stockwell, M. D. Shair, S. L. Schreiber, *J. Am. Chem. Soc.* **1999**, *121*, 9073–9087.
31. C. Chen, X. Li, S. L. Schreiber, *J. Am. Chem. Soc.* **2003**, *125*, 10174–10175.
32. J. M. Longmire, B. Wang, X. Zhang, *J. Am. Chem. Soc.* **2002**, *124*, 13400–13401.
33. P. Garner, H. Ümit Kaniskan, J. Hu, W. J. Youngs, M. Panzner, *Org. Lett.* **2006**, *8*, 3647–3650.
34. I. S. Young, M. A. Kerr, *Org. Lett.* **2004**, *6*, 139–141.
35. I. Uchida, T. Shigehiro, K. Hiroshi, K. Sumio, M. Hashimoto, T. Tada, K. Shigetaka, Y. Morimoto, *J. Am. Chem. Soc.* **1987**, *109*, 4108–4109.
36. (a) D. L. Hertzog, D. J. Austin, W. R. Nadler, A. Padwa, *Tetrahedron Lett.* **1992**, *33*, 4731–4734; (b) J. M. Mejia-Oneto, A. Padwa, *Org. Lett.* **2004**, *6*, 3241–3244; for review, see (c) M. H. Osterhout, W. R. Nadler, A. Padwa, *Synthesis* **1994**, 123–140; (d) A. Padwa, D. M. Weingarten, *Chem. Rev.* **1996**, *96*, 223–269.
37. H. Oguri, S. L. Schreiber, *Org. Lett.* **2005**, *7*, 47–50.
38. (a) M. T. Burger, P. A. Bartlett, *J. Am. Chem. Soc.* **1997**, *119*, 12697–12698; (b) Y.-K. Kim, M. A. Arai, T. Arai, J. O. Lamenzo, E. F. Dean, N. Patterson, P. A. Clemons, S. L. Schreiber, *J. Am. Chem. Soc.* **2004**, *126*, 14740–14745.
39. R. E. Looper, D. Pizzirani, S. L. Schreiber, *Org. Lett.* **2006**, *8*, 2063–2066.
40. E. Comer, E. Rohan, L. Deng, J. A. Porco, Jr., *Org. Lett.* **2007**, *9*, 2123–2126.
41. H. Li, Y. Wang, L. Tang, F. Wu, X. Liu, C. Guo, B. M. Foxman, L. Deng, *Angew. Chem. Int. Ed.* **2005**, *44*, 105–108.
42. L. A. Marcaurelle, C. Johannes, D. Yohannes, B. P. Tillotson, D. Mann, *Bioorg. Med. Chem. Lett.* **2009**, *19*, 2500–2503.
43. R. A. Brawn, M. Welzel, J. T. Lowe, J. S. Panek, *Org. Lett.* **2010**, *12*, 336–339.
44. L. A. Marcaurelle, E. Comer, S. Dandapani, J. R. Duvall, B. Gerard, S. Kesavan, M. D. Lee IV, H. Liu, J. T. Lowe, J.-C. Marie, C. A. Mulrooney, B. A. Pandya, A. Rowley, T. D. Ryba, B.-C. Suh, J. Wei, D. W. Young, L. B. Akella, N. T. Ross, Y.-L. Zhang, D. M. Fass, S. A. Reis, W.-N. Zhao, S. J. Haggarty, M. Palmer, M. A. Foley, *J. Am. Chem. Soc.* **2010**, *132*, 16962–16976.
45. A. R. Kelly, J. Wei, S. Kesavan, J.-C. Marié, N. Windmon, D. W. Young, L. A. Marcaurelle, *Org. Lett.* **2009**, *11*, 2257–2260.
46. G. Muncipinto, T. Kaya, J. A. Wilson, N. Kumagai, P. A. Clemons, S. L. Schreiber, *Org. Lett.* **2010**, *12*, 5230–5233.
47. (a) D. A. Horton, G. T. Bourne, M. L. Smythe, *Chem. Rev.* **2003**, *103*, 893–930; (b) Y. Che, B. R. Brooks, G. R. Marshall, *J. Comput. Aided Mol. Des.* **2006**, *20*, 109–130; (c) M. E. Welsch, S. A. Snyder, B. R. Stockwell, *Curr. Opin. Chem. Biol.* **2010**, *14*, 347–361.

48. S. F. Martin, B. Benage, L. S. Geraci, J. E. Hunter, M. P. Mortimore, *J. Am. Chem. Soc.* **1991**, *113*, 6161–6171.
49. J. R. Donald, S. F. Martin, *Org. Lett.* **2011**, *13*, 852–855.
50. S. Hardy, S. F. Martin, *Org. Lett.* **2011**, *13*, 3102–3105.
51. B. A. Granger, K. Kaneda, S. F. Martin, *Org. Lett.* **2011**, *13*, 4542–4545.
52. J. W. VanDyke, Jr., H. J. Havera, R. D. Johnson, *J. Med. Chem.* **1971**, *15*, 91–93.
53. (a) D. K. Mohapatra, P. K. Maity, M. Shabab, M. I. Khan, *Bioorg. Med. Chem. Lett.* **2009**, *19*, 5241–5245; (b) L. Bertelli, G. Biagi, I. Giorgi, O. Livi, C. Manera, V. Scartoni, C. Martini, G. Giannaccini, L. Trincavelli, P. L. Barili, *Farmaco* **1998**, *53*, 305–311.
54. A. Isidro-Llobet, T. Murillo, P. Bello, A. Cilibrizzi, J. T. Hodgkinson, W. R. J. D. Galloway, A. Bender, M. Welch, D. R. Spring, *Proc. Natl. Acad. Sci. U.S.A.* **2011**, *108*, 6793–6798.
55. (a) W. Horne, C. Olsen, J. Beierle, A. Montero, M. Gadhiri, *Angew. Chem. Int. Ed.* **2009**, *48*, 4718–4724; (b) V. D. Bock, D. Speijer, H. Hiemstra, J. H. Maarseveen, *Org. Biomol. Chem.* **2007**, *5*, 971–975.
56. R. F. Sweis, M. P. Schramm, S. A. Kozmin, *J. Am. Chem. Soc.* **2004**, *126*, 7442–7443.
57. Y. Zhou, A. B. Beeler, S. Cho, Y. Wang, S. G. Franzblau, J. K. Snyder, *J. Comb. Chem.* **2008**, *10*, 534–540.
58. B. L. Gray, X. Wang, W. C. Brown, L. Kuai, S. L. Schreiber, *Org. Lett.* **2008**, *10*, 2621–2624.
59. (a) G. Corfas, K. Roy, J. D. Buxbaum, *Nat. Neurosci.* **2004**, *7*, 575–580; (b) T. T. Junttila, M. Sundvall, M. Lundin, J. Lundin, M. Tanner, P. Härkönen, H. Joensuu, J. Isola, K. Elenius, *Cancer Res.* **2005**, *65*, 1384–1393; (c) L. Kuai, S. E. Ong, J. M. Madison, X. Wang, J. R. Duvall, T. A. Lewis, C. J. Luce, S. D. Conner, D. A. Pearlman, J. L. Wood, S. L. Schreiber, S. A. Carr, E. M. Scolnick, S. J. Haggarty, *Chem. Biol.* **2011**, *18*, 891–906.
60. H. Guo, Q. Xu, O. Kwon, *J. Am. Chem. Soc.* **2009**, *131*, 6318–6319.
61. R. A. Bauer, C. M. DiBiasi, D. S. Tan, *Org. Lett.* **2010**, *12*, 2084–2087.
62. (a) F. Ferreira, C. Botuha, F. Chemia, A. Perez-Luna, *Chem. Soc. Rev.* **2009**, *38*, 1162–1186; (b) D. Morton, R. A. Stockman, *Tetrahedron* **2006**, *62*, 8869–8905.
63. J. E. Robinson, E. W. Baum, A. N. Fazal, P. A. Evans, *J. Am. Chem. Soc.* **2002**, *124*, 8782–8783.
64. B. Witulski, T. Stengel, J. M. Fernandez-Hernandez, *Chem. Commun.* **2000**, 1965–1966.
65. Y. Yamamoto, K. Kinpara, T. Saigoku, H. Takagishi, S. Okuda, H. Nishiyama, K. Itoh, *J. Am. Chem. Soc.* **2005**, *127*, 605–613.
66. K. Tanaka, N. Suzuki, G. Nishida, *Eur. J. Org. Chem.* **2006**, *71*, 3917–3922.
67. K. Maeda, S. Saito, *Tetrahedron Lett.* **2007**, *48*, 3173–3176.
68. T. B. Samarakoon, M. Y. Hur, R. D. Kurtz, P. R. Hanson, *Org. Lett.* **2010**, *12*, 2182–2185.
69. A. Rolfe, T. B. Samarakoon, P. R. Hanson, *Org. Lett.* **2010**, *12*, 1216–1219.
70. (a) J. S. Panek, M. Yang, *J. Am. Chem. Soc.* **1991**, *113*, 6594–6600; (b) J. S. Panek, M. Yang, F. Xu, *J. Org. Chem.* **1992**, *57*, 5790–5792; (c) J. Wu, J. S. Panek, *Angew. Chem. Int. Ed.* **2010**, *49*, 6165–6168.
71. Y. Zhang, J. S. Panek, *Org. Lett.* **2009**, *11*, 3366–3369.
72. (a) S. Kesavan, Q. Su, J. Shao, J. A. Porco, Jr., J. S. Panek, *Org. Lett.* **2005**, *7*, 4435–4438; (b) Y. Chen, J. A. Porco, Jr., J. S. Panek, *Org. Lett.* **2007**, *9*, 1529–1532; (c) S. Kesavan, J. S. Panek, J. A. Porco, Jr., *Org. Lett.* **2007**, *9*, 5203–5206.

73. J. Wu, J. Becerril, Y. Lian, H. M. L. Davies, J. A. Porco, Jr., J. S. Panek, *Angew. Chem. Int. Ed.* **2011**, *50*, 5938–5942.

74. C. H. Arnaud, *Chem. Eng. News* **2009**, *87*, 16–19.

75. (a) G. L. Verdine, L. D. Walensky, *Clin. Cancer Res.* **2007**, *13*, 7264–7270; (b) D. J. Payne, M. N. Gwynn, D. J. Holmes, D. L. Pompilano, *Nat. Rev. Drug Discov.* **2007**, *6*, 29–40; (c) S. L. Schreiber, *Nature* **2009**, *457*, 153–154.

76. S. L. Schreiber, *ChemBioChem* **2009**, *10*, 26–29.

4

PHOSPHINE ORGANOCATALYSIS AS A PLATFORM FOR DIVERSITY-ORIENTED SYNTHESIS

ZHIMING WANG AND OHYUN KWON

4.1 INTRODUCTION

With the supposition that libraries of small organic molecules would expedite the identification of novel therapeutic lead compounds, combinatorial synthesis has been forged into one of the pillars of synthetic organic chemistry [1–3]. Whereas the first libraries generated were based on the combinatorial assembly of readily available synthetic building blocks (e.g., amino acids) to produce "oligomeric" molecules [4–8], it did not take long for the allure of synthesizing a large number of "drug-like" molecules to capture the imagination of synthetic organic chemists. Early combinatorial libraries were constructed around a single core scaffold with varying substituents [9,10]. Structural diversity in those libraries stemmed from the availability and reactivity of synthetic building blocks, which manifested themselves as variable substituent groups around the core structure in the final products. Although these "focused" libraries, in which known synthetic routes have been converted to the solid-phase or parallel synthesis format, still constitute the majority of chemical libraries, the idea of constructing combinatorial libraries that possess diversity in their core scaffolds has been discussed since the early formative years of combinatorial chemistry.

In 1996, Gordon et al. constructed a library of heterocycles (4-thiazolidinones, β-lactams, pyrrolidines, and dihydropyridines) from imines derived from aldehydes and resin-bound amino acids [11]. In another early example, Marx et al. prepared a library

Diversity-Oriented Synthesis: Basics and Applications in Organic Synthesis, Drug Discovery, and Chemical Biology, First Edition. Edited by Andrea Trabocchi.

of diverse pyrrolidine-fused multicyclic compounds through dipolar cycloadditions of azomethine ylides [12]. The same year, Tempest and Armstrong proposed the idea of employing squaric acid derivatives as a fluid core system to generate a multiple-core structure library (MCSL) [13]. Contemporarily, Spaller et al. presented a thoughtful account of the synthetic strategies for constructing "prospecting" libraries, in comparison with the more prevalent focused libraries. The design of prospecting libraries relied on identifying clusters of available reactions for a selected set of building blocks to generate structurally novel drug-like molecules [14].

Whereas these aforementioned examples and discussions failed to gain traction or a following, diversity-oriented synthesis (DOS), since its introduction, has expounded the need for and design of libraries of structurally novel "natural product–like" and drug-like molecules and has succeeded in providing a platform upon which more-recent endeavors in library synthesis could be expanded and elaborated. *Although early DOS libraries also relied on the adaptation of known chemistry* [15], *the development of new reactions with the goal of generating diverse libraries of novel compounds distinguished DOS from the early examples of libraries with various core scaffolds* [16]. It is in this vein that we embarked on developing new phosphine catalysis reactions to maximize structural novelty and diversity in the construction of our libraries.

One of the major goals in the field of DOS is the development of efficient synthetic methods for the construction of libraries encompassing the maximum amount of chemical space [17–21]. A key factor in addressing this challenge is establishing the maximum amount of skeletal diversity of small-molecule collections to improve the probability and efficiency of discovering new biological probes and potential therapeutic lead compounds [22–25]. Although DOS library approaches to the discovery of small-molecule enzyme inhibitors or receptor ligands are well established [26–36], many reactions continue to pose challenges when applied to solid-phase synthesis for the generation of more-extensive analog libraries. Indeed, combinatorial chemistry encompasses a wide range of technical aspects, including soild-phase synthesis [37,38], split synthesis [39], encoding, high-throughput screens, and deconvolution of hits. Nevertheless, in this chapter we focus mainly on synthetic strategies for building libraries of structurally diverse compounds having novel structures. The strategies for preparing DOS libraries require different dimensions of creativity and intuition from those of target-oriented synthesis (TOS).

In the early examples of pre-DOS libraries, one can identify a functional group (e.g., imines, squaric acid derivatives) known to be applicable in a series of reactions that would be identified to provide a structurally diverse sets of compounds. Based on the difficulty, illustrated in Armstrong's example, of translating known chemistry into the efficient solid-phase split-and-pool syntheses of resin-bound squaric acid derivatives, we envisioned developing a novel platform that would be even more streamlined for the generation of structural diversity and novelty. Ideally, the synthetic sequence would be short; each set of reaction conditions would be mild, efficient, and amenable to diverse sets of readily available building blocks; and each product would be diverse in both its core scaffold and its substituents. We anticipated that multiple bond-forming reactions that produce nonaromatic heterocycles of low molecular weight would be an ideal platform. Nucleophilic phosphine catalysis of allenoates and propiolates presented such an opportunity [40–45]. When performing phosphine catalysis–based DOS and related chemical biological studies, our research activities fit into three stages. The first phase involved the development of new reactions for allenoates and propiolates under nucleophilic phosphine catalysis conditions as well as compiling and testing a pilot library of diverse small molecules. The second stage involved converting the newly developed reactions to a format for solid-phase synthesis; allenoic acids and propiolic acids could be loaded onto resin-bound alcohols simply through ester linkages, and the fact that our phosphine-catalyzed reactions worked best in nonpolar organic solvents forecasted minimal effort in this translational stage. Upon identification of specific scaffold of compounds as biomodulators, the third phase involved solid-phase combinatorial synthesis of analogs of the hits, potentially facilitating related biological studies and the development of therapeutic leads; more conventional medicinal chemistry could further expedite the progression of therapeutic leads, when appropriate. The efficient preparation of small molecules and their analogs would boost research and development in all of these stages.

In the first stage of our DOS research, we demonstrated that various reaction modalities could be controlled through appropriate choice of the phosphine catalysts as well as structural variations in the starting materials. For the allenes, we performed one-step ring-closing reactions using either commercially available or otherwise readily available imines, maleimides, aldehydes, aziridines, electron-deficient olefins, azomethine imines, and dinucleophiles, resulting in distinctive carbo- and heterocyclic scaffolds, including tetrahydropyridines, pyrrolines, pyrones, dihydropyrones, dioxanes, cyclohexenes, coumarins, bicyclic succinimides, tetrahydropyrazolopyrazolones, tetrahydropyrazolopyridazinones, tetrahydropyrazolodiazepinones, tetrahydropyrazolodiazocinones, hexahydropyrazoloisoquinolines, hexahydrodiazepinoisoquinolines, benzimidazolines, benzoxazolines, benzothiazolines, benzodioxoles, benzoxathioles, and benzodithioles [46–58]. For electron-deficient acetylenes, one-step double-Michael reactions with readily available dinucleophiles were effected when using a diphenylphosphinopropane catalyst, providing several heterocyclic frames, including oxazolidines, thiazolidines, pyrrolidines, indolines, dihydropyrrolopyridines, benzimidazolines, tetrahydroquinolines, tetrahydroisoquinolines, dihydrobenzo-3,1-oxazines, and dihydrobenzo-1,4-oxazines [59–61]. At the same time, phosphine-catalyzed Michael addition followed by Heck cyclization occured with *o*-iodobenzyl alcohols and electron-deficient acetylenes, giving phthalans [62]. We were especially pleased to find that the phosphine catalysis reactions are extremely compatible with polystyrene resin-bound allenoates, allowing the efficient construction of focused libraries through split-and-pool synthesis [63,64]. In several bioassays we tested our DOS libraries prepared through phosphine organocatalysis, resulting in the identification of geranylgeranyltransferase type I (GGTase-I) inhibitors [63,65–67], activators of endothelium-driven immunity [64], the cholesterol-lowering compound aplexone [68], and antimigratory compounds [69].

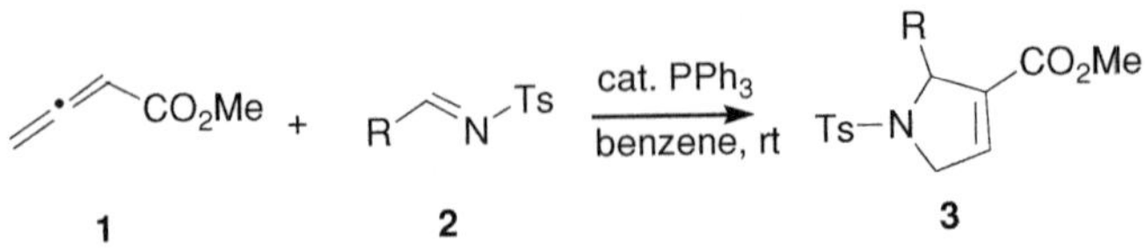

SCHEME 4.1 Xu and Lu's [3 + 2] annulation with *N*-tosylimines.

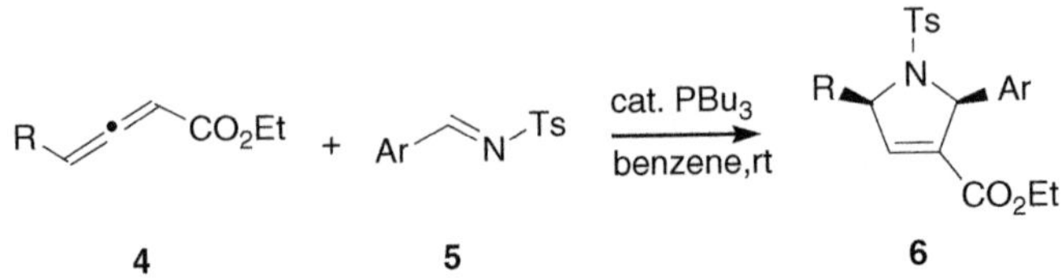

SCHEME 4.2 Synthesis of tetra-substituted dihydropyrroles.

4.2 DOS USING PHOSPHINE ORGANOCATALYSIS

4.2.1 Phosphine Organocatalysis of Allenes with Imines

4.2.1.1 [3 + 2] Annulation with N-Tosylimines to Form Dihydropyrroles In 1997, Xu and Lu established the construction of dihydropyrroles **3** as a single annulation product when using *N*-tosylimines **2** as reaction partners in triphenylphosphine-catalyzed [3 + 2] annulations (Scheme 4.1) [70]. As part of a program to synthesize a library of heterocycles, we decided to employ γ-substituted allenoates **4** to maximize the diversity in the dihydropyrrole products. The mechanism of the [3 + 2] annulation dictates that the γ-substituted allenoates **4** should provide tetra-substituted dihydropyrrole products **6** and, therefore, increase the versatility of the reaction (Scheme 4.2 and Table 4.1) [46].

When we subjected γ-ethyl allenoate to Xu and Lu's original reaction conditions with *N*-tosylbenzaldimine, we obtained only 44% of the desired 2,5-disubstituted dihydropyrrole as a 15 : 1 mixture of diastereoisomers. We hypothesized that the sterically more-hindered γ-substituted allenoate would require a more nucleophilic

TABLE 4.1 Tetra-Substituted Dihydropyrroles

Entry	R	Ar	Yield (%)	cis : trans
1	Me	Ph	89	91 : 9
2	Et	Ph	99	95 : 5
3	*n*-Pr	Ph	98	96 : 4
4	*i*-Pr	2-F-Ph	97	cis only
5	*i*-Pr	4-CF_3-Ph	96	cis only
6	Ph	3-Cl-Ph	99	cis only
7	Ph	4-OMe-Ph	99	cis only
8	*t*-Bu	3,4-Cl_2-Ph	>99	cis only
9	*t*-Bu	1-Naphthyl	>99	cis only

phosphine for its initial conjugate addition. Indeed, dramatic increases in the efficiency of the reaction occurred when using more nucleophilic phosphines. With tributylphosphine, various aryl imines **5** furnished 89 to 99% yields of the tetrasubstituted dihydropyrroles desired, favoring the cis isomers, when using γ-methyl, γ-ethyl, γ-propyl, γ-isopropyl, γ-*tert*-butyl, or γ-phenyl allenoate as the reaction substrate. Employing allenoates with bulky γ-substituents led to the exclusive production of *cis*-dihydropyrroles.

In our [3 + 2] annulation reaction, tributylphosphine behaved as a nucleophilic organocatalyst to facilitate the formation of 3-carbethoxyl-2,5-disubstituted-3-dihydropyrroles in excellent yields with high diastereoselectivities. This method has many attractive features, including short reaction times at room temperature and a simple nonaqueous workup. In addition, the reactants were readily prepared in one step from commercially available materials using known procedures. Gratifyingly, the reaction efficiencies and selectivities meet the high standards for the construction of DOS libraries.

4.2.1.2 [4 + 2] Annulation with N-Tosylimines to Form Functionalized Guvacines

In the [3 + 2] annulation with imines described above, unsubstituted or γ-substituted allenoates **4** reacted readily with *N*-sulfonylimines in the presence of phosphine organocatalysts to generate functionalized pyrrolines. To expand the scope of the phosphine-catalyzed annulations of imines, we employed α-substituted allenoates as starting materials. Our efforts led to the development of unprecedented [4 + 2] annulations with *N*-tosylimines, procuring highly functionalized guvacines in high efficiency [47]. The reaction began with the 1,2-addition of intermediate **11** to the imine **8a**, producing the vinylphosphonium species **12**, which underwent two proton transfer steps to form the allylic phosphonium species **14**. Intramolecular conjugate addition of the amide unit to the enoate moiety in **14** yielded the final product **9a** with the release of PBu_3, completing the catalytic cycle (Scheme 4.3). The 2-methylbuta-2,3-dienoates readily served as a four-carbon synthon in this transformation.

The [4 + 2] annulation reaction works well for a variety of *N*-tosylimines and α-substituted allenoates in the presence of 20 mol% tributylphosphine in dichloromethane. Several characteristics of this transformation are noteworthy. Excellent yields were obtained when electron-donating *N*-tosylimines were used, but a slight decrease in yield was discerned when strongly electron-withdrawing groups were attached to the imines. Interestingly, *N*-tosylpivalaldimine afforded the desired product in good yield when we employed sodium carbonate as an additive. When we applied 2-benzyl-2,3-butadienoates as reaction partners, we obtained 2,6-diarylguvacine esters in almost quantitative yields with good to excellent diastereoselectivities favoring cis isomers (Scheme 4.4 and Table 4.2).

The robust [4 + 2] annulation with *N*-tosylimines can be employed to generate tetrahydropyridines on large scales [71], and we have demonstrated their application to natural product synthesis through formal syntheses of (±)-alstonerine and (±)-macroline [72] and the total synthesis of hirsutine [73]. Catalytic asymmetric synthesis of guvacine derivatives through the [4 + 2] annulation of imines with allenes was described by Wurz and Fu [74] and Xiao et al. [75], respectively.

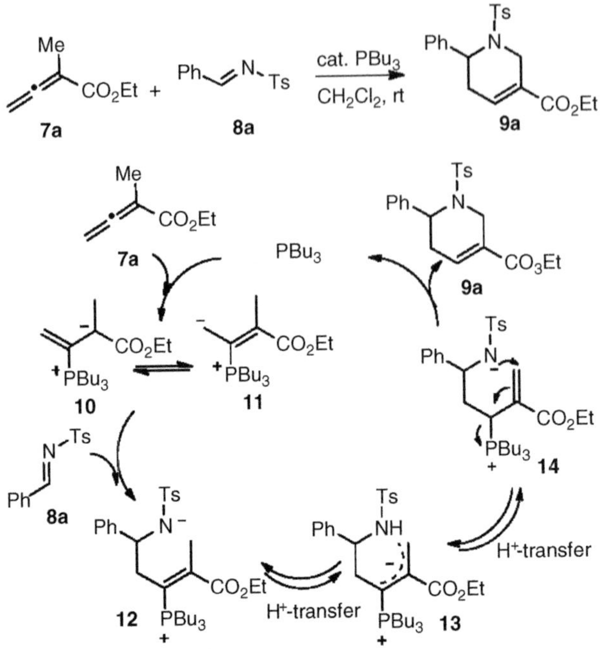

SCHEME 4.3 Phosphine-catalyzed [4 + 2] annulation with *N*-tosylimines. (Adapted from [47], with permission; copyright © 2003 American Chemical Society.)

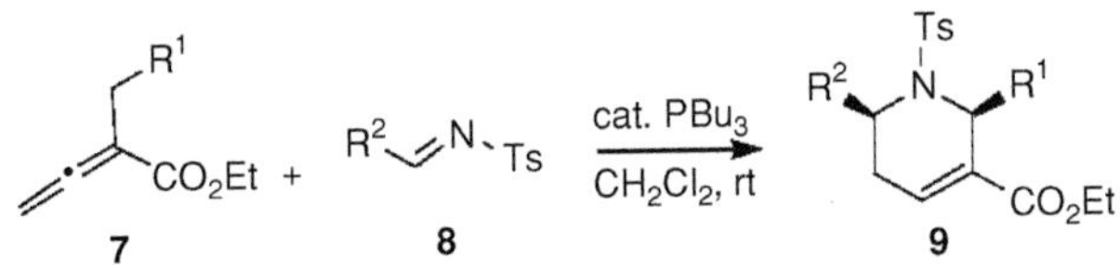

SCHEME 4.4 Synthesis of highly functionalized guvacines.

TABLE 4.2 Tetra-Substituted Tetrahydropyridine

Entry	R^1	R^2	Yield (%)	cis : trans
1	H	Ph	98	
2	H	4-OMe-Ph	99	
3	H	4-Me-Ph	95	
4	H	2-CF_3-Ph	98	
5	H	4-NO_2-Ph	86	
6	H	2-Furyl	97	
7	H	4-Py	92	
8	H	*t*-Bu	86[a]	
9	Ph	Ph	99	98 : 2
10	4-CN-Ph	3-Cl-Ph	99	98 : 2
11	3-OMe-Ph	2-Cl-Ph	98	83 : 17
12	3-OMe-Ph	4-Me-Ph	99	98 : 2

[a] 3 equiv of Na_2CO_3 was added.

4.2.2 Phosphine Organocatalysis of Allenes with Azomethine Imines

Azomethine imines **15** are readily accessible, stable compounds that have been employed recently as efficient 1,3-dipoles in various metal-catalyzed and organocatalytic cycloadditions [76–78]. To expand the scope of our phosphine-catalyzed annulations of allenes, we tested azomethine imines as the reaction partners. To our delight, the phosphine-catalyzed [3 + 2], [3 + 3], [4 + 3], and [3 + 2 + 3] annulations of azomethine imines and allenoates produced unusual dinitrogen-fused heterocycles with seven distinctive scaffolds, including tetrahydropyrazolopyrazolones, tetrahydropyrazolopyridazinones, tetrahydropyrazolodiazepinones, and tetrahydropyrazolodiazocinones (Scheme 4.5) [48]. The reaction outcome depends on the structure and chemical properties of the allenoates and the nature of the phosphine catalyst.

The [3 + 2] annulation occurred when we used an azomethine imine **15a** as the electrophilic partner with α-methyl allenoate **7a** in the presence of a trialkylphosphine (20 mol%). The first step in the mechanism proposed (Scheme 4.6) is the nucleophilic addition of tributylphosphine to the allenoate **7a** to give the zwitterion **11**. The addition of **11** to the azomethine imine **15a** generates the intermediate **23**, the conjugate addition of which would yield the final product **16a** through the release of PBu_3 from the intermediate **24**.

Azomethine imines readily served as a 1,3-dipole synthon in this annulation. Under optimized conditions, a variety of aryl or alkyl azomethine imines underwent [3 + 2] annulation, providing the anticipated products in moderate to excellent yields as single *E*-isomers. Both β′-alkyl- and β′-aryl-substituted allenoates were amenable to the [3 + 2] annulation, providing the corresponding tetrahydropyrazolopyrazolones in excellent yields (Scheme 4.7 and Table 4.3).

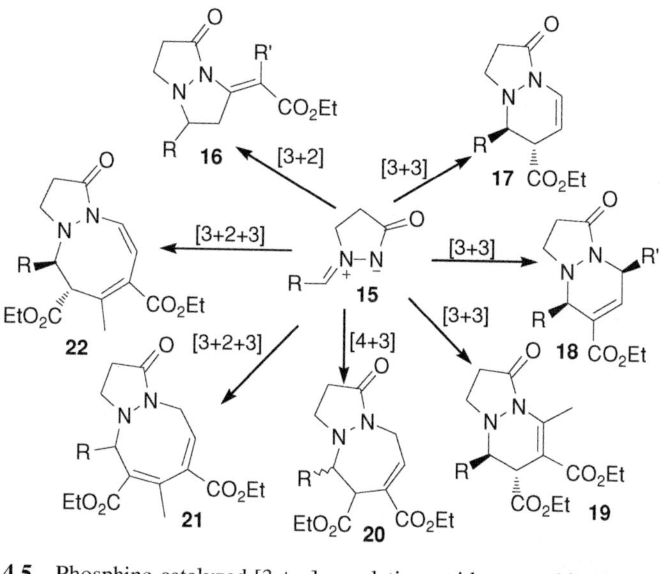

SCHEME 4.5 Phosphine-catalyzed [3 + *n*] annulations with azomethine imines. (Adapted from [48], with permission; copyright © 2011 American Chemical Society.)

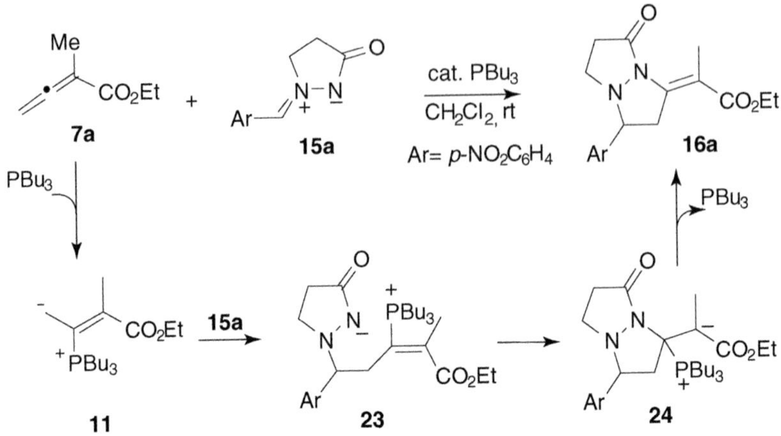

SCHEME 4.6 Phosphine-catalyzed [3 + 2] annulations with azomethine imine.

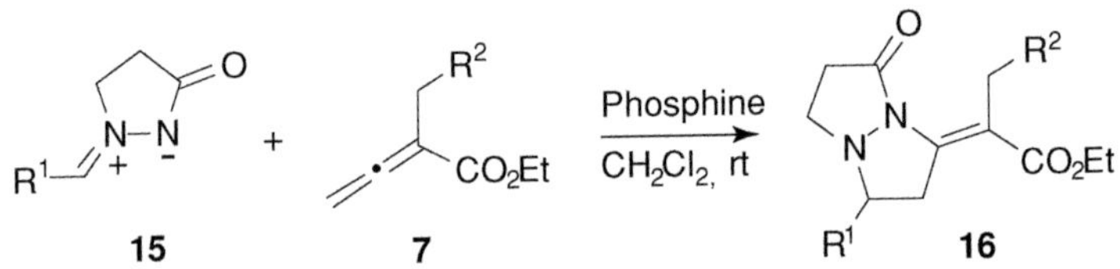

SCHEME 4.7 Synthesis of highly functionalized tetrahydropyrazolopyrazolones.

TABLE 4.3 Highly Functionalized Tetrahydropyrazolopyrazolones

Entry	R^1	R^2	Phosphine (20 mol%)	Time (h)	Yield (%)
1	Ph	H	PBu_3	24	97
2	4-OMe-Ph	H	PBu_3	30	85
3	4-CF_3-Ph	H	PBu_3	24	98
4	3-NO_2-Ph	H	PBu_3	24	97
5	2-Me-Ph	H	PBu_3	48	70
6	2-Naphthyl	H	PBu_3	24	96
7	4-Py	H	PBu_3	20	86
8	2-Furanyl	H	PBu_3	20	99
9	4-NO_2-Ph	Me	PMe_3	20	96[a]
10	4-NO_2-Ph	*t*-Bu	PMe_3	24	98[a]
11	4-NO_2-Ph	2-Me-Ph	PMe_3	20	95
12	4-NO_2-Ph	4-F-Ph	PMe_3	15	99

[a]At 40°C.

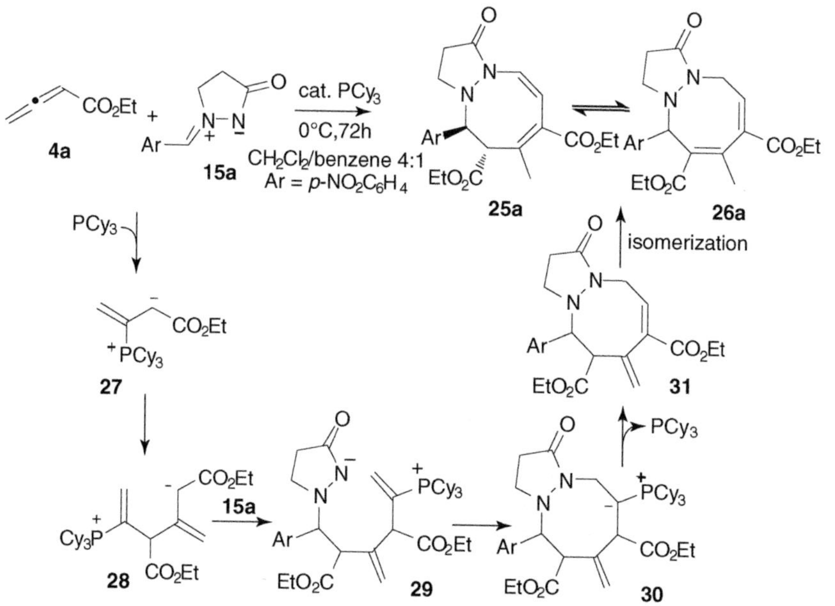

SCHEME 4.8 Phosphine-catalyzed [3 + 2 + 3] annulations with azomethine imines.

The [3 + 2 + 3] annulation occurred when we reacted the azomethine imines **15a** with ethyl 2,3-butadienoate **4a** in the presence of PCy_3 (20 mol%). In the mechanism proposed (Scheme 4.8), the key event is the formation of the trimeric zwitterionic intermediate **28**, which is generated through the conjugate addition of the phosphonium dienolate **27** to another molecule of the allenoate **4a**. With the introduction of the azomethine imine **15a**, the intermediate **28** generates **29**, which gives compound **31** after cyclization, proton transfer, and phosphine elimination. The isomerization of the olefin unit in intermediate **31** yields the final products **25a** and **26a** as a mixture of tautomers. The trimeric zwitterionic intermediate, which forms from two molecules of ethyl 2,3-butadienoate and a molecule of phosphine, acts as a 1,5-dipole in the cycloaddition reaction. We employed a mixed solvent of CH_2Cl_2 and benzene for the [3 + 2 + 3] annulation because CH_2Cl_2 aids dissolution of the azomethine imine and benzene favors formation of the trimeric zwitterionic intermediate **28**.

Under the optimized reaction conditions, both electron-rich and electron-deficient aryl azomethine imines, as well as a polyaromatic azomethine imine, were compatible substrates. Among the many azomethine imine substrates tested, 2-halophenyl azomethine imines provided higher yields, and 2-nitrophenyl azomethine imine gave a better ratio of one isomer over the other (Scheme 4.9 and Table 4.4).

4.2.3 Phosphine Organocatalysis of Allenes with Electron-Deficient Olefins

4.2.3.1 [3 + 2] Annulation with Maleimides to Form Bicyclic Succinimides

In 1995, Zhang and Lu described the phosphine-catalyzed [3 + 2] annulation of

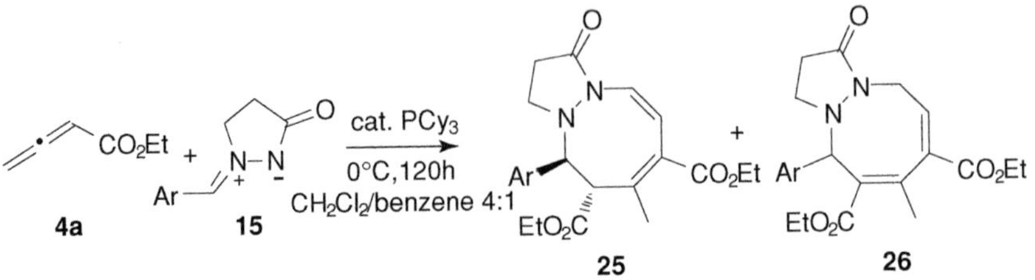

SCHEME 4.9 Synthesis of highly functionalized tetrahydropyrazolodiazocinones. (Adapted from [48], with permission; copyright © 2011 American Chemical Society.)

TABLE 4.4 Highly Functionalized Tetrahydropyrazolodiazocinones

Entry	Ar	Yield (%)	**25 : 26**
1	Ph	81	34 : 66
2	4-Me-Ph	64	39 : 61
3	4-OMe-Ph	76	44 : 56
4	4-Br-Ph	77	29 : 71
5	3-NO_2-Ph	77	17 : 83
6	2-Br-Ph	92	28 : 72
7	2-Cl-Ph	90	24 : 76
8	2-F-Ph	88	25 : 75
9	2-NO_2-Ph	73	9 : 91
10	2-Naphthyl	76	32 : 68

2,3-butadienoates and electron-deficient olefins to form functionalized cyclopentenes [79]. Unlike the reaction of allenoates with imines, the reaction of allenoates with electron-deficient olefins typically furnishes a mixture of the two regioisomeric products **33** and **34** (Scheme 4.10).

To alleviate this regioscrambling, we explored the dipolar [3 + 2] cycloadditions of the allenoate **4a** and C2-symmetric cyclic electron-deficient olefins (e.g., maleimides). When we mixed *N*-phenylmaleimide with ethyl allenoate **4a** in the presence of triphenylphosphine (10 mol%) in benzene at 60°C, the reaction mixture smoothly underwent the desired [3 + 2] annulation to form the bicyclic succinimide **36** in 99% isolated yield (Scheme 4.11).

To increase the versatility of the reaction and the diversity of the reaction products, we also investigated the reactions of γ-substituted allenoates with other maleimides.

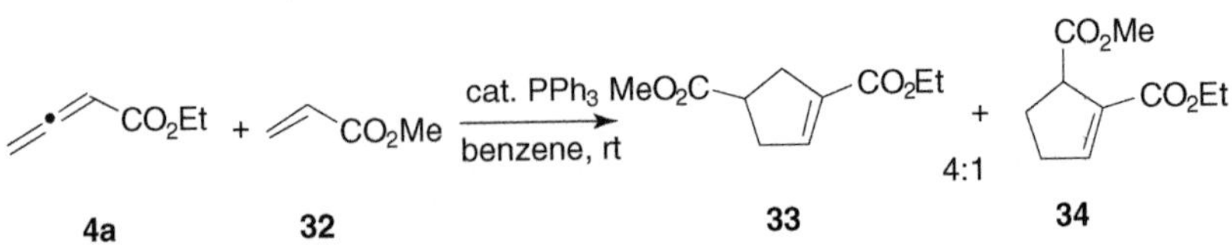

SCHEME 4.10 Zhang and Lu's [3 + 2] annulation with an electron-deficient olefin.

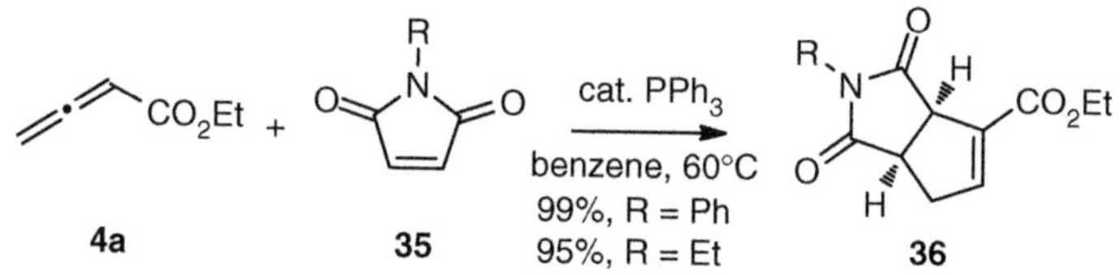

SCHEME 4.11 Allene-maleimide [3 + 2] annulations.

This highly efficient reaction has also worked well for resin-bound allenoates, allowing us to use it to generate a library of 145 bicyclic succinimides.

4.2.3.2 Intramolecular [3 + 2] Annulation to Form Coumarins Another way to avoid regioscrambling, and thereby obtain a single product from the allene–alkene [3 + 2] annulation, is to perform the reaction in an intramolecular setting. The geometric constraint inherent in the starting material can control the reaction mode so that only one pathway will dominate. We synthesized compounds **37** from salicylaldehydes through Wittig reactions followed by coupling with 3-butynoic acid under the influence of Mukaiyama's reagent. Treatment of **37** with 20 mol% of tributylphosphine in THF at room temperature for 6 h produced the tricyclic dihydrocoumarins **38** in good to excellent yields as a single diastereoisomers. Salicylaldehydes featuring either electron-donating or electron-withdrawing substituents were readily converted into the coumarin derivatives **38** (Scheme 4.12) [49].

From an extensive investigation of the role played by the activating substituents on the olefin moiety, we found that the reaction of the nitrostyrenyl derivative **39** with triphenylphosphine in benzene provided the nitronate **40** in 58% yield. The nitronate **40** underwent 1,3-dipolar cycloadditions with a variety of commercially available alkenes, providing highly functionalized tetracyclic coumarins **41** in excellent yields (83 to 97%) with high exo selectivity (exo/endo = 9–11 : 1) and exclusive facial selectivity (Scheme 4.13).

4.2.3.3 [4 + 2] Annulation with Arylidenemalononitriles to Form Cyclohexenes In the imine [4 + 2] annulation, α-alkyl allenoates reacted with *N*-tosylimine to form tetrahydropyridines. The reaction works well for a variety of *N*-sulfonylimine electrophiles [47]. To expand the utility of the [4 + 2] annulation, we became interested in

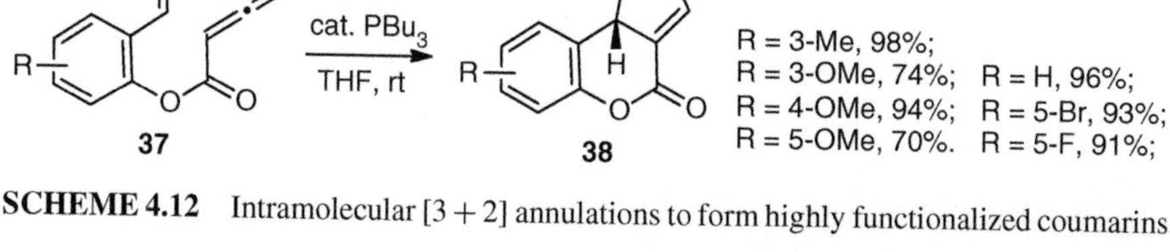

SCHEME 4.12 Intramolecular [3 + 2] annulations to form highly functionalized coumarins.

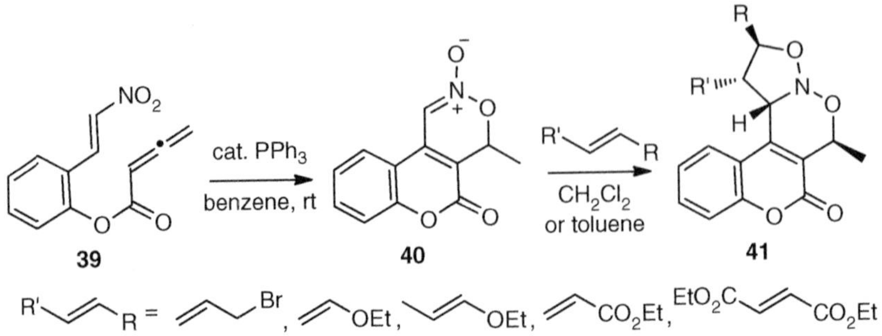

SCHEME 4.13 Intramolecular [3 + 2] annulations and 1,3-dipolar cycloadditions.

broadening the scope of the electrophilic partner. When we used the electron-deficient olefin **42a** as the electrophile, the [4 + 2] annulation of the α-methyl allenoate **7a** afforded functionalized cyclohexenes (Scheme 4.14) [50].

Notably, two distinct cyclohexene regioisomers were formed selectively, depending on the electronic nature of the phosphine catalyst. When we used hexamethylphosphorus triamide (HMPT) as the catalyst, the reaction produced the cyclohexene **44a** via the phosphonium dienolate intermediate **43**. On the other hand, when we employed a more electron-withdrawing triarylphosphine, such as *tris*-(*p*-fluorophenyl)phosphine, as the catalyst, the phosphonium dienolate **43** isomerized into the vinylogous ylide intermediate **45**, leading to the reaction producing the alternative cyclohexene regioisomer **46a**.

Under the optimized reaction conditions, both electron-deficient and electron-rich arylidenes can be reacted with the α-methyl allenoate **7a** to provide the desired cyclohexene products in high yields. Notably, these conditions also worked well for activated heteroarylidenes, furnishing the corresponding functionalized cyclohexenes. The reaction tolerated a wide range of allenylic β′-alkyl and β′-aryl substituents on the allenoates, providing the cyclohexenes in excellent isolated yields with good to excellent diastereoselectivities favoring the cis isomers (Scheme 4.15 and Table 4.5).

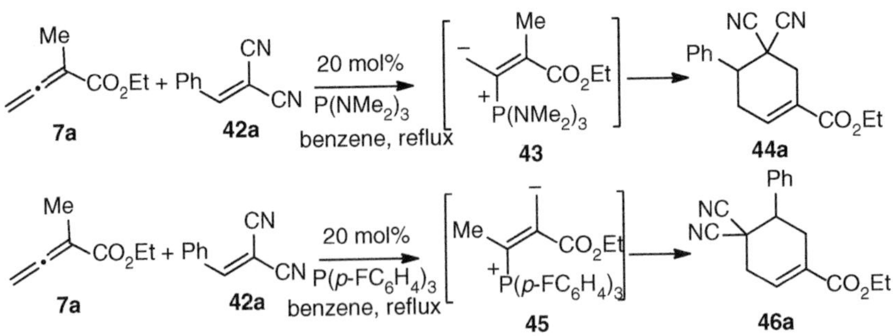

SCHEME 4.14 Phosphine-catalyzed [4 + 2] annulations with arylidenemalononitriles.

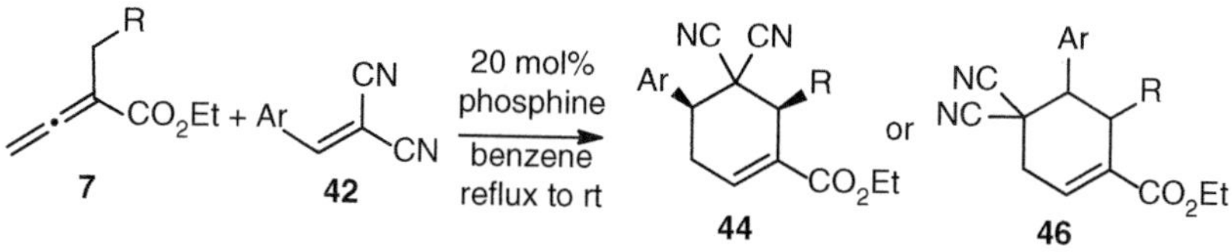

SCHEME 4.15 Synthesis of highly functionalized cyclohexenes.

4.2.4 Phosphine Organocatalysis of Allenes with Aldehydes

Relative to the extensive studies of annulation reactions between allenoates and various imines or electron-deficient olefins described above, examples of phosphine-catalyzed reactions between allenes and aldehydes are rather limited. To expand the synthetic scope of phosphine-catalyzed annulations to construct heterocycles beyond azacycles, we explored the reactivity of allenoates with aldehydes (Scheme 4.16) [51–54].

Following the logic demonstrated for the [3 + 2] annulations between the allenoate **1** and *N*-sulfonylimines to form dihydropyrroles, we expected to harvest dihydrofuran adducts. Instead, the phosphine-catalyzed reactions between allenoates and aldehydes yielded dioxanes, pyrones, and dihydropyrones, with selectivity depending on the nature of the phosphine catalyst and the reaction conditions. When the phosphine was small (e.g., trimethylphosphine), the intermediate dienolate **48** having s-trans geometry was favored, due to electrostatic association between the dienolate oxygen anion and the phosphonium cation, leading to formation of the dioxane **2**. When a sterically demanding phosphine was used or when a hydrogen-bond donor was present, the s-cis-phosphonium dienolate intermediate was dominant and the reaction provided pyrones **55** (bulky phosphine) or dihydropyrones **56** (in the presence of protic solvents).

After investigating the steric and electronic effects of the allenoate and phosphine on the reaction efficiency, we selected isopropyl 2,3-butadienoate **47a** as the starting material and trimethylphosphine as the catalyst for the preparation of various dioxanylidenes. Both pyridyl aldehydes and benzaldehydes bearing electron-withdrawing

TABLE 4.5 Highly Functionalized Cyclohexenes

Entry	R	Ar	Phosphine	Product	Yield (%)	cis : trans
1	H	Ph	HMPT	**44**	98	
2	H	4-OMe-Ph	HMPT	**44**	94	
3	H	4-Br-Ph	$P(4\text{-F-Ph})_3$	**46**	86	
4	H	2-Furyl	$P(4\text{-F-Ph})_3$	**46**	88	
5	H	*N*-Me-2-indolyl	$P(4\text{-F-Ph})_3$	**46**	91	
6	Ph	Ph	HMPT	**44**	93	82 : 18
7	CO_2Et	Ph	HMPT	**44**	96	66 : 33
8	Et	Ph	HMPT	**44**	98	92 : 8
9	$CH{=}CH_2$	Ph	HMPT	**44**	94	91 : 9

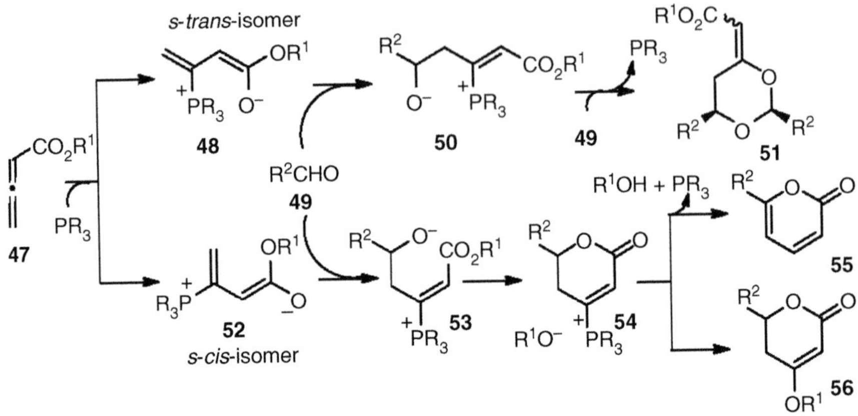

SCHEME 4.16 Phosphine organocatalysis of allenes with aldehydes. (Adapted from [53], with permission; copyright © 2008 American Chemical Society.)

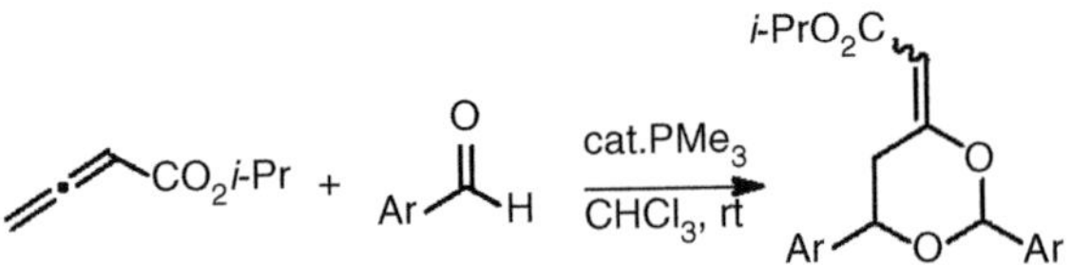

SCHEME 4.17 Synthesis of 1,3-dioxan-4-ylidenes.

groups afforded the desired products in good to excellent yields with good to excellent stereoselectivities favoring the *E*-isomers. Less reactive electron-rich benzaldehydes afforded moderate reaction yields (Scheme 4.17 and Table 4.6).

When using relatively bulky tricyclopentylphosphine as the catalyst, the phosphine-catalyzed reactions of allenoates and aldehydes provided 2-pyrones as products. Ethyl 2,3-butadienoate provided yields greater than those of other 2,3-butadienoates. Various aromatic aldehydes bearing electron-withdrawing and

TABLE 4.6 1,3-Dioxan-4-ylidenes

Entry	Ar	Yield (%)	*E* : *Z*
1	4-Py	99	8 : 1
2	3-Py	96	9 : 1
3	4-CF_3-Ph	99	7 : 1
4	3-NO_2-Ph	97	7 : 1
5	2-Cl-Ph	64	8 : 1
6	Ph	54	only *E*
7	3-OMe-Ph	47	only *E*

TABLE 4.7 Synthesis of 6-Substituted 2-Pyrones

Entry	R^2	$PCyp_3$ (mol%)	Yield (%)
1	Ph	10	60
2	3-OMe-Ph	20	69
3	3-Cl-Ph	10	91
4	2-Naphthyl	30	66
5	2-Furyl	30	61
6	*n*-Pr	30	34

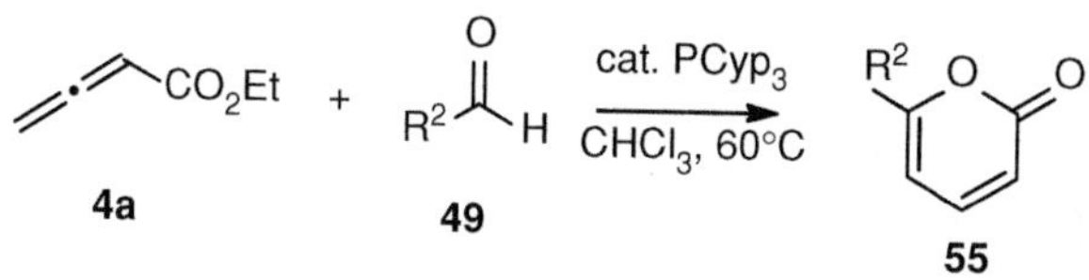

SCHEME 4.18 Synthesis of 6-substituted 2-pyrones.

electron-donating groups, including 2-naphthaldehyde and 2-furaldehyde, afforded 6-aryl-2-pyrones in yields greater than 60%. Although the yield was moderate when using an aliphatic aldehyde, the reaction provided a valuable product—specifically, 6-*n*-propyl-2-pyrone (Table 4.7, entry 6), which possesses a sweet, creamy, coumarin-like herbal flavor [80] and is used as an additive to alter the organoleptic properties of tobacco [81–83]—in only a single step from a commercially available aldehyde (Scheme 4.18 and Table 4.7).

When using methanol and *n*-butyllithium as additives in the presence of trimethylphosphine, the annulations of allenoates with aldehydes produced the dihydro-2-pyrones as products. The presence of methanol as a hydrogen-bond donor induced the formation of dihydropyrones; the presence of *n*-butyllithium as an additive suppressed formation of the noncyclized product. Benzaldehydes possessing a variety of electron-withdrawing substituents provided the desired dihydropyrones in good yields. Aromatic substitution in the meta position was optimal, but substitution at the ortho position, the most sterically susceptible site, was also tolerated. Similar to the formation of dioxanes and pyrones, the use of more electron-rich aromatic aldehydes rendered lower yields (Scheme 4.19 and Table 4.8). Nevertheless, the convenience of this one-step synthesis of dihydro-2-pyrones from two commercially

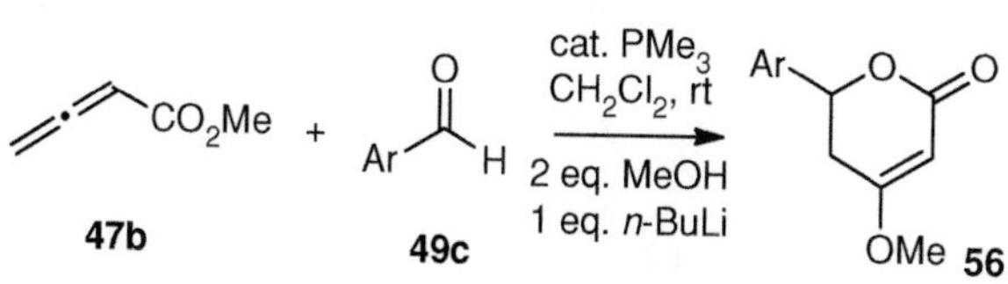

SCHEME 4.19 Synthesis of functionalized dihydro-2-pyrones.

TABLE 4.8 Functionalized Dihydro-2-pyrones

Entry	R^2	Yield (%)
1	3-CN-Ph	83
2	4-CF_3-Ph	63
3	3-CF_3-Ph	74
4	2-CF_3-Ph	58
5	3-Me-Ph	36

available starting materials was recognized immediately: Pollastri and his colleagues employed our method to prepare dihydro-2-pyrone suppressors of TNFα [84].

4.2.5 Phosphine Organocatalysis of Allenes with Aziridines

Aziridines are extremely versatile synthetic building blocks in modern synthetic chemistry [85]. By incorporating them in the phosphine organocatalysis of allenes, we developed a novel [3 + 3] annulation to generate highly functionalized tetrahydropyridines [55]. In the mechanism proposed (Scheme 4.20), addition of triphenylphosphine to the allenoate forms the corresponding phosphonium zwitterion **60**, which undergoes proton transfer to provide the vinylogous ylide **61**. The addition of the intermediate **61** to the phenyl aziridine occurs to form **62**, which is converted to the phosphonium dienolate **63** through proton transfer. Intramolecular

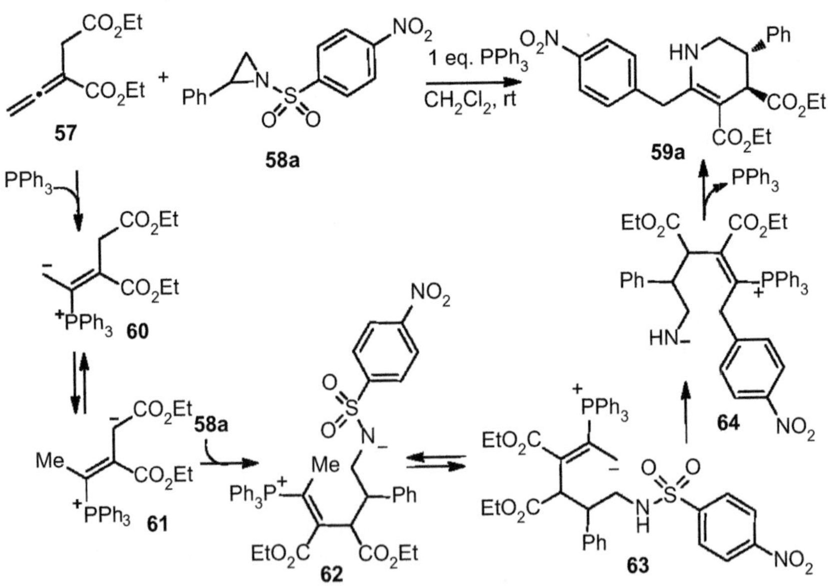

SCHEME 4.20 Phosphine-catalyzed [3 + 3] annulations with aziridines.

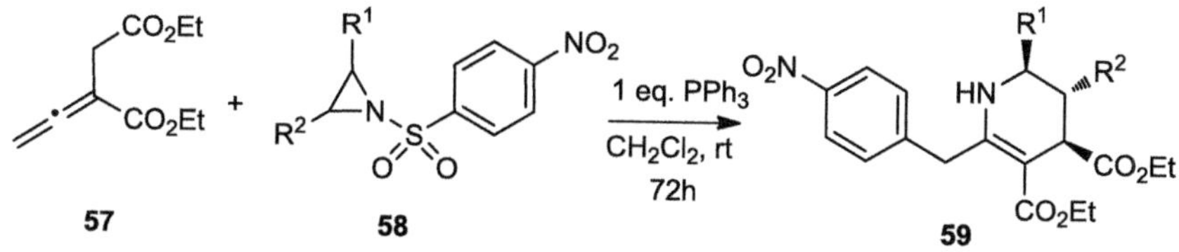

SCHEME 4.21 Synthesis of highly functionalized tetrahydropyridines.

nucleophilic aromatic substitution and desulfonylation affords **64**; intramolecular conjugate addition of the amide **64** and β-elimination of the phosphine yields the final product **59a**.

Although the reaction is catalytic in nature, 1 equiv of triphenylphosphine can be used to expedite the transformation. Under the optimized reaction conditions, we tested a range of aziridine derivatives as partners for the [3 + 3] annulations. Phenyl aziridines featuring both electron-rich and electron-deficient substituents at the ortho, meta, and para positions are amenable to the reaction, giving products in good to excellent yields. With aryl-substituted aziridines, the reaction produced tetrahydropyridines favoring 1,2-trans, whereas alkyl-substituted aziridines produced 1,3-cis isomers preferably (Scheme 4.21 and Table 4.9).

4.2.6 Phosphine Organocatalysis of Allenes with Dinucleophiles

Whereas phosphines catalyzes Michael additions of nucleophiles onto activated alkenes [86], nucleophiles undergo γ-umpolung addition onto activated allenes [87–89] and β′-umpolung additions onto α-alkyl allenoates [90,91] in the presence of phosphine catalysts. Considering that the γ-umpolung addition product of a nucleophile onto an allenoate is a α,β-enoate, we envisioned the possibility of tandem γ-umpolung-/Michael addition of a dinucleophile onto an allenoate [92]. Indeed, we isolated the benzomorpholine **66** in 88% yield when using *N*-tosyl-2-aminophenol **65a** as the dinucleophile reaction partner in the presence of triphenylphosphine (Scheme 4.22). Switching the catalyst to trimethylphosphine, however, led to

TABLE 4.9 Highly Functionalized Tetrahydropyridines

Entry	R^1	R^2	Yield (%)	trans : cis
1	H	2-Me-Ph	88	90 : 10
2	H	3-Me-Ph	82	86 : 14
3	H	4-Me-Ph	64	89 : 11
4	H	2,5-Me_2-Ph	98	92 : 8
5	H	4-F-Ph	76	88 : 12
6	H	3-Cl-Ph	86	83 : 17
7	Me	H	66	41 : 59
8	H	H	37	

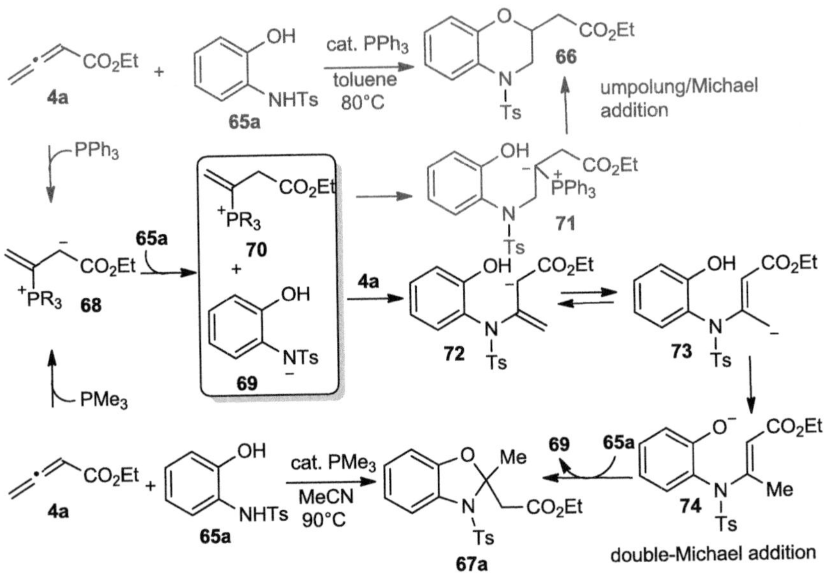

SCHEME 4.22 Phosphine organocatalysis of an allenoate with dinucleophiles.

production of the double-Michael product **67a** in 92% yield. In the mechanism proposed, the phosphonium enolate **68** (R = Me) formed upon addition of trimethylphosphine to the allenoate **4a** acts as a general base and promotes formation of the double-Michael product **67a** [56].

Under optimized conditions, our phosphine-catalyzed double-Michael reactions could produce six distinct heterocyclic scaffolds—benzimidazoline, benzoxazoline, benzothiazoline, 1,3-benzodioxole, 1,3-benzoxathiole, and 1,3-benzodithiole—in high efficiency when using various ortho-substituted phenols, anilines, and thiophenols as the dinucleophiles. We observed comparable reactivity for both α- and γ-substituted allenoates (Scheme 4.23 and Table 4.10).

4.2.7 Phosphine Organocatalysis of Acetylenes with Dinucleophiles

A different class of phosphine-catalyzed double-Michael reaction occurs when employing acetylenes as the electrophilic partner of the dinucleophiles. Using amino

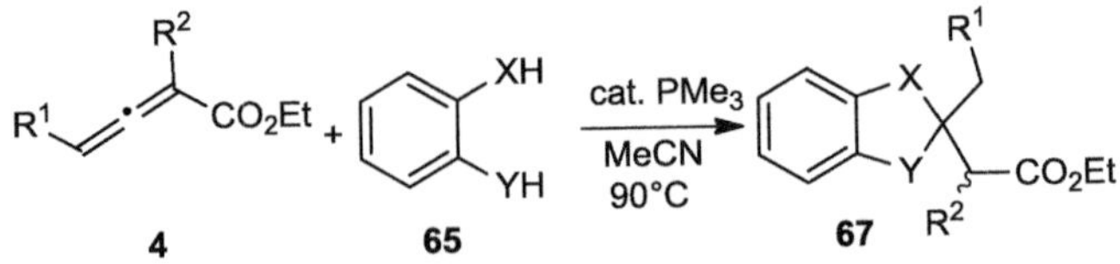

SCHEME 4.23 Synthesis of benzannulated heterocycles.

TABLE 4.10 Benzannulated Heterocycles

Entry	X	Y	R^1	R^2	Yield (%)
1	O	S	H	H	93
2	O	O	H	H	80
3	S	S	H	H	74
4	S	NTs	H	H	68
5	NTs	NTs	H	H	79
6	O	NTs	H	H	84
7	O	S	Ph	H	86
8	O	O	Bn	H	89
9	O	O	H	Me	89
10[a]	O	NTs	H	CH_2CO_2Et	84[b]

[a] NaOAc (50 mol%) was added.
[b] dr = 1.2 : 1, based on the ^{1}H NMR spectrum of the crude reaction mixture.

acid–derived amino alcohols, thiols, malonates, and malononitriles as Michael donors and electron-deficient acetylenes as Michael acceptors provided efficient access to enantiomerically pure oxazolidines, thiazolidines, pyrrolidines, and octahydroindoles in excellent yields with high diastereoselectivities [59,61]. In the mechanism proposed, the bisphosphine [e.g., diphenylphosphinopropane (DPPP)] stabilizes the intermediate phosphonium ions and improves the reaction efficiency (Scheme 4.24).

Switching to aromatic dinucleophiles, we also prepared seven distinctive heterocyclic compounds—indolines, dihydropyrrolopyridines, benzimidazolines, dihydrobenzo-3,1-oxazines, benzomorpholines, tetrahydroquinolines, and

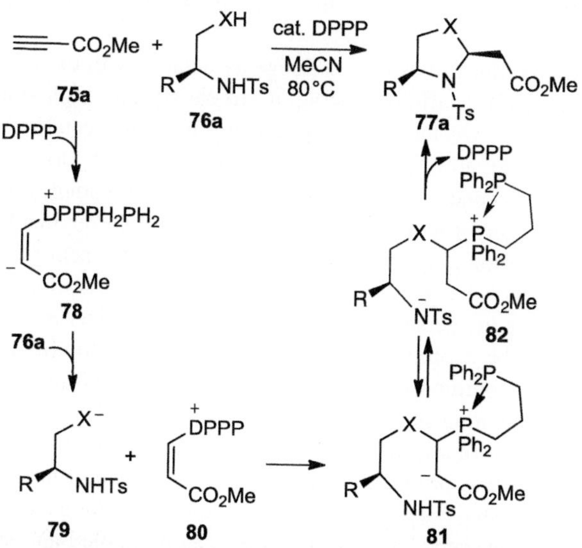

SCHEME 4.24 Phosphine-catalyzed double-Michael reaction.

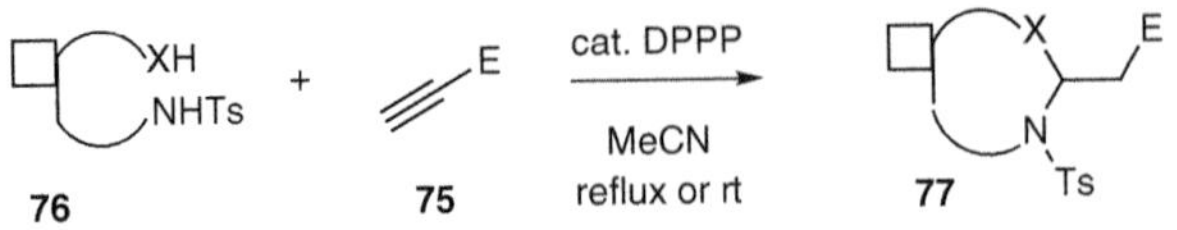

SCHEME 4.25 Synthesis of 11 different heterocyclic scaffolds.

tetrahydroisoquinolines—through the mixed double-Michael strategy. In some cases, the use of acetic acid and sodium acetate as additives expedited the proton transfer steps and, consequently, the overall reaction (Scheme 4.25 and Table 4.11) [60]. In the realm of phosphine organocatalysis, the mixed double-Michael reaction of acetylenes and dinucleophiles appears to be the most versatile annulation process for generating heterocyclic compounds.

4.3 SKELETAL DIVERSITY BASED ON A PHOSPHINE CATALYSIS/COMBINATORIAL SCAFFOLDING STRATEGY

The algorithm of DOS (i.e., in split-and-pool format) manifests two contradicting principles in operation: maximizing structural diversity and maintaining common reactivity. For the sake of diversity, one must generate structures that are as different as possible. This diversity requirement alone is not particluarly challenging and quite manageable [26–36]. Examples have appeared in the literature since the 1990s, such as Armstrong's proposal that squaric acid could be used as a fluid core system to generate a multiple-core structure library. The result could be eight drastically different core structures (Scheme 4.26) [13]; these eight compounds, however, cannot be pooled and subjected to the next split step, because they do not share any common chemical functionality to exploit. In this case, maintaining common reactivity has been sacrificed for the sake of diversity.

This quandary demonstrates the challenge associated with more advanced forms of DOS: combinatorial scaffold generation. To satisfy this high standard of combinatorial scaffolding, we have developed new phosphine catalyses of allenoates to produce a diverse array of compounds, including dihydropyrroles, tetrahydropyridines, cyclohexenes, bicyclic succinimides, dioxanylidenes, tetrahydropyrazolopyrazolones tetrahydropyrazolopyridazinones, tetrahydropyrazolodiazepinones, and tetrahydropyrazolodiazocinones [46–48,50,51,53]. This set of compounds is as diverse as Armstrong's, yet, by design, only one unit of unsaturation in the allenoates is consumed, leaving the versatile α,β-unsaturated ester moiety as a common chemical functionality to be exploited in the next split step after pooling all 11 scaffolds. Exploration of this common chemical functionality (α,β-unsaturated ester) in the second scaffold-generating step (Michael addition or Tebbe/Diels–Alder reaction) is one of the key features of our library design: combinatorial scaffolding. In combinatorial scaffolding, diverse products generated from a set of phosphine organocatalysis reactions in one split step are programmed to contain a common chemical functional group (e.g., α,β-unsaturated ester) so that they can be pooled and subjected to the next scaffold-generating split step.

TABLE 4.11 Eleven Different Heterocyclic Scaffolds

Entry	NuH	E	Product	Yield (%)
1	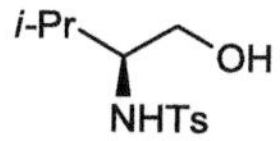	Ts	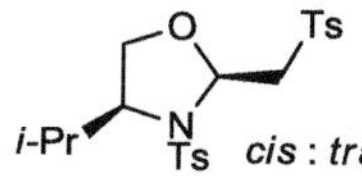 *cis* : *trans* = 97 : 3	87
2	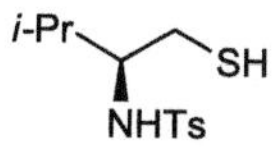	COMe	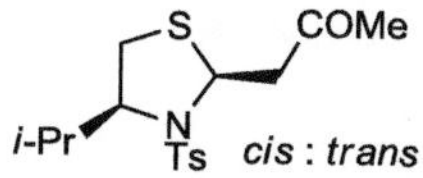 *cis* : *trans* = 95 : 5	93
3	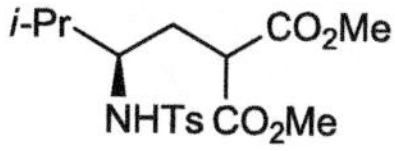	CO_2Me	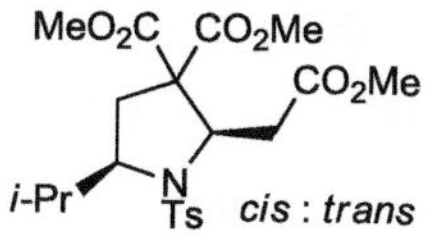 *cis* : *trans* = 94 : 6	82
4	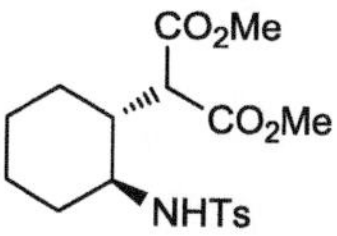	COMe	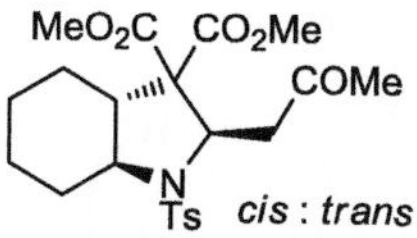 *cis* : *trans* = 100 : 0	80
5[a]	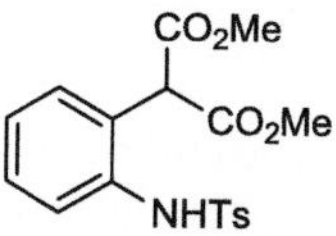	CO_2Me	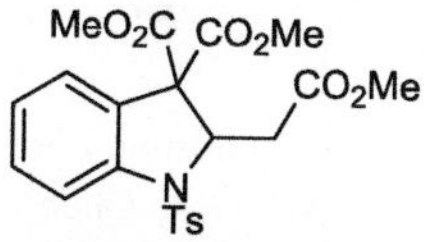	87
6[a]	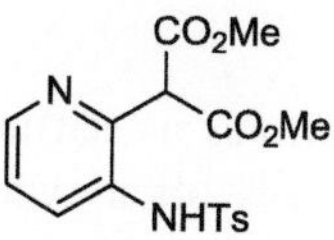	COPh	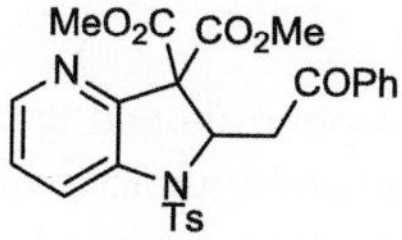	96
7[a]	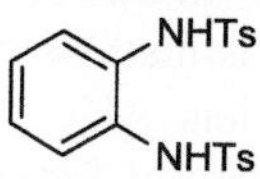	COMe	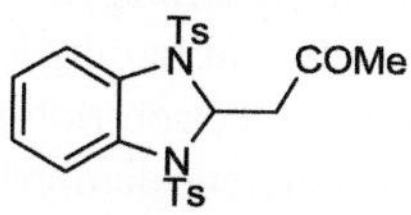	81
8[a]	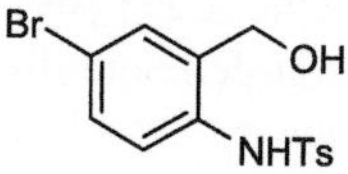	COMe	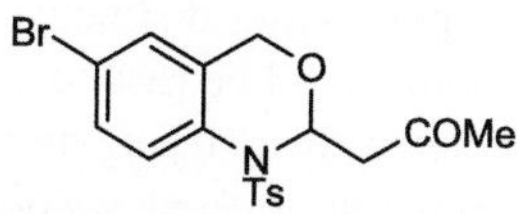	88
9	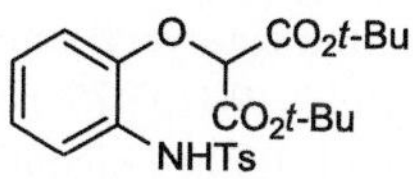	COPh	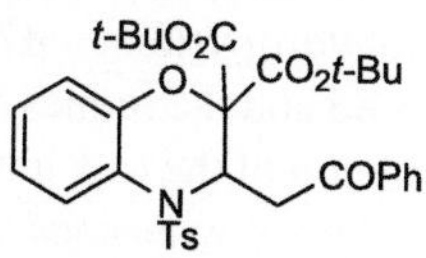	80
10	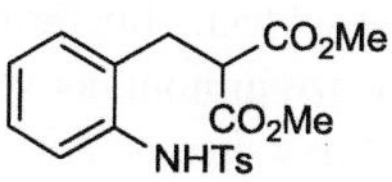	COPh	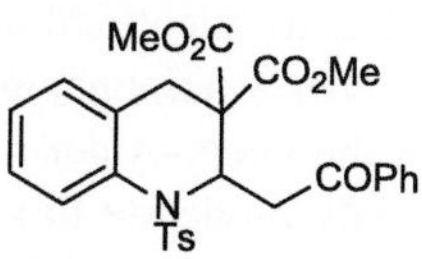	92
11	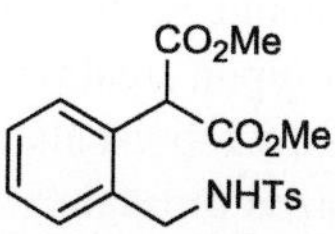	COMe	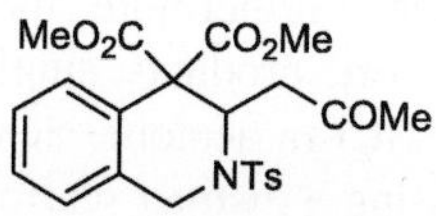	83

[a]50 mol % of AcOH and NaOAc were added.

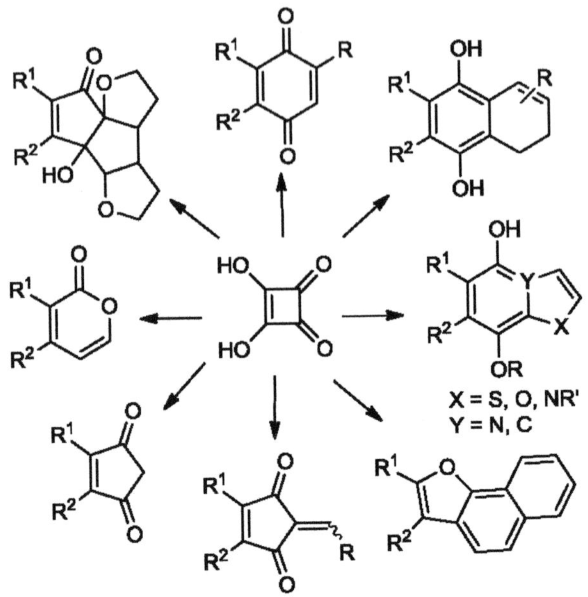

SCHEME 4.26 Armstrong's multiple-core structure library. (Adapted from [13], with permission; copyright © 1997 American Chemical Society.)

We chose the dihydropyrrole and tetrahydropyridine scaffolds as starting points to explore the generation of a library with high skeletal diversity library when using the combinatorial scaffolding strategy. When the C=O moiety of an α,β-unsaturated ester in the dihydropyrrole and tetrahydropyridine compounds is methylenated, the resulting dienol ether (an electron-rich diene) can undergo Diels–Alder reactions with different electron-deficient dienophiles to generate a variety of fused heterocyclic compounds possessing distinctive frameworks (Scheme 4.27) [69]. Indeed, Diels–Alder reactions of Tebbe reaction products and electron-deficient dienophiles provided a key skeleton-diversifying branch in our DOS pathway.

To begin exploring this concept, we prepared, as starting materials, the dihydropyrroles **85** and the tetrahydropyridines **87** through phosphine-catalyzed annulations between the allenoates **83** and the imines **84**, in excellent yields (90 to 97%) on multigram scales. Methylenation of the α,β-unsaturated esters with Tebbe reagent gave the corresponding ethoxy dienes with good reaction efficiencies (50 to 82%). Through screening of dienophiles, we identified maleimides, *N*-phenyl triazolinedione, tetracyanoethylene, imines, benzoquinone, and 2,6-dichlorobenzoquinone as very good reaction substrates for the Diels–Alder reactions of the ethoxy dienes. Although the reaction yields were only moderate to good (35 to 85%), the reaction stereoselectivities were excellent. Furthermore, the enol ether units could be hydrolyzed to the corresponding ketone products, single diastereoisomers, upon treatment with a solution of aqueous HCl in acetone. Scheme 4.28 displays 16 representative compounds, each possessing a distinct scaffold that we synthesized through a sequence

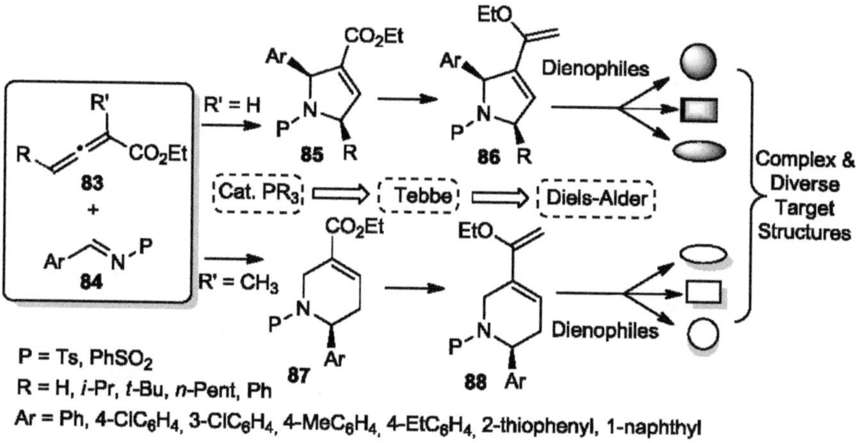

SCHEME 4.27 Skeletal diversity generated using a phosphine organocatalysis/Tebbe/Diels–Alder reaction sequence.

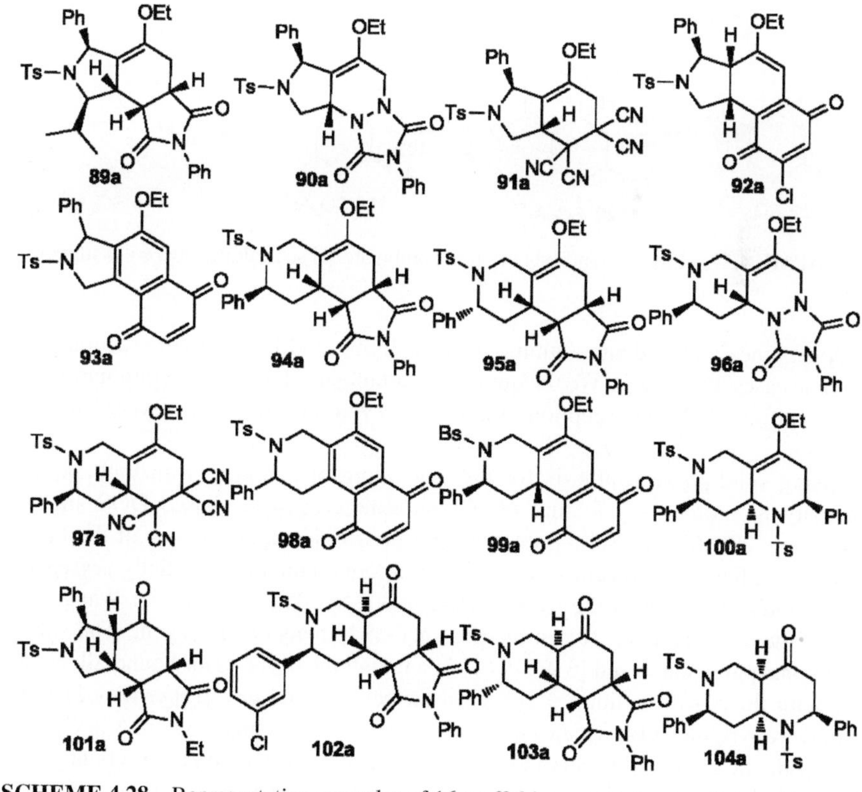

SCHEME 4.28 Representative examples of 16 scaffolds generated using the combinatorial scaffolding strategy. (From [69], with permission of John Wiley & Sons; copyright © 2011 John Wiley & Sons.)

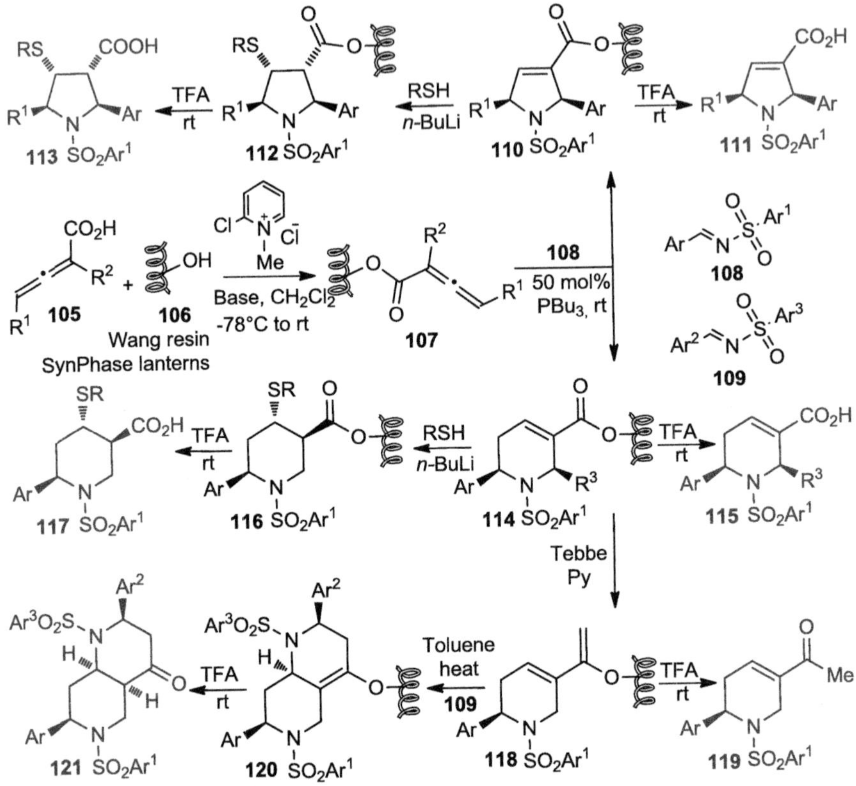

SCHEME 4.29 Library synthesis through combinatorial scaffolding on a solid support.

of phosphine-catalyzed annulation reactions, Tebbe, and Diels–Alder reactions and, in some cases, hydrolysis. We unequivocally established the relative configuration of each scaffold, with the exception of compound **99a**, through x-ray crystallographic analysis.

Using solid-phase synthesis for the generation of a library with high skeletal diversity is another key feature of our phosphine catalysis–based DOS strategy (Scheme 4.29) [63,64]. Split–pool synthesis on a solid support is one of the fastest and most efficient approaches to generating a large number of spatially segregated compounds. We coupled the hydroxyl groups of SynPhase lanterns of Wang resin **106** to the allenoic acids **105** in the presence of Mukaiyama's reagent. The subsequent phosphine-catalyzed [3 + 2]/[4 + 2] annulations of the resin-bound allenoates **107** with the *N*-sulfonylimines **108** gave the resin-bound dihydropyrroles **110** and tetrahydropyridines **114**, which we cleaved from the resin using 2.5% TFA in CH_2Cl_2 to provide the carboxylic acids **111** and **115**, respectively, in good to excellent yields with high diastereoselectivities after chromatographic purification.

We employed the common chemical functional group (α,β-unsaturated ester) of the heterocycles **110** and **114** in the next scaffold-generating split step. The Michael additions of thiols to **110** and **114** gave **112** and **116**, respectively, which, upon TFA-mediated cleavage, yielded the highly functionalized pyrrolidines **113** and piperidines **117**, respectively, as single diastereoisomers in good to excellent yields. We treated the resin-bound tetrahydropyridines **114** with Tebbe reagent and anhydrous pyridine to obtain the dienol ethers **118**, which give the tetrahydropyridine enones **119** in overall yields (over four steps) of up to 53% with excellent purity (>95%, ^{1}H NMR). Furthermore, endo-selective Diels–Alder reactions with the *N*-sulfonylimines **109** in toluene produced the octahydro-1,6-naphthyridines **120**, which we hydrolyzed from the resin using 2.5% TFA in CH_2Cl_2 to yield, after chromatographic purification, the octahydro-1,6-naphthyridin-4-ones **121** in overall yields (over five steps) of up to 39% with high diastereoselectivities. Notably, this strategy for construction of our combinatorial library is very efficient in terms of the consumption of the building blocks, because we used the *N*-sulfonylimines in both the phosphine-catalyzed [3 + 2]/[4 + 2] annulations and the Diels–Alder reactions. Employing 19 allenoic acids, 46 *N*-sulfonylimines, and 32 thiols as building blocks, we synthesized a library of 4388 heterocyclic compounds, featuring five distinct scaffolds, through our combinatorial scaffolding strategy on a solid support.

4.4 A DOS LIBRARY BASED ON PHOSPHINE ORGANOCATALYSIS: BIOLOGICAL SCREENING, ANALOG SYNTHESIS, AND STRUCTURE–ACTIVITY RELATIONSHIP ANALYSIS

Although many examples exist of collections of structurally diverse small organic molecules serving as reservoirs of therapeutic lead compounds, we were cognizant of the need to identify multiple biomodulators from our phosphine catalysis–driven library. Although we did not build our library with any specific target or known leads in mind, it exhibited intuitively desirable traits: heterocyclic compounds with low molecular weights and conformational constraints; complexity; novelty; and diversity. In this context, we were delighted to find that the library did, indeed, allow the discovery of many biomodulators of therapeutic potential. Combining phosphine organocatalysis of allenes and acetylenes with the combinatorial scaffolding strategy allowed us to prepare a designed DOS library of up to 4600 carbo- and heterocyclic compounds, featuring 55 distinctive scaffolds (Scheme 4.30). As a pilot library, we arrayed 642 model compounds, selected from our DOS library, in nine 96-well microtiter plates as 5 mM solutions in DMSO and disseminated them to our biologist collaborators. Four different types of biological systems were used in the bioassays of our pilot library compounds: (1) in vitro enzymatic assays by the Tamanoi group in the Department of Microbiology, Immunology and Molecular Genetics at UCLA; (2) endothelial cell activation bioassays by the Cruz lab in the Department of Medicine, Division of Cardiology at UCLA; (3) cancer cell migration assays by the Fenteany group in the Department of Chemistry at the University of Connecticut; and (4) zebrafish developmental screens by the Chen group in the Department of Molecular, Cell, and Development Biology at UCLA. In the following sections we provide a summary of the results from these assays.

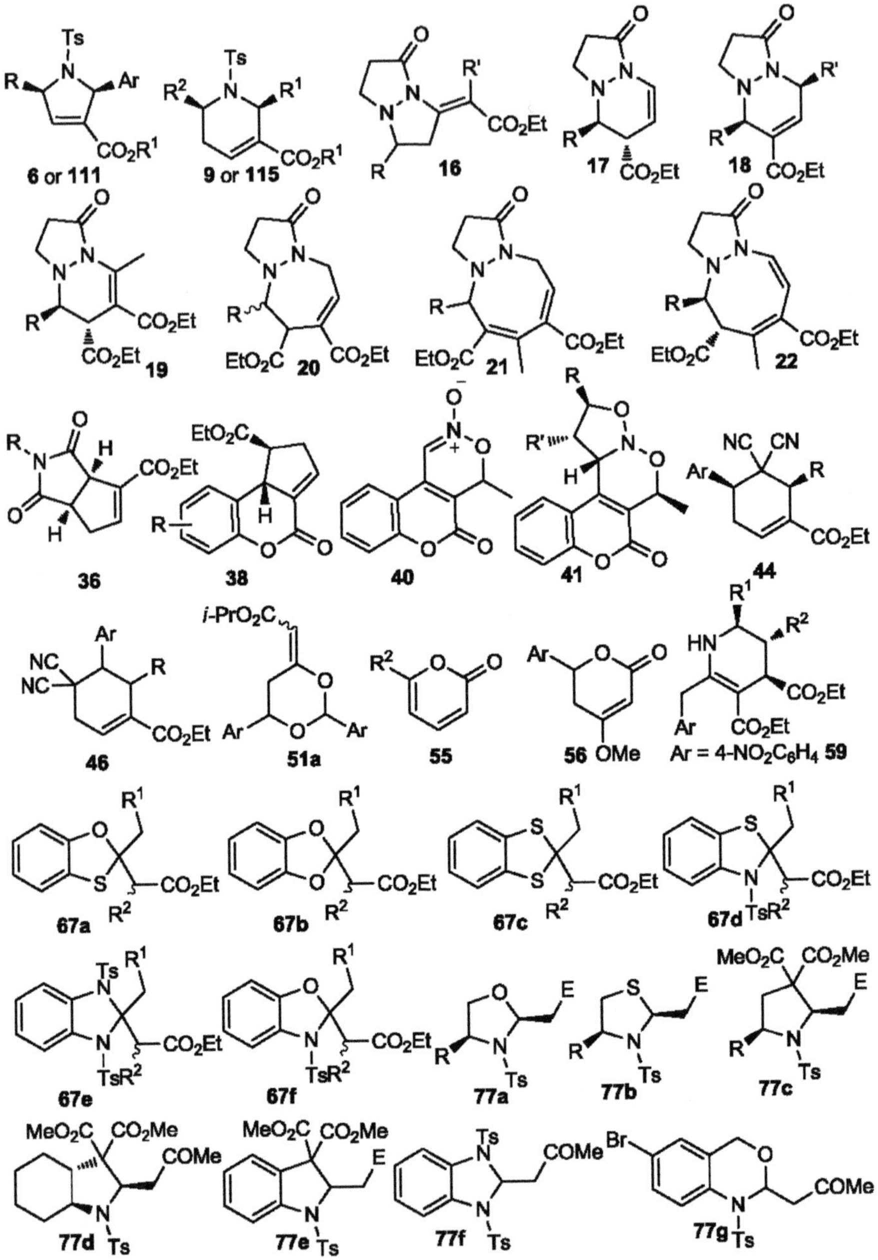

SCHEME 4.30 DOS library, featuring 55 distinct scaffolds, prepared using phosphine organocatalysis.

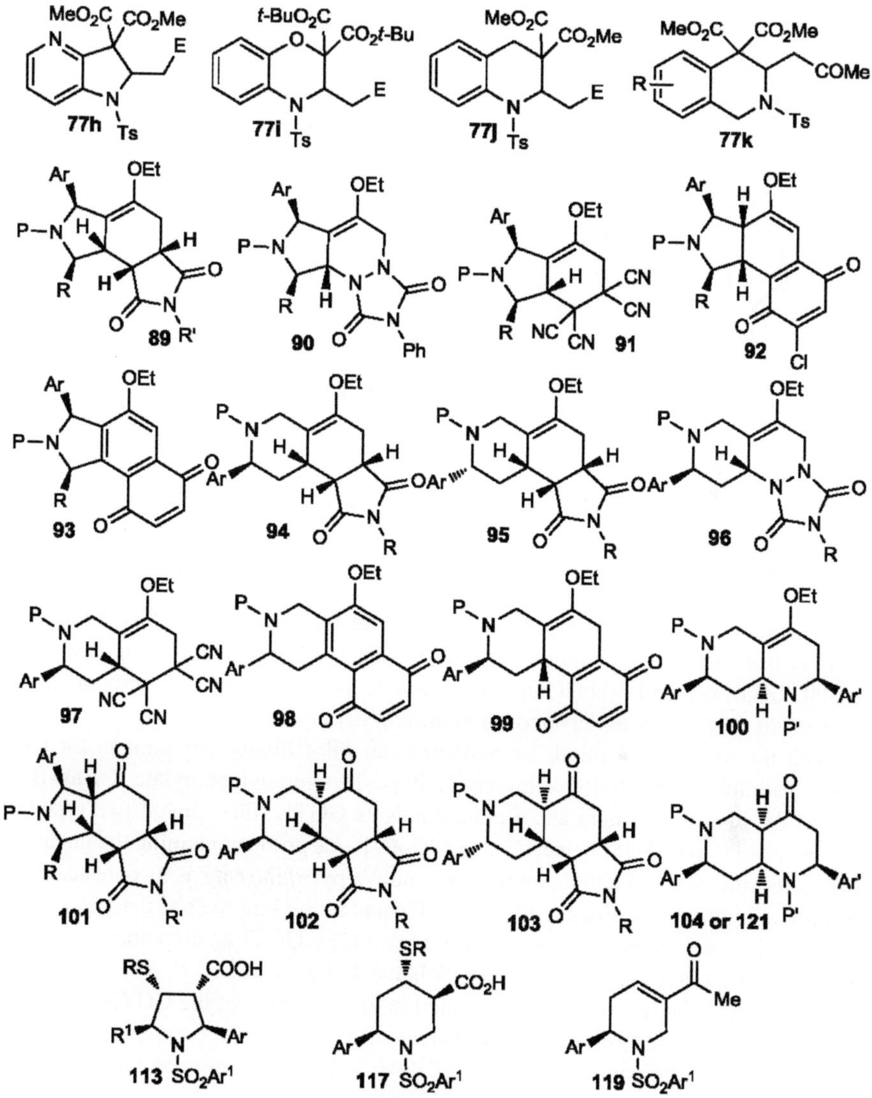

SCHEME 4.30 (*Continued*)

4.4.1 Protein Geranylgeranyltransferase Type I and Rab Inhibitors

Protein geranylgeranyltransferase type I (GGTase-I) and Rab geranylgeranyltransferase (RabGGTase) catalyze protein geranylgeranylation, a lipid modification that is critical for the function of a number of proteins, including RhoA, Rac, and Rab [93,94]. Because these prenyltransferases are implicated in a diverse number of human disorders, including cancer and cardiovascular diseases, small-molecule inhibitors are expected to play pivotal roles in the treatment of related diseases. Although Sebti,

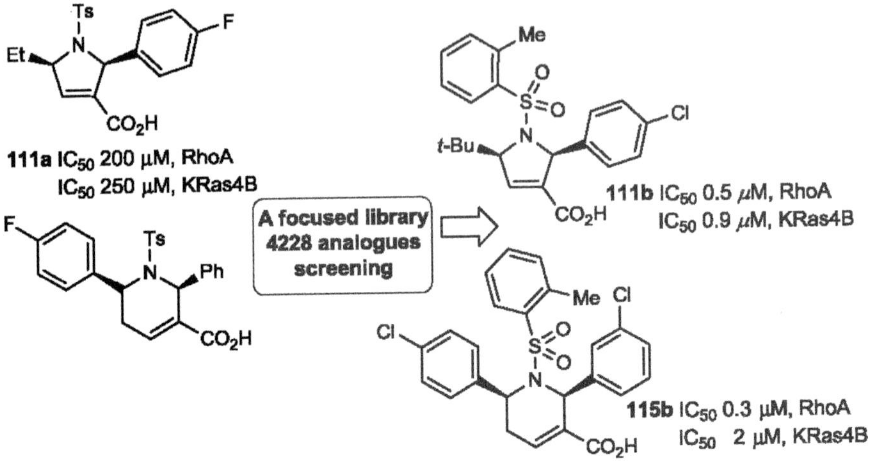

SCHEME 4.31 Representative GGTI compounds obtained from a DOS library based on phosphine catalysis.

Lerner, and co-workers prepared peptidomimetic inhibitors of GGTase-I (GGTIs) derived from the *CAAL* peptide [95,96], further development of compounds that inhibit GGTase-I and RabGGTase is desirable because they will provide novel reagents to deepen our understanding of protein geranylgeranylation.

With this premise in mind, we screened our pilot library compounds for their ability to inhibit the activity of human GGTase-I to geranylgeranylate K-Ras4B or RhoA. We identified a number of compounds as GGTIs; they shared two types of unique novel scaffolds (Scheme 4.31) [63,65–67]: one group containing the guvacine ring as its core scaffold and the other containing a pyrroline ring as its core scaffold. To further improve the efficacy of the GGTIs and to explore their structure–activity relationships, we prepared a focused library of 4288 GGTI analogs through solid-phase synthesis in a split–pool manner (Scheme 4.29).

Screening of the focused library resulted in identification of the GGTIs **111b** and **115b**, which exhibited submicromolar values of IC_{50}. Notably, compound **115b** did not induce any significant inhibition of FTase or RabGGTase, even at concentrations of 50 μM. At the same time, we found that coupling the carboxyl group of the dihydropyrrole ring of compound **111b** with phenylalanine derivatives significantly improved the cellular activity (Scheme 4.32). Furthermore, to examine the specificity of GGTase-I inhibition by compounds **111b** and **115b**, we also tested their inhibition behavior toward two closely related enzymes: FTase and RabGGTase. Surprisingly, compound **111b** induced slight inhibition of RabGGTase at concentrations greater than 10 μM. With this trail, the rescreening of the focused library resulted in identification of inhibitors of RabGGTase (RabGGTIs). As indicated in Table 4.12, compounds **113a** to **e** (Scheme 4.33) inhibited RabGGTase with IC_{50} values of 2 to 5 μM, whereas their inhibition of GGTase-I and FTase required concentrations greater than 50 μM and greater than 100 μM, respectively. Compounds **113a** to **e** are the first potentially useful inhibitors of RabGGTase [97–99].

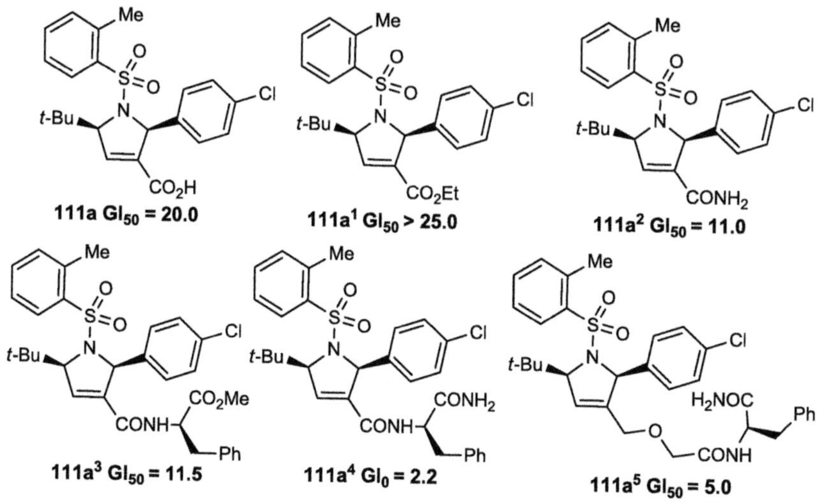

SCHEME 4.32 Values of GI_{50} for modified GGTIs toward K562 cells (μM).

TABLE 4.12 IC_{50} Values (μM) of Preferential RabGGTIs

RabGGTI	GGTase-I	RabGGTase	FTase
113a	>50	4.5	>100
113b	>50	3.6	>100
113c	>50	4.8	>100
113d	>50	2.1	>100
113e	>50	2.2	>100

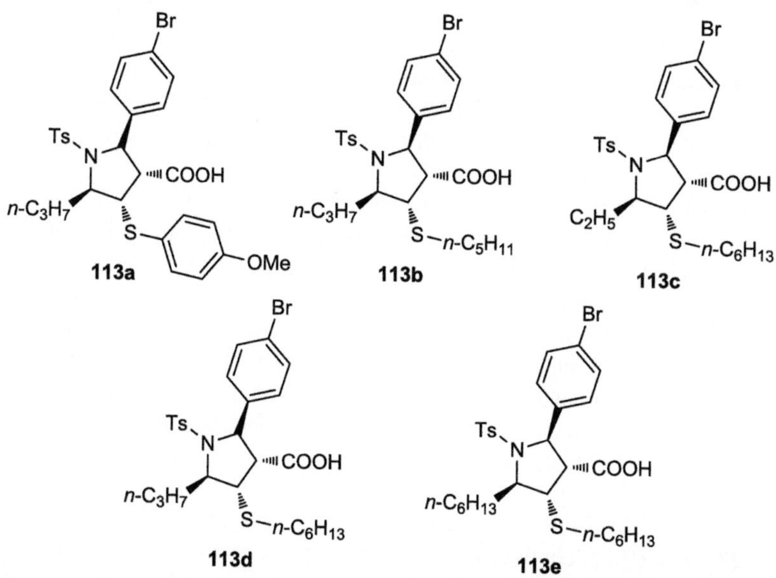

SCHEME 4.33 Molecular structures and IC_{50} values (μM) of preferential RabGGTIs.

4.4.2 Activators of Endothelium-Driven Innate Immunity of Macrophages

The endothelium plays a critical role in promoting inflammation in cardiovascular diseases and other chronic inflammatory conditions; accordingly, many small-molecule screens have sought to identify agents to prevent endothelial cell activation. Conversely, an augmented immune response can be protective against microbial pathogens and in cancer immunotherapy. Nevertheless, small-molecule screens to identify agents that induce endothelial cell (EC) activation had not been reported prior to our studies. The identification of a family of chemical probes that augment innate immunity through EC activation would provide a framework for understanding the gene networks involved in endothelial inflammation, as well as the development of novel endothelium-driven immunotherapeutic agents [64].

Activation of the endothelium is linked intricately to innate immunity. For example, ECs activated by IFNγ were shown to trigger the production of macrophage inflammatory protein 1 alpha (MIP1α) from primary human monocytes [97]. Using the designed EC-monocyte coculture system, we screened our pilot library of 642 carbo- and heterocycles, featuring 55 distinctive scaffolds, for their ability to activate human ECs, which, in turn, trigger monocyte activation. We identified seven distinct scaffolds (**89**, **90**, **94** to **96**, **100**, and **104**; Scheme 4.30) among 35 compounds derived from the sequence of phosphine-catalyzed annulation, Tebbe methylenation, Diels–Alder reaction, and sometimes hydrolysis (Scheme 4.27) for their capacity to bestow upon the endothelium the ability to trigger MIP1α and MIP1β production from previously quiescent monocytes [64].

The octahydro-1,6-naphthyridine framework was particularly interesting in that it exhibited the highest hit rate (all 10 of the octahydro-1,6-naphthyridin-4-ones in our pilot library were active), and because the synthesis of octahydro-1,6-naphthyridin-4-one analogs could be performed from resin-bound allenoates on a solid support. For further analysis of structure–activity relationships, we synthesized 100 ($1 \times 10 \times 10$) octahydro-1,6-naphthyridin-4-one analogs using the tagging and split-and-pool combinatorial techniques; as building blocks, we used 2-methyl-2,3-butadienoic acid and 10 different *N*-sulfonylimines (Scheme 4.29). In addition, we prepared several other octahydro-1,6-naphthyridine analogs, with different functional groups, through solution-phase syntheses (Scheme 4.34).

After completing the syntheses of the analogs in the solid phase and the solution phase, we dissolved the products in DMSO and analyzed their degrees of EC activation. The focused library of 96 analogs produced compounds with distinct capacities for EC activation, revealing some essential structural features for the biological activity of the octahydro-1,6-naphthyridin-4-ones. In particular, the aryl groups on the N1 and N6 arylsulfonly groups were indispensible; both N1 mesyl (Ms) and N6 mesyl naphthyridinone compounds (**104c** and **d**, Scheme 4.35) lost their activity. Based on the fact that the mesylated naphthyridinone at N1 was inactive, it was not surprising that the derivative **129** featuring a pyridyl ring was also inactive. On the other hand, removal of the aryl group at C7 (analog **125**) did not affect the naphthyridinone's ability to induce EC activation. At the same time, the alcohol **126** and its ethyl ether **127** were active, confirming that the ketone group was less essential for the naphthyridinone's EC-activating effect.

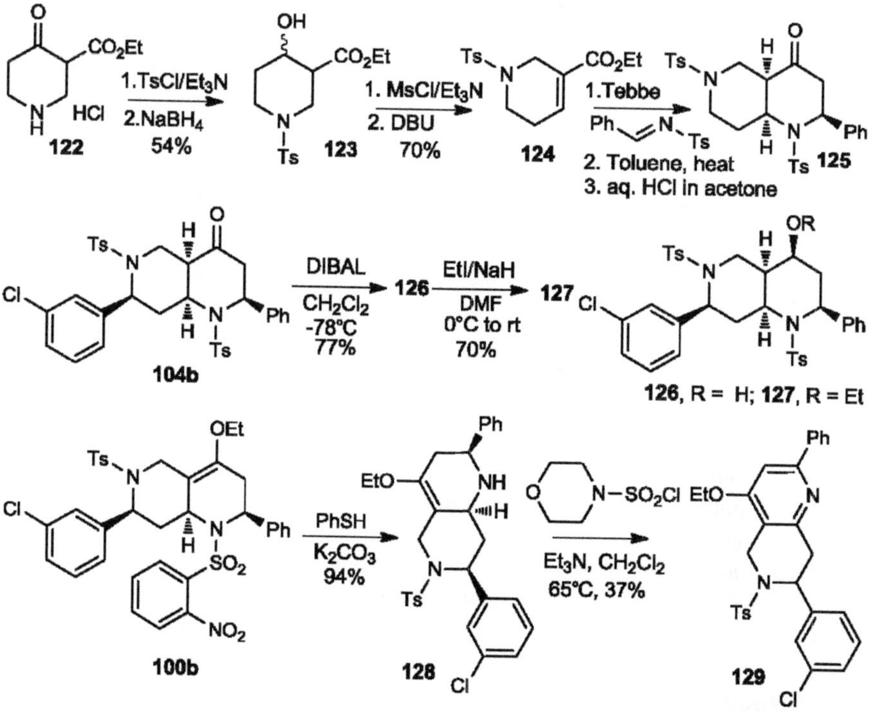

SCHEME 4.34 Solution-phase synthesis of naphthyridines.

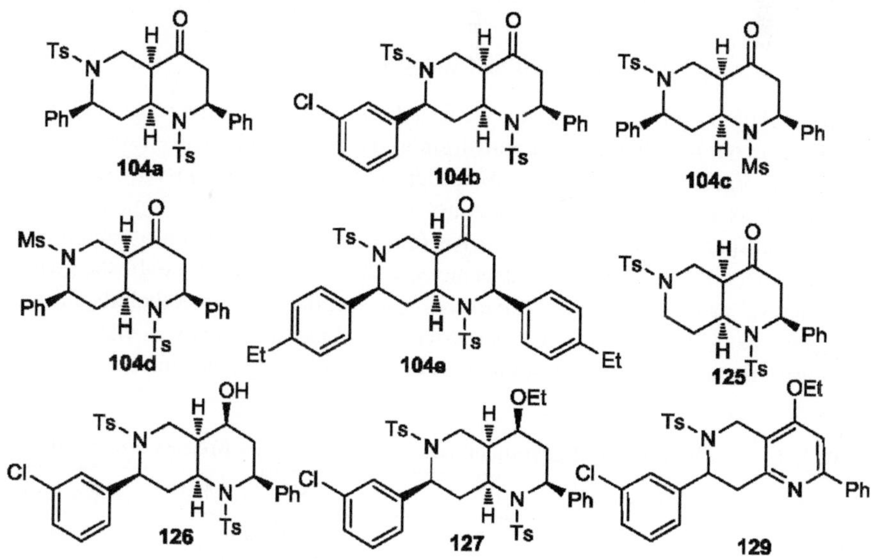

SCHEME 4.35 Structures of selected active (**104a**, **b**, and **e**, **125**, **126**, **127**) and inactive (**104c** and **d**, **129**) naphthyridine analogs.

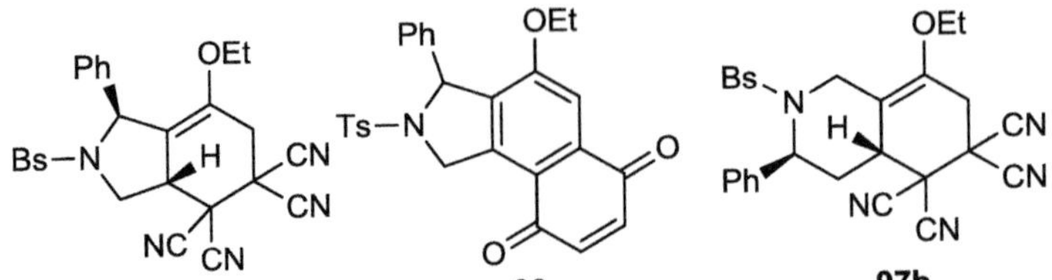

SCHEME 4.36 Selected inhibitors of MDA-MB-231 human breast cancer cells. (From [69], with permission of John Wiley & Sons; copyright © 2011 John Wiley & Sons.)

4.4.3 Cancer Cell Migration Assays

Cell migration is a basic biological phenomenon in several normal and disease processes, including cancer cell invasion and metastasis. Despite the relatively small number of compounds in our library, we were optimistic about the prospect of identifying compounds exhibiting antimigratory activity because of the broad chemical space covered by the products generated from the phosphine organocatalysis of allenes and acetylenes in conjunction with combinatorial scaffolding strategy. We screened the pilot libarry in a medium-throughput wound closure assay to identify molecules that inhibited the migration of MDA-MB-231 human breast cancer cells. To our delight, compounds **91b**, **93a**, and **97b** exhibited subtoxic antimigratory activity in the wound closure assay (Scheme 4.36 and Table 4.13). Their values of IC_{50} for inhibition of cell migration were comparable with that of the natural product migrastatin, which has an IC_{50} of 29 μM and has served as a lead for the subsequent development of more potent analogs [69].

4.4.4 Aplexone Decreases Cellular Cholesterol Level

In collaboration with Chen's group, we tested the biological activities of 171 compounds from our pilot library on zebrafish embryos. The zebrafish was chosen because this model organism provides an unprecedented opportunity for in vivo screens as a result of its small size (allowing assays in 96-well format), its high fecundity, and the transparent bodies of its embryos (allowing ready detection of subtle phenotypic defects). In addition, a transgenic zebrafish line has been established using the VEGFR2 promoter to drive GFP expression, specifically in endothelial cells and their precursors. These transgenic fish provide a unique opportunity to visualize the cardiovasculature development of the live embryos, allowing efficient screening of a

TABLE 4.13 Inhibition of Migration of MDA-MB-231 Human Breast Cancer Cells

Compound	MIC (μM)	IC_{50} (μM)	95% CI (μM)	MLC (μM)
91b	5	43.1	39.9–46.6	150
93a	5	14.8	13.1–16.7	50
97b	5	21.6	19.1–24.5	100

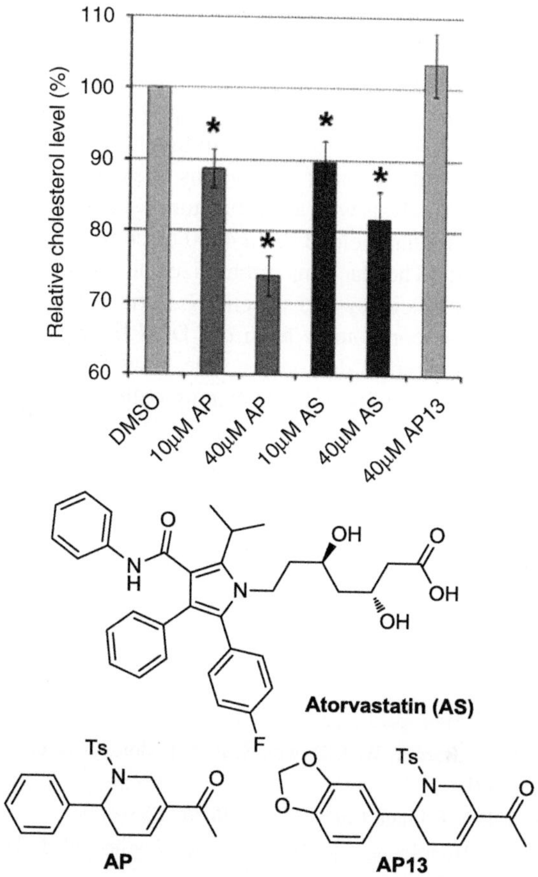

SCHEME 4.37 Cellular cholesterol levels in zebrafish embryos.

large number of compounds for their impacts on the formation and function of the heart and its vessels. We expected to observe effects on several phenotypes, including changes in the populations of cardiac and endothelial precursors, abnormal patterning of the heart and the vasculature, and abnormal cardiac functions, including arrhythmia. Using this model, we identified a novel compound, aplexone, which affects the transcription of the HMG-CoA reductase pathway and differentially regulates arteriovenous angiogenesis in zebrafish and human endothelial cells. Aplexone was also effective in lowering the cellular cholesterol levels—even more efficiently than atorvastatin (Scheme 4.37) [68].

4.5 CONCLUSIONS

We have established collections of compound libraries, based on phosphine catalysis of allenes and acetylenes, as common starting points for chemical biological studies and the discovery of potential therapeutic leads. The development of efficient methods for the construction of libraries featuring structural diversity that encompasses the maximum amount of chemical space is a particularly challenging task for organic chemists. DOS entails the development of pathways leading to the efficient syntheses of collections of small molecules exhibiting rich skeletal and stereochemical diversity. Here, the combination of annulations of electron-deficient allenes and alkynes under phosphine catalysis with a combinatorial scaffolding strategy provided a branching DOS pathway that yielded up to 4600 discrete compounds, comprising 55 distinct cyclic scaffolds. The reactions summarized herein meet the high standards of DOS in terms of their efficiency and selectivity. Furthermore, we have identified several biologically active compounds from our DOS library based on phosphine organocatalysis. These outcomes demonstrate the powerful premise behind DOS: The greater the structural diversity in the screening collection, the higher the probability of discovering small-molecule biomodulators. Indeed, the implementation of nucleophilic phosphine catalysis resulted in a compound library featuring rich structural diversity, making it well suited to probe critical biological questions and for creative new applications. The basic transformations described herein will undoubtedly instigate further expansions of the scope and utility of nucleophilic phosphine catalysis in DOS.

REFERENCES

1. E. M. Gordon, R. W. Barrett, W. J. Dower, S. P. A. Fodor, M. A. Gallop, *J. Med. Chem.* **1994**, *37*, 1385–1401.
2. L. A. Thompson, J. A. Ellman, *Chem. Rev.* **1996**, *96*, 555–600.
3. R. E. Dolle, B. Le Bourdonnec, K. Worm, G. A. Morales, C. J. Thomas, W. Zhang, *J. Comb. Chem.* **2010**, *12*, 765–806.
4. S. P. Fodor, J. L. Read, M. C. Pirrung, L. Stryer, A. T. Lu, D. Solas, *Science* **1991**, *251*, 767–773.
5. K. S. Lam, S. E. Salomon, E. M. Hersh, V. J. Hruby, W. M. Kazmierski, R. J. Knapp, *Nature* **1991**, *354*, 82–84.
6. R. A. Houghten, C. Pinilla, S. E. Blondelle, J. R. Appel, C. T. Dooley, J. H. Cuervo, *Nature* **1991**, *354*, 84–86.
7. G. Jung, A. G. Beck-Sickinger, *Angew. Chem. Int. Ed.* **1992**, *31*, 367–383.
8. M. A. Gallop, R. W. Barrett, W. J. Dower, S. P. A. Fodor, E. M. Gordon, *J. Med. Chem.* **1994**, *37*, 1233–1333.
9. B. A. Bunin, J. A. Ellman, *J. Am. Chem. Soc.* **1992**, *114*, 10997–10998.
10. M. J. Plunkett, J. A. Ellman, *J. Am. Chem. Soc.* **1995**, *117*, 3306–3307.
11. E. M. Gordon, M. A. Gallop, D. V. Patel, *Acc. Chem. Res.* **1996**, *29*, 144–154.
12. M. A. Marx, A.-L. Grillot, C. T. Louer, K. A. Beaver, P. A. Bartlett, *J. Am. Chem. Soc.* **1997**, *119*, 6153–6167.
13. P. A. Tempest, R. W. Armstrong, *J. Am. Chem. Soc.* **1997**, *119*, 7607–7608.
14. M. R. Spaller, M. T. Burger, M. Fardis, P. A. Bartlett, *Curr. Opin. Chem. Biol.* **1997**, *1*, 47–53.

15. O. Kwon, S. B. Park, S. L. Schreiber, *J. Am. Chem. Soc.* **2002** *124*, 13402–13404.
16. J. E. Biggs-Houck, A. Younai, J. T. Shaw, *Curr. Opin. Chem. Biol.* **2010**, *14*, 371–382.
17. S. L. Schreiber, *Science* **2000**, *287*, 1964–1969.
18. D. R. Spring, *Org. Biomol. Chem.* **2003**, *1*, 3867–3870.
19. G. W. Bemis, M. A. Murcko, *J. Med. Chem.* **1996**, *39*, 2887–2893.
20. M. D. Burke, E. M. Berger, S. L. Schreiber, *Science* **2003**, *302*, 613–618.
21. W. R. J. D. Galloway, S. Isidro-Llobet, D. R. Spring, *Nat. Commun.* **2010**, *1*, 80–86.
22. T. E. Nielsen, S. L. Schreiber, *Angew. Chem. Int. Ed.* **2008**, *47*, 48–56.
23. D. P. Walsh, Y.-T. Chang, *Chem. Rev.* **2006**, *106*, 2476–2530.
24. R. E. Dolle, B. L. Bourdonnec, A. J. Goodman, G. A. Morales, C. J. Thomas, W. Zhang, *J. Comb. Chem.* **2009**, *11*, 739–790.
25. D. S. Tan, *Nat. Chem. Biol.* **2005**, *1*, 74–84.
26. J. M. Mitchell, J. T. Shaw, *Angew. Chem. Int. Ed.* **2006**, *45*, 1722–1726.
27. E. Comer, E. Rohan, L. Deng, J. A. Porco, *Org. Lett.* **2007**, *9*, 2123–2126.
28. H. An, S.-J. Eum, M. Koh, S. K. Lee, S. B. Park, *J. Org. Chem.* **2008**, *73*, 1752–1761.
29. G. L. Thomas, R. J. Spandl, F. G. Glansdorp, M. Welch, A. Bender, J. Cockfield, J. A. Lindsay, C. Bryant, D. F. J. Brown, O. Loiseleur, H. Rudyk, M. Ladlow, D. R. Spring, *Angew. Chem. Int. Ed.* **2008**, *47*, 2808–2812.
30. D. Morton, S. Leach, C. Cordier, S. Warriner, A. Nelson, *Angew. Chem. Int. Ed.* **2008**, *48*, 104–109.
31. W. R. J. D. Galloway, A. Bender, M. Welch, D. R. Spring, *Chem. Commun.* **2009**, 2446–2462.
32. A. Kumar, R. A. Maurya, P. Ahmad, *J. Comb. Chem.* **2009**, *11*, 198–201.
33. P. Mukherjee, S. J. S. Roy, T. K. Sarkar, *Org. Lett.* **2010**, *12*, 2472–2475.
34. J. Cui, J. Hao, O. A. Ulanovskaya, J. Dundas, J. Liang, S. A. Kozmin, *Proc. Natl. Acad. Sci. U.S.A.* **2011**, *108*, 6763–6768.
35. A. Isidro-Llobet, T. Murillo, P. Bello, A. Cilibrizzi, J. T. Hodgkinson, W. R. J. D. Galloway, A. Bender, M. Welch, D. R. Spring, *Proc. Natl. Acad. Sci. U.S.A.* **2011**, *108*, 6793–6798.
36. A. W. Hung, A. Ramek, Y. Wang, T. Kaya, J. A. Wilson, P. A. Clemons, D. W. Young, *Proc. Natl. Acad. Sci. U.S.A.* **2011**, *108*, 6799–6804.
37. J. M. J. Frechet, *Tetrahedron* **1981**, *37*, 663–683.
38. P. Hodge, in: *Syntheses and Separation Using Functional Polymers*, D. C. Sherrington, P. Hodge, Eds., Wiley, New York, **1988**, pp. 43–122.
39. A. Furka, F. Sebestyen, M. Asgedom, G. Dibo, *Int. J. Pept. Protein Res.* **1991**, *37*, 487–493.
40. X. Lu, C. Zhang, Z. Xu, *Acc. Chem. Res.* **2001**, *34*, 535–544.
41. J. L. Methot, W. R. Roush, *Adv. Synth. Catal.* **2004**, *346*, 1035–1050.
42. L.-W. Ye, J. Zhou, Y. Tang, *Chem. Soc. Rev.* **2008**, *37*, 1140–1152.
43. B. J. Cowen, S. J. Miller, *Chem. Soc. Rev.* **2009**, *38*, 3102–3116.
44. A. Marinetti, A. Voituriez, *Synlett* **2010**, 174–194.
45. Y. C. Fan, O. Kwon, in: *Science of Synthesis: Asymmetric Organocatalysis 1*, Vol. 2011/5, *Lewis Base and Acid Catalysts*, B. List, Ed., Georg Thieme, Stuttgart, Germany, **2012**, pp. 723–782.

46. X.-F. Zhu, C. E. Henry, O. Kwon, *Tetrahedron* **2005**, *61*, 6276–6282.
47. X.-F. Zhu, J. Lan, O. Kwon, *J. Am. Chem. Soc.* **2003**, *125*, 4716–4717.
48. R. Na, C. Jing, Q. Xu, J. Shi, H. Jiang, X. Wu, J. Zhong, M. Wang, D. Benitez, W. A. Goddard, H. Guo, O. Kwon, *J. Am. Chem. Soc.* **2011**, *133*, 13337–13348.
49. C. E. Henry, O. Kwon, *Org. Lett.* **2007**, *9*, 3069–3072.
50. Y. S. Tran, O. Kwon, *J. Am. Chem. Soc.* **2007**, *129*, 12632–12633.
51. X. Zhu, C. E. Henry, J. Wang, T. Dudding, O. Kwon, *Org. Lett.* **2005**, *7*, 1387–1390.
52. X. Zhu, A. Schaffner, R. C. Li, O. Kwon, *Org. Lett.* **2005**, *7*, 2977–2980.
53. G. S. Creech, O. Kwon, *Org. Lett.* **2008**, *10*, 429–432.
54. G. S. Creech, X. Zhu, B. Fonovic, D. Travis, O. Kwon, *Tetrahedron* **2008**, *64*, 6935–6942.
55. H. Guo, Q. Xu, O. Kwon, *J. Am. Chem. Soc.* **2009**, *131*, 6318–6319.
56. J. Szeto, V. Sriramurthy, O. Kwon, *Org. Lett.* **2011**, *13*, 5420–5423.
57. Y. S. Tran, T. J. Martin, O. Kwon, *Chem. Asian J.* **2011**, *6*, 2101–2106.
58. C. Jing, R. Na, H. Liu, L. Zhang, J. Liu, M. Wang, J. Zhong, O. Kwon, H. Guo, *Adv. Synth. Catal.* **2012**, *354*, 1023–1034.
59. V. Sriramurthy, G. A. Barcan, O. Kwon, *J. Am. Chem. Soc.* **2007**, *129*, 12928–12929.
60. V. Sriramurthy, O. Kwon, *Org. Lett.* **2010**, *12*, 1084–1087.
61. Y. C. Fan, O. Kwon, *Molecules* **2011**, *16*, 3802–3825.
62. Y. C. Fan, O. Kwon, *Org. Lett.* **2012**, *14*, 3264–3267.
63. S. Castellano, H. D. G. Fiji, S. S. Kinderman, M. Watanabe, P. De Leon, F. Tamanoi, O. Kwon, *J. Am. Chem. Soc.* **2007**, *129*, 5843–5845.
64. D. Cruz, Z. Wang, J. Kibbie, R. Modlin, O. Kwon, *Proc. Natl. Acad. Sci. U.S.A.* **2011**, *108*, 6769–6774.
65. M. Watanabe, H. D. G. Fiji, L. Guo, L. Chan, S. S. Kinderman, D. J. Slamon, O. Kwon, F. Tamanoi, *J. Biol. Chem.* **2008**, *283*, 9571–9579.
66. J. Lu, L. Chan, H. D. G. Fiji, R. Dahl, O. Kwon, F. Tamanoi, *Mol. Cancer Ther.* **2009**, *8*, 1218–1226.
67. L. N. Chan, H. D. G. Fiji, M. Watanabe, O. Kwon, F. Tamanoi, *PloS ONE* **2011**, *6*, e26135.
68. J. Choi, K. Mouillesseauex, Z. Wang, H. D. G. Fiji, S. S. Kinderman, G. W. Otto, R. Geisler, O. Kwon, J.-N. Chen, *Development* **2011**, *138*, 1173–1182.
69. Z. Wang, S. Castellano, S. S. Kinderman, C. E. Argueta, A. B. Beshir, G. Fenteany, O. Kwon, *Chem. Eur. J.* **2011**, *17*, 649–654.
70. Z. Xu, X. Lu, *Tetrahedron Lett.* **1997**, *38*, 3461–3464.
71. K. Lu, O. Kwon, *Org. Synth.* **2009**, *86*, 212–224.
72. Y. S. Tran, O. Kwon, *Org. Lett.* **2005**, *7*, 4289–4291.
73. R. A. Villa, Q. Xu, O. Kwon, *Org. Lett.* **2012**, *14*, 4634–4637.
74. R. P. Wurz, G. C. Fu, *J. Am. Chem. Soc.* **2005**, *127*, 12234–11235.
75. H. Xiao, Z. Chai, H.-F. Wang, X.-W. Wang, D.-D. Cao, W. Liu, Y.-P. Lu, Y.-Q. Yang, G. Zhao, *Chem. Eur. J.* **2011**, *17*, 10561–10564.
76. C. Perreault, S. R. Goudreau, L. E. Zimmer, A. B. Charette, *Org. Lett.* **2008**, *10*, 689–692.
77. M. Keller, A. S. S. Sido, P. Pale, J. Sommer, *Chem. Eur. J.* **2009**, *15*, 2810–2817.
78. T. Hashimoto, Y. Maeda, M. Omote, H. Nakatsu, K. Maruoka, *J. Am. Chem. Soc.* **2010**, *132*, 4076–4077.

79. C. Zhang, X. Lu, *J. Org. Chem.* **1995**, *60*, 2906–2908.
80. A. O. Pittet, E. M. Klaiber, *J. Agric. Food Chem.* **1975**, *23*, 1189–1195.
81. J. F. Vinals, A. D. Quinn, A. O. Pittet, U.S. Patent 3,861,403, **1975**.
82. J. F. Vinals, A. D. Quinn, A. O. Pittet, U.S. Patent 3,890,981, **1975**.
83. E. J. Shuster, A. O. Pittet, U.S. Patent 3,996,170, **1976**.
84. M. P. Pollastri, A. Whitty, J. C. Merrill, X. Tang, T. D. Ashton, S. Amar, *Chem. Biol. Drug Des.* **2009**, *74*, 121–128.
85. G. S. Singh, M. D'hooghe, N. De Kimpe, *Chem. Rev.* **2007**, *107*, 2080–2135.
86. D. A. White, M. M. Baizer, *Tetrahedron Lett.* **1973**, *14*, 3579–3600.
87. H.-J. Christau, J. Viala, H. Christol, *Tetrahedron Lett.* **1982**, *23*, 1569–1572.
88. B. M. Trost, C.-J. Li, *J. Am. Chem. Soc.* **1994**, *116*, 3167–3168.
89. X. Lu, C. Zhang, *Synlett* **1995**, 645–646.
90. T. J. Martin, V. G. Vakhshori, Y. S. Tran, O. Kwon, *Org. Lett.* **2011**, *13*, 2586–2589.
91. X.-Y. Guan, Y. Wei, M. Shi, *Eur. J. Org. Chem.* **2011**, 2673–2677.
92. C. Lu, X. Lu, *Org. Lett.* **2002**, *4*, 4677–4679.
93. K. F. Leung, R. Baron, M. C. Seabra, *J. Lipid Res.* **2006**, *47*, 467–475.
94. H. Zhang, M. C. Seabra, J. Deisenhofer, *Structure* **2000**, *8*, 241–251.
95. S. M. Sebti, A. A. Adjei, *Semin. Oncol.* **2004**, *31*, 28–39.
96. E. C. Lerner, Y. Qian, A. D. Hamilton, S. M. Sebti, *J. Biol. Chem.* **1995**, *270*, 26770–26773.
97. N. W. Lukacs, *Blood* **1994**, *83*, 1174–1178.

5

DOMINO REACTIONS IN LIBRARY SYNTHESIS

MATTHEW G. LAPORTE, JOHN R. GOODELL, SAMMI TSEGAY, AND PETER WIPF

5.1 INTRODUCTION

Domino reactions have been defined by Tietze et al. as "sequences in which a bond formation (or a bond-breaking process) is combined with the formation of a new functionality" [1]. In the purest version of a domino sequence, the functionalities would be prearranged for further transformations without the need for subsequent reagents, catalysts, or altered conditions. There are many elegant examples in total synthesis that satisfy these stipulations [2]. The arrival of diversity-oriented synthesis (DOS) is providing a powerful further incentive for the invention of new domino reactions. Concise new reaction methodologies leading to the rapid synthesis of novel chemotypes are the most effective source of structurally and stereochemically diverse libraries for biological screening. The use of domino processes can offer several benefits (financial, atom efficiency, environmental) compared to multistep syntheses containing discrete workups and product isolations [1].

In this chapter we utilize the domino reaction classification system put forth by Tietze et al. [1]. A modified interpretation of this definition also allows for the inclusion of consecutive conversion(s) without the necessity for isolation of intermediates in the sequence [3]. The examples selected are categorized transformation-dependent as either cationic, anionic, pericyclic, oxidative–reductive, transition metal–mediated, or radical reactions. A total of 14 case studies are described, with the largest group using the powerful Diels–Alder cycloaddition process.

Diversity-Oriented Synthesis: Basics and Applications in Organic Synthesis, Drug Discovery, and Chemical Biology, First Edition. Edited by Andrea Trabocchi.

Now more than ever, there is a need for the development of structurally novel chemical entities to keep pace with the ever-evolving field of biomedical sciences [4]. Over the past several decades, many technological advances in synthetic organic chemistry have contributed to a more ambitious design as well as the practical generation of diverse chemical libraries [5]. For example, single-mode microwave heating has moved into the mainstream synthetic laboratory, and flow chemistry is rapidly following suit. This chapter is not intended to be either a comprehensive review of domino reactions or a chemical library synthesis but, rather, a compilation of some recent applications of domino reaction sequences in the design of unique chemical libraries.

5.2 PERICYCLIC DOMINO REACTIONS

Based on a previously reported resin-linked two-component-domino-Knoevenagel-ene reaction [6], Tietze et al. reported the solid-phase synthesis of 3,4-dihydropyrans **3** and **4** via a domino-Knoevenagel-hetero-Diels–Alder sequence (Scheme 5.1) [7]. This intermolecular three-component reaction was performed similar to a one-pot reaction without the requirement of workup and purification of intermediates. The utilization of solid-supported chemistry allowed for the use of reagent excesses that were required for driving the equilibrium to reaction completion. Knoevenagel condensation of the polymer-bound acetoacetate **1** with aliphatic aldehydes (5 equiv) in the presence of catalytic piperidinium acetate provided polymer-bound oxabutadienes **2**. Treatment with enol ethers (10 equiv) and heating in pressure flasks for three days was required to achieve the inverse electron-demand hetero-Diels–Alder cycloaddition. Cleavage from the resin utilized a basic trans esterification with sodium methoxide to afford methyl-3,4-dihydro-2*H*-pyran-5-carboxylates **3** and **4**

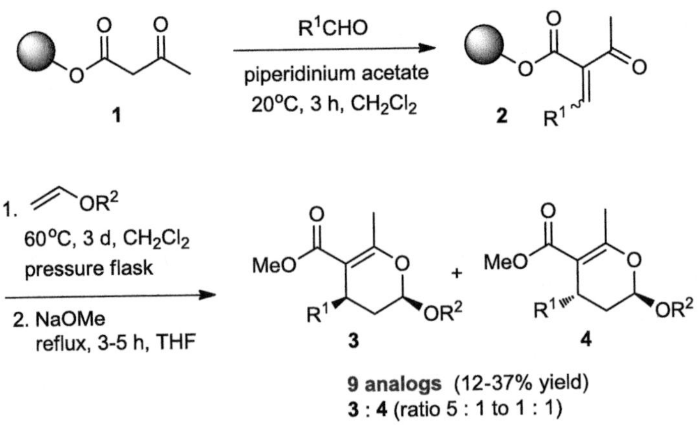

SCHEME 5.1 Synthesis of 3,4-dihydropyrans **3** and **4** via a domino-Knoevenagel-hetero-Diels–Alder sequence.

with modest diastereoselectivity. The crude cycloadducts were isolated in about 90% purity.

Based on this methodology, a small prototype library of nine compounds was synthesized with overall yields ranging from 12 to 37%. Diastereoselectivities ranged from 1 : 1 to 5 : 1 as a result of endo/exo cycloaddition modes, as shown in Table 5.1. Since a large variety of aldehydes and a more limited selection of enol ethers are available commercially, this one-pot three-component domino sequence should be amendable to the synthesis of more extensive and quite diverse libraries.

The Aubé group made use of a Diels–Alder-acylation sequence to assemble an *N*-alkyloctahydroisoquinlin-1-one-8-carboxamide library (Scheme 5.2) [8]. Although the mechanistic pathway for this domino sequence is still undefined, it is believed to involve either the Diels–Alder reaction followed by an intramolecular acylation, or the reverse sequence (i.e., acylation, then cycloaddition). The optimal cycloaddition conditions for the aminoethyl-tethered butadienes **5** and maleic anhydride involved microwave irradiation at 165°C in 1,2-dichloroethane. All isoquinolinone–carboxylic acids **7** were obtained as single stereoisomers in good yields (68 to 80%). Six structurally diverse heterocyclic building blocks were further elaborated in a parallel format to generate a 72-membered compound library through conversion to amides **8** (Scheme 5.2 and Table 5.2).

The Aubé group also reported an elegant Diels–Alder–Schmidt domino reaction approach to the tricyclic core ring systems of various *Stemona* alkaloids (Scheme 5.3) [9]. This synthetic strategy not only provided the four contiguous stereocenters of the *Stemona* core but also proved to be sufficiently efficient for total synthesis applications [10]. Several of these library members have recently been shown to have potent binding efficiencies to numerous GPCR receptors (i.e., sigma 1/2, adrenergic alpha 1/2) [11].

The treatment of azidoethyl-tethered butadienes **9** with cyclohexenenones (or cyclopentenones) in the presence of stoichiometric $BF_3{\cdot}OEt_2$ exclusively provided the endo-cycloadduct ring-expanded lactam products **11**. It was also found that $SnCl_4$ could promote this reaction sequence to afford high to exclusive ratios of endo isomers. In many instances, the cycloadducts **10** were not isolated, as these intermediates underwent a subsequent intramolecular Schmidt reaction under Diels–Alder reaction conditions. Based on this methodology, a series of six-tricyclic templates were prepared in sufficient quantities to allow for further diversifications (Scheme 5.4).

The carbonyl group containing intermediates **11** were converted further to either phenylamines or 4-phenylpiperazine derivatives via imine formation with $TiCl_4$/$Ti(O\textit{i}\text{-}Pr)_4$ followed by reduction with $NaBH_3CN$ (Scheme 5.4). In most cases, these aminations were unselective and gave inseparable mixtures of diastereomers, with the exception of phenylamine **12**. The configuration in this series of analogs was firmly established by the x-ray structure analysis of a dichlorobenzamide derivative. The diastereomerically pure α- and β-phenylamines **12** were converted to a set of structurally diverse tertiary amines **16** by a second reductive amination (Scheme 5.5).

A stereochemically diverse collection of carbamates **13** was readily assembled from the corresponding alcohols derived from the pivotal ketone intermediates **11**

TABLE 5.1 Preparation of Methyl-3,4-dihydro-2*H*-pyran-5-carboxylates 3/4

Entry	Aldehyde	Enol Ether	Product	Yield (%)[a]	3 : 4[b]
1	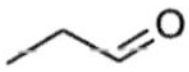	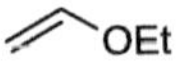	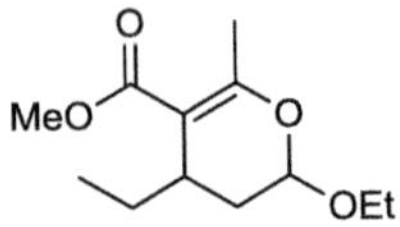	**3a/4a**, 13	3.5 : 1
2	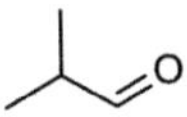	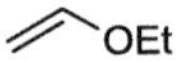	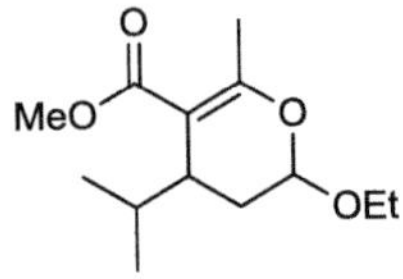	**3b/4b**, 13	1 : 1
3	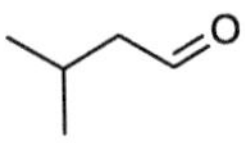	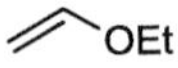	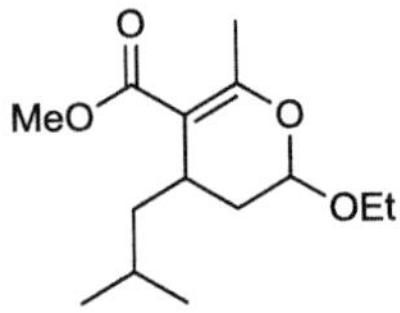	**3c/4c**, 37	3.5 : 1
4	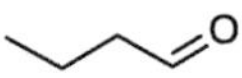	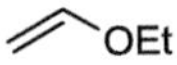	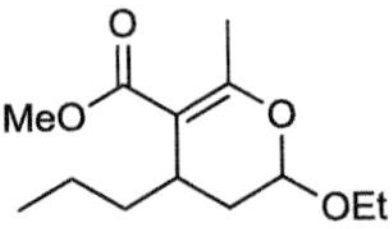	**3d/4d**, 21	4 : 1
5	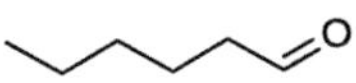	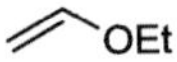	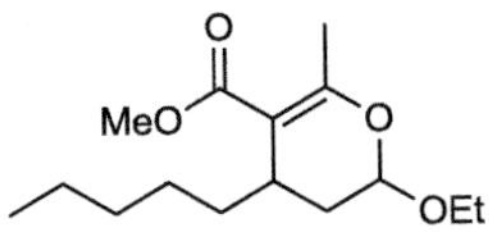	**3e/4e**, 22	5 : 1
6	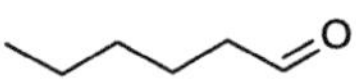	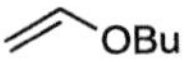	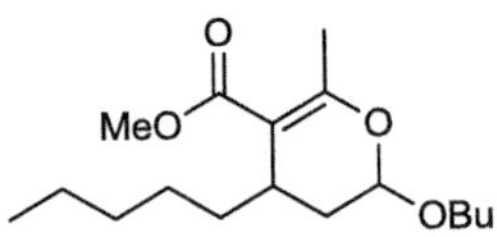	**3f/4f**, 32	5 : 1
7	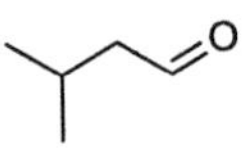	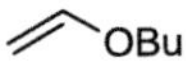	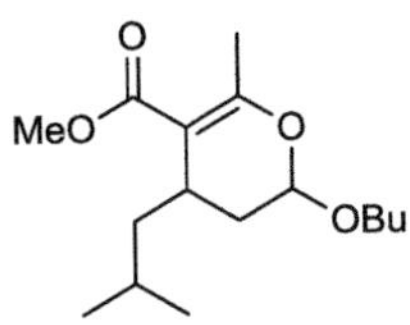	**3g/4g**, 26	3.6 : 1
8	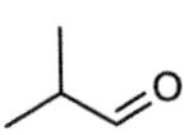	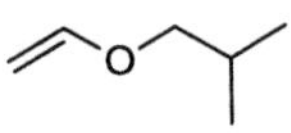	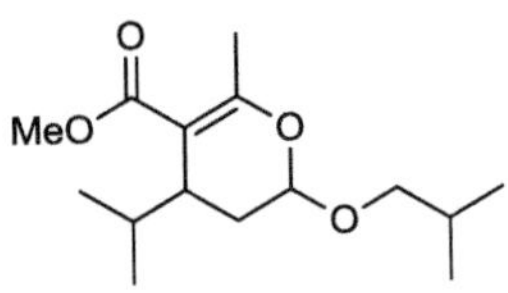	**3h/4h**, 20	3.3 : 1
9	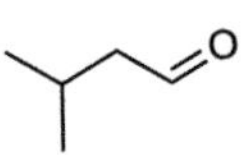	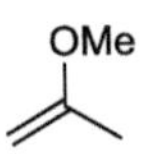	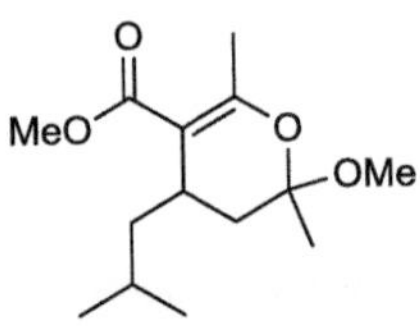	**3i/4i**, 12	2.8 : 1

[a]Isolated yields.
[b]Ratios determined by NMR spectroscopy and GC analysis.

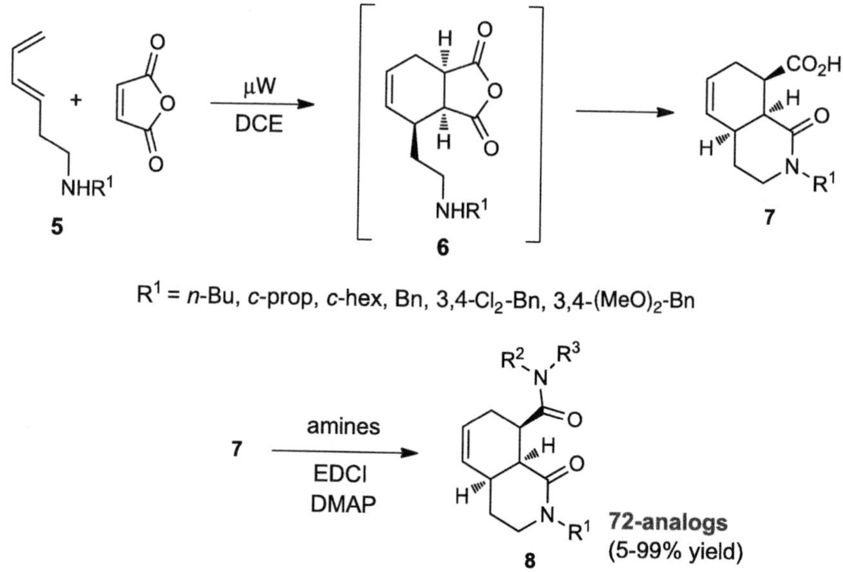

SCHEME 5.2 Synthesis of an isoquinolone library.

(Scheme 5.6). A 6 × 4 (alcohol × isocyanate) matrix produced 22 new analogs (2 to 98% yields, Table 5.3). Several of these scaffolds were manipulated further to incorporate fused indole or quinoline functionalities that are present in many biologically active substrates (Table 5.4) [12]. For example, Fischer indolization procedures were successfully applied to the scaffold **11** to give the pentacycles **14**, whose configurations were elucidated by x-ray crystallography.

The fused quinolines **18** were prepared in good yields (48 to 86%) using a modified Friedländer synthesis (Scheme 5.7) [13]. In these cases, the ketones were treated with the substituted 2-aminobenzaldehydes, generated in situ, in the presence of $ZnCl_2$ and molecular sieves to provide five fused heterocycles **18**.

Janvier et al. followed a guiding principle combining the use of multicomponent reactions and domino processes to develop a highly efficient synthesis of a combinatorial library [14]. They coupled a three-component reaction successfully with an acylation/IMDA/retro-Michael cycloreversion triple domino sequence to afford pyrrolo[3,4-*b*]pyridin-5-ones **23** (Scheme 5.8) [14].

Initial studies toward the formation of library **23** via a three-component condensation of an aldehyde, an amine, and the α-isocyanoacetamide **19** in methanol were focused on 5-aminooxazole **20**. The use of α-isocyanoacetamides **19** in place of commonly used α-isocyanoacetate in this Ugi type of transformation facilitated the desired cyclocondensations in good overall yields (60 to 96%). Acylation of the secondary amine to amide **21** followed by an intramolecular Diels–Alder cycloaddition produced the bridged tricycles **22**. Finally, a base-mediated retro-Michael cycloreversion afforded the desired pyrrolopyridines **23** (Table 5.5).

TABLE 5.2 Select Members of an Isoquinolone–Carboxamide Library Based on Scaffold 8

Entry	R^1	Amine	Product, Yield (%)
1	*n*-Bu	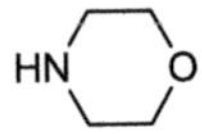	**8a**, 82
2	*n*-Bu	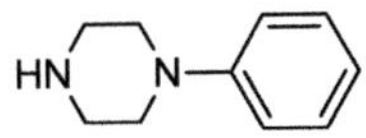	**8b**, 88
3	*c*-Pr	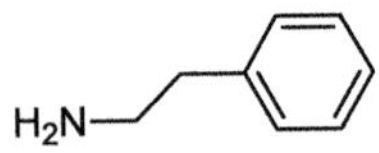	**8c**, 50
4	*c*- Pr	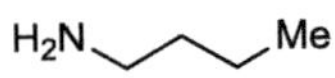	**8d**, 67
5	*c*-Hex	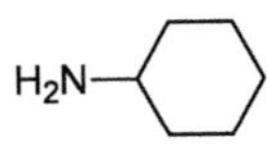	**8e**, 83
6	*c*-Hex	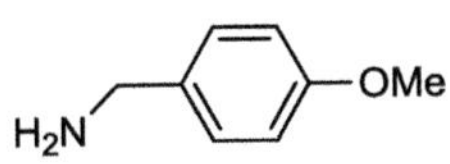	**8f**, 90
7	Bn	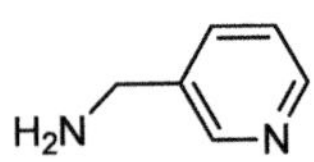	**8g**, 54
8	Bn	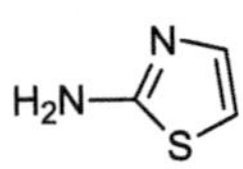	**8h**, 90
9	3,4-Cl_2-Bn	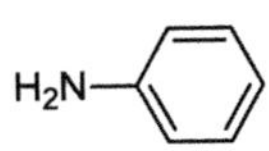	**8i**, 77
10	3,4-Cl_2-Bn	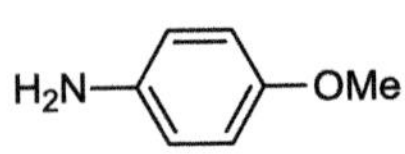	**8j**, 75
11	3,4-$(OMe)_2$-Bn	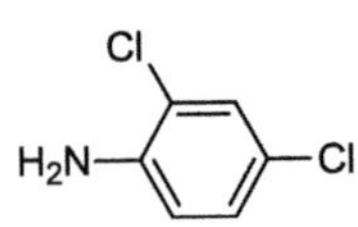	**8k**, 39
12	3,4-$(OMe)_2$-Bn	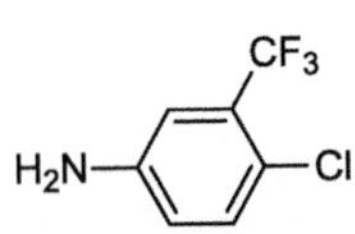	**8l**, 61

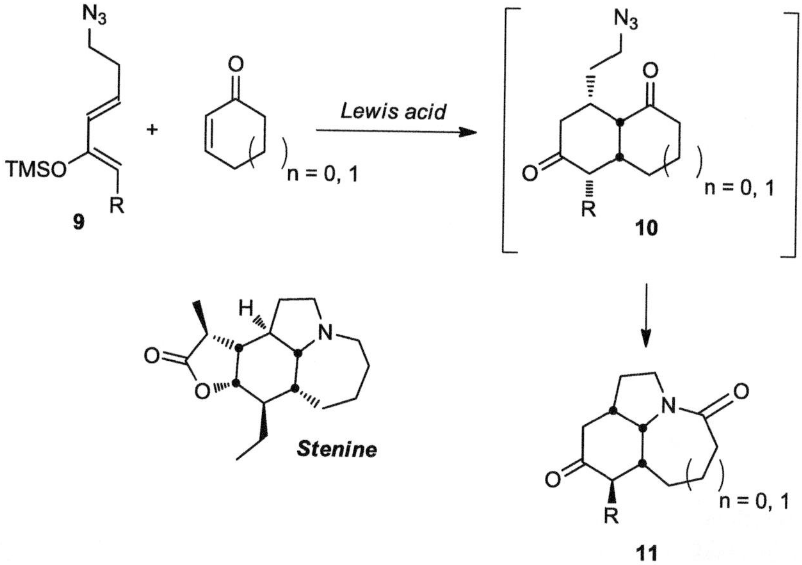

SCHEME 5.3 Diels–Alder–Schmidt domino sequence toward the *Stemona* alkaloid core structure.

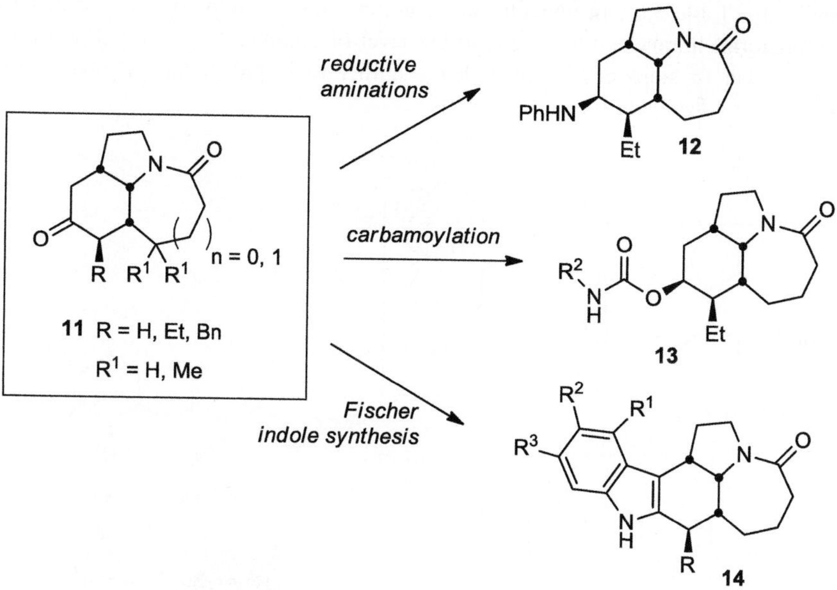

SCHEME 5.4 Diversification of the *Stemona* core structure.

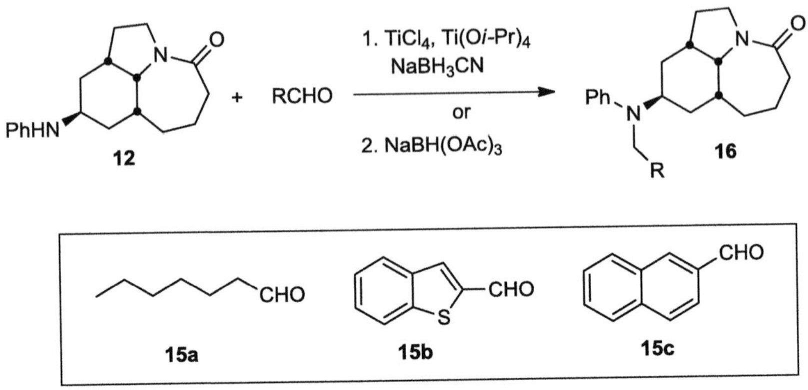

SCHEME 5.5 Reductive aminations of tricyclic *Stemona* alkaloid analogs.

This domino reaction sequence was combined into a one-pot four-component process by conducting the reactions in toluene in the presence of ammonium chloride (Scheme 5.9). The addition of ammonium chloride is believed to facilitate imine formation and activation. Toluene was the solvent of choice, as other solvents, such as THF, MeCN, and benzene, resulted in lower overall yields of the desired pyrrolopyridines **23**. Utilizing this methodology, 12 additional pyrrolopyridines **23** were synthesized with slightly better overall yields than those of the protocol described previously (Table 5.6). In general, this reaction sequence is particularly suited for combinatorial library synthesis due to the level of complexity generated in a minimal number of steps combined with the ability to install greater diversity from commercially available starting materials.

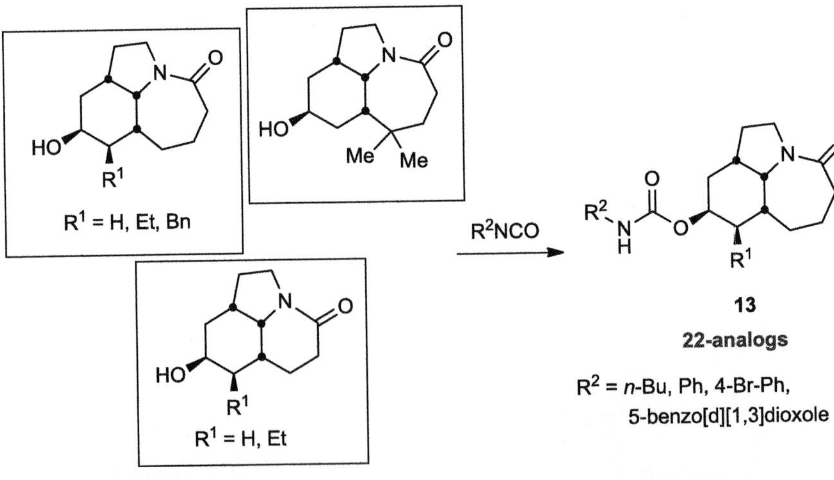

SCHEME 5.6 Carbamate analogs of the tricyclic *Stemona* alkaloid core.

TABLE 5.3 Carbamate Library (13)

Entry	Product	Yield (%)	Entry	Product	Yield (%)
1	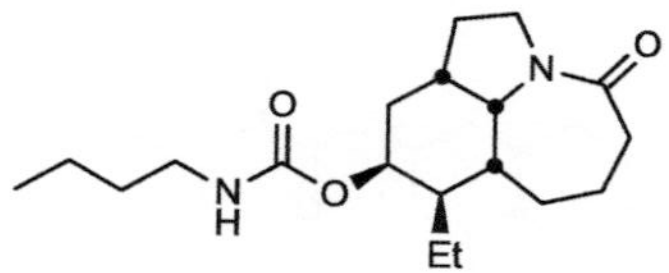	**13a**, 25	7	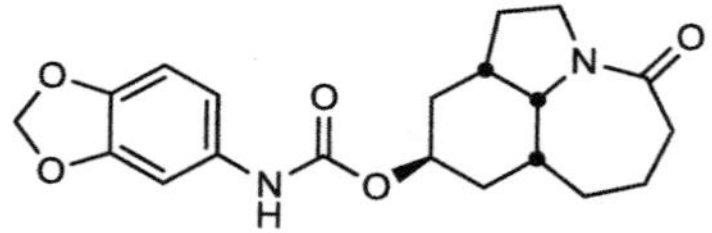	**13g**, 30
2	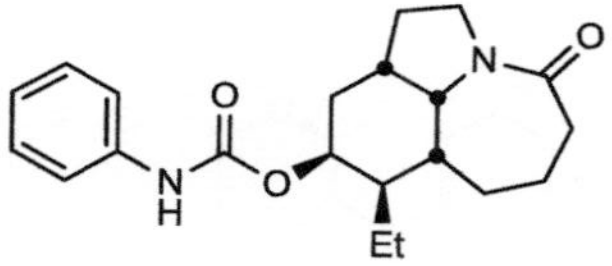	**13b**, 67	8	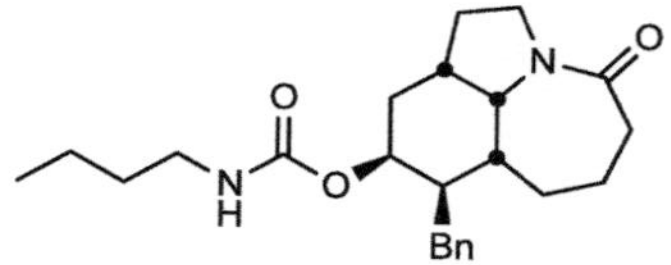	**13h**, 35
3	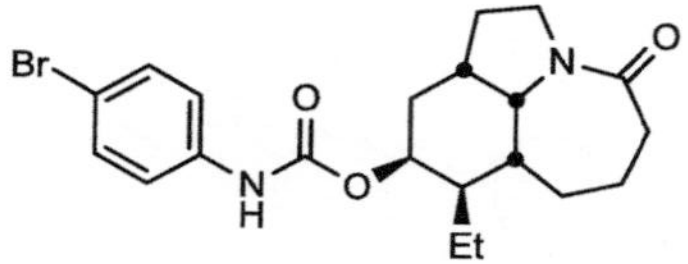	**13c**, 25	9	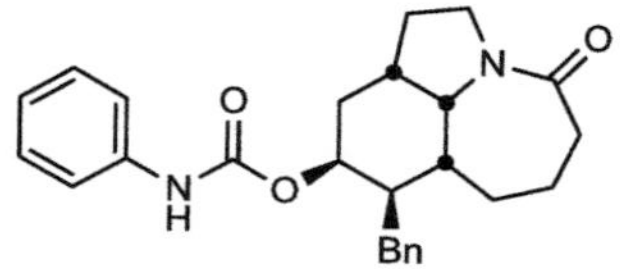	**13i**, 28
4	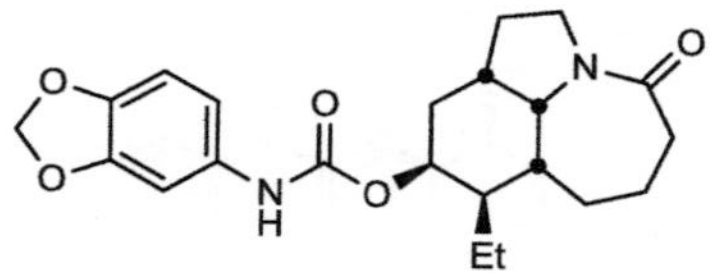	**13d**, 43	10	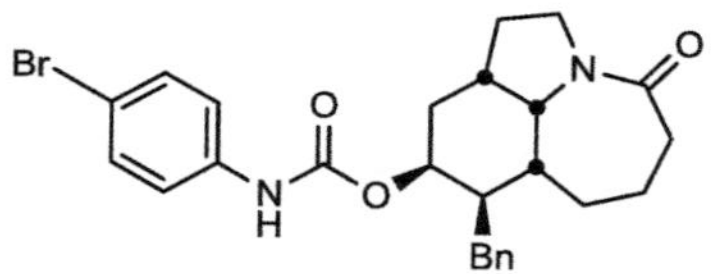	**13j**, 2
5	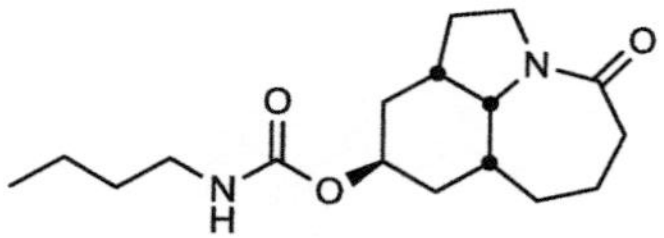	**13e**, 90	11	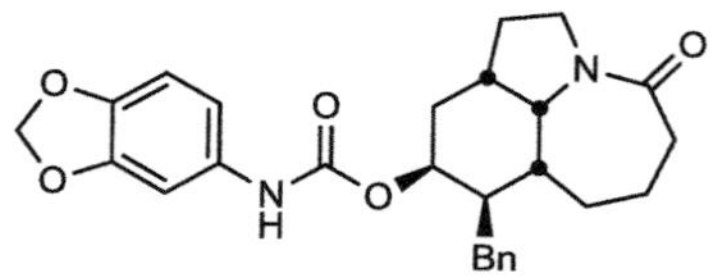	**13k**, 33
6	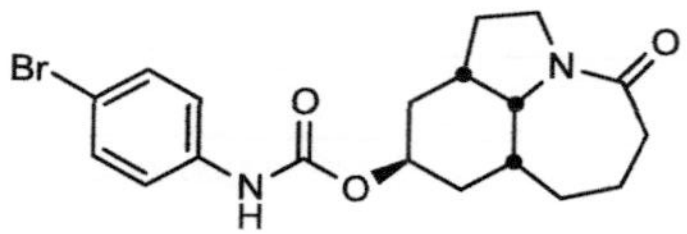	**13f**, 5	12	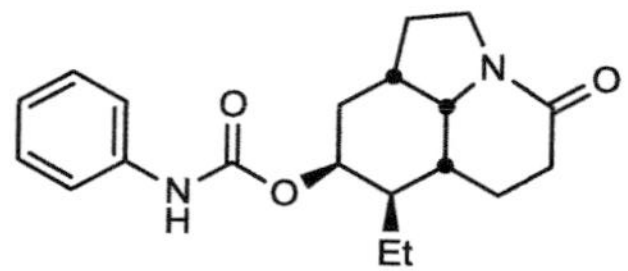	**13l**, 19

(continued)

TABLE 5.3 (*Continued*)

Entry	Product	Yield (%)	Entry	Product	Yield (%)
13	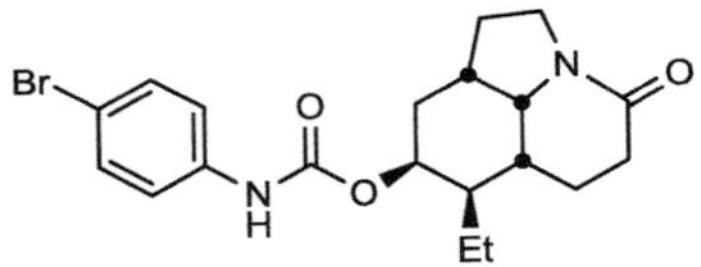	**13m**, 37	18	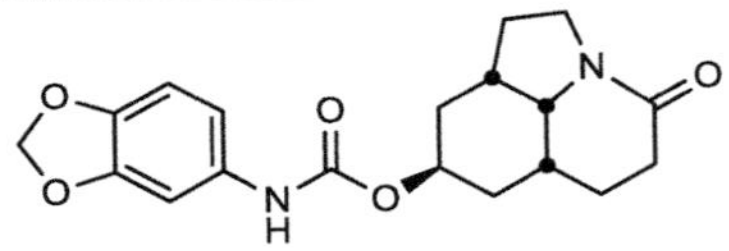	**13r**, 34
14	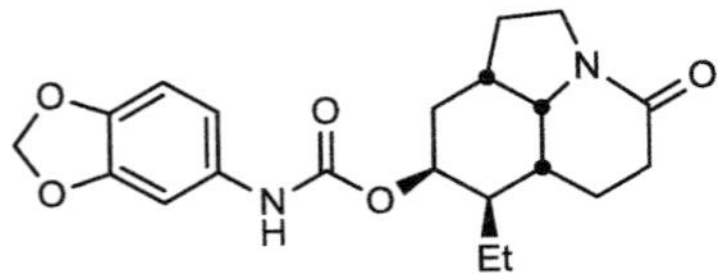	**13n**, 26	19	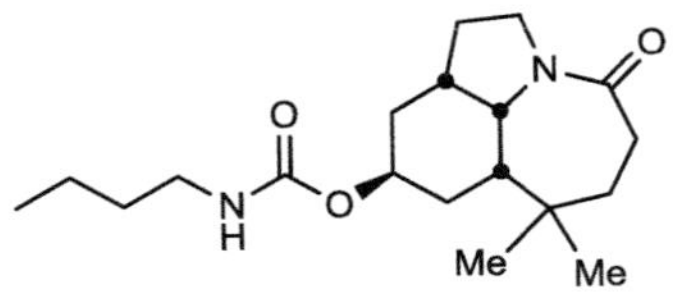	**13s**, 52
15	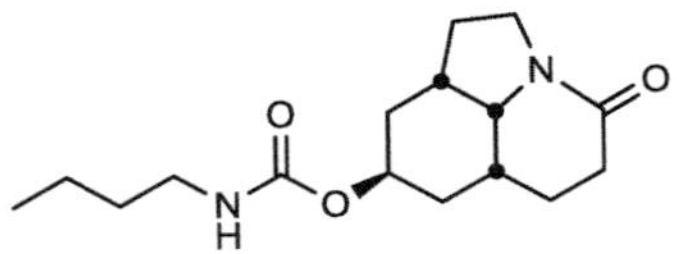	**13o**, 49	20	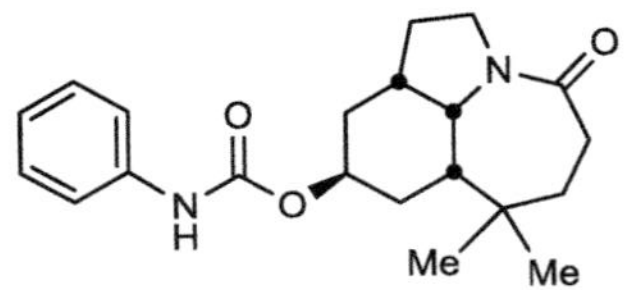	**13t**, 82
16	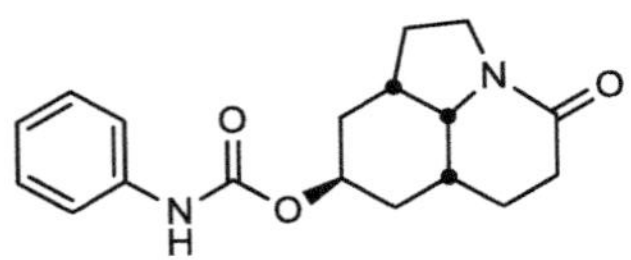	**13p**, 45	21	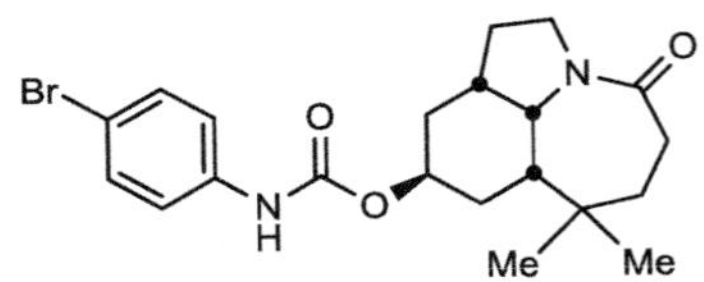	**13u**, 20
17	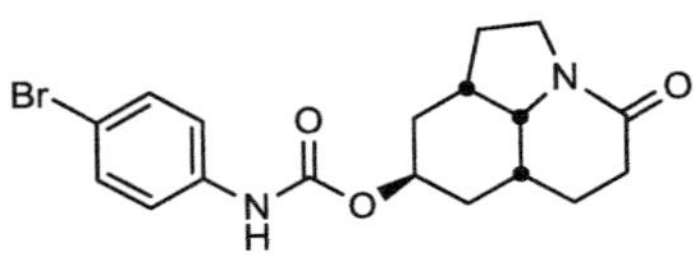	**13q**, 15	22	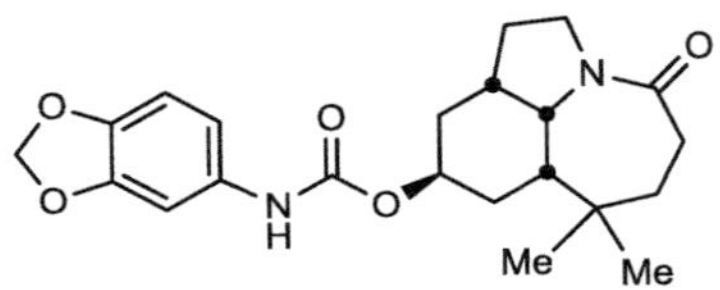	**13v**, 98

TABLE 5.4 Fused Indole Analogs 14 of the Tricyclic *Stemona* Alkaloid Core

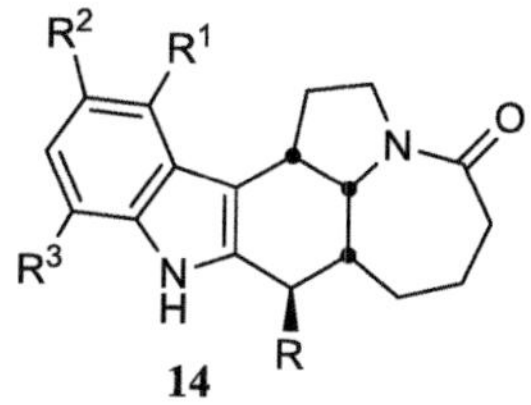

14

Entry	R	R^1	R^2	R^3	Yield (%)
1	H	H	H	H	**14a**, 76[a]
2	H	H	*i*-Pr	H	**14b**, 24[b]
3	H	H	OMe	H	**14c**, 18[b]
4	H	H	H	Cl	**14d**, 37[b]
5	H	Cl	H	Cl	**14e**, 37[b]
6	Et	H	*i*-Pr	H	**14f**, 26[c]
7	Et	H	OMe	H	**14g**, 77[c]

[a] $ZnCl_2$/AcOH.
[b] TsOH/EtOH.
[c] EtOH.

In an effort to develop an efficient method for the synthesis of pyrimidine-fused heterocycles, Yang et al. envisioned a domino reaction sequence involving the condensation of *N*-allylaminopyrimidine aldehydes **24** with anilines **25**, followed by an acid-catalyzed intramolecular inverse electron-demand hetero-Diels–Alder reaction (Scheme 5.10) [15]. As expected, this reaction sequence afforded the desired cis-configured tetracycles **27**, consistent with the proposed concerted intramolecular Diels–Alder cycloaddition.

Twenty tetracyclic analogs **27** were prepared using this methodology, with yields ranging from 49 to 98% (Table 5.7). Whereas this domino sequence demonstrates a suitable reaction scope, secondary aryl amines required harsher conditions to obtain the desired products (Table 5.7, entries 13 and 14). Several attempts

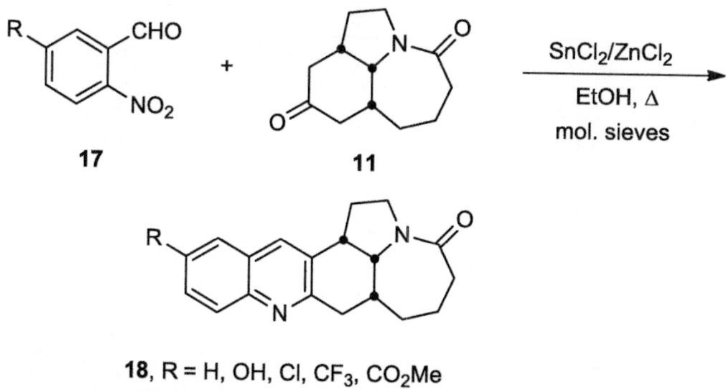

SCHEME 5.7 Quinoline analogs of the tricyclic *Stemona* alkaloid core.

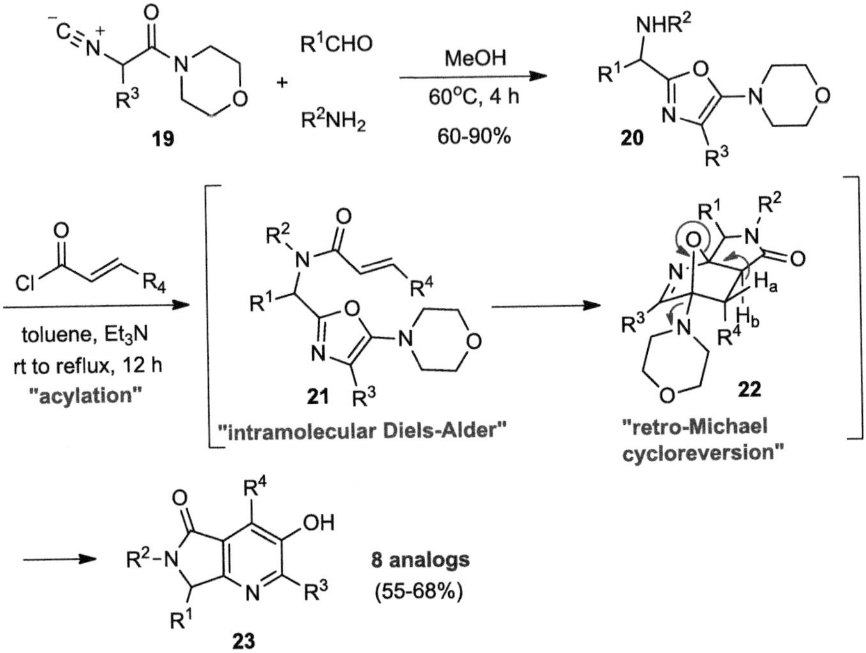

SCHEME 5.8 Three-component/triple domino synthesis of pyrrolo[3,4-*b*]pyridin-5-ones **23**.

TABLE 5.5 Pyrrolopyridines 23 Derived from a Three-Component/Triple Domino Reaction Sequence

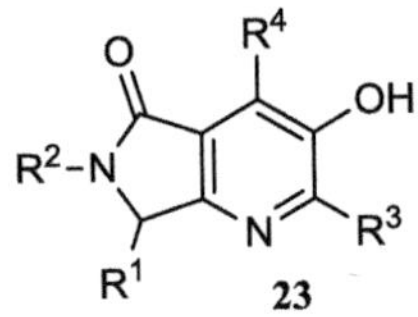

Entry	R^1	R^2	R^3	R^4	Yield (%)[a]
1	*n*-Hex	2-[3′,4′-$(OMe)_2$]Ph$(CH_2)_2$	Bn	CO_2Et	**23a**, 58
2	*n*-Pr	4-OMe-Bn	Bn	CO_2Et	**23b**, 57
3	*n*-Pr	3,5-$(OMe)_2$-Bn	Bn	CO_2Et	**23c**, 68
4	*c*-Oct	4-F-Bn	Bn	CO_2Et	**23d**, 66
5	*n*-Pr	3,4-dioxolanyl-Ph	Bn	CO_2Et	**23e**, 58
6	*n*-Hex	2-[3′,4′-$(OMe)_2$]Ph$(CH_2)_2$	Bn	4-NO_2-Ph	**23f**, 63
7	*n*-Hex	2-[3′,4′-$(OMe)_2$]Ph$(CH_2)_2$	Bn	4-OMe-Ph	**23g**, 55
8	*n*-Hex	*n*-Bu	*i*-Pr	4-NO_2-Ph	**23h**, 59

[a]Isolated yields.

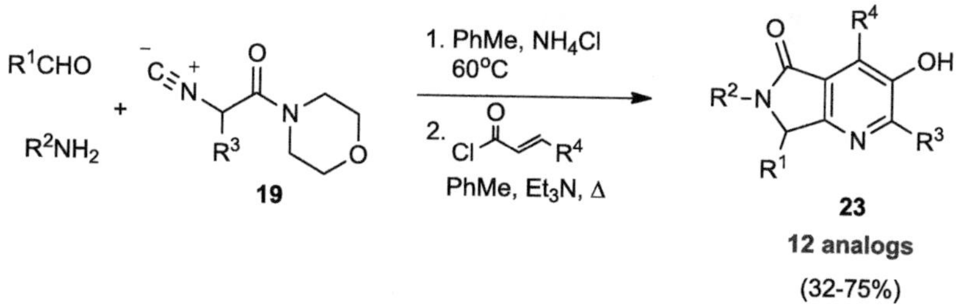

SCHEME 5.9 One-pot four-component synthesis of pyrrolo[3,4-*b*]pyridin-5-ones **23**.

TABLE 5.6 Pyrrolo[3,4-*b*]pyridin-5-ones 23

Entry	R^1	R^2	R^3	R^4	Yield (%)[a]
1	*n*-Hex	$(CH_2)_5CO_2Me$	Bn	CO_2Et	**23i**, 50
2	4-MeO-Ph	*n*-Bu	Bn	CO_2Et	**23j**, 55
3	*n*-Hex	(*R*)-$CH(PhCH_2)CO_2Me$	Bn	CO_2Et	**23k**, 54
4	*n*-Hex	(*S*)-$CH(PhCH_2)CO_2Me$	Bn	CO_2Et	**23l**, 48
5	*n*-Hex	(*S*)-$CH(CH_2CO_2Me)CO_2Me$	Bn	CO_2Et	**23m**, 40
6	4-Cl-Ph	*n*-Bu	Bn	4-NO_2-Ph	**23n**, 32
7	*n*-Hex	$(CH_2)_5CO_2Me$	Bn	4-NO_2-Ph	**23o**, 75
8	*n*-Hex	*n*-Bu	Bn	4-NO_2-Ph	**23p**, 71
9	*n*-Hex	2-[3′,4′-$(OMe)_2$]$Ph(CH_2)_2$	Bn	4-NO_2-Ph	**23q**, 70
10	Ph	*n*-Bu	Bn	4-NO_2-Ph	**23r**, 70
11	Ph	$(CH_2)_5CO_2Me$	Bn	4-NO_2-Ph	**23s**, 35
12	*n*-Hex	$(CH_2)_5CO_2Me$	Bn	4-OMe-Ph	**23t**, 65

[a]Isolated yields.

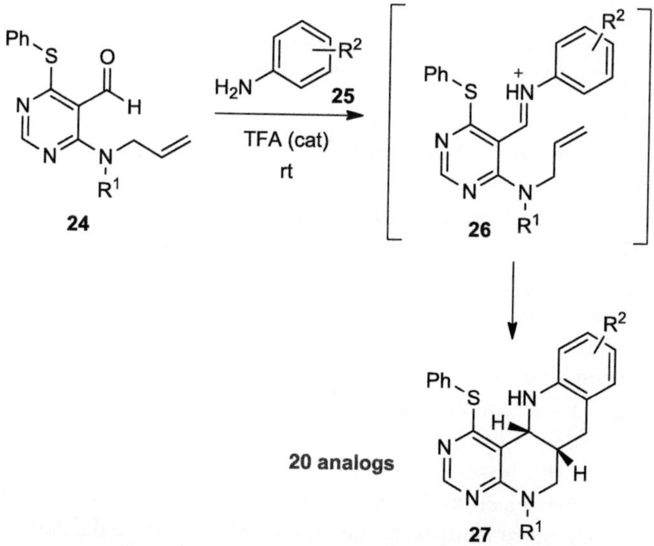

SCHEME 5.10 Synthesis of tetracyclic pyrimidines.

TABLE 5.7 Representative Pyrimidine-Fused Tetracycles 27

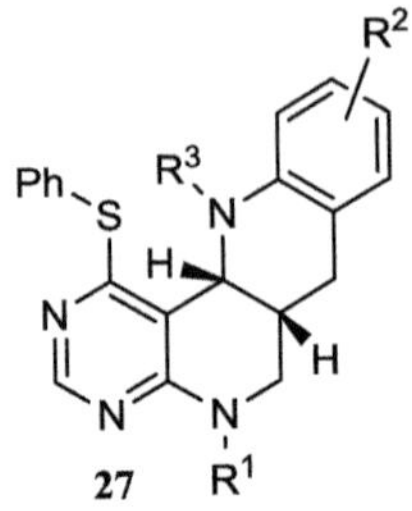

27

Entry	R^1	R^2	R^3	Time (h)	Yield (%)[a]
1	Me	Ph	H	11	**27a**, 94
2	Me	3-NO_2-Ph	H	4	**27b**, 92[b]
3	Me	4-NO_2-Ph	H	7	**27c**, 87
4	Me	4-CO_2H-Ph	H	6	**27d**, 86
5	Me	2,4-F_2-Ph	H	23	**27e**, 83
6	Me	3-Me-Ph	H	10	**27f**, 88 (3 : 1)[c]
7	Me	2,4-Me_2-Ph	H	27	**27g**, 78
8	Me	4-OH-Ph	H	20	**27h**, 98
9	Me	1,3-Bn-dioxol-5-yl	H	27	**27i**, 85[d]
10	Me	1-Naph	H	15	**27j**, 83
11	Bn	Ph	H	3	**27k**, 98
12	CH_2CO_2Et	Ph	H	7	**27l**, 95
13	Me	Ph	Me	33[e]	**27m**, 78
14	Me	Ph	Et	85[e]	**27n**, 49

[a]Isolated yields.
[b]Obtained as a single regioisomer (cyclized at C2 of aniline).
[c]Obtained as a mixure of regioisomers.
[d]Obtained as a single regioisomer [cyclized at C6 of 3,4-(methylenedioxy)aniline].
[e]Anilines (1.05 equiv), TsOH (0.05 equiv), in toluene at reflux with a Dean–Stark trap.

to expand the generality of this methodology further (i.e., by *O*-allylation or *N*-propargylation) were unproductive, possibly due to the insufficient reactivity of the dienophile. Nonetheless, the ease of formation and further functionalization of tetracycles **27** renders this methodology well suited for the synthesis of an expanded library of diverse drug discovery candidates. For example, a small sublibrary was readily assembled through nucleophilic aromatic displacement of the sulfoxide **28** (Scheme 5.11 and Table 5.8).

While investigating the oxidation of primary amines **31/32** to imines **33** using electrogenerated 3,4-azaquinone catalysts $\mathbf{30_{ox}}$ (derived from *o*-aminophenols $\mathbf{30_{red}}$), Largeron, Blattes, Xu, and co-workers observed that this catalytic cycle ceased after several turnovers [16]. It was determined that the 3,4-azaquinones $\mathbf{30_{ox}}$ underwent a regioselective inverse electron-demand Diels–Alder cycloaddition with the simultaneously electrogenerated tautomeric ene-amines **34** to produce the densely functionalized 1,4-benzoxazines **35** (Scheme 5.12).

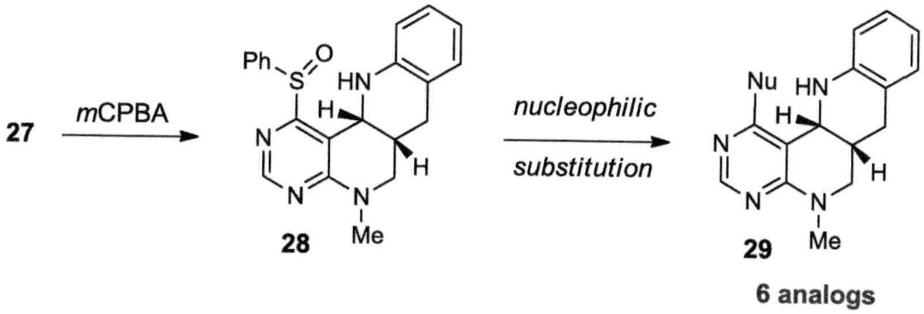

SCHEME 5.11 Nucleophilic substitutions of sulfoxide **28** to give heterocycles **29**.

TABLE 5.8 Tetracyclic Library 29

Entry	Nu	Yield (%)
1	Pyrrolidine	**29a**, 98
2	Aniline	**29b**, 60
3	SEt	**29c**, 55
4	SBn	**29d**, 87
5	O*n*-Bu	**29e**, 80
6	OPh	**29f**, 60

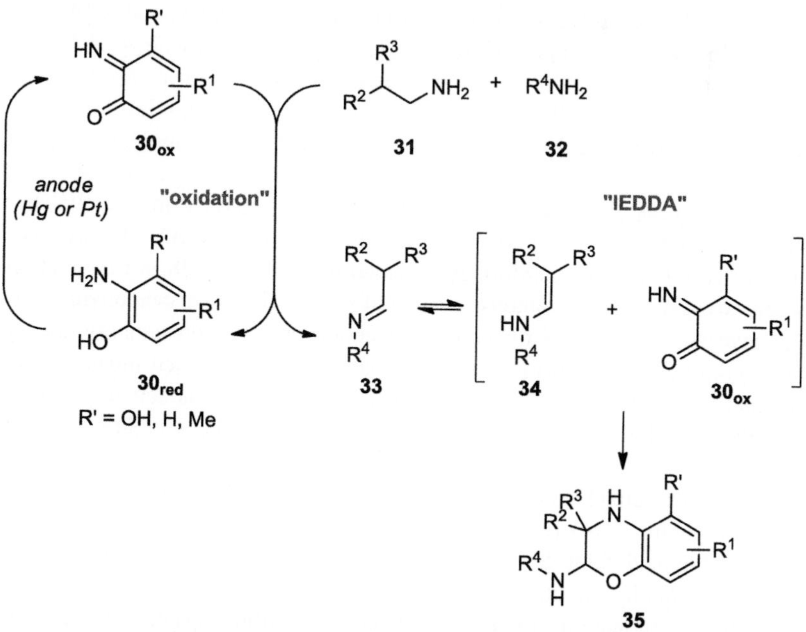

SCHEME 5.12 Oxidation/IEDDA domino sequence to produce of 1,4-benzoxazines **35**.

A special feature of this domino sequence is the in situ generation of both cycloaddition components at room temperature under metal-free conditions. Subsequent exploratory studies to extend the scope of this domino process lead to the synthesis of 22 1,4-benzoxazines using 3,4-azaquinone **$30a_{ox}$** and various aliphatic amines, with yields ranging from 5 to 80% (Table 5.9). A high degree of tolerance for the amine component was demonstrated in this reaction sequence. Interestingly, the initial cycloadducts derived from phenethylamine building blocks (Table 5.9, entries 4 and 5) underwent further oxidation to afford the stable imines **35d** and **e**. It is also noteworthy that single regioisomers are formed from a mixture of two different amines (Table 5.9, entries 6 to 8).

An alternative one-pot tandem oxidation/IEDDA sequence was developed for aminophenols **30_{red}** ($R' = H$, Me) and cyclohexyl-enamine **34**, using a platinum anode with a higher anodic decomposition potential (Table 5.10). With this procedure, eight additional 1,4-benzoxazines **35** were prepared (10 to 65%), further expanding the generality of the reaction sequence (Table 5.10).

5.3 ANIONIC DOMINO REACTIONS

In 2007, Wan et al. reported a multicomponent protocol for the conversion of nitriles into acyl aminals through a domino reaction sequence comprising nitrile hydrozirconation, acylation, and nucleophilic additions [17]. The scope of this methodology was later elaborated to include a Lewis-acid-promoted intramolecular Friedel–Crafts annulation of the acylimine **38** to give various bicyclic β-alkoxyamides **39** (Scheme 5.13) [18]. In general, this domino sequence proceeds efficiently with good diastereoselectivities, making it ideal for library synthesis. Ultimately, a 35-membered bicyclic-benzyloxy amide library **39** was constructed and provided a high level of structural novelty compared to the National Institutes of Health small-molecule repository (SMR).

This reaction cascade is initiated by the treatment of cyanohydrin **36** with Schwartz reagent [$Cp_2Zr(H)Cl$] to give the metalloimine intermediate **37** [19]. Subsequent treatment with various acid chlorides provided the acylimine **38**, which cyclized to **39** when exposed to a stoichiometric amount of $ZnCl_2$. The use of cyanohydrin-benzyl ethers gave improved yields of the desired amido-bicycles, presumably due to the lower acylimine tautomerization rates compared to unsubstituted nitrile substrates or the increased rate of nitrile hydrozirconation due to the inductive effect of the benzylether. The latter functionality was used to enhance the structural diversity of this library.

Cyanohydrin benzyl ethers **36a** to **c** were readily prepared using established protocols (Scheme 5.14) [20]. The library design included a set of 13 commercially available acid chlorides comprised of aliphatic, aromatic, and heteroaromatic derivatives. Cyanohydrin ether **36a** afforded a series of tetrahydronaphthyl amides after the hydrozirconation–acylation–cyclization reaction sequence (Table 5.11, 16 to 73%). The resulting stereoisomers were separated using mass-directed HPLC with subsequent ^{1}H NMR analysis indicating the major diastereomers to be the syn

TABLE 5.9 Representative Set of 1,4-Benzoxazines Illustrating the Scope of Functionalized Amines[a]

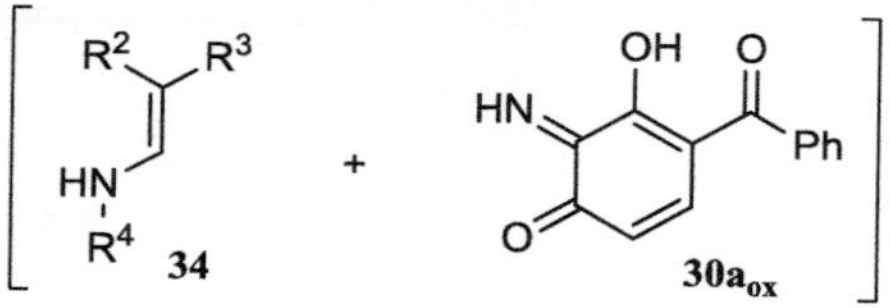

Entry	Amine	R^4NH_2	Product	Yield (%)[b]
1	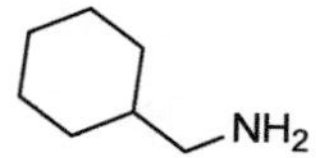 **31a**	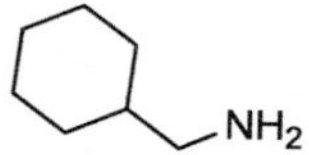 **31a**	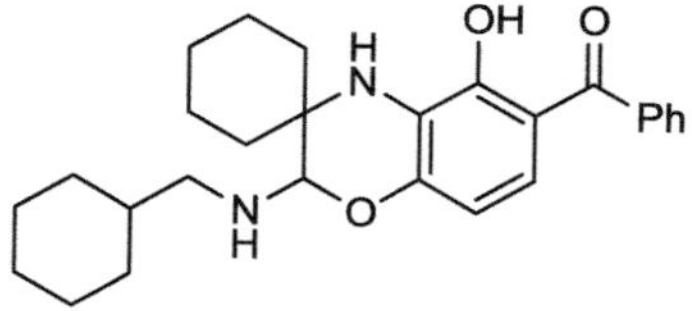	**35a**, 77
2	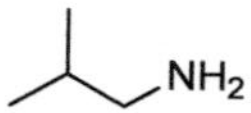 **31b**	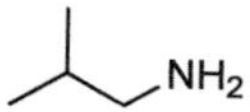 **31b**	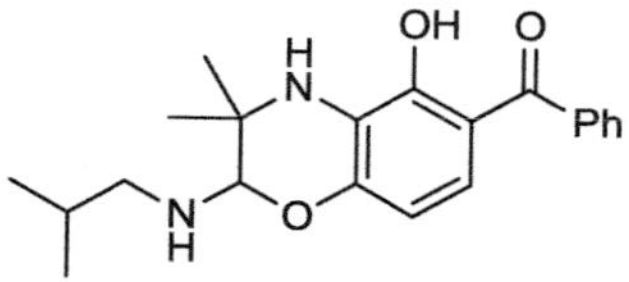	**35b**, 71
3	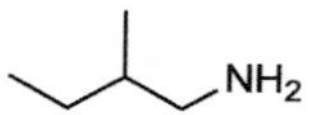 **31c**	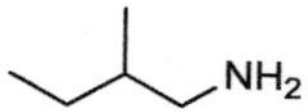 **31c**	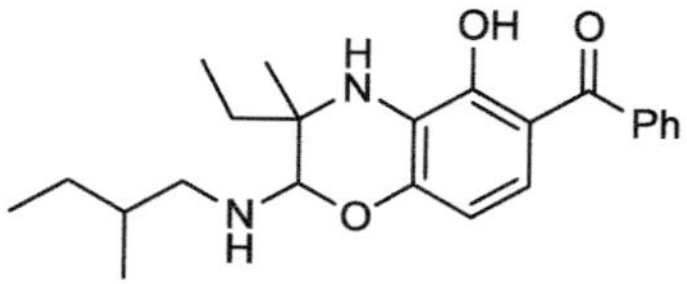	**35c**, 70[c]
4	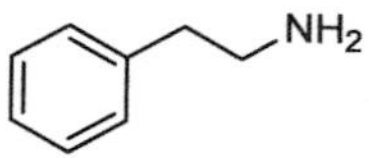 **31d**	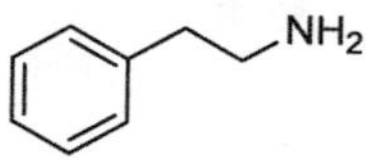 **31d**	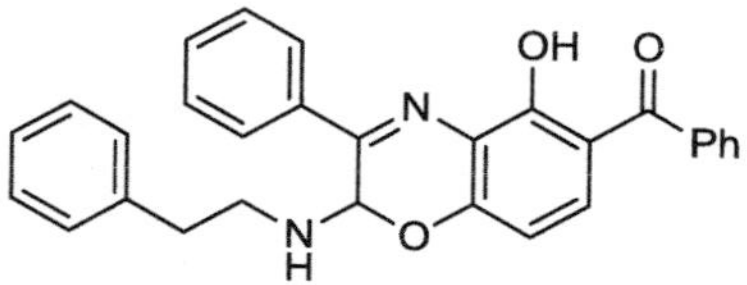	**35d**, 58

(continued)

TABLE 5.9 (*Continued*)

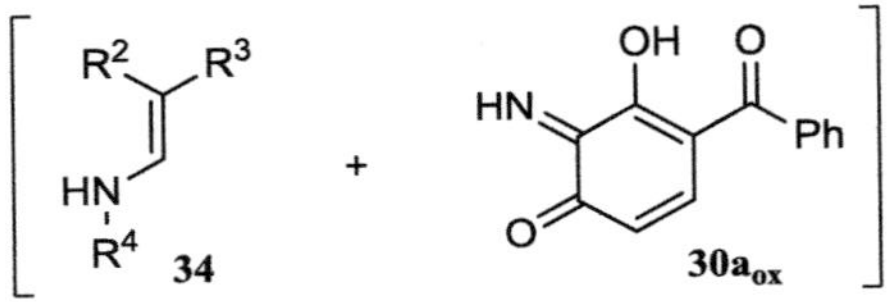
34 + $30a_{ox}$

Entry	Amine	R^4NH_2	Product	Yield (%)[b]
5	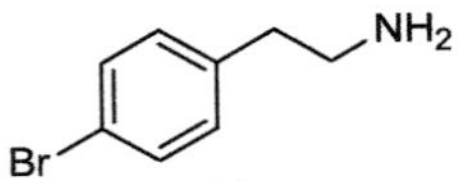 **31e**	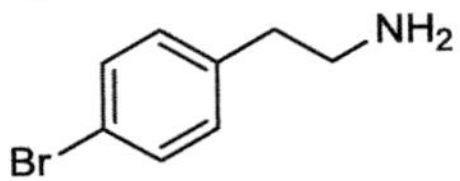 **31e**	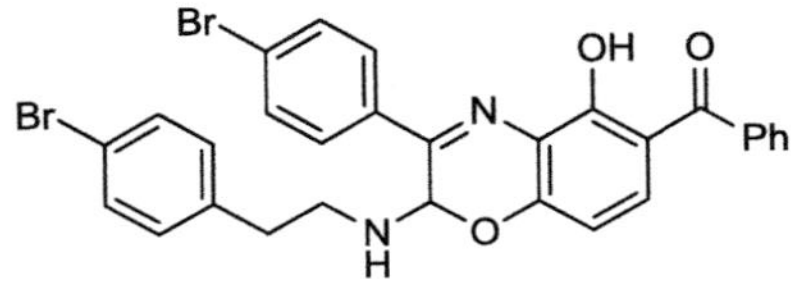	**35e**, 76
6	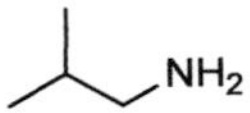 **31b**	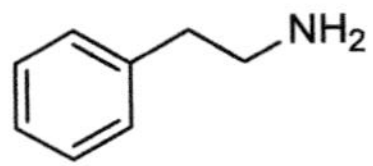 **31d**	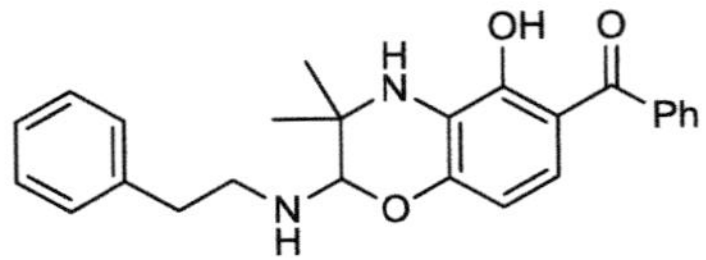	**35f**, 60
7	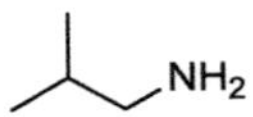 **31b**	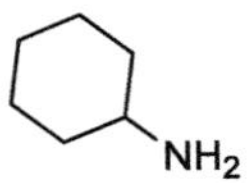 **32a**	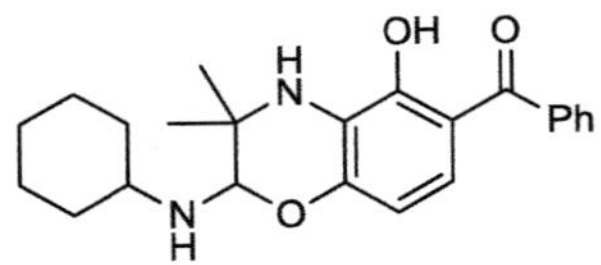	**35g**, 50
8	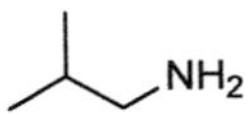 **31b**	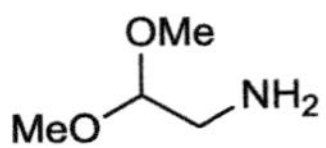 **32b**	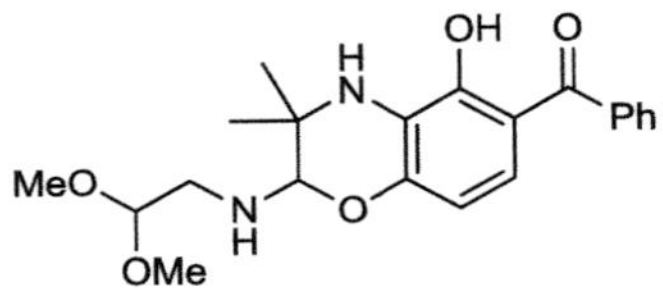	**35h**, 68

[a]Reaction conditions: (**$30a_{red}$**) = 2 mM, amine, **31** = R^4NH_2 = 20 mM, MeOH, rt, Hg anode (E = +50 mV vs. SCE), 6 to 10 h.
[b]Yields refer to chromatographically pure isolated products.
[c]Obtained as a mixture of two unassigned diastereoisomers (ca. 1 : 1 ratio).

TABLE 5.10 Representative Set of 1,4-Benzoxazines 35 Demonstrating the Scope with *o*-Aminophenols (R′ = H, Me)[a]

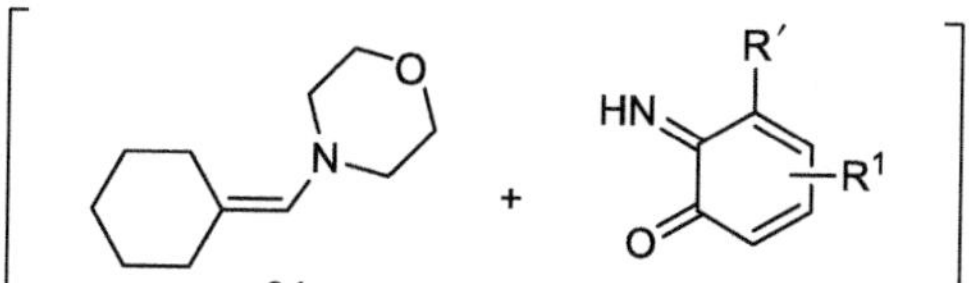

34 + **30_{ox}**

Entry	Substrate	Product	Yield (%)[b]
1	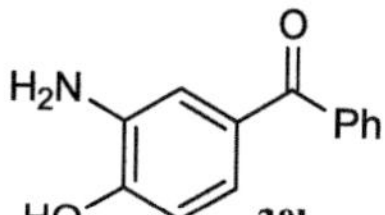 **$30b_{red}$**	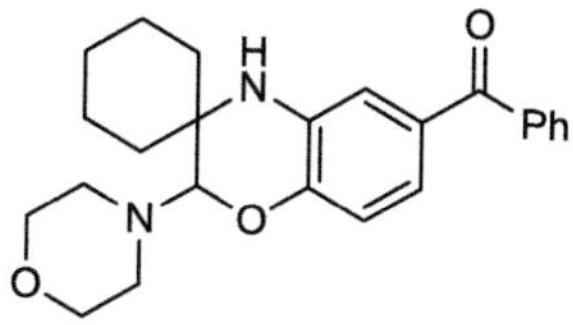	**35i**, 56
2	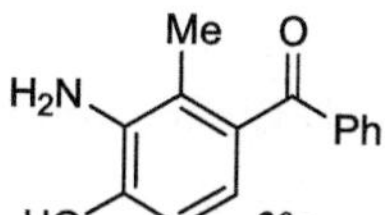 **$30c_{red}$**	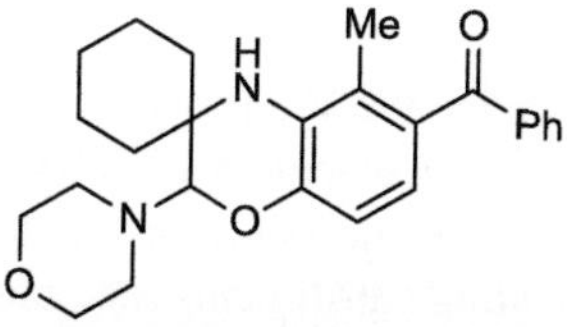	**35j**, 60
3	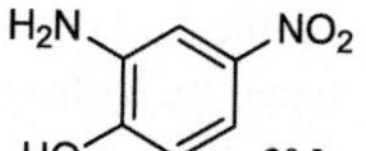 **$30d_{red}$**	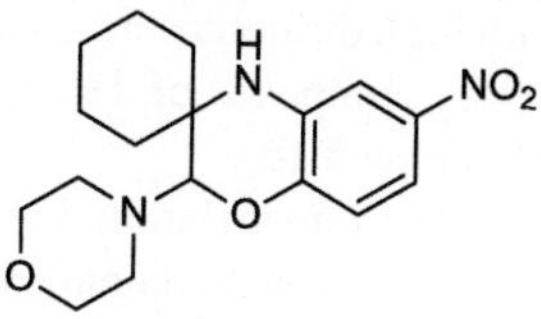	**35k**, 65
4	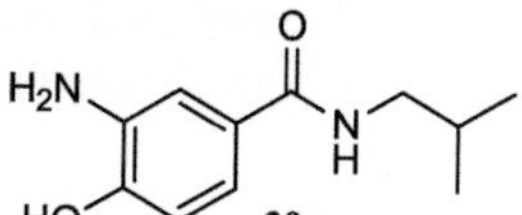 **$30e_{red}$**	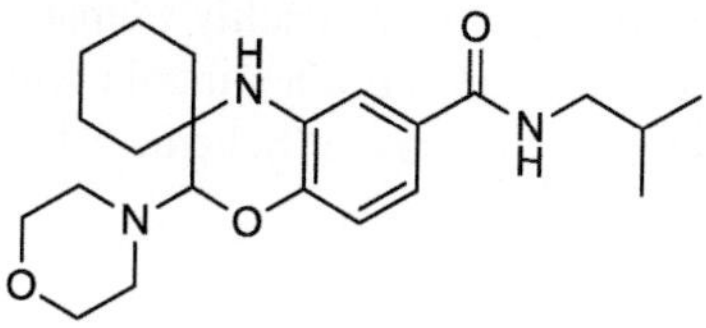	**35l**, 26
5	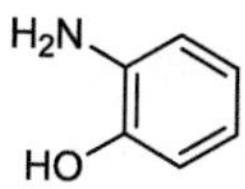 **$30f_{red}$**	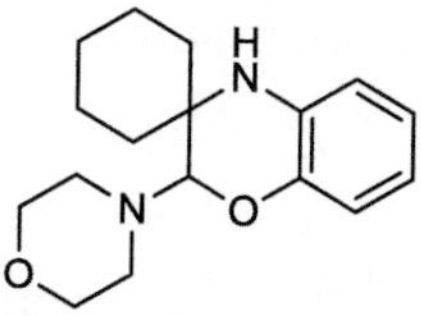	**35m**, 10

[a]Reaction conditions: ($\mathbf{30_{red}}$) = 2 mM, enamine **34** = 10 mM, MeOH, rt, Pt anode (E = +450 mV vs. SCE), 4 h; 1 equiv of morpholine was added to the reaction to produce $\mathbf{30_{ox}}$.
[b]Isolated yields.

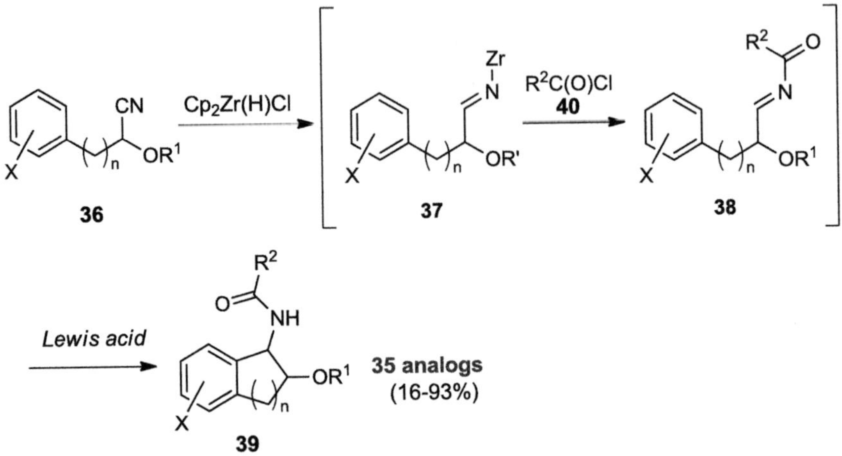

SCHEME 5.13 Indanolamide synthesis.

isomers. The syn isomers **39a** are probably formed via a chelation-controlled five-membered transition state (Scheme 5.15, **TS1**).

The formation of cyclopentyl bicycles via the cascade sequence were slower and required electron-donating substituents (i.e., methoxyarenes, **36b** or an indole group, **36c**). Unlike the tetrahydronaphthyl-containing examples, these derivatives gave anti isomers (**39b** and **c**) with high diastereoselectivities (>10 : 1, Table 5.12). In addition, a β-hydroxy amide library composed of **41a** and **b** was prepared by hydrogenolysis of the benzyl groups (Scheme 5.16).

The nitrile hydrozirconation–acylation–Friedel–Crafts alkylation protocol was expanded further to produce a stereochemically diverse series of spirooxindoles **43** and **44** (Scheme 5.17) [21]. The pivotal 2-chloro- and 2-triisopropylsiloxyindoles, **42a** and **b**, respectively, were readily prepared from the 3-substituted indole **42c**, which is available through a Fischer indole synthesis from *N*-benzylphenylhydrazine and the corresponding aldehyde (Scheme 5.18).

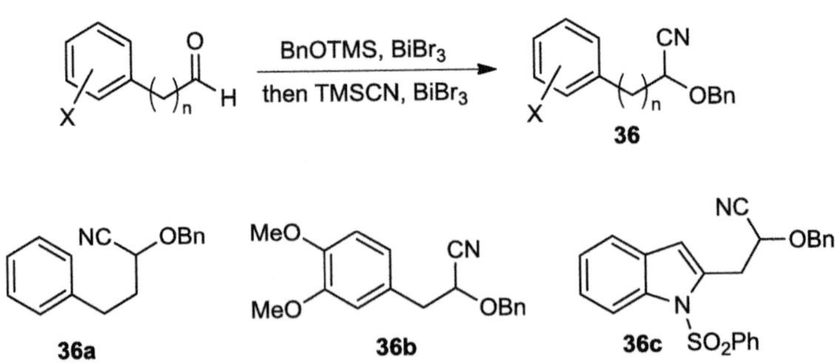

SCHEME 5.14 Preparation of cyanohydrins. (Adapted from [18], with permission; copyright © 2009 American Chemical Society.)

TABLE 5.11 Synthesis of a Tetrahydronaphthylamide Library (39)

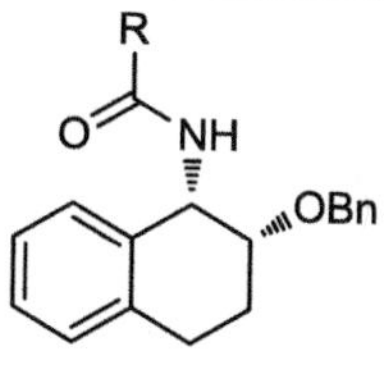

syn-**39a** (*major*)

Entry	R (40)	Yield (%)[a]	dr[b]	Purity[c]
1	*i*-Pr	52	3 : 1	100
2	HC=CHMe	59	3 : 1	92
3	$MeOCH_2$	34	3 : 1	100
4	$PhOCH_2$	73	3 : 1	100
5	$BnOCH_2$	39	4 : 1	100
6	4-OMe-Bn	41	3 : 1	100
7	Bn	34	3 : 1	100
8	4-F-Bn	50	3 : 1	100
9	$Ph(CH_2)_2$	54	3 : 1	100
10	$4\text{-}CF_3\text{-}Ph$	28	3 : 1	100

[a]Isolated yield.
[b]Syn : anti.
[c]UV 214 nM.

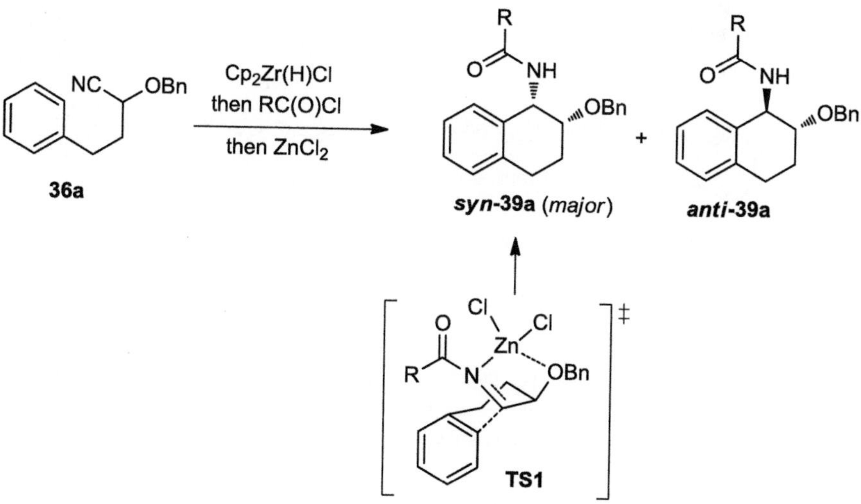

SCHEME 5.15 Bicyclic amide formation from **36a**. (Adapted from [18], with permission; copyright © 2009 American Chemical Society.)

TABLE 5.12 Synthesis of Indanyl- (39b) and Cyclopentaindolyl Amides (39c)

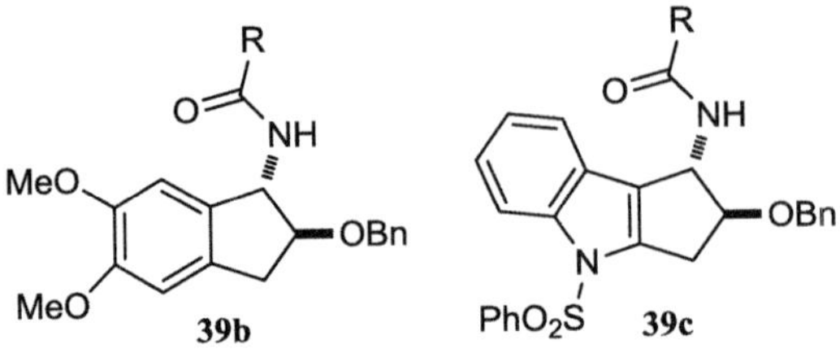

39b 39c

Entry	Nitrile	R (40)	Yield (%)[a,b]	Purity[c]
1	**36b**	*i*-Pr	40	100
2	**36b**	$CH_3(CH_2)_8$	64	100
3	**36b**	$MeOCH_2$	18	100
4	**36b**	$PhOCH_2$	33	100
5	**36b**	4-OMe-Bn	40	100
6	**36b**	4-F-Bn	17	99
7	**36b**	$Ph(CH_2)_2$	54	100
8	**36b**	4-CF_3-Ph	27	100
9	**36c**	*i*-Pr	14	99
10	**36c**	4-OMe-Bn	19	100
11	**36c**	$Ph(CH_2)_2$	14	96

[a]Isolated yield.
[b]>10 : 1, anti : syn.
[c]UV 214 nM.

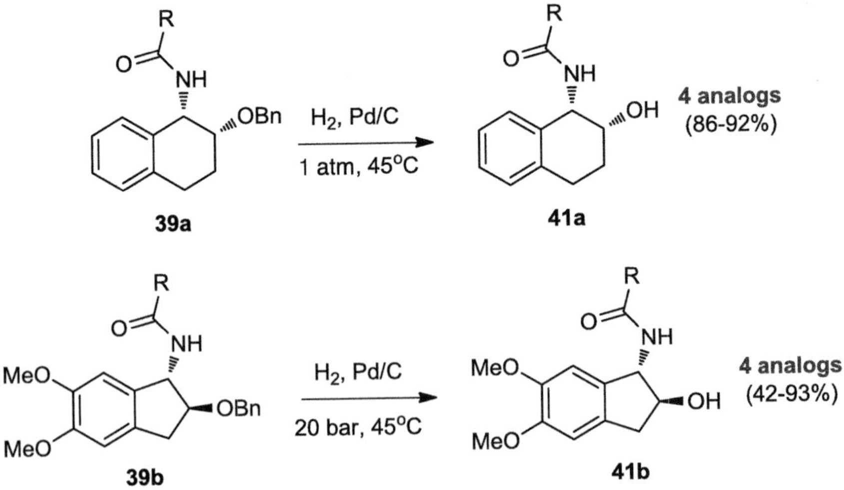

SCHEME 5.16 Representative β-hydroxyamides **41a** and **b**. (Adapted from [18], with permission; copyright © 2009 American Chemical Society.)

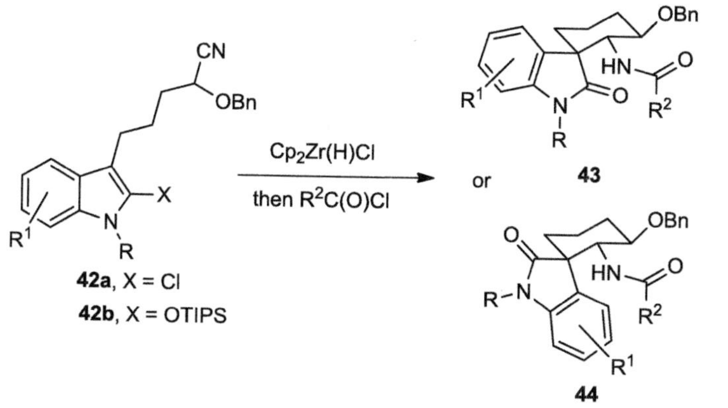

SCHEME 5.17 Spirooxindole formation. (Adapted from [21], with permission; copyright © 2010 American Chemical Society.)

Hydrozirconation of nitriles **42a** and **b** with Schwartz reagent provided metalloimines that were subjected to acylation by treatment with hydrocinnamoyl chloride to provide building blocks **43** and **44.** The configuration of the Friedel–Crafts cyclization step was controlled by the C2 indole substitution, with the less hindered chlorosubstituent favoring a pseudoaxial conformation (**A**), whereas the inverted conformation (**B**) would be preferred with larger groups at C2 (i.e., OTIPS) (Scheme 5.19). Following hydrolysis of the chloroiminium intermediate (not shown), the spirooxindoles **43** and **44** were isolated in 83% yield, with a dr = 7.3 : 1, from the 2-chloroindole substrate. In the presence of $Sc(OTf)_3$, 2-siloxyindole **42b** also gave high selectivities. In this case, the spirooxindole isomer **44** was the major product (dr = 5 : 1) formed in a slightly lower ratio of trans : cis amide to benzyloxy groups

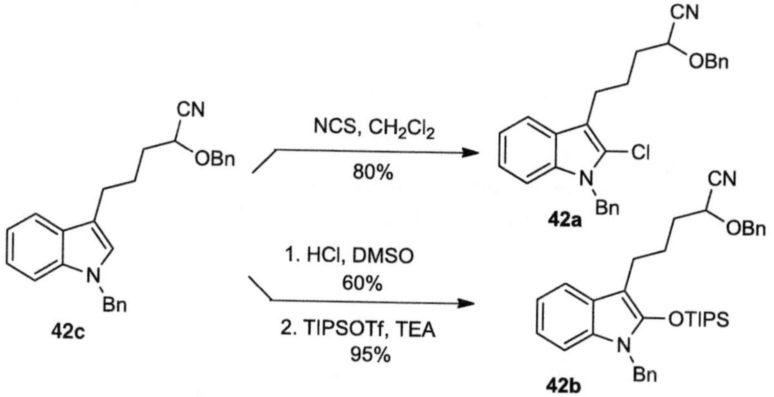

SCHEME 5.18 Preparation of cascade precursors.

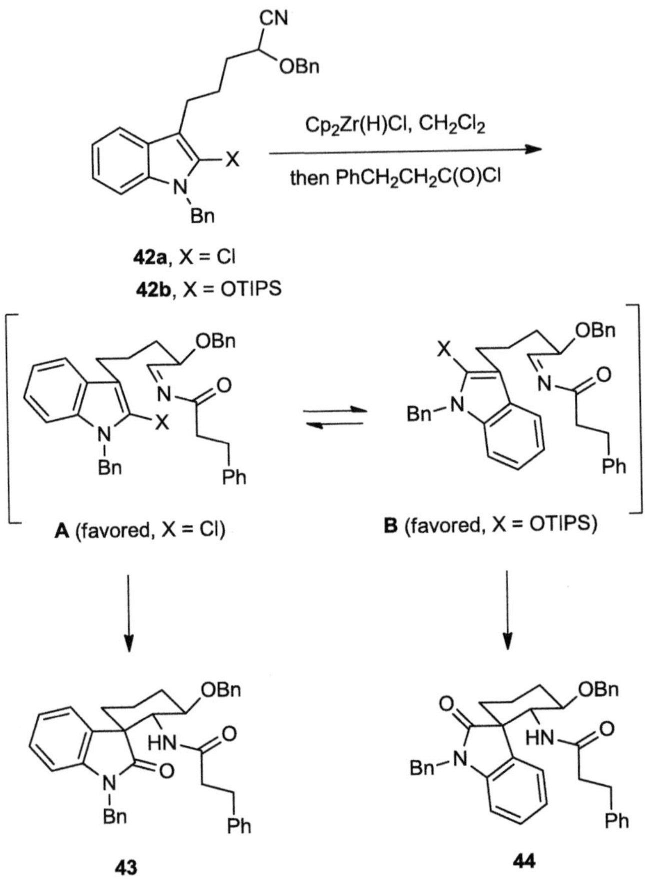

SCHEME 5.19 Domino process to oxindoles. (Adapted from [21], with permission; copyright © 2010 American Chemical Society.)

compared to the 2-chloroindole example. This convenient access to polyfunctional oxindole substrates facilitated the construction of a stereochemically diverse library.

An alternative process to generate 2-aminoindole-3-carboxamides **49** has recently been reported by Wang et al. (Scheme 5.20 and Table 5.13) [22]. Their approach combines a nucleophilic aromatic substitution reaction with a one-pot reductive cyclization without the isolation of intermediates. The cyanoacetamide starting materials **46** were obtained via amidations of cyanoacetic acid methyl ester (**45**) under neat conditions [23].

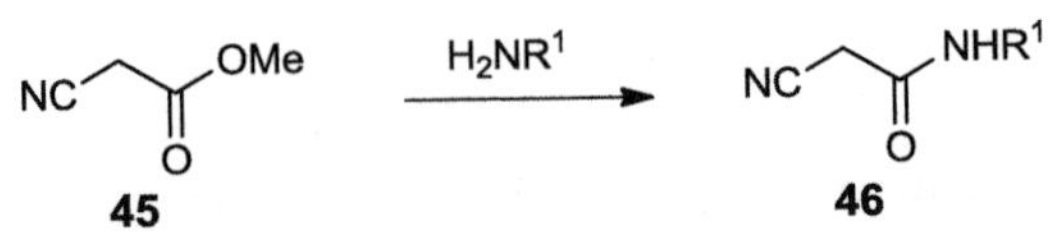

(5.1)

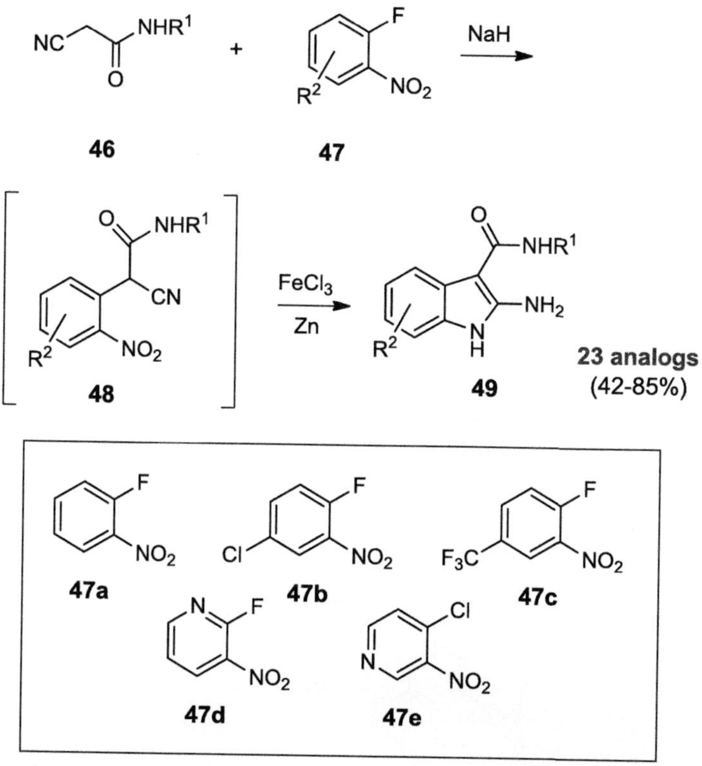

SCHEME 5.20 Formation of 2-aminoindole-3-carboxamides **49**.

Deprotonation of cyanoacetamides **46** with NaH and treatment with various 2-halonitrobenzenes (or pyridines, **47**), followed by an acidic quench, led to crude reaction mixtures that were subjected to a combination of $FeCl_3$ (3 equiv) and Zn dust (10 equiv) at 100°C for 1 h (Scheme 5.20). These Fe/Zn molar ratios were essential for the formation of the desired 2-aminoindole products **49**.

5.4 TRANSITION-METAL-MEDIATED DOMINO REACTIONS

While investigating selective chemical reactions of trienyl-containing compounds, Brummond et al. found that by utilizing Davies chemistry [24] to react triene **50** with (*E*)-diethyl-4-diazo-2-pentene dioate and Rh(II) acetate hydroazulenone, **52** was obtained as a single diastereomer (Scheme 5.21) [25]. This formal [4 + 3] cycloaddition proceeds through the selective cyclopropanation of the vinyl group to give *cis*-divinylcyclopropane intermediate **51**, which subsequently undergoes a strain-induced Cope rearrangement to produce racemic cycloheptadiene **52**.

TABLE 5.13 Representative Indole Products 49

Entry	Product Structure	Product, Yield (%)
1	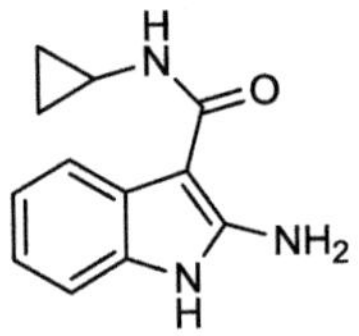	**49a**, 76
2	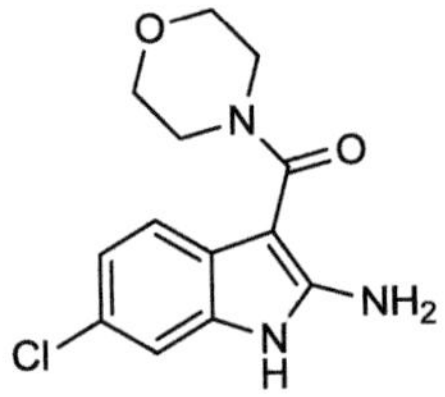	**49b**, 55
3	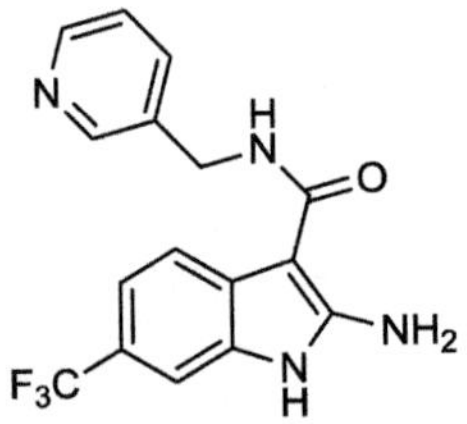	**49c**, 71
4	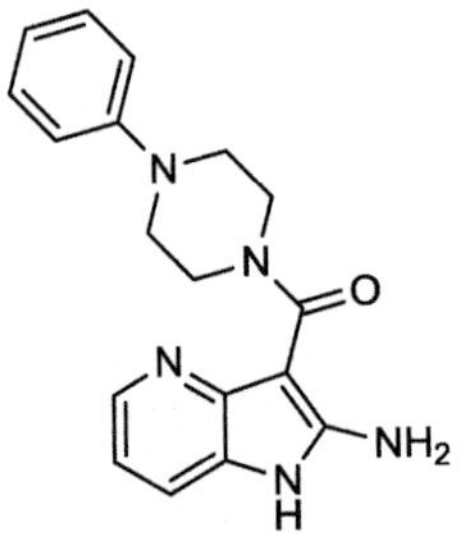	**49d**, 54
5	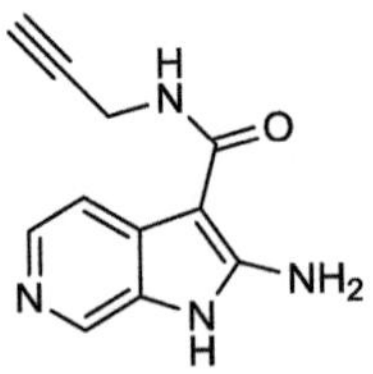	**49e**, 42

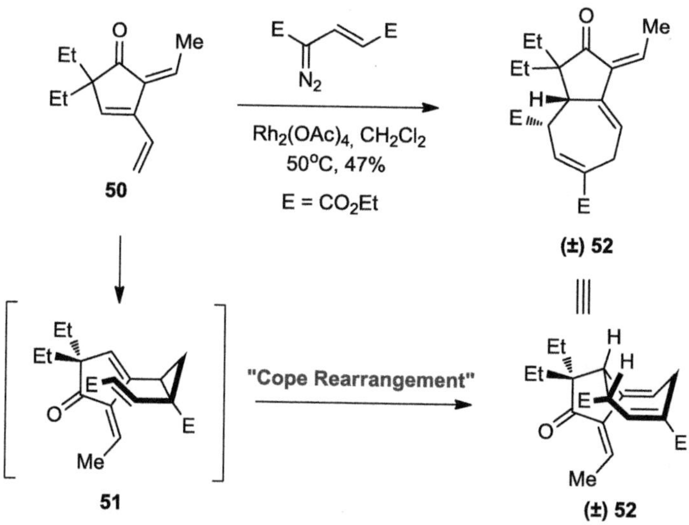

SCHEME 5.21 Domino cyclopropanation/Cope rearrangement of trienone **50**. (Adapted from [26], with permission; copyright © 2009 American Chemical Society.)

After reaction optimization, this domino cyclopropanation/Cope rearrangement sequence could be applied to the benzamide-containing trienone **53** to afford a separable mixture of diastereomers **54a** : **54b** (1 : 1.4) in good overall yield (Scheme 5.22). Based on this methodology, a library of hydroazulenoisoindoles was prepared (Scheme 5.23) [25b]. Eight trienone analogs **55**, prepared as reported previously [25a], were subjected to the domino cyclopropanation/Cope rearrangement sequence and yielded a total of 10 cycloheptadiene analogs **56** (44 to 75%). Further structural diversity was introduced via Diels–Alder cycloadditions with six readily available dienophiles, affording 40 hydroazulenoisoindoles **57** as single diastereomers (Table 5.14).

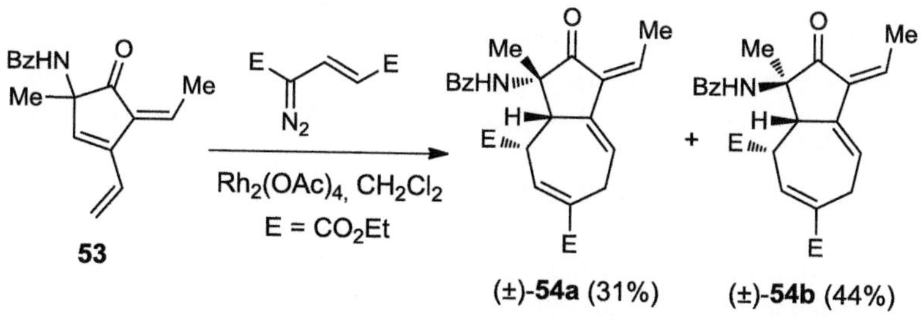

SCHEME 5.22 Domino cyclopropanation/Cope rearrangement of trieneone **53**. (Adapted from [26], with permission; copyright © 2009 American Chemical Society.)

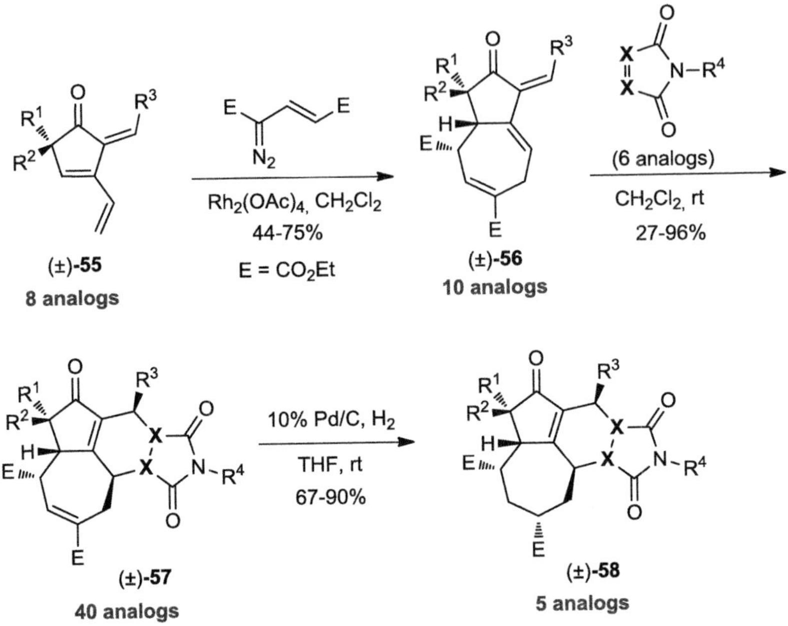

SCHEME 5.23 Synthesis of hydroazulenoisoindoles.

Symmetrical dienophiles were utilized for this library to avoid the formation of mixtures of regioisomers. In all cases, the endo mode of cycloaddition was observed and confirmed by x-ray crystallographic analysis of select compounds. Additionally, five hydroazulenoisoindoles **57** were converted into the saturated fused cycloheptanes **58** through diastereoselective hydrogenation.

The Hanson group has published extensively on the preparation of various substituted sultams (cyclic sulfonamides) [26]. Most recently, a 1,1-dioxido-1,2-benzothiazoline-3-acetic acid library **63** was prepared using the domino Heck-aza-Michael (HaM) sequence (Scheme 5.24). The versatility of this protocol is reflected by sultam formations with three variable points of diversity, originating from the three starting material components [2-bromobenzylsulfonyl chlorides (**59**) amines (**60**) and Michael acceptors (**61**)] (Scheme 5.24 and Table 5.15).

The utility of the HaM process was further enhanced by the assembly of a sultam-containing sublibrary that utilized an additional aryl halogen (**64**), which under the reaction conditions allowed for a double-Heck alkylation to give analogs **65** (Scheme 5.25 and Table 5.16).

In an effort to provide an alternative, milder route toward benzoxazoles that allowed for a greater starting material compatibility, Viirre et al. developed a copper-catalyzed domino annulation using microwave-accelerated conditions [27]. Previous reports highlighted the intramolecular copper-catalyzed cross-coupling reaction of a 2-haloanilide (isolated **68**) to afford benzoxazole **69** [28], and the authors now applied two one-pot domino strategies (Scheme 5.26).

TABLE 5.14 Hydroazulenoisoindole Library 57

Entry	R[1]	R[2]	R[3]	R[4]	X	Yield (%)[a]
1	Me	Me	Me	Ph	N	**57a**, 55
2	Me	Me	Me	Me	N	**57b**, 42
3	Me	Me	Me	Ph	CH	**57c**, 40
4	Me	Me	*n*-Bu	Ph	N	**57d**, 62
5	Me	Me	*n*-Bu	Me	N	**57e**, 50
6	Me	Me	Ph	Ph	CH	**57f**, 52
7	Me	Me	Ph	Me	CH	**57g**, 59
8	Et	Et	Me	Ph	N	**57h**, 92; **58h**, 88
9	Et	Et	Me	Me	N	**57i**, 93; **58i**, 90
10	Et	Et	Me	Ph	CH	**57j**, 85; **58j**, 81
11	Et	Et	Me	CH_2CO_2Me	CH	**57k**, 44
12	Et	Et	Me	Me	CH	**57l**, 66; **58l**, 84
13	Et	Et	Me	Et	CH	**57m**, 30
14	Et	Et	Me	Bn	CH	**57n**, 36
15	Et	Et	Me	H	CH	**57o**, 59
16	$-(CH_2)_4-$	—	Me	Ph	N	**57p**, 70
17	$-(CH_2)_4-$	—	Me	Me	N	**57q**, 84
18	$-(CH_2)_4-$	—	Me	CH_2CO_2Me	CH	**57r**, 61
19	Me	NHBz	Me	Ph	N	**57s**, 70; **58s**, 67
20	Me	NHBz	Me	Me	N	**57t**, 77
21	Me	NHBz	Me	Ph	CH	**57u**, 80
22	Me	NHBz	Me	Me	CH	**57v**, 76
23	Me	NHBz	Me	H	CH	**57w**, 65
24	Me	NHBz	Me	CH_2CO_2Me	CH	**57x**, 50
25	Me	NHBz	Me	Bn	CH	**57y**, 77
26	Me	NHBz	Ph	Ph	N	**57z**, 93
27	Me	NHBz	Ph	Me	CH	**57aa**, 92
28	Me	NHBz	Ph	Ph	CH	**57bb**, 85
29	Me	NHBz	Ph	H	CH	**57cc**, 96
30	NHBz	Me	Me	Ph	N	**57dd**, 54
31	NHBz	Me	Me	Me	N	**57ee**, 51
32	NHBz	Me	Me	Me	CH	**57ff**, 54
33	NHBz	Me	Me	H	CH	**57gg**, 60
34	NHBz	Me	Me	CH_2CO_2Me	CH	**57hh**, 56
35	$-CH_2CH(Me)_2$	NHBz	Me	Ph	N	**57ii**, 50
36	$-CH_2CH(Me)_2$	NHBz	Me	Me	N	**57jj**, 37
37	$-CH_2CH(Me)_2$	NHBz	Me	CH_2CO_2Me	CH	**57kk**, 27
38	NHBz	$-CH_2CH(Me)_2$	Me	Ph	N	**57ll**, 63
39	NHBz	$-CH_2CH(Me)_2$	Me	Me	N	**57mm**, 76

Source: Adapted from [25b], with permission; copyright © 2009 American Chemical Society.

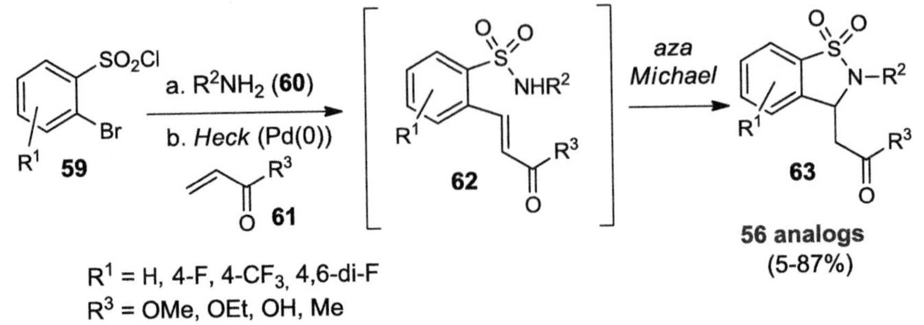

SCHEME 5.24 Heck-aza-Michael domino sequence.

TABLE 5.15 Representative Sultam Analogs 63 Obtained from the Reaction Sequence Shown in 5.24

Entry	R^1	R^2	R^3	Yield (%)
1	H	*i*-Pr	OMe	**63a**, 44
2	H	*c*-Hex	Me	**63b**, 29
3	H	Ph	OH	**63c**, 35
4	H	3-OMe-Ph	OEt	**63d**, 22
5	F	Bn	OMe	**63e**, 75
6	F	*n*-Oct	OEt	**63f**, 54
7	F	*c*-Pent	OEt	**63g**, 87
8	F	*t*-Bu	OEt	**63h**, 46
9	CF_3	*i*-Pr	OEt	**63i**, 50
10	CF_3	*c*-Hex	OMe	**63j**, 51
11	CF_3	*c*-Hex	OH	**63k**, 46
12	CF_3	Bn	OH	**63l**, 44

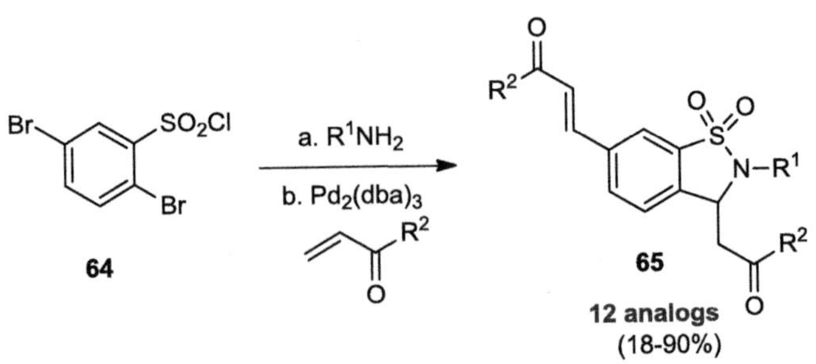

SCHEME 5.25 Double-Heck coupling derived sultams **65**.

TABLE 5.16 Representative Sultam Analogs 65 Obtained from the Reaction Sequence Shown in Scheme 5.25

Entry	R^1	R^2	Yield (%)
1	4-OMe-Bn	OH	**65a**, 62
2	4-OMe-Bn	Me	**65b**, 86
3	*n*-Oct	OEt	**65c**, 90
4	2-OMe-Ph	OMe	**65d**, 78

The 2-haloanilide **68** can either be formed in situ using a copper-catalyzed cross-coupling of a corresponding 1,2-dihaloarene **66** and a primary amide **67** (method A), or via acylation of a 2-haloamine **70** and an acid chloride **71** (method B). The limited commercial availability of the 1,2-dihaloarenes as well as lower regioselectivities render the first procedure less versatile in terms of library design. While optimizing reaction conditions, Viirre et al. found that a one-pot domino acylation/cross-coupling approach proceeded more efficiently under microwave irradiation conditions. Reaction times of 15 min in the microwave were obtained compared to 24 h under standard heating conditions. It should be noted that the microwave irradiations were carried out on a smaller scale and under higher dilutions due to the stirring limitations with the insoluble base. A library of 24 analogs was constructed using four 2-bromoanilines and six acid chlorides, and the optimized microwave conditions provided isolated yields ranging from 21 to 97% (Figure 5.1).

5.5 RADICAL DOMINO REACTIONS

With a particular interest in exploring the synthetic utility of the high π-character of the central C–C bond of bicyclo[1.1.0]butanes [29] **70**, Wipf and Walczak developed an *N*-allylation/Alder-ene and an *N*-allylation/formal [2 + 2] domino reaction

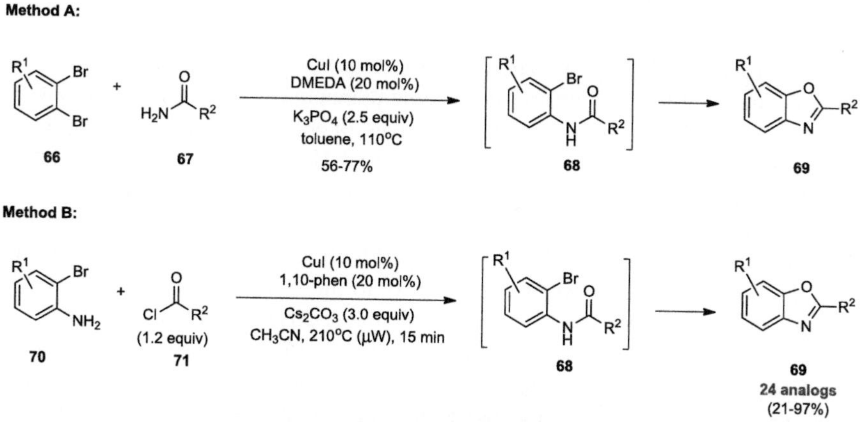

SCHEME 5.26 Synthesis of benzoxazoles.

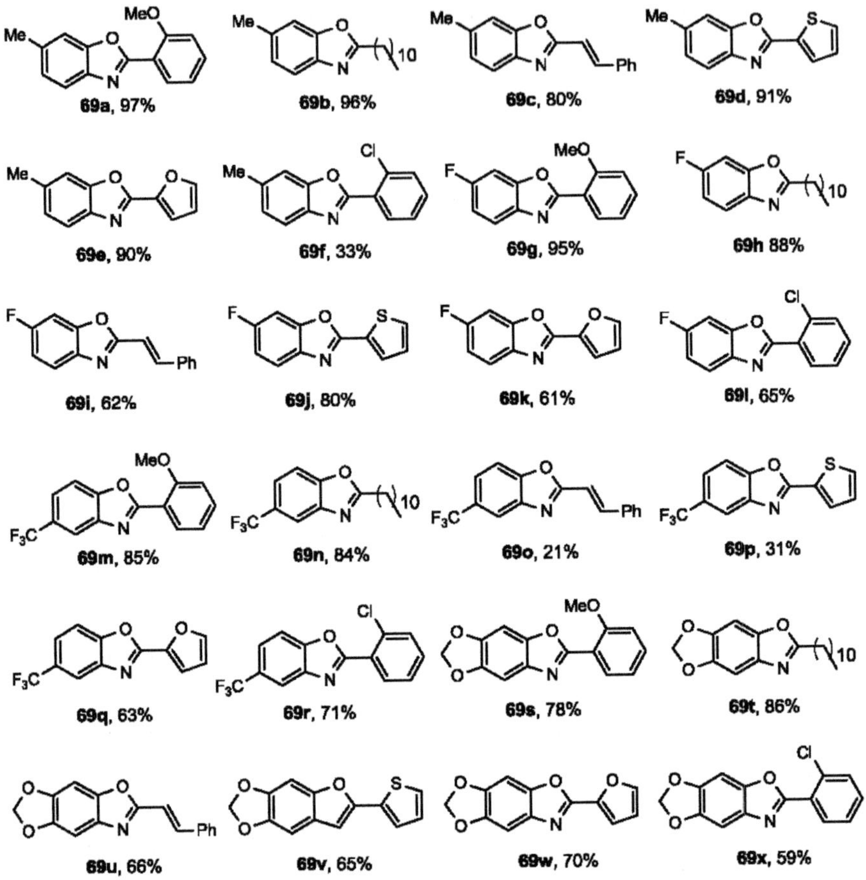

FIGURE 5.1 Benzoxazole library **69**.

sequence (Scheme 5.27) [30]. The intramolecular cycloadditions of these strained ring systems with tethered alkene and alkyne dienophiles led to novel heterocyclic scaffolds. *N*-Allylations of bicyclo[1.1.0]butanes **70** proceeded smoothly under modified phase-transfer conditions (allyl bromide, Bu_2NHSO_4, 50% NaOH aq., toluene), and the alkylated intermediates spontaneously underwent further transformations that were dependent on the electronic nature of the allylic substituent (Scheme 5.27). When bicyclo[1.1.0]butane **70** was substituted with alkyl groups (R^3 = alkyl), the reaction proceeded via an Alder-ene pathway to the spirocyclic butenes **72**. Interestingly, under the same reaction conditions, the *N*-allyl bicyclobutane with R^3 = aryl substitution gave the unprecedented 3-azatricyclo[5.1.1.0]nonanes **73** by means of a formal [2 + 2] reaction pathway. It was also observed that only bicyclobutanes conjugated to aromatic rings (R^1 = aryl) participated in these domino reaction sequences.

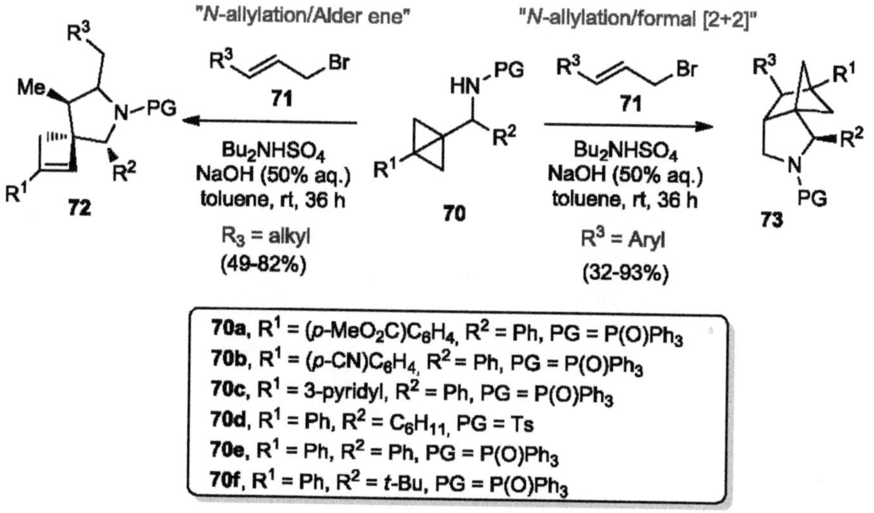

SCHEME 5.27 *N*-Allylation/Alder-ene and *N*-allylation/formal [2 + 2] domino reactions.

The *N*-allylated intermediates **74** could undergo a homolytic cleavage of the central bicyclobutane C–C σ-bond (possessing high π-character), affording the putative biradical intermediates **75** (Scheme 5.28). The resonance-stabilized biradical **75** could recombine to afford the 3-azatricyclo[5.1.1.0]nonanes **73**. This mechanistic hypothesis was supported by the observation of multiple radical species in ESR-spin trapping experiments [31].

With this straightforward access to the structurally unique azatricycles, a 5 × 3 matrix (bicyclobutanes × allyl bromide, Figure 5.2 and Table 5.17) library was constructed [32]. A larger, tertiary amine containing sublibrary **77** (75 analogs)

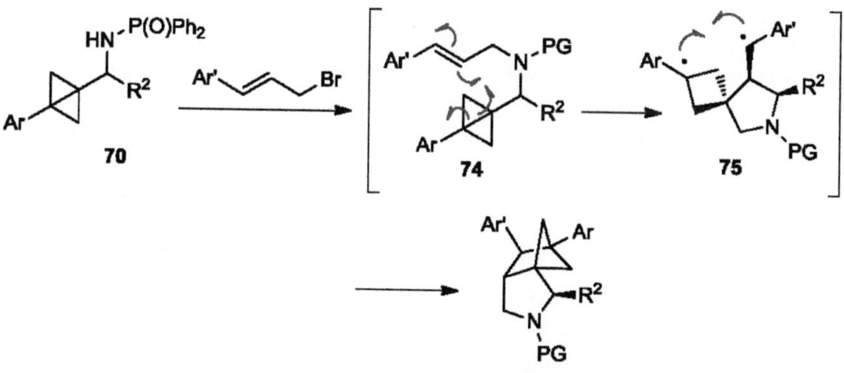

SCHEME 5.28 Proposed mechanism for *N*-allylation/formal [2 + 2] domino reactions.

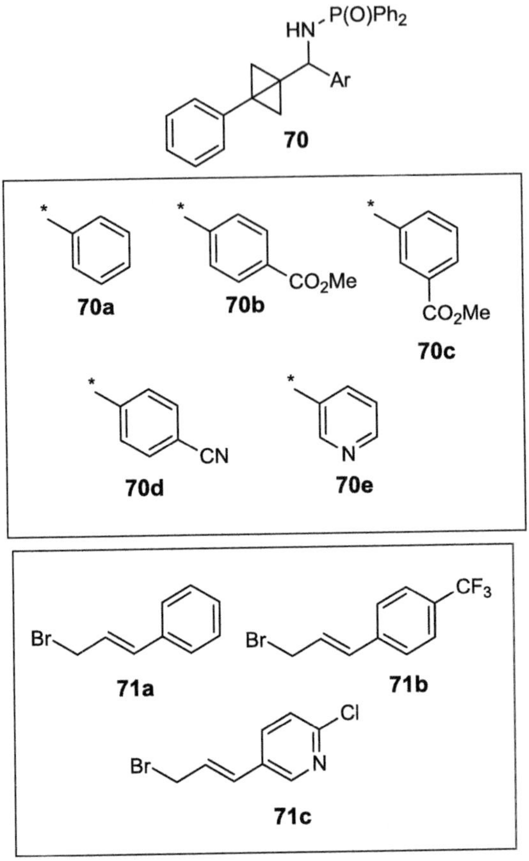

FIGURE 5.2 Bicyclobutane and allylbromide building blocks.

was completed in a parallel format from reductive aminations of the corresponding pyrrolidine substrates (Scheme 5.29 and Table 5.18). High-throughput biological assay screening of this library identified a series of inhibitors of the malaria parasite, *Plasmodium falciparum* [33].

Frutos and Curran expanded previously reported studies of a radical cascade process en route to the synthesis of the camptothecin class of antitumor agents and generated a structurally related mappicine and mappicine ketone library of 112 compounds using established parallel synthesis techniques (Scheme 5.30) [34]. This study highlighted a late-stage cascade radical annulation approach for the assembly of the polycyclic core, using readily available building blocks in a relatively short reaction sequence (six or seven steps) (Scheme 5.30). As a second point of diversification, *N*-alkylation of lactams **78** with the desired propargyl bromides (**79**, four variants) afforded the cyclization precursors **80**. Iodopyridones **80** underwent the radical cascade annulation when combined with phenyl isonitriles **81** (four variants) and

TABLE 5.17 3-Azatricyclo[5.1.1.0]nonanes 73 Obtained in the Reaction Sequence Shown in Scheme 5.28

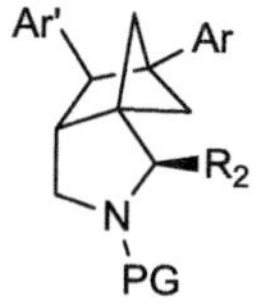

73

Entry	70	71	Product, Yield (%)
1	**70a**	**71a**	**73a**, 80
2	**70a**	**71b**	**73b**, 95
3	**70a**	**71c**	**73c**, 55
4	**70b**	**71a**	**73d**, 68
5	**70b**	**71b**	**73e**, 84
6	**70b**	**71c**	**73f**, 76
7	**70c**	**71a**	**73g**, 84
8	**70c**	**71b**	**73h**, 57
9	**70c**	**71c**	**73i**, 71
10	**70d**	**71a**	**73j**, 75
11	**70d**	**71b**	**73k**, 72
12	**70d**	**71c**	**73l**, 70
13	**70e**	**71a**	**73m**, 69
14	**70e**	**71b**	**73n**, 79
15	**70e**	**71c**	**73o**, 74

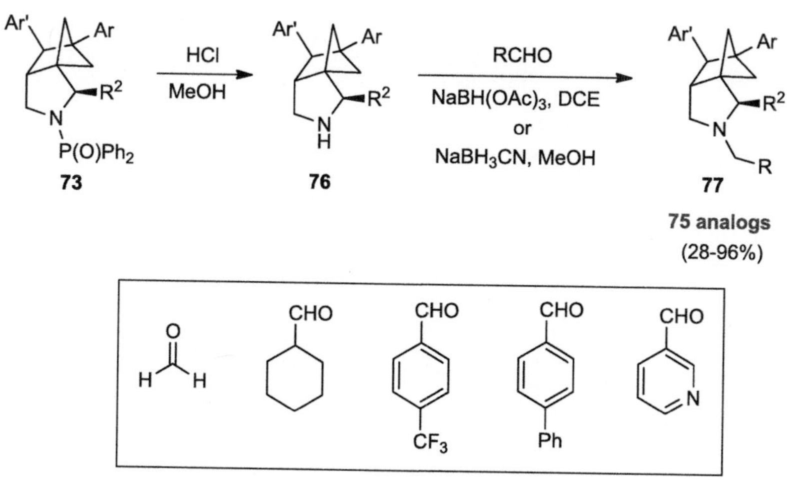

SCHEME 5.29 Preparation of a tricyclic amine library.

TABLE 5.18 Tertiary Amine Library 77 Obtained from the Reaction Sequence Shown in Scheme 5.29

Entry	Product Structure	Yield (%)	Purity[a]
1	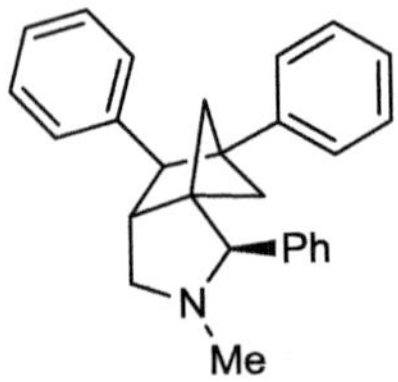	93	100
2	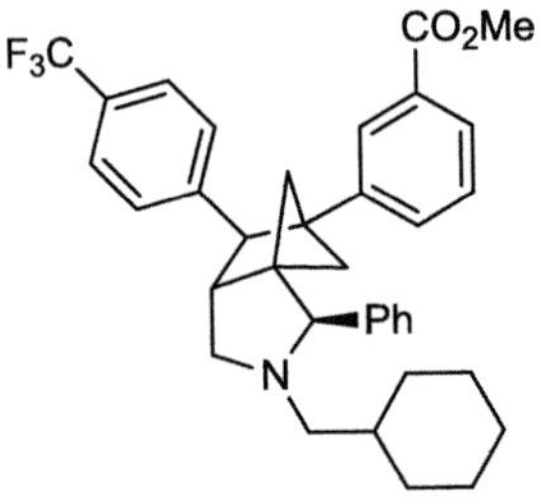	76	100
3	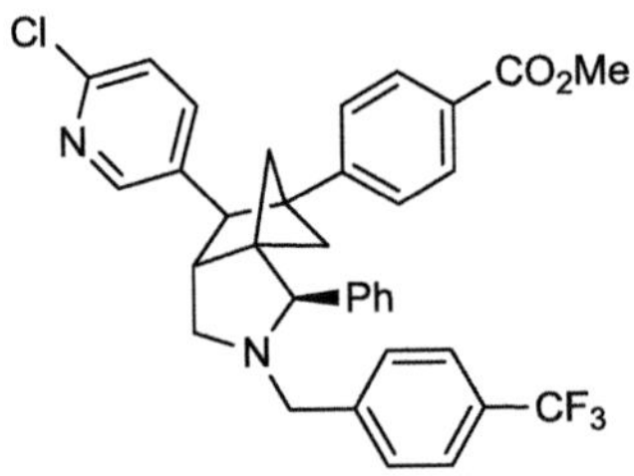	55	100
4	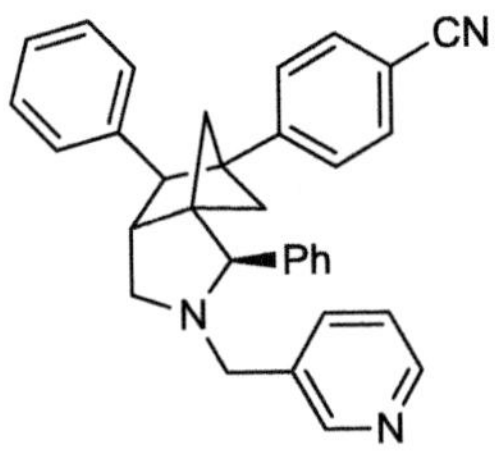	47	94
5	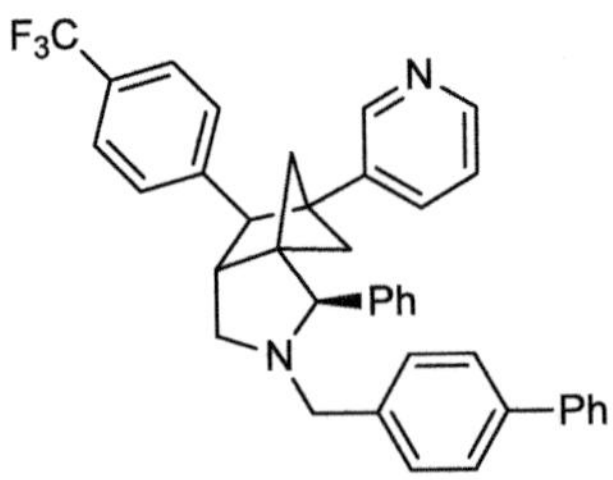	93	100

[a]Purity by LC-MS-ELSD analysis.

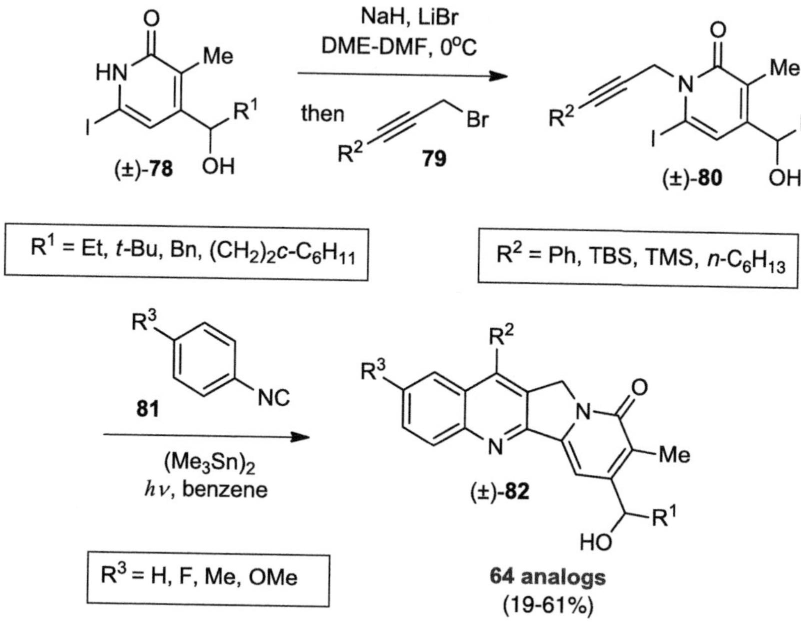

SCHEME 5.30 Synthesis of a mappicine library.

hexamethylditin under sunlamp irradiation conditions. Product mixtures were first purified by solid-phase extraction (SPE), followed by flash chromatography to afford the mappicine analogs **82** (19–61%) (Table 5.19). A second-generation library consisting of mappicine ketones was completed through the use of solid-supported oxidants, with resin-bound chromic acid being the most effective.

With the exception of the benzyl-substituted analogs (R^1 = benzyl), all mappicine alcohols **82** cleanly provided the desired ketone products (Scheme 5.31 and Table 5.20). This radical annulation cascade was also applied to the camptothecin class of anticancer agents (Scheme 5.32) [35]. The radical cascade precursors **86** were again assembled by *N*-propargylation of iodopyridones **84.** In a parallel format, the iodopyridones **84** were exposed to a set of isonitriles (seven analogs), hexamethylditin, and light.

The streamlined reaction workup and purifications involved automated solid-phase extractions followed by serial reversed-phase chromatographies to provide 91 homosilatecan analogs **88** in yields ranging from 3 to 64% (Scheme 5.32 and Table 5.21). A subset of the 91 pentacyclic analogs were deprotected to give the amino- and hydroxy substituted homosilatecans (Scheme 5.33, **89** and **90**, respectively). Overall, the use of the late-stage cascade radical annulation approach for generating the polycyclic core of camptothecin-related compounds allowed the synthesis of several analog libraries.

TABLE 5.19 Representative Mappicine Analogs 82

Entry	Product Structure	Prodict, Yield (%)
1	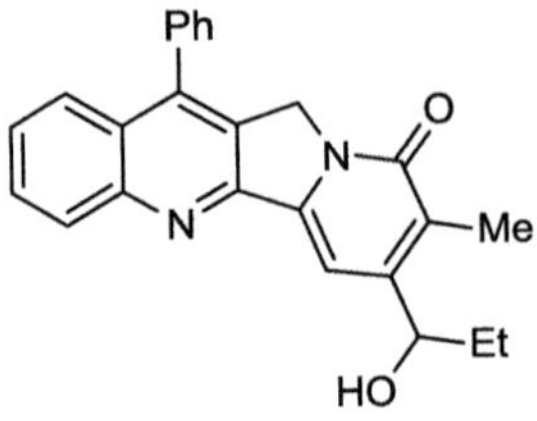	**82a**, 58
2	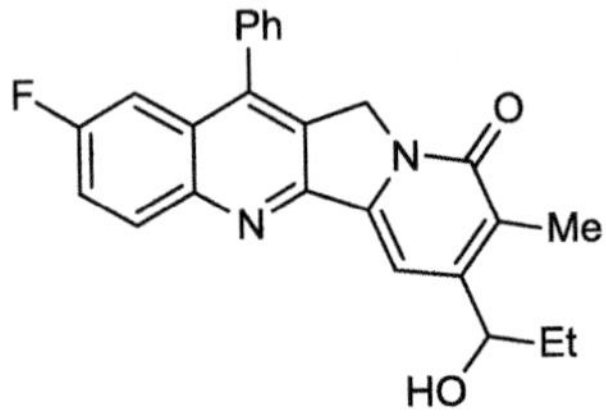	**82b**, 61
3	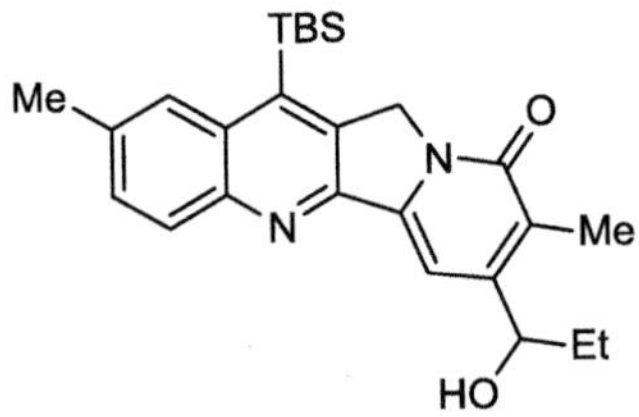	**82c**, 56
4	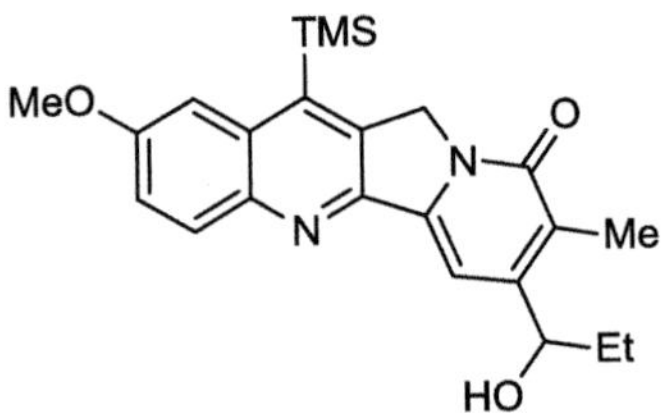	**82d**, 59
5	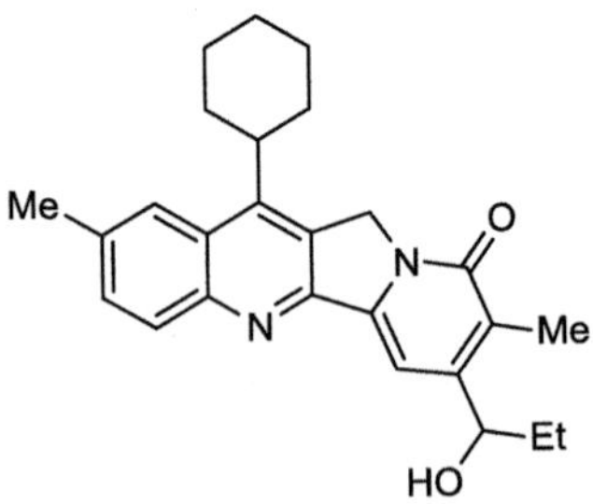	**82e**, 58

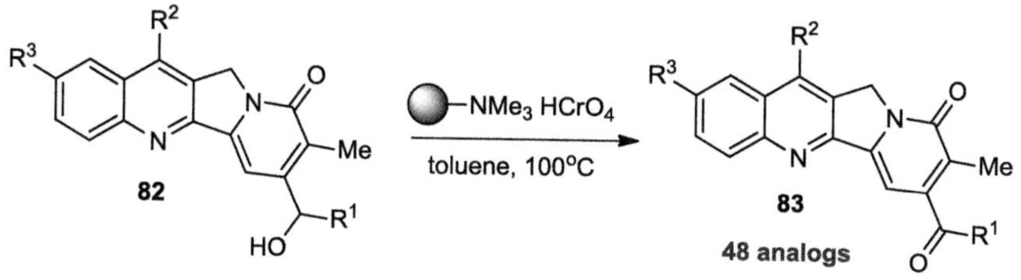

R^1 = Et, *t*-Bu, CH_2CH_2c-C_6H_{11}
R^2 = Ph, TBDMS, TMS, *n*-Hexyl
R^3 = H, F, Me, OMe

SCHEME 5.31 Preparation of mappicine ketones **83**.

TABLE 5.20 Representative Members of Library 83

Entry	R^1	R^2	R^3	Yield (%)
1	Et	Ph	F	**83a**, 86
2	Et	TBS	OMe	**83b**, 91
3	Et	*n*-Hex	F	**83c**, 88

[a]Yields of ketone formation.

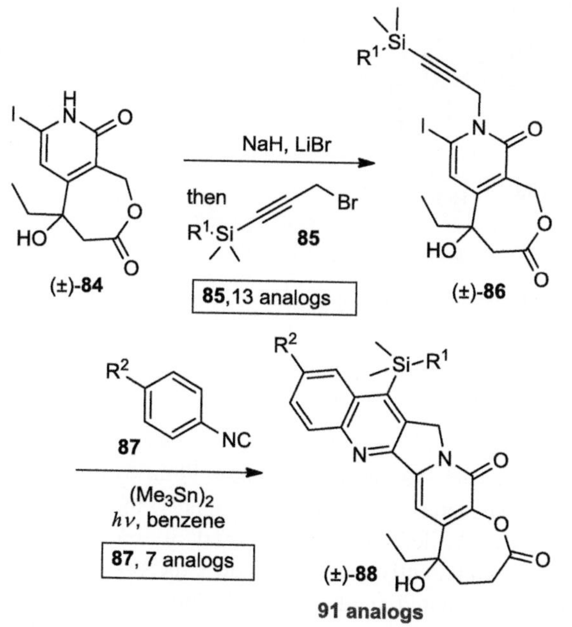

SCHEME 5.32 Construction of a (13 × 7) homosilatecan library **88**.

TABLE 5.21 Representative Examples of Homosilatecan Library 88

Entry	R^1	R^2	Product, Yield (%)
1	$(CH_2)_2CF_3$	H	**88a**, 42
2	*n*-Bu	F	**88b**, 35
3	Me	CF_3	**88c**, 49
4	Me	OAc	**88d**, 29
5	*n*-Bu	NHBoc	**88e**, 26

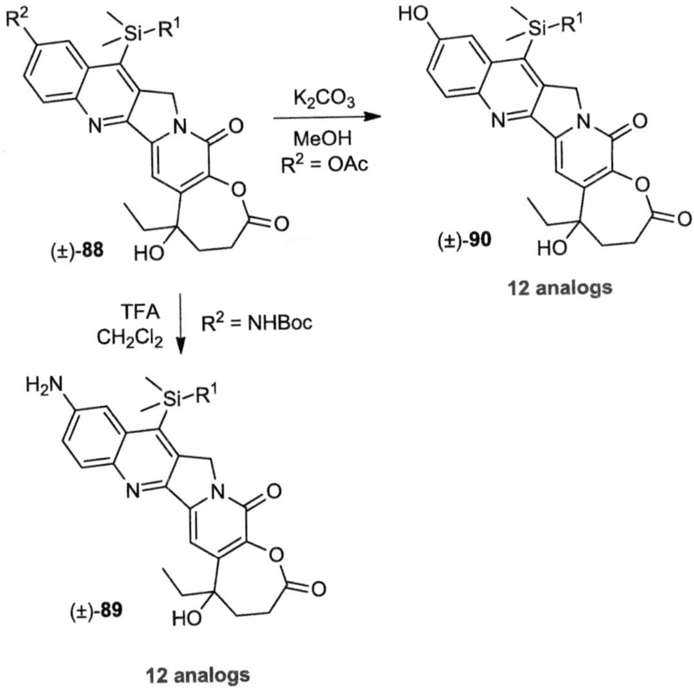

SCHEME 5.33 Synthesis of 24 homosilatecan analogs by parallel deprotections.

5.6 CONCLUSIONS

Over the past decade, the synthesis of chemical libraries has matured from a pure enumeration to a consideration of compound quality, including chemical diversity and scaffold novelty. While libraries composed of individual, pure samples vs. mixture libraries are inherently more difficult, expensive, and slower to prepare, steady advances in automated synthesis, purification, and analysis are still greatly expediting access to high-quality compound collections. This includes technologies such as microwave reactors, flow chemistry, solid-phase and split–pool synthesis strategies, as well as polymer-linked reagents and dynamic combinatorial library strategies. The literature continues to suggest a shift in chemical library design to smaller and more focused structurally diverse libraries [36]. It remains to be seen if these efforts will result in the accelerated discovery of new therapeutic agents, but it is clear that in a streamlined discovery approach that starts with the high-throughput screen of an unbiased library, the outcome is greatly influenced by the quality of the screening collection. The goal of DOS is the rapid assembly and manipulation of natural product–derived or rationally designed, stereochemically rich scaffolds [37]. DOS inspires the discovery of new reaction methodologies and their application in the synthesis of novel scaffolds [32,38]. Imaginative domino reaction sequences are ideally suited to create these versatile chemotypes and structurally and stereochemically diverse libraries. The combination of reactions performed in sequence without the need for isolation is an ideal solution to the paradigm to cover more chemical space in fewer chemical steps.

Acknowledgments

The authors thank the NSF and NIH for financial support. The NIGMS Center for Chemical Methodologies and Library Development program (P50-GM067082) is gratefully acknowledged. Maciej M. M. Walczak and Zhenglai Fang in our laboratories at the University of Pittsburgh were valuable contributors to the synthetic work on 3-azatricyclo[5.1.1.0]nonanes summarized in this review.

REFERENCES

1. (a) L. F. Tietze, G. Brasche, K. M. Gericke, *Domino Reactions in Organic Synthesis*, Wiley-VCH, Weinheim, Germany, **2006**; (b) L. F. Tietze, M. E. Lieb, *Curr. Opin. Chem. Biol.* **1998**, *2*, 363–371.
2. For recent reviews of total syntheses using cascade reactions, see (a) K. C. Nicolaou, J. S. Chen, *Chem. Soc. Rev*. **2009**, *38*, 2993–3009; (b) K. C. Nicolaou, D. J. Edmonds, P. G. Bulger, *Angew. Chem. Int. Ed*. **2006**, *45*, 7134–7186.
3. S. E. Denmark, A. Thorarensen, *Chem. Rev.* **1996**, *96*, 137–165.
4. S. L. Schreiber, *Proc. Nat. Acad. Sci. U.S.A.* **2011**, *108*, 6699–6702.
5. For an updated survey of chemical libraries for drug discovery and chemical biology, see R. E. Dolle, B. Le Bourdonnec, K. Worm, G. A. Morales, C. J. Thomas, W. Zhang, *J. Comb. Chem.* **2010**, *12*, 765–806.
6. L. F. Tietze, A. Steinmetz, *Angew. Chem. Int. Ed.* **1996**, *35*, 651–652.
7. L. F. Tietze, T. Hippe, A. Steinmetz, *Synlett* **1996**, 1043–1044.
8. K. J. Frankowski, E. E. Hirt, Y. Zeng, B. Neuenswander, D. Fowler, F. Schoenen, J. Aubé, *J. Comb. Chem.* **2007**, *9*, 1188–1192.
9. K. J. Frankowski, J. E. Golden, Y. Zeng, Y. Lei, J. Aubé, *J. Am. Chem. Soc.* **2008**, *130*, 6018–6024.
10. K. J. Frankowski, B. Neuenswander, J. Aubé, *J. Comb. Chem.* **2008**, *10*, 721–725.
11. K. J. Frankowski, V. Setola, J. M. Evans, B. Neuenswander, B. L. Roth, J. Aubé, *Proc. Natl. Acad. Sci. U.S.A.* **2011**, *108*, 6727–6732.

12. P. Ertl, S. Jelfs, J. Mühlbacher, A. Schuffenhauer, P. Selzer, *J. Med. Chem.* **2006**, *49*, 4568–4573.
13. B. R. McNaughton, B. L. Miller, *Org. Lett.* **2003**, *5*, 4257–4259.
14. P. Janvier, X. Sun, H. Bienaymé, J. Zhu, *J. Am. Chem. Soc.* **2002**, *124*, 2560–2567.
15. F. Yang, L. Zheng, J. Xiang, Q. Dang, X. Bai, *J. Comb. Chem.* **2010**, *12*, 476–481.
16. (a) M. Largeron, A. Neudorffer, M. Vuilhorgne, E. Blattes, M.-B. Fleury, *Angew. Chem. Int. Ed.* **2002**, *41*, 824–827; (b) E. Blattes, M.-B. Fleury, M. Largeron, *J. Org. Chem.* **2004**, *69*, 882–890; (c) D. Xu, A. Chiaroni, M.-B. Fleury, M. Largeron, *J. Org. Chem.* **2006**, *71*, 6374–6381.
17. S. Wan, M. E. Green, J.-H. Park, P. E. Floreancig, *Org. Lett.* **2007**, *9*, 5385–5388.
18. L. Zhang, Q. Xiao, C. Ma, X.-Q. Xie, P. E. Floreancig, *J. Comb. Chem.* **2009**, *11*, 640–644.
19. For review of organochlorozirconocene complexes, see P. Wipf, H. Jahn, *Tetrahedron* **1996**, *52*, 12853–12910.
20. N. Komatsu, M. Uda, H. Suzuki, T. Takahashi, T. Domae, M. Wada, *Tetrahedron Lett.* **1997**, *38*, 7215–7218.
21. C. Lu, Q. Xiao, P. Floreancig, *Org. Lett.* **2010**, *12*, 5112–5115.
22. K. Wang, E. Herdtweck, A. Dömling, *ACS Comb. Sci.* **2011**, *13*, 140–146.
23. K. Wang, K. Nguyen, Y. Huang, A. Dömling, *J. Comb. Chem.* **2009**, *11*, 920–927.
24. H. M. L. Davies, *Tetrahedron* **1993**, *49*, 5203–5223.
25. (a) K. M. Brummond, D. Chen, T. O. Painter, S. Mao, D. D. Seifried, *Synlett* **2008**, 759–764; (b) K. M. Brummond, S. Mao, S. N. Shinde, P. J. Johnston, B. W. Day, *J. Comb. Chem.* **2009**, *11*, 486–494.
26. A. Rolfe, K. Young, K. Volp, F. Schoenen, B. Neuenswander, G. H. Lushington, P. R. Hanson, *J. Comb. Chem.* **2009**, *11*, 732–738.
27. R. D. Viirre, G. Evidar, R. A. Batey, *J. Org. Chem.* **2008**, *73*, 3452–3459.
28. G. Evindar, R. A. Batey, *J. Org. Chem.* **2006**, *71*, 1802–1808.
29. P. Wipf, C. R. J. Stephenson, K. Okumura, *J. Am. Chem. Soc.* **2003**, *125*, 14694–14695.
30. P. Wipf, M. A. A. Walczak, *Angew. Chem. Int. Ed.* **2006**, *45*, 4172–4175.
31. M. A. A. Walczak, B.-K. Shin, P. Wipf, S. Saxena, *Org. Biomol. Chem.* **2009**, *7*, 2363–2366.
32. P. Wipf, Z. Fang, L. Ferrié, M. Ueda, M. A. A. Walczak, Y. Yan, M. Yang, *Pure Appl. Chem.* http://dx.doi.org/10.1351/PAC-CON-12-09-03.
33. PubChem Project Homepage, http://pubchem.ncbi.nlm.nih.gov; accessed Mar. 29, 2012; AID 504834.
34. O. de Frutos, D. P. Curran, *J. Comb. Chem.* **2000**, *2*, 639–649.
35. A. E. Gabarda, D. P. Curran, *J. Comb. Chem.* **2003**, *5*, 617–624.
36. D. J. Newman, G. M. Cragg, *J. Nat. Prod.* **2012**, *75*, 311–335, and references therein.
37. M. D. Burke, S. L. Schreiber, *Angew. Chem. Int. Ed.* **2004**, *43*, 46–58.
38. (a) P. Wipf, *DOS* **2012**, 6–10; (b) E. M. Skoda, G. C. Davis, P. Wipf, *Org. Process Res. Dev.* **2012**, *16*, 26–34; (c) P. Wipf, J. Xiao, C. R. J. Stephenson, *Chimia* **2009**, *63*, 764–775.

6

DIVERSITY-ORIENTED SYNTHESIS OF AMINO ACID–DERIVED SCAFFOLDS AND PEPTIDOMIMETICS: A PERSPECTIVE

ANDREA TRABOCCHI

6.1 INTRODUCTION

Protein–protein interactions (PPIs) play a pivotal role in many biological mechanisms, such as proteases complexation of substrates, the binding of peptide hormones to receptors, and protein–protein interplay to produce signal transduction. Nevertheless, the discovery of modulators of PPIs has been limited to structurally characterized targets. As a consequence of interaction with their membrane-bound receptors, bioactive peptides influence cell–cell communication and control a series of vital functions, and a huge number of biologically active peptides have been discovered and characterized, including hormones, vasoactive peptides, and neuropeptides. Thus, they are of great interest in the biomedical field, and the number of native and modified peptides used as therapeutics is ever increasing. Many bioactive peptides have been prepared on a large scale and tested both in pharmacology and in the clinic, thus allowing the development of new therapies for a variety of pathologies.

Nevertheless, the use of peptides as therapeutics is limited due to some factors, including low metabolic stability toward proteolysis in the gastrointestinal tract, poor absorption after oral ingestion, rapid excretion through liver and kidneys, and undesired effects due to interaction of flexible peptides with several receptors [1]. Moreover, the conformational flexibility of most natural peptides and the high dependence

Diversity-Oriented Synthesis: Basics and Applications in Organic Synthesis, Drug Discovery, and Chemical Biology, First Edition. Edited by Andrea Trabocchi.

of their conformation on the environment are relevant issues that make it difficult to use such entities as templates for drug discovery.

In addition to all these limitations, biomedical research is constantly oriented toward the development of new therapeutics based on peptides and proteins, by introducing both structural and functional specific modifications and maintaining the features responsible for the biological activity. These requirements are all matched in the development of *peptidomimetics* [2–4]. In this approach, peptides and proteins are considered as leads for the discovery of other classes of compounds. Peptidomimetics play a prominent role as candidate compounds to induce phenotypic effects on biological systems. In fact, side-chain recognition dominates the biological interactions of almost all cellular processes; thus, peptidomimetics, developed initially for their property of preventing degradation and improving oral bioavailability of peptide-based drugs, have been envisaged as a tool for perturbing such interactions and identifying protein functions. Small peptide-based agents have attracted wide interest as anticancer and anti-infective agents, and there is a continuous need to develop new high-affinity and high-specificity peptidomimetic or small-molecule ligands in such widespread pathologies. Success in this area depends on the ability to create novel complex molecular structures of peptidomimetic nature as tools for probing protein–protein interactions.

A peptidomimetic compound may be defined as a molecule having a secondary structure, besides other structural features similar to the parent peptide, such that it binds to enzymes or receptors with higher affinity than the starting peptide. As an overall result, the native peptide effects are inhibited (antagonist) or increased (agonist). Apart from being much more selective and efficient than native peptides, thus resulting in fewer side effects, peptidomimetics show greater oral bioavailability, and biological activity is prolonged due to lowered enzymatic degradation [5,6].

The generation of peptidomimetic combinatorial libraries was reported initially as modified peptides, taking advantage of the chemical diversity of the amino acids and the sequence diversity. As an example, Dooley and Houghten described the generation of a wide range of combinatorial libraries, from peptides to low-molecular-weight heterocyclic compounds, which have been screened successfully in assays specific for opioid receptors, thus enabling the identification of new ligands for these receptors [7]. During the past decade, the combinatorial concept has evolved to diversity-oriented synthesis (DOS) in view of expanding the chemical diversity of the central scaffold, as well as exploiting the appendage diversity on a fixed cyclic chemical entity, which has been the most popular approach in classic combinatorial libraries. Very recently, several reports on DOS chemical methods encompassing peptidomimetics and amino acid–derived scaffolds appeared in the literature, possibly opening a perspective on expansion of the chemical diversity around the peptidomimetic concept, in an effort to address “undruggable” protein–protein interactions that have been identified as potential therapeutic targets for relevant diseases. Accordingly, selected contributions addressing the generation of peptidomimetic and amino acid–derived scaffolds following a DOS approach are presented in this chapter, with the aim of giving a perspective view of what may be a promising way to merge two powerful concepts to achieve brand new privileged structure carriers of high chemical diversity.

6.2 DEFINITION AND CLASSIFICATION OF PEPTIDOMIMETICS

The standard approach to the generation of peptidomimetics is focused on knowledge of the electronic and conformational features of the native peptide and its receptor or active site of an enzyme. Thus, the development of peptidomimetics as compounds possessing biological activity must take account of some basic principles [8], including:

- Replacement of the peptide backbone with a nonpeptide framework: If an amide bond does not influence the biological activity, or amide bonds are not exposed to the active site, the template may be designed to eliminate such peptide bonds.
- Preservation of the side-chain functionalities involved in biological activity, as they constitute the pharmacophore. In the development of second-generation mimetics, several modifications should be taken into account so as to improve the biological activity, including chain-length modification, introduction of constraints, cyclopeptide bond replacement with a covalent binding, and the introduction of isosteric replacements [9].
- Maintenance of flexibility in first-generation peptidomimetics: If a biological activity is observed for a flexible mimetic, the introduction of elements of rigidity to side chains is a rational approach to improving the preliminary activity observed. This can be done by incorporating amino acids, which can adopt only a very limited number of different conformations, or by introducing cycles in the molecule.
- Selection of proper targets based on a pharmacophore hypothesis; knowledge of the structure–activity relationship or the three-dimensional structure of the bioactive conformation is an important issue in achieving the best compound rapidly without generating an enormous number of compounds with poor biological activity.

Peptidomimetics may be subdivided into three classes, depending on their structural and functional characteristics [10]:

1. Structural mimetics or type I mimetics. These compounds show the strict analogy of a local topography with the native substrate, and they carry all the functionalities responsible for interaction with an enzyme or a receptor in a well-defined spatial orientation.
2. Functional mimetics or type II mimetics. In these molecules the analogy with the native compound is based on interaction with the target receptor or enzyme, without apparent structural analogies.
3. Structural–functional mimetics or type III mimetics. This class of peptidomimetics is commonly represented by a scaffold having a structure different from the substrate, in which all the functional groups necessary for biological interactions are mounted in a well-defined spatial orientation.

A typical example of the third class of peptidomimetics is reported in the literature for thyrotropin-releasing hormone (TRH) mimetics, which are based on a central

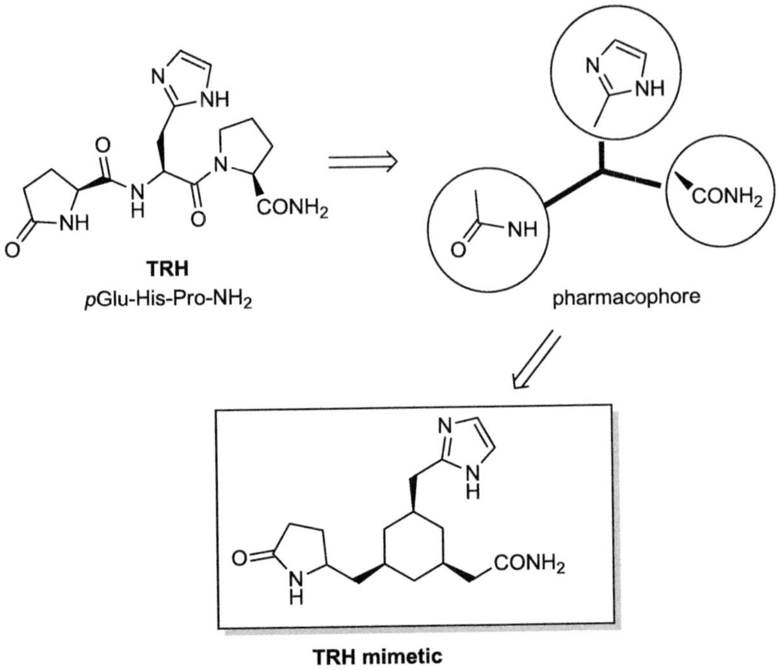

FIGURE 6.1 Cyclohexane-based thyrotropin-releasing hormone mimetic.

cyclohexane scaffold [6]. In this case, the cyclohexane ring replaces the peptide backbone, and the three functional groups that constitute the pharmacophore are placed on the scaffold with the same spatial orientation of amino acid side chains found in TRH (Figure 6.1). Moreover, by conferring additional rigidity on the system, cyclohexane determines an increase in the biological activity, although the side-chain rotation is still maintained.

6.3 EARLY COMBINATORIAL APPROACHES TO PEPTIDOMIMETIC SCAFFOLDS

Several reports on application of the concepts of combinatorial chemistry in the field of peptidomimetics appeared in 1996, in an effort to develop molecules resembling peptide pharmacophores [11], including the generation of peptidomimetics containing the phosphinic group as a transition-state isostere in neurolysin inhibitors [11b] and the development of a library of molecules as dipeptide isosteres to expand the side-chain functionality of the central core scaffold through consecutive amide alkylations [11d]. Successively, many research groups turned their attention to the generation of scaffold-based chemical libraries possessing a more "drug-like" profile, still taking into account amino acid moieties as a source of chemical diversity. Accordingly, early combinatorial libraries were conceived around a readily available

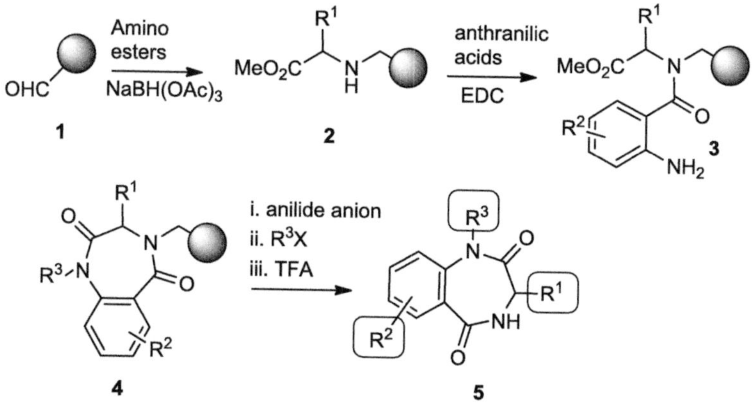

SCHEME 6.1 Solid-phase synthesis of a library of 1,4-benzodiazepine-2,5-diones using α-amino acid derivatives as the building blocks.

scaffold capable of being decorated with varying substituents in order to expand the chemical diversity around it.

Other entries to peptidomimetic combinatorial libraries encompassed known classes of enzyme inhibitors as powerful tools in rapidly profiling novel protease drug targets and identifying most effective pharmacophores at their inhibition. Specifically, the syntheses of phosphonyl acid–based pharmacophore libraries were proposed with the aim of facilitating the discovery of novel metalloprotease inhibitors of medicinal interest; phosph-inyl (-onyl) acids were selected as transition-state analogs and metal chelators for inclusion in peptidomimetic inhibitors of metalloproteases such as thermolysin and angiotensin-converting enzyme [12].

Among early examples of small-molecule libraries containing amino acid moieties, Ellman and co-workers reported a solid-phase synthesis of 1,4-benzodiazepine-2,5-diones from three commercially available components: namely, anthranilic acids, α-amino esters, and alkylating agents (Scheme 6.1). Moreover, a library of 1,4-benzodiazepine-2,5-diones was prepared from 11 alkylating agents, 12 anthranilic acids, and a total of 19 α-amino esters coming from nine sets of enantiomeric pairs and glycine methyl ester hydrochloride, to give 2508 total compounds, or 1320 spatially separate compounds [13].

Also, another interesting report by Ellman's group is represented by the generation of a β-turn mimetic library (Scheme 6.2) [14], taking advantage of a facile intramolecular–cyclative thiol S_N2 displacement and the simultaneous cleavage of the molecule from resin to create nine- and ten-membered rings.

In 1996, Gordon et al. reported on a library of heterocycles (4-thiazolidinones, β-lactams, pyrrolidines, and dihydropyridines) from imines derived from resin-bound amino acids, following the principles of skeletal diversity from a common starting substrate [15]. The approach to heterocyclic diversity was conceived by exploiting the imine chemistry as a useful synthetic strategy and using amino acid building blocks to give a convenient starting point for exploring the scope of solid-phase organic

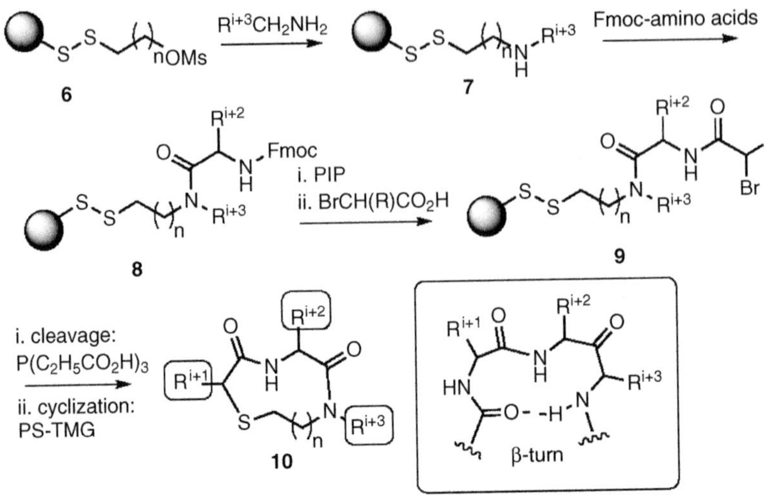

PS-TMG: polymer-supported tetramethylguanidine
PIP: piperidine

SCHEME 6.2 Generation of a β-turn mimetic library using an intramolecular-cyclative thiol S_N2 displacement.

synthesis by conventional peptide activation chemistries. The formation of imines via condensation of the amino group of resin-bound amino acids with aldehydes appeared a particularly appropriate reaction, followed by cycloadditions, condensation reactions, and other nucleophilic additions to give diverse nitrogen heterocycles (Scheme 6.3).

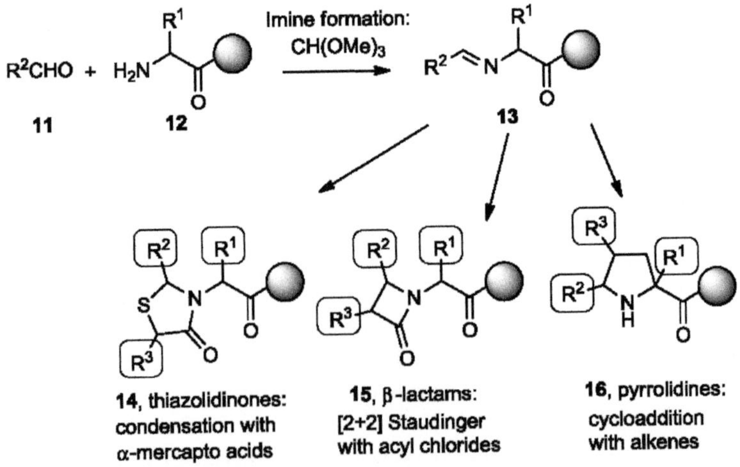

SCHEME 6.3 Skeletal diversity from the chemistry of imines derived from resin-bound amino acids.

6.4 AMINO ACID–DERIVED SCAFFOLDS

A special interest in the development of suitable peptidomimetics for targeting peptide–protein and protein–protein interactions has been focused on the use of amino acids as building blocks. Side chains of natural amino acids are of great importance for the activity of proteins, due to their functional group diversity, which allows them to interact with other peptides or proteins; thus, amino acids and their derivatives have been taken into account as building blocks to enable the generation of peptidomimetic scaffolds possessing side-chain conformational restriction. Interest in the use of easily accessible and versatile proteinogenic amino acids as a chiral pool for synthesis of optically active heterocycles has been growing rapidly [16] because of their versatility in the panorama of chemical transformations and for their chemical diversity. Many reports focused on the generation of cyclic scaffolds using proline derivatives by virtue of the conformational restriction of the pyrrolidine ring and took advantage of the chemical and stereochemical diversity around this scaffold. Also, a significant contribution to the generation of amino acid–derived peptidomimetic scaffolds came out of the exploitation of the chemistry of amino aldehydes, particularly as related to iminium chemistry, and the inclusion of α-amino acid derivatives into cyclic scaffolds.

6.4.1 Scaffolds from α-Amino Acids

Nefzi et al. reported the application of α-amino acid–derived ethylenediamines on solid phase to achieve an array of cyclic scaffolds differing by ring size and functional groups [17]. The starting diamine deriving from N-acylated amino acid was treated with an array of reactants such as carbonyldiimidazole, thiocarbonyldiimidazole, and cyanogen bromide to give the corresponding five-membered ring imidazolidin-2-one, imidazolidine-2-thione, and imidazolidin-2-imine. Six-membered heterocycles consisted of diketopiperazines from the reaction with oxalyldiimidazole, and seven-membered ring diazepinediones were achieved from the reaction with malonyl dichloride (Scheme 6.4).

Recent applications of α-amino acids as building blocks for the diversity-oriented synthesis of heterocyclic scaffolds were reported by Mishra and Panda. A first report described the synthesis of benzannulated medium-ring heterocyclic scaffolds employing inter- and intramolecular Mitsunobu reactions in enantiomerically pure form [18].

Seven- and eight-membered rings bearing nitrogen, oxygen, and sulfur atoms were synthesized following the Mitsunobu approach for the formation of the carbon–heteroatom bond as a convenient route to the synthesis of benzannulated medium-ring heterocycles in an optically pure form, containing the naturally occurring α-amino acids (Scheme 6.5). The general synthetic strategy consisted of two Mitsunobu reactions to install the amino acid moiety on the aromatic scaffold. The intermolecular Mitsunobu reaction on the amino or carboxylic end was followed by functional group interconversions enabling a second intramolecular Mitsunobu reaction to achieve the final bicyclic scaffolds. This simple protocol is quite straightforward and may give

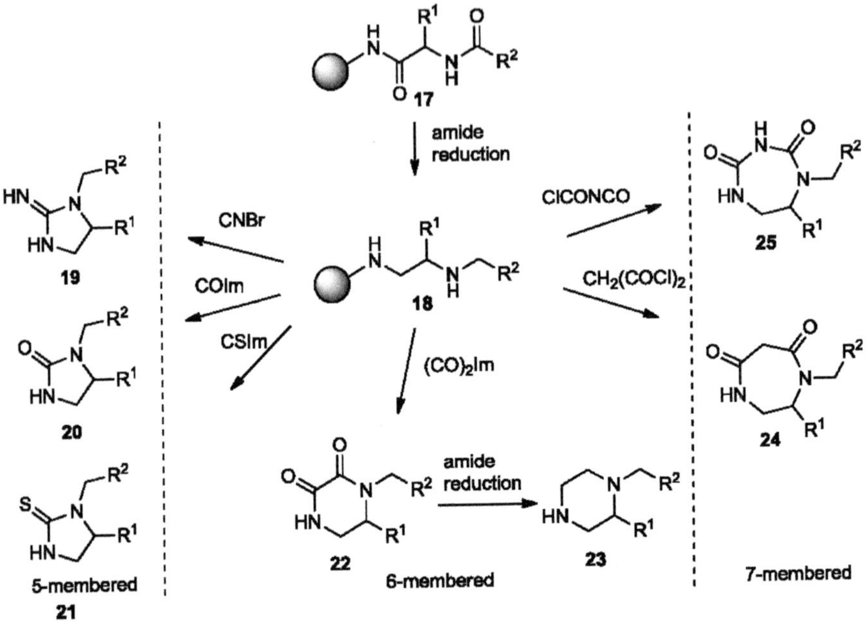

SCHEME 6.4 Skeletal diversity from α-amino acid–derived ethylenediamines on solid phase.

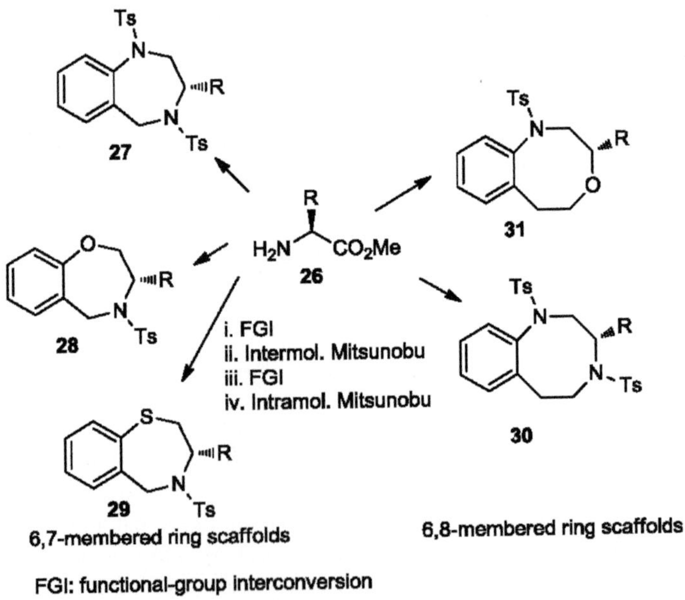

SCHEME 6.5 Synthesis of benzannulated medium-ring heterocycles containing α-amino acid moieties through a Mitsunobu approach for formation of the carbon–heteroatom bond.

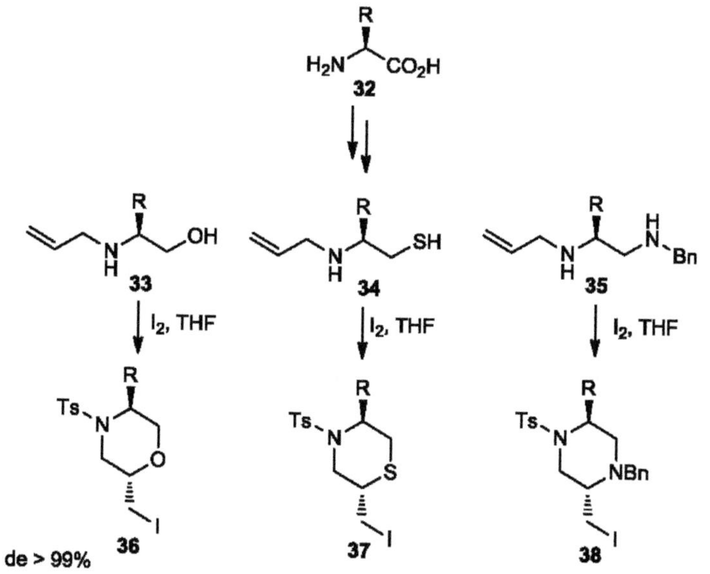

SCHEME 6.6 Synthetic approach to diastereoselective 2,5-disubstituted diverse morpholines, piperazines, and thiomorpholines from α-amino acid derivatives and molecular iodine.

access to further skeletal diversity by extending the process to other unnatural amino acid–based building blocks.

In a recent paper, Bera and Panda described the synthesis of disubstituted six-membered ring heterocycles taking advantage of the chemistry of alkenes with I_2 [19]. This simple and powerful synthetic route provided access to diastereoselective 2,5-disubstituted diverse morpholines, piperazines, and thiomorpholine starting from commercially available α-amino acid–derived synthetic intermediates (Scheme 6.6). The key step involved iodine-mediated cyclization under mild reaction conditions to give six-membered ring heterocycles. Such compounds may be elaborated further in several ways, such as by nucleophilic substitution on the rings as well as by incorporating a substituent at the 5-position from amino acids constituents.

6.4.2 Scaffolds Containing the Pyrrolidine Ring

Proline has a special place as a template for the generation of peptidomimetic scaffolds because as the cyclic proteinogenic amino acid, it has a role in influencing the biological behavior of peptides due to the secondary structure's inducing and stabilizing properties. In fact, the cyclic structure of proline is capable of restricting the conformational space of the peptide chain, and many papers have appeared in the literature focusing on the strategic decoration of the pyrrolidine ring to generate suitable proline analogs for medicinal chemistry applications [20]. Also, early reports on combinatorial approaches described the generation of collections of proline-containing

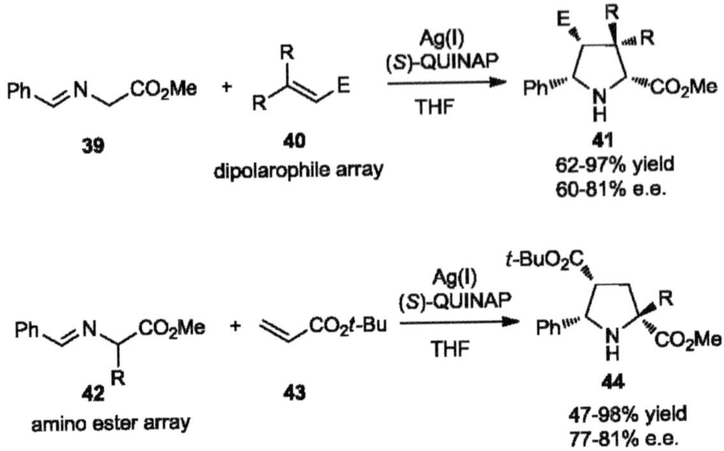

SCHEME 6.7 Enantioselective 1,3-dipolar cycloadditions of azomethine ylides with electron-deficient olefins in the divergent pathways for the generation of highly substituted proline analogs.

scaffolds [21]. Thus, in view of expanding the array of peptidomimetics according to DOS principles, proline is a privileged template as the starting point for the construction of skeletal diversity of complex scaffolds from the chiral pool or taking advantage of asymmetric processes and the extensive amount of chemistry focusing on the generation of pyrrolidine-based scaffolds.

Chen et al. described a procedure for the synthesis of highly substituted proline analogs, taking advantage of the use of enantioselective 1,3-dipolar cycloadditions of azomethine ylides with electron-deficient olefins in the divergent pathways of diversity-oriented synthesis [22]. The stereospecificity of the reaction enabled stereochemical diversification of up to four tetrahedral centers on pyrrolidine rings, using a silver(I) acetate/QUINAP catalyst system as being most effective for the reaction of azomethine ylides derived from benzaldehyde and several α-amino esters, with 3 mol% catalyst loading of various dipolarophiles (Scheme 6.7)

A very interesting entry to peptidomimetic scaffolds containing the proline ring according to DOS principles was recently reported by Hung et al. [23]. In this paper, a build/couple/pair (B/C/P) strategy was devised to achieve small fused bicyclic and spirocyclic three-dimensional fragments consisting of nonaromatic, high-sp^3 content, and chiral fragments. The pathways were constructed to generate small molecules according to the *fragment rule of 3* [24], which is a structural criterion for molecular compounds to limit the molecular weight to less than 300 g/mol and to allow for a maximum of three hydrogen-bond donors or acceptors and a maximum of three rotatable bonds.

The strategy was exemplified by selecting three proline analogs as the building blocks and applying coupling reactions with alkene derivatives to access the final

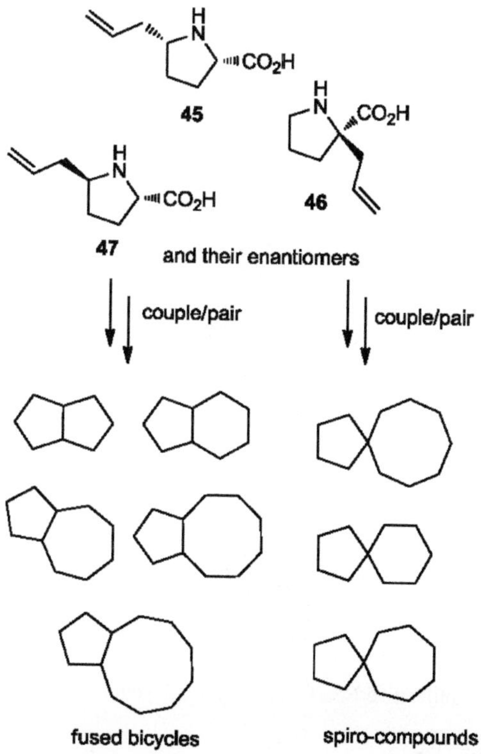

SCHEME 6.8 Selected proline analogs as building blocks for the generation of skeletal diversity through RCM reaction in the B/C/P approach.

pairing phase with ring-closing metathesis (RCM) reactions (Schemes 6.8 and 6.9). The B/C/P strategy devised allowed for the generation of a small library of fragments possessing diverse skeletal structures and rich in stereochemical variation, thanks to the use of different stereoisomers as starting building blocks for synthesis.

6.4.3 Scaffolds from Amino Aldehyde Intermediates

The field of peptidomimetics has taken advantage of iminium chemistry for the generation of suitable scaffolds possessing amino acid moieties, taking advantage of the amino group and the carbonyl resulting from manipulation of the carboxylic acid of the amino acids. In particular, resin-bound amino acids have attracted considerable attention for their use in synthetic methods consisting of functionalization of the amine moiety, followed by cyclative cleavage to access the target scaffolds.

Accordingly, an interesting paper by Scott et al. described an entry to skeletal diversity by exploiting the chemistry of resin-bound amino aldehydes deriving from α-amino acids [25]. In this paper, the authors reported the preparation, via solid-phase

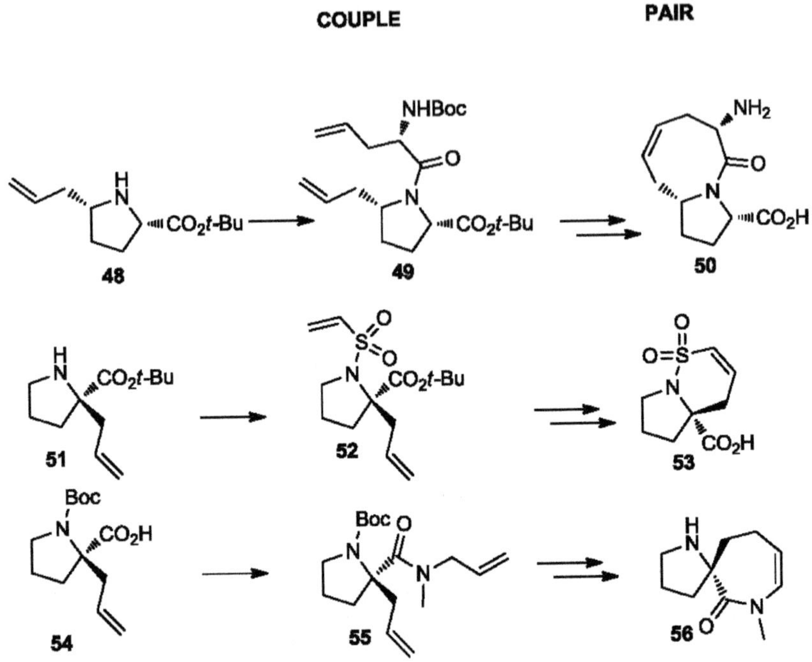

SCHEME 6.9 Exemplificative B/C/P approach to bicyclic scaffolds using selected proline analogs.

synthesis, of a wide variety of highly substituted lactam-containing peptidomimetic scaffolds through the synthesis and use of a remarkably stable and versatile class of resin-bound aldehyde intermediates. The amino aldehyde was easily prepared by ozonolysis of a α-amino acid containing an allyl group as the side chain. The skeletal diversity coming from this approach was demonstrated by preparing a representative series of five peptidomimetic scaffolds, consisting of lactones, lactams, bicyclics, tricyclics, and tetracyclics (Scheme 6.10). In every case, the release of final products from resin is by a cyclative cleavage process, which also removed the need for resin-linker handles that remained on the products. As representative examples, five-membered lactams were formed by reductive amination, followed by cyclative cleavage to form the amide bond. Similarly, cysteine was used to react with the aldehyde group to produce a thiazolidine ring, followed by generation of the final bicyclic scaffold by cyclative cleavage to generate the amide bond (Scheme 6.11).

Similarly, García-Cuadrado et al. reported a paper on the generation of skeletal diversity using the chemistry of amino aldehydes according to a couple/cyclization process, and leading to diverse heterocycles in four steps containing up to five points of diversity [26]. Specifically, a concise synthetic strategy consisting of a Petasis three-component reaction followed by a tandem aza-Cope–Mannich cyclization was developed. Alternative iminium cyclization based on a Pictet–Spengler reaction or aminal

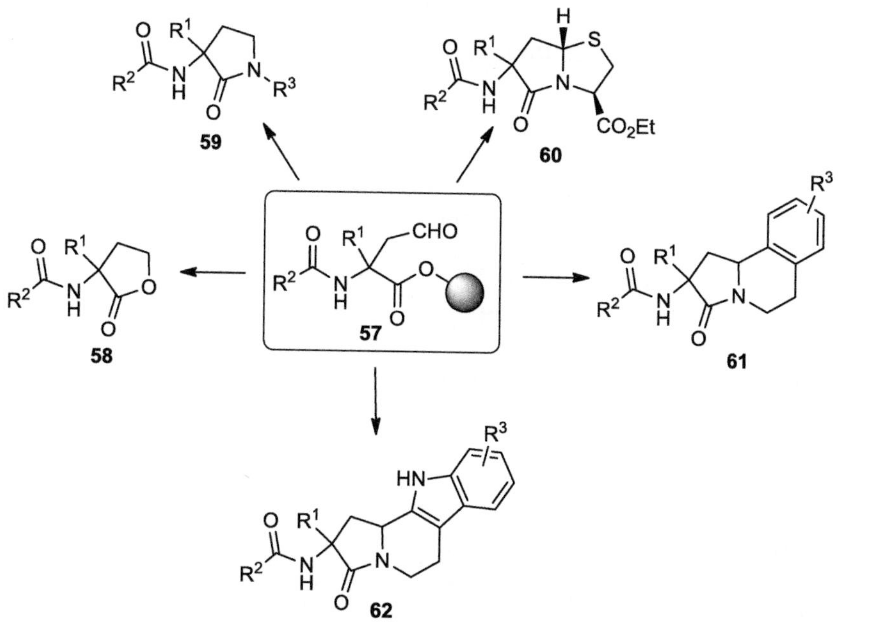

SCHEME 6.10 Highly substituted peptidomimetic scaffolds containing the lactam moiety through the application of resin-bound aldehyde intermediates.

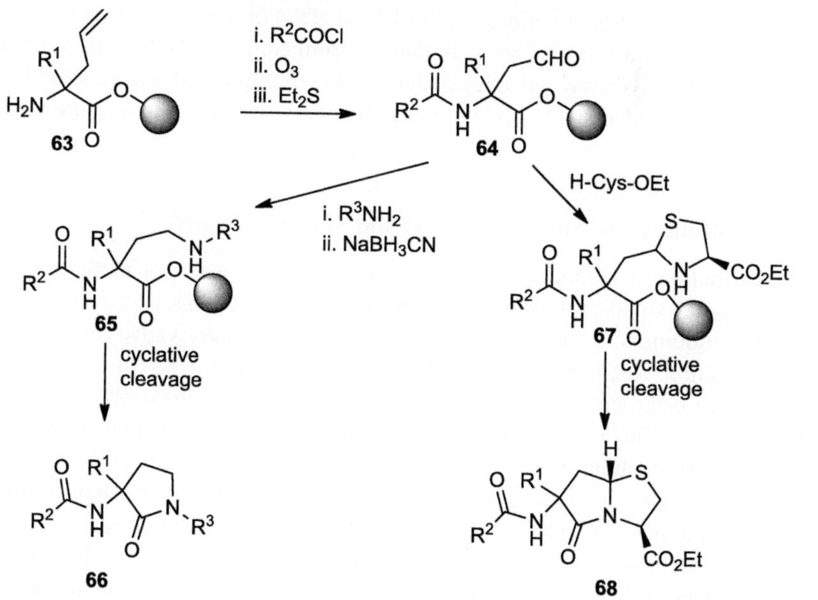

SCHEME 6.11 Representative examples of peptidomimetic scaffolds obtained from resin-bound α-amino aldehydes through a cyclative cleavage to form the lactam ring.

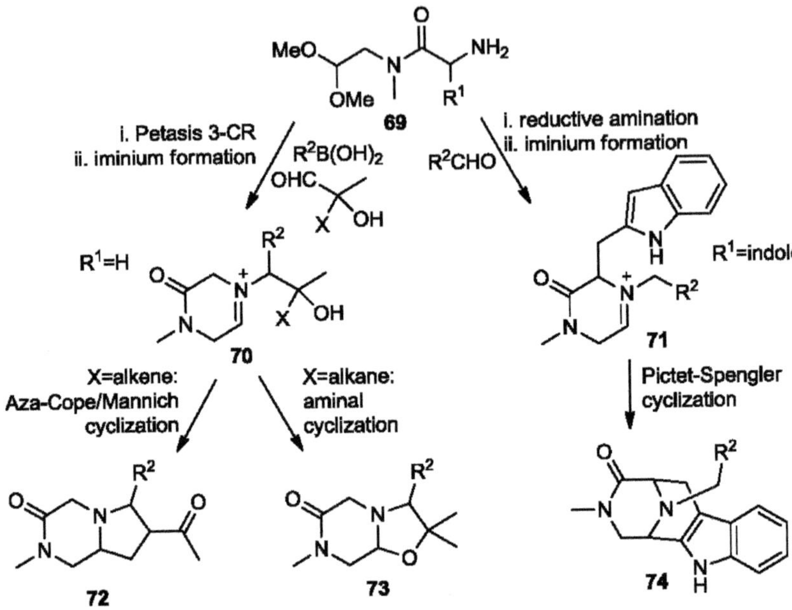

SCHEME 6.12 Skeletal diversity using the chemistry of amino aldehydes according to a couple/cyclization process.

formation led to divergent pathways affording skeletal diversity (Scheme 6.12). The starting materials for this process were derived from amino acids, boronic acids, and readily available hydroxyl aldehydes, and the use of supported reagents facilitated isolation of the products, thus allowing this chemistry to be open to high-throughput synthesis.

6.4.4 Scaffolds from Amino Carbonyl and Sugar Derivatives

The contribution of Guarna and Trabocchi to the development of methods for the identification of small-molecule modulators of protein function focused on the generation of peptidomimetic libraries from amino acid and sugar derivatives in a diversity-oriented fashion. Accordingly, the chemistry of amino carbonyls was applied in the generation of heterocycles, taking advantage of amino acid and sugar derivatives as building blocks. This is a powerful approach to accessing chemical and geometrical diversity, given the high number of stereocenters and the multifunctionality of such compounds, as also demonstrated by literature reports on the use of such substrates to access combinatorial libraries of heterocyclic compounds [27].

We envisaged the application of selected building blocks from the chiral pool to develop a DOS strategy to access constrained and stereochemically-rich morpholine-based heterocycles according to the build/couple/pair approach. The morpholine ring is of special interest, as it represents a common motif among the various structures employed by medicinal chemists [28]. In fact, the morpholine moiety is embedded in

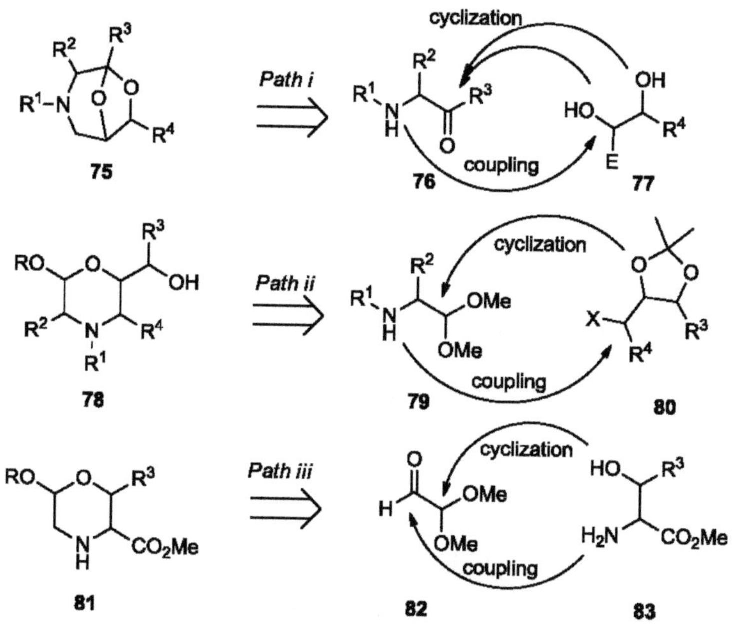

SCHEME 6.13 Diversity pathways to morpholine scaffolds through the application of α-amino acid derivatives to a couple/pair approach. (From [34], with permission of John Wiley & Sons; copyright © 2009 John Wiley & Sons.)

several bioactive molecules, such as MMP and TNF inhibitors [29], and is found in the structure of tricyclic benzodiazepines [30] and of 8,6-fused bicyclic peptidomimetic compounds such as interleukin-1β-converting enzyme inhibitors [31]. The strategic approach to the generation of morpholine scaffolds encompassed the application of α-amino acid derivatives such as amino carbonyls (path i) and amino acetals (path ii), or using β-amino alcohols (path iii) to be coupled with suitable building blocks possessing complementary functionalities to achieve structurally diverse heterocycles after the cyclization step (Scheme 6.13).

6.4.4.1 Bicyclic Peptidomimetics During the last decade we reported several papers on the generation of bicyclic compounds from a combination of sugar or tartaric acid and amino acid derivatives. The synthetic process consisted of two key couple/pair steps: the coupling of two components from the chiral pool, followed by an intramolecular cyclization to achieve the bicyclic structure [32].

This class of bicyclic compounds showed structural similarity to dipeptides, allowing for their application as suitable molecular platforms for peptidomimetic chemistry. Morever, a wide array of molecules bearing such a bicyclic moiety was achieved with skeletal and stereochemical diversity by tuning the starting compounds from the chiral pool and the cyclization process, and was obtained as readily available constrained amino acids for insertion in peptide structures (Figure 6.2).

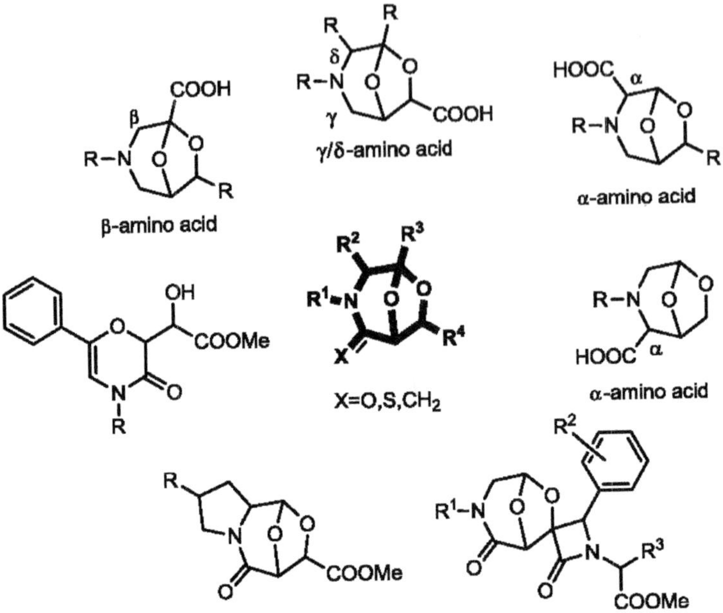

FIGURE 6.2 Skeletal and stereochemical diversity to bicyclic compounds by tuning the starting compounds from the chiral pool and the cyclization process.

A collection of such constrained dipeptide isosteres were used as probes to dissect their effects on a panel of *Saccharomyces cerevisiae* strains as a model to screen the activity of these compounds toward cell growth [33]. Specifically, we conceived the screening of wild-type strains and selected deletants involved in multidrug resistance using cell growth as the phenotype of study. This selection process through phenotypic screening on yeast strains enabled the selection of library member **089**, which was identified as the molecule inducing the most intense cell growth decrease.

6.4.4.2 DOS of Morpholine-Containing Scaffolds In a second application of a library of peptidomimetics in chemical genetics studies, we generated a new set of morpholine-based compounds to be screened toward yeast strains. Following the diversity-oriented synthesis of morpholine scaffolds, and taking advantage of a two-step process involving the combination of amino acid derivatives as building blocks, a high degree of chemical diversity was achieved around such heterocycles. Specifically, the reactivity of the acetal moiety and the amino and carboxylic groups of the parent amino acid component was exploited with a series of acid-mediated reactions (Scheme 6.14) [34]. Specifically, constrained mono-, bi-, and tricyclic scaffolds were obtained from a linear densely functionalized intermediate taking advantage of lactone or lactam formation, and of trans-acetalization reactions.

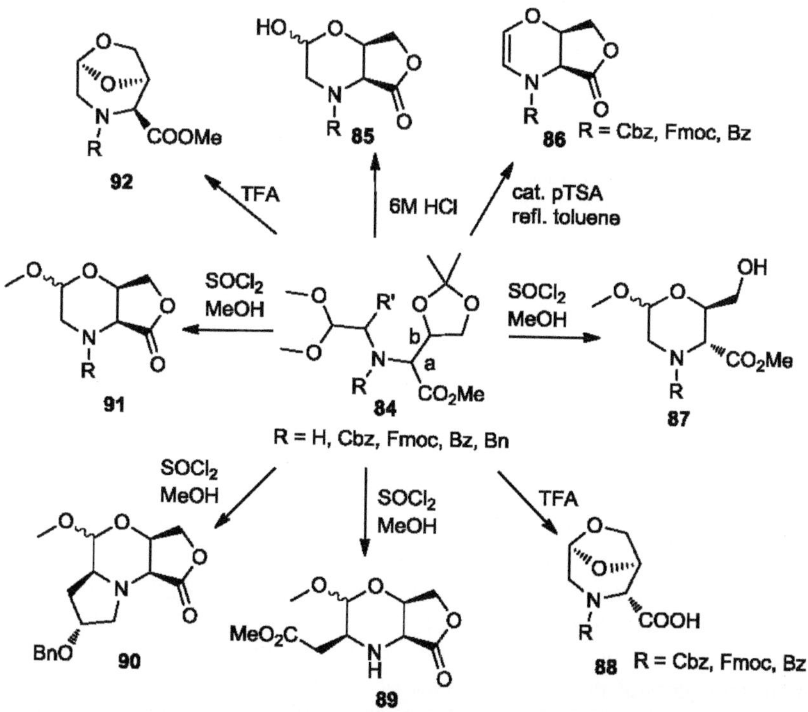

SCHEME 6.14 Chemical diversity of morpholine-based scaffolds through the reactivity of the acetal moiety and the amino and carboxylic groups of the parent amino acid component with a series of acid-mediated reactions. (From [34], with permission of John Wiley & Sons; copyright © 2009 John Wiley & Sons.)

We applied the reactivity of some morpholine acetals to give access to a library of bicyclic scaffolds containing either the ketopiperazine or the 2,5-diketopiperazine moieties, which are considered privileged scaffolds in medicinal chemistry, owing to the wide number of bioactive natural products and drugs containing such chemical entities. Such a library of new chemical entities was employed in the systematic exploration of yeast deletants to classify deletant strains as a function of bioactive molecules. According to the synthetic strategy, skeletal diversity was accessed by applying morpholine acetals in a one-pot multistep process giving diverse bicyclic scaffolds as a function of the reaction methodology; the one-pot process, consisting of a couple/cyclization sequence, enabled the generation of 2,5-diketopiperazine-containing bicyclic compounds as a consequence of intramolecular amide bond formation following Fmoc deprotection. The stepwise route, also consisting of a intermediate workup step, produced a different bicyclic compound as a consequence of the reactivity of the acetal moiety with the Fmoc-protected amino group, followed by rearrangement [35] (Scheme 6.15).

The effects of 48 morpholine-based library members on cell growth were tested in a phenotypic screening toward *S. cerevisiae* at 0.3 mM concentration on the BY4742 wild-type strain. Compounds responsible for induction variations in cell generation

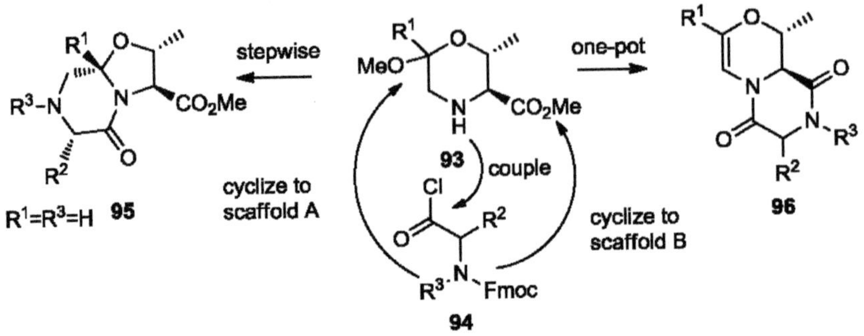

SCHEME 6.15 Stepwise and one-pot processes to address skeletal diversity around the starting morpholine ring. (From [35], with permission; copyright © 2010 Elsevier.)

time, which is the time interval required for a yeast cell to divide, or in cell growth at the stationary phase, or both, were selected for further characterization [36].

Screening of the wild-type strain allowed for the selection of 21 molecules inducing a phenotypic effect higher than 10% or other interesting effects (e.g., a decrease in yeast fitness). In a second round of phenotype screening, eight molecules were selected and tested for mitochondrial membrane potential activation and peroxisomal proliferation. Both mitochondrial membrane activation and peroxisomal proliferation are indicators of the metabolic state of a cell, disclosing respiratory and fatty acid metabolism, respectively, and ultimately resulting in the identification of new chemotypes involved in mitochondria metabolism and respiration events.

6.5 MACROCYCLIC PEPTIDOMIMETIC SCAFFOLDS

The field of macrocycles in diversity-oriented synthesis is emerging, and it is a very promising research area in addressing the need for peptidomimetic compounds targeting extended protein–protein interactions (see Chapter 8 for a comprehensive review of the topic). In a series of publications, Wessjohann and Ruijter described the successful generation of macrocyclic peptide-based structures taking advantage of multicomponent reactions to address both chemical diversity and macrocyclization [37]. Specifically, they applied an expanded strategy using the potential of multicomponent reactions not only for diversity generation, but also for the macrocyclization reaction. This approach allowed for building a complex macrocycle in one step from simple precursors, taking advantage of at least one bifunctional building block as the main requirement.

In case of a single Ugi reaction forming the ring, two of the four R groups were replaced by a tether joining two functionalities. This approach required three components to achieve macrocyclization through a single Ugi reaction, one being an unsymmetrically bifunctional building block (see Scheme 6.16). When employing more than one Ugi reaction to construct macrocycles, several bifunctional building blocks were required to carry out multiple-multicomponent macrocyclizations.

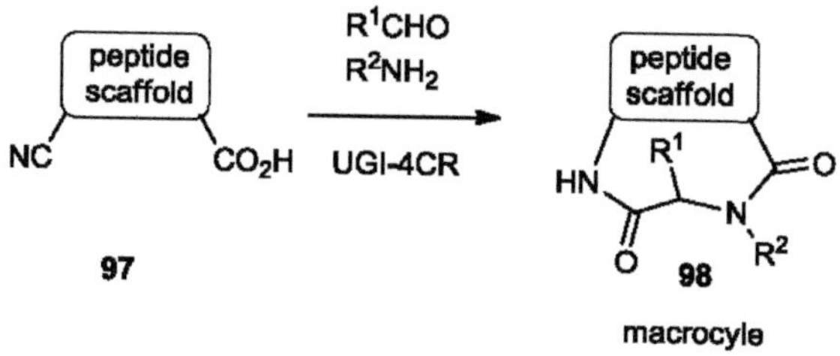

SCHEME 6.16 Ugi 4CR to achieve macrocycles through bifunctional peptidomimetic building blocks.

Several subtypes of such reactions conceived the diversity-oriented character of this synthetic approach by varying both the type and the number of MCRs and components involved, and changing the directionality of the amide bonds formed as well. In this paper, the authors reported the Ugi 4CR to form the macrocycle desired in either a uni- or a bidirectional way, which was referred to as having the resulting dipeptide moieties running in the same (parallel) or the opposite (counter) direction (the N- to C-terminal direction), respectively (Figure 6.3).

The generation of peptide-like macrocycles via multiple multicomponent macrocyclizations using bifunctional building blocks allowed for achieving macrocycles with diverse properties, such as ring size, flexibility, rigidity, core size, shape, or polarity distribution. The same approach was applied successively to hybrid macrocycles containing steroid–peptoid moieties [38]. This paper proposed the generation

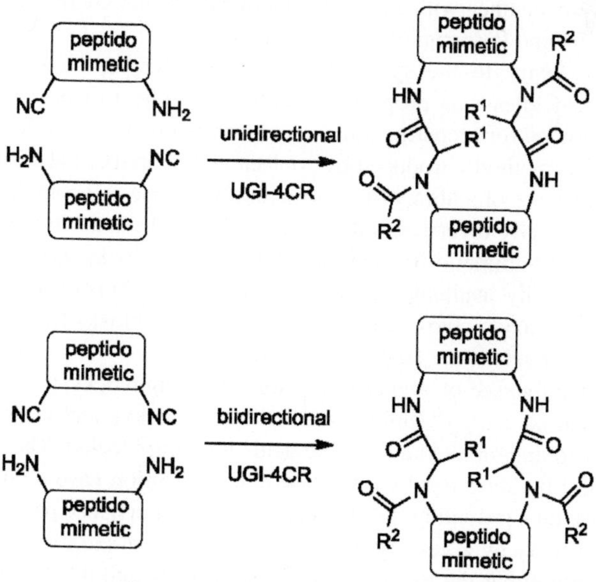

FIGURE 6.3 Uni- and bidirectional approaches to macrocyclization through Ugi 4CR.

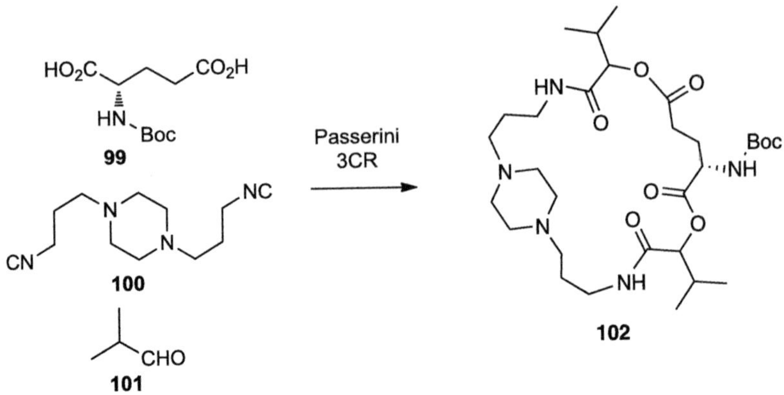

SCHEME 6.17 Classic Passerini-3CR approach to peptidomimetic macrocycles.

of hybrid systems as an entry toward a nonrepetitive type of macrocycle resembling the structural diversity of those naturally occurring, which are usually endowed with unique structural complexity. Specifically, the straightforward strategy was based on multicomponent reactions to generate a collection of very large chimeric peptoid macrocycles, specifically containing steroid moieties in this report.

Successively, Leon et al. reported additional examples of multicomponent macrocyclizations using both Staudinger and Passerini three-component reactions [39]. In a specific report, the versatility and wide scope of the macrocylization methodology were shown to achieve complex macrocyclic structures in one pot and in a straightforward manner with multiple functional groups encompassing multiple Staudinger-3CRs and Passerini-3CRs (Scheme 6.17).

A very nice entry to the synthesis of macrocyclic peptidomimetics according to DOS principles was proposed recently by Isidro-Llobet et al. [40], who reported the strategy for preparing amino acid–containing macrocycles based on the build/couple/pair method introduced by Nielsen and Schreiber [41]. The study consisted of the synthesis of a library of 14 structurally diverse and complex compounds containing structurally diverse small molecules based around biologically relevant macrocyclic peptidomimetic frameworks. All building blocks were prepared from simple, commercially available amino acid derivatives. Specifically, azido-amine building blocks, α-azido acids and alkyne-acid building blocks from phenylalanine and lysine, respectively, were taken into account.

The build step focused on two types of chiral building blocks: the first containing a free amine and an azide (“azido-amine” building blocks), and the second having a carboxylic acid and an alkyne (“alkyne-acid” building blocks). The coupling step of three of these building blocks via amide bond formation gave a range of tripeptide derivatives, thus providing the stereochemical diversity. The pair phase allowed for scaffold diversity by accessing two cyclization steps in the formation of macrocycles containing either the triazole or both the triazole and diketopiperazine rings (Scheme 6.18).

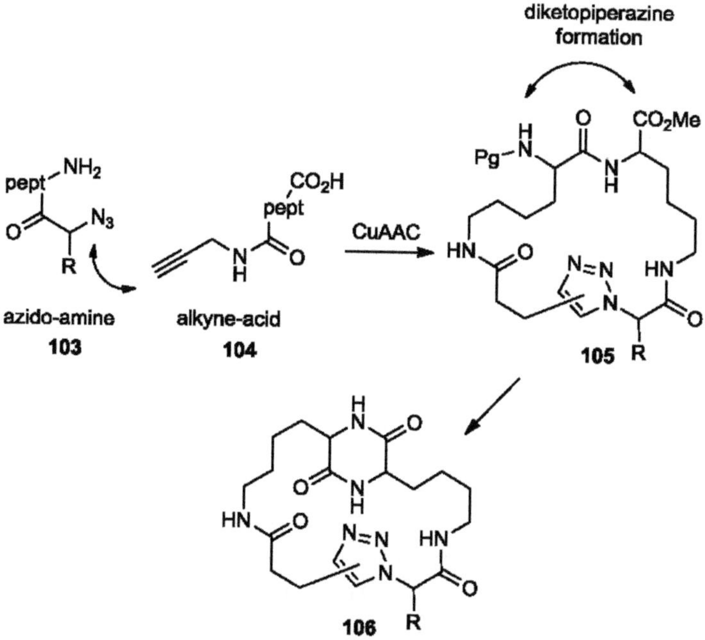

SCHEME 6.18 Couple/pair approach to macrocyclic peptidomimetics using azido-amines and alkyne-acids for the CuAAC process.

The major contribution to stereochemical and scaffold diversity was recognized to be from the build phase, taking advantage of the wide variety of natural and unnatural α-, β-, and γ-amino acids amenable for the build step of the synthesis. Using principal components analysis (PCA), the authors demonstrated that the library of the 12 synthesized analogs covered a distinct and broad chemical space compared to a set of 657 approved drugs.

6.6 CONCLUSIONS

The combinatorial concept is evolving to diversity-oriented synthesis (DOS) with the aim of expanding the chemical diversity around the molecular scaffolds and addressing structural complexity, in addition to the appendage diversity on a fixed cyclic scaffold traditionally taken into account in classic combinatorial chemistry. Peptidomimetics may represent a privileged set of compounds, due to the importance of peptide–protein and protein–protein in biological processes, and the diversity-oriented synthesis of small-molecule peptidomimetics may give access to collections of new chemotypes bearing a high level of structural diversity. Accordingly, recent papers on DOS approaches in the field of peptidomimetics have appeared in the literature, suggesting an increasing interest in developing chemical diversity around the amino acid scaffolds and peptidomimetics. This is of primary importance in view of addressing "undruggable" protein–protein interactions that have been identified as key players in cell cycle events or of a still unknown function, thus paving the way toward the development of new therapeutic strategies for relevant diseases. Taking advantage of peptidomimetics as privileged structures in medicinal chemistry, the combination of interests in the fields of DOS and peptidomimetics may open a new avenue toward important molecular probes and hit compounds for biomedical applications.

REFERENCES

1. A. Giannis, T. Kolter, *Angew. Chem. Int. Ed.* **1993**, *32*, 1244–1267.
2. J. Gante, *Angew. Chem. Int. Ed.* **1994**, *33*, 1699–1720.
3. C. Adessi, C. Soto, *Curr. Med. Chem.* **2002**, *9*, 963–978.
4. A. Grauer, B. König, *Eur. J. Org. Chem.* **2012**, 5099–5111.
5. R. M. J. Liskamp, *Recl. Trav. Chim. Pays-Bas* **1994**, *113*, 1–19.
6. G. L. Olson, D. R. Bolin, M. P. Bonner, M. Bös, C. M. Cook, D. C. Fry, B. J. Graves, M. Hatada, D. E. Hill, M. Kahn, V. S. Madison, V. K. Rusiecki, R. Sarabu, J. Sepinwall, G. P. Vincent, M. E. Voss, *J. Med. Chem.* **1993**, *36*, 3039–3049.
7. C. T. Dooley, R. A. Houghten, *Biopolymers*, **1999**, *51*, 379–390.
8. P. S. Farmer, in: *Drug Design*, E. J. Ariëns, Ed., Academic Press, New York, **1980**, Vol. X, pp. 119–143.
9. This concept has been referred to a *hierarchical approach to peptidomimetic design*; see G. R. Marshall, *Tetrahedron* **1993**, *49*, 3547–3558.
10. A. S. Ripka, D. H. Rich, *Curr. Opin. Chem. Biol.* **1998**, *2*, 441–452.
11. (a) M. E. Houston, Jr., A. Wallace, E. Bianchi, A. Pessi, R. S. Hodges, *J. Mol. Biol.* **1996**, *262*, 270–282; (b) J. Jiráček, B. Vincent, F. Checler, V. Dive, *J. Biol. Chem.* **1996**, *271*, 19606–19611; (c) K. Müller, F. O. Gombert, U. Manning, F. Grossmüller, P. Graff, H. Zaegler, J. F. Zuber, F. Freuler, C. Tschopp, G. Baumann, *J. Biol. Chem.* **1996**, *271*, 16500–16505; (d) B. Dörner, G. M. Husar, J. M. Ostresh, R. A. Houghten, *Bioorg. Med. Chem.* **1996**, *4*, 709–715; (e) A. Rockwell, M. Melden, R. A. Copeland, C. P. Decicco, W. F. De Grado, *J. Am. Chem. Soc.* **1996**, *118*, 10337–10338.
12. (a) P. A. Bartlett, C. K. Marlowe, *Biochemistry* **1987**, *26*, 8553–8561; (b) B. P. Morgan, J. M. Scholtz, M. D. Ballinger, I. D. Zipkin, P. A. Bartlett, *J. Am. Chem. Soc.* **1991**, *113*, 297–307; (c) D. S. Karanewsky, M. C. Badia, S. W. Cushman, J. M. DeForrest, T. Dejneka, M. J. Loots, M. G. Perri, E. W. Petrillo, J. R. Powell, *J. Med. Chem.* **1988**, *31*, 204–212.
13. (a) B. A. Bunin, J. A. Ellman, *J. Am. Chem. Soc.* **1992**, *114*, 10997–10998; (b) M. J. Plunkett, J. A. Ellman, *J. Am. Chem. Soc.* **1995**, *117*, 3306–3307.
14. C. Haskell-Luevano, A. Rosenquist, A. Souers, K. C. Khong, J. A. Ellman, R. D. Cone, *J. Med. Chem.* **1999**, *42*, 4380–4387.
15. E. M. Gordon, M. A. Gallop, D. V. Patel, *Acc. Chem. Res.* **1996**, *29*, 144–154.
16. For a review on amino acid–derived heterocycles, see F. J. Sardina, H. Rapport, *Chem. Rev.* **1996**, *96*, 1825–1872.

17. A. Nefzi, J. M. Ostresh, J. Yu, R. A. Houghten, *J. Org. Chem.* **2004**, *69*, 3603–3609.
18. J. K. Mishra, G. Panda, *J. Comb. Chem.* **2007**, *9*, 321–338.
19. S. Bera, G. Panda, *ACS Comb. Sci.* **2012**, *14*, 1–4.
20. For a review, see M. I. Calaza, C. Cativiela, *Eur. J. Org. Chem.* **2008**, 3427–3448.
21. (a) A. Bianco, C. P. Sonksen, P. Roepstorff, J. P. Briand, *J. Org. Chem.* **2000**, *65*, 2179–2187; (b) A. L. Vergnon, R. S. Pottorf, M. R. Player, *J. Comb. Chem.* **2004**, *6*, 91–98.
22. C. Chen, X. Li, S. L. Schreiber, *J. Am. Chem. Soc.* **2003**, *125*, 10174–10175.
23. A. W. Hung, A. Ramek, Y. Wang, T. Kaya, J. A. Wilson, P. A. Clemons, D. W. Young, *Proc. Natl. Acad. Sci. U.S.A.* **2011**, *108*, 6799–6804.
24. M. Congreve, R. Carr, C. Murray, H. Jhoti, *Drug Discov. Today* **2003**, *8*, 876–877.
25. W. L. Scott, J. G. Martynow, J. C. Huffman, M. J. O'Donnell, *J. Am. Chem. Soc.* **2007**, *129*, 7077–7088.
26. D. García-Cuadrado, S. Barluenga, N. Winssinger, *Chem. Commun. 2008*, 4619–4621.
27. (a) M. Ghosh, R. G. Dulina, R. Kakarla, M. J. Sofia, *J. Org. Chem.* **2000**, *65*, 8387–8390; (b) T. M. Chapman, S. Courtney, P. Hay, B. G. Davis, *Chem. Eur. J.* **2003**, *9*, 3397–3414; (c) J. G. Lewis, P. A. Bartlett, *J. Comb. Chem.* **2003**, *5*, 278–284; (d) R. P. Trump, P. A. Bartlett, *J. Comb. Chem.* **2003**, *5*, 285–291; (e) A. Nefzi, J. M. Ostresh, J. Yu, R. A. Houghten, *J. Org. Chem.* **2004**, *69*, 3603–3609; (f) T. E. Nielsen, M. Meldal, *J. Comb. Chem.* **2005**, *7*, 599–610; (g) W. Meutermans, G. T. Le, B. Becker, *ChemMedChem* **2006**, *1*, 1164–1194; (h) G. Cervi, F. Peri, C. Battistini, C. Gennari, F. Nicotra, *Bioorg. Med. Chem.* **2006**, *14*, 3349–3367; (i) A. Trabocchi, D. Scarpi, A. Guarna, *Amino Acids* **2008**, *34*, 1–24.
28. R. Wijtmans, M. K. S. Vink, H. E. Schoemaker, F. L. van Delft, R. H. Blaauw, F. P. J. T. Rutjes, *Synthesis* **2004**, 641–662.
29. N. G. Almstead, R. S. Bradley, S. Pikul, B. De, M. G. Natchus, Y. O. Taiwo, F. Gu, L. E. Williams, B. A. Hynd, M. J. Janusz, C. M. Dunaway, G. E. Mieling, *J. Med. Chem.* **1999**, *42*, 4547–4562.
30. J. M. Matthews, A. B. Dyatkin, M. Evangelisto, D. A. Gauthier, L. R. Hecker, W. J. Hoeckstra, F. Liu, B. L. Poulter, K. L. Sorgi, B. E. Maryanoff, *Tetrahedron: Asymmetry* **2004**, *15*, 1259–1267.
31. S. V. O'Neil, Y. Wang, M. C. Laufersweiler, K. A. Oppong, D. L. Soper, J. A. Wos, C. D. Ellis, M. W. Baize, G. K. Bosch, A. N. Fancher, W. Lu, M. K. Suchanek, R. L. Wang, B. De, T. P. Demuth, Jr., *Bioorg. Med. Chem. Lett.* **2005**, *15*, 5434–5438.
32. (a) A. Guarna, A. Guidi, F. Machetti, G. Menchi, E. G. Occhiato, D. Scarpi, S. Sisi, A. Trabocchi, *J. Org. Chem.* **1999**, *64*, 7347–7364; (b) A. Trabocchi, G. Menchi, F. Guarna, F. Machetti, D. Scarpi, A. Guarna, *Synlett* **2006**, 331–353.
33. I. Stefanini, A. Trabocchi, E. Marchi, A. Guarna, D. Cavalieri, *J. Biol. Chem.* **2010**, *285*, 23477–23485.
34. C. Lalli, A. Trabocchi, F. Sladojevich, G. Menchi, A. Guarna, *Chem. Eur. J.* **2009**, *15*, 7871–7875.
35. L. Ciofi, M. Morvillo, F. Sladojevich, A. Guarna, A. Trabocchi, *Tetrahedron Lett.* **2010**, *51*, 6282–6285.
36. A. Trabocchi, I. Stefanini, M. Morvillo, L. Ciofi, D. Cavalieri, A. Guarna, *Org. Biomol. Chem.* **2010**, *8*, 5552–5557.
37. L. A. Wessjohann, E. Ruijter, *Mol. Divers.* **2005**, *9*, 159–169.

TABLE 7.1 Factors Attributed to the Decision Process

Criteria	Solution Phase	Solid Phase
Reaction condition optimization	None or fast	*Critical and tedious*
Additional synthetic step	No	*Anchoring and release*
Excess of reagents	None or small	*High*
Limits on chemistry	No	*Some*
Analysis of intermediates	Simple	*Requires release*
Isolation of intermediates	*Time consuming*	Fast and simple
Integration/automation	*Complex*	Simple
Solvent removal	*Evaporation*	Washing (DMF, DMSO okay)
Compound loss due to handling	*Some*	None
Pooling of intermediates	*No*	Yes

thus benefiting from the effort to develop their synthesis (c.f. page 220). In the following pages we provide useful coordinates for those considering solid-phase chemistry a valid alternative to solution-based synthetic strategies.

We used diversity classification as suggested by Burke and Schreiber and divided our chapter into the following diversity sections: skeletons, stereochemistry, and appendages [1]. Skeletal diversity involves the formation of several discrete scaffolds and it was approached via either branching or folding processes (for more information, see Section 7.2). Stereochemical diversity represents different three-dimensional spatial arrangements of identical two-dimensional structures (different *S/R* orientation) and thus influences the behavior (e.g., binding, affinity) of a given compound. Finally, appendage diversity is achieved by incorporation of various substructures (e.g. heterocycles) on one core unit using several chemistries and building blocks. The build/couple/pair (B/C/P) [2] approach to chemical diversity includes all three categories mentioned above. In addition, chemical diversity can be achieved by a scaffold hopping process.

This chapter is devoted to diversity-oriented solid-phase synthesis with the primary emphasis on skeletal diversity. Several syntheses that qualify for both skeletal and stereochemical diversity are listed in the skeletal section; only a reference is given in the stereochemical section. One particular scaffold decorated with a set of various building blocks does not fulfill our criteria for diversity-oriented synthesis (DOS), even though the combination of several building blocks undoubtedly increases the diversity and thus induces an effect on the biological activity of the particular heterocycle.

There are a limited number of realistic chemical reactions; however, the diversity of compounds is practically unlimited. Thus, several reactions are recurrent throughout this chapter, some of them bearing an inherent propensity to serve as diversity creators. Among them, cycloadditions, including Diels–Alder reactions and 1,3-dipolar additions, and iminium chemistry with subsequent nucleophilic addition are transformations that are used frequently.

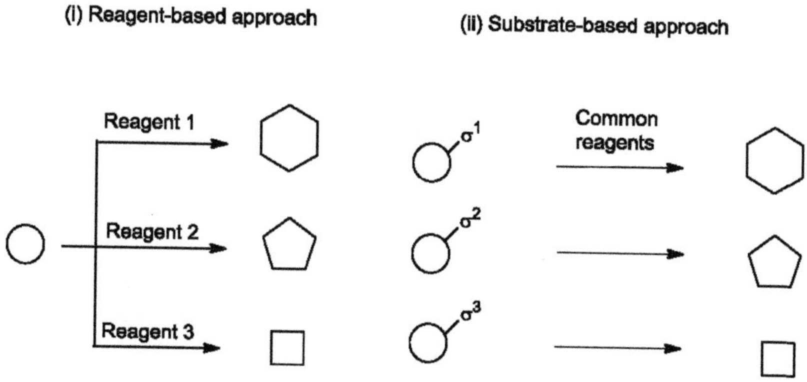

SCHEME 7.1 Two different approaches to generate skeletal diversity.

7.2 SKELETAL DIVERSITY

Skeletal diversity has generally been achieved by two different strategies: (1) a common intermediate was converted into diverse skeletons under different reaction conditions (referred to as a reagent-based approach or a differentiating or branching process), and (2) substrates bearing different appendages were prepared that preencoded skeletal information (called σ-elements), referred to as a substrate-based approach or folding process (Scheme 7.1) [1]. However, in numerous instances the approaches are combined, resulting in increased diversity of compounds. One older example of skeletal diversity is the synthesis of an α,β,γ-peptide library [3] that included β- and γ-peptide bonds in addition to traditional α-peptides. Side-chain amino groups of diaminopropionic and diaminobutyric acids formed the β- and γ-peptide bonds of the backbone.

7.2.1 Reagent-Based Strategy: Branching Process

As mentioned above, reagent-based strategy utilizes a common precursor that is transformed into diverse skeletons as a result of treatment with various reagents. Due to the diversity of structures and applied chemistries that fall into this category, we divided individual diversity routes according to the number of core structures and the type of heterocycle. Most, if not all, chemical routes reported describe the syntheses of heterocyclic compounds.

7.2.1.1 One Core Structure In this section we describe the DOS of heterocyclic compounds that share a core structure. This core molecule is converted into several different derivatives; however, all of them contain the same central core structure [4]. In medicinal chemistry, core molecules possessing a range of biological activities are termed *privileged structures*. The term was introduced by Evans et al. in 1988 [5]. Privileged structures are capable of binding to multiple unrelated classes of protein receptors as high-affinity ligands [5]. Medicinal chemists used this concept to design inhibitors of a given enzyme around a substructure (privileged structure), which is known to interact with specific binding sites [6]. In traditional medicinal chemistry a core structure is decorated with various groups at particular positions. Such chemistries are not designed to prepare diverse sets of compounds and are not discussed in this section. In the one-core-structure section, we include syntheses characterized by the formation of additional rings (fused or connected) that result in compounds with skeletal diversity (topography).

7.2.1.1.1 Oxygen Heterocycles Three examples of the DOS of natural product–like compounds embedded in benzopyran structures are described. The benzopyran moiety is contained in many natural products, such as flavonoids, coumarins, rotenoids, stilbenoids, and chromene glycosides [7]. The synthesis of coumarin-based libraries begins from tetrasubstituted 2,2-dimethylbenzopyran **2** attached to polystyrene-based selenenyl bromide resin (Scheme 7.2) [7–9]. Key intermediate **2** underwent different reactions, including annulation (path A), glycosidation (path B), condensation (path C), and coupling reactions (path D and E), to afford diverse natural product–based libraries containing seven different bi-, tri-, and tetracyclic scaffolds **3** to **9**.

Parallel diversity-oriented solid-phase synthesis leading to a library of compounds containing the benzopyran privileged structure has subsequently been reported [10,11]. A benzopyranyl intermediate is transformed into bi-, tri-, and tetracyclic benzopyran libraries. Syntheses involve treatment of either polystyrene resin–supported vinyl triflate **11** (Scheme 7.3) [10] or vinyl bromide **16** (Scheme 7.4) [11], with different reagents leading to four [10] and six [11] discrete core skeletons embedded with a benzopyran ring (Schemes 7.3 and 7.4). Chemical libraries contained 434 [10] and 284 [11] members, respectively.

Chemical transformation of vinyl triflate **11** involved four pathways, resulting in target scaffolds (Scheme 7.3) [10]. The first one was based on palladium-mediated Suzuki coupling leading to bicycle **12** (path A), and the second reaction route applied Stille-type vinylation followed by Diels–Alder reaction to yield tetracycle **13** (path B), which was either cleaved from the support or underwent further DDQ oxidation and resulted in product **14** (path C). Finally, Negishi-type alkynylation and subsequent Huisgen 1,3-dipolar [3 + 2] cycloaddition led to target triazoles **15** (path D).

Similar synthetic routes were also applied to polymer-supported vinyl bromide **16** (Scheme 7.4) [11]. Vinyl bromide **16** was converted into **17** by Suzuki coupling (path A). Suzuki coupling–based vinylation led to intermediate **18** (path B), which underwent Diels–Alder reaction with two different dienophiles to afford heterocyclic products **19** and **20** (path C). Tetracycle **20** was either isolated as a final compound or was subsequently oxidized to yield derivative **21** (path D). Finally, products **23** and **24** were prepared via copper(I)- and ruthenium(II)-catalyzed regioselective cycloaddition of compound **22** (path F), obtained by Stille coupling–based alkynylation (path E) with two different azides. All 10 prepared unique scaffolds [10,11] were visualized in chemical space using principal components analysis.

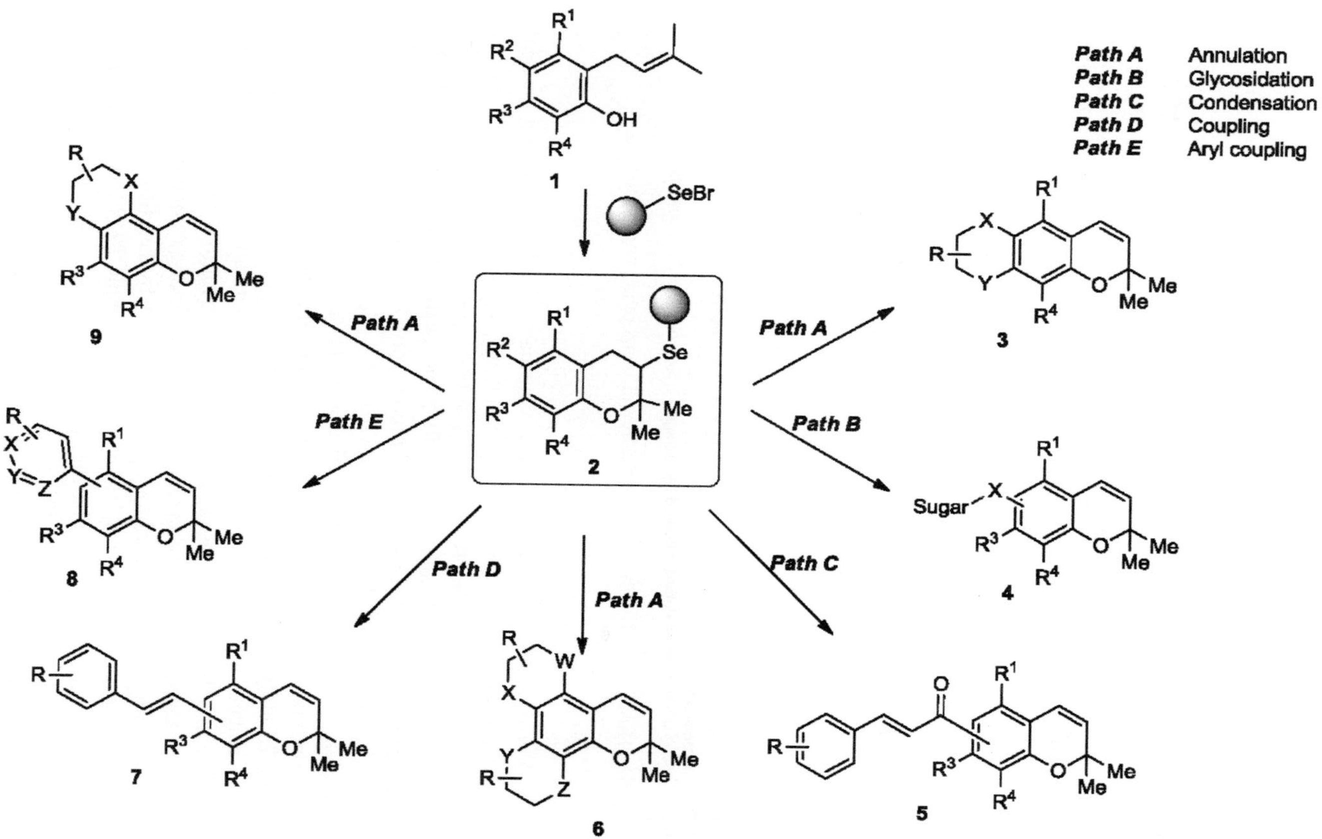

SCHEME 7.2 Synthesis of seven different skeletons containing a benzopyran core molecule. (From [7], with permission of John Wiley & Sons; copyright (©) 2000 John Wiley & Sons.)

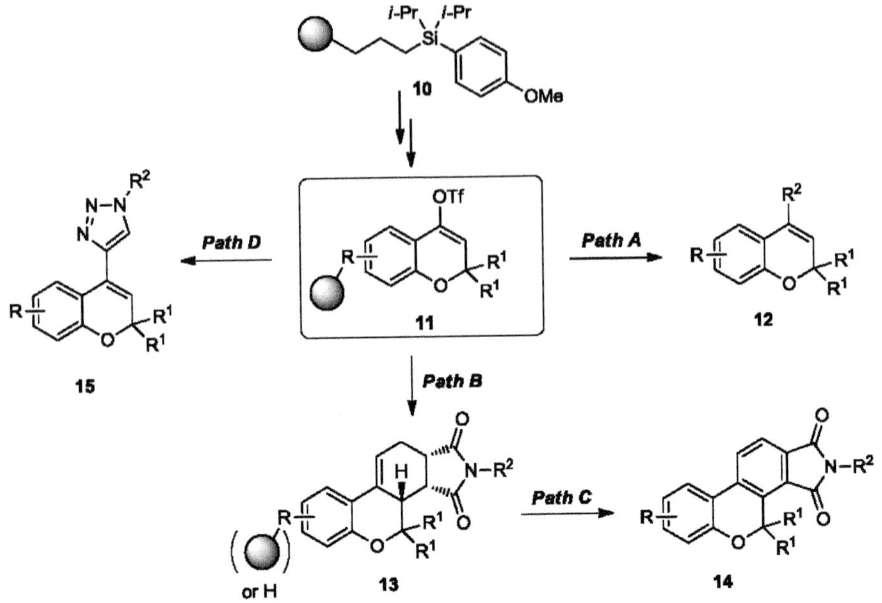

Path A Suzuki coupling
Path B Stille-type vinylation coupling followed by Diels-Alder reaction
Path C Aromatization
Path D Negishi-type alkynylation followed by Huisgen 1,3-dipolar [3 + 2] cycloaddition

SCHEME 7.3 Synthesis of four different benzopyran core skeletons.

7.2.1.1.2 Nitrogen Heterocycles Diverse pharmaceutically interesting nitrogenous heterocyclic compounds were synthesized using selenium resin [12]. Polymer-supported pyrrolidine **25** was prepared by a three-step reaction sequence (Scheme 7.5). This key precursor was subjected to different reaction conditions to yield 2,5-dihydro-1*H*-pyrroles **26**, 1,3-dioxo-2,3,5,7a-tetrahydro-1*H*-pyrrolo[1,2-*c*]imidazoles **28**, and 1,4-dioxo-1,2,3,4,6,8a-hexahydropyrrolo[1,2-*a*]pyrazines **30**, respectively.

First, polymer-supported pyrrolidine **25** was acylated with acyl chlorides to yield the corresponding amide, which was cleaved from the resin with 30% H_2O_2 to afford final 2,5-dihydro-1*H*-pyrroles **26**. Alternatively, treatment of the precursor **25** with isocyanates afforded polymer-supported urea derivatives **27**. Subsequently, the fused five-membered ring was closed under basic conditions and the final product **28** was released from the resin under the same cleavage conditions. The last nitrogen-containing heterocycle reported was obtained by reaction of pyrrolidine **25** with *o*-chloroacetyl chloride to yield chloromethylamide **29** chlorine was substituted with amines to afford, after cleavage from resin, final 1,4-dioxo-1,2,3,4,6,8a-hexahydropyrrolo[1,2-*a*]pyrazines **30**.

The next example illustrates the synthesis of spirocompounds, spirohydantoins, and spiro-2,5-diketopiperazines, which are pharmacologically important molecules [13]. Spirohydantoins were shown to have antiepileptic [14], antidepressant,

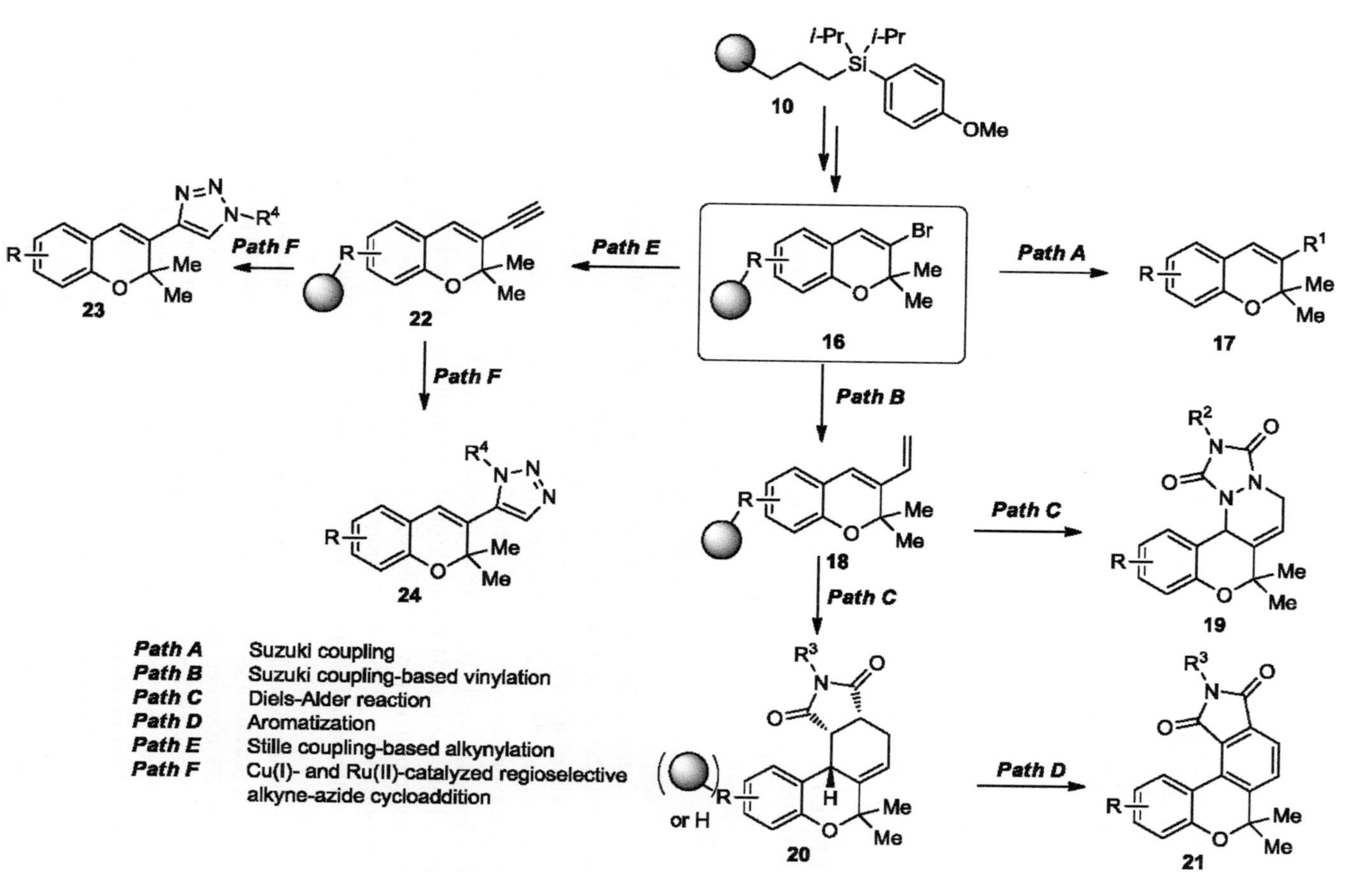

SCHEME 7.4 Synthesis of six different benzopyran core skeletons.

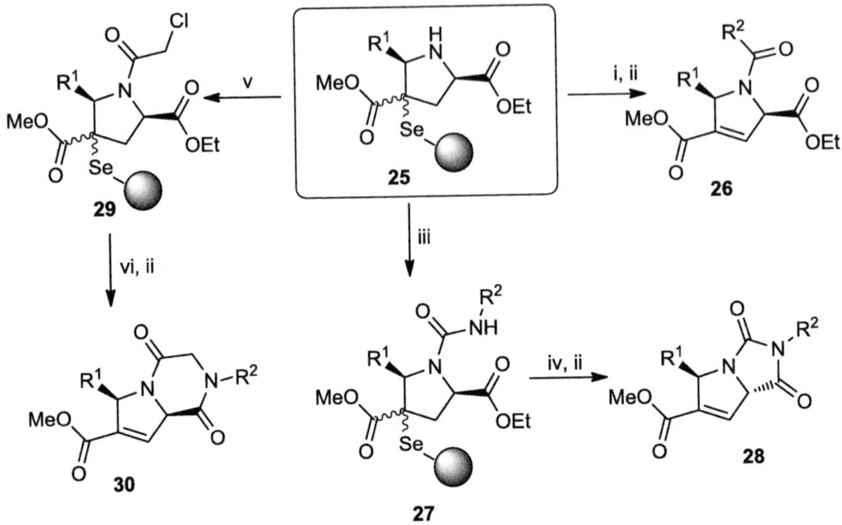

SCHEME 7.5 Synthesis of 2,5-dihydro-1*H*-pyrroles and their fused derivatives. Reagents and conditions: (i) R^2COCl, TEA, DCM, rt, 24 h; (ii) 30% H_2O_2, THF, rt, 1 h; (iii) R^2NCO, toluene, rt, 12 h; (iv) DBU, 80°C, 6 h; (v) $ClCH_2COCl$, TEA, DCM, rt, 24 h; (vi) R^2NH_2, TEA, MeOH/THF, reflux, 10 h.

anxyolytic, and antipsychotic properties [15], while spirodiketopiperazines exhibited antiproliferative and antiinflammatory activities [16].

The synthetic route started with a diastereomeric mixture of Merrifield resin-bound amino ester **31** (Scheme 7.6). This precursor was reacted with either phenyl isocyanate or Fmoc-amino acid. Reaction with phenyl isocyanate yielded derivative **32**, which was cyclized to final spirohydantoin **33**. Acylation with Fmoc-alanine afforded intermediate **34**. Cleavage of a Fmoc-protecting group triggered spontaneous cyclization to final spiro-2,5-diketopiperazines **35** and cleavage from the resin. After evaluation of the scope and limitations of the synthetic route, two libraries of 56 spirohydantoins and 56 spiro-2,5-diketopiperazines were prepared in a combinatorial manner.

Other work described a DOS of skeletally diverse alkaloid-like compounds [17,18]. A precursor of the distinct products was the polymer-supported dihydroisoquinoline **38** together with dihydropyridine. Only the route involving the dihydroisoquinoline skeleton is outlined here. 7-Hydroxyisoquinoline was attached to macrobeads to yield compound **36**, which was alkylated with *o*-bromobenzyl bromide to afford iminium salt **37** (Scheme 7.7). Addition of vinyl magnesium bromide yielded the target precursor dihydroisoquinoline **38**. It is worth noting that polymer-supported dihydroisoquinolines (as well as dihydropyridines) were stable compared to those not attached to the solid support. This enamine was subjected to different reactions; selective reduction with $NaBH_3CN$ in acidic conditions led to 1,2,3,4-tetrahydroisoquinoline **39**. The second route involved reaction with

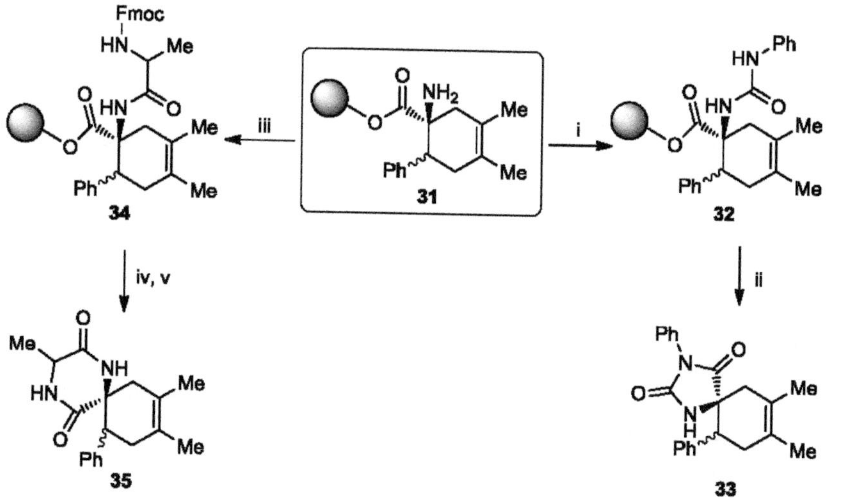

SCHEME 7.6 Synthesis of spirohydantoins and spiro-2,5-diketopiperazines. Reagents and conditions: (i) PhNCO, DCM, rt, 18 h; (ii) *t*-BuOK, THF, rt, 16 h; (iii) Fmoc-Ala-OH, PyBOP/HOBt, rt, 4 h; (iv) 20% piperidine/DMF, rt, 20 min; (v) 5% AcOH, toluene, reflux, 4 h.

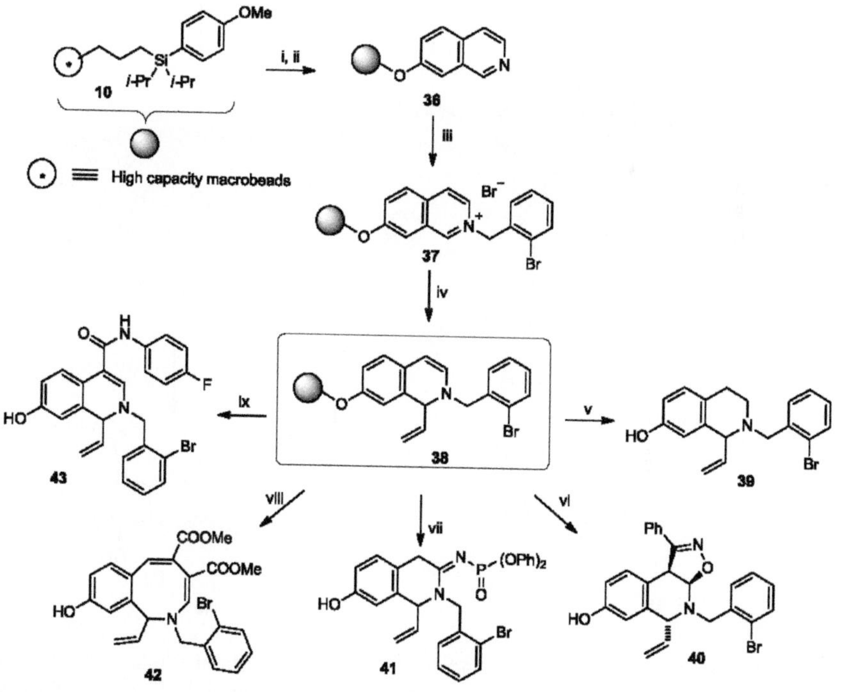

SCHEME 7.7 Synthesis of five various heterocycles derived from the dihydroisoquinoline core skeleton. Reagents and conditions: (i) TfOH, DCM; (ii) 7-hydroxyisoquinoline, 2,6-lutidine; (iii) *o*-bromobenzyl bromide; (iv) CH_2=CH-MgBr, −78°C → 0°C; (v) $NaBH_3CN$, TFA/DCM; (vi) benzohydroximinoyl chloride; (vii) $N_3PO(OPh)_2$; (viii) DMAD; (ix) *p*-fluorophenyl isocyanate.

benzohydroximinoyl chloride and yielded cycloadduct **40**, while [3 + 2] cycloaddition of precursor **38** with electron-deficient azide resulted in the release of molecular nitrogen and concomitant formation of product **41**. The fourth diversity route involved [2 + 2] cycloaddition with DMAD, followed by ring expansion and afforded 6,8-fused adduct **42**. Finally, the last transformation of precursor **38** leading to skeletal diversity involved reaction with *p*-fluorophenyl isocyanate and produced conjugated amide **43**. The authors also calculated a low-energy conformation for each compound prepared.

The last example of DOS of complex heterocycles containing one core structure utilized ring-closing metathesis and resembled indoline alkaloid natural product–like tricyclic compounds [19] and included seven- and eight-membered unsaturated lactam rings.

The synthesis started from polystyrene macrobead-supported aminoindoline **44** with orthogonally protected amino groups (Scheme 7.8). Aminoindoline was

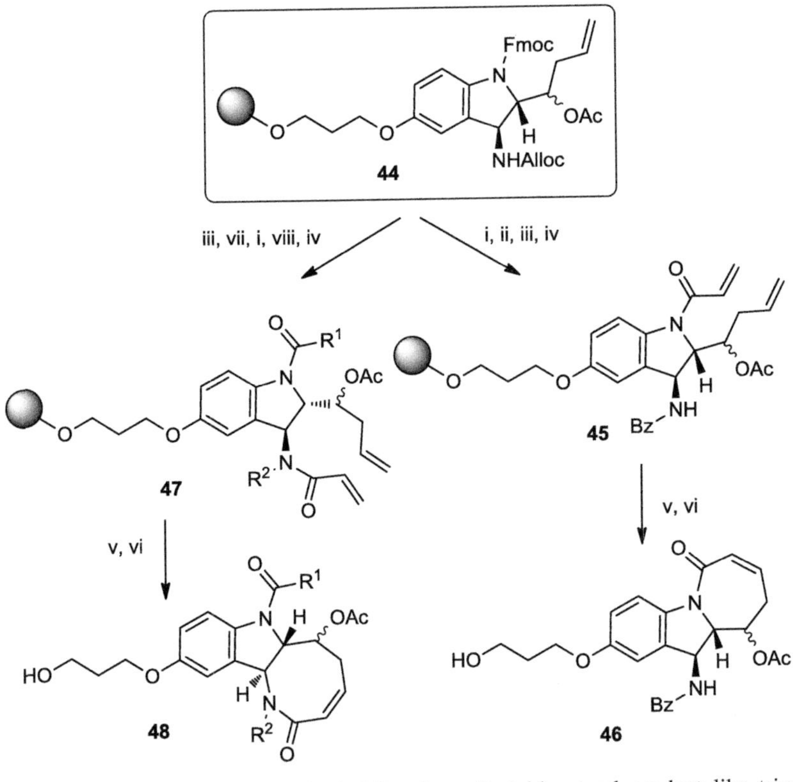

SCHEME 7.8 Conversion of aminoindoline into alkaloid natural product–like tricyclic derivatives by a ring-closing metathesis. Reagents and conditions: (i) Pd(0), PPh_3, *N*-methylmorpholine, AcOH, DCM; (ii) BzCl, 2,6-collidine, DCM; (iii) 20% piperidine/DMF; (iv) acryloyl chloride, 2,6-collidine; (v) Grubbs II (40 to 50 mol%), DCM, 40°C; (vi) HF·pyridine, THF; (vii) isobutyroyl chloride, 2,6-collidine, DCM; (viii) isobutyraldehyde, $NaBH_3CN$, MeOH, AcOH, TMOF.

immobilized deliberately via an alkylsilyl linker through a three-carbon linker, which served as a spacer, utilized a primary hydroxyl group for immobilization, and also enabled printing the small-molecule library onto glass slides and conducting protein-binding studies with small-molecule microarrays [20].

The Alloc group was cleaved from the aminoindoline precursor **44**, and the resin-bound amine was acylated with BzCl. After Fmoc-protecting group removal, the intermediate was coupled with acryloyl chloride to yield compound **45**. The key step of this reaction sequence, the ring-closing metathesis, yielded, after cleavage from the resin, the final tricyclic compound **46**.

Alternatively, the Fmoc-protecting group was removed from the resin-bound precursor **44**. Subsequent conversion of the amine to amide was followed by Alloc removal. Liberated amine was subjected to reductive alkylation to yield secondary amine. Acylation with acryloyl chloride led to intermediate **47**, which underwent ring-closing metathesis, yielding compound **48**.

7.2.1.1.3 Heterocycles with More Heteroatoms Several studies focused on solid-phase synthesis of five-membered heterocyclic structures containing nitrogen, oxygen, and sulfur heteroatoms. Synthesis of 1*H*-imidazoles, thiazoles, and oxazoles was reported on polystyrene sodium sulfinate resin [21]. Resin-bound sulfinic acid **49** afforded precursor **50** by condensation with aldehyde and amide. Cleavages of the products by one-pot elimination–cyclization reaction yielded three five-membered heterocycles: **51** to **53** (Scheme 7.9). Polymer-supported syntheses of

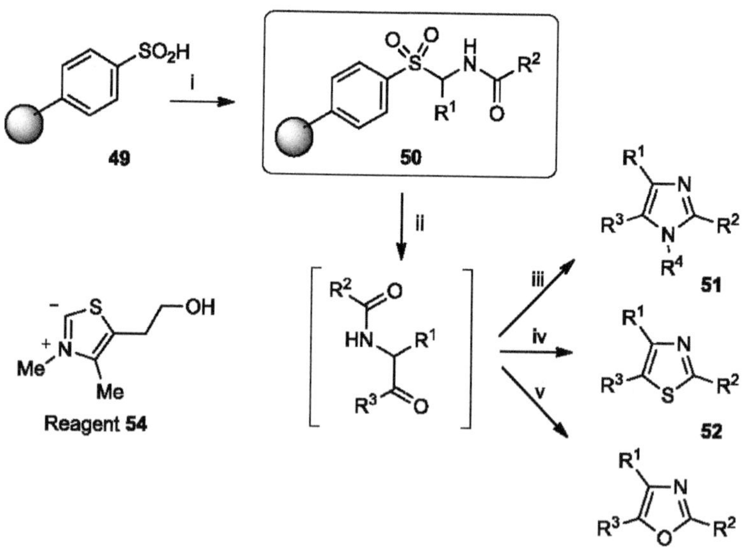

SCHEME 7.9 Synthesis of three different five-membered heterocycles. Reagents and conditions: (i) R^1CHO, R^2CONH_2, TMSCl, MeCN/toluene, 80°C; (ii) reagent **54** (20 mol%), TEA, DCM, R^3CHO; (iii) EtOH, R^4NH_2, AcOH, reflux; (iv) Lawesson's reagent; (v) PPh_3, I_2.

their isomers, such as 1*H*-pyrazoles, isoxazoles, and 1*H*-1,2,4-triazoles, were also described [22–24]. Fused derivatives (i.e., benzimidazoles, benzoxazoles, and benzothiazoles) have been prepared from Wang resin-bound esters **55** by condensation under microwave irradiation [25].

The DOSs of various nitrogen- and oxygen-containing heterocyclic scaffolds—furo[3,4-*d*]pyrimidines, pyrrolo[3,4-*d*]pyrimidines, and pyridazino[4,5-*d*]pyrimidines—were prepared on hydroxymethylpolystyrene resin [26]. The key common intermediate **56** was prepared from resin-bound 4-chloroacetoacetate precursor **55** by three-component Biginelli-type condensations utilizing urea and various aromatic aldehydes (Scheme 7.10). Precursor **56** afforded furo[3,4-*d*]pyrimidines **57** by cyclative cleavage. Pyrrolo[3,4-*d*]pyrimidines **59** were obtained by reaction with primary amines followed by cyclative cleavage. Pyridazino[4,5-*d*]pyrimidine derivatives **61** were achieved by substitution with hydrazines following cyclative

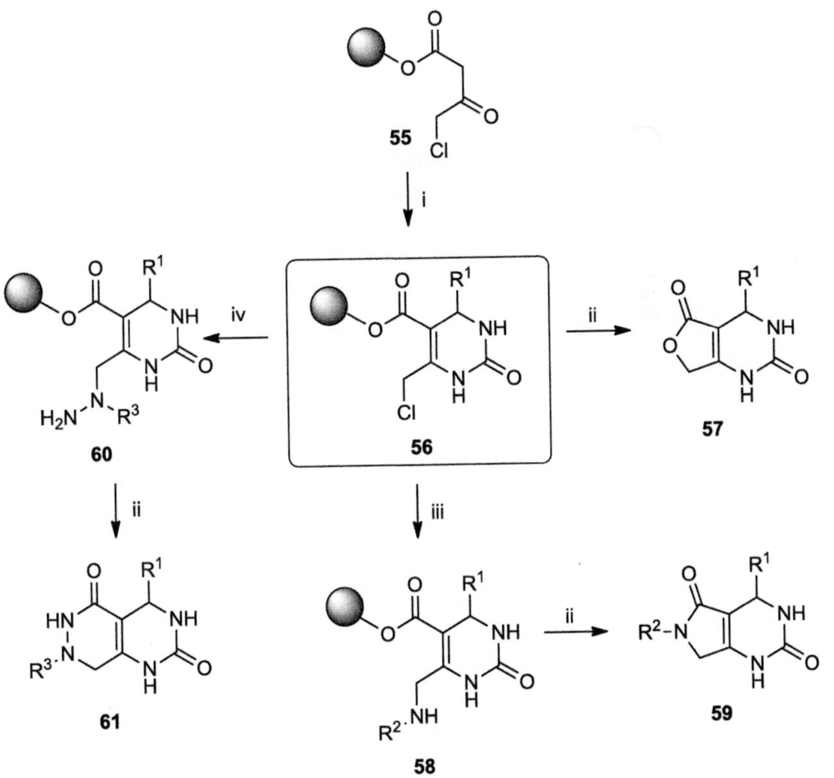

SCHEME 7.10 Synthesis of three various heterocycles embedded with 3,4-dihydropyrimidin-2(1*H*)-one core skeleton by cyclative cleavage. Reagents and conditions: (i) R^1CHO (3 equiv), urea (3 equiv), dioxane, HCl (cat.), 70°C, on, then conc. HCl, rt, 5 min; (ii) DMF, μw (150 or 200°C), 10 min; (iii) R^2NH_2 (5 equiv), DMF, rt, 50 or 70°C, on; (iv) R^3NHNH_2 (5 equiv), DMF, rt, 30 min.

cleavage. Synthetic transformations afforded bicyclic heterocycles in good to moderate yields, and they were therefore applicable for synthesis of combinatorial libraries for HTS.

DOS of nitrogen- and sulfur-containing heterocycles enabled the synthesis of 30 to 50 examples for each heterocyclic scaffold, including thiazolo[4,5-*b*]pyridine, thiazolo[4,5-*d*]pyrimidine-5,7(4*H*,6*H*)-dione, and 1*H*-thiazolo[4,5-*c*][1,2]thiazin-4(3*H*)one 2,2-dioxide, compounds with a wide range of important biological properties [27–30]. The precursor for all those compounds was thiazole-modified resin **62** (Scheme 7.11).

The synthesis of thiazolo[4,5-*b*]pyridine derivatives **65** involved conversion of thiazole precursor **62** into the corresponding thiazolo[4,5-*b*]pyridine resin **63** under Friedländer conditions. Subsequent oxidation of sulfides to sulfones **64** followed by nucleophilic substitution with various (aliphatic/cyclic/aromatic/heterocyclic) amines afforded 50 derivatives of thiazolo[4,5-*b*]pyridine **65** [27]. The second route led to thiazolo[4,5-*d*]pyrimidine-5,7(4*H*,6*H*)-dione **68** via thiazolourea **66** from thiazole amino ester **62** ($R^1 = OEt$). The ester underwent one-pot cyclization/*N*-alkylation and afforded derivative **67** after oxidation. Finally, nucleophilic substitution with amines gave 48 derivatives of the target heterocyclic scaffold **68** [28]. The last skeleton from the polymer-supported thiazole precursor **62** was prepared by a five-step reaction sequence. Treatment with benzylsulfonyl chlorides yielded sulfonamide derivative **69**, and subsequent Mitsunobu alkylation with benzyl alcohol followed by cyclization and oxidation led to sulfone **70**. Final substitution with various amines yielded 34 derivatives of the target 1*H*-thiazolo[4,5-*c*][1,2]thiazin-4(3*H*)one 2,2-dioxide skeleton **71** [29].

Stereo- and regioselective tandem acylation/1,3-dipolar cycloaddition of epoxycyclohexenol carboxylic acid with nitrones afforded polymer-supported tetracycle **72** (Scheme 7.12) [31]. Treatment of tetracycle **72** with $Yb(OTf)_3$ and MeCN or PhCN afforded derivatives **73**. Lactone **72** was also reacted with primary amine to yield γ-hydroxyamides **74**, which underwent aminolysis to form compound **75**. Alternatively, the liberated hydroxyl- group of γ-hydroxyamides **74** was reacted with PhNCO and afforded tricyclic compounds **76**.

Ring-closing metathesis was used not only for the DOS of complex tricyclic indoline-based compounds (see Section 7.2.1.1.2), but also for the synthesis of complex heterocycles containing more heteroatoms [32]. In an analogous manner, orthogonally protected tetrahydroquinoline **77** was immobilized to alkylsilylated macrobeads through a three-carbon spacer (Scheme 7.13).

The tetrahydroquinoline precursor **77** underwent three parallel reactions to yield bridged 10- and 12-membered rings **79** and **81** or trans-fused 12-membered ring-based polycyclic derivative **83**. The reaction sequences started with *O*-pentenoylation of the free hydroxyl group, *N*-Fmoc-protecting group removal and *N*-acylation with *trans*-crotonoyl chloride, and 4-pentenoyl chloride or BzCl. Subsequent *N*-Alloc-protecting group removal followed by *N*-acylation with either BzCl or 4-pentenoyl chloride afforded intermediates **78**, **80**, and **82**, respectively. These derivatives were subjected to ring-closing metathesis, and tricyclic heterocycles **79**, **81**, and **83** were obtained after cleavage from the resin.

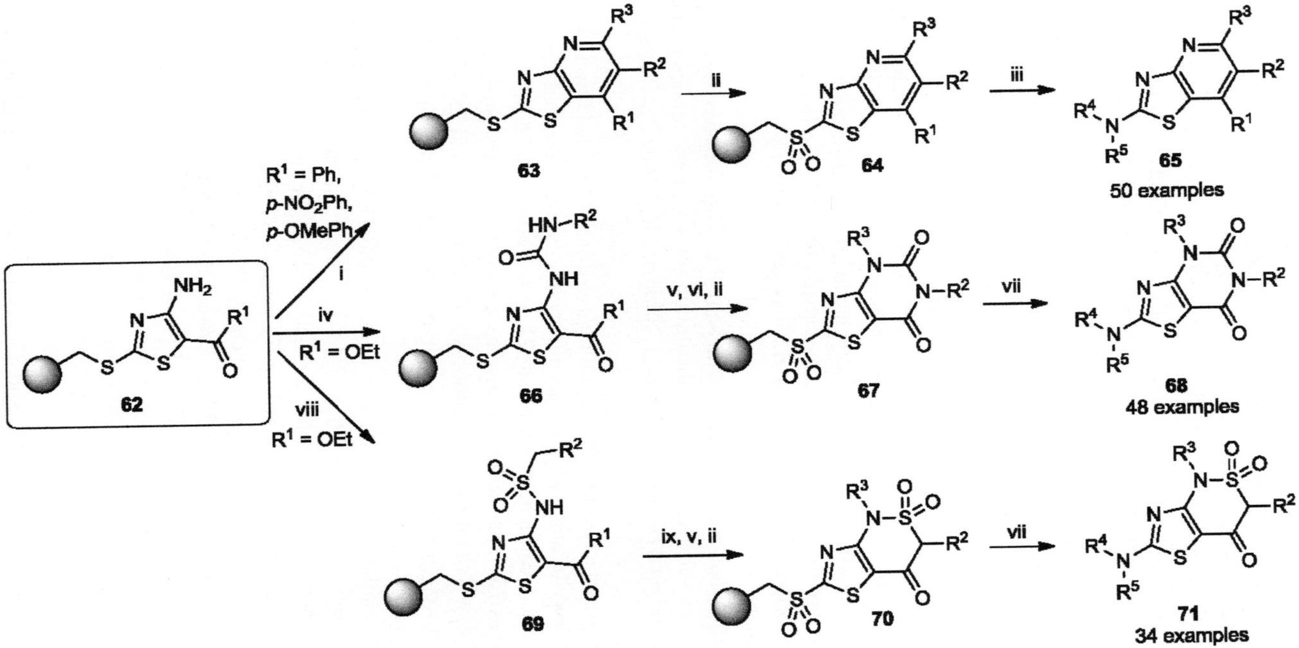

SCHEME 7.11 Synthesis of three different skeletons containing thiazole core structures. Reagents and conditions: (i) $R^2CH_2COR^3$, $AlCl_3$, μw, MeCN, 150°C; (ii) *m*CPBA, DCM, rt; (iii) R^4R^5NH, TEA, THF, 60°C; (iv) R^2NCO, DIEA, μw, DMSO, 150°C; (v) NaH, DMF, rt; (vi) R^3X; (vii) R^4R^5NH, TEA, DCM, 40°C; (viii) $R^2CH_2SO_2Cl$, *t*-BuOLi, DCM, rt; (ix) R^3OH, DIAD, PPh_3, THF, rt.

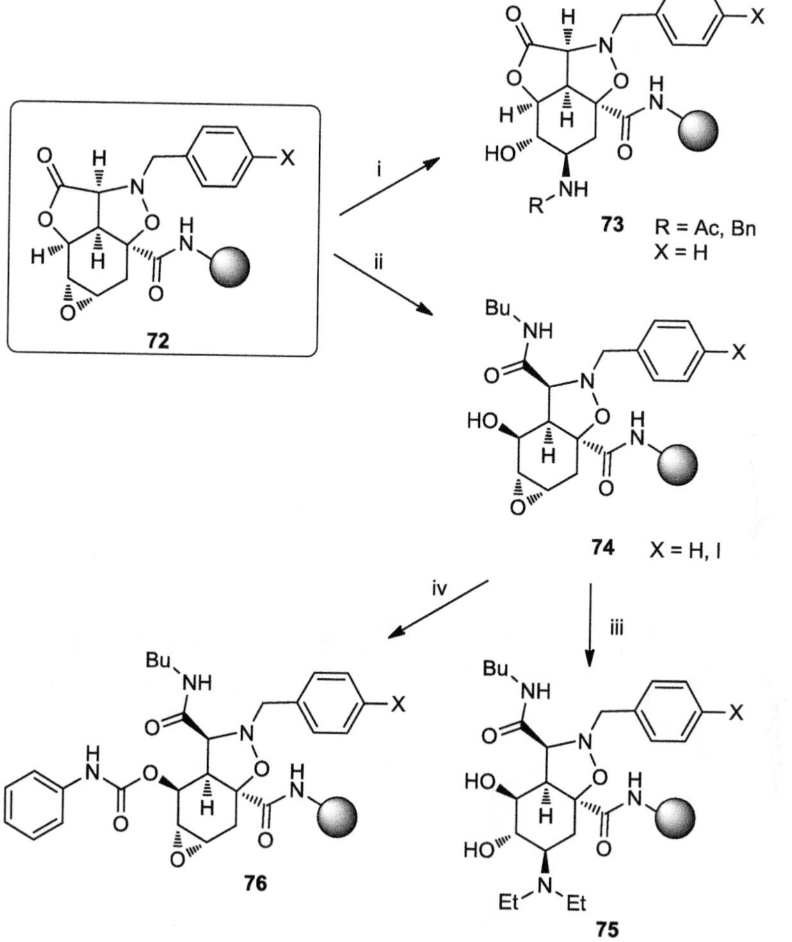

SCHEME 7.12 Synthesis of bi- and tricyclic heterocycles from tetracyclic template. Reagents and conditions: (i) $Yb(OTf)_3$, MeCN or PhCN, rt, 16 h; (ii) $BuNH_2$, THF, rt, 24 h; (iii) Et_2NH, $Yb(OTf)_3$, THF, rt, 16 h; (iv) PhNCO, toluene, rt, 16 h.

7.2.1.2 ***Multiple Core Structures*** This section is dedicated to the synthesis of multiple-core structures using a strategy based on transformation of a common precursor to structurally unrelated skeletons (multiple-core structures) in a few reaction steps. The diversity of such libraries is generally higher than in those based on one core structure.

7.2.1.2.1 Nitrogen Heterocycles DOS of three different nitrogenous heterocyclic compounds started from a sulfone linker which enabled traceless release of the products [33]. The synthesis started with S-alkylation of sulfinate resin yielding

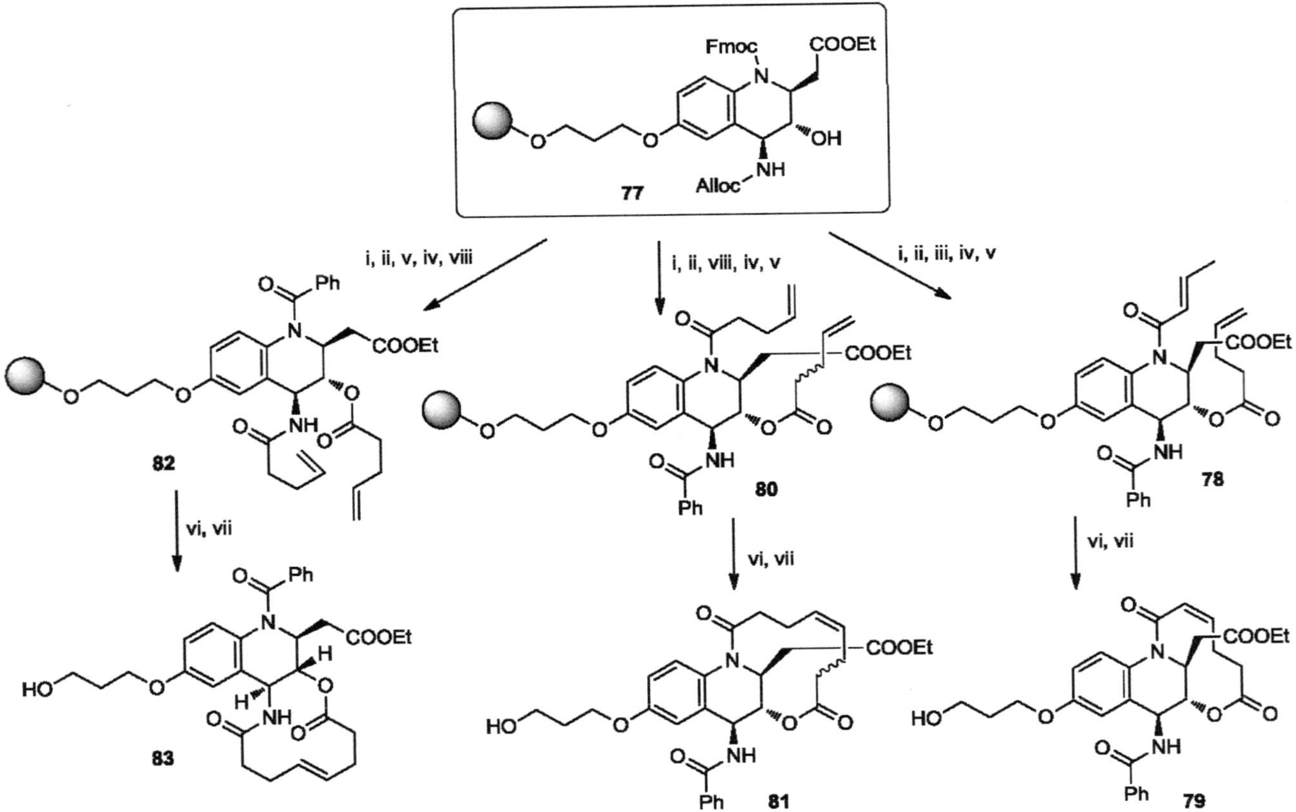

SCHEME 7.13 Conversion of tetrahydroquinoline into natural product–like tricyclic derivatives by a ring-closing metathesis. Reagents and conditions: (i) 4-pentenoic acid, DMAP, DIC; (ii) piperidine; (iii) *trans*-crotonoyl chloride, 2,4,6-collidine; (iv) $Pd(PPh_3)_4$, PPh_3, 4-methyl morpholine, AcOH; (v) BzCl, 2,4,6-collidine; (vi) Grubbs II; (vii) HF·pyridine; (viii) 4-pentenoyl chloride, 2,4,6-collidine.

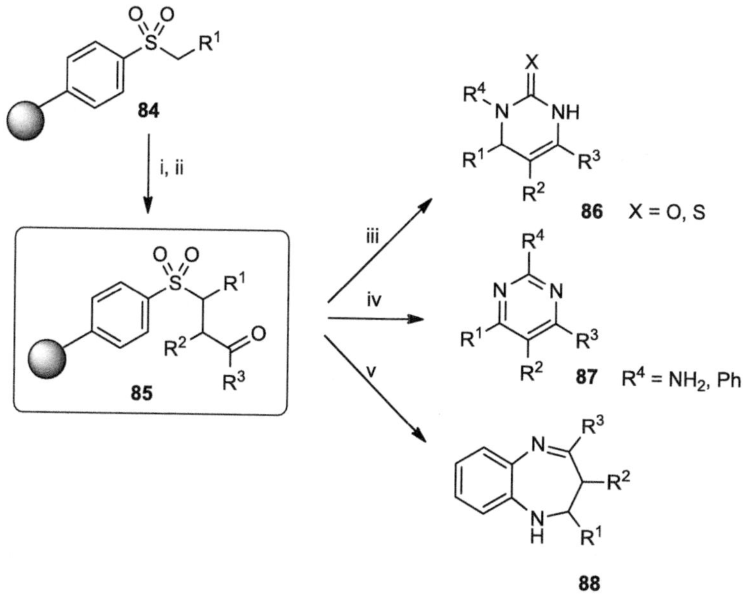

SCHEME 7.14 Transformation of γ-ketosulfone into four diverse heterocycles. Reagents and conditions: (i) $LiCH_2SOMe$, 2,3-disubstituted oxirane; (ii) Jones reagent, acetone, 0°C; (iii) $N(R^4)$-urea or $N(R^4)$-thiourea; (iv) guanidine or benzamidine; (v) *o*-phenylenediamine.

polymer-supported sulfone, **84** (Scheme 7.14). Subsequent sulfone anion alkylation with an epoxide led to formation of γ-hydroxyl sulfone, which was oxidized to γ-ketosulfone **85**. Finally, one-pot elimination–cyclization led to final compounds pyrimidine-2-ones **86** (X = O), pyrimidine-2-thiones **86** (X = S), pyrimidines **87**, and benzo[*b*][1,4]diazepines **88**, respectively.

Five various nitrogenous nonplanar heterocycles including five- to seven-membered rings were prepared from solid-supported *N*-acylated amino acid amides **89** (Scheme 7.15) [34]. Two amides were reduced to afford the key precursor having two secondary amino groups, polymer-supported ethylenediamines **90**. Further reaction with different bifunctional reagents resulted in formation of heterocyclic scaffolds with therapeutic and diagnostic potential. Treatment of diamine precursor **90** with carbonyldiimidazole, thiocarbonyldiimidazole, or cyanogen bromide yielded imidazolidin-2-one **91** (X = O), imidazolidine-2-thione **91** (X = S), and imidazolidin-2-imine **91** (X = NH), respectively. Reaction of precursor **90** with oxalyldiimidazole led to formation of diketopiperazines **92**; further reduction afforded piperazines **93**. Reaction with malonyl dichloride gave diazepinediones **94**, and finally, treatment with carbonisocyanatidic chloride afforded triazinediones **95**.

Further, five- and six-membered bis-heterocyclic compounds were prepared from resin-bound orthogonally protected lysine **96** (Scheme 7.16) [34,35]. The reaction sequence was based on consequent selective cleavage of particular protecting groups and acylation of liberated primary amino groups. Subsequently, all amidic

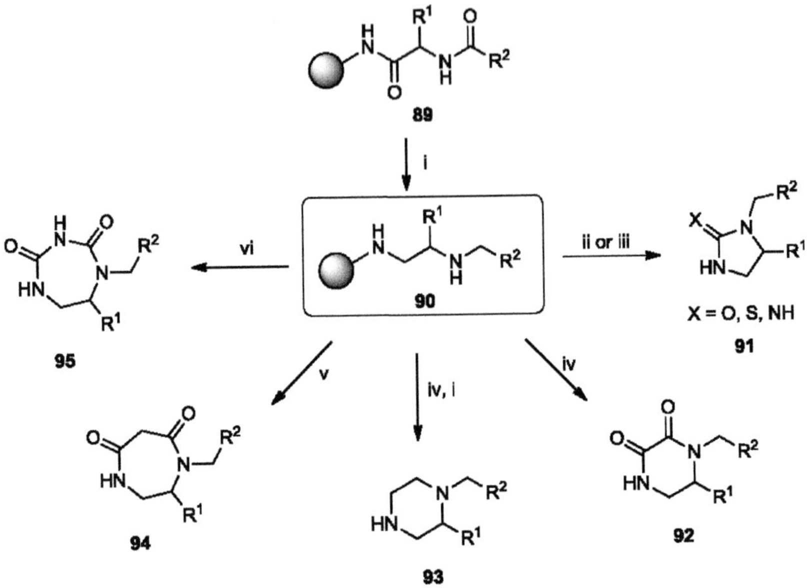

SCHEME 7.15 Ethylenediamines as precursors for preparation of di- and triazacyclic compounds. Reagents and conditions: (i) [H]; (ii) for X = NH: CNBr; (iii) for X = O,S: $CXIm_2$; (iv) $C_2O_2Im_2$; (v) $CH_2(COCl)_2$; (vi) ClCONCO.

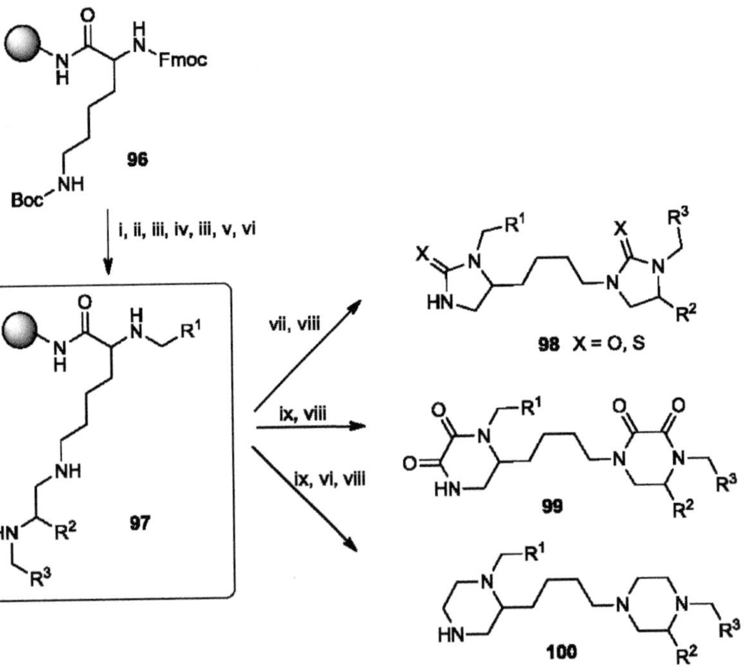

SCHEME 7.16 Synthesis of bis-heterocyclic compounds from orthogonally protected lysine. Reagents and conditions: (i) 25% piperidine/DMF; (ii) R^1COOH, DIC, HOBt, DMF; (iii) 50% TFA/DCM; (iv) Boc-X-amino acid, DIC, HOBt, DMF; (v) R^3COOH, DIC, HOBt, DMF; (vi) BH_3-THF, 65°C; (vii) $CXIm_2$ (X = O,S), DCM; (viii) HF/anisole; (ix) $C_2O_2Im_2$, DMF.

groups were reduced by borane in THF to yield appropriate secondary amines **97**. Following cyclization with various imidazoles and cleavage from the resin afforded the nonplanar bis-heterocycles derived from imidazolidin-2-one **98** (X = O), imidazolidine-2-thione **98** (X = S), piperazine-2,3-dione **99**, and piperazine **100**, respectively.

In addition to ethylenediamines, *o*-phenylenediamines were often used for the synthesis of a wide range of diverse heterocycles (Scheme 7.17) [36–39]. The synthetic route most frequently used included nucleophilic substitution of fluorine in 1-fluoro-2-nitrobenzenes by polymer-supported amines, either on Rink resin (resin **101**) [37] or on BAL resin after reductive amination (resin **102**) [36]. *o*-Phenylenediamine was also immobilized to selenium resin via one of the amino groups (resin **102**) [38]. Alternatively, the synthesis started from *o*-phenylenediamine attached to the resin-bound imidazole (resin **103**) [39].

Polymer-supported *o*-phenylenediamines **101** to **103** that have two diversity positions already built were further transformed into nitrogen-containing heterocyclic compounds, including benzimidazoles **104** [36,37] and **105** [39], benzoimidazolimine **106** [39], benzotriazoles **107** [37], quinoxalinones **108** [36,37], dihydroquinoxalinones **109** [36], quinoxalinedione **110** [39], benzodiazepinediones **111** [37], benzimidazopyrimidinones **112** [38], and finally, benzoimidazodiazepinones **113** [38]. Those chemistries were generally based on reduction of a nitro group, acylation, alkylation, and intramolecular cyclization.

Parallel solid-phase synthesis of a library containing 384 natural product–like diaza-bridged heterocycles was based on tandem acidolytic cleavage and in situ iminium formation, followed by Pictet–Spengler intramolecular cyclization (Scheme 7.18) [40]. The synthesis was carried out on Wang resin that was converted to bromoacetal resin **114**. The bromine was substituted with amines, yielding precursor **115**, which was acylated with Fmoc-Trp-OH and Fmoc(*O*-diTBS)DOPA to afford derivatives **116** and **117**. The Fmoc-protecting group was cleaved and the liberated amines were acylated with carboxylic acids to afford cyclization precursors **118** to **121**. The last step, Pictet–Spengler intramolecular cyclization, was performed in neat formic acid and led to final compounds **122** to **125**, respectively.

The last example of polymer-supported synthesis-generating nitrogenous multiple-core scaffolds represents stereoselective synthesis of skeletons having up to six stereocenters [41]. A library yielding 29,400 discrete compounds, including 10 distinct polycyclic skeletons, was synthesized using a split-and-pool technique. Many building blocks were evaluated for their compatibility with the synthetic route proposed. Selected building blocks were then used for the DOS. Specifically, 40 of the 64 hydroxyaldehydes tested were attached to the macrobeads through silylation of the hydroxyl group. For simplicity, only polymer-supported vanillin derivative **127** is portrayed [step (i), Scheme 7.19]. In the next combinatorial step, 41 of the 53 originally disubstituted cyclic dienophiles underwent double addition to afford derivatives **129** and **131** [step (ii)], while 22 of 44 tri- and tetrasubstituted cyclic dienophiles underwent monocycloaddition and yielded intermediates **130** [step (iii))].

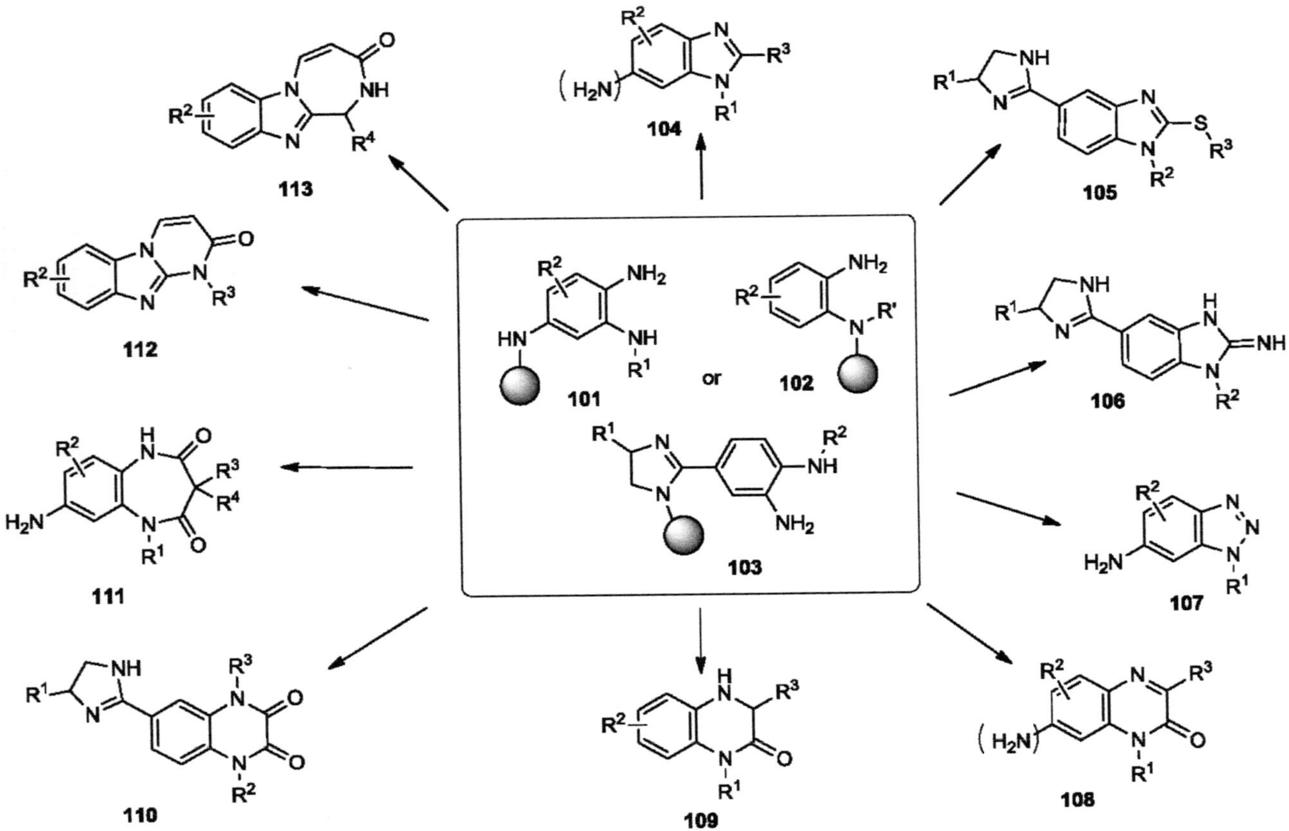

SCHEME 7.17 *o*-Phenylendiamines as precursors for preparation of several diverse skeletons.

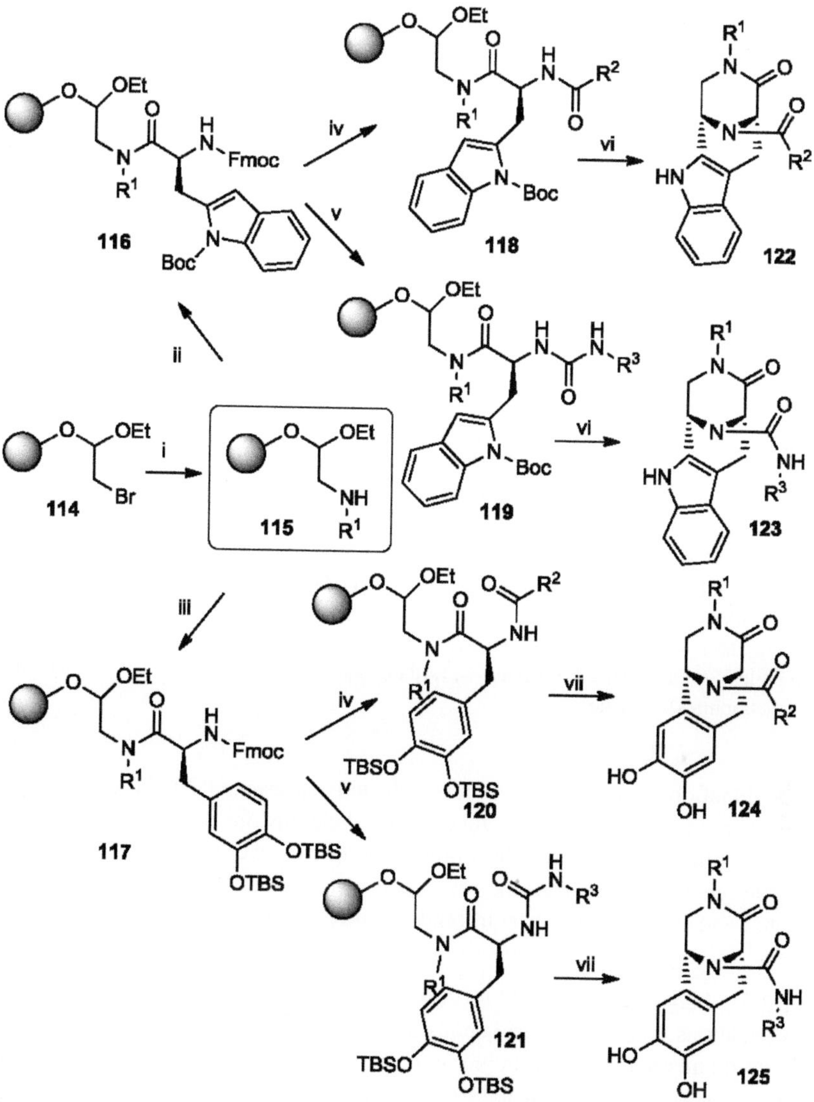

SCHEME 7.18 Synthesis of natural product–like diaza-bridged heterocycles through Pictet–Spengler intramolecular cyclization. Reagents and conditions: (i) R^1NH_2, DMSO, 60°C; (ii) Fmoc-Trp(Boc)-OH, HATU, DIEA, DMF, rt; (iii) Fmoc(*O*-diTBS)DOPA, HATU, DIEA, DMF, rt; (iv) (a) 25% piperidine, rt; (b) R^2COOH, DIC, HOBt, DIEA, rt; (v) R^3NCO, DIEA, DCE, rt; (vi) neat formic acid, rt; (vii) neat formic acid, 60°C. (Adapted from [40], with permission; copyright © 2006 American Chemical Society.)

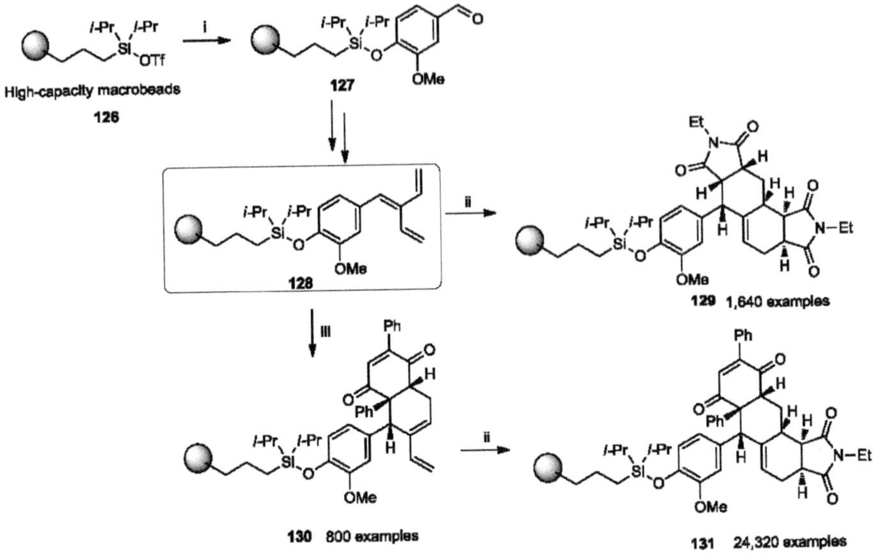

SCHEME 7.19 Synthesis of a library containing 29,400 distinct compounds. Reagents and conditions: (i) 4-hydroxy-3-methoxybenzaldehyde (as one of 40 hydroxy-substituted aromatic aldehydes), 2,6-lutidine, DCM; (ii) *N*-ethylmaleimide (as one of 41 disubstituted dienophiles), toluene; (iii) 2,5-diphenyl-[1,4]benzoquinone (as one of 22 tri- and tetrasubstituted dienophiles), toluene.

Combination of all building blocks selected represented 29,400 different polycyclic compounds. Generic structures of 10 skeletons are portrayed in Figure 7.1. After finishing the synthesis, the library macrobeads were arrayed into wells, target compounds cleaved from the resin, and a stock solution of library compounds prepared and submitted for protein-binding and phenotypic assays.

7.2.1.2.2 Heterocycles with More Heteroatoms DOS of seven structurally unrelated skeletons containing nitrogenous and oxygenous heterocycles having a four- to seven-membered ring and including both planar and nonplanar scaffolds utilized α-acylamino ketone **142** attached to the polystyrene resin (Scheme 7.20) [42]. The synthesis started with sulfonylation of polymer-supported primary amines **140** with 4-Ns-Cl followed by *N*-alkylation and cleavage of the 4-Ns-protective group. Subsequent acylation with different acids yielded intermediates **142**, which were further transformed to final heterocycles.

The first heterocycle, dihydropyrrolone **143**, was obtained by intramolecular Wittig reaction. The second, pyrazinone **144**, was achieved by using two different pathways. The first involved substitution of bromine (R^4 = Br) with $NH_2NH_2{\cdot}H_2O$ to afford hydrazinoderivative, which cyclized spontaneously during TFA-mediated cleavage. The second route involved cleavage of Fmoc-protective group (R^4 = NH-Fmoc) and spontaneous cyclization to yield pyrazinone **144**. The seven-membered heterocycle,

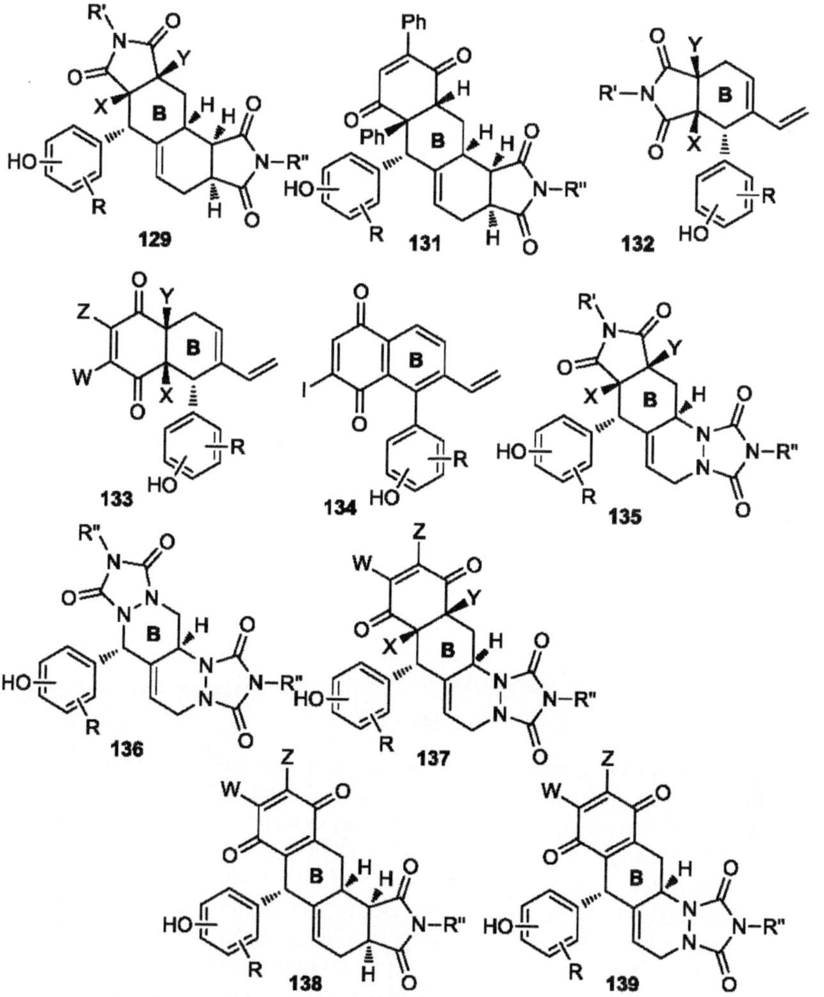

FIGURE 7.1 Generic structures of 10 skeletons prepared from vanillin-based precursors.

dihydrotriazepinone **145**, was prepared from bromoacylamino ketone **142** (R^4 = Br) by treatment with $BnNHNH_2 \cdot HCl$ and spontaneous cyclization during acid-mediated cleavage from the resin. The fourth reaction pathway resulted in formation of morpholinone **146**. BEMP-mediated cyclization of bromoacylamino ketones **142** (R^2 = aryl, R^4 = Br) afforded polymer-supported oxazinones that were released from the resin by TFA in DCM in the presence of triethylsilane (TES) as a reducing agent to afford morpholinones **146**. α-Acylamino ketones **142** were also utilized for synthesis of imidazoles **147**. In this case, α-acylamino ketones **142** were cyclized in acetic acid in the presence of $AcONH_4$ and yielded imidazole derivatives **147**. Treatment of precursor **142** with Schwesinger base (BEMP) caused cyclization to lactam that was subsequently released from the resin to get heterocycle **148**. The last heterocycle

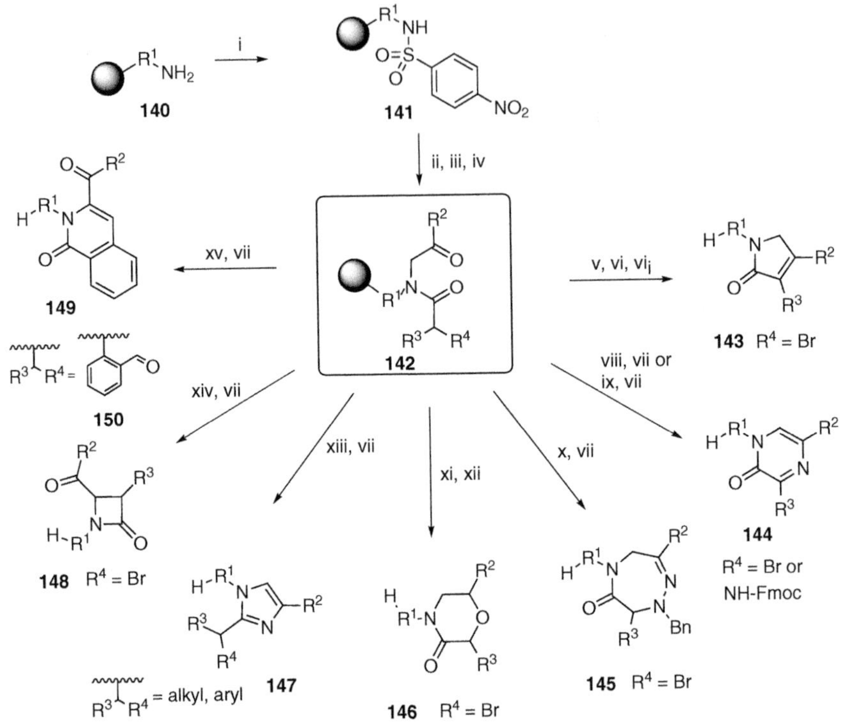

SCHEME 7.20 α-Acylamino ketones as a precursor for the synthesis of seven diverse heterocycles. Reagents and conditions: (i) 4-Ns-Cl, 2,6-lutidine, DCM, 16 h; (ii) bromoketone, DIEA, DMF, 16 h; (iii) 2-mercaptoethanol, DBU, DMF, 5 min; (iv) R^4 = Br: alpha-bromocarboxylic acid, DIC, DCM, 15 min, then DIEA, 4 h; R^4 = NH-Fmoc: Fmoc-alpha-amino acid, DIC, DCM/DMF, 16 h; R^4 = **150**: *o*-carboxybenzaldehyde, DIC, DMF, 16 h; (v) PPh_3, anh. NMP, 16 h; (vi) TEA, anh. NMP, 2 h; (vii) 50% TFA/DCM, 1 h; (viii) R^4 = Br: $NH_2NH_2 \cdot H_2O$, THF, 2 h; (ix) R^4 = NH-Fmoc: 50% piperidine/DMF, 15 min; (x) $BnNHNH_2 \cdot HCl$, DMF, 16 h; (xi) BEMP, anh. NMP, 15 min (for R^2 = aryl); (xii) TFA/TES/DCM (5:1:4), 1 h; (xiii) $AcONH_4$, AcOH, 100 °C, 16 h; (xiv) BEMP, anh. NMP, 30 min (for R^2 = OEt); (xv) DBU, DMF, 30 min.

from this series, isoquinolinone **149**, utilized polymer-supported alkylamines acylated with in situ–activated *o*-carboxybenzaldehyde that underwent intramolecular aldol condensation.

Stereoselective synthesis based on tandem *N*-acyl iminium ion cyclizations/nucleophilic addition chemistry enabled access to various fused heterocycles that also serve as peptidomimetics [43]. The key precursor, *N*-(2,2-dimethoxyethyl)amide **151**, was converted into a wide range of structurally unrelated scaffolds **152** to **159** bearing at least two sp^3-hybridized carbon atoms and creating one asymmetric carbon (Scheme 7.21). The chemistry is generally based on unmasking of the aldehyde that triggered iminium ion formation followed by nucleophilic addition with internal nucleophile. Synthesis of 1,2,10,10a-tetrahydrobenzo[4,5]imidazo[1,2-*a*]pyrazin-3(4*H*)-ones **157** has already been published [44].

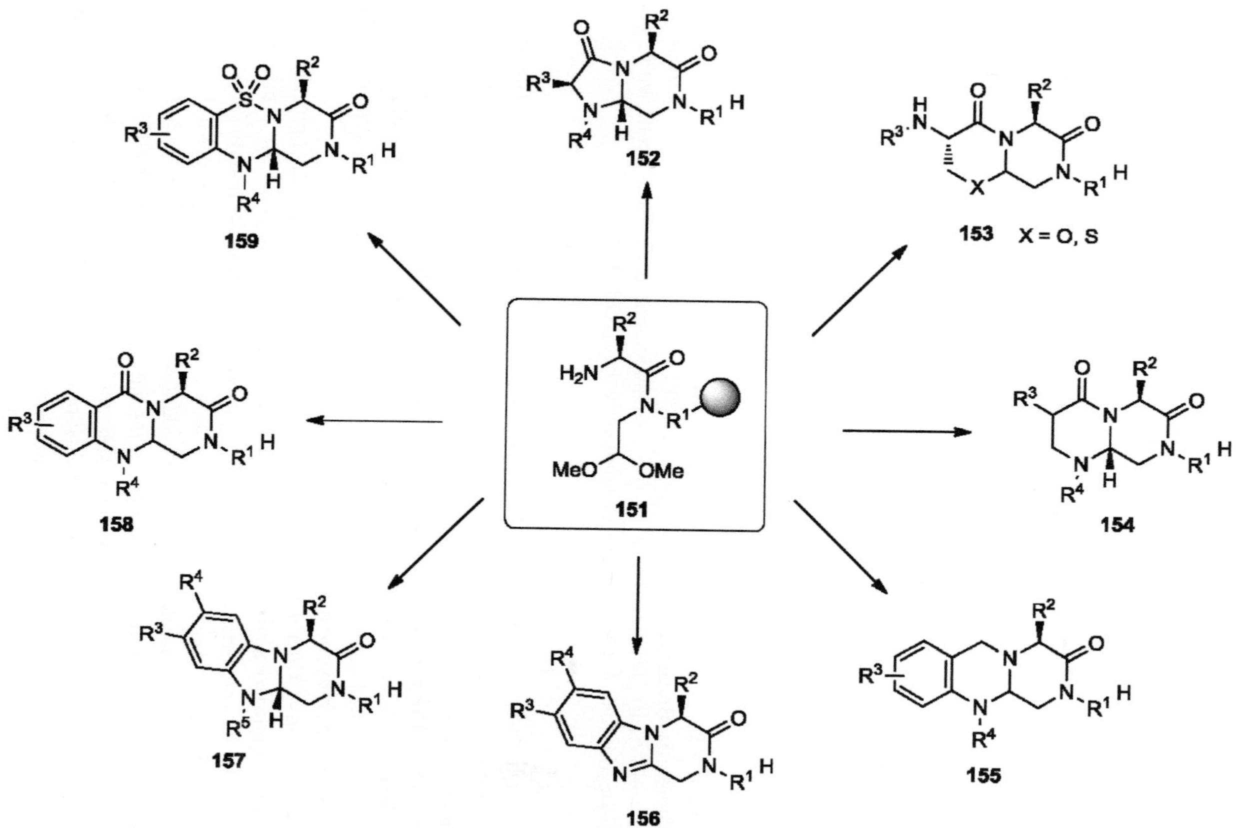

SCHEME 7.21 Synthetic routes leading to eight diverse heterocycles involving formation of iminium salts.

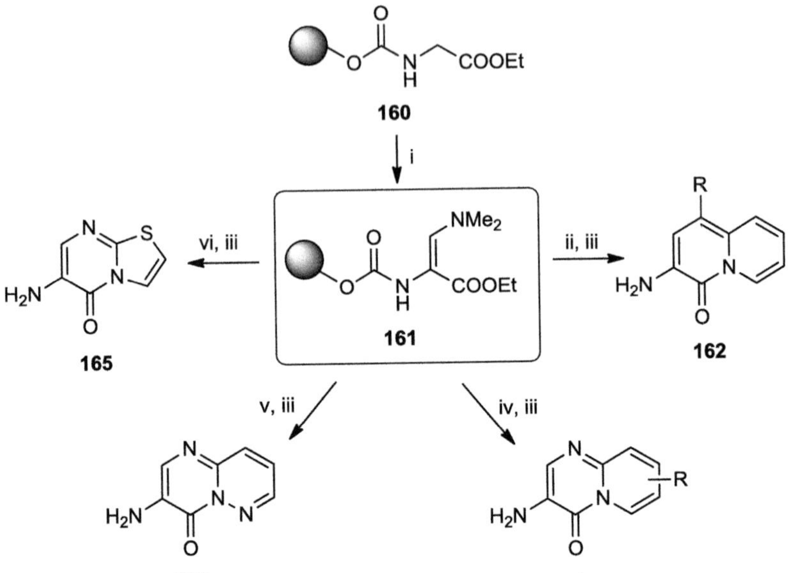

SCHEME 7.22 Conversion of propenoate into four different heterocycles. Reagents and conditions: (i) Bredereck's reagent, toluene, 100°C; (ii) pyridineacetic acid derivatives, AcOH, 90°C; (iii) 50% TFA/DCM, rt; (iv) 2-aminopyridine derivatives, AcOH, 90°C; (v) 3-aminopyridazine, AcOH, 90°C; (vi) 2-aminothiazole, AcOH, 90°C.

Other libraries of heterocyclic scaffolds were achieved from the key intermediate, resin-bound propenoate **161** prepared by reaction of Wang resin with ethyl isocyanatoacetate and subsequent treatment with Bredereck's reagent (Scheme 7.22) [45]. The intermediate **161** was then cyclized with different nucleophiles such as pyridineacetic acid derivatives [step (ii), Scheme 7.22], 2-aminopyridine derivatives [step (iv)], 3-aminopyridazine [step (v)], and 2-aminothiazole [step (vi)] to form, after the cleavage from the resin, bicyclic heterocyclic scaffolds **162** to **165**.

Hetero-Diels–Alder reaction of the diene-containing natural product piperine with polymer-supported acyl- and arylnitroso dienophiles was applied for synthesis of structurally diverse heterocycles [46]. An attractive feature of the hetero-Diels–Alder reaction of diene-containing natural products with acyl- and arylnitroso derivatives is its potential to generate versatile cycloadducts (evolvable scaffolds) and provided opportunities for "Modular enhancement of nature's diversity" (MEND) [47]. Conversion of cycloadduct intermediate **166** into four diverse products is shown in Scheme 7.23. The synthesis utilized diisopropylsilyl linker **I** and Rink linker **II**. Cleavage of the precursor **166** from the resin with 50% TFA in DCM led to 1-(2-nitrophenyl)-1*H*-pyrrol-2-yl(piperidin-1-yl)methanone **167**, while reduction of nitro group prior to the cleavage from the solid support afforded 1-(benzo[*d*][1,3]dioxol-5-yl)pyrrolo[1,2-*a*]quinoxalin-4(5*H*)-one **168**. Reduction of the nitro group to an amino group and subsequent cleavage from the silicon linker opened access

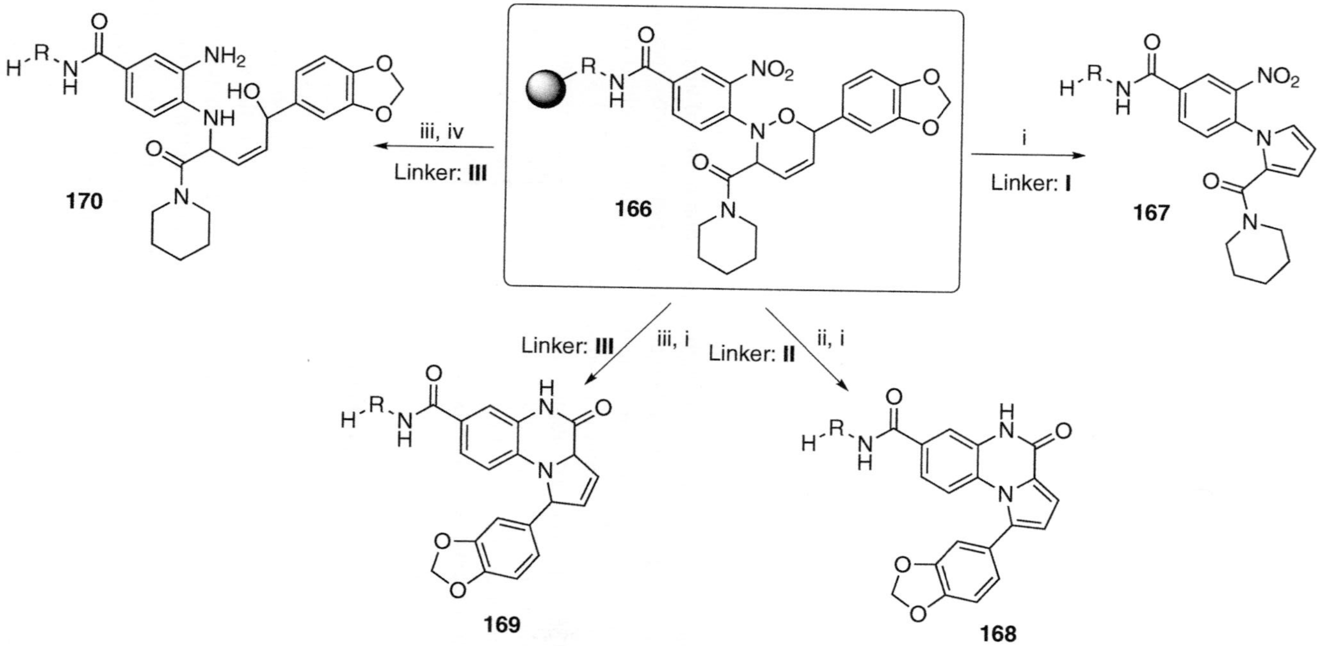

SCHEME 7.23 Conversion of cycloadduct precursor into four acyclic and heterocyclic derivatives. Reagents and conditions: (i) 50% TFA/DCM, rt, 30 min; (ii) $SnCl_2{\cdot}2H_2O$, DIEA, NMP, rt, 2 h; (iii) $SnCl_2{\cdot}2H_2O$, NMP, rt, 16 h; (iv) 0.01 M TBAF, THF, rt, 30 min.

to two different products, 1-(benzo[*d*][1,3]dioxol-5-yl)-3a,5-dihydropyrrolo[1,2-*a*]quinoxalin-4(1*H*)-one **169** and an acyclic product **170**, depending on the cleavage conditions.

Related work described the Diels–Alder reaction of resin-bound acyl- and arylnitroso derivatives with dienes and yielded oxazine derivatives that were transformed further [48]. The synthesis was carried out using different linkers: Rink amide **I**, diisopropylsilyl **II**, Wang **III**, and dual linker **IV** (diisopropylsilyl linker and Wang resin), respectively (Scheme 7.24). Polymer-supported *N*-hydroxycarbamates **171** oxidized in the presence of cyclic and linear dienes including 1,3-cyclohexadiene, 2,4-hexadiene, cycloheptatriene and 9,10-dimethylanthracene resulted in diverse oxazine-based derivatives **172** to **180**. Diels–Alder reactions of arylnitroso compounds were also investigated [49].

Extensive research focused on the discovery of novel compounds with anti-MRSA (methicillin-resistant *Staphylococcus aureus*) activity [50,51]. 242 Derivatives of 18 discrete natural product–like scaffolds were prepared in four steps. The synthesis was carried out on the resin **181** bearing phosphonate moieties and permitted the stereoselective formation of α,β-unsaturated acyl imidazolidinones (Scheme 7.25). In addition, to expand compound diversity, the imidazolidinone linker could be further derivatized, hydrolyzed, esterified, reduced, and amidated.

Resin **182** underwent [2 + 3] cycloaddition, dihydroxylation, or Diels–Alder cycloaddition to yield intermediates **183** to **185**. Pyrrolidine intermediate **183** was further either acylated or underwent reductive amination to afford final compounds **186** and **187**. Diol **184** was transformed into four different products. Treatment with $SOCl_2$, NaN_3, and DMAD led to triazole **188**. O-Alkylation yielded diether **189**, while condensation with carbonyl derivatives afforded ketals **190** and acetals **191**. Cycloadduct **185** underwent oxidation to bridged scaffold **192** or metathesis to yield cyclopentane derivative **193**. Alternatively, it was oxidized to derivative **194**, which was further converted into four diverse scaffolds **197** to **200**. Finally, intermediate **185** was transformed into products of metathesis **195** and **196**.

7.2.2 Substrate-Based Strategy: Folding Process

The folding approach is based on the transformation of a common precursor bearing various appending substituents (σ-elements) into different skeletons under identical reaction conditions. An illustrious example of this approach is the work by Burke et al. [52]. They have prepared three precursors having different σ-elements in a few reaction steps using macrobeads as a solid support. Furan precursors **201** to **203** (Scheme 7.26) were exposed to the same reaction conditions: treatment with NBS, followed by reaction with PPTS. As a result, three diverse skeletons, which included 1260 individual compounds, were synthesized. Compounds were prepared in all possible combinations of building blocks and encompassed skeletal diversity.

While precursors **201** and **203** afforded the anticipated products [3.2.1]bicyclic ketal **204** and (*E*)-enedione **206**, precursor **202** underwent an initial oxidative ring expansion to cyclic hemiketal followed by unexpected, acid-mediated dehydration to yield pyranyl derivative **205**.

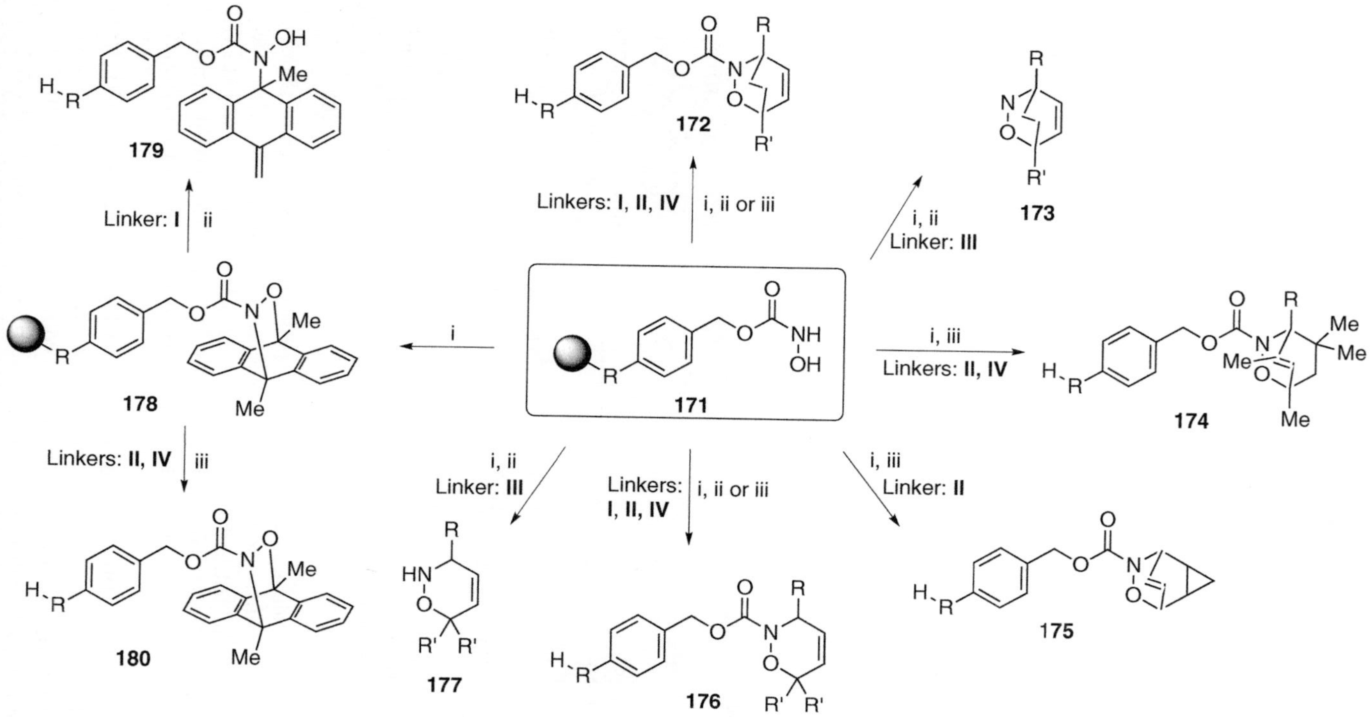

SCHEME 7.24 Synthesis of several diverse compounds by Diels–Alder reaction. Reagents and conditions: (i) diene, 0.1 M n-Bu_4NIO_4, DCM, rt, 1 h; (ii) 50% TFA/DCM, rt, 30 min; (iii) 0.1 M TBAF, THF, rt, 30 min.

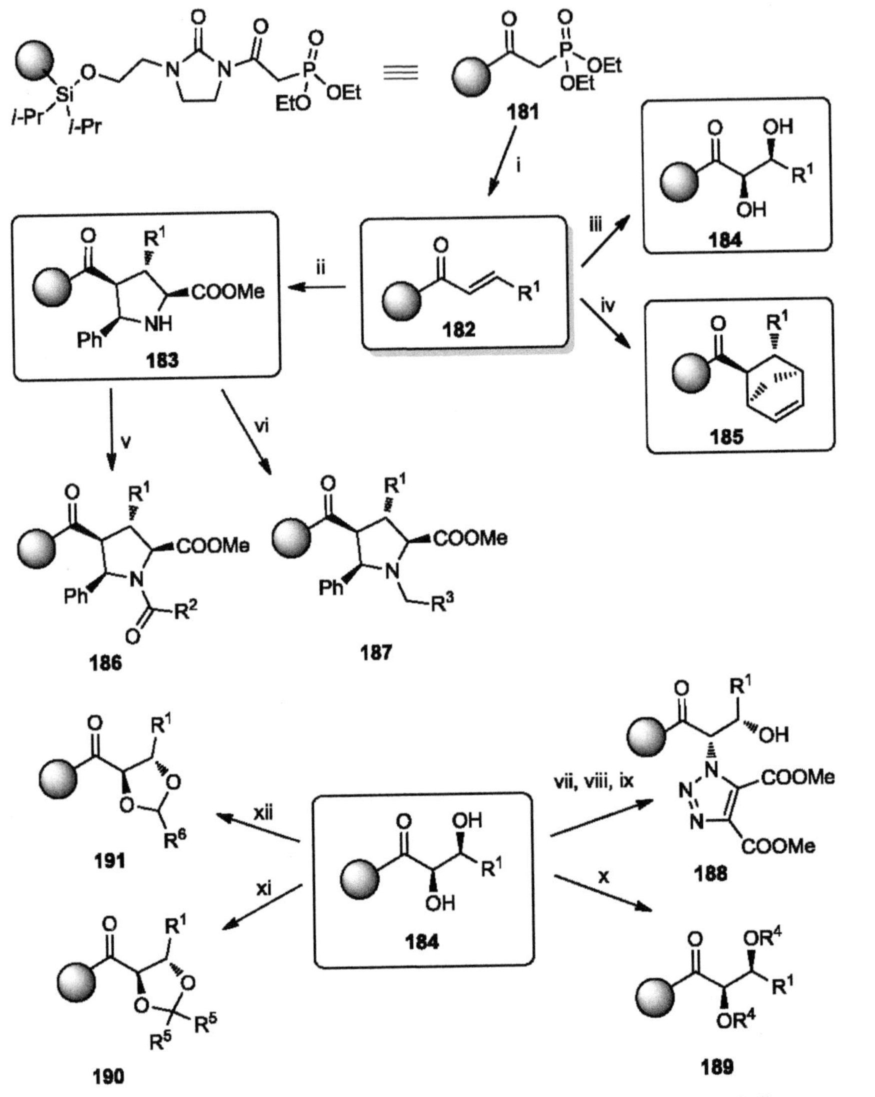

SCHEME 7.25 Synthesis of a library containing 242 compounds based on 18 discrete natural product–like scaffolds. Reagents and conditions: (i) R^1CHO, LiBr, DBU, MeCN; (ii) (*R*)-QUINAP, AgOAc, DIEA, THF, −78°C → rt; (iii) AD-mix, (DHQD)2-PHAL, THF/H_2O (1 : 1); (iv) chiral bis(oxazoline), $Cu(OTf)_2$, DCM, C_5H_6; (v) R^2COCl, DMAP, pyridine, DCM; (vi) R^3CHO, BH_3·pyridine, MeOH; (vii) $SOCl_2$, pyridine, DCM, 40°C; (viii) NaN_3, DMF, 100°C; (ix) DMAD, toluene, 65°C; (x) R^4Br, Ag_2O, DCM, 40°C; (xi) $(R^5)_2CO$, TsOH, DMF, 65°C; (xii) R^6CHO, TsOH, DMF, 65°C; (xiii) *m*CPBA, DCM; (xiv) MeOH, 65°C; (xv) $CH_2{=}CHCOOBn$, Grubbs II, $CH_2{=}CH_2$, toluene, 120°C; (xvi) OsO_4, NMO, acetone/H_2O (10 : 1); (xvii) RNH_2, Me_2AlCl, toluene, 120°C; (xviii) NaH, $R^{11}X$, DMF/THF; (xix) Grubbs II, $CH_2{=}CH_2$, toluene, 120°C; (xx) $NaIO_4$, THF/H_2O (1 : 1); (xxi) R^7NH_2, $NaBH(OAc)_3$, DCM; (xxii) $R^{8a}R^{8b}NH$, $NaBH(OAc)_3$, DCM; (xxiii) R^9CHO, TsOH, DMF, 65°C; (xxiv) $(R^{10})_2CO$, TsOH, DMF, 65°C.

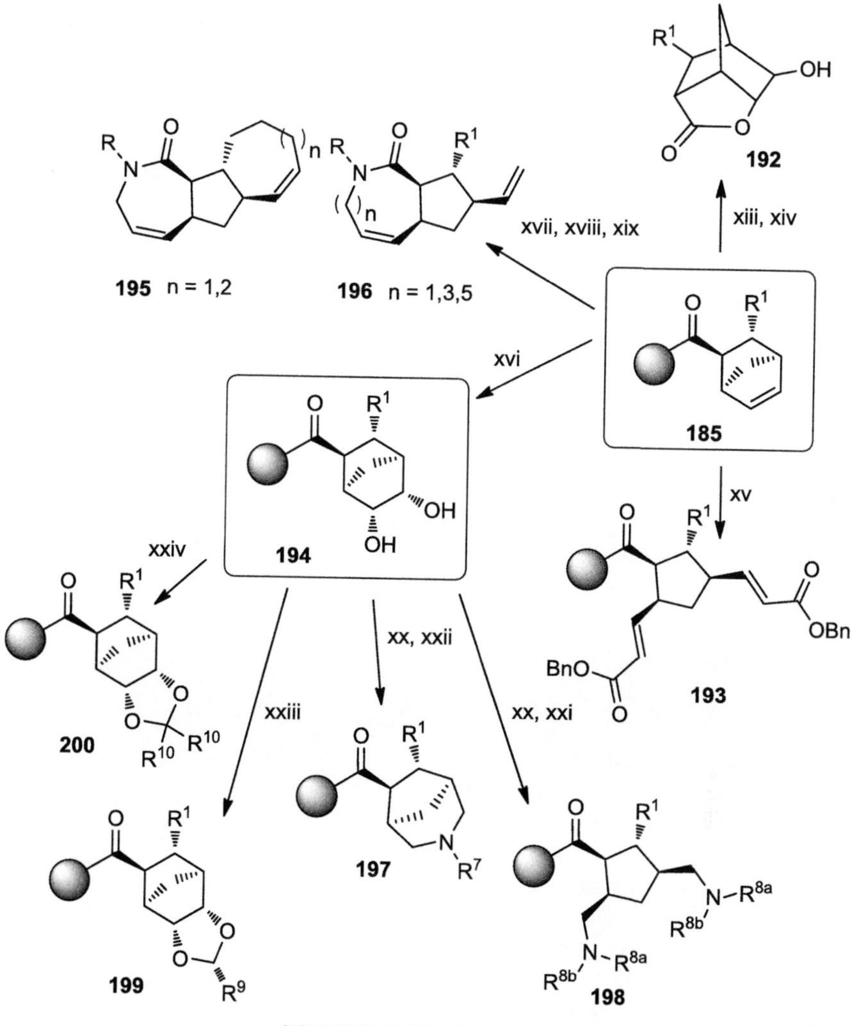

SCHEME 7.25 (*Continued*)

Synthesis of fused isoindolinones started with oxidative cleavage of 2-vinylbenzamides [53]. Isoindolinones are known as core structures of complex alkaloids [54–56]. During the folding process, 2-vinylbenzamide derivatives, **207** to **209**, underwent $OsO_4/NaIO_4$/DABCO-mediated oxidative cleavage followed by TFA-triggered tandem *N*-acyl iminium ion cyclizations/nucleophilic addition to products **210** to **212** (Scheme 7.27).

The same authors [53] performed the synthesis of different nitrogenous heterocycles. The reaction sequence started from TBDMS-protecting building blocks which were attached to three different solid-supported peptide residues containing

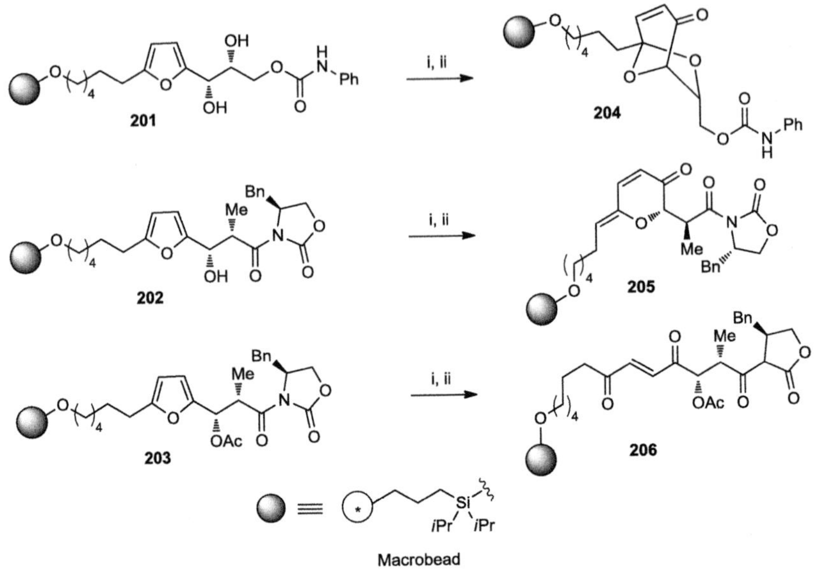

SCHEME 7.26 Transformation of furan precursors into three diverse scaffolds. Reagents and conditions: (i) NBS, $NaHCO_3$, NaOAc, THF/H_2O (4 : 1), rt, 1 h; (ii) PPTS, DCM, 40°C, 20 h.

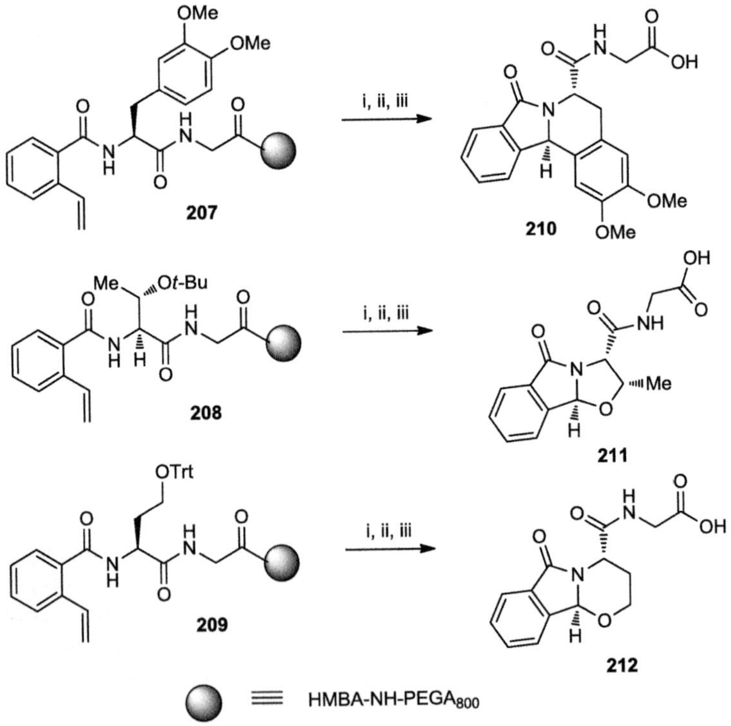

SCHEME 7.27 Conversion of 2-vinylbenzamides to fused isoindolinones. Reagents and conditions: (i) OsO_4/$NaIO_4$/DABCO, THF/H_2O (1 : 1); (ii) 50% TFA/DCM; (iii) 0.1 M NaOH (aq.).

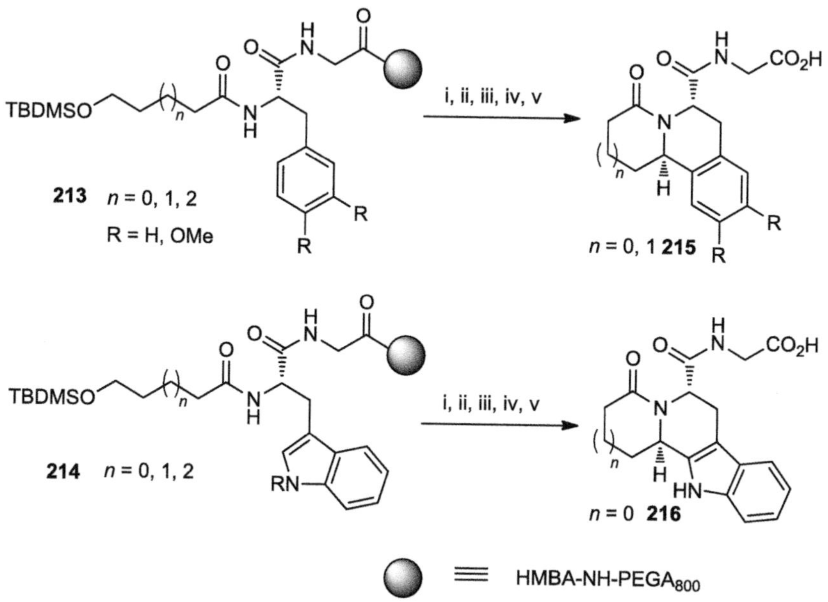

SCHEME 7.28 Conversion of silylethers precursors to tetrahydropyrrolo- and tetrahydropyrido-isoquinolinones and tetrahydroindolizino-indolones. Reagents and conditions: (i) TBAF, AcOH, THF; (ii) Dess–Martin periodinane, DCM; (iii) 10% TFA (aq); (iv) 50% TFA/DCM; (v) 0.1 M NaOH (aq.).

phenylalanine (R = H), (3,4-dimethoxyphenyl)alanine (R = OMe), and tryptophan to form precursors **213** (R = H, OMe), and **214** suitable for the following folding process (Scheme 7.28). The length of a side chain, as well as the amino acid residue, preencoded the character of the final products.

Deprotection of hydroxyl groups was followed by Dess–Martin periodinane-mediated oxidation. However, final intramolecular *N*-acyl iminium Pictet–Spengler reaction only led to products **215** (n = 0,1) and **216** (n = 0); formation of (7 + 6)-fused ring systems (for compounds **215** and **216**) and (6 + 6)-fused ring system (for compound **216**) failed.

Other examples of the use of a folding process to approach diversity scaffolds reported by the same authors [53] are discussed in Section 7.5. The folding process is synonymous for the pair phase, the last step of the B/C/P approach (Scheme 7.39).

An example of a folding process based on regioselectivity has also been described [57]. Heating of hydrazide precursors **217** and **218** in basic conditions at 80°C resulted in the formation of either 3-aminohydantoins **219** or 1,2,4-triazine-3,6-diones **220**, depending on the substitution of two nitrogen atoms of the hydrazide moiety (Scheme 7.29). Resins **217** afforded 3-aminohydantoins **219**. In contrast, when the hydrogen atom was attached to terminal nitrogen, 1,2,4-triazine-3,6-diones **220** were obtained instead.

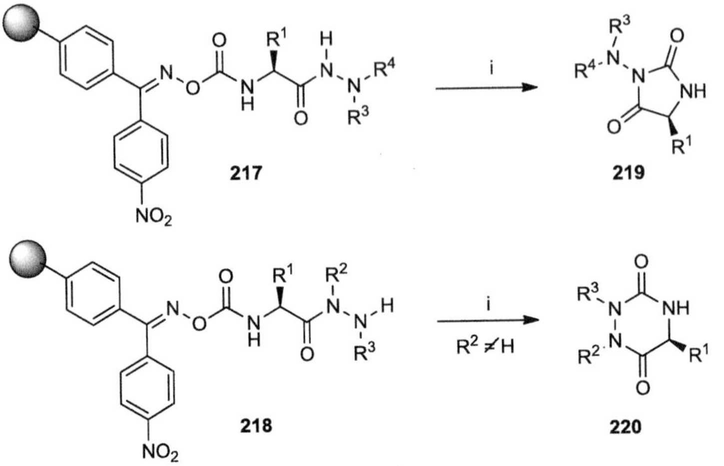

SCHEME 7.29 Conversion of hydrazide precursors to 3-aminohydantoins and 1,2,4-triazine-3,6-diones. Reagents and conditions: (i) DIEA, DMF, 80°C.

7.3 STEREOCHEMICAL DIVERSITY

An intriguing approach to chemical diversity is the preparation of compounds differing in stereochemistry. These molecules have different three-dimensional spatial arrangements but identical two-dimensional structures. The most illustrious example of stereochemical diversity is a combinatorial synthesis of peptides using both L- and D-amino acids, applied successfully long before the term DOS was introduced [58]. In an analogous manner, using amino acid–based building blocks, stereochemical diverse spiro-ladder tetramers were synthesized [59]. All of the tetramers were a permutation of four *pro4* monomers (Scheme 7.30). To synthesize a particular oligomer, monomers **225** were coupled in appropriate sequence to Rink amide resin **221**. Final tetramers **222** to **224** were cleaved from the resin with 8% TFMSA in TFA, precipitated, and dissolved in 20% piperidine in NMP to trigger closure of the second amide bond between adjacent residues.

Another example focusing exclusively on stereochemical diversity is the synthesis of pyrrolo[3,4-*c*]pyrrole and pyrrolidine derivatives from Merrifield resin-supported α-silylimines [60,61]. α-Silylimines **226** were treated with dipolarophiles to yield two heterocyclic cores in two and four diastereomeric forms, respectively (Scheme 7.31).

First, the Ph derivative (R = Ph) was examined. When the reaction proceeded at 180°C for 6 h, the two diastereoisomers **227** and **228** were obtained in the ratio 36 : 64 (the stereoselectivity was rather low under those conditions). As expected, when the reaction was carried out at 40°C for 48 h, the ratio changed in favor of diastereoisomer **227** (the ratio was 93 : 7). The cycloaddition was then carried out with other para-substituted phenyl derivatives. Even though the ratio was influenced by the substitution, more than 74% of cycloadduct **227** was formed in all cases.

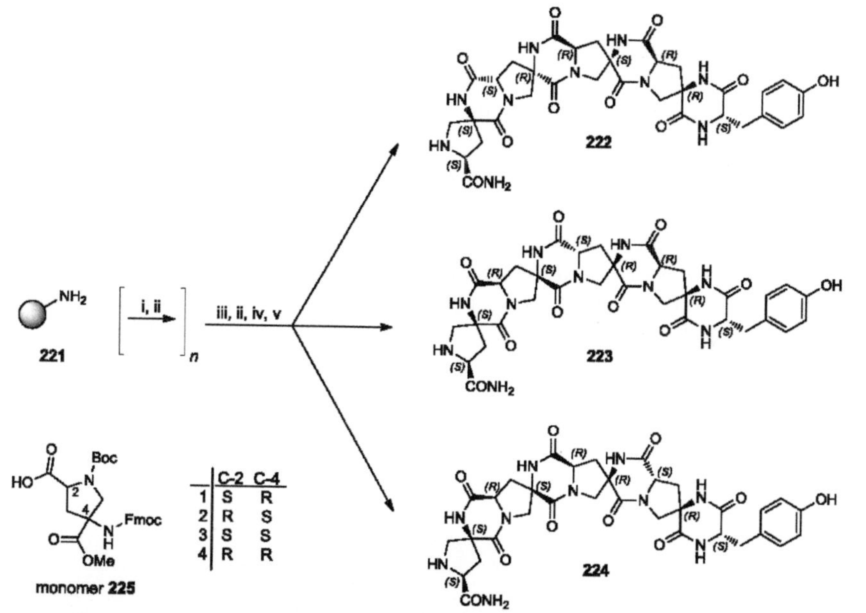

SCHEME 7.30 Synthesis of diastereoisomeric tetramers. Reagents and conditions: (i) monomer **225**, HATU, DIEA, DMF/DCM; (ii) 20% piperidine/DMF; (iii) Fmoc-L-Tyr, HATU, DIEA, DMF/DCM; (iv) 8% TFMSA/TFA, ice bath, 1.5 h; (v) precipitation from Et_2O, then 20% piperidine/NMP, 48 h.

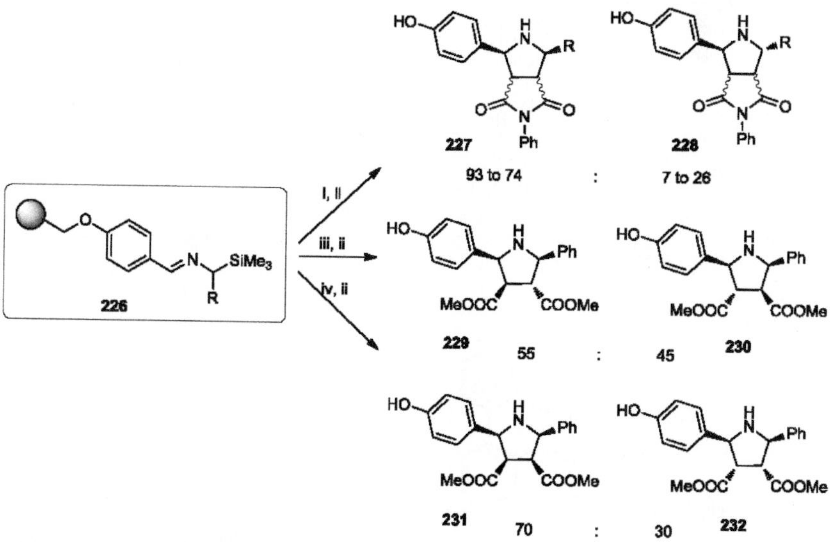

SCHEME 7.31 Synthesis of diastereoisomers of two different cycloadducts. Reagents and conditions: (i) *N*-phenylmaleimide, F_3SiPh, toluene, 40 or 60°C; (ii) TFA/DCM, rt, 24 h; (iii) dimethyl fumarate, F_3SiPh, toluene, 40°C, 48 h; (iv) dimethyl maleate, F_3SiPh, toluene, 40°C, 48 h.

The last two reaction sequences involved treatment of precursor **226** with two dipolarophiles, dimethyl fumarate and dimethyl maleate, to yield diastereoisomers **229**, **230** and **231**, **232**, respectively. Although the stereoselectivities of those reactions were not satisfactory, retention of the fumarate and maleate stereochemistry was observed in both cases.

The next example illustrates the introduction of chirality during diversity-oriented stereoselective synthesis leading to complex small molecules with different three-dimensional structures [62,63]. An aldehyde was immobilized on high-capacity macrobeads **10** and subjected to reductive amination with an amino alcohol. Subsequent reductive alkylation of the secondary amine with *o*-bromobenzaldehyde followed by O-alkylation resulted in polymer-supported O-benzylated intermediate **233** (Scheme 7.32).

Intermediate **233** was subjected further to metalation with *t*-BuLi or *i*-$PrBu_2MgLi$, followed by transmetallation with CuCN·2LiBr. Oxidation of the cuprate by 1,3-DNB

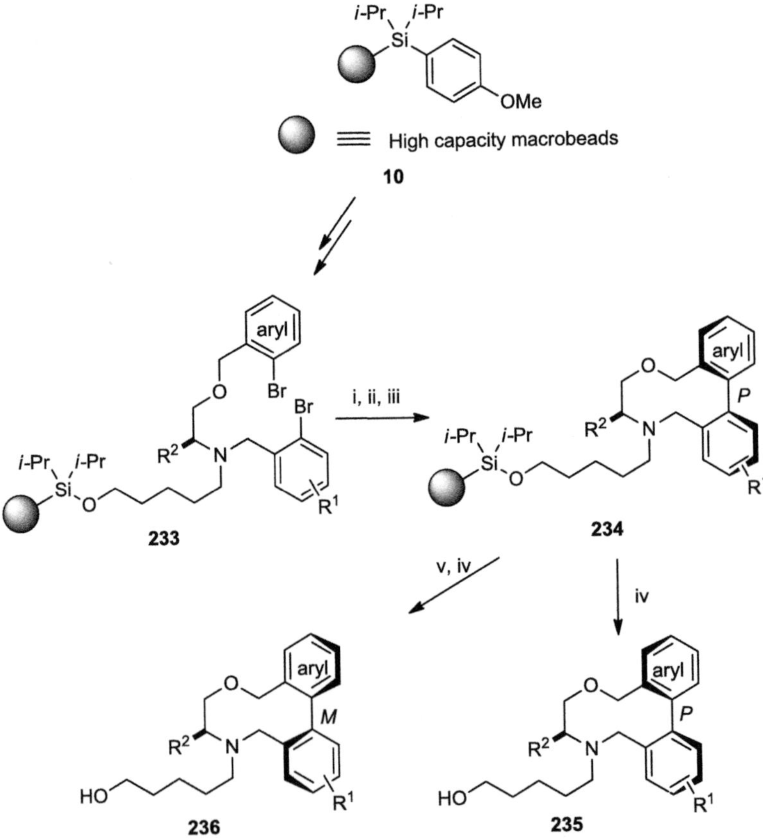

SCHEME 7.32 Stereoselective synthesis of a library containing natural product–like scaffolds. Reagents and conditions: (i) *t*-BuLi, benzene or *i*-$PrBu_2MgLi$ (2.2 equiv); (ii) CuCN·2LiBr, 2-MeTHF; (iii) 1,3-DNB; (iv) HF·pyridine; (v) 150°C, 24 h.

yielded derivative **234**. The final product **235** [dr = 6 : 1 (*P* : *M*)] was released from the resin by treatment with HF·pyridine. Alternatively, the precursor **234** was heated prior to cleavage from the support to reverse the atropdiastereomeric ratio, leading to product **236** [dr = 1:7 (*P* : *M*)]. The number of library compounds was 1412. The products were submitted for numerous phenotypic and protein-binding assays.

N-Acyl iminium ion cyclizations/nucleophilic addition allowed formation of stereochemically diverse skeletons based on tetrahydro-2*H*-oxazolo[3,2-*a*]pyrazin-5(3*H*)-ones and tetrahydro-2*H*-thiazolo[3,2-*a*]pyrazin-5(3*H*)-ones (Scheme 7.33) [64]. The conversion was based on unmasking of the aldehyde that underwent intramolecular condensation with an amide to yield iminium ions. The cyclization proceeded toward the peptide carboxy terminus (called *eastbound cyclization*). The iminium ions were attacked immediately by internal nucleophiles (oxygen or sulfur) to afford final bicyclic heterocycles **238** to **241**. Depending on the stereochemistry of the α-amino acid used (introducing an R^2 substituent), diastereoisomers **238** or **240** were created

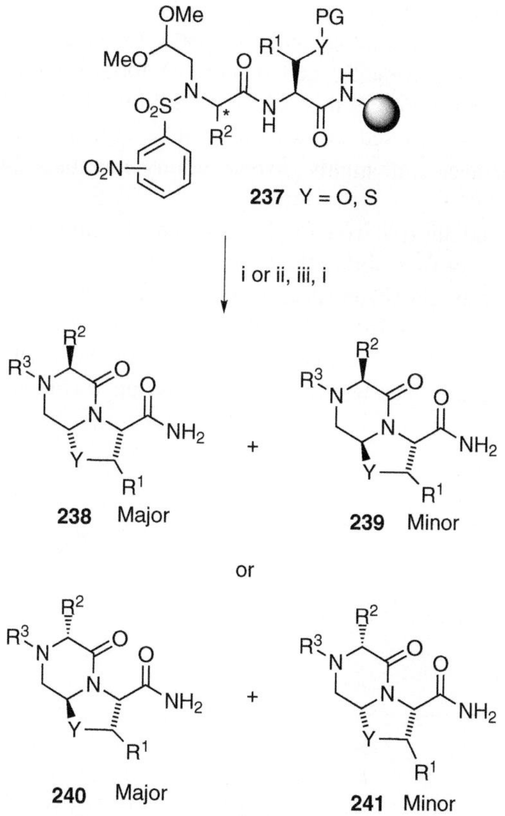

SCHEME 7.33 Stereoselective synthesis of eight diastereoisomers based on two types of heterocycles. Reagents and conditions: (i) 50% TFA/DCM, rt, 2 to 3 h; (ii) 2-mercaptoethanol, DBU, DMF, rt, 10 min; (iii) R^3SO_2Cl, 2,6-lutidine, DCM, rt, on.

as the major products. The iminium chemistry that enabled synthesis of heterocycles with skeletal diversity also proceeded with full control of their stereochemistry (Scheme 7.21) [43].

7.4 APPENDAGE DIVERSITY

Appendage diversity is achieved by incorporation of various substructures onto a common-core unit. To enhance the diversity with an array of compounds, the substructural elements need to possess an inherent diversity and contribute to the topographic variability of individual compounds. As an example we included a combinatorial library of bis-heterocyclic structures. The diversity is achieved by incorporation of the second heterocycle (the appendage) into three different positions of the parent molecule.

The synthesis of bis-heterocyclic compounds, where two heterocyclic cores were connected by a variable spacer, was carried out in polypropylene-reaction vessels by using split-and-split methodology and was made "by hand" without sophisticated instrumentation [65]. Traditionally, generic combinatorial libraries contain one heterocyclic scaffold, and therefore their diversity is limited. The library compounds herein presented are composed of two different heterocyclic rings connected by a spacer that gave access to highly diverse library members **243**, **245**, and **247** (Scheme 7.34).

Benzimidazole and thiazole were used as core units because of their drug-likeness. The building blocks for assembling the benzimidazole core were selected to enable the introduction of functionalized spacers in any of three diversity positions of benzimidazole. The spacer in position 1 of the benzimidazole ring was introduced in the first combinatorial step by reductive amination of the aldehyde linker with a protected amino alcohol. The terminal hydroxyl group was then converted into an

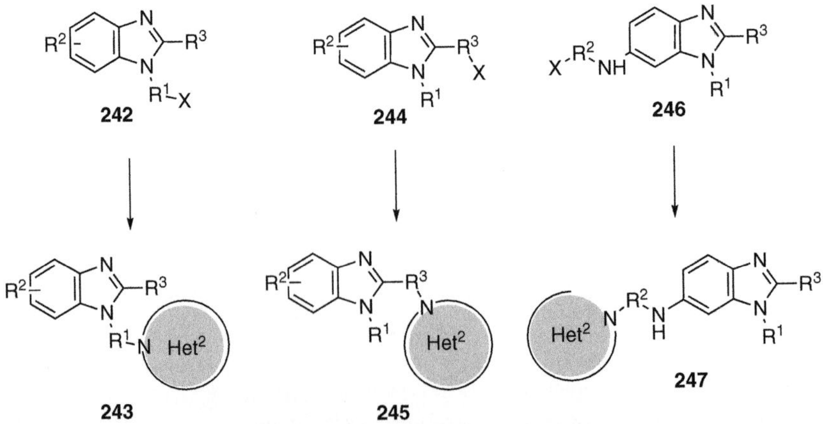

SCHEME 7.34 Bis-heterocyclic library members.

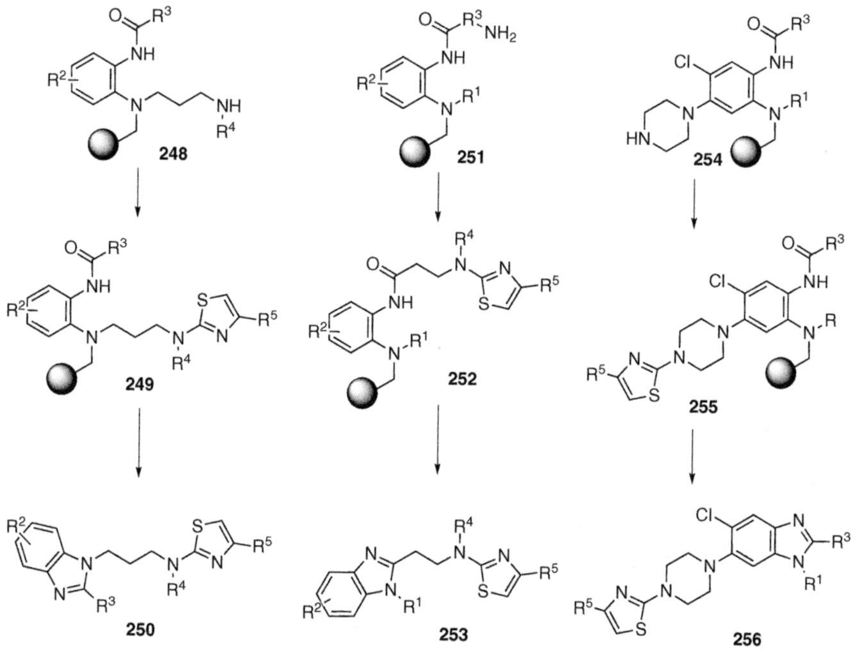

SCHEME 7.35 Benzimidazole connected to thiazole core at three positions by different spacers.

amino group via mesylation to afford structure **248** (after introduction of an *o*-aminobenzene moiety) (Scheme 7.35). Alternatively, acylation of polymer-supported *o*-phenylenediamines with Fmoc-amino acid introduced the spacer in position 2 (structure **251**). To introduce the spacer at position 6 of benzimidazole ring, a 1,2-dichloro-4-fluoro-5-nitrobenzene was used. First, the fluorine was substituted with polymer-supported amines and the chlorine in the para position with respect to the nitro group was subsequently displaced by a secondary diamine (piperazine) to afford structure **254** (after acylation of aniline nitrogen). As an example, we included the reaction sequence for the synthesis of one sublibrary (Scheme 7.36). The reaction sequence involved reduction, acylation, alkylation, and cyclization reactions; over 100 derivatives were prepared.

7.5 BUILD/COUPLE/PAIR STRATEGY

In 2008, Nielsen and Schreiber introduced a new term for a general approach to synthesis of stereochemically and skeletally diverse small molecules, referred to as the *build/couple/pair* (B/C/P) *strategy* (Scheme 7.37) [2]. The B/C/P approach describes the synthesis of small molecules having diverse skeletons, stereochemistry, and appendages [66]. The B/C/P pathway starts with the build phase, in which building blocks with preexisting stereogenic centers are synthesized [67]. Building blocks also bear functionalities suitable for further coupling reactions. The following

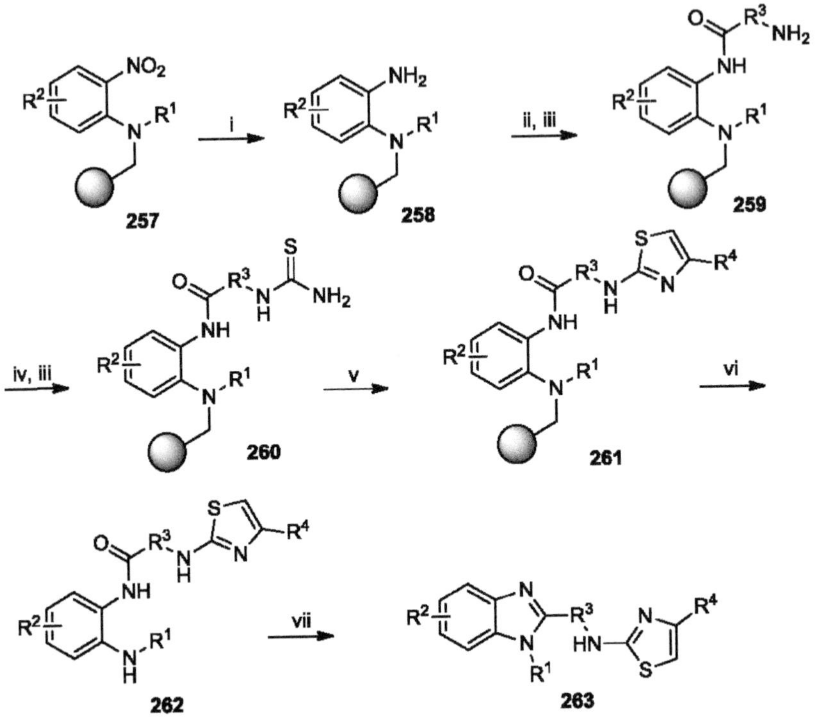

SCHEME 7.36 Synthesis of a library of bis-heterocyclic compounds. Reagents and conditions: (i) $SnCl_2 \cdot 2H_2O$, DIEA, DMF, rt, 2 h to on; (ii) Fmoc-amino acid, DIC, THF (for temperature and time, see original paper); (iii) 50% piperidine/DMF, rt, 30 min; (iv) Fmoc-NCS, DCM, rt, 1 h; (v) haloketone, DCM, rt, on; (vi) 50% TFA/DCM, rt, 30 min; (vii) AcOH, 60°C (for reaction time, see original paper), then 5% NaOH/MeOH, 60°C, on (when amino alcohols were used in the first combinatorial step).

couple phase involves intermolecular coupling reactions. At this stage building blocks are joined in all stereochemical combinations and yield functionalized molecules ready to enter into the last stage, the pair phase. In this last step, intramolecular coupling reactions (functional group-pairing reactions) between appendages within the molecule lead to diverse scaffolds [2]. The simplest illustrious example of the B/C/P strategy is the synthesis of cyclic peptides using L- and D-amino acids.

The synthetic strategies of several DOSs that fall into more than one category were described in previous sections. For example, in 2006, two years before the B/C/P pathway was introduced [2], Mitchell and Shaw described the synthesis of a diverse library of polycyclic lactams that can be regarded as a B/C/P approach [68]. The synthesis began with the build step, which involved the preparation of polymer-supported methoxy-substituted oxazole **264** and of various aromatic aldehydes as well as benzyl halides decorated with azide in the ortho position (both prepared in solution) (Scheme 7.38). In the next step, the couple phase [step (i)], enantioselective

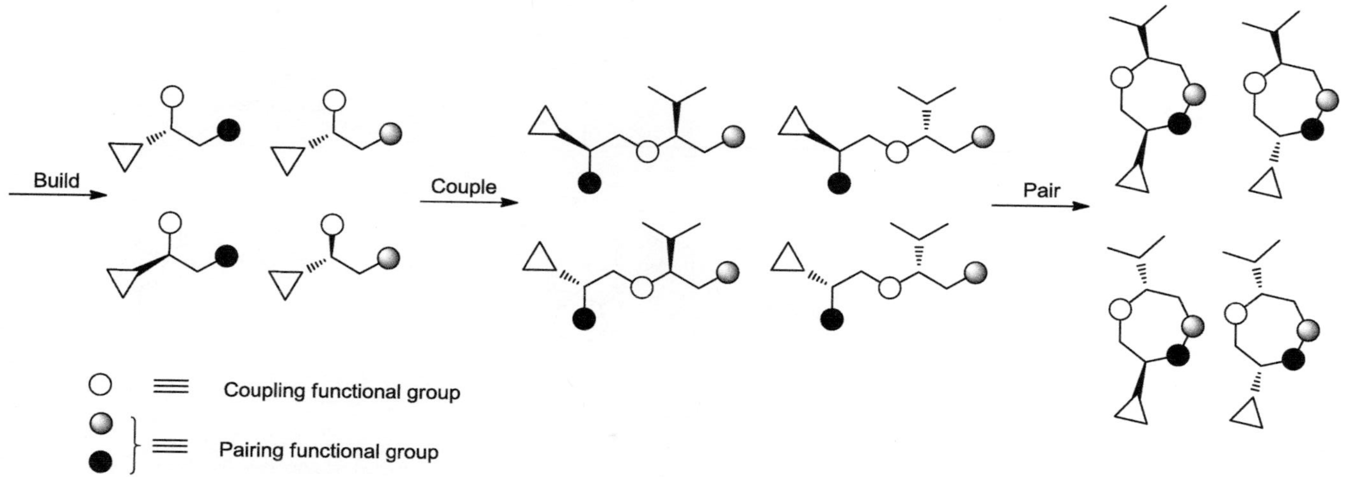

SCHEME 7.37 B/C/P approach to generate skeletal diversity.

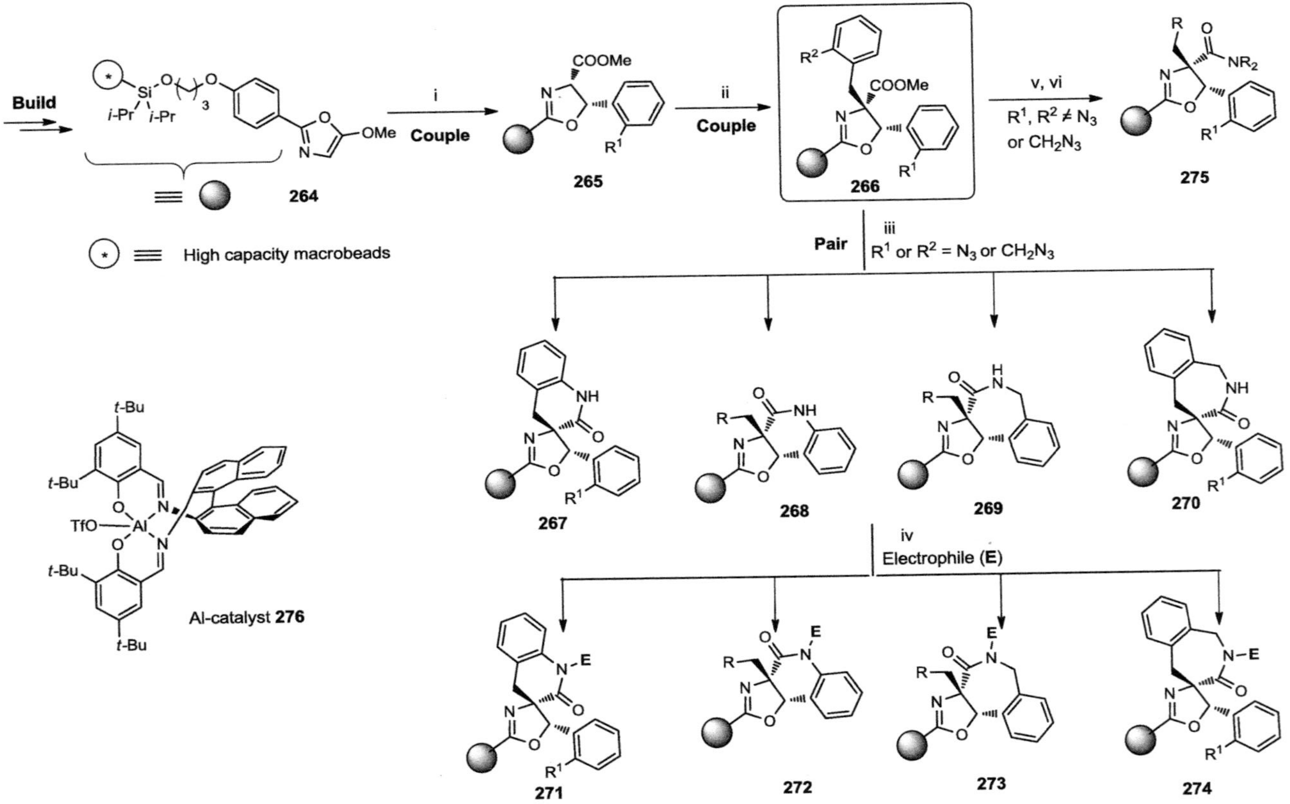

SCHEME 7.38 Catalytic enantioselective Suga–Ibata reactions and diastereoselective enolate alkylation (couple phase) and subsequent Staudinger-type reductive cyclizations (pair phase). Reagents and conditions: (i) *o*-azide-substituted benzaldehydes, Al-complex **276**, 3 Å MS, DCM; (ii) *o*-azide-substituted benzyl bromides, BEMP or BTPP, tetrabutylammonium iodide, NMP; (iii) PMe_3, DBU, toluene/H_2O/dioxane; (iv) BEMP or BTPP, electrophiles; (v) TMSOK, THF; (vi) PyBrOP, R^4NH, DIEA, DCM.

Suga–Ibata reaction of oxazole **264** proceeded with aromatic aldehydes to afford intermediates **265**. Subsequent alkylation with benzylhalides yielded compounds bearing azide moiety **266**, which were ready for the pair phase [step (iii)], Staudinger-type reduction, which resulted in spontaneous cyclization to appropriate spirocyclic and fused lactams **267** to **270**. Lactams were alkylated to the target products **271** to **274** [step (iv)]. In addition, methyl esters **266** were converted into the corresponding amides **275** to complement the complexity of the library. The pair phase of the B/C/P pathway is very often connected narrowly with the folding pathway. For example, the pair phase in the synthesis discussed above (Scheme 7.38) [68] resulted in either spirocyclic or fused lactams **267** to **270**, depending on the type of azide moiety (R^1 and R^2) attached to the intermediates **266**.

The following syntheses included the folding process in the pair phase [2,53,69–71]. Various α-amino acids were coupled by traditional solid-phase peptide synthesis to afford masked peptide aldehydes **280** (Scheme 7.39). In the pair phase the aldehyde was liberated and reacted immediately with the amide backbone to yield *N*-acyl iminium intermediates, which were stabilized by internal nucleophiles to provide a wide range of diverse heterocyclic scaffolds **281** to **289**.

7.6 SCAFFOLD HOPPING

Last but not least, access to chemical diversity represents the *scaffold hopping approach*, a term introduced in 1999 [72]. For medicinal chemists this method represents an important drug-design strategy. The aim of scaffold hopping is to discover structurally novel compounds starting from known active compounds by modifying the central core structure of the molecule [73]. Scaffold hopping can be used not only for improving the selectivity/specificity of a lead compound, but also for replacing a component of a molecule that is undesirable (e.g., for its toxicity or insolubility) or to avoid intellectual property infringements [74]. For example, replacement of the 1*H*-pyrazolo[4,3-*d*]pyrimidin-7(6*H*)-one core in Sildenafil to imidazo[5,1-*f*][1,2,4]triazin-4(3*H*)-one led to Vardenafil [75].

Scaffold hopping was classified into four categories: heterocycle replacement, ring opening and closure, peptidomimetics, and topology-based scaffold hopping [76]. Only a few examples of scaffold hopping were reported on solid phase.

An example of the ring opening and closure category is a conversion of polymer-supported *N*-alkyl-2-nitro-*N*-(2-oxo-2-arylethyl)benzenesulfonamides **290** into 2*H*-indazole 1-oxides and quinazolines, respectively (Scheme 7.40) [77]. The presence of an "activated" methylene derived from glycine was critical for the transformation to quinazolines. Exposure of *N*-alkyl-2-nitro-*N*-(2-oxo-2-arylethyl)benzenesulfonamides **290** to a base led to formation of polymer-supported 2*H*-indazole 1-oxides **291**. While following cleavage from the support, derivatives not bearing the methylene afforded indazoles **292** [78], the presence of methylene activated by the neighboring carbonyl group had a crucial influence on further transformations of indazoles **291** to quinazolines **293**. An exposure to DBU in DMF led

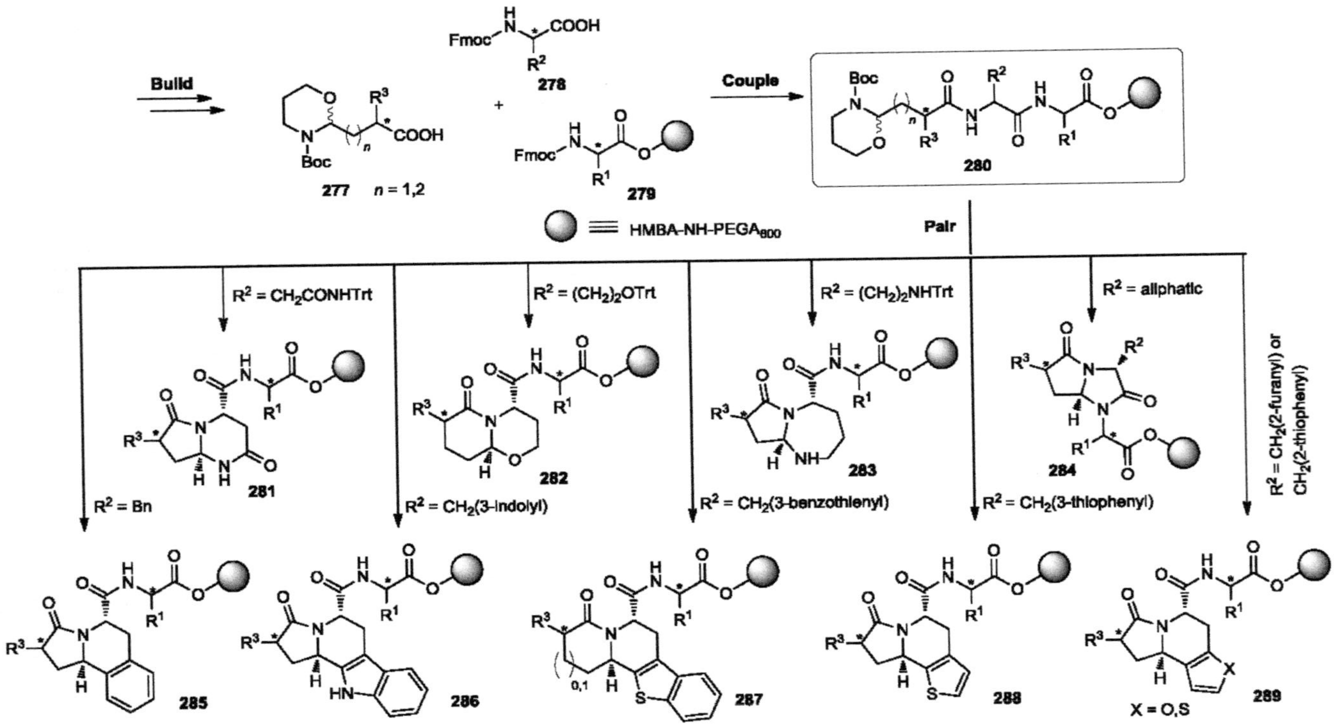

SCHEME 7.39 Peptide synthesis (couple phase) and subsequent aldehyde–amide condensation followed by addition of a nucleophile to iminium ion intermediates (pair phase). (From [2], with permission of John Wiley & Sons; copyright © 2008 John Wiley & Sons.)

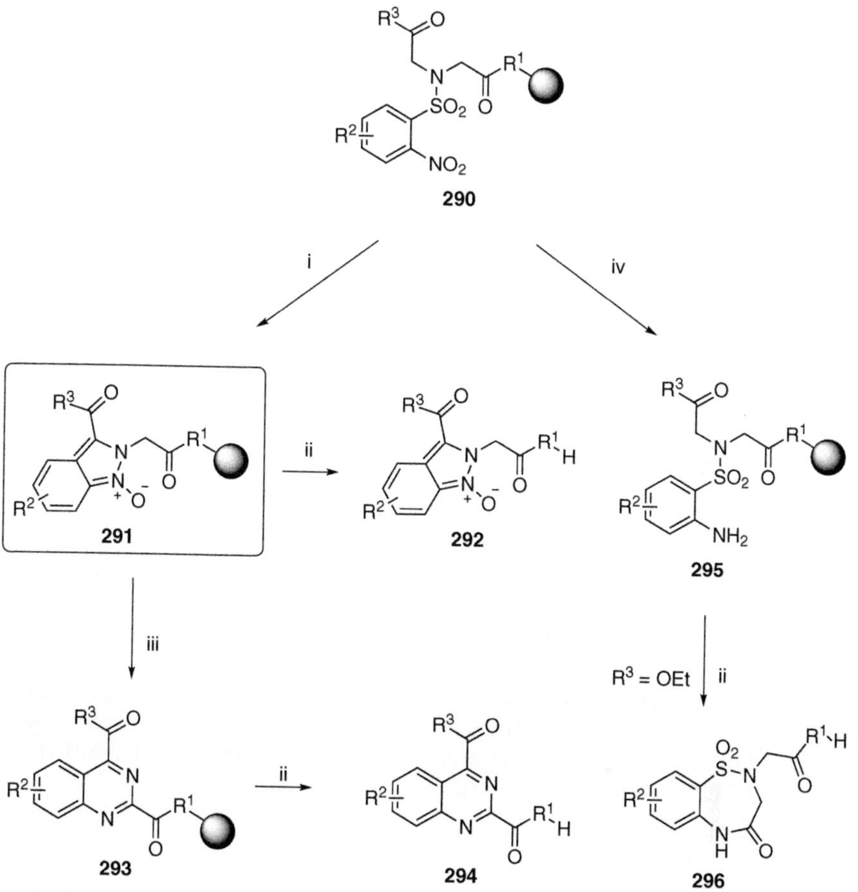

SCHEME 7.40 2*H*-Indazole 1-oxides-ring expansion to quinazolines and to 2,3-dihydrobenzo[*f*][1,2,5]thiadiazepine 1,1-dioxides. Reagents and conditions: (i) 0.2 M DBU, DMF, rt, 30 min; (ii) 50% TFA/DCM, rt, 1 h; (iii) 0.1 M to 0.2 M DBU, DMF, rt, 10 min to 16 h; (iv) $SnCl_2{\cdot}2H_2O$, DIEA, DMF, rt, on.

to rearrangement involving 2*H*-indazole 1-oxide core opening followed by quinazoline **293** ring closure. *N*-Alkyl-2-nitro-*N*-(2-oxo-2-arylethyl)benzenesulfonamides **290** can also serve for the synthesis of (6 + 7)-membered heterocycle **296**. However, this route does not represent a scaffold hopping process.

γ-Lactones, γ-lactams, and cyclobutane derivatives were synthesized from Merrifield resin-supported cyclobutanone precursors **298** [61,79]. [2 + 2] Cycloaddition was used to prepare cyclobutanone iminium species **297** that was hydrolyzed to the ketones **298** and further subjected to various reaction routes, leading to either four- or five-membered rings (Scheme 7.41).

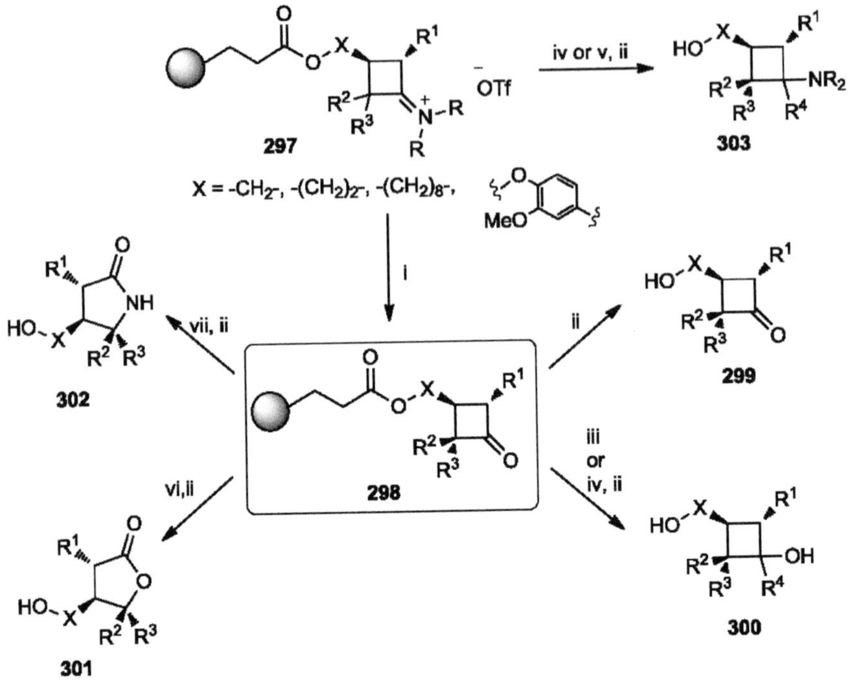

SCHEME 7.41 Synthesis of cyclobutanes and ring expansion to lactone and lactam derivatives. Reagents and conditions: (i) THF, $NaHCO_3$ (aq.); (ii) TMSOK, MeOH, DCM or pyrrolidine; (iii) $LiBH_4$, MeOH, THF; (iv) MeMgCl, THF, −78°C → −10°C; (v) $Me_4NBH(OAc)_3$, DCM; (vi) *m*CPBA, DCM; (vii) *O*-mesitylenesulfonylhydroxylamine, DCM, rt.

The polymer-supported cyclobutanone was cleaved from the resin directly by treatment with TMSOK to afford liberated cyclobutanone **299**. Cyclobutane with hydroxyl group **300** was prepared via two routes. The first route involved nucleophilic addition with Grignard reagent and subsequent cleavage from the resin. The second involved reduction with $LiBH_4$ to yield tertiary alcohol **300** (R^4 = Me) and secondary alcohol **300** (R^4 = H), respectively. Alternatively, iminium ions **297** were converted into amino derivatives **303**. In addition to four-membered ring hydrocarbons, ring expansion to five-membered lactons and lactams was also performed. Baeyer–Villiger ring expansion of ketones **298** afforded γ-lactones **301**, while Beckmann rearrangement yielded γ-lactams **302**.

The next example outlines the synthesis of a pyranone-derived scaffold. Polymer-supported furyl alcohols **304** were oxidized to 2*H*-pyran-3(6*H*)-ones **305**, having many reactive sites and functionalities and thus amenable to further transformation into diverse pharmacologically important heterocyclic compounds by selective pairing (Scheme 7.42) [4].

The key precursor **305** was converted into carbamate **306** that was either treated with alcohols to give ethers **307** or afforded oxazolediones **308** under basic

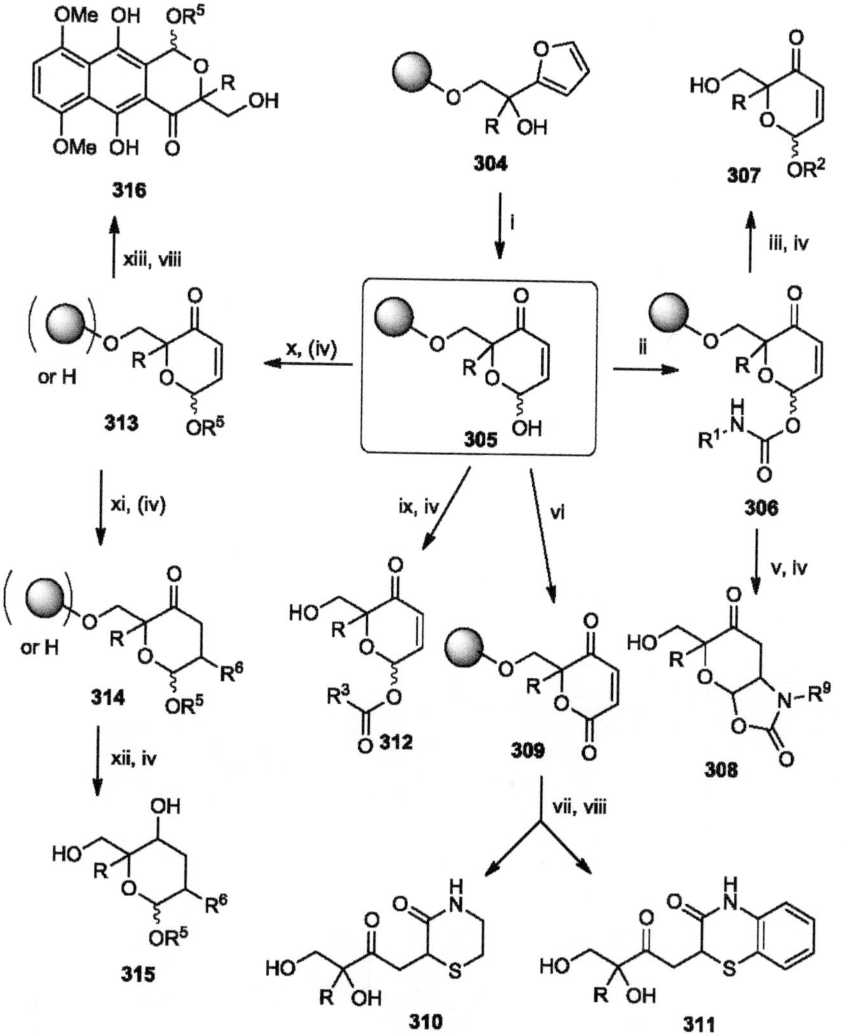

SCHEME 7.42 Conversion of furyl alcohols to 2*H*-pyran-3(6*H*)-ones and their further derivatization. Reagents and conditions: (i) NBS, THF/H_2O (4 : 1), 0 → 25°C, 1 h; (ii) R^1NCO, TEA, 0 → 25°C, 2 h; (iii) R^2OH, THF, 70% $HClO_4$ (cat.), 25°C, 1 h; (iv) DDQ, DCM/H_2O (18:2), 25°C, 2 to 3 h, then ascorbic acid; (v) DBU, DCM, 25°C, 12 h; (vi) Dess–Martin reagent, DCM, 25°C, 12 h; (vii) binucleophile, DCM, 25°C, 10 h; (viii) DCM/TFA (6 : 1), 25°C, 8 to 10 h; (ix) R^3COCl, TEA, 25°C, 12 h; (x) $CH(OR^5)_3$, $BF_3{\cdot}Et_2O$, THF, 0°C, 1 h; (xi) R^6SH, TEA, DCM, 25°C, 2 h or NaN_3 (aq.), AcOH, pH 5, 25°C, 1 h; (xii) $NaBH_4$, THF/H_2O (1 : 4), 25°C, 0.5 h; (xiii) 4,7-dimethoxy-3-oxodihydroisobenzofuran-1-carbonitrile, LDA, THF, −78 → 0°C, 4 h.

conditions. In both cases the final products were released from the support by treatment with DDQ. Alternatively, oxidation with Dess–Martin reagent afforded pyrones **309**, which were treated with binucleophiles to afford thiazinones **310** or benzothiazinones **311**. These products were cleaved from the resin under acidic conditions. Precursor **305** was also esterified or etherified. While ester **312** was released from the resin with DDQ, ether **313** was either cleaved under the same conditions or was further derivatized to deoxysugar derivative **314**. It was either cleaved from the resin as a final compound or was reduced prior to the cleavage to yield tetrahydro-2*H*-pyran-2,5-diol **315**. Finally, intermediate **313** was also treated with 4,7-dimethoxy-3-oxodihydroisobenzofuran-1-carbonitrile to afford naphthoquinone **316** under acidic cleavage from the solid support.

Alternatively, scaffold hopping was utilized for the preparation of two diverse scaffolds by Diels–Alder reaction. The key precursor **319** was achieved by extraction from several plants, such as *Piper nigrum*, *Eugenia caryophyllata*, and *Pimenta dioica*, in organic solvent in the presence of resin-bound nitroso compound **317** (Scheme 7.43) [80]. The resin selectively sequestered only one of the substances extracted, the β-caryophyllene **318**. Resin-bound ene product **319** was converted to two structurally

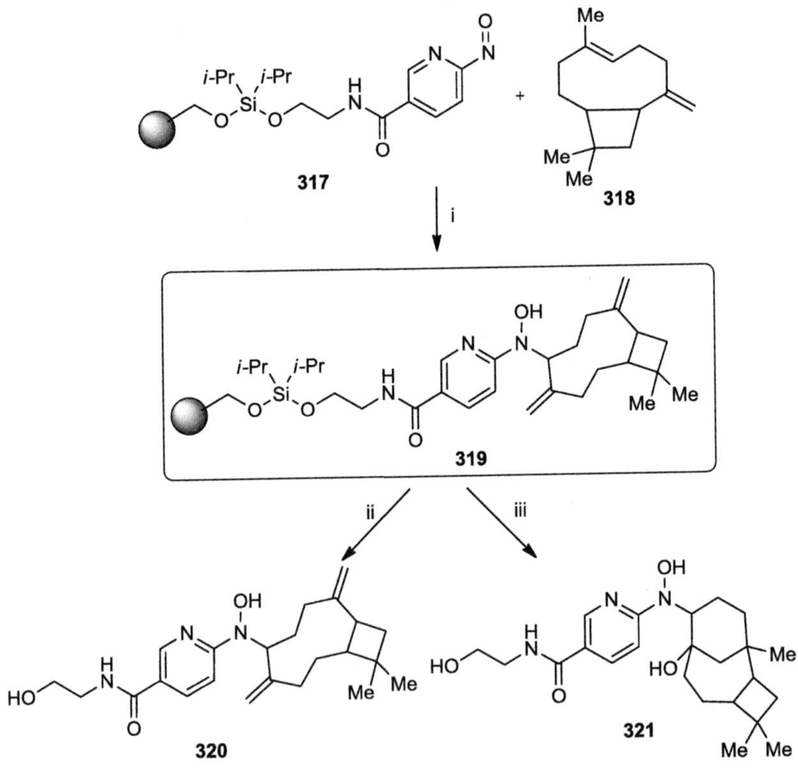

SCHEME 7.43 Synthesis of diverse scaffolds from natural product extracts. Reagents and conditions: (i) DCM, 1 h; (ii) 0.1 M TBAF, THF, rt, 30 min (iii) 50% TFA/DCM, rt, 1 h.

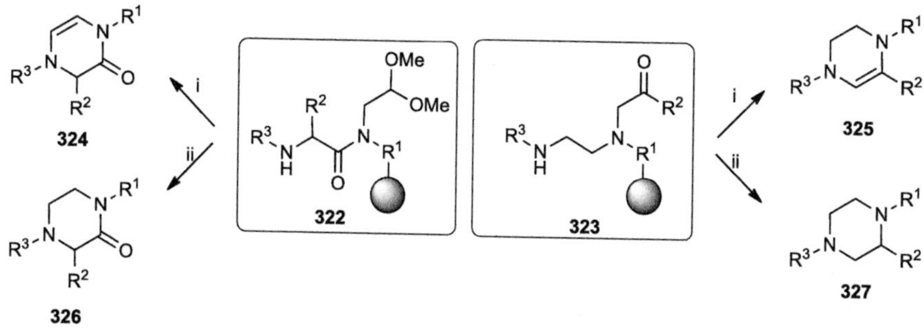

SCHEME 7.44 Synthesis of trisubstituted piperazinones, piperazines, and their unsaturated analogs. Reagents and conditions: (i) 50% TFA/DCM, rt, 30 min; (ii) TFA/TES/DCM (5 : 1 : 4), rt, 30 min.

unrelated carbocycles. While treatment of the precursor **319** with TBAF-cleavage cocktail resulted in ene product **320**, cleavage in TFA led to rearrangement and formation of tricycle **321**, which was also obtained using Wang resin and TFA-mediated cleavage. Both isolated compounds were screened for antibacterial activity assay.

An analogous strategy of scaffold hopping portrays the conversion of resin-bound precursors **322** or **323** into nitrogenous heterocycles, depending on cleavage conditions. TFA solution in DCM and solution of TFA with addition of a reducing agent (TES) afforded different heterocycles (Scheme 7.44) [81]. Iminium ions were transformed to 3,4-dihydropyrazin-2(1*H*)-ones **324** and 1,2,3,4-tetrahydropyrazines **325** (in the case of TFA-mediated cleavage) or by formation of piperazin-2-ones **326** or piperazines **327** (in the case of TFA/TES-mediated cleavage). Piperazinones and piperazines are known to be important pharmacophores [82].

7.7 CONCLUSIONS

In this chapter we have described numerous approaches to arriving at diverse compounds using polymer-supported synthesis, including skeletal-, stereochemical-, and appendage diversity-generating processes. We focused mainly on skeletal diversity, due to its great potential to create diverse heterocyclic compounds, although a combination of strategies such as those described in the B/C/P method provides the most promising potential for diversity creation. In our opinion the critical advantage of the use of solid-phase chemistry for DOS resides in a potential to build (and synthesize) a large number of very diverse compounds in a relatively short period of time together with the operational simplicity of the process (remember, isolation of intermediates after each reaction step takes about 5 minutes!). Obviously, the chemistry needs to be optimized and robust enough to afford reasonable purity of target compounds. Because solid-phase chemistry does not allow the purification of intermediates, the purity of crude products depends on the entire sequence of individual reaction steps. However, one should not be discouraged if one reaction in the entire sequence does not provide high-purity intermediates; the entire sequence of reaction steps may still be amenable to solid-phase synthesis. Purification of crude compounds is typically carried out by reversed-phase HPLC.

Acknowledgments

Writing of this chapter was supported by the Department of Chemistry and Biochemistry, University of Notre Dame, by projects P207/12/0473 from GACR and CZ.1.07/2.3.00/20.0009 from the European Social Fund.

REFERENCES

1. M. D. Burke, S. L. Schreiber, *Angew. Chem. Int. Ed.* **2004**, *43*, 46–58.
2. T. E. Nielsen, S. L. Schreiber, *Angew. Chem. Int. Ed.* **2008**, *47*, 48–56.
3. V. Krchňák, A. S. Weichsel, D. Cabel, M. Lébl, *Pept. Res.* **1995**, *8*, 198–205.
4. E. A. Couladouros, A. T. Strongilos, *Angew. Chem. Int. Ed.* **2002**, *41*, 3677–3680.
5. B. E. Evans, K. E. Rittle, M. G. Bock, R. M. DiPardo, R. M. Freidinger, W. L. Whitter, G. F. Lundell, D. F. Veber, P. S. Anderson, R. S. L. Chang, V. J. Lotti, D. J. Cerino, T. B. Chen, P. J. Kling, K. A. Kunkel, J. P. Springer, J. Hirshfieldt, *J. Med. Chem.* **1988**, *31*, 2235–2246.
6. J. S. Mason, I. Morize, P. R. Menard, D. L. Cheney, C. Hulme, R. F. Labaudiniere, *J. Med. Chem.* **1999**, *42*, 3251–3264.
7. K. C. Nicolaou, J. A. Pfefferkorn, G. Q. Cao, *Angew. Chem. Int. Ed.* **2000**, *39*, 734–739.
8. K. C. Nicolaou, J. A. Pfefferkorn, A. J. Roecker, G. Q. Cao, S. Barluenga, H. J. Mitchell, *J. Am. Chem. Soc.* **2000**, *122*, 9939–9953.
9. D. G. Hall, S. Manku, F. Wang, *J. Comb. Chem.* **2001**, *3*, 125–150.
10. S. Oh, H. J. Jang, S. K. Ko, Y. Ko, S. B. Park, *J. Comb. Chem.* **2010**, *12*, 548–558.
11. M. Zhu, B. J. Lim, M. Koh, S. B. Park, *ACS Comb. Sci.* **2011**, *14*, 124–134.
12. X. Huang, J. F. Xu, *J. Comb. Chem.* **2009**, *11*, 350–354.
13. G. J. T. Kuster, L. W. A. van Berkom, M. Kalmoua, A. van Loevezijn, L. A. J. M. Sliedregt, B. J. van Steen, C. G. Kruse, F. P. J. T. Rutjes, H. W. Scheeren, *J. Comb. Chem.* **2006**, *8*, 85–94.
14. H. U. Stilz, W. Guba, B. Jablonka, M. Just, O. Klingler, W. Koenig, V. Wehner, G. Zoller, *J. Med. Chem.* **2001**, *44*, 1158–1176.
15. E. Naydenova, N. Pencheva, J. Popova, N. Stoyanov, M. Lazarova, B. Aleksiev, *Farmaco* **2002**, *57*, 189–194.
16. J. Vinsova, K. Kosar, E. Kasafirek, *Collect. Czech. Chem. Commun.* **1993**, *58*, 2987–2993.
17. S. J. Taylor, A. M. Taylor, S. L. Schreiber, *Angew. Chem. Int. Ed.* **2004**, *43*, 1681–1685.
18. D. S. Tan, *Nat. Chem. Biol.* **2005**, *1*, 74–84.
19. P. T. Reddy, S. Quevillon, Z. Gan, N. Forbes, D. M. Leek, P. Arya, *J. Comb. Chem.* **2006**, *8*, 856–871.
20. D. Barnes-Seeman, S. B. Park, A. N. Koehler, S. L. Schreiber, *Angew. Chem. Int. Ed.* **2003**, *42*, 2376–2379.

21. W. Li, Y. Lam, *J. Comb. Chem.* **2005**, *7*, 644–647.
22. W. Ma, B. Peterson, A. Kelson, E. Laborde, *J. Comb. Chem.* 2009, *11*, 697–703.
23. L. De Luca, G. Giacomelli, A. Porcheddu, M. Salaris, M. Taddei, *J. Comb. Chem.* **2003**, *5*, 465–471.
24. J. Y. Hwang, H. S. Choi, D. H. Lee, S. E. Yoo, Y. D. Gong, *J. Comb. Chem.* **2005**, *7*, 136–141.
25. H. J. Lim, D. Myung, I. Y. C. Lee, M. H. Jung, *J. Comb. Chem.* **2008**, *10*, 501–503.
26. R. Pérez, T. Beryozkina, O. I. Zbruyev, W. Haas, C. O. Kappe, *J. Comb. Chem.* **2002**, *4*, 501–510.
27. T. Lee, D. Lee, I. Y. Lee, Y. D. Gong, *J. Comb. Chem.* **2010**, *12*, 95–99.
28. T. Lee, J. H. Park, D. H. Lee, Y. D. Gong, *J. Comb. Chem.* **2009**, *11*, 495–499.
29. T. Lee, J. H. Park, M. K. Jeon, Y. D. Gong, *J. Comb. Chem.* **2009**, *11*, 288–293.
30. Y. D. Gong, T. Lee, *J. Comb. Chem.* **2010**, *12*, 393–409.
31. D. S. Tan, M. A. Foley, B. R. Stockwell, M. D. Shair, S. L. Schreiber, *J. Am. Chem. Soc.* **1999**, *121*, 9073–9087.
32. M. Prakesch, U. Sharma, M. Sharma, S. Khadem, D. M. Leek, P. Arya, *J. Comb. Chem.* **2006**, *8*, 715–734.
33. K. H. Kong, Y. Chen, X. Ma, W. K. Chui, Y. Lam, *J. Comb. Chem.* **2004**, *6*, 928–933.
34. A. Nefzi, J. M. Ostresh, J. Yu, R. A. Houghten, *J. Org. Chem.* **2004**, *69*, 3603–3609.
35. A. Nefzi, M. A. Giulianotti, R. A. Houghten, *J. Comb. Chem.* **2001**, *3*, 68–70.
36. V. Krchňák, J. Smith, J. Vágner, *Collect. Czech. Chem. Commun.* **2001**, *66*, 1078–1106.
37. S. Křupková, P. Funk, M. Soural, J. Hlaváč, *ACS Comb. Sci.* **2013**, *15*, 20–28.
38. X. Huang, J. Cao, J. Huang, *J. Comb. Chem.* **2009**, *11*, 515–518.
39. A. N. Acharya, J. M. Ostresh, R. A. Houghten, *J. Comb. Chem.* **2002**, *4*, 214–222.
40. S. C. Lee, S. B. Park, *J. Comb. Chem.* **2006**, *8*, 50–57.
41. O. Kwon, S. B. Park, S. L. Schreiber, *J. Am. Chem. Soc.* **2002**, *124*, 13402–13404.
42. N. Pudelová, V. Krchňák, *J. Comb. Chem.* **2009**, *11*, 851–859.
43. A. La Venia, B. Vaňková, N. Cankařová, V. Krchňák, manuscript in preparation.
44. N. Cankařová, V. Krchňák, *J. Org. Chem.* **2012**, *77*, 5687–5695.
45. P. Čebašek, D. Bevk, S. Pirc, B. Stanovnik, J. Svete, *J. Comb. Chem.* **2006**, *8*, 95–102.
46. V. Krchňák, K. R. Waring, B. C. Noll, U. Moellmann, H. M. Dahse, M. J. Miller, *J. Org. Chem.* **2008**, *73*, 4559–4567.
47. F. Li, B. Yang, M. J. Miller, J. Zajíček, B. C. Noll, U. Moellmann, H. M. Dahse, P. A. Miller, *Org. Lett.* **2007**, *9*, 2923–2926.
48. V. Krchňák, U. Moellmann, H. M. Dahse, M. J. Miller, *J. Comb. Chem.* **2008**, *10*, 94–103.
49. V. Krchňák, U. Moellmann, H. M. Dahse, M. J. Miller, *J. Comb. Chem.* **2008**, *10*, 104–111.
50. G. L. Thomas, R. J. Spandl, F. G. Glansdorp, M. Welch, A. Bender, J. Cockfield, J. A. Lindsay, C. Bryant, D. F. Brown, O. Loiseleur, H. Rudyk, M. Ladlow, D. R. Spring, *Angew. Chem. Int. Ed.* **2008**, *47*, 2808–2812.
51. R. E. Dolle, B. L. Bourdonnec, A. J. Goodman, G. A. Morales, C. J. Thomas, W. Zhang, *J. Comb. Chem.* **2009**, *11*, 739–790.
52. M. D. Burke, E. M. Berger, S. L. Schreiber, *J. Am. Chem. Soc.* **2004**, *126*, 14095–14104.
53. S. T. Le Quement, T. E. Nielsen, M. Meldal, *J. Comb. Chem.* **2007**, *9*, 1060–1072.

54. I. Takahashi, E. Hirano, T. Kawakami, H. Kitajima, *Heterocycles*. **1996**, *43*, 2343–2346.
55. R. Alonso, L. Castedo, D. Dominguez, *Tetrahedron Lett*. **1985**, *26*, 2925–2928.
56. C. J. Moody, G. J. Warrellow, *Tetrahedron Lett*. **1987**, *28*, 6089–6092.
57. Y. Hamuro, W. J. Marshall, M. A. Scialdone, *J. Comb. Chem*. **1999**, *1*, 163–172.
58. K. S. Lam, M. Lébl, V. Krchňák, *Chem. Rev*. **1997**, *97*, 411–448.
59. C. G. Levins, Z. Z. Brown, C. E. Schafmeister, *Org. Lett*. **2006**, *8*, 2807–2810.
60. H. Okada, T. Akaki, Y. Oderaotoshi, S. Minakata, M. Komatsu, *Tetrahedron*. **2003**, *59*, 197–205.
61. L. Feliu, P. Vera-Luque, F. Albericio, M. Alvarez, *J. Comb. Chem*. **2007**, *9*, 521–565.
62. D. R. Spring, S. Krishnan, H. E. Blackwell, S. L. Schreiber, *J. Am. Chem. Soc*. **2002**, *124*, 1354–1363.
63. R. E. Dolle, *J. Comb. Chem*. **2003**, *5*, 693–753.
64. A. La Venia, B. Dolenský, V. Krchňák, *ACS Comb. Sci*. **2013**, *15*, 162–167.
65. M. Soural, I. Bouillon, V. Krchňák, *J. Comb. Chem*. **2008**, *10*, 923–933.
66. T. Luo, S. L. Schreiber, *J. Am. Chem. Soc*. **2009**, *131*, 5667–5674.
67. T. Uchida, M. Rodriquez, S. L. Schreiber, *Org. Lett*. **2009**, *11*, 1559–1562.
68. J. M. Mitchell, J. T. Shaw, *Angew. Chem. Int. Ed*. **2006**, *45*, 1722–1726.
69. T. E. Nielsen, M. Meldal, *J. Org. Chem*. **2004**, *69*, 3765–3773.
70. T. E. Nielsen, M. Meldal, *J. Comb. Chem*. **2005**, *7*, 599–610.
71. T. E. Nielsen, S. Le Quement, M. Meldal, *Org. Lett*. **2005**, *7*, 3601–3604.
72. G. Schneider, W. Neidhart, T. Giller, G. Schmid, *Angew. Chem. Int. Ed*. **1999**, *38*, 2894–2896.
73. H. J. Böhm, A. Flohr, M. Stahl, *Drug Discov. Today Technol*. **2004**, *1*, 217–224.
74. S. H. Fitzgerald, M. Sabat, H. M. Geysen, *J. Comb. Chem*. **2007**, *9*, 724–734.
75. Y. S. Tung, M. S. Coumar, Y. S. Wu, H. Y. Shiao, J. Y. Chang, J. P. Liou, P. Shukla, C. W. Chang, C. Y. Chang, C. C. Kuo, T. K. Yeh, C. Y. Lin, J. S. Wu, S. Y. Wu, C. C. Liao, H. P. Hsieh, *J. Med. Chem*. **2011**, *54*, 3076–3080.
76. H. Sun, G. Tawa, A. Wallqvist, *Drug Discov*. Today. **2012**, *17*, 310–324.
77. S. Křupková, G. A. Slough, V. Krchňák, *J. Org. Chem*. **2010**, *75*, 4562–4566.
78. I. Bouillon, J. Zajíček, N. Pudelová, V. Krchňák, *J. Org. Chem*. **2008**, *73*, 9027–9032.
79. R. C. D. Brown, J. Keily, R. Karim, *Tetrahedron Lett*. **2000**, *41*, 3247–3251.
80. V. Krchňák, J. Zajíček, P. A. Miller, M. J. Miller, *J. Org. Chem*. **2011**, *76*, 10249–10253.
81. B. Vaňková, L. Brulíková, B. Wu, V. Krchňák, *Eur. J. Org. Chem*. **2012**, *26*, 5075–5084.
82. C. J. Dinsmore, D. C. Beshore, *Org. Prep. Proc. Int*. **2002**, *34*, 367–404.

8

MACROCYCLES AS TEMPLATES FOR DIVERSITY GENERATION IN DRUG DISCOVERY

Eric Marsault

8.1 INTRODUCTION

Macrocycles occupy a distinct segment of the chemical space compared to traditional small molecules. They have been described as an intermediate class between small molecules and larger biomolecules, borrowing features of both classes [1]. At one end of the spectrum, small molecules have been the main objective of drug discovery for several decades, owing primarily to their favorable PK-ADME (pharmacokinetics, absorption, distribution, metabolism, and excretion) properties and synthetic accessibility. Small molecules are the target of numerous and versatile synthetic strategies as well as the object of massive structural and computational support [2]. Numerous molecular descriptors have been refined in the attempt to predict their properties, starting with Lipinski's rule of 5 [3]. The elucidation of their mode of interaction with biological targets has been the object of numerous studies, and they have been the objective of most diversity-oriented synthesis (DOS) approaches within their chemical space (see also Chapters 10, 15–18) [4]. At the other end of the spectrum are found biological drugs, which have witnessed steady growth for the past two decades. Proteins occupy the lion's share of this chemical space, particularly antibodies [5]. Biomolecules generally possess large interacting surfaces with their targets, often featuring distant epitopes, and have proven their high potential for difficult targets such as protein–protein interactions [6]. Nevertheless, their molecular size generally prevents them from crossing biological membranes, makes them susceptible to

Diversity-Oriented Synthesis: Basics and Applications in Organic Synthesis, Drug Discovery, and Chemical Biology, First Edition. Edited by Andrea Trabocchi.

enzymatic cleavage, and generally limits them to indications compatible with parenteral administration.

Macrocycles occupy a unique position between these two chemical spaces in terms of molecular properties [1a], with molecular weights ranging from 400 to 1500 Da. At equivalent molecular weight, they generally possess fewer degrees of freedom and are more conformationally restricted than their linear congeners, which represents an advantage for oral absorption, other parameters aside [3,7]. This is controlled by several factors, such as ring size, ring strain, the orientation of exocyclic substituents and transannular interactions. In the context of a biological target, this translates into a reduced entropic loss upon binding, provided that preorganization mimics a binding conformation. Owing to their ability to display distant pharmacophores on conformationally restricted templates, macrocycles have the potential to recapitulate the large interaction surfaces of biological molecules such as antibodies. This is demonstrated convincingly by the isolation of the first antagonist of the IL-17 receptor, which exemplifies the ability of this class to mimic an antibody–protein interface [8]. Macrocycles have successfully provided drug candidates for most biological target classes; however, their use in drug discovery is limited by the lack of broadly applicable synthetic methods capable of generating the diversity required for lead optimization in medicinal chemistry [1a and c], despite great advances in the total synthesis of complex macrocyclic natural products of various chemical classes [9]. Indeed, as opposed to total synthesis, which favors convergent approaches, lead optimization requires the generation of multiple analogs and several points of synthetic divergence, in order to generate the necessary diversity that will deliver a development candidate endowed with the optimal balance of properties. As it stands today, no fewer than 100 drugs are macrocycles, most of which are peptidic or natural product derivatives [10]. The predominance of these two categories largely reflects synthetic accessibility: natural products are available as such and can generally be biosynthesized or extracted from natural sources, whereas peptides benefit from robust and automated production methods [9e,10,11].

In summary, macrocycles have the potential to tackle targets that represent high-hanging fruits in terms of druggability; however, they also constitute high-hanging fruit in terms of synthetic challenge and diversity generation. Technologies for macrocycle synthesis and diversification have given rise to several platforms that constitute the foundation of companies incorporated in the last decade [12].

In the following paragraphs, we first discuss some general issues inherent to macrocycles and their synthesis, then we cover diversity generation strategies that have either demonstrated their potential for drug discovery or that appear very promising among different chemical classes of macrocycles. The reader is referred to earlier reviews for additional information and examples [1c,9c and d,10,13].

8.2 CHALLENGES ASSOCIATED WITH MACROCYCLES

8.2.1 Synthetic Challenge

Synthetically, the main hurdle in the path toward macrocycles is the need to overcome entropy and favor an intramolecular reaction over the competing intermolecular

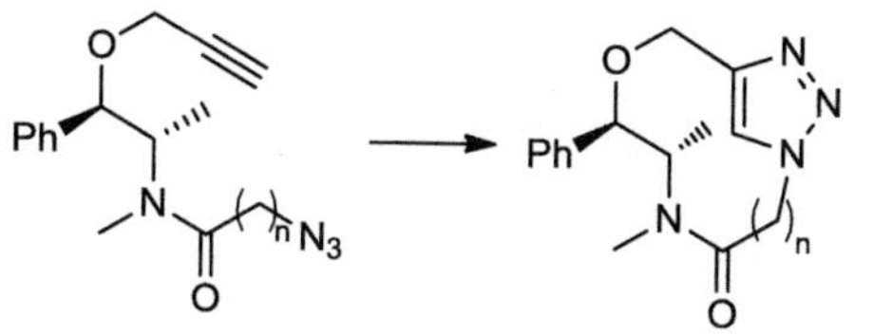

1 (n = 1-5)

	ring size (n)	yield
2	10 (n=1)	< 1%
3	11 (n=2)	20%
4	12 (n=3)	73%
5	13 (n=4)	66%
6	14 (n=5)	80%

SCHEME 8.1 Macrocyclization mediated by copper-catalyzed azide–alkyne cycloaddition (CuAAC).

reactions that produce dimers and higher oligomers. In this context, conformational control in the macrocyclization precursor is critical, as reviewed by Blankenstein and Zhu [14]. A traditional way to circumvent this limitation is to perform ring closure at low concentrations, typically millimolar or lower. While this generally leads to high yields and low proportions of multimers, it is particularly inefficient in terms of scale-up, an obligatory step in drug development. The monomer/oligomer ratio is governed by ring strain and transannular interactions at the transition state of the macrocyclization step, a situation that becomes very acute in small macrocycles (ring size 13 or below). Indeed, in the more strained structures, the desired monomer is often simply not observed, because ring strain increases the activation barrier so high that dimerization and oligomerization become predominant [15].

Bogdan et al. recently reported a systematic study of the macrocyclization and conformational flexibility of small (i.e., 10- to 14-membered) macrocycles using a combination of x-ray crystallography, ^{1}H NMR, and molecular modeling (Scheme 8.1) [16]. Macrocyclization mediated by copper-catalyzed azide–alkyne cycloaddition (CuAAC) was conducted using flow chemistry in a heated copper tube [17]. The authors demonstrated that both experimental yield and geometrical distortions correlated with the ring strains calculated. The most strained 10-membered ring (**2**), which could not be observed experimentally, was associated with a ring strain of 25.1 kcal/mol. Severe geometrical distortions to amide bond and triazole bond dihedral angles were also observed in the smaller rings. The authors concluded that macrocyclization is unlikely to be practical when the calculated strain energy of the product approaches about 20 kcal/mol.

A productive macrocyclization step relies on a subtle balance between enthalpic and entropic factors. It is difficult to rigorously separate the contribution of both factors, as reported by Zhang and Tam, who studied silver ion–catalyzed orthogonal cyclization of unprotected peptide thioesters in the formation of lactones and lactams [18]. The authors demonstrated the critical role of silver ions in interacting with both ends of the peptide chain, enabling spatial proximity between the two extremities and subsequent ring closing. Silver may also act as a Lewis acid able to coordinate oxygen atoms in the linear precursor, templating chain folding. This was also observed by Marsault et al. in the macrolactamization of peptidomimetics [19].

Macrolactamization and macrolactonization are probably the most widely used methods for large ring closure, as exemplified in the synthesis of numerous natural

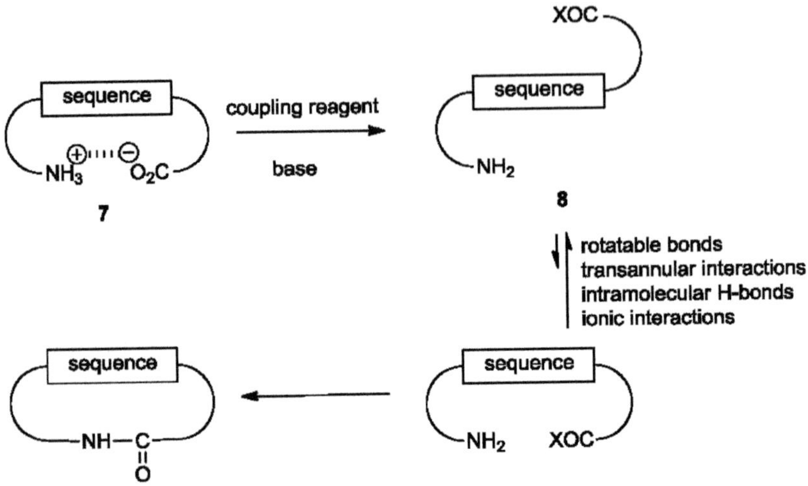

SCHEME 8.2 Macrolactamization using coupling reagents and basic conditions.

products and macrocyclic peptides [9d,20]. In the latter case, macrolactamization usually makes use of coupling reagents and basic conditions, which annihilate a critical polar interaction that otherwise favors linear chain folding as arecently articulated by White and Yudin (Scheme 8.2) [21].

Indeed, linear zwitterionic precursor **7** benefits from a polar interaction between the ammonium and carboxylate moieties, which contributes to stabilizing a fold that brings extremities into close proximity. Upon activation with a coupling reagent and basic conditions, the macrocyclization precursor exists under multiple conformations (schematized as **8** and **9**), among which only **9** is productive to yield macrocycle **10**. The preorganization of the chain and the population of correctly folded precursor **9** depend on the accessible secondary structures, hence the sequence. This duality was exploited by Yudin and co-workers, who developed amphoteric reagents such as aziridine carboxaldehyde **11** (Scheme 8.3). This peculiar reagent mediates the three-component-mediated cyclization of six-membered and larger rings, such as 12-membered ring **12**, in concentrations as high as 0.2 M [22], whereas medium-sized macrocycles are usually extremely difficult to synthesize. To rationalize this high efficiency, the authors proposed that a stabilizing polar interaction is maintained throughout every intermediate step of the three-component reaction with the help of a solvent such as trifluoroethanol, as opposed to classical coupling reagents, which destroy the critical ionic interaction (**7** vs. **8**, Scheme 8.2). This new approach demonstrated its potential in the synthesis of medium-size macrocycles otherwise difficult to access, and provided some insight to improve macrocyclization [21–23].

The most widely used methods for macrocyclization involve ring-closing metathesis (RCM) [24], "click" [3 + 2] cycloaddition of alkyne and azide [25], macrolactonization [9d], macrolactamization [20a] and organometallic cross-couplings [1c,26].

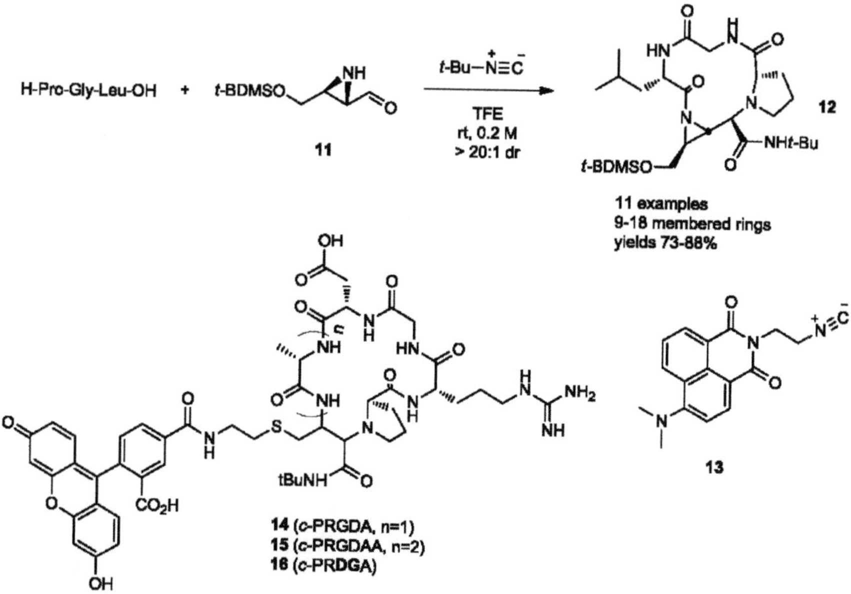

SCHEME 8.3 Multicomponent reaction using aziridine carboxaldehyde **11** as an amphoteric reagent to achieve macrocycles.

8.2.2 Assessment of Diversity

Molecular diversity in the chemical space of macrocycles is a parameter that has not been analyzed quantitatively, as opposed to extensive studies on small molecules [27]. Molecular diversity in macrocycles can be coarsely divided between the chemical functionality displayed by the scaffold on the one hand, which will often be critical in establishing direct interactions with a biological target, and on the other hand the topological diversity that defines the overall shape and volume occupancy of the molecule. These two parameters are intertwined with respect to target interactions and PK-ADME profile. To better characterize diversity in macrocyclic natural products, Wessjohann, Frank, Lachance, and others have analyzed naturally occurring macrocycles using data mining and cheminformatics [10,28]. This was done from a structural point of view based on molecular weight distribution (up to 2559 g/mol), ring size (13- to 72-membered) and the occurrence of common substructures, and from a production point of view taking into account biosynthetic pathways (polyketide pathways, isoprenoid metabolism, peptide biosynthesis and sugar condensation) [10,13a]. Macrocycles, which constitute about 3% of all natural products, possess very complex and diversified structures, as well as high levels of biological activities. Ranging from FK-506 (**17**, Figure 8.1) to erythromycin A (**18**), epothilone B (**19**), vancomycin (**20**), and amphotericin B (**21**), the authors observed the predominance of 14-membered rings and the frequent assembly, in the same molecule, of an apolar, hydrophobic side and a polar, hydrophilic side, or the presence of at least one small heterocycle. The diversity in naturally occurring macrocycles is further

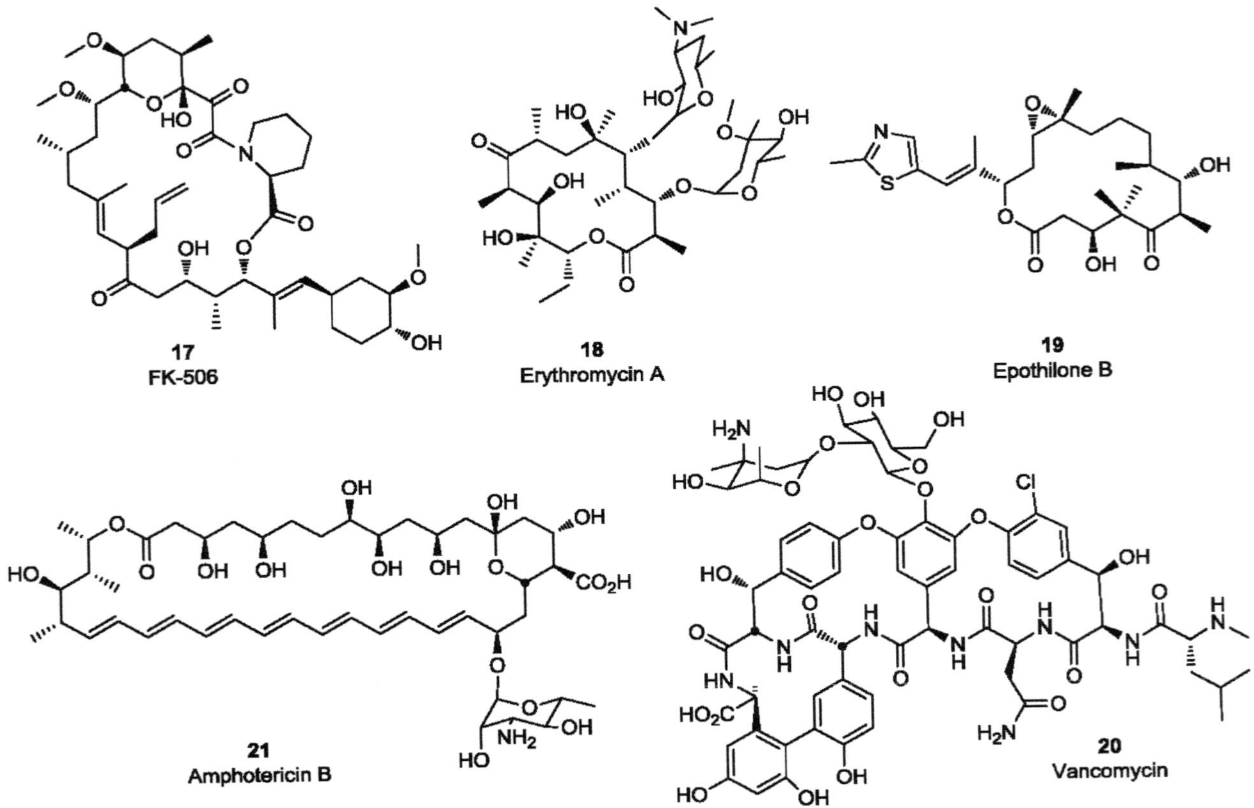

17 FK-506

18 Erythromycin A

19 Epothilone B

20 Vancomycin

21 Amphotericin B

FIGURE 8.1 Selected macrocyclic natural products, including FK-506 (**17**), erythromycin A (**18**), epothilone B (**19**), vancomycin (**20**), and amphotericin B (**21**).

characterized by a broad topological diversity, controlled again by ring size, transannular interactions, and the presence of conformational restrictions [10].

8.2.3 Macrocycles in Drug Discovery

The role and potential of macrocycles in drug discovery has recently been the object of several reviews [1a and c,13b,29]. Despite challenges inherent in such structures, macrocycles have provided drugs or clinical lead compounds for most target classes, including GPCRs, enzymes, DNA, and protein–protein interactions as well as otherwise highly challenging targets [1c,29]. Such is the case for the HCV NS3/4A protease, which presents a very shallow and featureless surface that escaped strategies based on small molecules [30], or immunosuppressant drugs such as FK-506 or rapamycin, macrolide antibiotics, and an antagonist of IL-17 [1a and c].

A common requirement of medicinal chemistry programs is the necessity to generate analogs with a sufficient diversity in order to optimize simultaneously the multiple parameters associated with a drug candidate profile. As a result, access to synthetic methods that enable diversity generation is a *sine qua non* component of a successful medicinal chemistry program.

8.3 MACROCYCLIC PEPTIDES

Macrocyclic peptides have been known for decades and been initially isolated from natural sources [31]. They are present in all kingdoms of life in sizes ranging from 10 to over 80 amino acids and possess a wide variety of natural and unnatural amino acid composition, functioning generally as defense against predators [31]. Macrocyclization, predominantly head to tail or via disulfide bonds, plays a major role in conferring on the molecule a high level of chemical and biological stability compared to its linear congeners [32]. This favorable characteristic was recognized early by drug discovery researchers, and hence has been the subject of numerous efforts to stabilize the topology of drug candidates to endow them with added chemical and biological stability [9e,11d and e,33]. Structure–permeation relationships and the improvement in the oral bioavailability of peptides by way of macrocyclization has been studied extensively by several groups [11b and c,34]. In particular, amide *N*-methylation, as exemplified in cyclosporine, the influence of side-chain substituents, and lipophilicity on permeation have all been scrutinized in several classes of macrocyclic peptides. The assembly of macrocyclic peptides is also facilitated by four decades of automated peptide synthesis and the ready availability of a broad diversity of natural and unnatural amino acids. The following section reports selected examples of technologies which have been developed for the production of large libraries of macrocyclic peptides.

8.3.1 Split-and-Pool Synthesis of Macrocyclic Peptides

The split-and-pool approach is an efficient avenue to generate large libraries (see also Chapter 1). Pei and co-workers have recently reported the synthesis of a 10^7-membered library of macrocyclic peptides using a carefully engineered method

(Scheme 8.4) [35]. The strategy hinges on the differentiation between the outer layer and the inner layer of TentaGel resin beads as reported by Lam and co-workers (e.g., **22**) [36]. First, resin beads were functionalized with a β-Ala–β-Ala–Arg–Met handle using standard Fmoc peptide synthesis. The two β-Ala residues serve as a spacer, the Arg residue functions as a constant ionization handle for subsequent MALDI-MS analysis, and the Met residue is critical to effect CNBr cleavage in order to analyze bead contents during synthesis and after screening. Subsequently, resin beads were soaked in water and reacted with Fmoc-Glu(OSu)-OAll. This simple step, which exploits the resin's physicochemical properties, partially shrinks the bead and leaves only the outer-layer amino groups available for reaction in water with the succinimidyl side chain of Fmoc–Glu(OSu)–OAll [36].

Subsequent Fmoc peptide chemistry in organic solvents allowed the introduction of Fmoc–Glu(O*t*-Bu)–OH on the inner layer only. The outer layer was subsequently functionalized into the macrocycle, and the inner layer (**23** to **26**) served to introduce a linear tag sequence identical to that of the macrocycle for MALDI-MS sequence determination after CNBr cleavage and partial Edman degradation [37]. Subsequently, the authors elaborated the library using a split-and-pool technique. After introduction of additional residues to provide the diversity elements, the allyl ester was cleaved and the octapeptide macrocyclized using PyBOP reagent on the outer-layer-anchored sequence, whereas the inner-layer sequence, which does not contain a free carboxylic acid, could not macrocyclize. Finally, all side-chain protections were cleaved with TFA. This approach allowed the refinement of the one bead–one compound approach to new ligands [37,38].

The screening of the above and similar libraries on several targets led to the identification of ligands of the prolactin receptor (K_d = 2.0 μM) [35], inhibitors of the calcineurin/NFAT interaction (K_d = 0.7 μM) [39], inhibitors of the Grb2/SH2 interaction (K_d = 45 to 900 nM) [40], as well as tyrocidine A analogs with antibacterial properties [41]. The approach exemplifies how a judicious exploitation of the resin's physical properties allows the development of a robust tagging methodology, which facilitates subsequent deconvolution steps.

8.3.2 Synthesis of Small-to-Medium-Sized Macrocycles Using Amphoteric Reagents

Yudin and co-workers reported an innovative three-component coupling reaction for the synthesis of medium-sized peptidic macrocycles based on the amphoteric aziridine carboxaldehyde reagent **11** (Scheme 8.2) [22,23]. This class of macrocycles is very difficult to synthesize, owing to ring strain, and as a result, exploitation of its chemical space has been limited by synthetic accessibility. Macrocyclization in this series was performed at room temperature and at 0.2 M concentration, as opposed to the usual millimolar or lower range. This remarkably high concentration was attributed to the conservation of an ionic pair throughout every intermediate step of the three-component coupling, which favored the folded precursor (**9**, Scheme 8.1) and contributed to overcoming the entropic penalty associated with ring closure. This

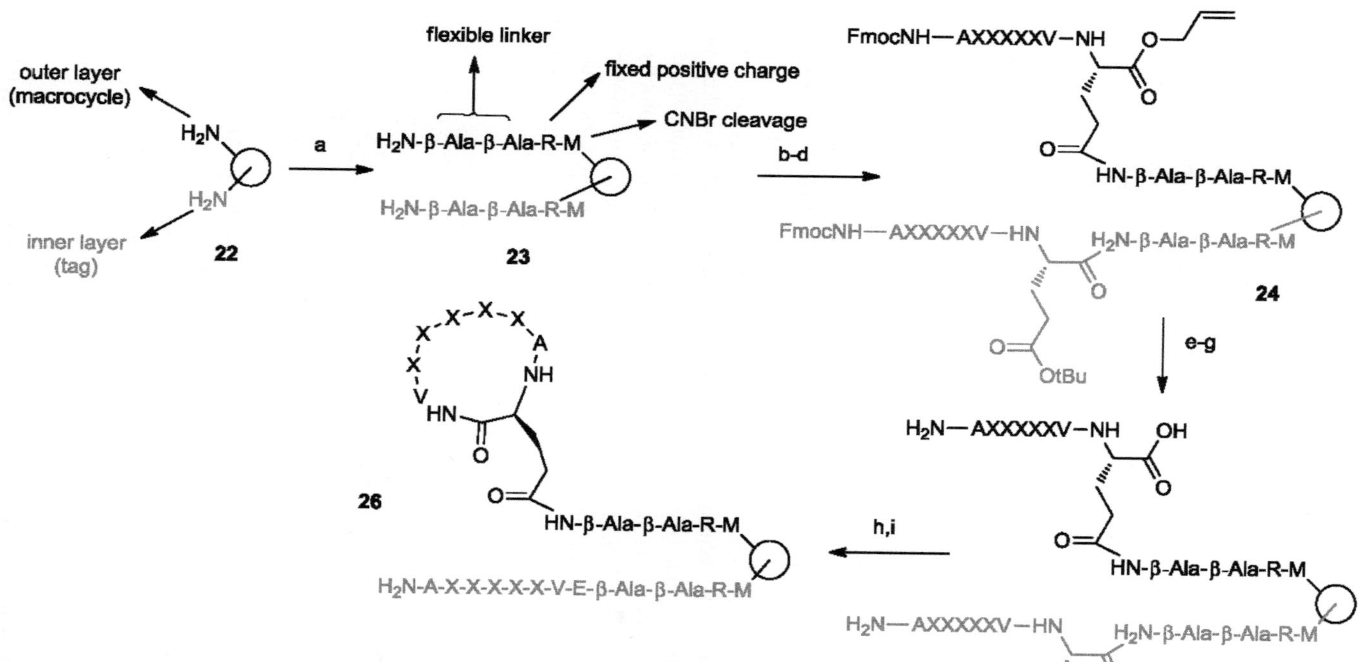

a. Fmoc chemistry; b. water soak; c. Fmoc-Glu(OSu)-OAll; d. Fmoc-Glu(O*t*-Bu)-OH, HBTU; e. split-and-pool Fmoc peptide synthesis; f. $Pd(PPh_3)_4$; g. piperidine; h. PyBOP; i. TFA;

X = Arg, Asp, Gln, Gly, His, Ile, Lys, Pro, Ser, Thr, Trp, Tyr, Phe(4-F), Nle, Orn, Phg, D-Ala, D -Asn, D-Glu, D-Leu, D-Phe, D-Val, NMe-Ala, NMe-Leu, NMe-Phe, Sar

diversity = 26^5 = 1.19 x 10^7

SCHEME 8.4 Synthesis of a library of 10^7 members exploiting the outer and inner layers of TentaGel resin beads.

technology was applied in microfluidics settings [42], and led to the design of isocyanide **13**, which enabled the synthesis of solvatochromic, conformation-responsive macrocycles and ionophores [43]. Subsequently, they synthesized a macrocycle capable of imaging mitochondria selectively [43]. Finally, the same authors demonstrated that fluorescently labeled 18- and 21-membered macrocycles **14** (*c*-PRGDA) and **15** (*c*-PRGDAA) were both able to bind the $\alpha_v\beta_3$ integrin receptor in a cell adhesion assay with affinities of 4.2 and 11.1 μM, respectively, as opposed to scrambled analog **16** (*c*-PRDGA) corroborating computational predictions [44]. This technology platform holds promise to support conformational fine-tuning in subsequent optimization [45].

8.3.3 DNA-, RNA-, and Phage-Templated Synthesis of Peptidic Macrocycles

The translation of nucleic acids into synthetic compounds is a recent achievement that has been pioneered largely by Liu and co-workers [46]. The authors have exploited DNA-templated chemistry in several directions, providing an efficient solution to the problem of broad diversity generation in massive libraries while exploiting one of the most robust encoding approaches available. This technology was applied to the synthesis of macrocyclic peptide-like structures (Scheme 8.5) [47]. Using a set of DNA codons able to support large library synthesis, the group constructed a library of 14,000 macrocycles of general structure **28**. Macrocyclization hinges around a biocompatible version of the Wittig reaction, which transformed DNA-linked precursor **27** into DNA-linked macrocycle **28**. Of particular interest, reactions were run at nanomolar concentrations, which minimized dimerization and building block consumption. In this application, DNA templating provides voth high effective molarities and high dilution ideal for monomer macrocyclization. Using this approach, libraries were produced and tested as mixtures in solution. Upon isolation of an active product,

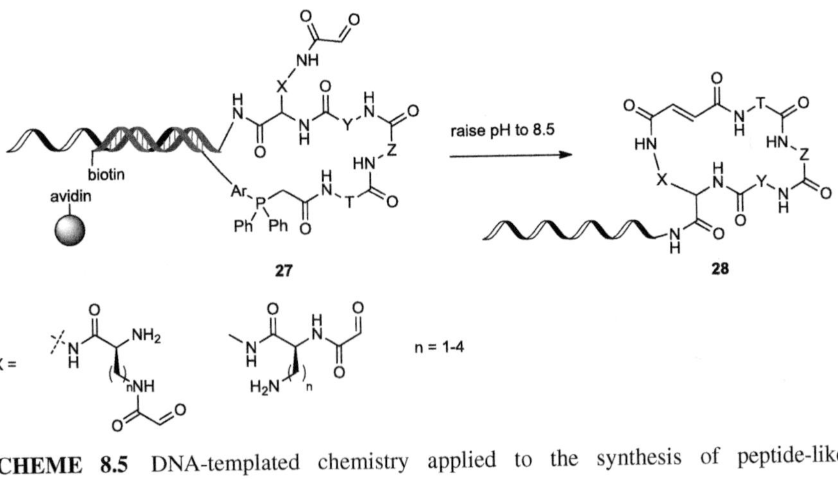

SCHEME 8.5 DNA-templated chemistry applied to the synthesis of peptide-like macrocycles.

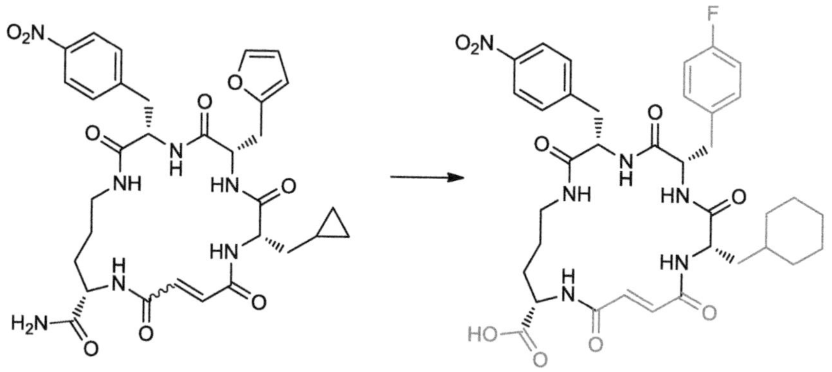

29 IC_{50} (Src) 0.68 $\propto$M

30 IC_{50} (Src) < 4 nM

FIGURE 8.2 Activators of the VEGFR2 kinase and inhibitors of the Akt3, MAPKAPK2, p38α-Src, and Pim1 kinases, as exemplified by **29** and the optimized analog **30**.

DNA amplification allowed identification of the active sequence. In terms of diversity, scaffolds X represented various chain lengths (Scheme 8.5), whereas building blocks Y, Z, and T (**27**) cover aliphatic, branched, aromatic, heteroaromatic, α- and γ-amino acids, and turn inducers, with a large representation of unnatural amino acids [46,47].

Screening of the library was conveniently performed in one test tube per target using in vitro affinity selection. The authors reported the identification of activators of the VEGFR2 kinase and inhibitors of the Akt3, MAPKAPK2, p38α-Src, and Pim1 kinases [48], as exemplified by **29** (Figure 8.2), which possesses $IC_{50} = 0.68$ μM against the Src kinase and excellent selectivity against a panel of 44 kinases, including related kinases such as Abl, Aurora A and B, RET, Blk, Fgr, Frk, Fyn, Hck, Lck, and LynA and B. This is noteworthy in light of the difficulties in reaching selectivity for that particular target. The authors tentatively attribute selectivity to library design, which emphasizes rigid macrocyclic structures, and to the larger molecular size of this class of inhibitors compared to classical ATP-competitive inhibitors. In this case, macrocycles have the unique capacity to interact simultaneously with the ATP binding pocket and reach less-conserved residues outside the ATP binding pocket. In a subsequent study [49], the activity of macrocycle **29** was further refined to provide analog **30** with $IC_{50} < 4$ nM. Using molecular modeling, they demonstrated that **30** behaves as a bisubstrate-competitive Src inhibitor with high potency and exceptional selectivity, owing to interactions with both the ATP binding pocket and the substrate-binding region. They demonstrated further that the level of peptide substrate competition could be modulated to inhibit preferentially the phosphorylation of strongly binding peptides over weaker substrates, opening the way to biasing the signaling of the Src kinase. In the present case, the ability of this scaffold to overlap with both binding pockets while retaining high affinity and high selectivity reflects the ability of the macrocyclic structure to display distant and precisely oriented pharmacophoric groups.

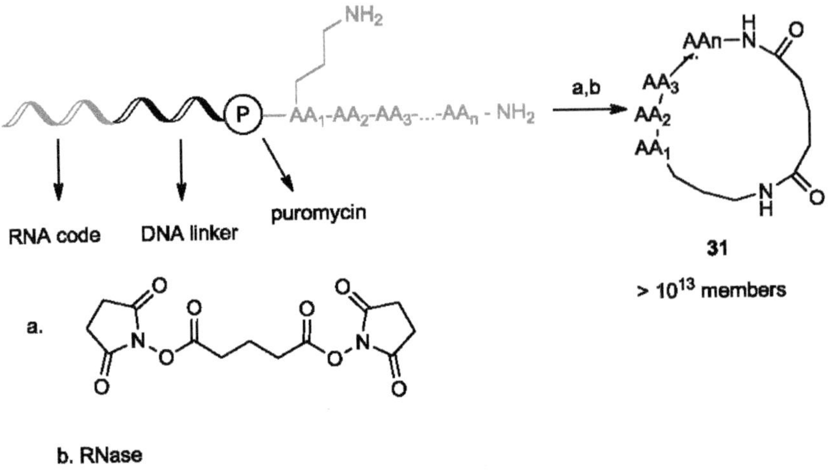

SCHEME 8.6 Complementary approach based on an *m*RNA display to achieve the synthesis of a library of 10^{13} macrocycles consisting of the general structure **31**.

A complementary approach based on *m*RNA display (Scheme 8.6) [50] involved anchoring of the peptide to puromycin and subsequent elongation using ribosomal translation. Subsequently, the side chain of lysine was linked to the N-terminal amine of the peptide using bis-succinimidyl glutarate, followed by RNA cleavage to complete the synthesis of macrocycles **31**. The authors reported the synthesis of a massive library of more than 10^{13} members using this approach. Recently, Yamagishi et al. reported the synthesis of a ribosome-expressed library of N-methylated macrocyclic peptides, which delivered an inhibitor of ubiquitin ligase [51].

Phage display has also been used to provide massive libraries of cyclic peptides closed via a disulfide linkage. Recently, Heinis et al. reported the synthesis of a library of more stable bicyclic macrocycles by phage display [52]. Compared to the chemically labile disulfide-linked macrocycles, these bicyclic macrocycles incorporate a stable 1,3,5-trimethylbenzene bridge (Figure 8.3) that simultaneously links three Cys residues. This library led to the isolation of a new class of inhibitors of kallikrein (**32**), with an IC_{50} value of 20 nM.

In summary, several methods are available for the generation of diversity in macrocyclic peptides, ranging from small to very large rings. Macrocyclic peptides are generally associated with higher chemical and biological stability than that of their linear analogs. The accessible diversity is limited only by the availability of natural and unnatural amino acids. The technologies discussed above are amenable to the production of massive libraries, such as those performed on phage or oligonucleotides, provided that both synthetic schemes and biological testing are compatible with the presence of such entities. Similarly, robust encoding methods have significantly simplified the deconvolution required compared with early diversity-based methodologies (see also Chapter 11). Nevertheless, the peptidic character of these types of macrocycles will probably remain a limiting factor for their direct use as

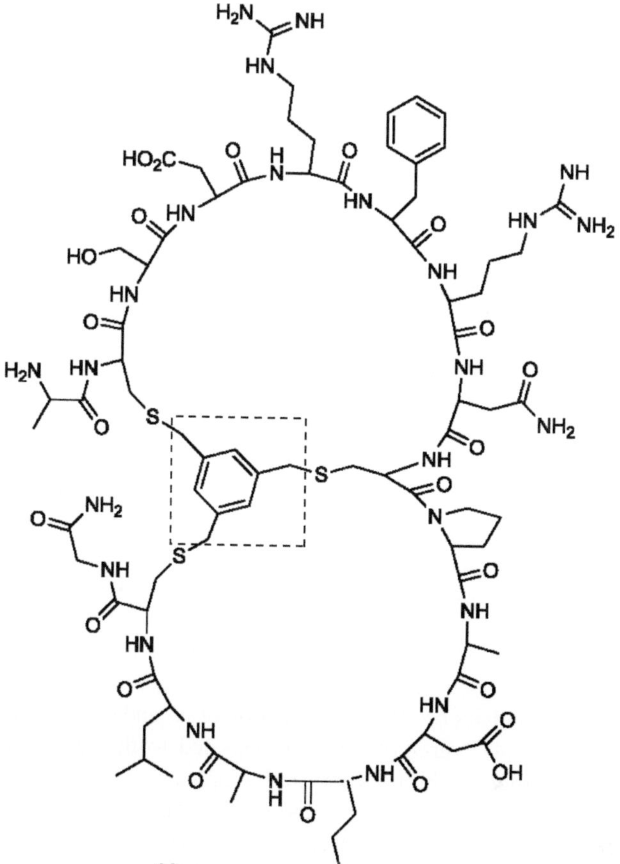

32
IC_{50} 20 nM (kallikrein)

FIGURE 8.3 Bicyclic macrocycle obtained by phage display and incorporating a stable 1,3,5-trimethylbenzene bridge.

drug leads when oral bioavailability is needed. Thus, there is a need for technologies to support diversity generation of less peptidic and nonpeptidic scaffolds, which is the focus of the next sections.

8.4 PEPTIDOMIMETIC MACROCYCLES

8.4.1 Mimics of Peptide Secondary Structures

Proteases recognize the β-strand conformation in their inhibitors and substrates [11d,53]. However, most substrate-based protease inhibitors are linear and conformationally flexible. Enzyme/inhibitor binding is highly cooperative, whereby small changes to an inhibitor structure and conformation induce changes to the

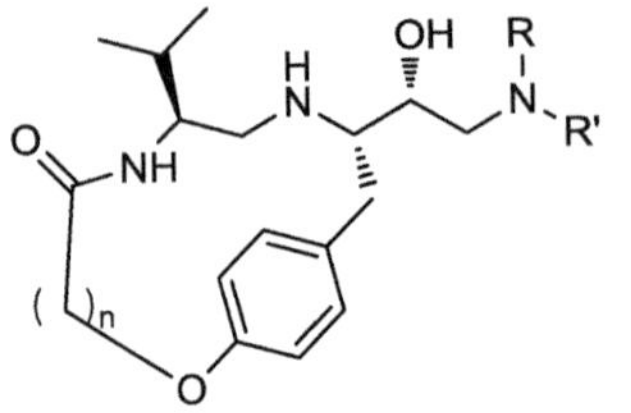

33

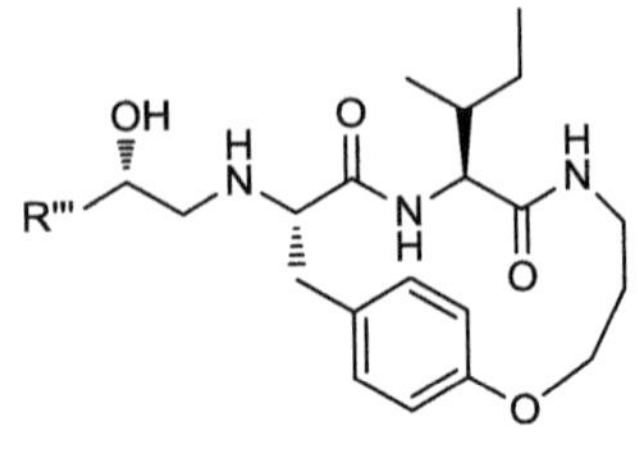

34

FIGURE 8.4 Conformationally constrainted macrocyclic tripeptide mimetics.

enzyme conformation at the binding site or at remote sites, a reciprocal phenomenon termed *induced fit*. Reciprocal conformational adjustment of both the enzyme and the inhibitor is common, which makes the prediction of enzyme/inhibitor geometry more difficult. Reid et al. reported the use of conformational constraints in the form of macrocyclic tripeptide mimetics, with the goal of minimizing the conformational rearrangement of the inhibitor upon binding (Figure 8.4) [54]. Macrocyclic templates **33** and **34** were designed to achieve this goal, based on the analysis of x-ray structures of protease-bond peptide inhibitors. **33** and **34** mimic the extended conformation of bound linear inhibitors, giving them an entropic advantage upon binding. Additionally, these macrocycles are more stable to proteolytic degradation and generally possess higher bioavailability then that of their linear congeners [55].

Accordingly, the authors produced a collection of N- and C-terminal macrocycles (**33** and **34**, Figure 8.4). Variations of R groups led to the identification of low-nanomolar HIV-1 protease inhibitors. Crystallographic analysis of protease/inhibitor complexes revealed a high level of conservation in residues surrounding the inhibitors and among contacts between inhibitors and enzyme. This suggests that induced fit is reduced among this series of macrocyclic HIV-1 inhibitors and opens new avenues for inhibitors of additional proteases as well.

Park and Burgess reported several classes of macrocyclic peptidomimetics designed to mimic protein β-turns [56]. Starting from solid-supported precursor **35** (Scheme 8.7), the macrocyclization relied on a key S_NAr reaction to yield macrocycle **36**. Oxygen, nitrogen, and sulfur nucleophiles all performed well in the S_NAr reaction, which was facilitated by the presence of the adjacent iodine atom [57] and could subsequently be diversified by cross-coupling. The same group reported the synthesis of semipeptoid analogs using an analogous approach [58]. As shown in Scheme 8.7, starting from solid-supported precursor **37**, macrocyclic semipeptoids were obtained by S_NAr to provide macrocycles **38**. Similarly, precursor **39** was cyclized by S_N2 on the benzylic bromide to provide, after cleavage from the resin, macrocycles **40**. Molecular modeling and conformational analysis by solution NMR indicated that these compounds adopted the desired β-turn-like conformations. This class of macrocycles led to the identification of TrkA-dependent NGF mimics and TrkC-dependent NT3 mimics **41** and **42** [59].

A related approach was reported by Jefferson, Swayze, and co-workers involving a combinatorial library of about 12,000 macrocycles based on scaffold **43**, which

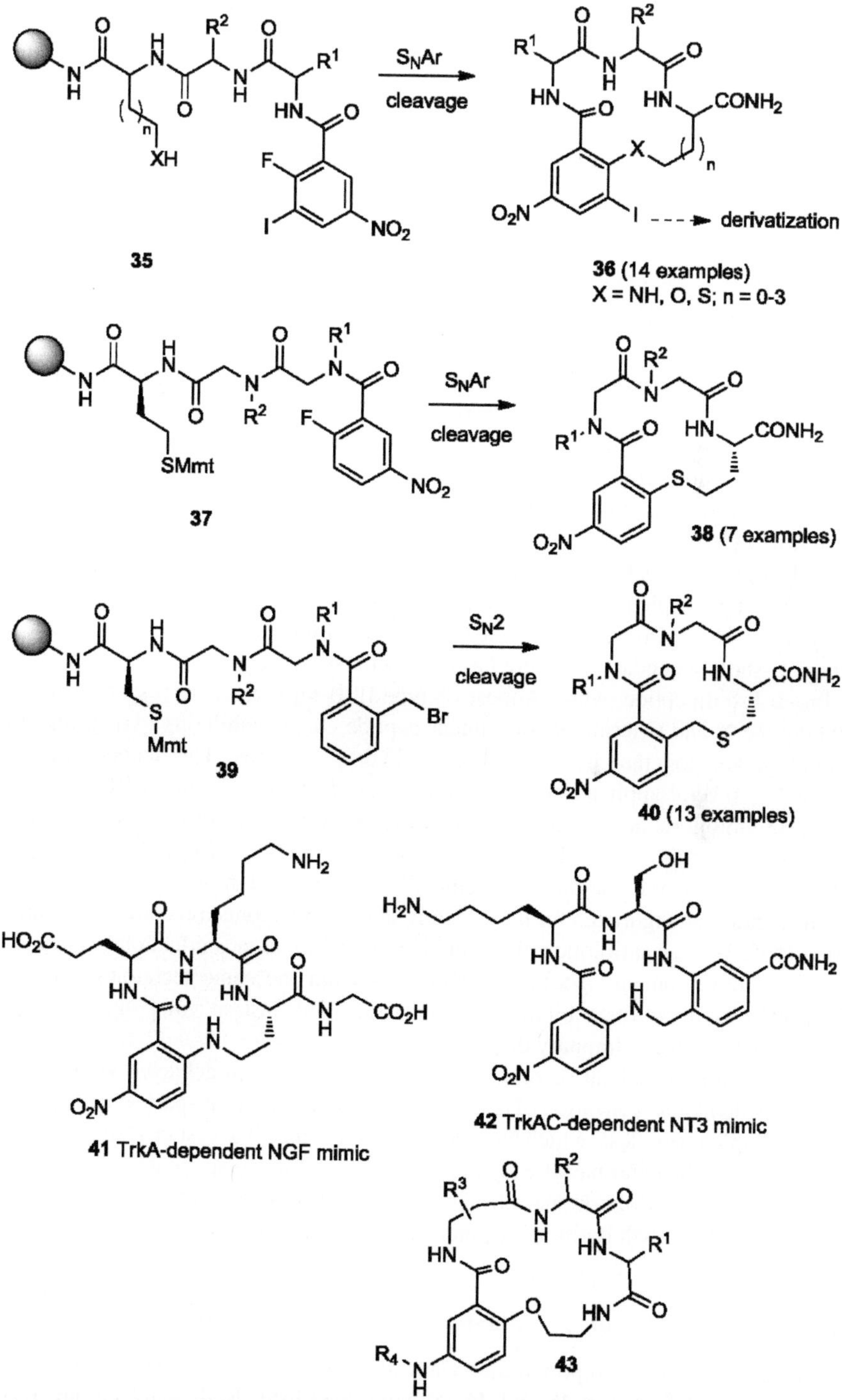

SCHEME 8.7 Macrocyclic compounds as β-turn mimetics obtained on solid phase through key S_NAr and S_N2 macrocyclization reactions.

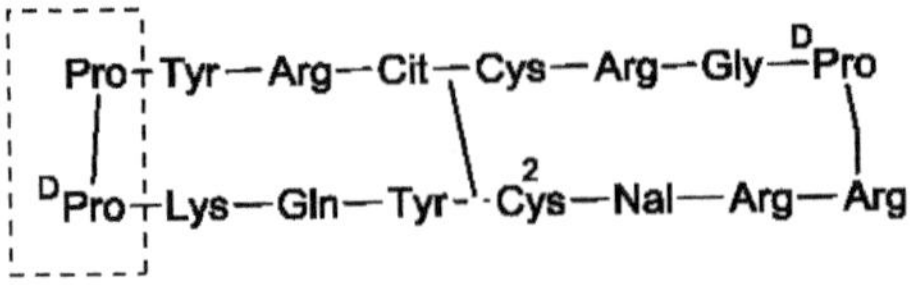

POL3026
(Ca^{2+}) flux IC_{50} 1.2 nM
SDF-1α displacement IC_{50} 1.2 nM

FIGURE 8.5 β-Hairpin mimetic possessing the L-Pro–D-Pro dipeptide motif: structure of the CXCR4 agonist precursor **POL3026**.

features 14-membered rings formed by amino acids selected to bind RNA targets, such as those found in naturally occurring macrocyclic antibiotics targeting bacterial ribosomal RNA. This library led to the identification of several compounds able to inhibit translation. However, none of the actives identified demonstrated minimal inhibitory concentrations against *Escherichia coli* or *Staphylococcus aureus* strains at 100 μM [60].

β-Hairpin mimics have been designed and synthesized using larger macrocycles, following a principle reminiscent of the secondary structure imposed to gramicidin S by two D-Phe–L-Pro residues at the turn regions (Figure 8.5) [61a–c]. Stabilization of the β-sheet secondary structure has been achieved successfully via the use of the L-Pro–D-Pro dipeptide, which imposes a type II′ β-turn which serves as a template linking the N and C terminals of a linear peptide chain, stabilizing two antiparallel strands linked together by intramolecular hydrogen bonds. This technology [62] has led to several applications, such as mimics of the TAT bound to HIV-1 RNA, protease inhibitors, and cationic host defense peptides [63]. So far it has delivered two clinical candidates. **POL6326** is a competitive inhibitor of CXCR4 currently in clinical phase II for hematopoietic stem cell transplant and in phase I for cancer and tissue repair (see Figure 8.5 for the structure of CXCR4 agonist precursor **POL3026**) [64]. **POL7080**, an antibiotic efficacious against *Pseudomonas* infections, is in phase I clinical development. Additional congeners from the same technology include **POL6014**, a competitive inhibitor of human neutrophil elastase with potential against chronic obstructive pulmonary disease [61c].

Stabilization of helical secondary structure has also been achieved via the side-chain to side-chain cyclization of residues located one helix turn apart, in an approach called "stapled peptides," which have been applied to multiple systems (see for example [61d–h]). The latter has already delivered a first clinical candidate, ALRN-5281, a long-acting growth hormone-releasing hormone agonist for the treatment of conditions such as growth hormone deficiency and HIV-related lipodystrophy [61d–h].

8.4.2 Diversity-Oriented Synthesis of Macrocyclic Peptidomimetics

Spring and co-workers reported the synthesis of macrocyclic peptidomimetics possessing general formulas **45** and **46** (Scheme 8.8) [65]. Both scaffolds are readily diversified, which is highly desirable for the optimization of potentially active

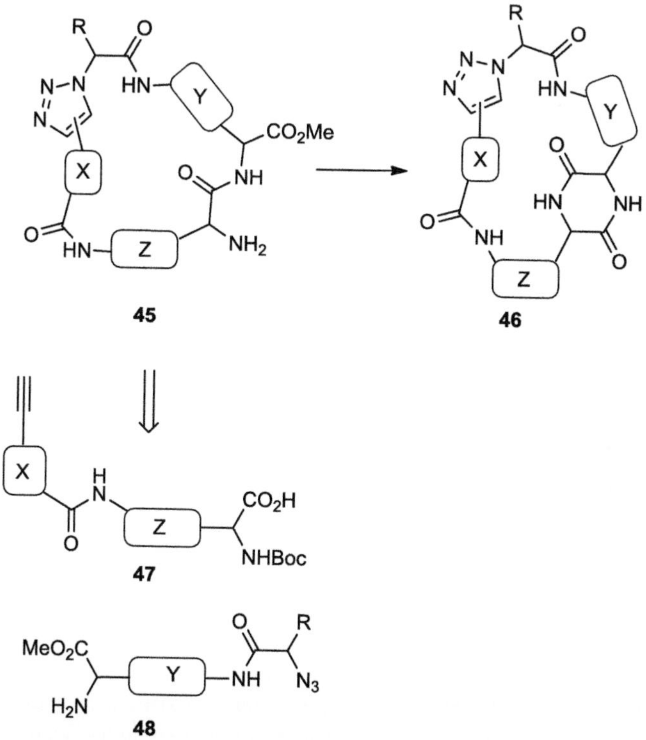

SCHEME 8.8 Diversity-oriented synthesis of macrocyclic peptidomimetics using "click chemistry."

compounds. Interestingly, the approach provides two synthetically related classes of macrocycles that differ by their degree of rigidity and topology. Thus, **46** can be obtained from **45** by diketopiperazine formation. Diversity stems from the choice of building blocks **47** and **48**, which governs the orientation and chemical nature of appendages as well as ring size. Building block assembly used standard peptide coupling, whereas macrocyclization was performed by "click" chemistry following the build/couple/pair (B/C/P) scheme developed by Nielsen and Schreiber [66]. Using principal components analysis (PCA), the authors demonstrated that this collection of 12 analogs covered a distinct and broad chemical space compared to a set of 657 approved drugs.

8.4.3 Macrocyclic Peptoid Libraries

Peptoids (N-alkylated glycine oligomers) are known to be proteolytically stable and can be accessed easily with broad diversity. They have been associated with increased cell permeability compared to native peptides [67]. Grubbs and co-workers recently reported the optimization of a synthetic method for solid-phase synthesis of cyclic peptoids (Scheme 8.9). The authors reported high RCM yields of precursor **49** to yield resin-supported macrocycle **50** using the Hoveyda–Grubbs

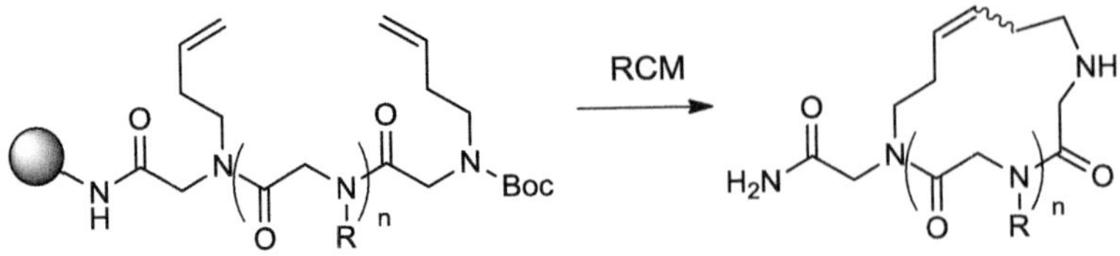

SCHEME 8.9 Solid-phase synthesis of cyclic peptoids using RCM as the macrocyclization reaction.

second-generation catalyst [68]. This follows a number of reports on the use of RCM for the solid-phase synthesis of surrogates of α-helices [69], cyclic peptides [70], and macrocyclic semipeptides targeting RNA [71].

8.4.4 Semipeptidic Macrocycles

Several groups have focused on the synthesis of semipeptidic macrocycles incorporating both peptidic and nonpeptidic fragments, with the goal of improving the physicochemical and PK-ADME profile of the resulting molecules by reducing its peptidic character while conserving the preorganized structure of the macrocycle and its concomitant advantages in terms of target potency and selectivity. Marsault et al. reported the synthesis of > 20,000 macrocycles using the strategy shown in Scheme 8.10 [72]. In this approach, 14- to 22-membered macrocycles were synthesized using a solid-phase approach [19]. The tripeptide group was anchored to the resin via a semilabile thioester linker, which prevented the use of Fmoc chemistry. Accordingly, Ddz was chosen as a highly acid-labile protective group for the α-amino group of amino acids, allowing side-chain protection with acid-labile *t*-Bu ethers, esters, and Boc carbamates. The third amino acid was protected as a benzothiazolyl sulfonamide (Bts) to give peptidic fragment **51**, which enabled attachment of the nonpeptidic tether component under mild Fukuyama–Mitsunobu conditions to provide macrocyclization precursor **52**. Deprotection of the Ddz group was followed by concomitant macrolactamization release, which delivered crude compounds of general formula **53** with acceptable purity since any uncyclized materials remained on the resin and were removed by filtration [73]. In this case, macrocyclization benefited from on-resin pseudo-high dilution [74]. Finally, Bts removal using a solid-supported thiophenoxide followed by side-chain acidolysis delivered crude macrocycles which were tested as such by high-throughput screening (HTS). Not surprisingly, macrolactamization yields were found to be very sensitive to tripeptide sequence, ring size, and the chirality of the amino acid side chains. Presumably due to easier chain folding, heterochiral sequences cyclized more readily than do their homochiral counterparts, and the use of silver(I) activation was found to be beneficial for both yield and monomer/dimer ratio [19]. Alternative approaches, such as RCM-mediated cyclative

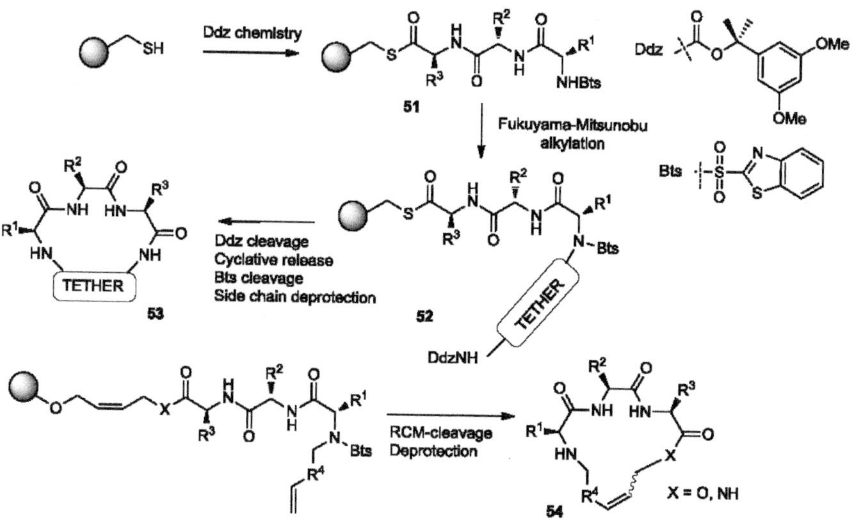

SCHEME 8.10 Solid-phase synthesis of a library of tethered 14- to 22-membered semipeptidic macrocycles using a semilabile thioester linker for cyclative cleavage as the macrocyclization reaction.

release (**54**), were also developed [75]. Using this technology, the authors produced a library of about 20,000 macrocycles with broad diversity in terms of side chains, stereochemistry, and types of amino acids, ring size, and the chemical nature of the tether [19]. This collection of macrocycles delivered nanomolar-level hits on several peptidergic G-protein-coupled receptors directly from HTS, which constituted an enviable starting point for multiple drug discovery programs.

Agonists of the ghrelin receptor identified via this technology are now in advanced clinical development, as exemplified by ulimorelin (TZP-101, **55**, Figure 8.6) [76], which was developed as an i.v. candidate for the treatment of postoperative ileus [77]. Although administered i.v. for this particular indication, the compound possesses 24% oral bioavailability with a low clearance (8.8 mL/min/kg) [76a]. A thorough analysis of structure–activity and structure–pharmacokinetic relationships disclosed the critical role of specific amino acids of the tripeptide portion and the nonpeptidic tether, revealing a subtle interplay between different parts of the molecule in the control of its pharmacodynamic and pharmacokinetic profiles. A second-generation ghrelin agonist (TZP-102) is currently in clinical phase II as an oral candidate for the treatment of diabetic gastroparesis [72a]. As a remarkable example of efficacy switch, this scaffold also delivered a novel series of antagonists of the ghrelin receptor exemplified by **56**, demonstrating how minute modifications of exocyclic substituents can affect biological activity dramatically [1c]. The same library also led to the discovery of potent and selective motilin antagonists exemplified by **57** [75,78].

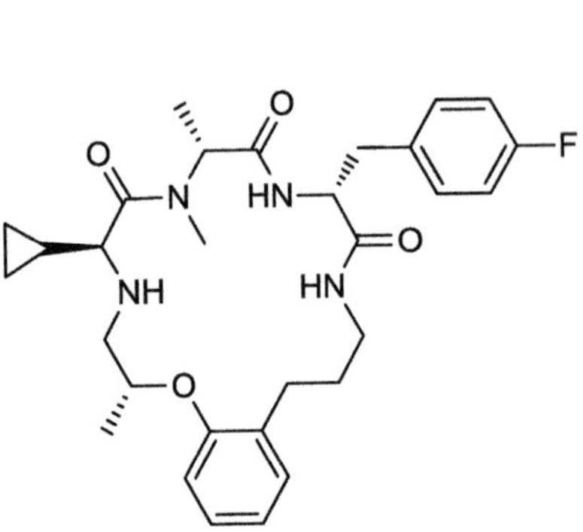

55 Ulimorelin (TZP-101)
K_i (*h*GHS-R1a) 16 nM
EC_{50} 22 nM

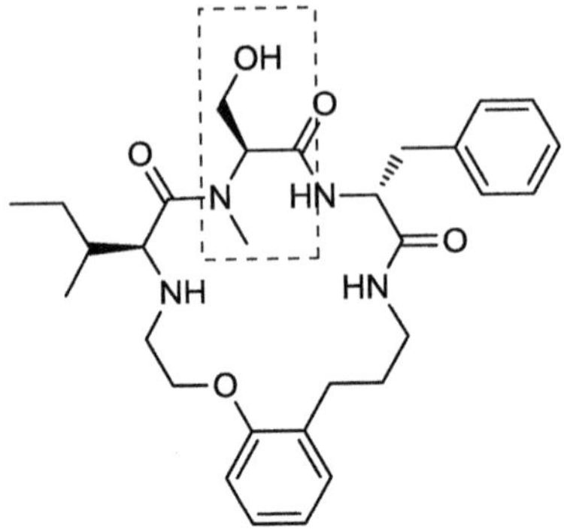

56
K_i (*h*GHS-R1a) 100 nM
EC_{50} NA
IC_{50} 500 nM

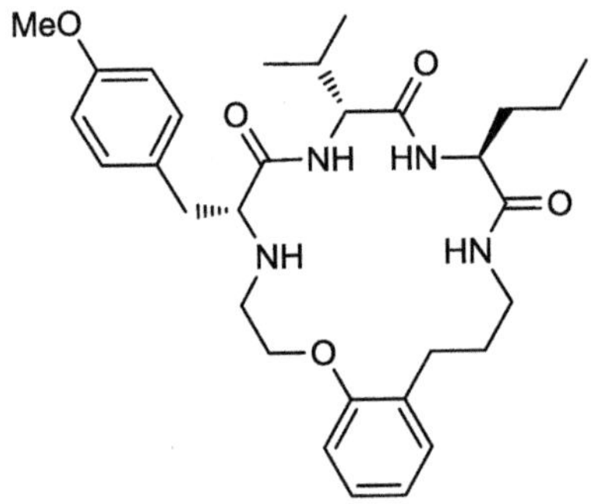

57 K_i (*h*MOT-r) 8 nM

FIGURE 8.6 Macrocyclic agonists of the ghrelin receptor.

The above macrocycles possess an interesting compromise between peptidic and nonpeptidic molecules, which retain the molecular recognition and selectivity of peptides while exhibiting desirable PK-ADME profiles. In all cases, the corresponding linear analogs were devoid of target activity.

In summary, several strategies are available to generate diverse collections of peptidomimetic and semipeptidic macrocycles, which have demonstrated their potential against a wide range of biological targets. Of particular relevance to drug discovery, some of these subclasses have been associated with acceptable oral bioavailability and a good PK-ADME profile, providing multiple clinical development candidates. Reduction in the peptidic character of the final molecules has proven to be a judicious choice, although the synthetic challenge is higher than that of all-peptidic macrocycles.

8.5 DIVERSITY-ORIENTED STRATEGIES BASED ON NONPEPTIDIC NATURAL PRODUCT SCAFFOLDS

Natural products represent the largest segment of macrocyclic drugs and an important source of new drugs [79]. Their complex architectural structures have been refined through the process of evolution to address several selection criteria, among which: biological activity, acceptable chemical stability and bioavailability for the intended purpose. It should be noted that the therapeutic indication and biological target for which they are optimized in drug discovery is often remote from their natural purpose. Nevertheless, natural products have served as templates for the generation of diversified libraries, essentially using two main approaches (Figure 8.7). In the first approach, the natural product itself (**58**) is used as a precursor that is further functionalized on areas (X and Y) available for functional group modifications to modulate its properties (e.g., **59**). In the second, essential elements of molecular recognition are kept constant and diversification is effected using simpler fragments (e.g., **60**). Whereas the first approach lends itself to narrow structural optimization, the second is amenable to broader chemical space coverage and is an interesting avenue to retain the most critical features of the natural product while synthetically simplifying it considerably with accessible building blocks. For examples of the first approach, readers are referred to the comprehensive review by Arya and co-workers [80]. Examples of the second approach are elaborated in more detail in the following sections [80].

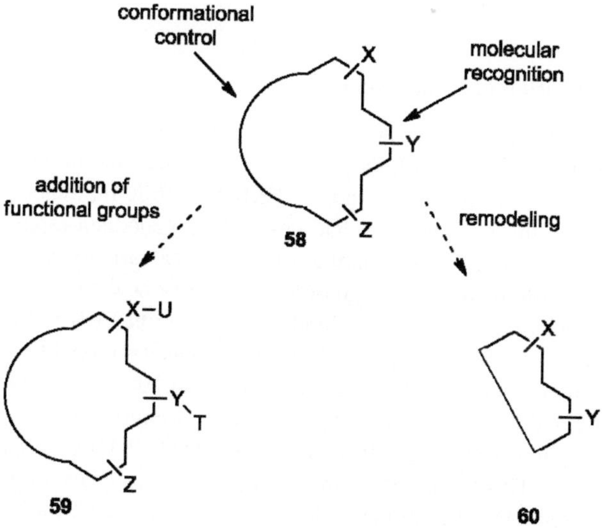

FIGURE 8.7 Natural products as templates for the generation of diversified libraries using two principal approaches: (1) the natural product is used as a precursor for further functionalization on areas (X and Y); and (2) essential elements of molecular recognition are kept constant and the diversification is achieved using simpler fragments.

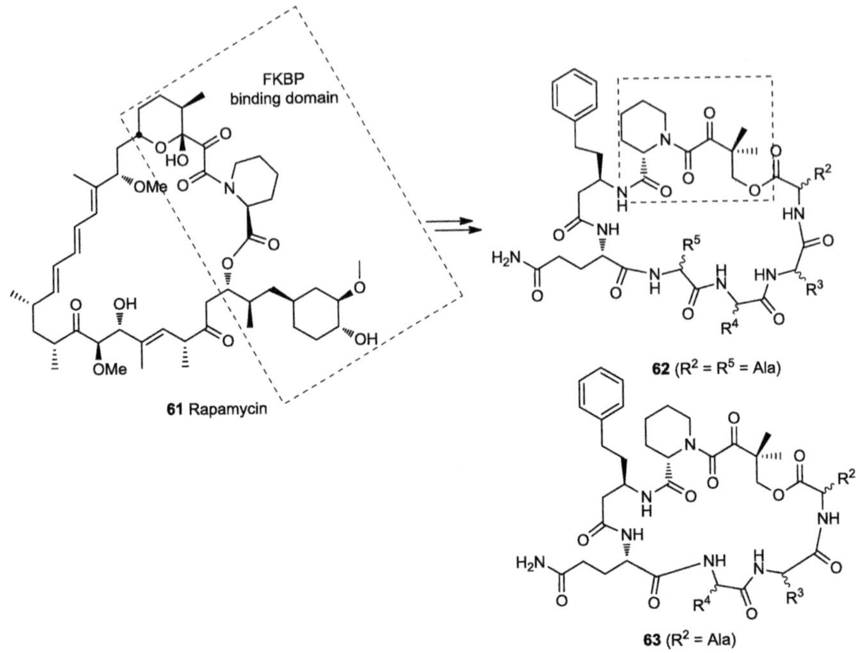

SCHEME 8.11 Rapamycin analogs combining a simplified FKBP binding domain and a variable peptidic region.

8.5.1 Diversification of Rapamycin

Rapamycin (**61**, Scheme 8.11) and FK506 are bifunctional molecules that pre-associate to their target FKBP12, then bind to calcineurin and mTOR, respectively. Part of the bifunctional molecule binds to FKBP12 while the other part serves as an effector on the third partner. Pei and co-workers generated rapamycin analogs that combine a simplified FKBP binding domain and a variable peptidic region, as exemplified by scaffolds **62** and **63**, in order to identify new partners of these complexes via the creation of composite binding surfaces on FKBP12 [81]. Building on the dimethyl-2-ketobutytyl-L-pipecolate fragment as a minimal FKBP12-binding fragment, the authors incorporated this fragment into a 200-membered library based on scaffolds **62** and **63** synthesized on solid phase. Compounds were subsequently tested for binding to FKBP12. Of 200 compounds tested, 163 bound to FKBP12 with $IC_{50} = 2$ to 93 μM. This library of rapamycin analogs, in complex with FKBP12, has the potential to serve as an enabling tool to design composite surfaces able to probe shallow protein–protein interactions or to explore binding sites for kinases outside the highly conserved ATP-binding site and open new avenues to fine-tune specific kinase pathways.

8.5.2 Diversification Strategies Based on Natural Macrolactones

Macrolactones represent an important segment of macrocyclic natural products possessing activity on numerous biological targets [10]. Schreiber and co-workers pioneered the synthesis of libraries of macrolactones based on scaffold **65** (Figure 8.8) [82]. Macrocycles were assembled by first anchoring monomer 1 on resin via a TIPS linker, followed by attachment of monomer 2 to give precursor **64**, which underwent macrocyclization using Yamaguchi's procedure [9d]. This library provided excellent stereochemical and conformational diversity; not surprisingly, yields varied widely (0 to 75%).

The same group then reported the synthesis of a 2070-membered library of macrolactones based on scaffold **67** assembled by RCM from precursor **66** (Scheme 8.12). Testing of this library for ligands of the Sonic Hedgehog N-terminal peptide identified robotnikinin **68**, which demonstrated its ability to bind the extracellular Sonic Hedgehog (Shh) protein and block Shh signaling in human cancer lines [82b,83]. Ring size and stereochemistry were found to greatly influence biological activity. This series of compounds represents an eloquent example of how to take advantage of synthetic diversification around natural structural motifs, which have been essentially "validated" through natural selection, to discover novel bioactive compounds [10].

Several natural products and structural analogs have been synthesized using solid-phase synthesis. Nicolaou et al. reported the synthesis of a library of analogs of the epothilones (**19**, Figure 8.1) using solid-phase synthesis (Scheme 8.13) [84]. Strategically, macrocycles were synthesized in microreactors on Merrifield resin functionalized as phosphonium salt **67**. Three building blocks were then added sequentially: aldehyde **68** ($n = 3$) via a Wittig reaction, ketone **69** ($n = 3$) via an aldol reaction, and alcohol **70** ($n = 5$) bearing an epothilone side chain via an esterification reaction. Macrocyclization was then performed as RCM-release, giving after side-chain deprotections epothilone analogs of general formula **71**. Subsequent epoxidation delivered congeners **72**. Using this strategy, $>$ 100 stereochemically diverse analogs bearing a variety of exocyclic appendages were synthesized. This library was used to probe the SAR of the epothilones systematically using rapidly accessible analog generation. The biological profile of analogs was tested in antitubulin polymerization and ovarian and breast carcinoma cell growth assays, delivering several analogs with activities similar or superior to those of natural products.

Using the fluorous mixture synthesis strategy developed in their labs [85], Curran and co-workers synthesized the 16 diastereomers of the natural macrolactone Sch727654 (**79**, Scheme 8.14) [86]. Using only one fluorous and one nonfluorous tag (T2 and T1, respectively), the authors synthesized four mixtures of isomers from each enantiomer of alcohols **73** and **74**, following esterification with tagged carboxylic acids **75** and **76** in the cis and trans series. Subsequent RCM followed by hydrogenation delivered four mixtures of isomers (**77**), which were independently subjected to preparative fluorous HPLC that allowed tag identification. Final detagging led to the 16 diastereomers (e.g., **78**) of Sch727654. Additional applications of the fluorous mixture synthesis to the synthesis of collections of macrocycles include

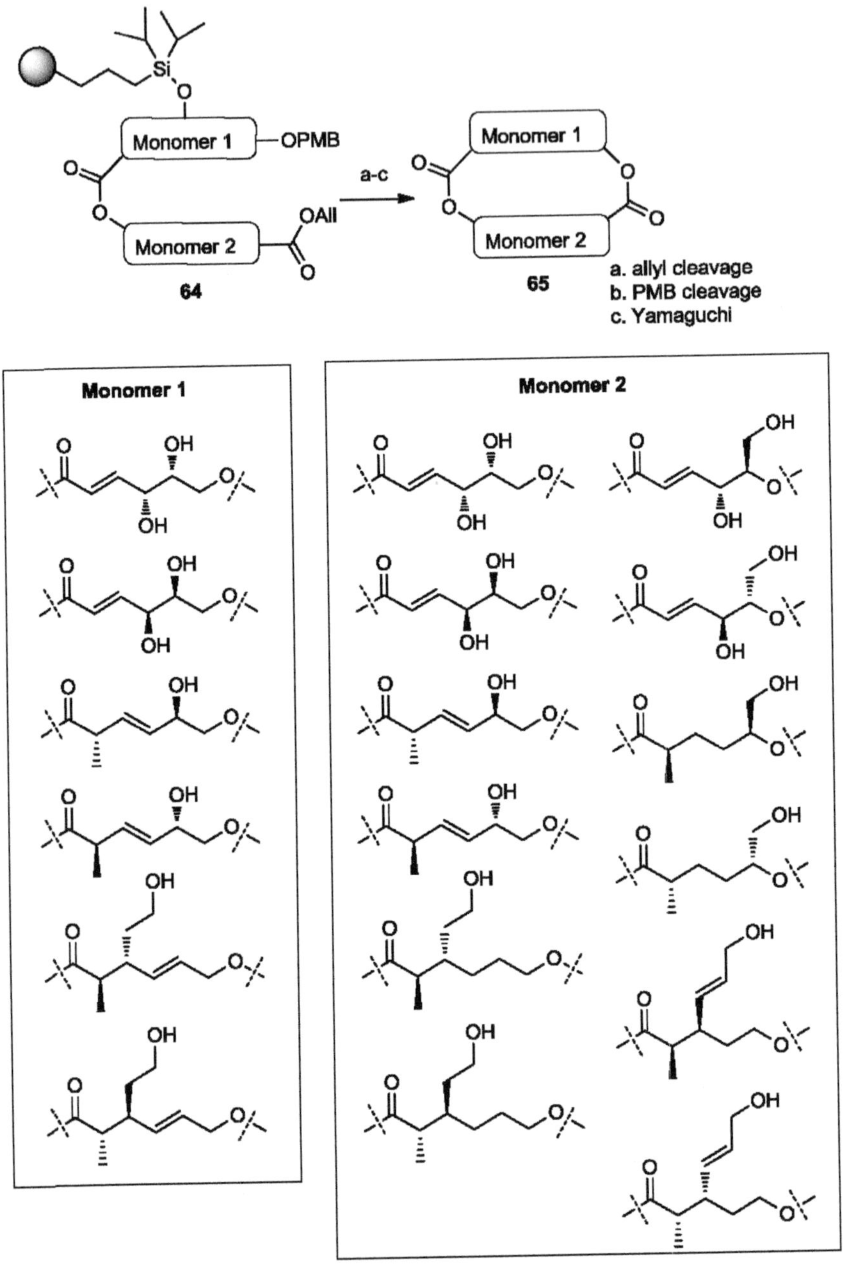

FIGURE 8.8 Library of macrolactones based on scaffold **65**.

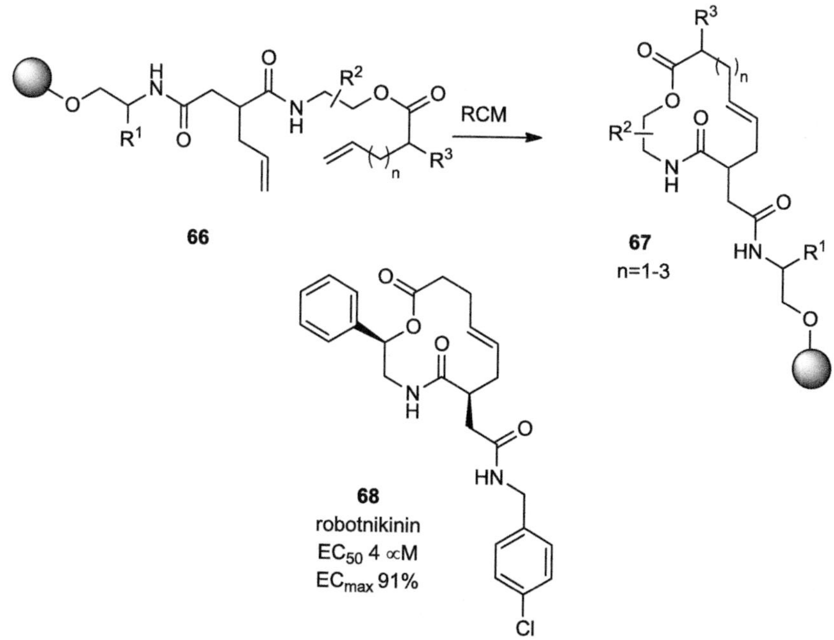

SCHEME 8.12 Synthesis of a 2070-member library of macrolactones based on scaffold **67** and assembled by RCM from precursor **66**.

the synthesis of a library of resorcylic acid lactones analogues of radicicol, pochonin C with kinase inhibitory activity [87], macrosphelides [88], and dictyostatin [89].

8.5.3 Diversification on Macrolactam Scaffolds

Marcaurelle and co-workers reported the structurally divergent exploitation of the eight stereoisomers of amino alcohol **80** (Scheme 8.15) in the construction of four types of macrocycles, 81–84, with the goal of building a library of stereochemically and chemically diverse structures. Scaffolds were obtained using four

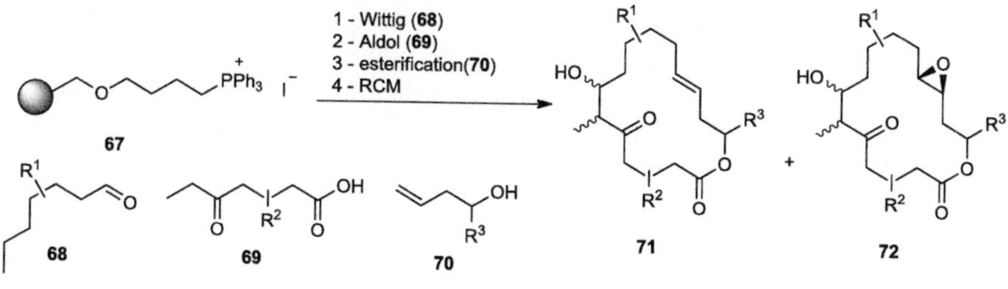

SCHEME 8.13 Synthesis of a library of analogs of the epothilones using solid-phase chemistry.

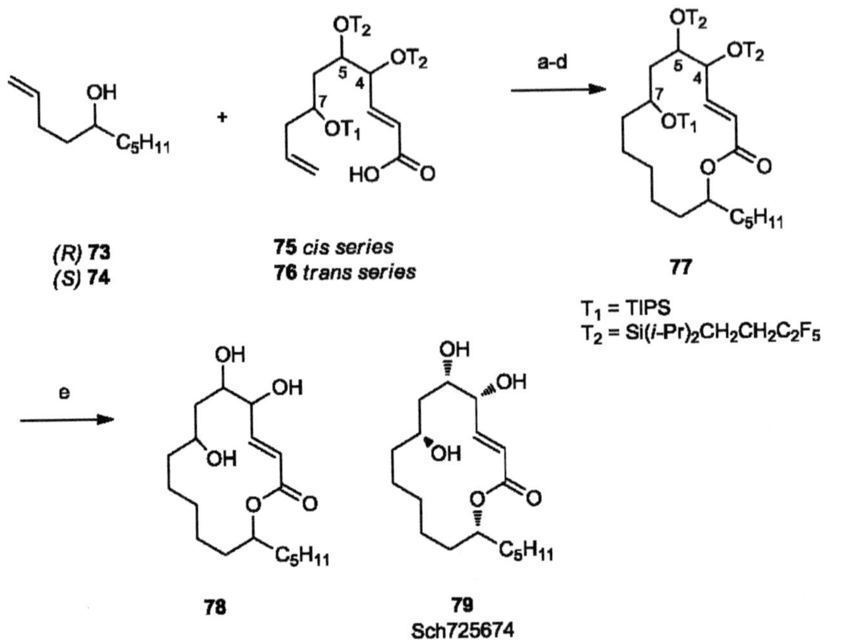

a. esterification (Yamaguchi);
b. ring-closing metathesis (Grubbs 2nd generation);
c. hydrogenation (H_2, Pd/$SrCO_3$);
d. preparative demixing (fluorous HPLC);
e. detagging (TBAF)

SCHEME 8.14 Diastereomeric library of the natural macrolactone Sch727654 using a fluorous mixture synthesis strategy.

macrocyclization strategies: S_NAr, Huisgen [3 + 2] cycloaddition (click), RCM, and macrolactamization [90]. These scaffolds combine some of the features of natural products and small molecules, such as contiguous chiral centers, endocyclic amides, or secondary alcohols. Libraries in the order of 15,000 members were produced using this approach, which have reportedly yielded leads on two targets. On the one hand, macrocycle **85** was found to be a low-micromolar inhibitor of histone deacetylases 1, 2, and 3 [90a]. On the other hand, macrocycle **86** inhibited the growth of the malaria parasite *Plasmodium falciparum*, with an IC_{50} value of 0.51 nM [91]. Although the target has not been identified, the compound displayed no toxicity toward erythrocytes.

8.5.4 Multicomponent Macrocyclization

Multicomponent reactions have been used extensively to generate diverse libraries of small molecules (see also Chapter 2). They have also been applied to the

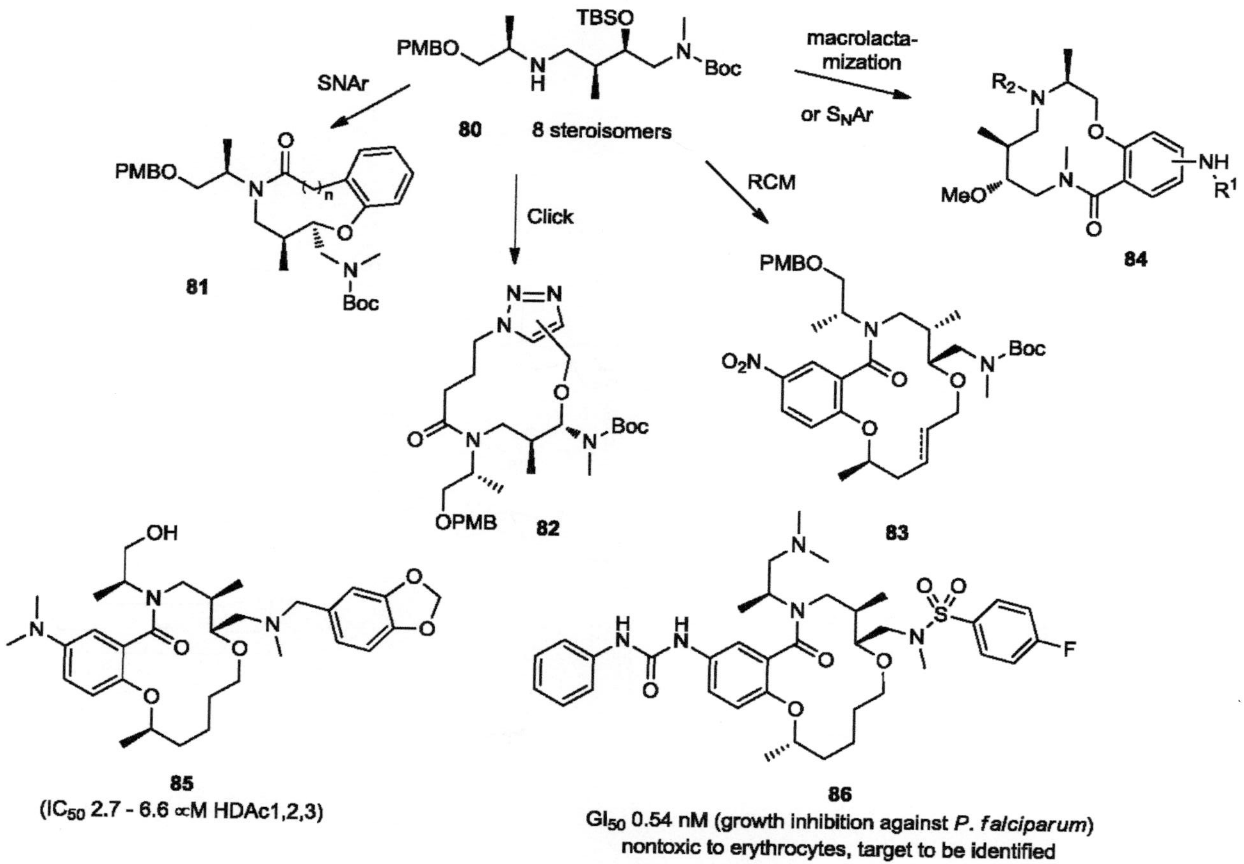

SCHEME 8.15 Amino alcohol **80** as the substrate for the diversity-oriented synthesis of four types of macrocycles, **81** to **84**.

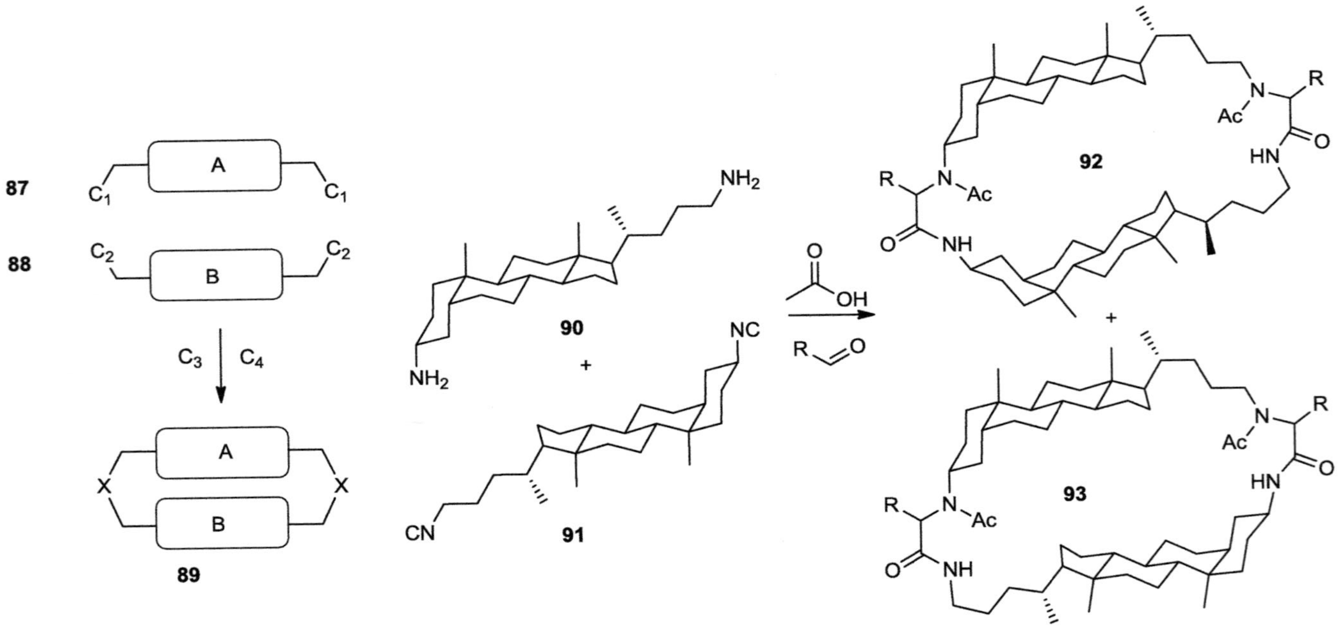

SCHEME 8.16 MiB approach (multicomponent macrocyclization including bifunctional building blocks) to address hybrid macrocycles.

generation of diversity in macrocycles via two approaches: through the assembly of linear precursors followed by macrocyclization mediated by one of the well-established methods, such as RCM, click, or macrolactamization [92], or as a point of diversification at the macrocyclization step. The latter approach represents the basis of the diversity-oriented strategy called MiB (multicomponent macrocyclization including bifunctional building blocks, Scheme 8.16; see also the works of Yudin and co-workers, Scheme 8.3) [93]. In this context, the Ugi four-component reaction between a primary amine, a carboxylic acid, a carbonyl, and an isocyanide has been used extensively [94]. The underlying principle in such an approach is to create complex macrocycles from simple building blocks with maximal diversity and versatility in terms of design. Wessjohann and co-workers have explored this avenue in several directions, one of which is presented in Scheme 8.16. The concept relies on the reaction between two bifunctional components **87** and **88**, reacted with additional components C3 and C4, together containing an amine, an aldehyde, a carboxylic acid, and an isocyanide. Two sequential Ugi four-component reactions led first to dimerization, then to ring closure.

The success of the reaction relies on the rigidity of scaffolds A and B, which prevent chain folding. For example, the reaction of diamine **90** with bis-isocyanide **91** gave, in the presence of acetic acid and an aldehyde, two macrocycles, **92** and **93**, macrocyclized head to head and head to tail. The same group has reported multiple variations around this theme, demonstrating the potential of this approach to generate diverse libraries of macrocycles based on natural and unnatural building blocks [93b].

In summary, natural product–inspired macrocycles definitely constitute an area worthy of further investment. They provide distinct, yet complementary coverage of chemical space compared to other structures, and probably distinct PK-ADME properties as well. By mimicking some of nature's secondary structures, they can be expected to benefit indirectly from the evolutionary pressure that has conferred natural macrocycles with their unique properties. Additionally, they will by design be more synthetically accessible and diversifiable than natural products, hopefully combining the best of both worlds.

8.6 CONCLUSIONS

Macrocycles represent a unique and very promising niche in the chemical space of druggable molecules. Structurally, they have the potential to reach targets that are very difficult to reach for traditional small molecules (high hanging fruits), as exemplified by the bioactive macrocycles developed for the HCV NS3/4A protease, or protein–protein interactions, such as those involved in immunomodulation or inflammation. This is due to macrocycles' ability to display distant pharmacophores on semirigid templates and provide large, yet conformationally defined interacting surfaces. Macrocycles are also considered challenging to synthesize and to diversify. More than ever, given the increasing number of targets and the limited chemical space covered in traditional libraries, medicinal chemistry needs access to novel and original scaffolds and chemical diversity. We believe that the multiple approaches reported in this chapter offer new avenues for exploiting and expanding macrocycle diversity. This ranges from cutting-edge technologies for their synthesis, such as split-and-pool production of large peptidic macrocycles or oligonucleotide-templated synthesis of peptidic macrocycles, to the efficient synthesis of small macrocycles or semipeptidic macrocycles. The past decade has also witnessed the emergence of large libraries of semipeptidic or nonpeptidic macrocycles that recapitulate some of the chemical motifs found in natural products and even some hybrid structures that borrow conserved recognition domains from macrocycles, such as rapamycin, further diversified by peptidic fragments [12]. These strategies represent a solid foundation on which to exploit the potential of macrocycles for drug discovery, and promise to deliver new macrocyclic compounds combining the best of two worlds and enabling broad coverage of the chemical space. In terms of molecular properties, macrocycles are often qualified as outliers relative to more traditional small molecules, which often makes them, at first glance, less attractive because more challenging. However, that same characteristic also endows them with a higher potential to identify novel drug candidates off the beaten path [95], as already demonstrated in numerous examples such as those reported in this chapter.

Acknowledgment

I am grateful to Dr. Mark L. Peterson for critically reviewing this manuscript.

REFERENCES AND NOTES

1. (a) E. M. Driggers, S. P. Hale, J. Lee, N. K. Terrett, *Nat. Rev. Drug. Discov.* **2008**, *7*, 608–624; (b) N. T. Terrett, *Drug Discov. Today Technol.* **2010**, *7*, e97–e104; (c) E. Marsault, M. L. Peterson, *J. Med. Chem.* **2011**, *54*, 1961–2004.
2. M. Colombo, I. Peretto, *Drug Discov. Today.* **2008**, *13*, 677–684.
3. C. A. Lipinski, *J. Pharmacol. Toxicol. Methods.* **2000**, *44*, 235–249.
4. (a) C. M. Dobson, *Nature.* **2004**, *432*, 824–828; (b) J. Zuegg, M. A. Cooper, *Curr. Top. Med. Chem.* **2012**, *12*, 1500–1513.
5. A. L. Nelson, E. Dhimolea, J. M. Reichert, *Nat. Rev. Drug Discov.* **2010**, *9*, 767–774.
6. O. Sperandio, C. H. Reynes, A. C. Camproux, B. O. Villoutreix, *Drug Discov. Today.* **2010**, *15*, 220–229.
7. (a) T. Rezai, B. Yu, G. L. Millhauser, M. P. Jacobson, R. S. Lokey, *J. Am. Chem. Soc.* **2006**, *128*, 2510–2511; (b) D. F. Veber, S. R. Johnson, H. Y. Cheng, B. R. Smith, K. W. Ward, K. D. Kopple, *J. Med. Chem.* **2002**, *45*, 2615–2623.
8. http://www.ensemblediscovery.com/pipeline/index.html.
9. (a) M. S. Butler, A. D. Buss, *Biochem. Pharmacol.* **2006**, *71*, 919–929; (b) M. S. Butler, *Nat. Prod. Rep.* **2005**, *22*, 162–195; (c) A. Gradillas, J. Perez-Castells, *Angew. Chem. Int. Ed. Engl.***2006**, *45*, 6086–6101; (d) A. Parenty, X. Moreau, J. M. Campagne, *Chem. Rev.* **2006**, *106*, 911–939; (e) C. Adessi, C. Soto, *Curr. Med. Chem.* **2002**, *9*, 963–978.
10. L. A. Wessjohann, E. Ruijter, D. Garcia-Rivera, W. Brandt, *Mol. Divers.* **2005**, *9*, 171–186.
11. (a) C. Gilon, D. Halle, M. Chorev, Z. Selinger, G. Byk, *Biopolymers* **1991**, *31*, 745–750; (b) D. Shan, M. G. Nicolaou, R. T. Borchardt, B. Wang, *J. Pharm. Sci.* **1997**, *86*, 765–767;

(c) P. S. Burton, R. A. Conradi, N. F. Ho, A. R. Hilgers, R. T. Borchardt, *J. Pharm. Sci.* **1996**, *85*, 1336–1340; (d) R. P. McGeary, D. P. Fairlie, *Curr. Opin. Drug Discov. Dev.* **1998**, *1*, 208–217; (e) B. Yoo, S. B. Shin, M. L. Huang, K. Kirshenbaum, *Chem. Eur. J.* **2010**, *16*, 5528–5537.

12. For a list of companies that developed platforms to exploit macrocycles, see C. Cain, *Biocentury*, **2012**, Sept. 17, A7–A12.
13. (a) L. A. Wessjohann, E. Ruijter, in: *Topics in Current Chemistry*, vol. 243, J. H. Mulzer, Ed., Springer-Verlag, Heidelberg, Germany, **2004**, pp. 137–184; (b) C. M. Madsen, M. H. Clausen, *Eur. J. Org. Chem.* **2011**, 3107–3115; (c) N. K. Terrett, *Drug Discov. Today Technol.* **2010**, *7*, e97–e104.
14. J. Z. Blankenstein, J. Zhu, *Eur. J. Org. Chem.* **2005**, 1949–1964.
15. W. D. Meutermans, G. T. Bourne, S. W. Golding, D. A. Horton, M. R. Campitelli, D. Craik, M. Scanlon, M. L. Smythe, *Org. Lett.* **2003**, *5*, 2711–2714.
16. A. R. Bogdan, S. V. Jerome, K. N. Houk, K. James, *J. Am. Chem. Soc.* **2012**, *134*, 2127–2138.
17. A. R. Bogdan, K. James, *Chem. Eur. J.* **2010**, *16*, 14506–14512.
18. L. T. Zhang, J. P. Tam, *J. Am. Chem. Soc.* **1999**, *121*, 3311–3320.
19. E. Marsault, H. R. Hoveyda, R. Gagnon, M. L. Peterson, M. Vezina, C. Saint-Louis, A. Landry, J. F. Pinault, L. Ouellet, S. Beauchemin, S. Beaubien, A. Mathieu, K. Benakli, Z. Wang, M. Brassard, D. Lonergan, F. Bilodeau, M. Ramaseshan, N. Fortin, R. Lan, S. Li, F. Galaud, V. Plourde, M. Champagne, A. Doucet, P. Bherer, M. Gauthier, G. Olsen, G. Villeneuve, S. Bhat, L. Foucher, D. Fortin, X. Peng, S. Bernard, A. Drouin, R. Deziel, G. Berthiaume, Y. L. Dory, G. L. Fraser, P. Deslongchamps, *Bioorg. Med. Chem. Lett.* **2008**, *18*, 4731–4735.
20. (a) J. S. Davies, *J. Pept. Sci.* **2003**, *9*, 471–501; (b) F. Kopp, M. A. Marahiel, *Nat. Prod. Rep.* **2007**, *24*, 735–749; (c) A. El-Faham, F. Albericio, *Chem. Rev.* **2011**, *111*, 6557–6602.
21. C. J. White, A. K. Yudin, *Nat. Chem.* **2011**, *3*, 509–524.
22. R. Hili, V. Rai, A. K. Yudin, *J. Am. Chem. Soc.* **2010**, *132*, 2889–2891.
23. B. H. Rotstein, V. Rai, R. Hili, A. K. Yudin, *Nat. Protoc.* **2010**, *5*, 1813–1822.
24. (a) C. Miller, R. Kirchmair, J. Troger, A. Saria, W. W. Fleischhacker, R. Fischer-Colbrie, A. Benzer, H. Winkler, *Biol. Psychiatry* **1996**, *39*, 911–918; (b) P. Van de Weghe, J. Eustache, *Curr. Top. Med. Chem.* **2005**, *5*, 1495–1519; (c) W. H. Martin, S. Blechert, *Curr. Top. Med. Chem.* **2005**, *5*, 1521–1540.
25. (a) J. M. Holub, K. Kirshenbaum, *Chem. Soc. Rev.* **2010**, *39*, 1325–1337; (b) R. A. Turner, A. G. Oliver, R. S. Lokey, *Org. Lett.* **2007**, *9*, 5011–5014.
26. M. A. J. Duncton, G. Pattenden, *J. Chem. Soc. Perkin Trans. 1* **1997**, 1235–1246.
27. (a) P. Bonnet, D. K. Agrafiotis, F. Zhu, E. Martin, *J. Chem. Inf. Model.* **2009**, *49*, 2242–2259; (b) Y. K. Kim, M. A. Arai, T. Arai, J. O. Lamenzo, E. F. Dean 3rd, N. Patterson, P. A. Clemons, S. L. Schreiber, *J. Am. Chem. Soc.* **2004**, *126*, 14740–14745.
28. (a) A. T. Frank, N. S. Farina, N. Sawwan, O. R. Wauchope, M. Qi, E. M. Brzostowska, W. Chan, F. W. Grasso, P. Haberfield, A. Greer, *Mol. Divers.* **2007**, *11*, 115–118; (b) H. Lachance, S. Wetzel, K. Kumar, H. Waldmann, *J. Med. Chem.* **2012**, *55*, 5989–6001.
29. J. Mallinson, I. Collins, *Future Med. Chem.* **2012**, *4*, 1409–1438.
30. S. Venkatraman, F. G. Njoroge, *Expert Opin. Ther. Pathol.* **2009**, *19*, 1277–1303.
31. L. Thorstholm, D. J. Craik, *Drug Discov. Today Technol.* **2012**, *9*, e13–e21.

32. B. Laufer, A. O. Frank, J. Chatterjee, T. Neubauer, C. Mas-Moruno, G. Kummerlowe, H. Kessler, *Chem. Eur. J.* **2010**, *16*, 5385–5390.

33. (a) D. J. Craik, J. E. Swedberg, J. S. Mylne, M. Cemazar, *Expert Opin. Drug Discov.* **2012**, *7*, 179–194; (b) M. Katsara, T. Tselios, S. Deraos, G. Deraos, M. T. Matsoukas, E. Lazoura, J. Matsoukas, V. Apostolopoulos, *Curr. Med. Chem.* **2006**, *13*, 2221–2232.

34. (a) G. M. Pauletti, S. Gangwar, T. J. Siahaan, J. Aubé, R. T. Borchardt, *Adv. Drug Deliv. Rev.* **1997**, *27*, 235–256; (b) H. Ouyang, D. G. Vander Velde, R. T. Borchardt, T. J. Siahaan, *J. Pept. Res.* **2002**, *59*, 183–195; (c) W. Wang, G. Camenisch, D. C. Sane, H. Zhang, E. Hugger, G. L. Wheeler, R. T. Borchardt, B. Wang, *J. Control. Release* **2000**, *65*, 245–251; (d) B. Wang, S. Gangwar, G. Pauletti, T. Siahaan, R. T. Borchardt, *Methods Mol. Med.* **1999**, *23*, 53–69; (e) S. Gangwar, G. M. Pauletti, T. J. Siahaan, V. J. Stella, R. T. Borchardt, *Methods Mol. Med.* **1999**, *23*, 37–51; (f) J. G. Beck, J. Chatterjee, B. Laufer, M. U. Kiran, A. O. Frank, S. Neubauer, O. Ovadia, S. Greenberg, C. Gilon, A. Hoffman, H. Kessler, *J. Am. Chem. Soc.* **2012**, *134*, 12125–12133; (g) J. Chatterjee, C. Gilon, A. Hoffman, H. Kessler, *Acc. Chem. Res.* **2008**, *41*, 1331–1342; (h) J. Chatterjee, B. Laufer, H. Kessler, *Nat. Protoc.* **2012**, *7*, 432–444; (i) O. Ovadia, S. Greenberg, J. Chatterjee, B. Laufer, F. Opperer, H. Kessler, C. Gilon, A. Hoffman, *Mol. Pharm.* **2011**, *8*, 479–487; (j) T. R. White, C. M. Renzelman, A. C. Rand, T. Rezai, C. M. McEwen, V. M. Gelev, R. A. Turner, R. G. Linington, S. S. Leung, A. S. Kalgutkar, J. N. Bauman, Y. Zhang, S. Liras, D. A. Price, A. M. Mathiowetz, M. P. Jacobson, R. S. Lokey, *Nat. Chem. Biol.* **2011**, *7*, 810–817; (k) A. C. Rand, S. S. F. Leung, H. Eng, C. J. Rotter, R. Sharma, A. S. Kalgutkar, Y. Zhang, M. V. Varma, K. A. Farley, B. Khunte, C. Limberakis, D. A. Price, S. Liras, A. M. Mathiowetz, M. P. Jacobson, R.S. Lokey, *Med. Chem. Commun.* **2012**, *3*, 1282–1289.

35. T. Liu, S. H. Joo, J. L. Voorhees, C. L. Brooks, D. Pei, *Bioorg. Med. Chem.* **2009**, *17*, 1026–1033.

36. R. Liu, J. Marik, K. S. Lam, *J. Am. Chem. Soc.* **2002**, *124*, 7678–7680.

37. (a) S. H. Joo, Q. Xiao, Y. Ling, B. Gopishetty, D. Pei, *J. Am. Chem. Soc.* **2006**, *128*, 13000–13009; (b) D. Pei, *Chem. Biol.* **2010**, *17*, 3–4.

38. X. Chen, P. H. Tan, Y. Zhang, D. Pei, *J. Comb. Chem.* **2009**, *11*, 604–611.

39. T. Liu, Z. Qian, Q. Xiao, D. Pei, *ACS Comb. Sci.* **2011**, *13*, 537–546.

40. Y. Zhang, S. Zhou, A. S. Wavreille, J. DeWille, D. Pei, *J. Comb. Chem.* **2008**, *10*, 247–255.

41. Q. Xiao, D. Pei, *J. Med. Chem.* **2007**, *50*, 3132–3137.

42. M. J. Jebrail, A. H. Ng, V. Rai, R. Hili, A. K. Yudin, A. R. Wheeler, *Angew. Chem. Int. Ed.* **2010**, *49*, 8625–8629.

43. B. H. Rotstein, R. Mourtada, S. O. Kelley, A. K. Yudin, *Chem. Eur. J.* **2011**, *17*, 12257–12261.

44. A. Roxin, J. Chen, C. C. Scully, B. H. Rotstein, A. K. Yudin, G. Zheng, *Bioconjug. Chem.* **2012**, *23*, 1387–1395.

45. This technology forms the basis of Encycle Therapeutics.

46. Z. J. Gartner, B. N. Tse, R. Grubina, J. B. Doyon, T. M. Snyder, D. R. Liu, *Science* **2004**, *305*, 1601–1605.

47. B. N. Tse, T. M. Snyder, Y. Shen, D. R. Liu, *J. Am. Chem. Soc.* **2008**, *130*, 15611–15626.

48. R. E. Kleiner, C. E. Dumelin, G. C. Tiu, K. Sakurai, D. R. Liu, *J. Am. Chem. Soc.* **2010**, *132*, 11779–11791.

49. G. Georghiou, R. E. Kleiner, M. Pulkoski-Gross, D. R. Liu, M. A. Seeliger, *Nat. Chem. Biol.* **2012**, *8*, 366–374.

50. (a) S. W. Millward, T. T. Takahashi, R. W. Roberts, *J. Am. Chem. Soc.* **2005**, *127*, 14142–14143; (b) A. Frankel, S. Li, S. R. Starck, R. W. Roberts, *Curr. Opin. Struct. Biol.* **2003**, *13*, 506–512; (c) R. W. Roberts, J. W. Szostak, *Proc. Natl. Acad. Sci. U.S.A.* **1997**, *94*, 12297–12302.

51. Y. Yamagishi, I. Shoji, S. Miyagawa, T. Kawakami, T. Katoh, Y. Goto, H. Suga, *Chem. Biol.* **2011**, *18*, 1562–1570.

52. C. Heinis, T. Rutherford, S. Freund, G. Winter, *Nat. Chem. Biol.* **2009**, *5*, 502–507.

53. (a) J. D. Tyndall, D. P. Fairlie, *J. Mol. Recognit.* **1999**, *12*, 363–370; (b) D. P. Fairlie, J. D. Tyndall, R. C. Reid, A. K. Wong, G. Abbenante, M. J. Scanlon, D. R. March, D. A. Bergman, C. L. Chai, B. A. Burkett, *J. Med. Chem.* **2000**, *43*, 1271–1281; (c) J. D. Tyndall, D. P. Fairlie, *Curr. Med. Chem.* **2001**, *8*, 893–907; (d) J. D. Tyndall, T. Nall, D. P. Fairlie, *Chem. Rev.* **2005**, *105*, 973–999; (e) P. K. Madala, J. D. Tyndall, T. Nall, D. P. Fairlie, *Chem. Rev.* **2010**, *110*, PR1–PR31.

54. R. C. Reid, L. K. Pattenden, J. D. Tyndall, J. L. Martin, T. Walsh, D. P. Fairlie, *J. Med. Chem.* **2004**, *47*, 1641–1651.

55. (a) M. P. Glenn, L. K. Pattenden, R. C. Reid, D. P. Tyssen, J. D. Tyndall, C. J. Birch, D. P. Fairlie, *J. Med. Chem.* **2002**, *45*, 371–381; (b) R. C. Reid, M. J. Kelso, M. J. Scanlon, D. P. Fairlie, *J. Am. Chem. Soc.* **2002**, *124*, 5673–5683; (c) J. D. Tyndall, R. C. Reid, D. P. Tyssen, D. K. Jardine, B. Todd, M. Passmore, D. R. March, L. K. Pattenden, D. A. Bergman, D. Alewood, S. H. Hu, P. F. Alewood, C. J. Birch, J. L. Martin, D. P. Fairlie, *J. Med. Chem.* **2000**, *43*, 3495–3504.

56. C. Park, K. Burgess, *J. Comb. Chem.* **2001**, *3*, 257–266.

57. Y. Feng, M. Pattarawarapan, Z. Wang, K. Burgess, *Org. Lett.* **1999**, *1*, 121–124.

58. E. Nnanabu, K. Burgess, *Org. Lett.* **2006**, *8*, 1259–1262.

59. (a) S. Maliartchouk, Y. Feng, L. Ivanisevic, T. Debeir, A. C. Cuello, K. Burgess, H. U. Saragovi, *Mol. Pharmacol. 2000*, *57*, 385–391; (b) M. Pattarawarapan, M. C. Zaccaro, U. H. Saragovi, K. Burgess, *J. Med. Chem.* **2002**, *45*, 4387–4390; (c) M. C. Zaccaro, H. B. Lee, M. Pattarawarapan, Z. Xia, A. Caron, P. J. L'Heureux, Y. Bengio, K. Burgess, H. U. Saragovi, *Chem. Biol.* **2005**, *12*, 1015–1028.

60. (a) E. A. Jefferson, S. Arakawa, L. B. Blyn, A. Miyaji, S. A. Osgood, R. Ranken, L. M. Risen, E. E. Swayze, *J. Med. Chem.* **2002**, *45*, 3430–3439; (b) E. E. Swayze, E. A. Jefferson, K. A. Sannes-Lowery, L. B. Blyn, L. M. Risen, S. Arakawa, S. A. Osgood, S. A. Hofstadler, R. H. Griffey, *J. Med. Chem.* **2002**, *45*, 3816–3819.

61. (a) F. Freire, S. H. Gellman, *J. Am. Chem. Soc.* **2009**, *131*, 7970–7972; (b) R. J. Woods, J. O. Brower, E. Castellanos, M. Hashemzadeh, O. Khakshoor, W. A. Russu, J. S. Nowick, *J. Am. Chem. Soc.* **2007**, *129*, 2548–2558; (c) D. Obrecht, E. K. Chevalier, J. A. Robinson, *Drug Discov. Today Technol.* **2012**, *9*, e63–e69; (d) This forms the technological basis of Aileron Therapeutics (www.aileronrx.com); (e) A. L. Jochim, P. S. Arora, *ACS Chem Biol.* **2010**, *5*, 919–923; (f) H.K. Cui, B. Zhao, Y. Li, Y. Guo, H. Hu, L. Liu, Y.G. Chen, *Cell Res.* **2013**, *23*, 581–584; (g) Y.W. Kim, T.N. Grossmann, G.L. Verdine, *Nat. Protoc.* **2011**, *6*, 761–771; (h) T.L. Joseph, D.P. Lane, C.S. Verma, *PLoS One*, **2012**, *7*, e43985.

62. This forms the basis of the PEMFinder and MacroFinder platforms of Polyphor: http://www.polyphor.com.

63. (a) A. Davidson, T. C. Leeper, Z. Athanassiou, K. Patora-Komisarska, J. Karn, J. A. Robinson, G. Varani, *Proc. Natl. Acad. Sci. U.S.A.* **2009**, *106*, 11931–11936; (b) A. Descours, K. Moehle, A. Renard, J. A. Robinson, *ChemBioChem* **2002**, *3*, 318–323; (c) N. Srinivas, P. Jetter, B. J. Ueberbacher, M. Werneburg, K. Zerbe, J. Steinmann, B. Van der Meijden, F. Bernardini, A. Lederer, R. L. Dias, P. E. Misson, H. Henze, J. Zumbrunn,

F. O. Gombert, D. Obrecht, P. Hunziker, S. Schauer, U. Ziegler, A. Kach, L. Eberl, K. Riedel, S. J. DeMarco, J. A. Robinson, *Science* **2010**, *327*, 1010–1013; (d) J. A. Robinson, S. C. Shankaramma, P. Jetter, U. Kienzl, R. A. Schwendener, J. W. Vrijbloed, D. Obrecht, *Bioorg. Med. Chem.* **2005**, *13*, 2055–2064.

64. S. J. DeMarco, H. Henze, A. Lederer, K. Moehle, R. Mukherjee, B. Romagnoli, J. A. Robinson, F. Brianza, F. O. Gombert, S. Lociuro, C. Ludin, J. W. Vrijbloed, J. Zumbrunn, J. P. Obrecht, D. Obrecht, V. Brondani, F. Hamy, T. Klimkait, *Bioorg. Med. Chem.* **2006**, *14*, 8396–8404.
65. A. Isidro-Llobet, T. Murillo, P. Bello, A. Cilibrizzi, J. T. Hodgkinson, W. R. Galloway, A. Bender, M. Welch, D. R. Spring, *Proc. Natl. Acad. Sci. U.S.A.* **2011**, *108*, 6793–6798.
66. T. E. Nielsen, S. L. Schreiber, *Angew. Chem. Int. Ed.* **2008**, *47*, 48–56.
67. Y. U. Kwon, T. Kodadek, *J. Am. Chem. Soc.* **2007**, *129*, 1508–1509.
68. S. N. Khan, A. Kim, R. H. Grubbs, Y. U. Kwon, *Org. Lett.* **2011**, *13*, 1582–1585.
69. (a) R. N. Chapman, P. S. Arora, *Org. Lett.* **2006**, *8*, 5825–5828; (b) G. Dimartino, D. Wang, R. N. Chapman, P. S. Arora, *Org. Lett.* **2005**, *7*, 2389–2392.
70. (a) J. Illesinghe, C. X. Guo, R. Garland, A. Ahmed, B. van Lierop, J. Elaridi, W. R. Jackson, A. J. Robinson, *Chem. Commun.* **2009**, 295–297; (b) J. Elaridi, J. Patel, W. R. Jackson, A. J. Robinson, *J. Org. Chem.* **2006**, *71*, 7538–7545.
71. M. Krishnamurthy, K. Simon, A. M. Orendt, P. A. Beal, *Angew. Chem. Int. Ed.* **2007**, *46*, 7044–7047.
72. (a) http://www.tranzyme.com; (b) this strategy, named MaTCh (Macrocyclic Template Chemistry), constitutes the technological basis of the company Tranzyme Pharma.
73. (a) J. H. van Maarseveen, *Comb. Chem. High Throughput Screen.* **1998**, *1*, 185–214; (b) J. G. Lewis, P. A. Bartlett, *J. Comb. Chem.* **2003**, *5*, 278–284.
74. F. Albericio, R. P. Hammer, C. Garcia-Echeverria, M. A. Molins, J. L. Chang, M. C. Munson, M. Pons, E. Giralt, G. Barany, *Int. J. Pept. Protein Res.* **1991**, *37*, 402–413.
75. E. Marsault, H. R. Hoveyda, M. L. Peterson, C. Saint-Louis, A. Landry, M. Vezina, L. Ouellet, Z. Wang, M. Ramaseshan, S. Beaubien, K. Benakli, S. Beauchemin, R. Deziel, T. Peeters, G. L. Fraser, *J. Med. Chem.* **2006**, *49*, 7190–7197.
76. (a) H. R. Hoveyda, E. Marsault, R. Gagnon, A. P. Mathieu, M. Vezina, A. Landry, Z. Wang, K. Benakli, S. Beaubien, C. Saint-Louis, M. Brassard, J. F. Pinault, L. Ouellet, S. Bhat, M. Ramaseshan, X. Peng, L. Foucher, S. Beauchemin, P. Bherer, D. F. Veber, M. L. Peterson, G. L. Fraser, *J. Med. Chem.* **2011**, *54*, 8305–8320; (b) G. L. Fraser, K. Veňková, H. R. Hoveyda, H. Thomas, B. Greenwood-Van Meerveld, *Eur. J. Pharmacol.* **2009**, *604*, 132–137.
77. (a) J. M. Wo, N. Ejskjaer, P. M. Hellstrom, R. A. Malik, J. C. Pezzullo, L. Shaughnessy, P. Charlton, G. Kosutic, R. W. McCallum, *Aliment. Pharm. Ther.* **2011**, *33*, 679–688; (b) G. Bochicchio, P. Charlton, J. C. Pezzullo, G. Kosutic, A. Senagore, *World J. Surg.* **2012**, *36*, 39–45.
78. E. Marsault, K. Benakli, S. Beaubien, C. Saint-Louis, R. Deziel, G. Fraser, *Bioorg. Med. Chem. Lett.* **2007**, *17*, 4187–4190.
79. D. J. Newman, G. M. Cragg, *J. Nat. Prod.* **2007**, *70*, 461–477.
80. J. P. Nandy, M. Prakesch, S. Khadem, P. T. Reddy, U. Sharma, P. Arya, *Chem. Rev.* **2009**, *109*, 1999–2060.
81. X. Wu, L. Wang, Y. Han, N. Regan, P. K. Li, M. A. Villalona, X. Hu, R. Briesewitz, D. Pei, *ACS Comb. Sci.* **2011**, *13*, 486–495.

82. (a) D. R. Schmidt, O. Kwon, S. L. Schreiber, *J. Comb. Chem.* **2004**, *6*, 286–292; (b) L. F. Peng, B. Z. Stanton, N. Maloof, X. Wang, S. L. Schreiber, *Bioorg. Med. Chem. Lett.* **2009**, *19*, 6319–6325; (c) D. S. Lee, J. K. Sello, S. L. Schreiber, *J. Am. Chem. Soc.* **1999**, *121*, 10648–10649.

83. B. Z. Stanton, L. F. Peng, N. Maloof, K. Nakai, X. Wang, J. L. Duffner, K. M. Taveras, J. M. Hyman, S. W. Lee, A. N. Koehler, J. K. Chen, J. L. Fox, A. Mandinova, S. L. Schreiber, *Nat. Chem. Biol.* **2009**, *5*, 154–156.

84. K. C. Nicolaou, N. Winssinger, J. Pastor, S. Ninkovic, F. Sarabia, Y. He, D. Vourloumis, Z. Yang, T. Li, P. Giannakakou, E. Hamel, *Nature* **1997**, *387*, 268–272.

85. Z. Luo, Q. Zhang, Y. Oderaotoshi, D. P. Curran, *Science* **2001**, *291*, 1766–1769.

86. J. D. Moretti, X. Wang, D. P. Curran, *J. Am. Chem. Soc.* **2012**, *134*, 7963–7970.

87. (a) R. Jogireddy, P. Y. Dakas, G. Valot, S. Barluenga, N. Winssinger, *Chem. Eur. J.* **2009**, *15*, 11498–11506; (b) P. Y. Dakas, S. Barluenga, F. Totzke, U. Zirrgiebel, N. Winssinger, *Angew. Chem. Int. Ed.***2007**, *46*, 6899–6902.

88. D. P. Curran, M. K. Sinha, K. Zhang, J. J. Sabatini, D. H. Cho, *Nat. Chem.* **2012**, *4*, 124–129.

89. Y. Fukui, A. M. Bruckner, Y. Shin, R. Balachandran, B. W. Day, D. P. Curran, *Org. Lett.* **2006**, *8*, 301–304.

90. (a) L. A. Marcaurelle, E. Comer, S. Dandapani, J. R. Duvall, B. Gerard, S. Kesavan, M. D. Lee IV, H. Liu, J. T. Lowe, J. C. Marie, C. A. Mulrooney, B. A. Pandya, A. Rowley, T. D. Ryba, B. C. Suh, J. Wei, D. W. Young, L. B. Akella, N. T. Ross, Y. L. Zhang, D. M. Fass, S. A. Reis, W. N. Zhao, S. J. Haggarty, M. Palmer, M. A. Foley, *J. Am. Chem. Soc.* **2010**, *132*, 16962–16976; (b) A. R. Kelly, J. Wei, S. Kesavan, J. C. Marie, N. Windmon, D. W. Young, L. A. Marcaurelle, *Org. Lett.* **2009**, *11*, 2257–2260; (c) E. Comer, H. Liu, A. Joliton, A. Clabaut, C. Johnson, L. B. Akella, L. A. Marcaurelle, *Proc. Natl. Acad. Sci. U.S.A.* **2011**, *108*, 6751–6756; (d) B. Gerard, J. R. Duvall, J. T. Lowe, T. Murillo, J. Wei, L. B. Akella, L. A. Marcaurelle, *ACS Comb. Sci.* **2011**, *13*, 365–374; (e) L. B. Akella, L. A. Marcaurelle, *ACS Comb. Sci.* **2011**, *13*, 357–364; (f) S. Dandapani, J. T. Lowe, E. Comer, L. A. Marcaurelle, *J. Org. Chem.* **2011**, *76*, 8042–8048.

91. R. W. Heidebrecht, Jr., C. Mulrooney, C. P. Austin, R. H. Barker, Jr., J. A. Beaudoin, K. C. Cheng, E. Comer, S. Dandapani, J. Dick, J. R. Duvall, E. H. Ekland, D. A. Fidock, M. E. Fitzgerald, M. Foley, R. Guha, P. Hinkson, M. Kramer, A. K. Lukens, D. Masi, L. A. Marcaurelle, X. Z. Su, C. J. Thomas, M. Weiwer, R. C. Wiegand, D. Wirth, M. Xia, J. Yuan, J. Zhao, M. Palmer, B. Munoz, S. Schreiber, *ACS Med. Chem. Lett.* **2012**, *3*, 112–117.

92. (a) B. Beck, G. Larbig, B. Mejat, M. Magnin-Lachaux, A. Picard, E. Herdtweck, A. Dömling, *Org. Lett.* **2003**, *5*, 1047–1050; (b) J. K. Sello, P. R. Andreana, D. Lee, S. L. Schreiber, *Org. Lett.* **2003**, *5*, 4125–4127; (c) T. Pirali, G. C. Tron, J. Zhu, *Org. Lett.* **2006**, *8*, 4145–4148; (d) C. Bughin, G. Zhao, H. Bienaymé, J. Zhu, *Chem. Eur. J.* **2006**, *12*, 1174–1184; (e) M. Greef, S. Abeln, K. Belskasmi, A. Dömling, R. V. A. Orru, L. A. Wessjohann, *Synthesis* **2006**, *23*, 3997–4004.

93. (a) L. A. Wessjohann, E. Ruijter, *Mol. Divers.* **2005**, *9*, 159–169; (b) L. A. Wessjohann, D. G. Rivera, O. E. Vercillo, *Chem. Rev.* **2009**, *109*, 796–814.

94. (a) I. Ugi, C. Steinbrückner, *Chem. Ber.* **1961**, *94*, 734–742; (b) A. Dömling, *Chem. Rev.* **2006**, *106*, 17–89.

95. Y. L. Bennani, *Drug Discov. Today* **2012**, *17* Suppl., S31–S44.

PART II

CHEMICAL LIBRARIES AND DIVERSITY-ORIENTED SYNTHESIS

9

DIVERSITY-ORIENTED SYNTHESIS OF NATURAL PRODUCT–LIKE LIBRARIES

Mark Dow, Francesco Marchetti, and Adam Nelson

9.1 INTRODUCTION

Small molecules have made a tremendous contribution to our understanding of biological mechanisms [1] and our ability to treat disease [2,3]. The field of chemical genetics, in which small molecules are used to modulate the cellular function of biological macromolecules, is dependent on the availability of small-molecule libraries that target biologically relevant chemical space. In this chapter we focus on the application of diversity-oriented synthesis (DOS) in the synthesis of libraries of natural product–like small molecules [4]. Natural products necessarily reside in biologically relevant chemical space, since they bind to both their biosynthetic enzymes and their target macromolecules. Some natural products (such as brefeldin A [5] and cyclopamine [6], Figure 9.1) have been exploited directly as valuable tools in chemical genetic studies. It is therefore unsurprising that natural products have provided an inspiration for library design and drug discovery [4,7]. Although "natural product likeness" might be considered to be a rather subjective attribute, it has been quantified and then exploited in the prioritization of small-molecule libraries [8].

DOS is a synthetic approach that was conceived to facilitate the discovery of small molecules with novel biological functions [9]. In this chapter we focus on strategies for the synthesis of libraries that have been inspired by natural products to some extent (rather than synthetic approaches that can yield only close analogs of a specific natural product). It is essential that this inspiration from natural products

Diversity-Oriented Synthesis: Basics and Applications in Organic Synthesis, Drug Discovery, and Chemical Biology, First Edition. Edited by Andrea Trabocchi.

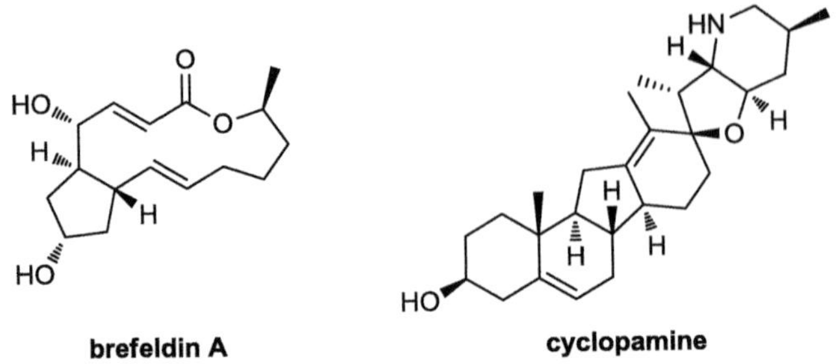

FIGURE 9.1 Examples of natural products exploited in chemical genetics.

does not inform DOS to the extent that compounds with novel functions are not identified. In some cases, the scaffold (or scaffolds) represented within a DOS library closely resemble those found in natural products. Inspiration from such scaffolds is, however, valuable because ligands based on similar natural product scaffolds have been shown to bind selectively to different ligand binding sites with similar folds [10]. In other examples, the relationship to natural products is much looser, with just some of the broad structural features of a natural product family being represented. Given that natural products necessarily have the structural features needed to recognize biological macromolecules, this level of inspiration is appropriate; indeed, it can allow the identification of ligands with functions that natural products have not yet been found to possess.

Although in this chapter we focus on the underpinning synthetic strategy, examples of valuable small molecules discovered using natural product–like libraries will be provided. In each scheme, generic structures for the synthetic intermediates are shown, generally with illustrative examples of specific final compounds. In Section 9.2 we focus on the synthesis of libraries that are inspired directly by a single (or a limited range of) natural products. A key challenge in the field has, however, been to vary the scaffolds of natural product–like molecules more widely. Some general strategies have been developed for the synthesis of skeletally diverse small molecules, including folding pathways (Section 9.3), branching pathways (Section 9.4), and oligomer-based approaches (Section 9.5).

9.2 LIBRARIES INSPIRED BY NATURAL PRODUCT SCAFFOLDS

Small-molecule scaffolds play a key role in guiding chemists' navigation of biologically relevant chemical space [11]. It has been proposed that biologically active small molecules may be focused on specific subfractions of chemical space [12]. Indeed, the field of biology-oriented synthesis (BIOS) [10d,13] seeks to target such "bioactivity islands" by designing libraries around scaffolds that have been validated biologically. Some drug scaffolds have been described as privileged [14] because representative compounds bind to different protein targets. The scaffolds found in natural products have been mapped comprehensively using hierarchical scaffold trees

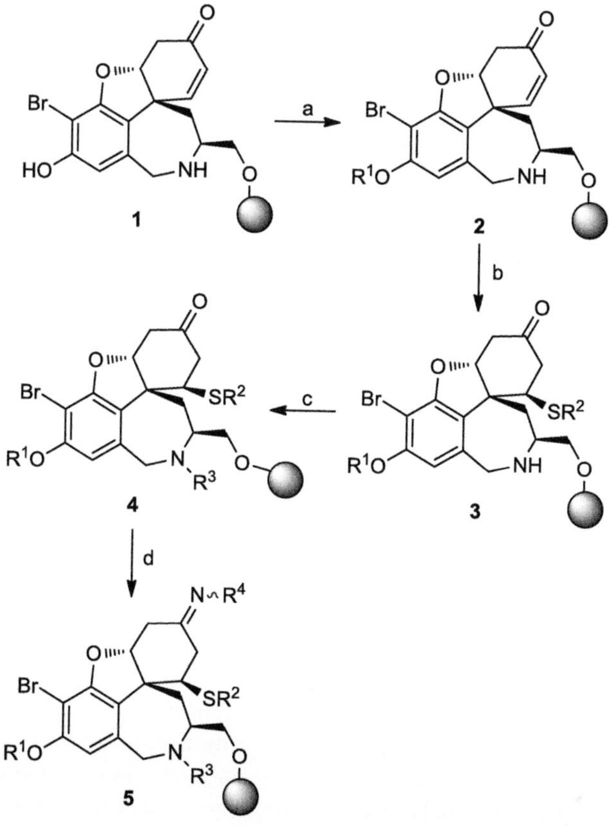

SCHEME 9.1 Pelish et al.'s Synthesis of galanthamine-like molecules with high appendage diversity: (a) R^1OH, PPh_3, DIAD, THF, 0°C; (b) R^2SH, 2,6-lutidine, *n*-BuLi, THF, $0 \rightarrow 40$°C; (c) (i) R^3CHO, AcOH, MeOH–THF; (ii) $NaBH_3CN$, MeOH or R^3COCl, 2,6-lutidine, CH_2Cl_2 or R^3NCO, CH_2Cl_2; (d) R^4NH_2, AcOH, MeOH–CH_2Cl_2.

[11c]. Such scaffold trees can facilitate navigation of the chemical space that is relevant to a particular biological function; specifically, analysis of such trees can help identify related, but previously unexplored, scaffolds that are worthy of investigation [11a and b].†

Scaffolds found in natural products have provided a direct inspiration in the design of small-molecule libraries. Pelish et al. prepared a substitutionally diverse library that was based on the scaffold of the alkaloid, galanthamine [15]. The scaffold **1** was prepared in five steps from a solid-supported tyrosine derivative. A sequence of reactions was then used to vary four of the appendages (R^1 to R^4) in the final compounds (Scheme 9.1). Functionalization of the phenol using a Mitsunobu reaction

†Scaffold Hunter is a software package that has been developed to facilitate the navigation of hierarchical scaffold trees [11a and b].

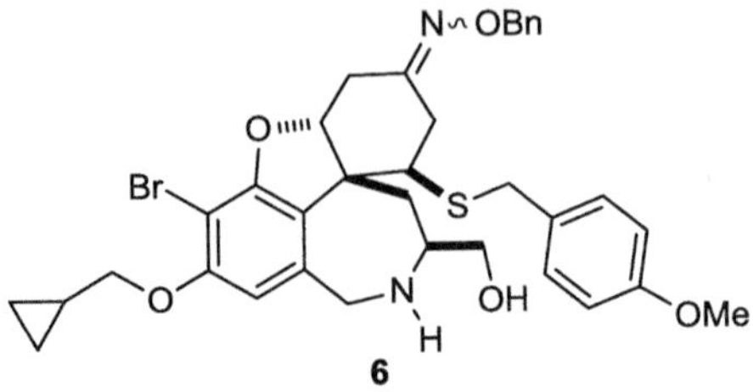

6

FIGURE 9.2 Chemical structure of secramine.

(→ **2**) was followed by conjugate addition of a thiol to the α,β-unsaturated ketone (→ **3**), and functionalization of the secondary amine (→ **4**). Finally, the ketones **4** were converted into imines or oxime ethers **5**. Following release from solid support, the successful formation of 2527 of 2946 possible compounds was confirmed by mass spectrometry. A cell-based phenotypic assay was used to identify secramine **6** (Figure 9.2) as a potent inhibitor of protein trafficking from the Golgi apparatus to the plasma membrane. The discovery of secramine emphasizes the validity of exploiting natural product scaffolds in the discovery of small molecules with novel functions; secramine was inspired by galanthamine but has an unrelated biological function.

Oh et al. have reported the diversity-oriented synthesis of a library that was inspired by benzopyran-containing natural products (see Figure 9.3 for examples of such natural products) [16]. The vinyl triflates **8**, prepared from the corresponding ketones **7**, were the key intermediates in the synthesis (Scheme 9.2). The final compounds **9** were prepared by Suzuki coupling, followed by release from the solid support. Alternatively, ethynylation, followed by a Cu-catalyzed "click" reaction, yielded the triazoles **10**. In a similar vein, Pd-catalyzed vinylation of the vinyl triflates **8**, followed by a Diels–Alder reaction, yielded the adducts **11**; the adducts **11** were either reduced (→ **12**) or aromatized (→ **13**) to yield final compounds with related scaffolds. Overall, the approach was used to prepare a 434-member polyheterocyclic benzopyran library. Nicolaou has described the synthesis of a member library of about 10,000 natural product–like benzopyran derivatives using a distinct synthetic approach [17]. Other libraries based on natural product–like scaffolds have exploited multicomponent aza-Diels–Alder chemistry [18], cyclizations of products of multicomponent reactions [19], and acetal formation [20].

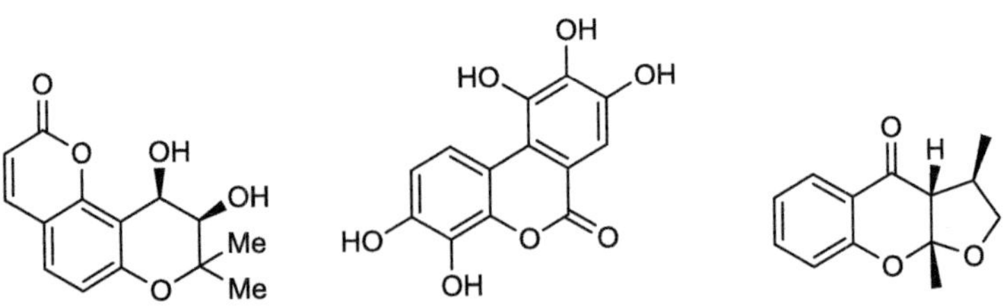

FIGURE 9.3 Examples of natural products with scaffolds related to benzopyrans.

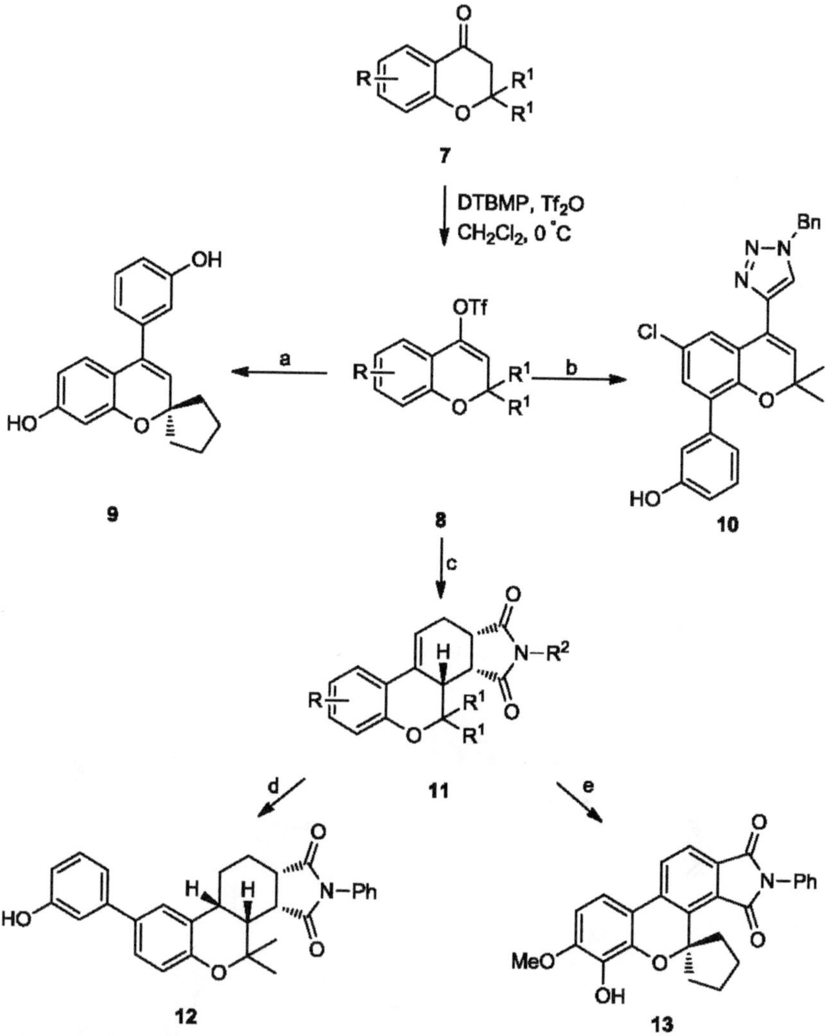

SCHEME 9.2 Illustrative syntheses of diverse molecules possessing a benzopyran core: (a) **9**, (i) $ArB(OH)_2$, 10 mol% $Pd(PPh_3)_4$, Na_2CO_3, 1,4-dioxane/H_2O, (ii) HF-pyridine, THF; (b) **10**, (i) ethynylmagnesium bromide, 10 mol% $Pd(PPh_3)_4$, $ZnCl_2$, 1,4-dioxane, (ii) R^2-N_3, $BrCu(PPh_3)_3$ (20%), DIPEA, toluene, (iii) HF-pyridine; (c) **11**, (i) tributylvinyltin, $Pd(PPh_3)_4$ (10%), 1,4-dioxane, (ii) *N*-phenylmaleimide, toluene; (d) **12**, (i) HF-pyridine, (ii) H_2, 10% Pd/C, MeOH–THF; (e) **13**, (i) DDQ, 1,4-dioxane, (ii) HF-pyridine. R- denotes a solid support.

Wilk et al. have prepared a library based on the scaffold of the alkaloid, macroline (Figure 9.4) [21]. Initially, the immobilized tryptophan derivative, **14**, was reductively aminated, and then subjected to a Pictet–Spengler reaction, to yield the trans-configured products **16** (Scheme 9.3). A Dieckmann cyclization was then used to prepare the bridged cycloocta[*b*]indole ring system found in macroline, thereby

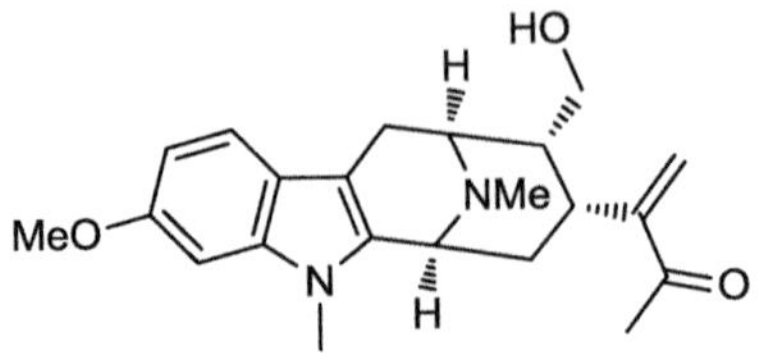

FIGURE 9.4 Chemical structure of macroline.

releasing the final compounds from the solid support. The substitutional diversity of the library was increased by exploiting functionalization reactions before release from the solid support.

Nören-Müller et al. exploited natural product–inspired chemical libraries in the discovery of novel inhibitors of protein phosphatases (see Figure 9.5) [13a]. Medium-sized compound libraries were prepared on the basis of simplified scaffolds related to those found in specific natural products. As an illustrative example, libraries inspired by yohombine, ajmalicine, and reserpine were prepared in which the complex polycyclic framework had been simplified. From these libraries, inhibitors of a range of protein phosphatases were discovered, including **18** to **20**, in which the core is bi-, tri-, or tetracyclic. For two specific phosphatases, the first inhibitors were thereby discovered (e.g., **18** to **20**, which inhibit MptpB). This overall approach demonstrates

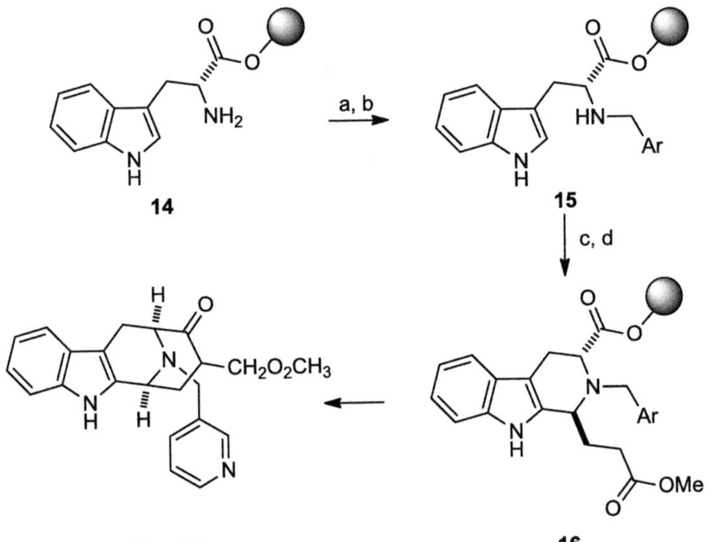

SCHEME 9.3 Synthesis of macroline-inspired compounds: (a) ArCHO, 1 : 1 $HC(OMe)_3$–CH_2Cl_2 (1 : 1); (b) $NaCNBH_3$, 20% AcOH in THF; (c) TFA in CH_2Cl_2 × 2; (d) methyl-4,4-dimethoxybutanoate, 15% TFA in CH_2Cl_2; (e) 15% Et_3N in CH_2Cl_2 × 2; (f) 1 M NaOMe, 1 : 1 MeOH-1,4-dioxane; (g) NaOMe, 10% MeOH in PhMe, reflux.

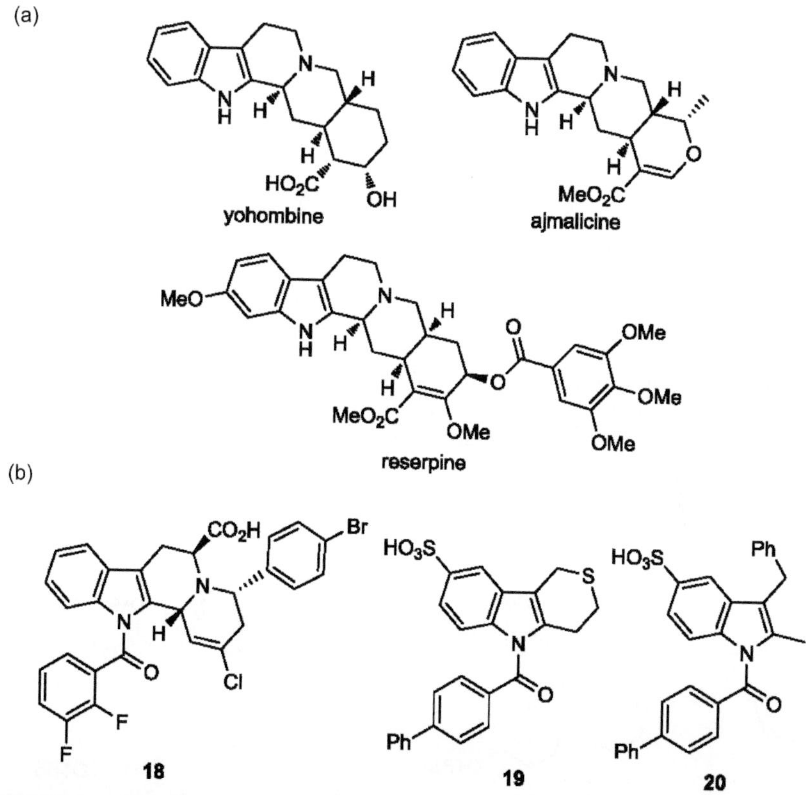

FIGURE 9.5 Discovery of protein phosphatase inhibitors inspired by scaffolds based on indole alkaloids: (a) chemical structures of yohombine, ajmalicine, and reserpine, which inhibit the dual specificity phosphatase Cdc25A (IC_{50}: 22, 32, and 64 μM, respectively); (b) chemical structures of inhibitors of the tyrosine phosphatase MptpB (IC_{50}: 18, 1.1 μM; 19, 0.36 μM; 20, 0.43 μM).

the value of hierarchical scaffold classification, which can facilitate effective navigation (between related molecular scaffolds) within biologically relevant chemical space.

9.3 FOLDING PATHWAYS IN THE SYNTHESIS OF NATURAL PRODUCT–LIKE LIBRARIES

The folding pathway [22] approach exploits common reaction conditions to transform a range of substrates into products with alternative molecular skeletons (Scheme 9.4). The substrates are encoded to “fold” into the alternative scaffolds through strategically placed appending groups, often referred to [22] as σ-elements. This strategy allows variation of the molecular scaffold using a single set of reaction conditions.

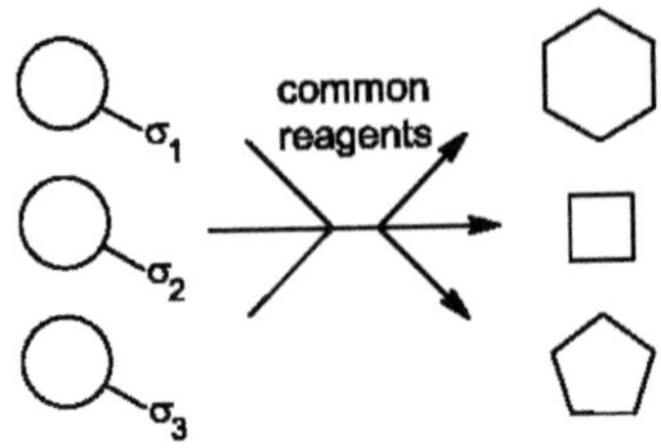

SCHEME 9.4 Overview of the folding pathway strategy in which alternative substrates are converted into alternative scaffolds under common reaction conditions.

Oguri and Schreiber exploited Rh-catalyzed tandem cyclizations/cycloaddition methodology in the synthesis of alternative indole alkaloid-like skeletons (Scheme 9.5) [23]. An α-diazo-β-keto-carbonyl group and an indole ring were strategically placed at different positions on the scaffolds of the starting materials (**21** to **23**). The substrates, **21** to **23**, were treated with a catalytic amount of rhodium(II)

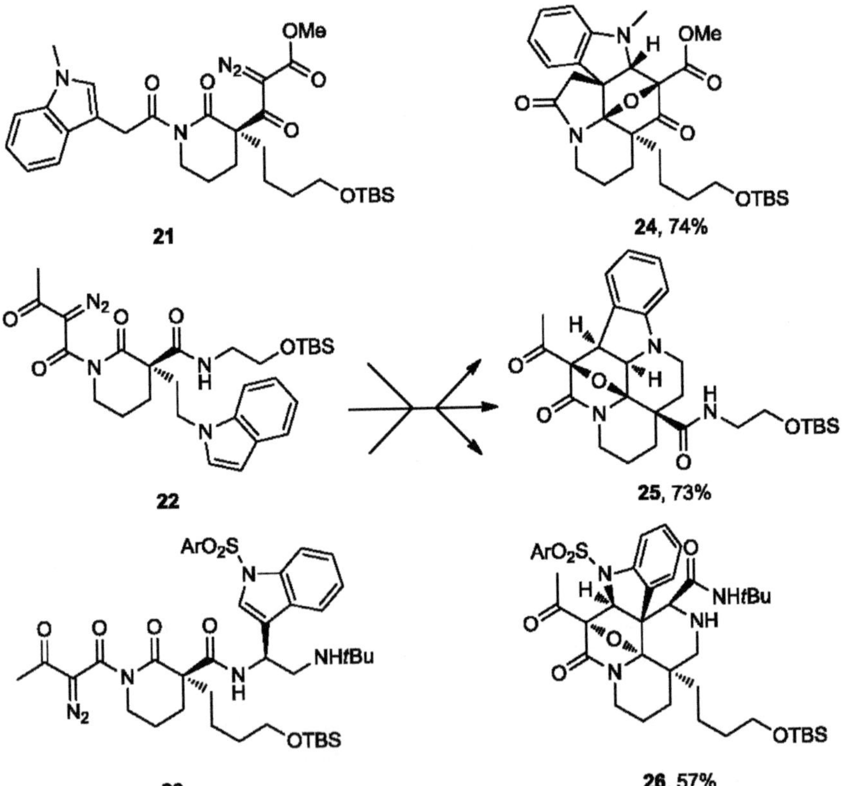

SCHEME 9.5 Folding pathway based on tandem cyclization–cycloaddition chemistry leading to skeletally diverse alkaloid-like small molecules. Conditions: Catalytic $Rh_2(O_2CC_7H_{15})_4$, benzene, 50°C.

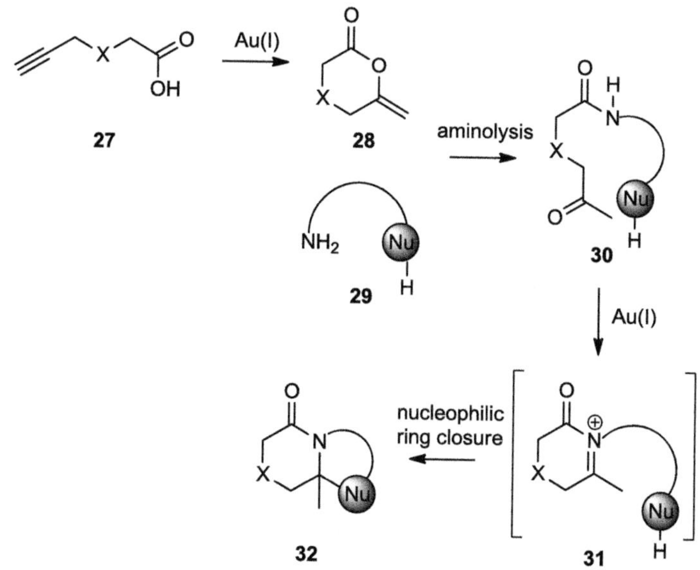

SCHEME 9.6 Overview of a Au-catalysed approach to the synthesis of skeletally diverse, bicyclic heterocycles. X = O or CH_2.

octanoate dimer in benzene at 50°C. Presumably, generation of a carbonyl ylid was followed by intramolecular 1,3-dipolar cycloaddition with the indole ring to give the alternative polycyclic skeletons, **24** to **26**.

Yang et al. have developed a one-pot gold-catalyzed cascade to prepare skeletally diverse alkaloid-like small molecules (Scheme 9.6) [24]. Initially, Au(I)-catalyzed cyclization of alkynyl carboxylic acids **27** gave the corresponding cyclic enol ethers **28**. Subsequently, attack of amine nucleophiles **29** on the cyclic enol ether generated ketoamides **30**. In other words, combination of the alkynyl carboxylic acid **27** and the amine **29** dictated the structure of the cyclization precursor **30**. Under these reaction conditions, the cyclization precursor was also converted into an *N*-acyl iminium ion **31**, which was then trapped by a tethered nucleophile (→ **32**). The scope of the methodology is summarized in Scheme 9.7. Using combinations of the alkynyl carboxylic acids (e.g., **33** or **34**) and the pyrrole- and indole-tethered amines (e.g., **35** or **36**), a remarkable range of alternative alkaloid-like scaffolds was prepared (e.g., **37** to **42**).

Dandapani et al. used radical cyclization chemistry to generate skeletally diverse products (Scheme 9.8) [25]. The tetrahydropyridine substrates **43** to **45** were prepared by stereoselective alkylation with a range of alkyl, alkenyl, and alkynyl halides. The skeletons of the products **46** to **48** were preencoded by the location of the bromine atoms and the unsaturated groups in the substrates **43** to **45**. The folding processes were triggered by treatment of the tetrahydropyridines **43** to **45** with an appropriate tin or silicon hydride and a substoichiometric amount of AIBN at 80°C. Abstraction of a strategically positioned bromine atom and cyclization yielded a range of distinct polycyclic alkaloid-like frameworks (e.g., **46** to **48**). Particularly high levels of skeletal

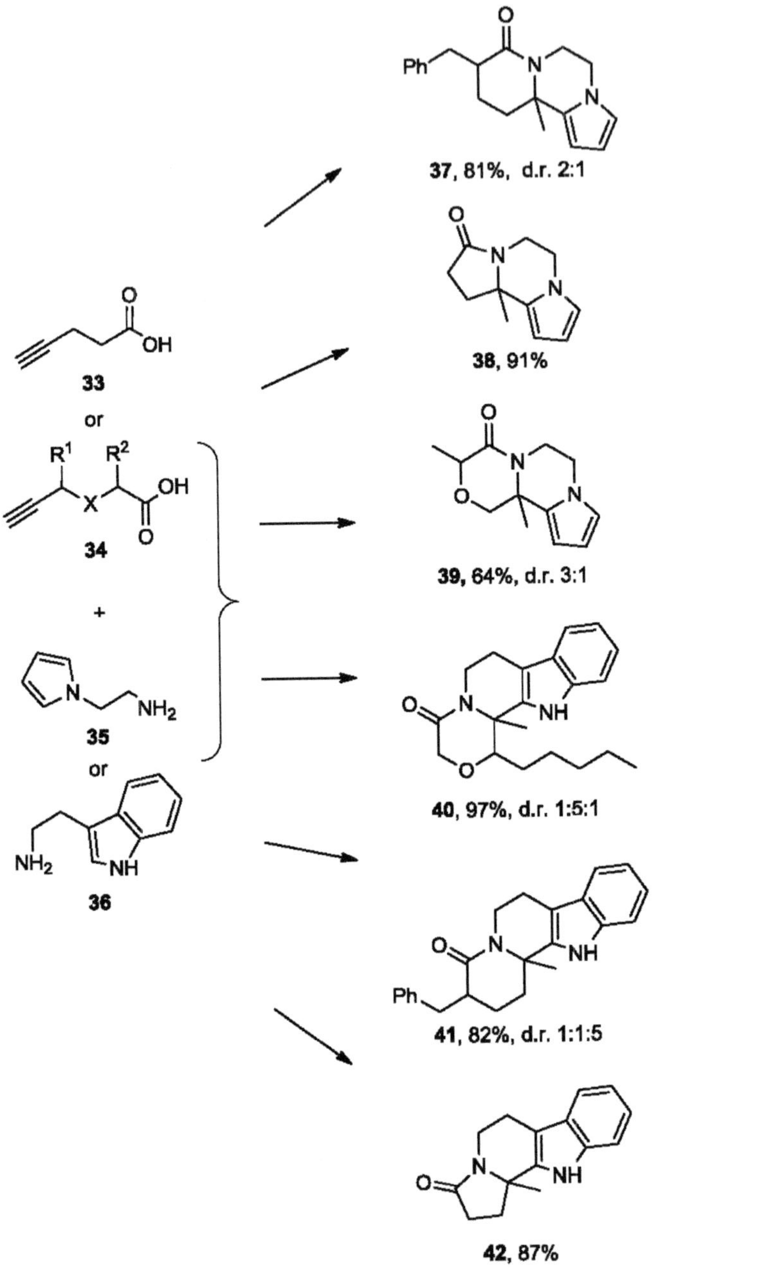

SCHEME 9.7 An Au(I)-catalysed cascade leading to alkaloid-like molecules: 1 mol% $AuPPh_3Cl/AgOTf$ (1%) toluene, reflux.

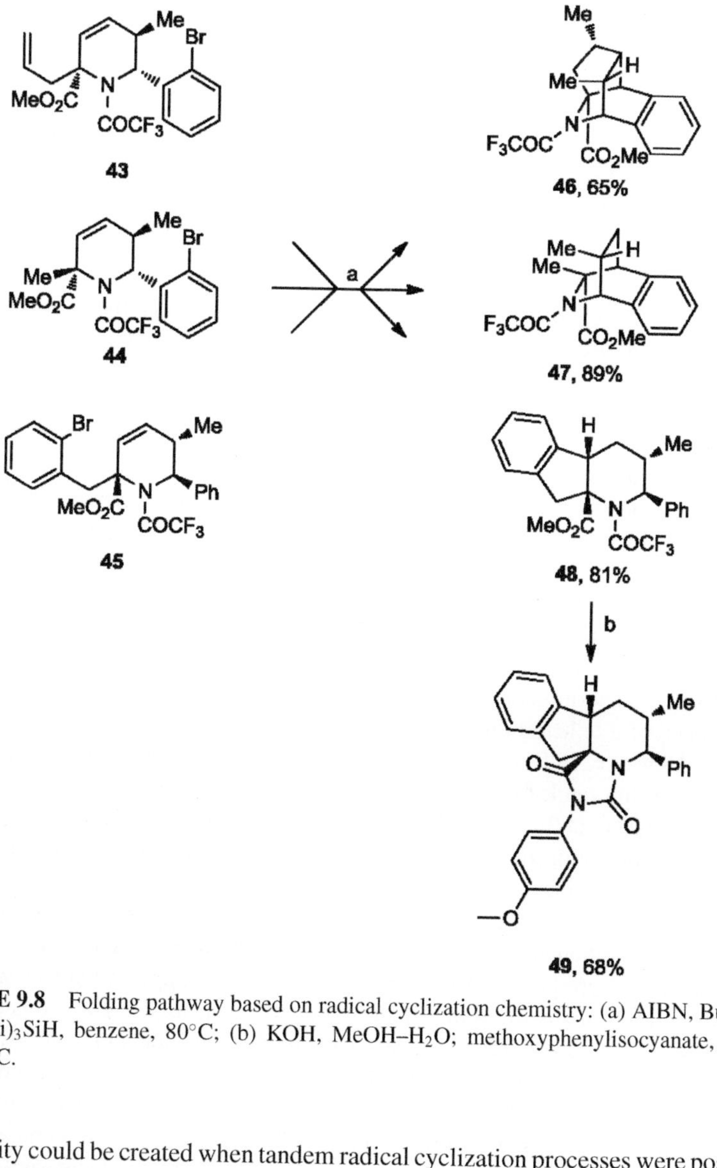

SCHEME 9.8 Folding pathway based on radical cyclization chemistry: (a) AIBN, Bu_3SnH or $(Me_3Si)_3SiH$, benzene, 80°C; (b) KOH, MeOH–H_2O; methoxyphenylisocyanate, Et_3N, THF, 45°C.

complexity could be created when tandem radical cyclization processes were possible (e.g., to yield **46**).

Oguri et al. developed a powerful approach in which metathesis was used to "fold" alternative substrates into polycyclic ring systems inspired by artemisinin (Figure 9.6) [26]. The folding substrates **51** to **56** were prepared in three or four steps from the α,β-unsaturated nitrile **50** (Scheme 9.9). Subsequent folding using ring-closing ene–yne–ene metathesis cascade chemistry generated the tricycles **57** to **62**. Further synthetic transformations yielded analogs of the sesquiterpene artemisinin (**63** to **65**), some of which exhibited antitrypanosomal activity.

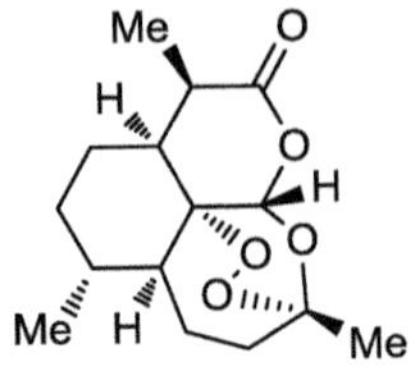

FIGURE 9.6 Chemical structure of artemisinin.

The efficiency of folding pathways may be increased if both the construction and the folding of substrates may take place in a single reaction. The Nelson group has exploited cascade chemistry [27], initiated by a three-component reaction, in the synthesis of a range of alkaloid-like scaffolds (Scheme 9.10) [28]. The approach could be used in combination with a second three-component reaction to control the substitutional diversity of the final compounds.

The reaction between secondary amines **66**, carbonyl compounds **67**, and triazines **68** yielded a range of skeletally diverse alkaloid-like compounds. Condensation between the secondary amines **66** and the carbonyl compounds **67** is believed to generate enamines **69**, which can undergo an inverse-electron-demand Diels–Alder reaction with the triazines **68** (→ **70**); expulsion of molecular nitrogen then yields 2-aza-dienes **71**, which are substrates for a subsequent folding pathway. The substituents in **71** determine the molecular scaffold that is ultimately obtained. Thus, in the absence of a tethered dienophile, the 2-aza diene **71** is itself the product of the reaction. However, with an appended dienophile in either the R^2 or the R^3 substituent, a subsequent intramolecular Diels–Alder reaction can yield, for example, either **72** [with R^2 = (*E*)-octa-3-enyl] or **75** (with R^3 = allyl). Finally, with R^3 = allyl, and a pendant nucleophile in R^2, intramolecular Diels–Alder reaction and subsequent cyclization are possible (→ **73**).

In many cases, the three-component cascade reaction could be followed by a second three-component reaction: a Joullié–Ugi reaction. Some selected examples of the alkaloid-like products that were accessible using the approach are shown. Thus, after the folding pathway, some of the products, such as the 2-aza diene **74** and the cyclic imines **72** and **75**, were reacted with an isocyanide and a carboxylic acid to give final products (such as **76** to **78**). By contrast, with an appropriate pendant nucleophile in place (such as the NHCbz group in **73**), a cyclization terminated the three-component cascade reaction. In total, 43 final products were prepared, which were in general derived from five separate components. The products were based on 28 distinct graph-node-level frameworks;† the high scaffold diversity of the products stemmed both from rings found in the individual components and from the power of the folding pathway.

†Frameworks defined at the graph-node level describe the connectivity and type of atoms (but not the bond types between atoms).

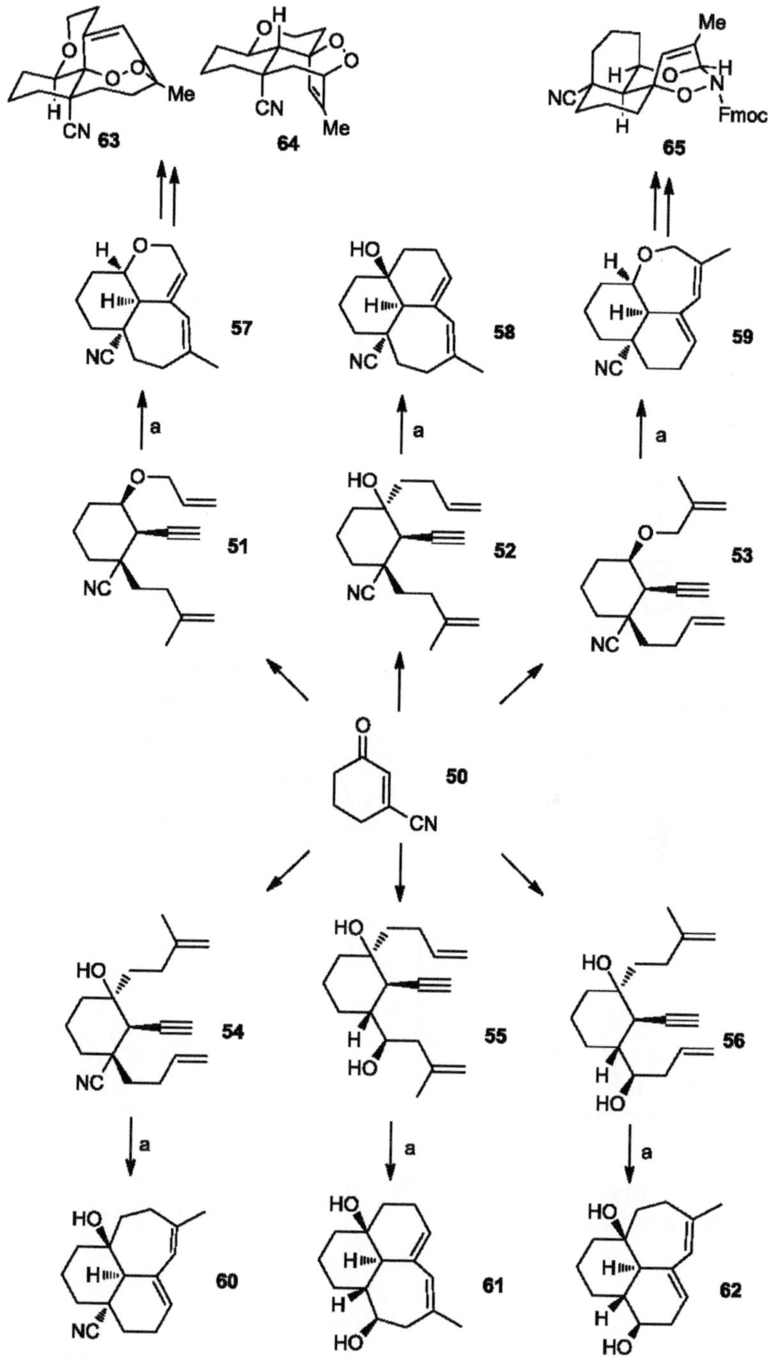

SCHEME 9.9 Folding pathway leading to the analogs of the sesquiterpene artemisin: (a) 10 to 66 mol% Grubbs's second-generation catalyst; yields from the folding step ranged between 39 and 90%.

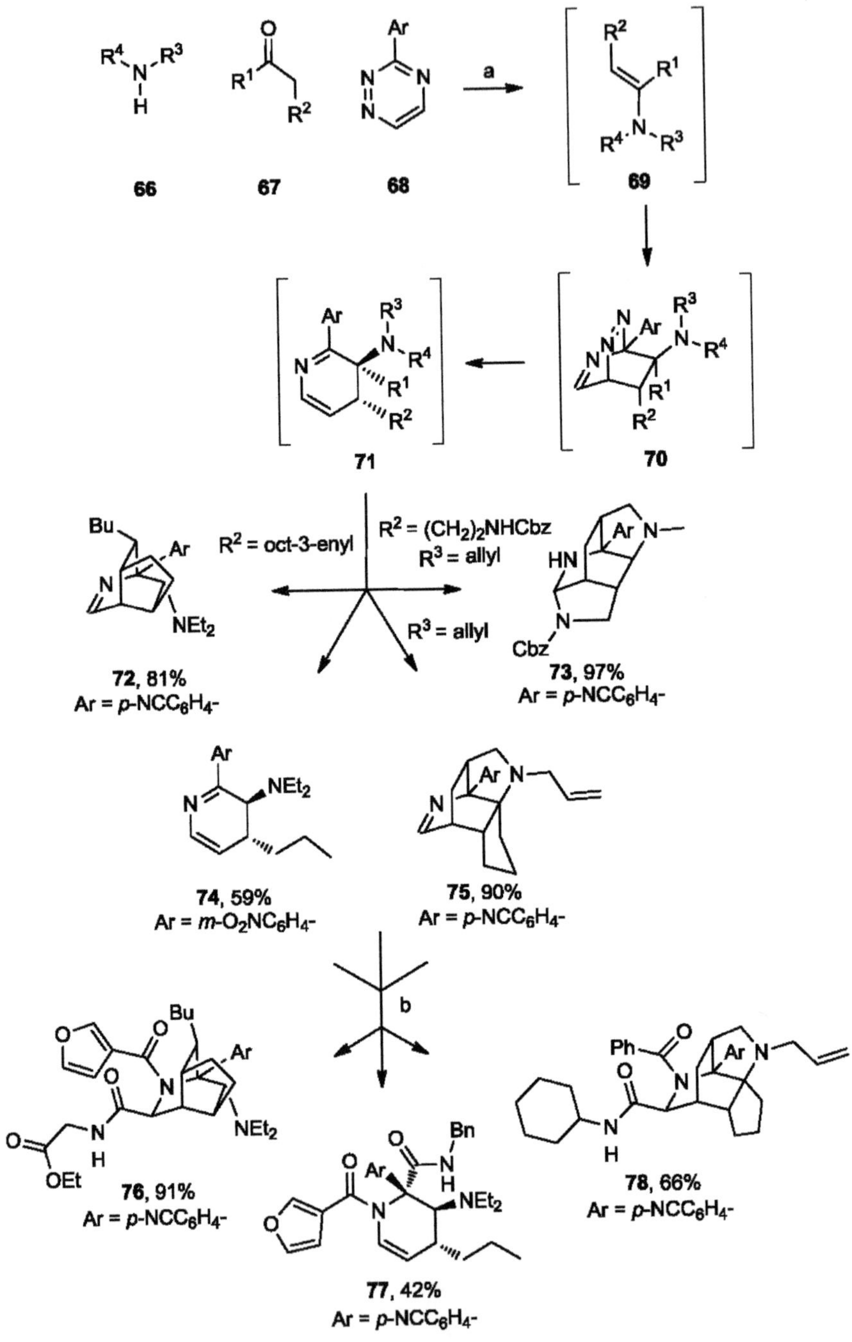

SCHEME 9.10 Selected syntheses of alkaloid-like compounds prepared using a synthetic approach involving two consecutive three-component reactions: (a) 4-Å molecular sieves, toluene, Δ; (b) imine, carboxylic acid, isocyanide, EtOH.

9.4 BRANCHING PATHWAYS IN THE SYNTHESIS OF NATURAL PRODUCT–LIKE LIBRARIES

The branching pathway strategy involves the conversion of common precursors into a range of distinct molecular scaffolds (Scheme 9.11). The development of a branching pathway may require considerable optimization of the individual skeletal transforming steps. It is important, however, to ensure that the ethos of diversity-oriented synthesis be retained: that is, that a library be prepared in a deliberate and simultaneous fashion. In some cases the implementation of an intricate branching pathway which exploits a very limited range of precursors may cause the parallel nature of diversity-oriented synthesis to be lost.

Complementary metal-catalyzed branching pathway which exploits complementary cyclization reactions (Scheme 9.12) [29]. A four-component Petasis condensation reaction was used to assemble flexible cyclization precursors (e.g., **79**). Alternative cyclization reactions were then used to yield products with distinct molecular skeletons: Pauson–Khand reaction (→ **80**); Pd-catalyzed cyclization (→ **81**); Ru-catalyzed cycloheptadiene formation (→ **82**); base-induced cyclization (→ **83**); Au-catalyzed cyclization of the alcohol onto the alkene (→ **84**); enyne metathesis (→ **85**); and Meisenheimer [2,3]-sigmatropic rearrangement (not shown).

Many of the initial cyclization products could be exploited in the synthesis of additional molecular scaffolds. For example, Diels–Alder reactions with 4-methyl-1,2,4-triazoline-3,5-dione converted dienes (such as **85**) into polycyclic products (such as **86**) (Scheme 9.12). In addition, four of the cyclization reactions could be used again to convert the enyne **83** into molecules with four additional skeletons (**87** to **90**) (Scheme 9.13). The key to this powerful synthetic approach lay in the design of precursors (e.g., **79**), which were effective substrates in a wide range of efficient and diastereoselective cyclization reactions.

Painter et al. prepared a library of alkaloid-like compounds designed to target a large region of biologically relevant chemical space as efficiently as possible (Scheme 9.14) [30]. The approach focused on the broad chemistry of derivatives of the allenylated tryptophan analogs **91**. A range of complementary transition metal–catalyzed reactions was used to convert the allene-ynes **92** into alternative scaffolds: for example, complementary Mo- and Rh-catalyzed cyclocarbonylations gave the regioisomeric cyclopentenones **94** and **95**, respectively, and thermal [2 + 2] cycloaddition gave the corresponding cyclobutenes **96**. Pictet–Spengler reactions were used

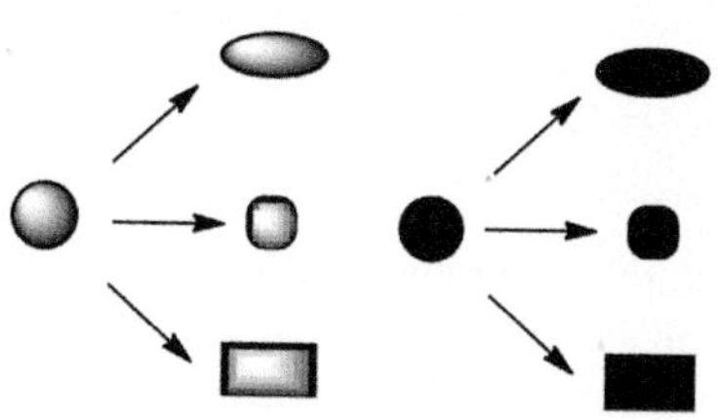

SCHEME 9.11 Overview of the branching pathway strategy in which common substrates are converted into alternative scaffolds under reagent control.

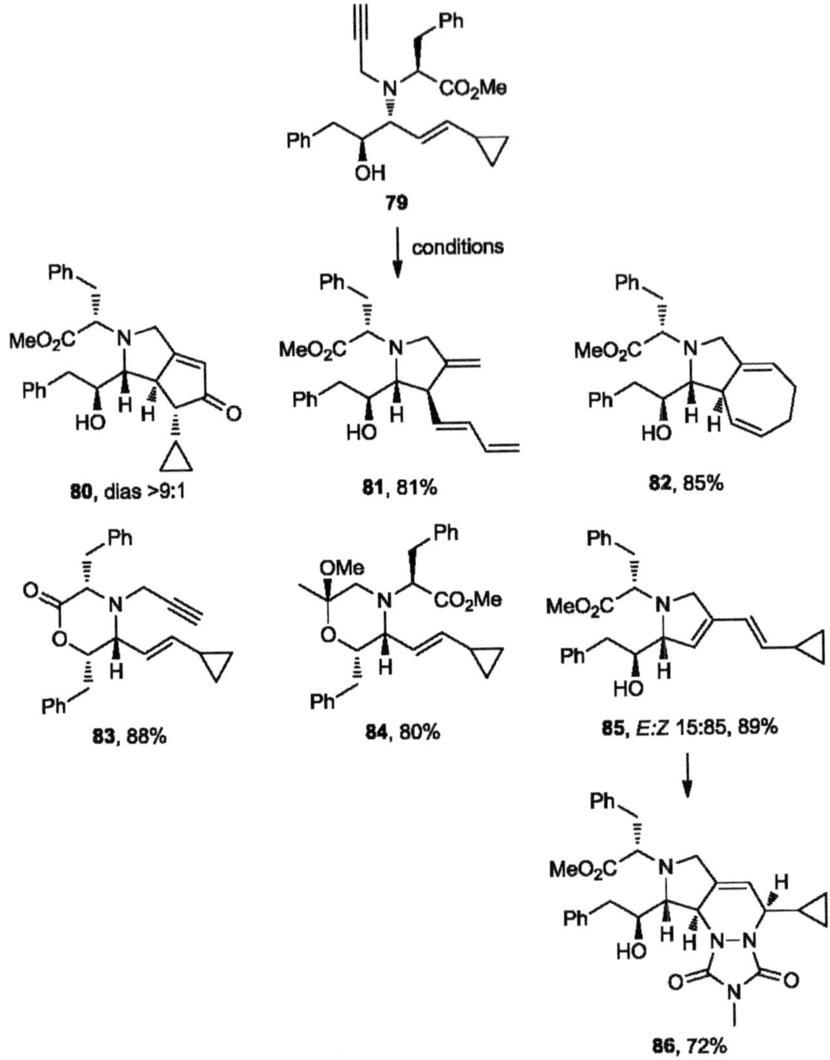

SCHEME 9.12 Complementary metal-catalyzed branching pathway: **80**, $Co_2(CO)_8$, $Et_3NO–NH_4Cl$, benzene; **81**, 10 mol% $Pd(PPh_3)_2(OAc)_2$, benzene, 80°C; **82**, 10 mol% $CpRu(MeCN)_3PF_6$, acetone; **83**, NaH, toluene; **84**, 10 mol% $NaAuCl_4$, MeOH; **85**, 10 mol% Hoveyda–Grubbs's second-generation catalyst, CH_2Cl_2; **86**, 4-methyl-1,2,4-triazoline-3,5-dione, CH_2Cl_2.

to prepare the allene-containing tetrahydro-β-carbolines **93**, which were subjected to Ag-catalyzed cyclization (→ **97**). Acylation gave the allenyl-ynes 99 which were subjected to cyclocarbonylation (→ **98**), and thermal [2 + 2] cycloaddition (→ **100**).

Comer et al. have developed a branching pathway that exploits reactions between alternative pairs of functional groups in the substrates (Scheme 9.15) [31]. The nitro

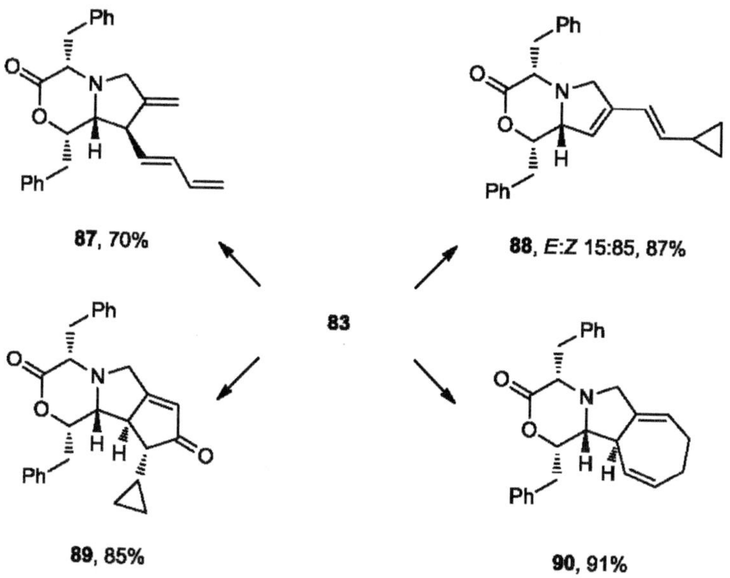

SCHEME 9.13 Exploitation of a common intermediate in a branching pathway: **87**, 10 mol% $Pd(PPh_3)_2(OAc)_2$, benzene, 80°C; **89**, $Co_2(CO)_8$, $Et_3NO–NH_4Cl$, benzene, rt; **90**, 10 mol% $CpRu(MeCN)_3PF_6$, acetone; **88**, 10 mol% Hoveyda–Grubbs's second-generation catalyst, CH_2Cl_2.

compounds **101** to **103** each contain additional functional groups: an alkyne, two methoxycarbonyl groups, and in the case of **103**, an alkene. Reduction of the nitro group in **101** triggered cyclization to yield the lactam **105**. Alternatively, conversion of the nitro group in **102** into a nitrile oxide resulted in a subsequent dipolar cycloaddition to give the isoxazole **106**. In contrast, ene–yne metathesis of **103** yielded a 1,3-diene substrate for a Diels–Alder reaction (→ **107**). The reactions of the nitro compound **104** illustrate how different cyclization reactions may be exploited to yield alternative scaffolds: metathesis and Diels–Alder reaction to yield **108**, Au-catalyzed cycloisomerization and lactamization to yield **109**, and Pauson–Khand reaction to yield the remarkable bridged cyclopentenone **110**.

Barjau et al. have developed a branching pathway based on the versatile chemistry of the polycycle **111**, prepared by electrooxidation of 2,4-dimethylphenol (Scheme 9.16) [32]. A wide range of reactions, several of which are rather remarkable, were used to prepare skeletally diverse polycycles (e.g., **112 to 121**). The reactions included Lewis acid–mediated substitution (→ **112**); stereochemically complementary Wagner–Meerwein rearrangements (→ **113** or **114**), reduction (→ **120**), and trapping of the open form of the hemiacetal (→ **121**). In addition, other products, such as **115** to **119**, were formed by more complex sequences of reactions. Although a wide range of interesting scaffolds was prepared, the approach does not allow independent variation of the substituents.

Cui et al. have developed a strategy for the preparation of a shape-diverse library of compounds with excellent physicochemical properties (Scheme 9.17) [33]. The

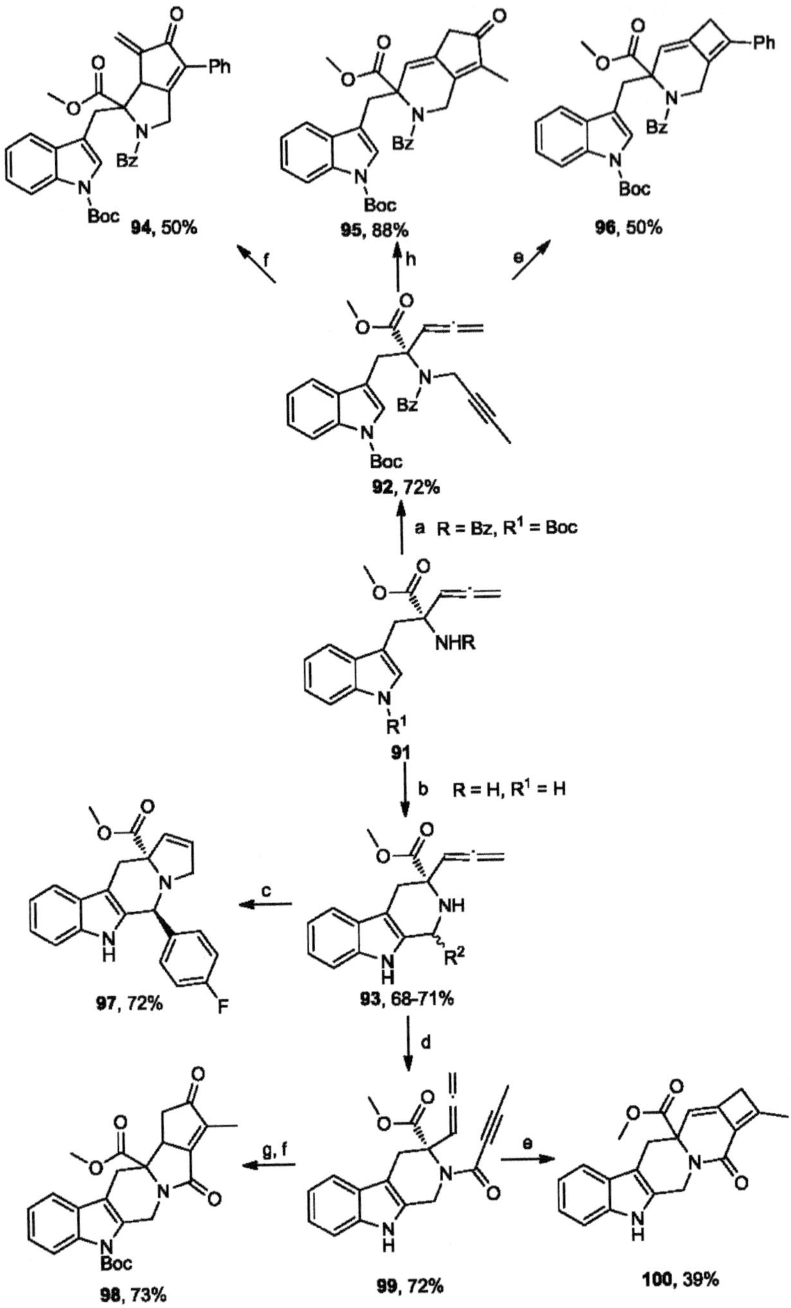

SCHEME 9.14 Branching pathways from an allene-containing tryptophan derivative: (a) NaH, 1-bromobut-2-eyne, DMF; (b) R_2CHO, TFA, CH_2Cl_2 or MeOH; (c) 20 mol% $AgNO_3$, acetone, 56 to 72%; (d) but-2-ynoic acid, isobutyl chloroformate, NMM, THF; (e) μw, 225°C, DMF; (f) $Mo(CO)_6$, DMSO, PhMe; (g) Boc_2O, DMAP; (h) 5 mol% $[Rh(CO)_2Cl]_2$, CO (1 atm), DCE.

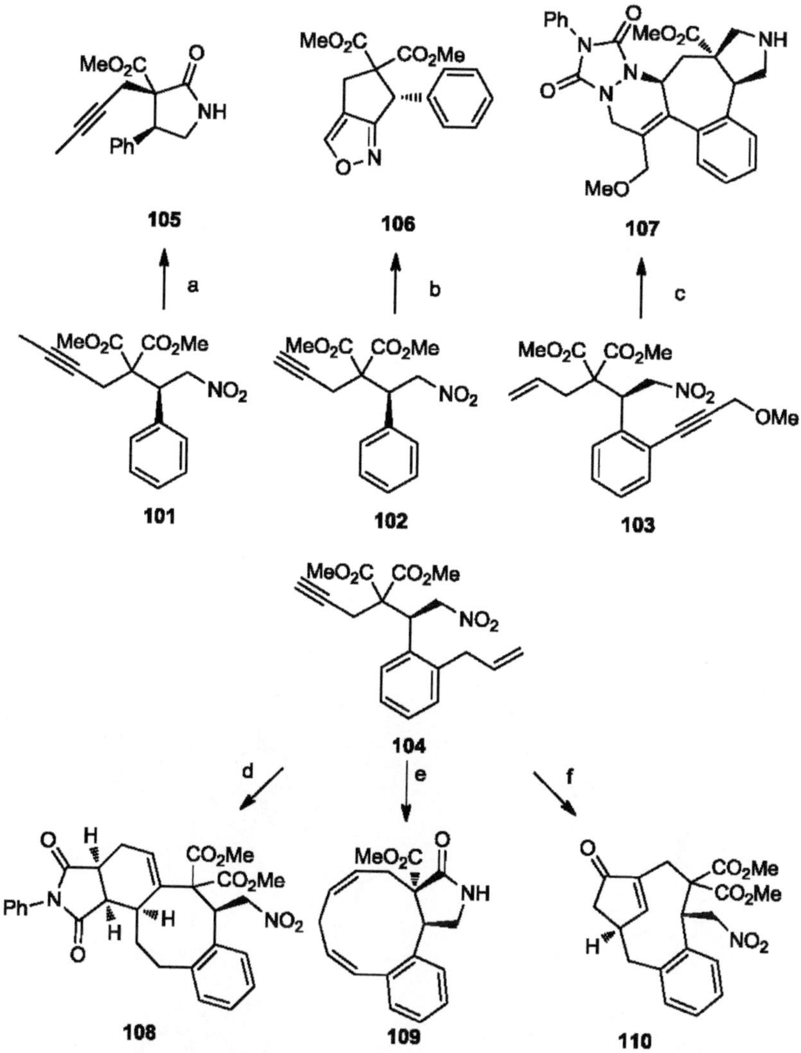

SCHEME 9.15 Branching pathway based on functional group pairing: (a) Zn, AcOH/THF; Na_2CO_3 (aq.), 92% (two steps); (b) Boc_2O, DMAP, toluene, 74%; (c) 10 mol% Grubbs's first-generation catalyst, ethylene, μw, 50°C, CH_2Cl_2; 4-phenyl-3*H*-1,2,4-triazole-3,5(4*H*)-dione, μw, 110°C, toluene, 86% (two steps); Zn, AcOH/THF, 90°C; Na_2CO_3 (aq.), 45% (two steps); (d) 10 mol% Grubbs's first-generation cataylst, ethylene, μw, 60°C, CH_2Cl_2; *N*-phenylmaleimide, μw, 160°C, toluene, 98% (two steps); (e) $AuCl(PPh_3)$, AgOTf, 50°C, toluene, 58%; Zn, AcOH–THF; NaOH/MeOH (aq.), 60% (two steps); (f) $Co_2(CO)_8$, μw, 80°C, CH_2Cl_2, 67%.

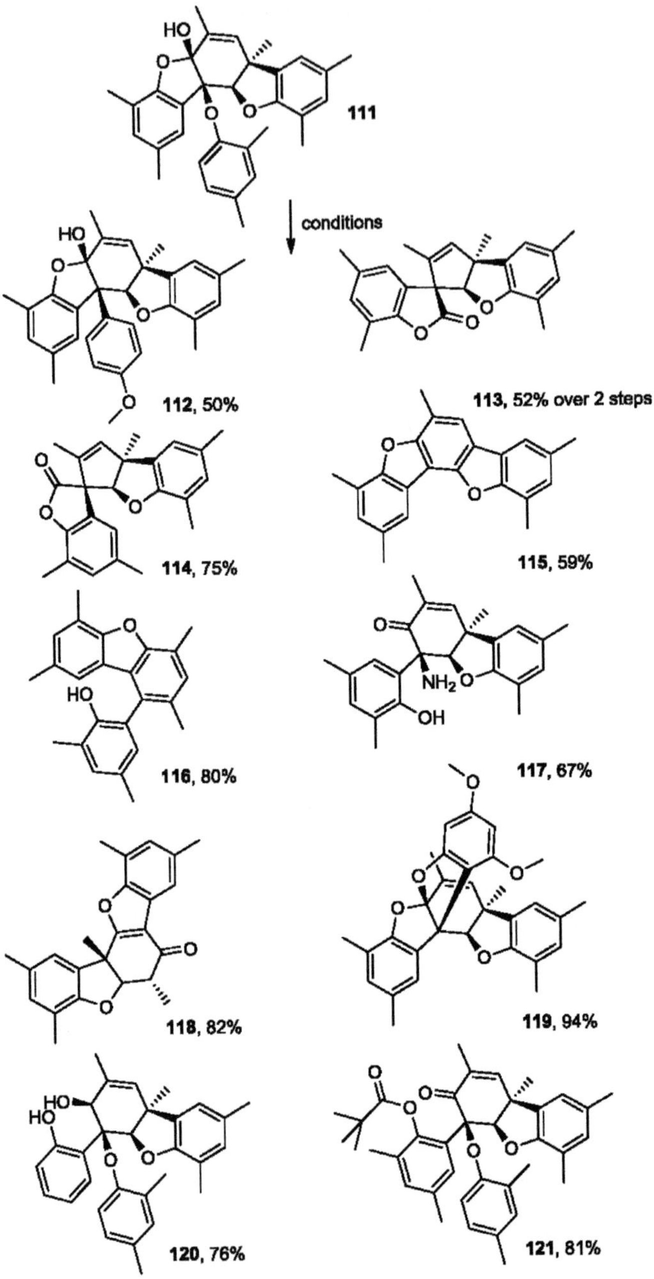

SCHEME 9.16 Distinct polycyclic scaffolds generated from a common substrate: **112**, anisole, $BF_3{\cdot}Et_2O$, CH_2Cl_2, $-78°C$; **113**, 2,4-lutidine, TIPSOTf, then $BF_3{\cdot}Et_2O$, CH_2Cl_2, 52%; **114**, conc. H_2SO_4, toluene, 75%; **115**, $SOCl_2$, CH_2Cl_2, 59%; **116**, Et_3SiH, TFA, CH_2Cl_2, 80%; **117**, NH_4F, CH_3CN, reflux, 67%; **118**, NaCN, CH_3CN, reflux, 82%; **119**, 3,5-dimethoxyphenol, $BF_3{\cdot}Et_2O$, CH_2Cl_2, $-78°C$, 94%; **120**, $LiAlH_4$, THF, 0°C, 76%; **121**, PvCl, 1 mol% DMAP, Et_3N, CH_2Cl_2, 81%.

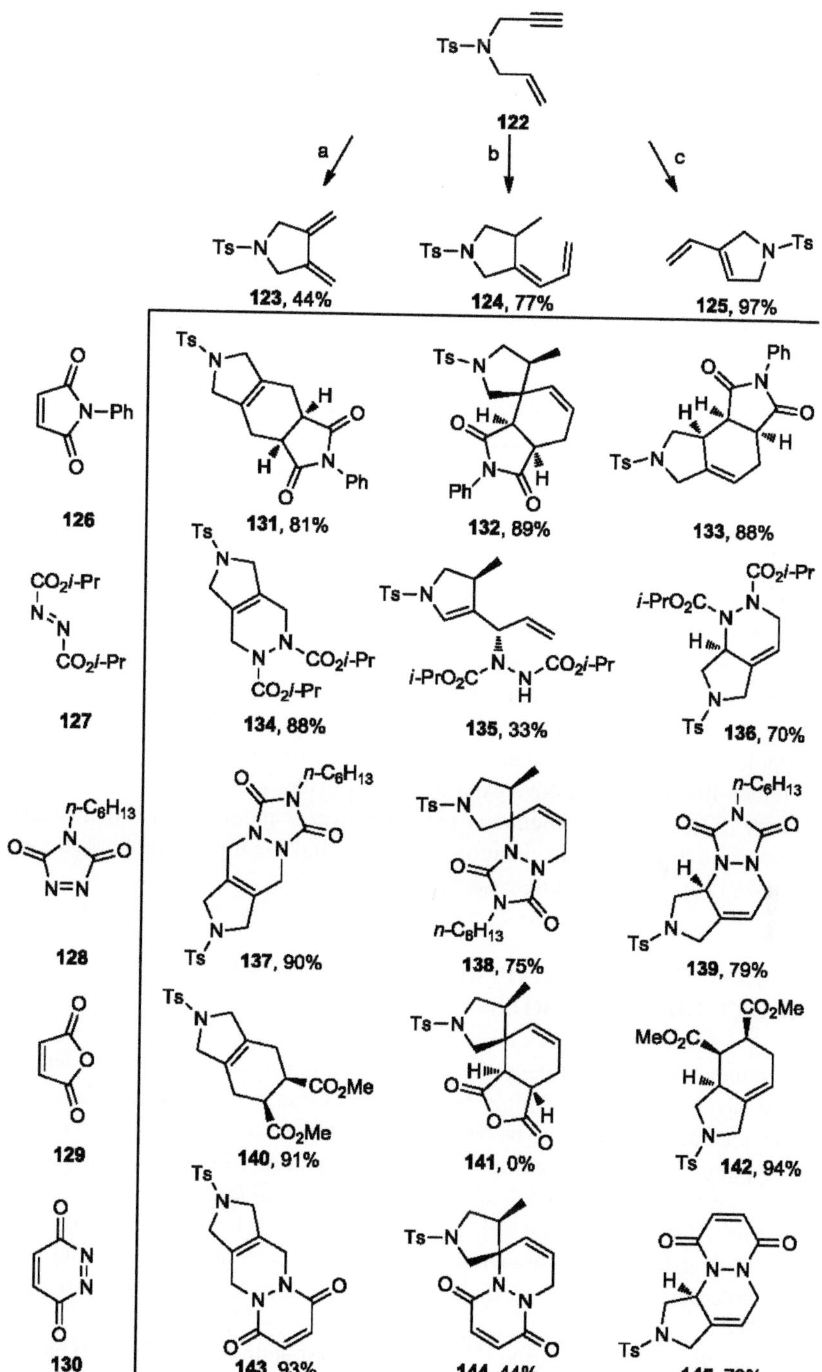

SCHEME 9.17 Synthesis of a skeletally-diverse library, exploiting metal-catalyzed cycloisomerizations of an acyclic enyne: (a) $Pd(PPh_3)_2(OAc)_2$ (5 mol%), C_6H_6, 44%; (b) $[Cp^*RuCl_2]_n$ (5 mol%), COD (5 mol%), ethene, toluene, 77%; (c) Grubbs's first-generation catalyst (3 mol%), ethylene, CH_2Cl_2, 97%.

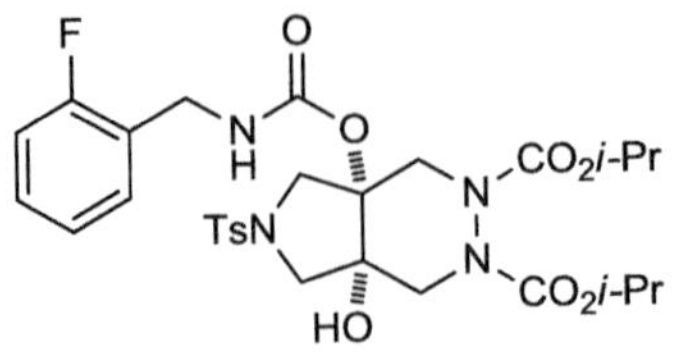

146

FIGURE 9.7 Chemical structure of 146, a suppressor of glycolytic ATP production.

branching approach utilized the common enyne **122**, which could be cycloisomerized to yield alternative 1,3-dienes (**123** to **125**). These were reacted with dienophiles to yield a diverse array of mono- and polycyclic compounds, which were then dihydroxylated and functionalized (not shown) to give a 191-member library. From this library, the compound **146** was identified as a suppressor of glycolytic ATP production in CHO-K1 cells (Figure 9.7).

Robbins et al. have prepared a wide range of natural product–like scaffolds (**148** to **159**) using an outstanding branching pathway (Scheme 9.18) [34,35]. The approach exploited the remarkable chemistry of the symmetrical ketone **147**, which has two appended unsaturated esters. The approach takes advantage of a wide range of reactions, including conjugate additions (e.g., of amines, enolates, and ketyl radicals) and dipole formation and cycloaddition. Other branching pathways leading to natural product–like molecules have exploited Diels–Alder reactions [36], boronic ester chemistry [37], diazoester chemistry [38], the diverse chemistry of α,β-unsaturated carbonyl compounds [39], cyclizations of *N*-allyl and *N*-propagyl propargylic amine derivatives [40], and branching cascade chemistry [41].

9.5 OLIGOMER-BASED APPROACHES TO NATURAL PRODUCT–LIKE LIBRARIES

An extremely powerful strategy for preparing small-molecule libraries with wide scaffold diversity exploits simple building blocks in combination. The approach requires the building blocks to be prepared ("built") and then connected ("coupled"). Finally, pairs of functional groups are reacted ("paired") intramolecularly to yield new ring systems in the final scaffolds. This *build/couple/pair strategy* which has been reviewed by Nielsen and Schreiber [9e], is extremely broad in scope. Indeed, some folding pathways and branching pathways can be considered to exemplify the build/couple/pair strategy (e.g., the synthetic approach described in Schemes 9.12 and 9.13).

Marcaurelle's group has used an oligomer-based approach has been exploited to prepare skeletally diverse macrocycles (Schemes 9.19 and 9.20) [42]. The approach involved the assembly of cyclization precursors from combinations of simple building blocks. For example, stereoisomeric amines **162** were prepared from the amines

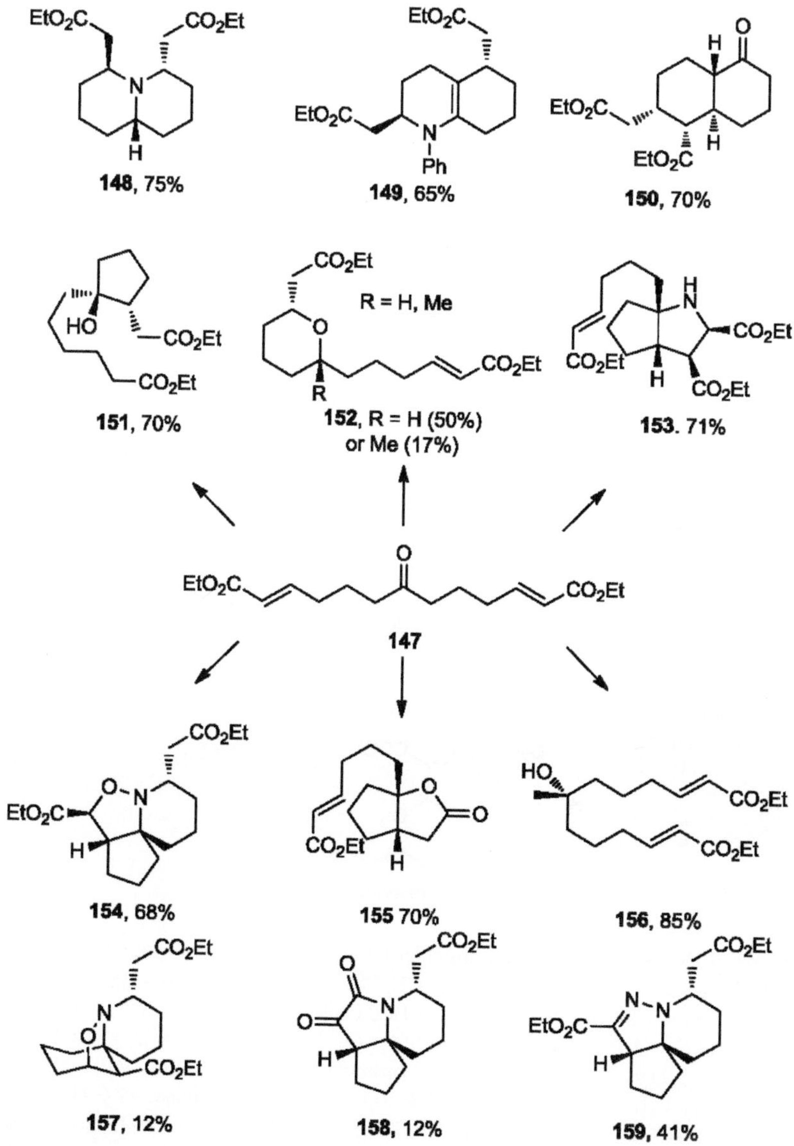

SCHEME 9.18 Branching pathway leading to skeletally diverse natural product–like compounds: **148**, $NaBH_4$, NH_3, EtOH, $Ti(OEt)_4$ then AcOH; **149**, $PhNH_2$, $TiCl_4$, CH_2Cl_2; **150**, NaH, THF; **151**, SmI_2 (5 equiv), THF/MeOH, −78°C; **152**, superhydride, THF or MeLi; **153**, glycine ethyl ester, DIPEA; **154**, $NH_2OH{\cdot}HCl$, NaOAc, MeOH/MeCN, 60°C; **155**, SmI_2 (2 equiv), THF–MeOH, −78°C; **156**, MeMgBr; **157**, $NH_2OH{\cdot}HCl$, NaOAc, MeCN then toluene, μw, 140°C or 39% PhCl, reflux from **154**; **158**, $NH_2OH{\cdot}HCl$, NaOEt, EtOH or 89% NaOEt from **154**; **159**, NH_2NHTs toluene, reflux.

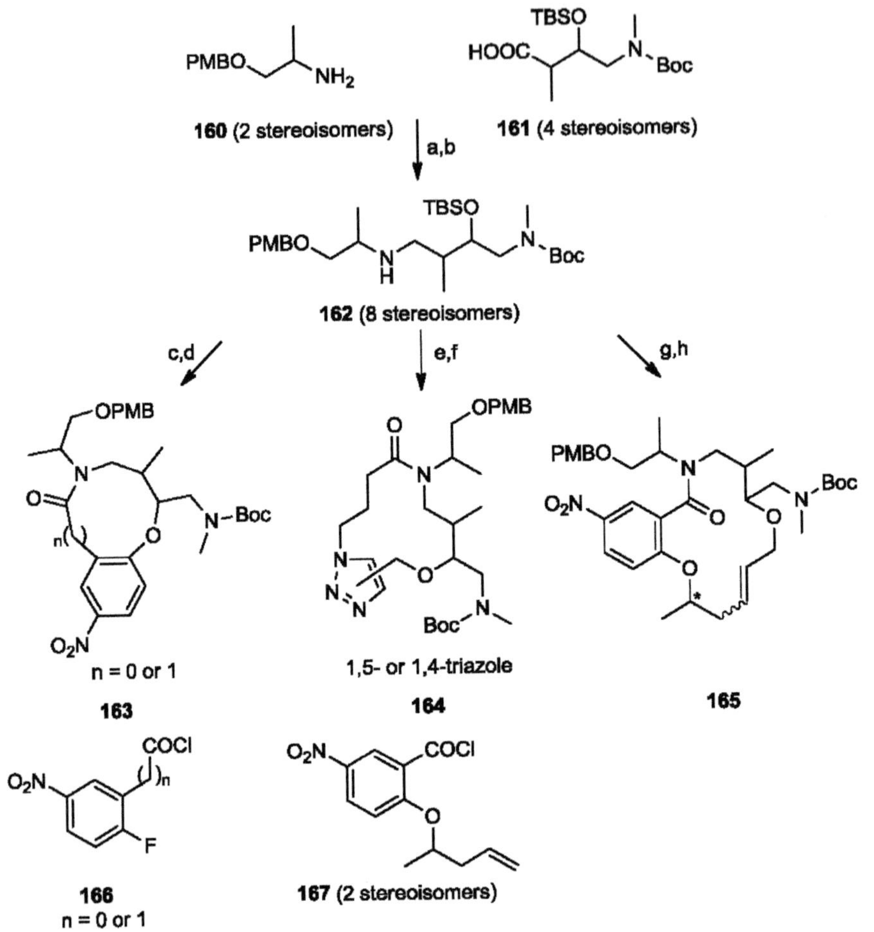

SCHEME 9.19 Application of an oligomer-based approach in the synthesis of skeletally diverse macrocycles: (a) PyBOP, DIPEA, CH_2Cl_2, 71 to 96%; b) BH_3–DMS, THF, 65°C; 10% Na/K tartrate, MeOH, 88 to 98%; (c) **166**, Et_3N, CH_2Cl_2, 88 to 99%; (d) CsF, DMF, 85°C; or TBAF, NH_4F; NaH, THF, 62 to 99%; (e) 3-azidobutanoic acid, PyBOP, DIPEA, CH_2Cl_2; TBAF, THF, 72 to 93%; propargyl bromide, NaHMDS, THF, DMF, −78°C, 91 to 96%; (f) $[Cp^*RuCl]_4$, toluene, 70°C (1,5-triazole) or PS-$CuPF_6$, toluene, 55°C (1,4-triazole), 56 to 78%; (g) **167**, DIPEA, CH_2Cl_2; TBAF, THF; NaH, allyl bromide, DMF, 50 to 77%; (h) Hoveyda–Grubbs's second-generation catalyst (10 mol%), 5 mol% benzoquinone, toluene, 40 to 60°C, 45 to 93%.

160 and the aldol products **161** by amide formation and subsequent reduction (Scheme 9.19) [42a]. The amines **162** were then combined with additional building blocks that were suitably armed to allow subsequent macrocyclization. Macrocyclization was then effected by nucleophilic aromatic substitution (e.g., → **163**), Huisgen [3 + 2] cycloaddition (e.g., → **164**), or ring-closing metathesis (→ **165**). The modular

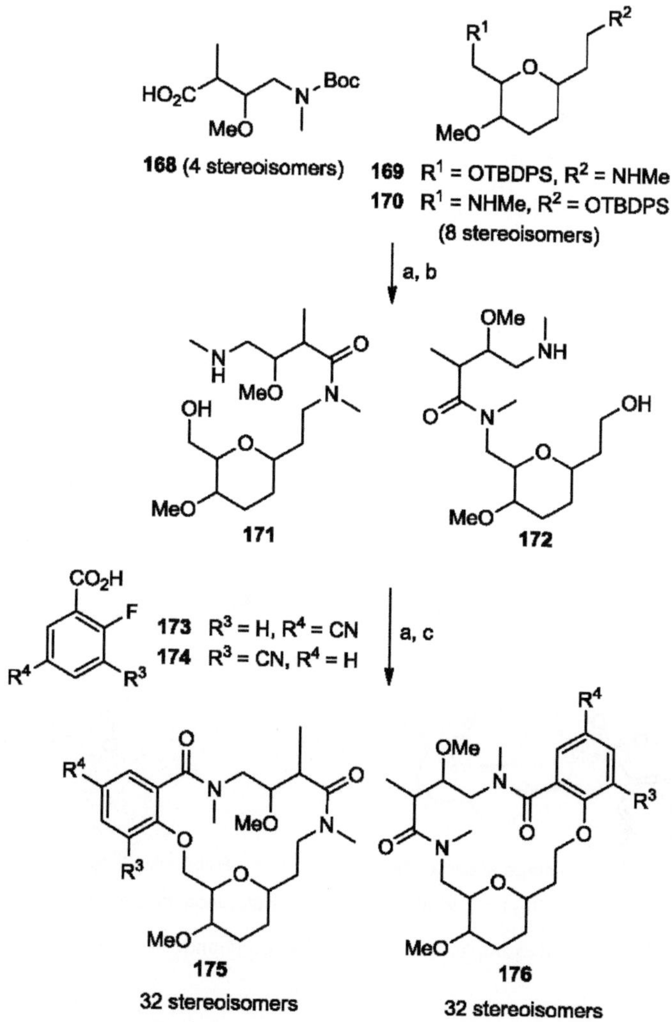

SCHEME 9.20 Parallel solution-phase synthesis of pyran-containing macrocycles: (a) Si–DCC, HOBt, DIPEA, CH_2Cl_2; Si–CO_3; b) TFA, CH_2Cl_2; Si–CO_3, 3 : 1 CH_2Cl_2–MeOH; (c) Cs_2CO_3, DMF, 110°C; Si–CO_3, CH_2Cl_2; 16 to 29%, yields over four steps.

approach also enabled the synthesis of pyran-containing macrocycles (Scheme 9.20) [42b]. By exploiting pyrans **169** and **170** as building blocks, a wide range of macrocycles (e.g., **175** and **176**) was prepared using nucleophilic aromatic substitutions in the macrocyclization step. The synthetic approach has enabled the discovery of a range of cyclic molecules with diverse biological functions (Figure 9.8) [42a,43]. Other researchers have also exploited oligomer-based approaches to prepare diverse macrocyclic scaffolds [44].

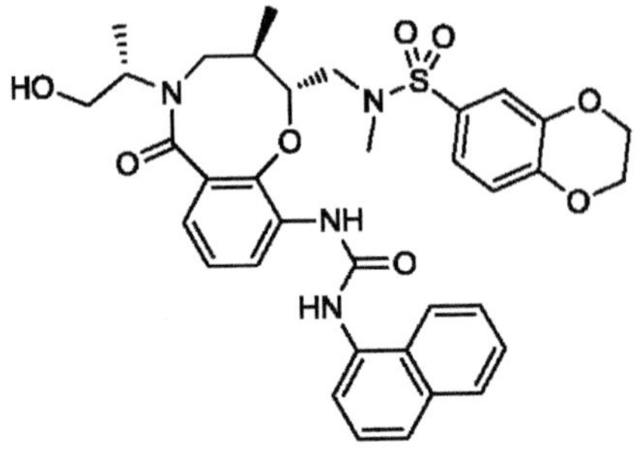

Suppressor of b-cell Apoptosis **177**

EC_{50} (μM) 0.78 ± 0.7

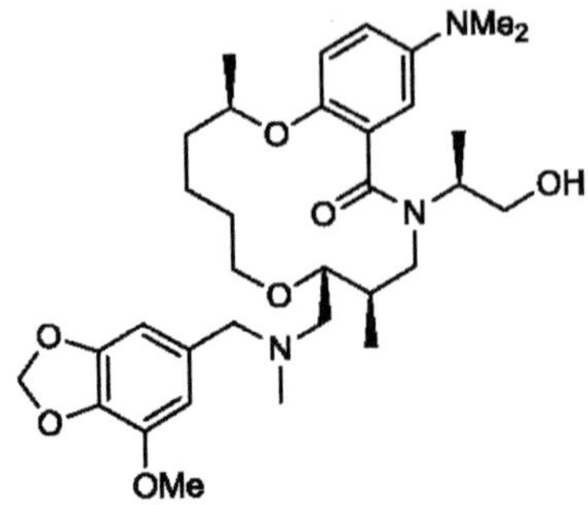

HDAC inhibitor **178**

HDAC1 IC_{50} (μM) 1.5 ± 0.7

HDAC2 IC_{50} (μM) 2.8 ± 0.9

HDAC3 IC_{50} (μM) 4.4 ± 2.0

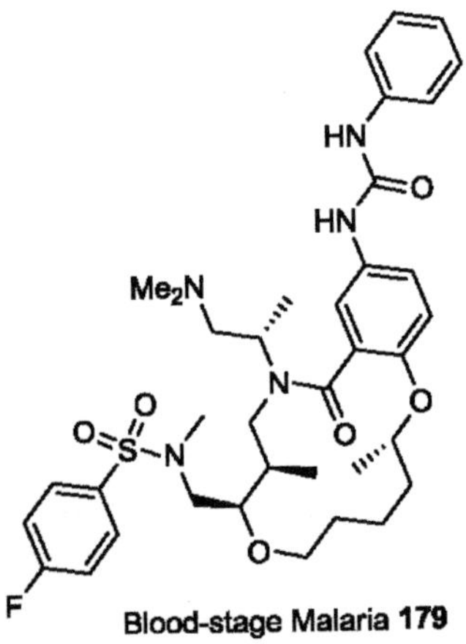

Blood-stage Malaria **179**

Dd2 (GI_{50}) = 0.54 nM

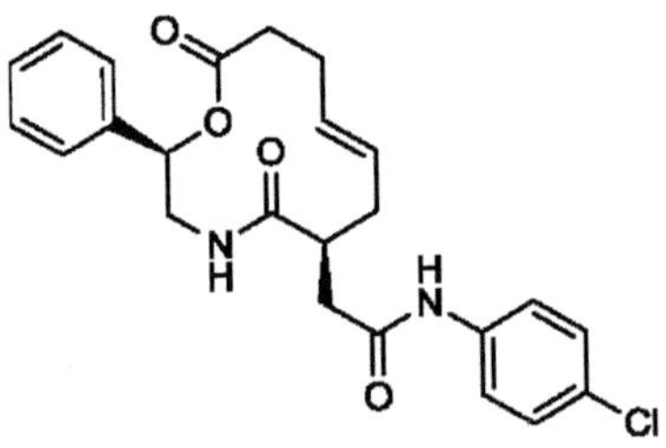

Robotnikinin **180**

Sonic Hedgehog Inhibitor, EC_{50} = 4 μM

FIGURE 9.8 Selected examples of diverse cyclic compounds with specific biological functions.

The cyclization reactions of *N*-acyl iminium ions have yielded a wide range of distinct small-molecule scaffolds (Scheme 9.21) [45]. Initially, peptide synthesis was used to connect a range of functionalized amino acid building blocks. Each of the resulting peptides, **181**, was designed to contain a masked aldehyde, a suitably positioned secondary amide, and a pendant nucleophile. Treatment of the peptides with acid triggered the release of an aldehyde, the formation of an *N*-acyl iminium ion, and cyclization to yield a final scaffold (e.g. **182** to **187**).

Metathesis cascades involving oligomeric substrates have underpinned the synthesis of skeletally diverse small-molecule libraries [46–48]. Within the Nelson group, for example, we prepared a wide range of oligomeric metathesis substrates by the iterative attachment of unsaturated building blocks to a fluorous-tagged linker (Scheme 9.22) [47]. Crucially, alternative attachment reactions were used such that

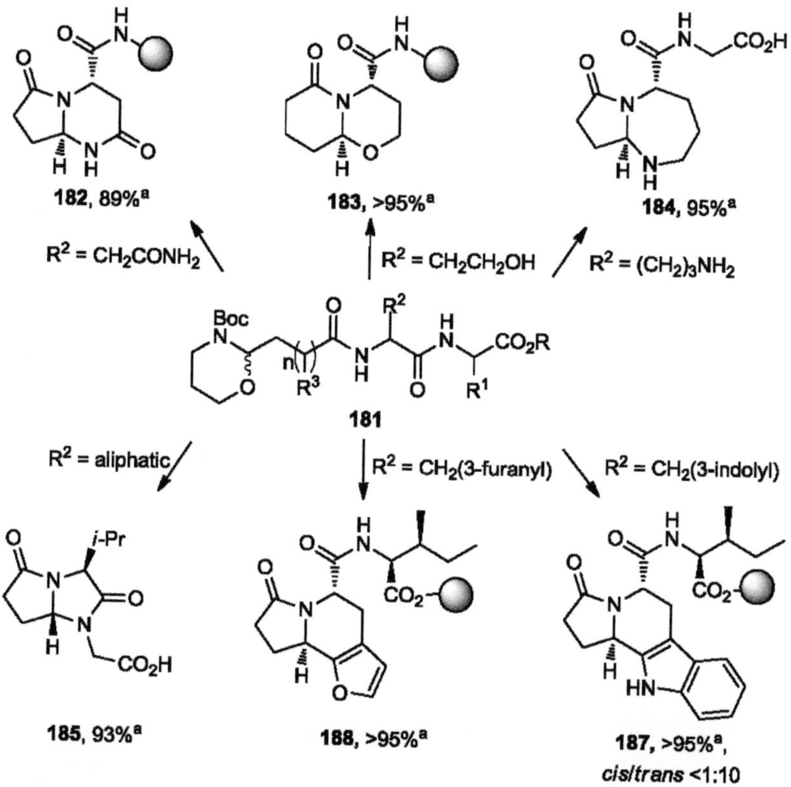

SCHEME 9.21 Oligomer-based approach exploiting *N*-acyl iminium chemistry. 10% TFA (aq). *a* HPLC purity.

the building blocks were connected through bonds that either did, or did not, remain as a vestige in the final products. Finally, metathesis cascades were used to "reprogram" the scaffolds and to release the products of successful cascade processes from the fluorous tag. The resulting library had unprecedented scaffold diversity (over 80 distinct molecular scaffolds).

The approach was extended by exploiting metathesis cascade chemistry in combination with either inter- or intramolecular Diels–Alder reactions (Scheme 9.23) [48]. The metathesis substrates (e.g., **197** to **199**) were assembled iteratively from the corresponding building blocks (in the case of **197** and **198**, using a fluorous tagged "safety catch" linker that had been developed earlier [49]). In the examples shown, the metathesis cascade resulted in the formation of a 1,3-diene whose reactivity was exploited in a Diels–Alder reaction. The fate of the 1,3-dienes **200** to **202** illustrates three distinct tactics that we employed: attachment of a dienophile and intramolecular Diels–Alder reaction (→ **203**); intermolecular Diels–Alder reaction (→ **204**); and intramolecular Diels–Alder reaction involving a tethered dienophile also formed in the cascade (→ **205**).

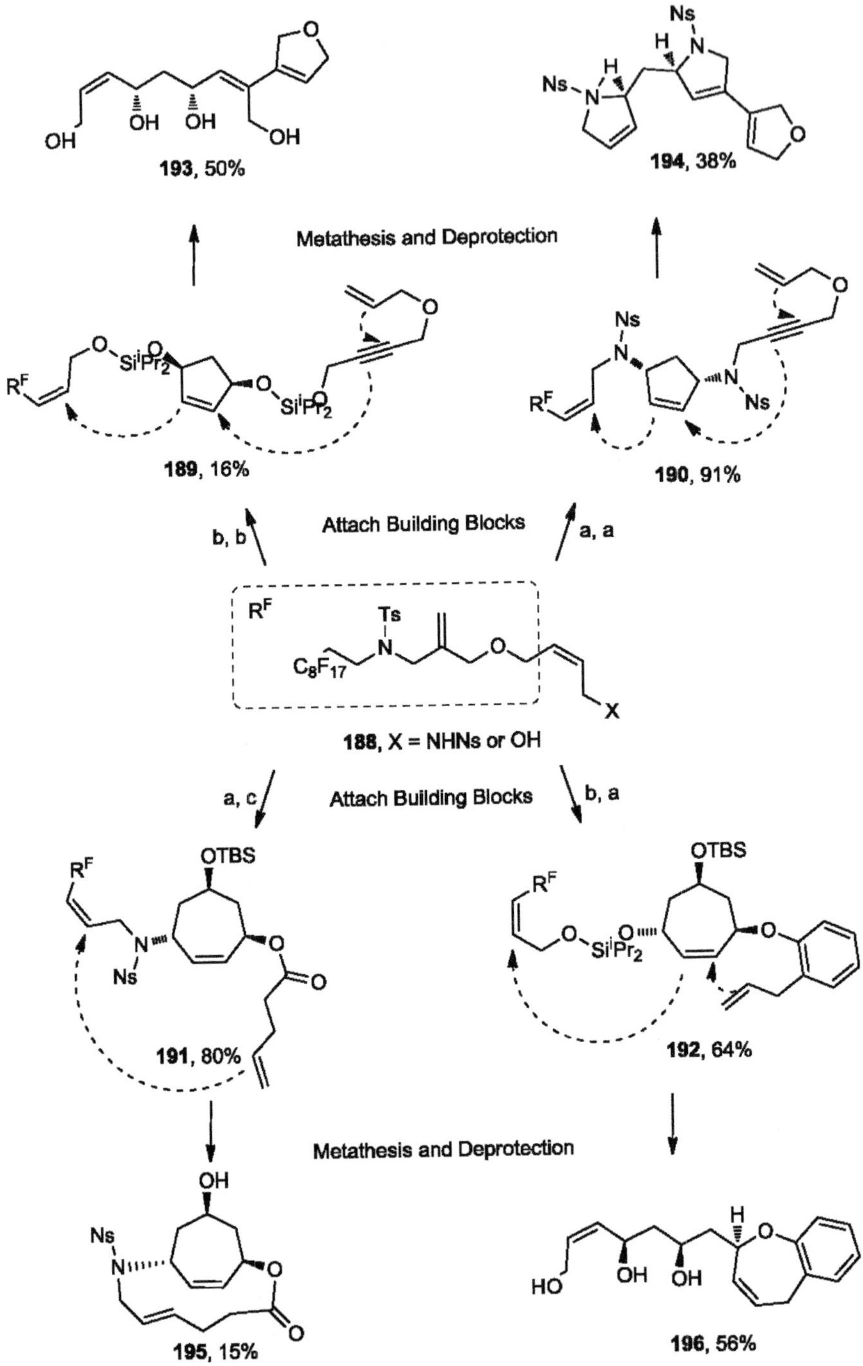

SCHEME 9.22 Selected examples of metathesis cascades leading to skeletally diverse products: (a) (i) alcohol, DEAD, PPh_3, THF, (ii) F-SPE; (b) (i) diisopropylsilyl ether, NBS, CH_2Cl_2, (ii) DMAP, Et_3N, (iii) F-SPE; (c) (i) acid, EDCI, DMAP, CH_2Cl_2, (ii) F-SPE.

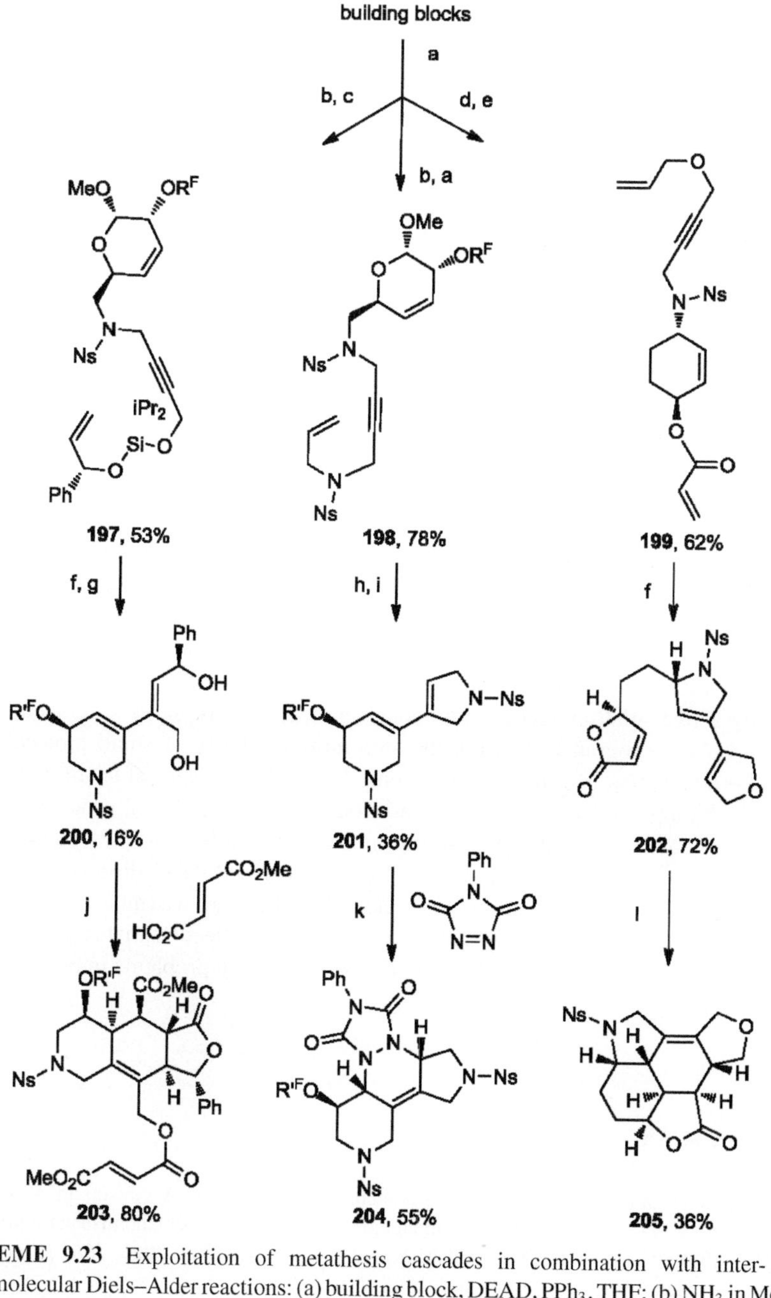

SCHEME 9.23 Exploitation of metathesis cascades in combination with inter- and intramolecular Diels–Alder reactions: (a) building block, DEAD, PPh_3, THF; (b) NH_3 in MeOH (sat.); building block; (c) NBS; CH_2Cl_2; (d) NaOMe, MeOH; (e) acryloyl chloride, *i*-Pr_2NEt, CH_2Cl_2; (f) 3 mol% Grubbs's first-generation catalyst, CH_2Cl_2, 45°C; Et_3N, $P(CH_2OH)_3$; (g) HF-pyridine, THF; (h) 3 mol% Hoveyda-Grubbs's second-generation catalyst, CH_2Cl_2, 45°C; Et_3N, $P(CH_2OH)_3$; (i) TFA, CH_2Cl_2; (j) methyl fumarate, DCC, DMAP, THF; MeCN, μw; (k) 4-phenyl-[1,2,4]-triazole-3,5-dione, CH_2Cl_2, TFA, CH_2Cl_2; (l) *p*-xylene, reflux. R^F and R'^F denote a fluorous tag.

9.6 SUMMARY

The historically uneven exploration of chemical space using synthesis presents a huge challenge to synthetic chemists: to develop synthetic approaches that allow chemical space to be probed more systematically. This challenge has required the development of synthetic approaches that allow molecule scaffolds to be varied combinatorially. In this chapter we have described some of the remarkable approaches that have been exploited in the diversity-oriented synthesis of natural product–like libraries.

Development of the branching and folding pathway strategies has enabled the synthesis of libraries based on up to about 30 scaffolds. These strategies are extremely ingenious; however, extension to many scores of alternative scaffolds is likely to be difficult (and very unlikely to be general). How feasible will it be, then, to devise reliable syntheses of molecules based on hundreds, or even thousands, of distinct scaffolds? Oligomer-based approaches provide the first glimpse into how this challenge might be met. The assembly of oligomeric substrates iteratively, using reliable reactions, may be combined with cyclization reactions of broad scope. A significant challenge will be to identify a broader range of reactions that have the broad scope and chemoselectivity needed to yield molecular scaffolds combinatorially. Nevertheless, it is at least possible that oligomer-based approaches will be harnessed in the future to yield libraries that target huge swathes of chemical space.

The success of diversity-oriented approaches to natural product-like molecules must ultimately be assessed in terms of their impact on the discovery of novel biologically active small molecules: both small-molecule tools and lead molecules for drug discovery programs. It is not the structural similarity of small molecules to natural products that is ultimately important—it is the discovery of small molecules that have biological functions that are not found in known natural products. Many diversity-oriented approaches to natural product–like molecules have yielded specific small molecules with exciting biological functions. The discovery of such molecules bodes well for the ultimate success of the field. To maximize the impact, it is essential that the collections of natural product–like molecules that result be made widely available to scientists addressing the most pressing problems in both biology and medicine.

REFERENCES

1. For reviews of the chemical genetic approach, see (a) M. Fisher, A. Nelson, in: *New Frontiers in Chemical Biology*, M. E. Bunnage, Ed., Royal Society of Chemistry, Cambridge, UK, **2011**, Chap. 1; (b) D. P. Walsh, Y. T. Chang, *Chem. Rev.* **2006**, *106*, 2476–2530; (c) S. L. Schreiber, *Nat. Chem. Biol.* **2005**, *1*, 64–66.
2. S. L. Schreiber, *Proc. Natl. Acad. Sci. U.S.A.* **2011**, *108*, 6699–6702.
3. (a) B. Munos, *Nature* **2009**, *8*, 959–968; (b) J. Drews, *Science* **2000**, *287*, 1960–1964.
4. (a) A. M. Boldi, *Curr. Opin. Chem. Biol.* **2004**, *8*, 281–286; (b) K. Kumar, H. Waldmann, *Angew. Chem. Int. Ed.* **2008**, *48*, 3224–3242; (c) C. Cordier, D. Morton, S. Murrison, A. Nelson, C. O'Leary-Steele, *Nat. Prod. Rep.* **2008**, *25*, 719–737.

5. (a) J. Lippincott-Schwartz, L. C. Yuan, J. S. Bonifacino, R. D. Klausner, *Cell* **1989**, *56*, 801–813; (b) R. D. Klausner, J. G. Donaldson, J. Lippincott-Schwartz, *J. Cell Biol.* **1992**, *116*, 1071–1080; (c) K. F. Xu, X. Shen, H. Li, G. Pacheco-Rodriguez, J. Moss, M. Vaughan, *Proc. Natl. Acad. Sci. U.S.A.* **2005**, *102*, 2784–2789.
6. J. Talpale, J. K. Chen, M. K. Cooper, B. Wang, R. K. Mann, L. Milenkovic, M. P. Scott, P. A. Beachy, *Nature* **2000**, *406*, 1005–1009.
7. (a) F. E Koehn, G. T. Carter, *Nat. Rev. Drug Discov.* **2005**, *4*, 206–220; (b) M. S. Butler, *J. Nat. Prod.* **2004**, *67*, 2141–2153; (c) J. Clardy, C. Walsh, *Nature* **2004**, *432*, 829–837; (d) U. Abel, C. Koch, M. Speitling, F. G, Hansske, *Curr. Opin. Chem. Biol.* **2002**, *6*, 453–458.
8. P. Ertl, S. Roggo, A. Schuffenhauer, *J. Chem. Inf. Model.* **2008**, *48*, 68–74.
9. (a) M. Dow, M. Fisher, T. James, F. Marchetti, A. Nelson, *Org. Biomol. Chem.* **2012**, *10*, 17–28; (b) M. D. Burke, S. L. Schreiber, *Angew. Chem. Int. Ed.* **2004**, *43*, 46–58; (c) W. R. J. D. Galloway, A. Isidro-Llobet, D. R. Spring, *Nat. Commun.* **2010**, *80*, 1–13; (d) S. Dandapani, L. A. Marcaurelle, *Curr. Opin. Chem. Biol.* **2010**, *14*, 362–370; (e) T. E. Nielsen, S. L. Schreiber, *Angew. Chem. Int. Ed.* **2008**, *47*, 48–56; (f) S. L. Schreiber, *Science* **2000**, *287*, 1964–1968; (g) D. S. Tan, *Nat. Chem. Biol.* **2005**, *1*, 74–84.
10. (a) M. A. Koch, L. O. Wittenberg, S. Basu, D. A. Jeyaraj, E. Gourzoulidou, K. Reinecke, A. Odermatt, H. Waldmann, *Proc. Natl. Acad. Sci. U.S.A.* **2004**, *101*, 16721–16726; (b) R. Breinbauer, I. R. Vetter, H. Waldmann, *Angew. Chem. Int. Ed.* **2002**, *41*, 2878–2890; (c) F. J. Dekker, O. Rocks, N. Vartak, S. Menninger, C. Hedberg, R. Balamurugan, S. Wetzel, S. Renner, M. Gerauer, B. Scholermann, M. Rusch, J. W. Kramer, D. Rauh, G. W. Coates, L. Brunsveld, P. I. H. Bastiaens, H. Waldmann, *Nat. Chem. Biol.* **2010**, *6*, 449–456; (d) S. Wetzel, R. S. Bon, K. Kumar, H. Waldmann, *Angew. Chem. Int. Ed.* **2011**, *50*, 10800–10826.
11. (a) S. Wetzel, K. Klein, S. Renner, D. Rauh, T. I. Oprea, P. Mutzel, H. Waldmann, *Nat. Chem. Biol.* **2009**, *5*, 581–583; (b) S. Renner, W. A. L. van Otterlo, M. D. Seoane, S. Möcklinghoff, B. Hofmann, S. Wetzel, A. Schuffenhauer, P. Ertl, T. I Oprea, D. Steinhilber, L. Brunsveld, D. Rauh, H. Waldmann, *Nat. Chem. Biol.* **2009**, *5*, 585–592; (c) M. A. Koch, A. Schuffenhauer, M. Scheck, S. Wetzel, M. Casaulta, A. Odermatt, P. Ertl, H. Waldmann, *Proc. Natl. Acad. Sci. U.S.A.* **2005**, *102*, 17272–17277; (d) A. A. Shelat, R. K. Guy, *Nat. Chem. Biol.* **2007**, *3*, 442–446; (e) Y. Hu, A. M. Wassermann, E. Lounkine, J. Bajorath, *J. Med. Chem.* **2010**, *53*, 752–758.
12. (a) H. Waldmann, *Nat. Chem. Biol.* **2009**, *5*, 76–77; (b) R. S. Bon, H. Waldmann, *Acc. Chem. Res.* **2010**, *43*, 1103–1114.
13. (a) A. Nören-Müller, I. Reis-Correa, H. Prinz, C. Rosenbaum, K. Saxena, H. J. Schwalbe, D. Vestweber, G. Cagna, S. Schunk, O. Schwarz, H. Schiewe, H. Waldmann, *Proc. Natl. Acad. Sci. U.S.A.* **2006**, *103*, 10606–10611; (b) W. Wilk, T. J. Zimmermann, M. Kaiser, H. Waldmann, *Biol. Chem.* **2010**, *391*, 491–497.
14. For the concept of a privileged scaffold, see B. E. Evans, K. E. Rittle, M. G. Bock, R. M. Di Pardo, R. M. Freidinger, W. L. Whitter, G. F. Lundell, D. F. Veber, P. S. Anderson, R. S. L. Chang, V. J. Lotti, D. J. Cerino, T. B. Chen, P. J. Kling, K. A. Kunkel, J. P. Springer, J. Hirshfieldt, *J. Med. Chem.* **1988**, *31*, 2235–2246.
15. H. E. Pelish, N. J. Westwood, Y. Feng, T. Kirchhausen, M. D. Shair, *J. Am. Chem. Soc.* **2001**, *123*, 6740–6741.
16. (a) S. Oh, H. J. Jang, S. K. Ko, Y. Ko, S. B. Park, *J. Comb. Chem.* **2010**, *12*, 548–558; (b) S. Oh, S. W. Cho, J. Y. Yang, H. J. Sun, Y. S. Chung, C. S. Shin, S. B. Park, *Med. Chem. Commun.* **2011**, *2*, 76–80.

17. (a) K. C. Nicolaou, J. A. Pfefferkorn, A. J. Roecker, G.-Q. Cao, S. Barluenga, H. J. Mitchell, *J. Am. Chem. Soc.* **2000**, *122*, 9939–9953; (b) K. C. Nicolaou, J. A. Pfefferkorn, H. J. Mitchell, A. J. Roecker, S. Barluenga, G.-Q. Cao, R. L. Affleck, J. E. Lillig, *J. Am. Chem. Soc.* **2000**, *122*, 9954–9967; (c) K. C. Nicolaou, J. A. Pfefferkorn, S. Barluenga, H. J. Mitchell, A. J. Roecker, G.-Q. Cao, *J. Am. Chem. Soc.* **2000**, *122*, 9968–9976; (d) K. C. Nicolaou, J. A. Pfefferkorn, G. -Q. Cao, *Angew. Chem. Int. Ed.* **2000**, *39*, 734–739; (e) K. C. Nicolaou, G.-Q Cao, J. A. Pfefferkorn, *Angew. Chem. Int. Ed.* **2000**, *39*, 739–743.
18. A. Kumar, S. Srivastava, G. Gupta, V. Chaturvedi, S. Sinha, R. Srivastava, *ACS Comb. Sci.* **2011**, *13*, 65–71.
19. J. D. Sunderhaus, C. Dockendorff, S. F. Martin, *Org. Lett.* **2007**, *9*, 4223–4226.
20. (a) J. S. Potuzak, S. B. Moilanen, D. S. Tan, *J. Am. Chem. Soc.* **2005**, *127*, 13796–13797; (b) S. B. Moilanen, J. S. Potuzak, D. S. Tan, *J. Am. Chem. Soc.* **2006**, *128*, 1792–1793; (c) L.-G. Milroy, G. Zinzalla, G. Prencipe, P. Michel, S. V. Ley, M. Gunaratnam, M. Beltran, S. Neidle, *Angew. Chem. Int. Ed.* **2007**, *46*, 2493–2496.
21. W. Wilk, A. Nören-Müller, M. Kaiser, H. Waldmann, *Chem. Eur. J.* **2009**, *15*, 11976–11984.
22. (a) M. D. Burke, E. M. Berger, S. L. Schreiber, *J. Am. Chem. Soc.* **2004**, *126*, 14095–14104; (b) M. D. Burke, E. M. Berger, S. L. Schreiber, *Science* **2003**, *302*, 613–618.
23. H. Oguri, S. L. Schreiber, *Org. Lett.* **2005**, *7*, 47–50.
24. T. Yang, L. Campbell, D. J. Dixon, *J. Am. Chem. Soc.* **2007**, *129*, 12070–12071.
25. S. Dandapani, M. Duduta, J. S. Panek, J. A. Porco, Jr., *Org. Lett.* **2007**, *9*, 3849–3852.
26. H. Oguri, T. Hiruma, Y. Yamagishi, H. Oikawa, A. Ishiyama, K. Otoguro, H. Yamada, S. Omura, *J. Am. Chem. Soc.* **2011**, *133*, 7096–7105.
27. (a) W. J. Bromley, M. Gibson, S. Lang, S. A. Raw, A. C. Whitwood, R. J. K. Taylor, *Tetrahedron* **2007**, *63*, 6004–6014; (b) S. A. Raw, R. J. K. Taylor, *J. Am. Chem. Soc.* **2004**, *126*, 12260–12261; (c) S. Murrison, D. Glowacki, C. Einzinger, J. Titchmarsh, S. Bartlett, B. McKeever-Abbas, S. Warriner, A. Nelson, *Chem. Eur. J.* **2009**, *15*, 2185–2189.
28. S. Murrison, S. K. Maurya, C. Einzinger, B. McKeever-Abbas, S. Warriner, A. Nelson, *Eur. J. Org. Chem.* **2011**, *12*, 2354–2359.
29. N. Kumagai, G. Muncipinto, S. L. Schreiber, *Angew. Chem. Int. Ed.* **2006**, *45*, 3635–3638.
30. T. O. Painter, L. Wang, S. Majumder, X. Q. Xie, K. M. Brummond, *ACS Comb. Sci.* **2011**, *13*, 166–174.
31. E. Comer, E. Rohan, L. Deng, J. A. Porco, Jr., *Org. Lett.* **2007**, *9*, 2123–2126.
32. J. Barjau, G. Schnakenburg, S. R. Waldvogel, *Angew. Chem. Int. Ed.* **2011**, *50*, 1415–1419.
33. J. Cui, J. Hao, O. A. Ulanovskaya, J. Dundas, J. Liang, S. A. Kozmin, *Proc. Natl. Acad. Sci. U.S.A.* **2011**, *108*, 6763–6768.
34. D. Robbins, A. F. Newton, C. Gignoux, J. C. Legeay, A. Sinclair, M. Rejzek, C. A. Laxon, S. K. Yalamanchili, W. Lewis, M. A. O'Connell, R. A. Stockman, *Chem. Sci.* **2011**, 2, 2232–2235.
35. For a related synthetic approach, see M. Díaz-Gavilán, W. R. J. D. Galloway, K. M. G. O'Connell, J. T. Hodkingson, D. R. Spring, *Chem. Commun.* **2010**, 776–778.
36. O. Kwon, S. B. Park, S. L. Schreiber, *J. Am. Chem. Soc.* **2002**, *124*, 13402–13404.
37. G. C. Micalizio, S. L. Schreiber, *Angew. Chem. Int. Ed.* **2002**, *41*, 152–154.
38. E. E. Wyatt, S. Fergus, W. R. J. D. Galloway, A. Bender, D. J. Fox, A. T. Plowright, A. S. Jessiman, M. Welch, D. R. Spring, *Chem. Commun.* **2006**, 3296–3298.

39. G. L. Thomas, R. J. Spandl, F. G. Glansdorp, M. Welch, A. Bender, J. Cockfield, J. A. Lindsay, C. Bryant, D. F. J. Brown, O. Loiseleur, H. Rudyk, M. Ladlow, D. R. Spring, *Angew. Chem. Int. Ed.* **2008**, *47*, 2808–2812.

40. G. Moura-Lettsa, C. M. DiBlasia, R. A. Bauerb, D. S. Tana, *Proc. Natl. Acad. Sci. U.S.A.* **2011**, *108*, 6745–6750.

41. W. Liu, V. Khedkar, B. Baskar, M. Schürmann, K. Kumar, *Angew. Chem. Int. Ed.* **2011**, *50*, 6900–6905.

42. (a) L. A. Marcaurelle, E. Comer, S. Dandapani, J. R. Duvall, B. Gerard, S. Kesavan, M. D. Lee IV, H. Liu, J. T. Lowe, J. C. Marie, C. A. Mulrooney, B. A. Pandya, A. Rowley, T. D. Ryba, B. C. Suh, J. Wei, D. W. Young, L. B. Akella, N. T. Ross, Y. L. Zhang, D. M. Fass, S. A. Reis, W. N. Zhao, S. J. Haggarty, M. Palmer, M. A. Foley, *J. Am. Chem. Soc.* **2010**, *132*, 16962–16976; (b) E. Comer, H. Liu, A. Joliton, A. Clabaut, C. Johnson, L. B. Akella, L. A. Marcaurelle, *Proc. Natl. Acad. Sci. U.S.A.* **2011**, *108*, 6751–6756.

43. (a) D. H. C. Chou, J. R. Duvall, B. Gerard, H. Liu, B. A. Pandya, B. C. Suh, E. M. Forbeck, P. Faloon, B. K. Wagner, L. A. Marcaurelle, *ACS Med. Chem. Lett.* **2011**, *2*, 698–702; (b) R. W. Heidebrecht, Jr., C. Mulrooney, C. P. Austin, R. H. Barker, Jr., J. A. Beaudoin, K. C. C. Cheng, E. Comer, S. Dandapani, J. Dick, J. R. Duvall, E. H. Ekland, D. A. Fidock, M. E. Fitzgerald, M. Foley, R. Guha, P. Hinkson, M. Kramer, A. K. Lukens, D. Masi, L. A. Marcaurelle, X. Z. Su, C. J. Thomas, M. Weïwer, R. C. Wiegand, D. Wirth, M. Xia, J. Yuan, J. Zhao, M. Palmer, B. Munoz, S. Schreiber, *ACS Med. Chem. Lett.* **2012**, *3*, 112−117; (c) L. F. Peng, B. Z. Stanton, N. Maloof, X. Wang, S. L. Schreiber, *Bioorg. Med. Chem. Lett.* **2009**, *12*, 6319–6325.

44. (a) A. Isidro-Llobet, T. Murillo, P. Bello, A. Cilibrizzi, J. T. Hodgkinson, W. R. J. D. Galloway, A. Bender, M. Welch, D. R. Spring, *Proc. Natl. Acad. Sci. U.S.A.* **2011**, *108*, 6793–6798; (b) T. Hu, R. Tannert, H. D. Arndt, H. Waldmann, *Chem. Commun.* **2007**, *38*, 3942–3944; (c) A. R. Bogdan, K. James, *Chem. Eur. J.* **2010**, *16*, 14506–14512; (d) G. Chouhan, K. James, *Org. Lett.* **2011**, *13*, 2754–2757; (e) A. R. Bogdan, K. James, *Org. Lett.* **2011**, *13*, 4060–4063; (f) E. Marsault, M. L. Peterson, *J. Med. Chem.* **2011**, *54*, 1961–2004.

45. (a) T. E. Nielsen, M. Meldal, *J. Org. Chem.* **2004**, *69*, 3765–3773; (b) T. E. Nielsen, M. Meldal, *J. Comb. Chem.* **2005**, *7*, 599–610; (c) T. E. Nielsen, S. L. Quement, M. Meldal, *Org. Lett.* **2005**, *7*, 3601–3604; (d) S. T. Le Quement, T. E. Nielsen, M. Meldal, *J. Comb. Chem.* **2007**, *9*, 1060–1072.

46. (a) D. A. Spiegel, F. C. Schroeder, J. R. Duvall, S. L. Schreiber, *J. Am. Chem. Soc.* **2006**, *128*, 14766–14767; (b) D. Lee, J. K. Sello, S. L. Schreiber, *Org. Lett.* **2000**, *2*, 709–712; (c) J. K. Sello, P. R. Andreana, D. Lee, S. L. Schreiber, *Org. Lett.* **2003**, *5*, 4125–4127; (d) D. Lee, J. K. Sello, S. L. Schreiber, *J. Am. Chem. Soc.* **1999**, *121*, 10648–10649; (e) R. J. Spandl, H. Rudyk, D. R. Spring, *Chem. Commun.* **2008**, 3001–3003.

47. D. Morton, S. Leach, C. Cordier, S. Warriner, A. Nelson, *Angew. Chem. Int. Ed.* **2009**, *48*, 104–109.

48. C. O'Leary-Steele, P. J. Pedersen, T. James, T. Lanyon-Hogg, S. Leach, J. Hayes, A. Nelson, *Chem. Eur. J.* **2010**, *16*, 9563–9571.

49. C. O'Leary-Steele, C. Cordier, J. Hayes, S. Warriner, A. Nelson, *Org. Lett.* **2009**, *11*, 915–918.

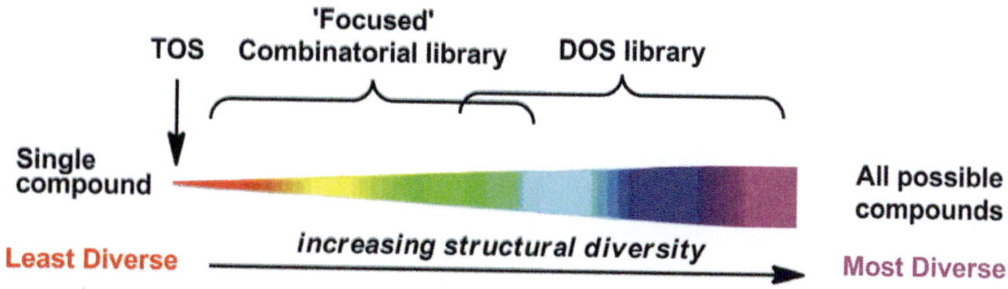

FIGURE 1.2 Molecular diversity spectrum: a representation of the relative degrees of molecular diversity achieved using TOS, focused library synthesis, and DOS. (From [19], with permission of The Royal Society of Chemistry.)

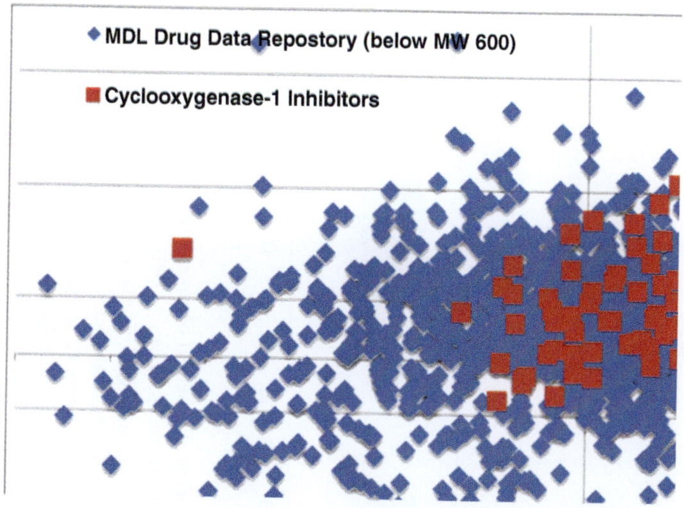

FIGURE 1.3 Chemical space analysis plot of cyclooxygenase-1 (COX-1) inhibitors (red squares) and MDDR compounds (blue diamonds), created using chemical descriptors and principal components analysis. The plot shows that COX-1 inhibitors occupy a wide range of chemical space.

Diversity-Oriented Synthesis: Basics and Applications in Organic Synthesis, Drug Discovery, and Chemical Biology, First Edition. Edited by Andrea Trabocchi.

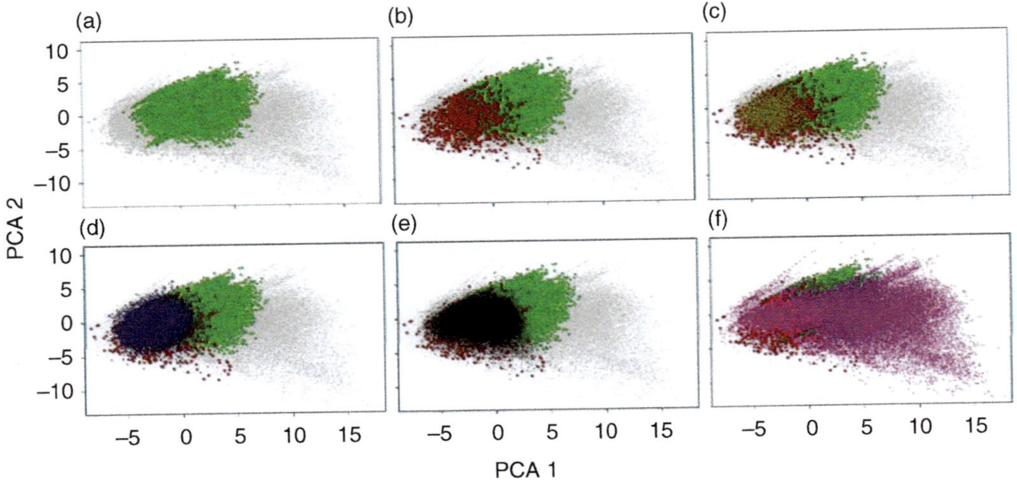

FIGURE 10.1 Application of principal components analysis (PCA) to generate a visual representation of the chemical space. The figure shows an ADME space representation of combinatorial libraries (green circles), approved drugs (red squares), diverse compounds from the National Cancer Institute (NCI) database (yellow triangles), commercial vendor compounds (blue diamonds), natural products from ZINC (black dots), and natural products from TCM (magenta dots). For better visualization, panels (a) to (f) show in the same graph one, two, or three libraries in color and the remaining libraries in gray for reference. In all panels, compounds are shown in the same coordinates. The first two principal components account for 74.8% of the variance.

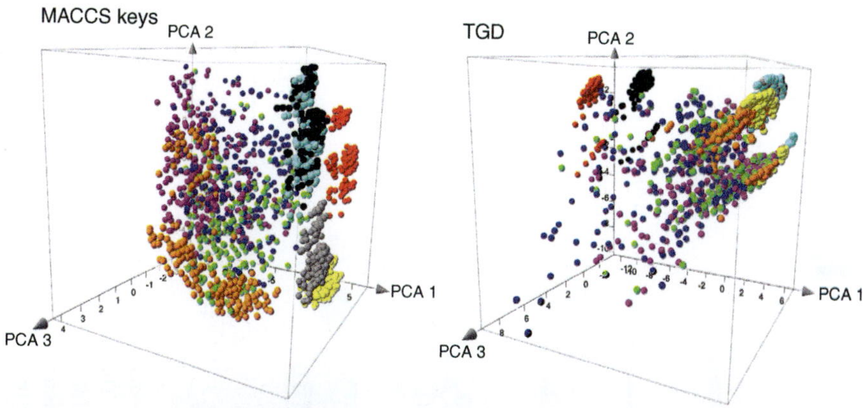

FIGURE 10.9 Dependence of chemical space with structure representation. Visual representation of the chemical space of a total of 2250 compounds selected from nine compound collections (250 compounds per data set). The data sets are natural products (green) and a general screening collection (magenta) from two commercial providers, five in-house combinatorial libraries (red, yellow, gray, black, and cyan), approved drugs (blue), and a set of compounds targeted to adenosine receptors (orange). The depiction of the chemical space was obtained by PCA of the similarity matrix computed by Tanimoto similarity and four different structure representations. The first three PCs account for 85.9% (MACCS) and 90.3% (TGD) of the variance, respectively.

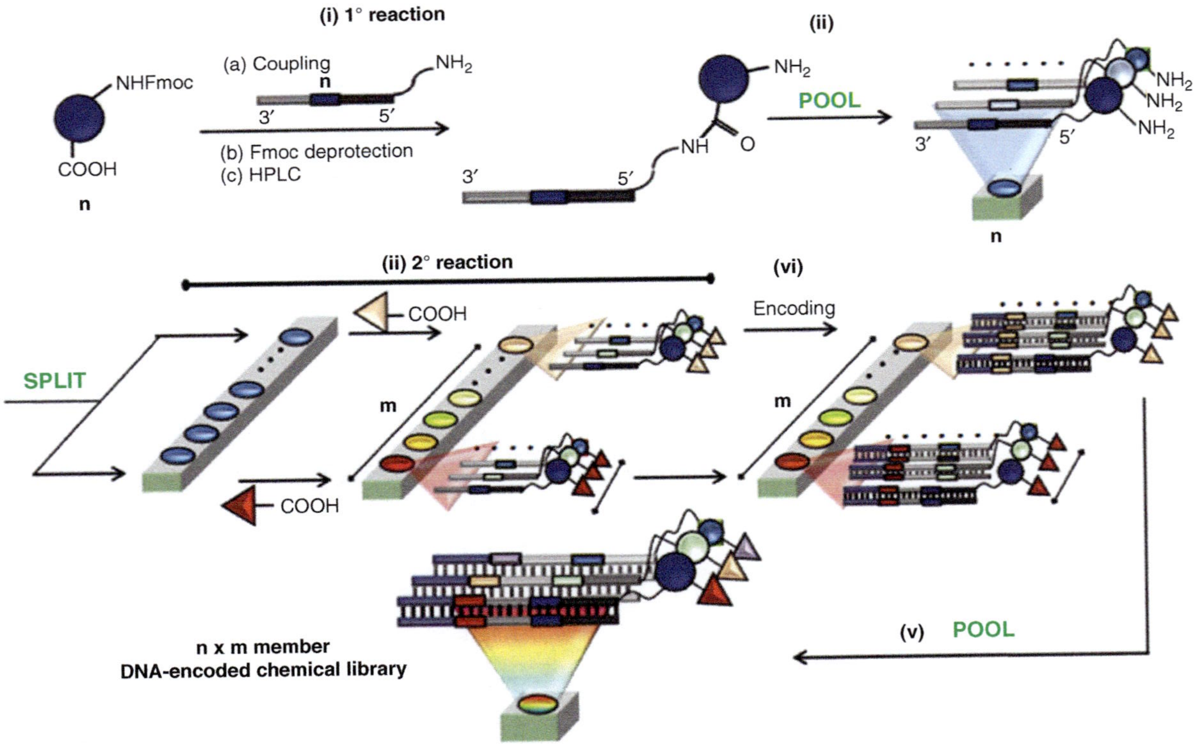

FIGURE 11.7 Schematic representation of the split–pool–split approach pioneered by Philochem for the construction of single-pharmacophore DNA-encoded chemical libraries based on two sets of building blocks. The strategies feature (i) conjugation of the first set of chemical compounds (e.g., *n* Fmoc amino acids) to distinct amino-modified oligonucleotides, followed by Fmoc removal and HPLC purification; (ii) pooling of the reactions and splitting into *m* separate reaction vessels, corresponding to the reactions for the insertion of the second building block; (iii) chemical incorporation of the second building block; (iv) encoding of the second building block (e.g., by hybridization of partially complementary oligonucleotides, followed by Klenow fragment-mediated DNA polymerization); (v) pooling of the encoded reactions to yield a DNA-encoded chemical library containing $n \times m$ molecules. (From [54]: copyright © 2008 National Academy of Sciences, U.S.A.)

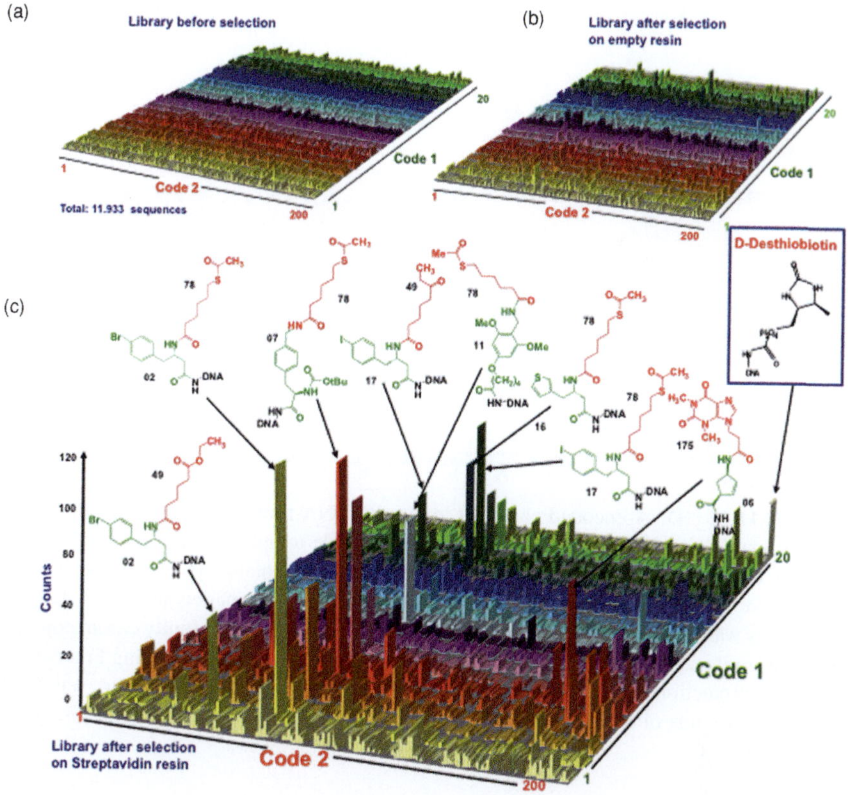

FIGURE 11.8 High-throughput-sequencing results before and after selection against streptavidin-coated Sepharose resin. (a–c) Histogram plots represent the 4000 library members in the *x*–*y* plane and the corresponding sequence counts on the *z*-axis. (a,b) Sequencing before selection and after selection on an empty resin showed equally distributed frequencies of library members; (c) streptavidin selection results exhibited a preferential enrichment of D-desthiobiotin (known streptavidin binder with nanomolar affinity spiked into the library prior to selection) and of additional structurally related compounds. The chemical structures of some of the most relevant straptavidin binders are indicated together with the library identification number. Building blocks used in the two synthetic steps are highlighted in green and red, respectively. Molecules identified with at least 30 counts exhibited after resynthesis preferential binding toward streptavidin, with K_d values ranging between 350 nM and 11 μM. In contrast, compounds tested with fewer than 10 counts did not exhibit appreciable binding to streptavidin ($K_d > 50$ μM). (From [54]; copyright © 2008 National Academy of Sciences, U.S.A.

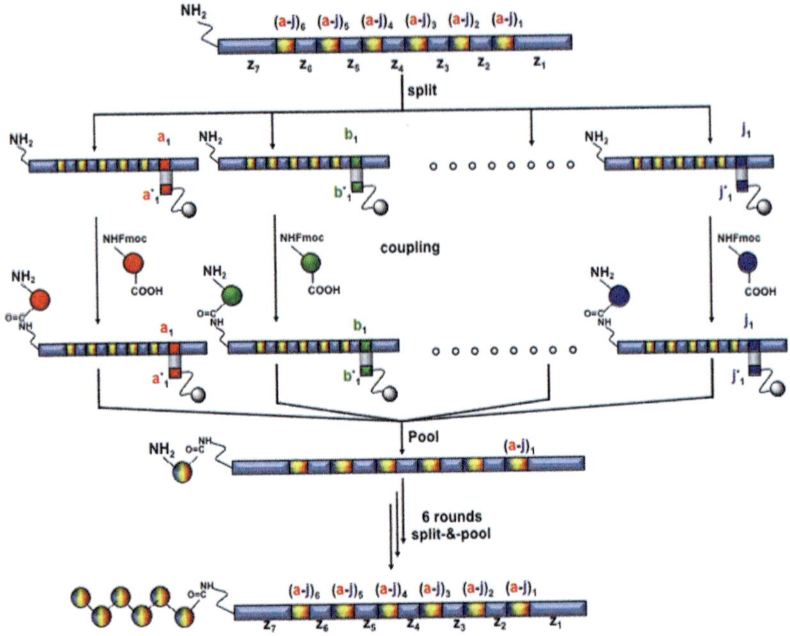

FIGURE 11.12 DNA-encoded chemical library by DNA routing. An initial combinatorial library comprising 10^6 different 340-mer-oligonucleotide templates was generated by PCR assembly of smaller oligonucleotides. Each oligonucleotide contains six coding regions, each encoding for 10 different amino acids [(a to j)$_{1\text{-}6}$], and seven constant domains ($z_{1\text{-}7}$). The template library was split using complementary oligonucleotides bound on affinity chromatography resins [(a* to j*)$_{1\text{-}6}$] (DAN routing). Following conjugation to the corresponding Fmoc-amino acid and deprotection, oligonucleotides were eluted and the initial template library was reconstituted. Six rounds of DNA routing yielded to the final library of 10^6 different N-acylated pentapeptides.

FIGURE 11.18 Vipergen's YoctoReactor technology. (a) A highly stable three-dimensional DNA complex (a DNA junction) makes it possible to confine DNA-conjugate reagents within a volume on the order 10^{-24} (yoctoliter). By means of the proximity effect, DTS reactions take place at the center of the molecular reactor, delivering the corresponding building blocks on the initial reaction site. (b) Synthesis of a three-way DNA-encoded chemical library. Individual building blocks are linked via cleavable or noncleavable linkers to distinct hairpin-shaped DNA oligonucleotides. *a*, building block 1, and *b*, building block 2, undergo the first DNA-templated reaction. Following encoding of building block 2 by enzymatic ligation, cleavage of the linker enables irreversible delivery of the compound to the initial reaction site. Iteration of the process enables incorporation of the third set of building blocks (*c*) and corresponding DNA-coding tags. After the reactions are completed, the hairpin-DNA reactor is converted into a linear oligonucleotide, yielding to the final library of small molecules displayed on double-stranded DNA templates. (c) Rolling translation: PCR-amplified DNA-coding tags selected in the first round of panning serve as yR templates for chemical translation of the next-generation library in a subsequent round of selection. Before entering the next round of translation and selection, the double-stranded DNA templates are site-specifically digested to expose single-stranded DNA codons, which effectively allow hybridization with the corresponding building block conjugates.

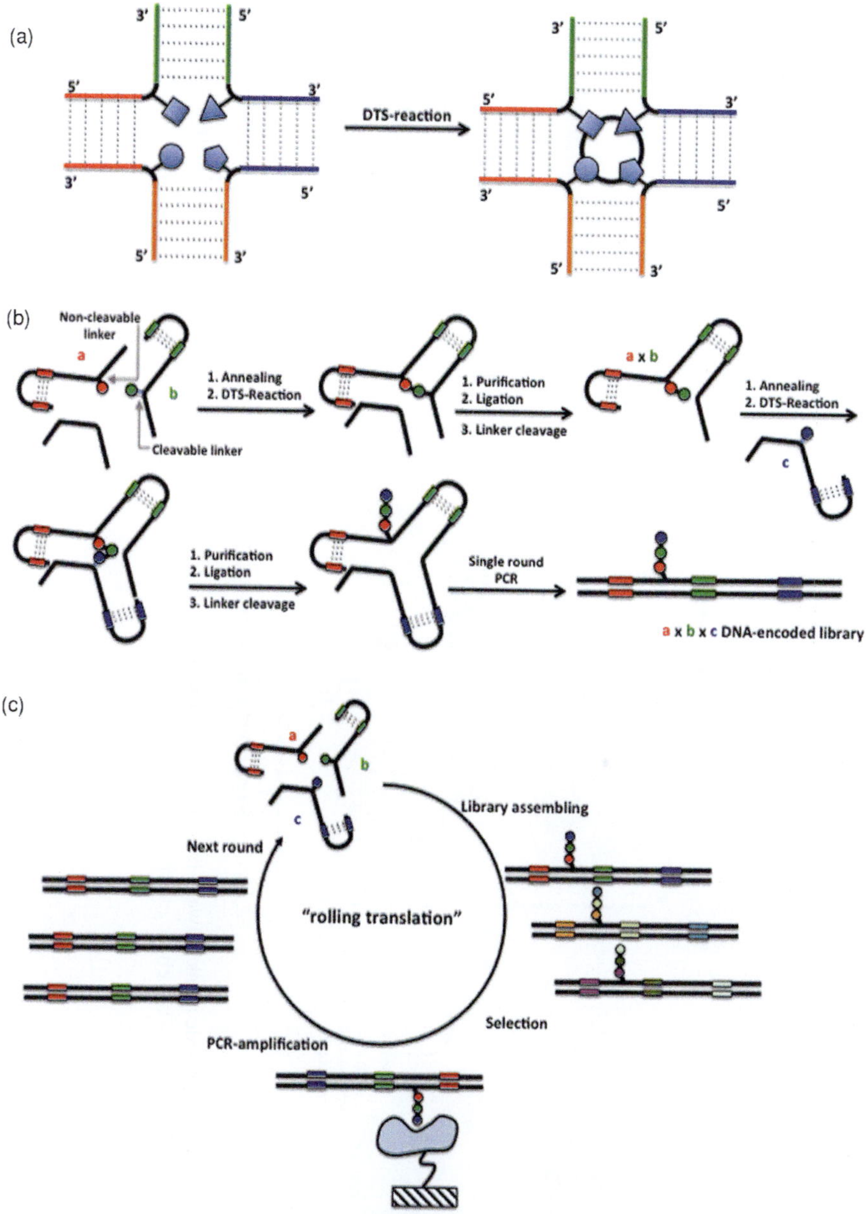
(a)
3′
5′
5′
3′
3′
5′
5′
3′
DTS-reaction
(b)
Non-cleavable linker
a
b
Cleavable linker
1. Annealing
2. DTS-Reaction
1. Purification
2. Ligation
3. Linker cleavage
a x b
1. Annealing
2. DTS-Reaction
c
1. Purification
2. Ligation
3. Linker cleavage
Single round PCR
a x b x c DNA-encoded library
(c)
a
b
c
Library assembling
Next round
"rolling translation"
Selection
PCR-amplification

FIGURE 11.18

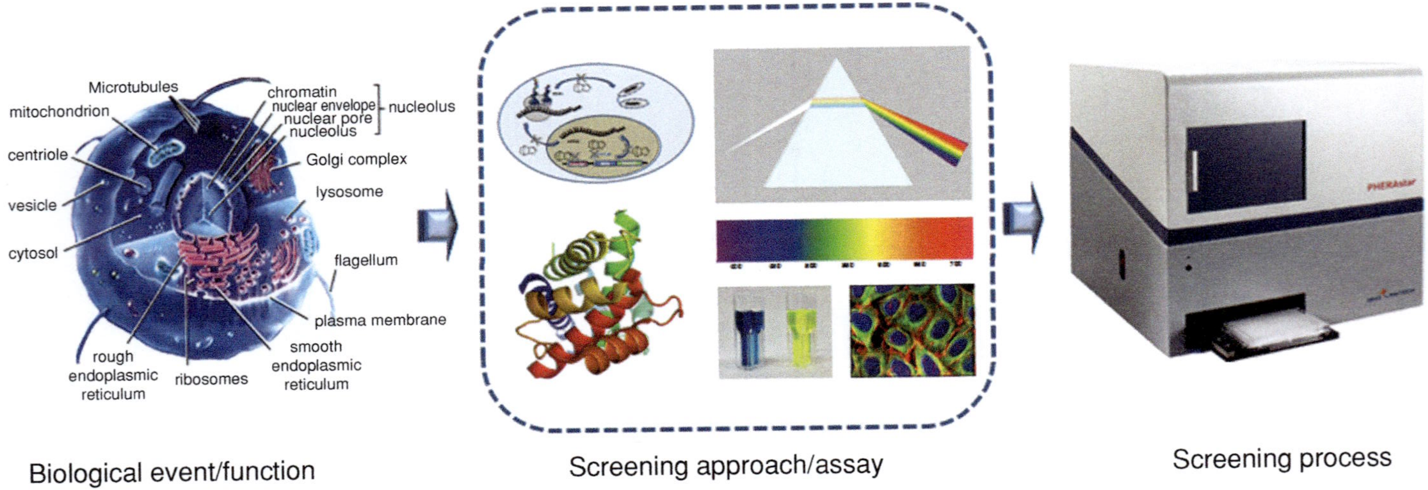

FIGURE 12.1 HTS assays provide a link between biological phenomena and a signal measured using special HTS instrumentation. (Adapted from Image of the 3D Cell is from http://www.ebi.ac.uk/microarray/biology_intro.html.)

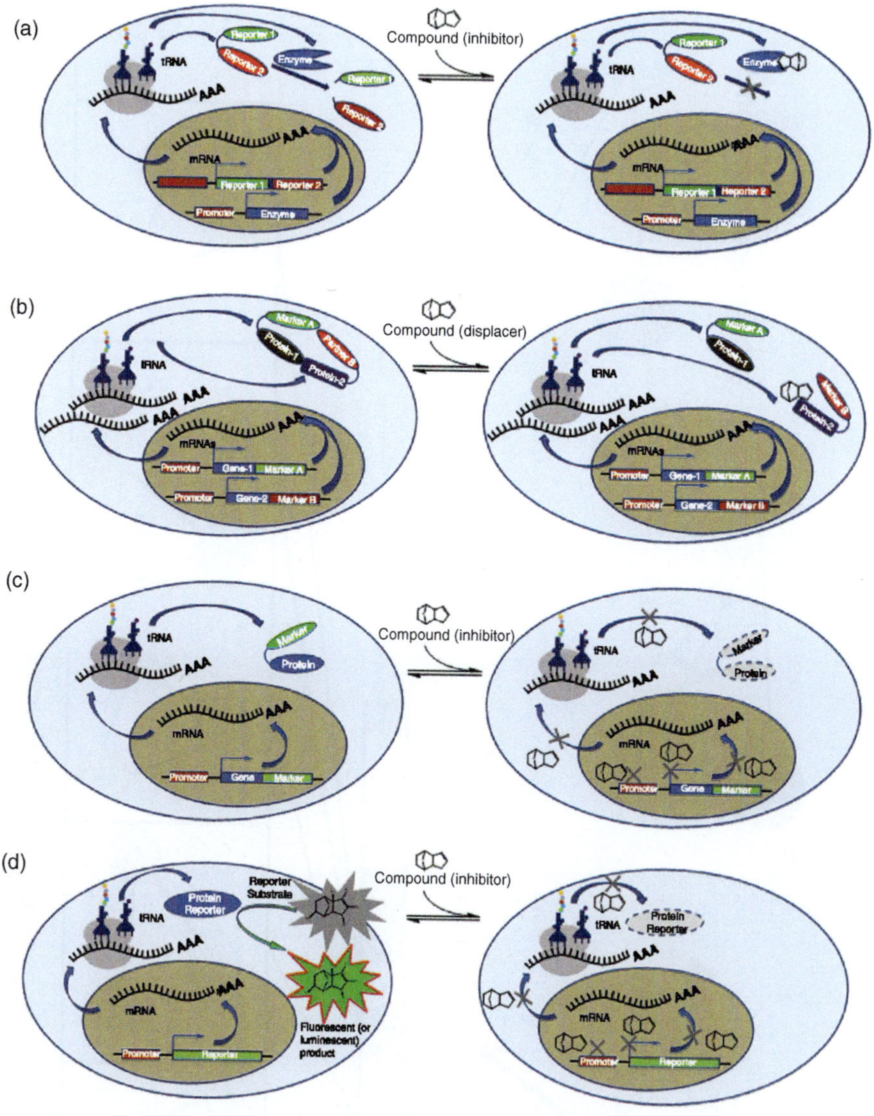

FIGURE 12.6 Cell-based assays: (a) functional cell-based assay; (b) binding cell-based assay; (c) reporter gene assay using a reporter with a directly measurable optical property; (d) reporter gene assay using a reporter with catalytic activity measured through fluorogenic or luminogenic substrates.

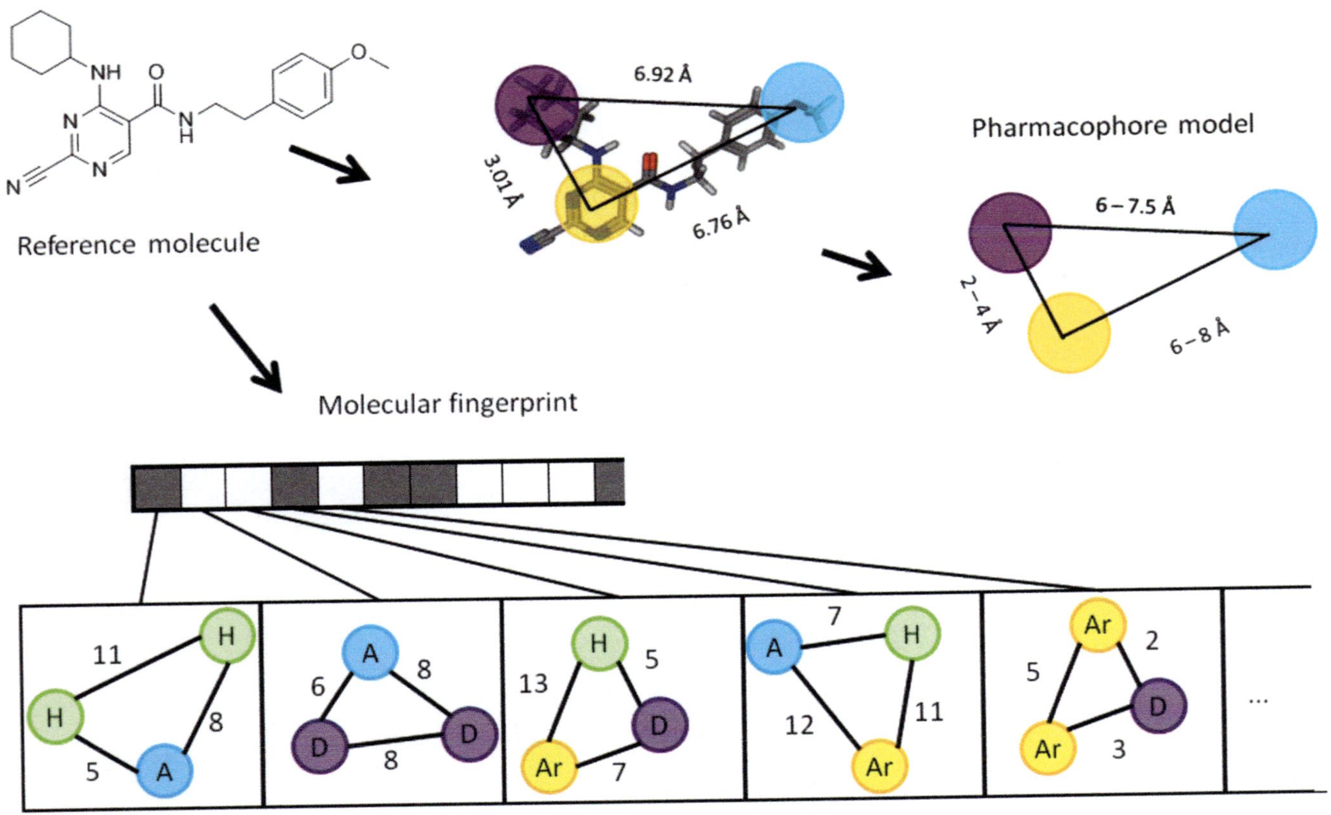

FIGURE 15.3 Pharmacophore representations. A three-point pharmacophore query derived from a single reference molecule and a three-point pharmacophore fingerprint of the same molecule are schematically represented.

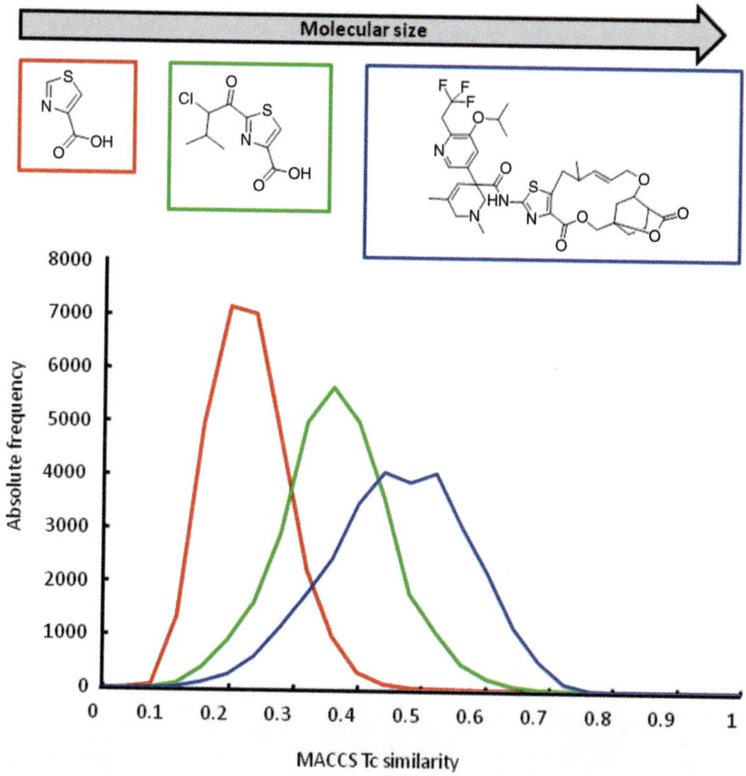

FIGURE 15.4 Biased similarity calculations. The influence of the molecular complexity effect on fingerprint calculations is illustrated. Three molecules of increasing size are shown (red, green, and blue, respectively). Each molecule is searched against the ZINC compound database [104] and the resulting distributions of MACCS T_c values are compared. As can be seen, apparent similarity to database compounds increases with molecular complexity and size.

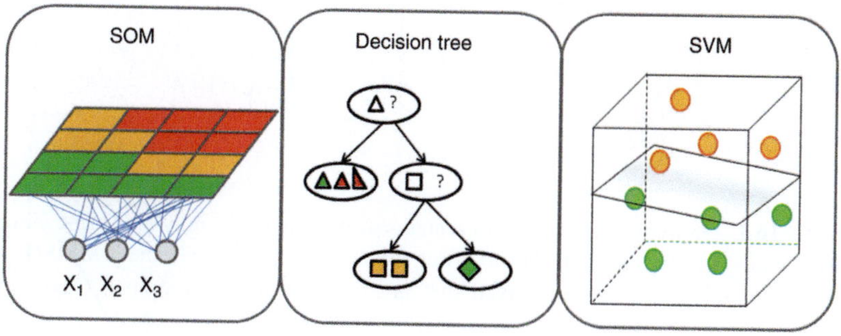

FIGURE 15.5 Representative machine learning approaches. From left to right, SOMs, decision trees, and SVMs are illustrated schematically.

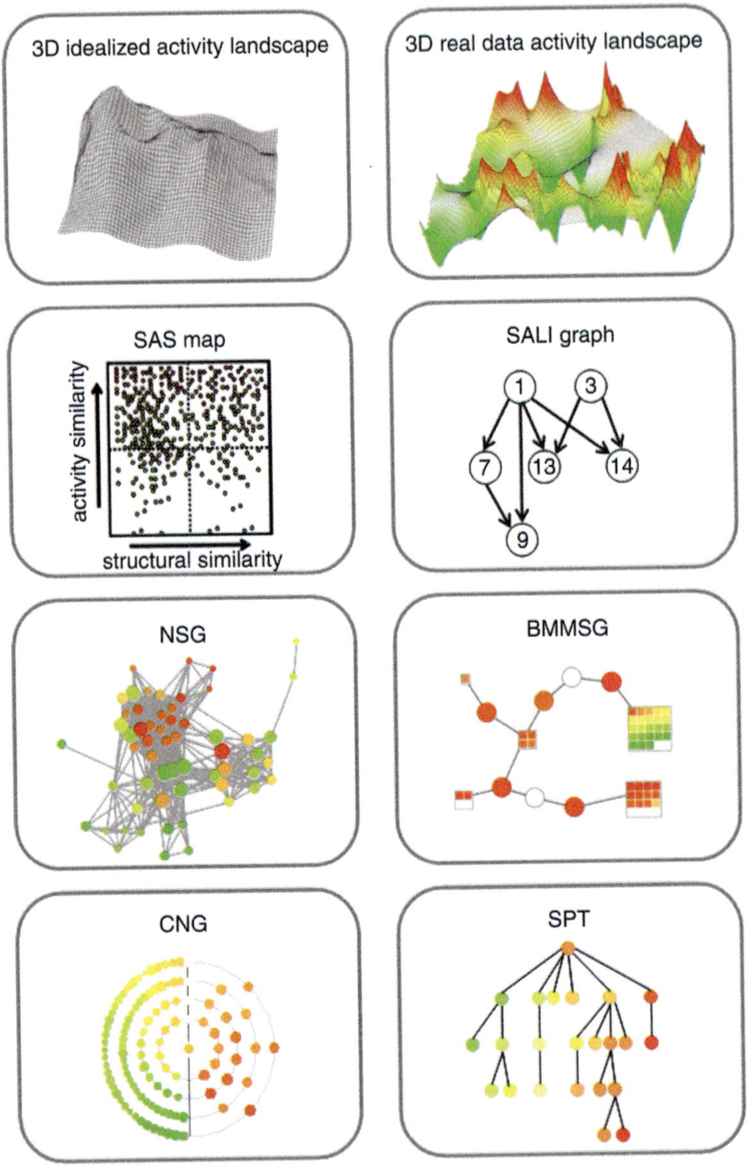

FIGURE 16.1 Activity landscape representations. Alternative activity landscape views are shown. Details are discussed in the text. For the SALI graph, NSG, and BMMSG, small representative sub-graphs are displayed that are taken from larger graphs of compound data sets.

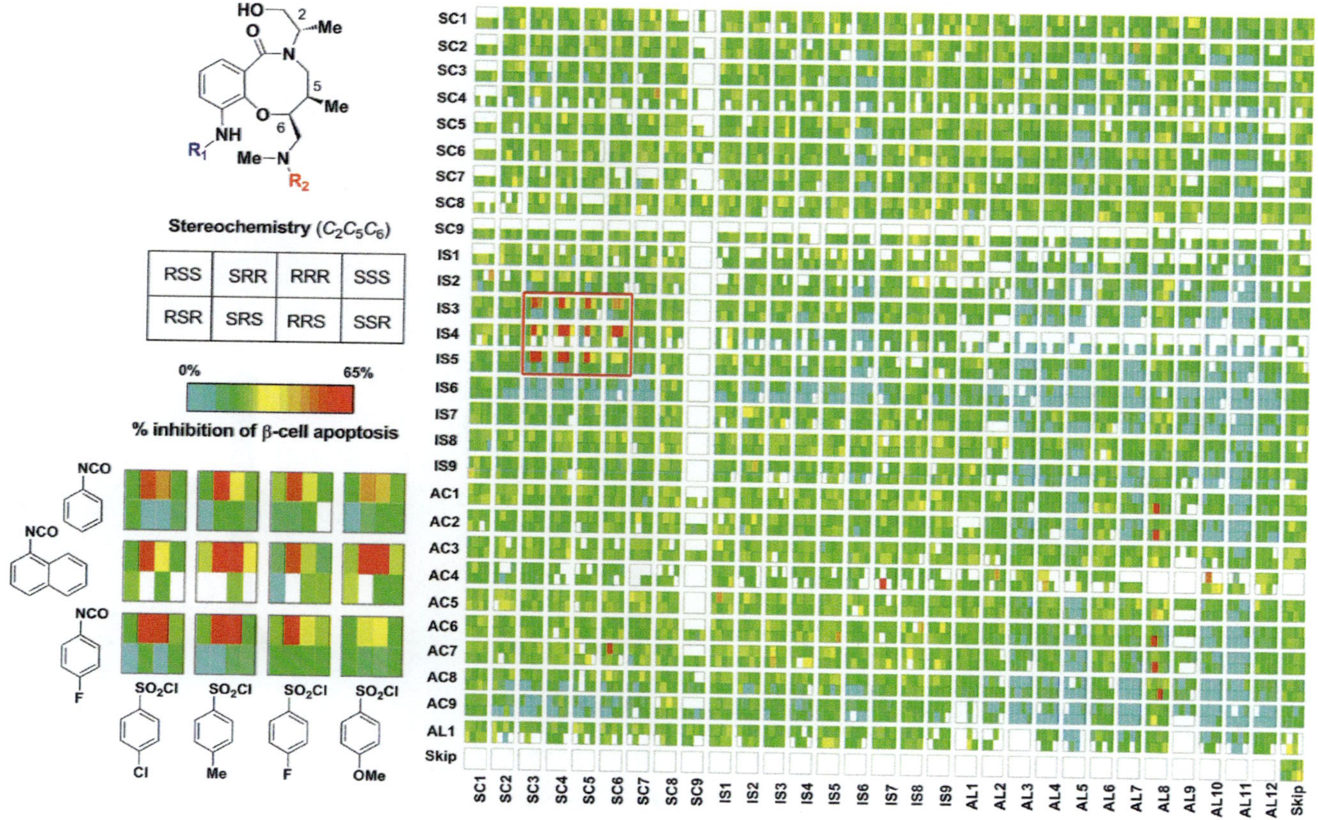

FIGURE 17.5 Primary screening data displayed as percent inhibition of β-cell apoptosis. The full matrix of building blocks is shown for R^1 (y-axis) and R^2 (x-axis). From [23], with permission; copyright © 2011 American Chemical Society.)

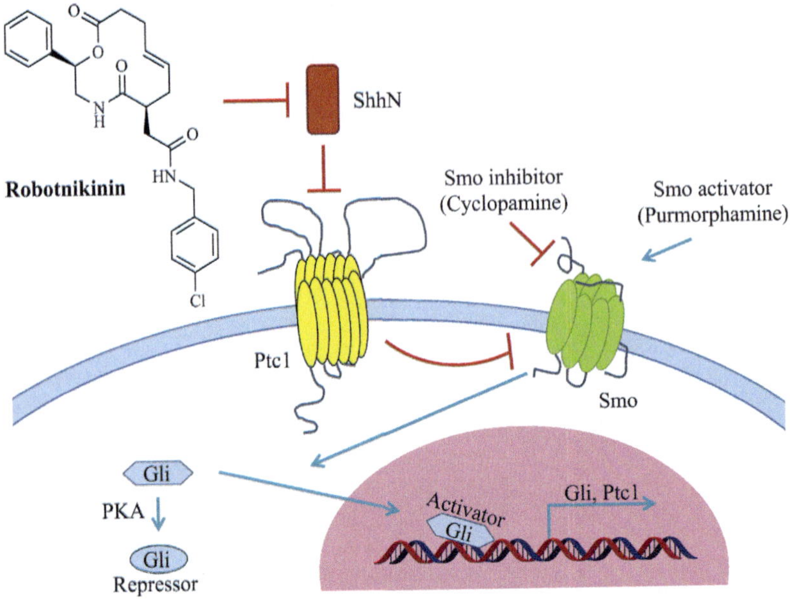

FIGURE 18.1 Robotnikinin inhibits Shh signaling by blocking Shh ligand binding to Ptc.

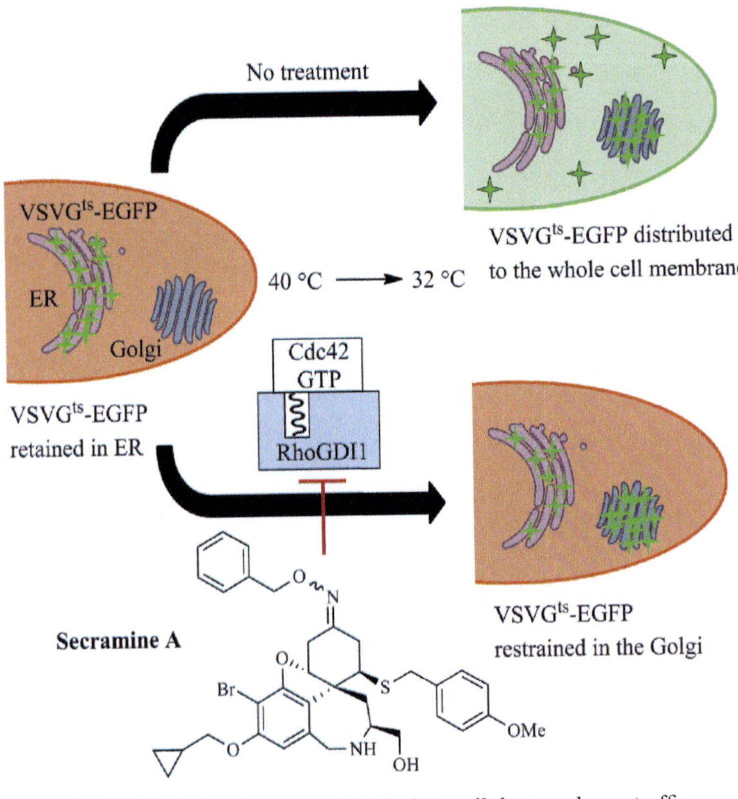

FIGURE 18.2 Secramine inhibits intracellular membrane traffic.

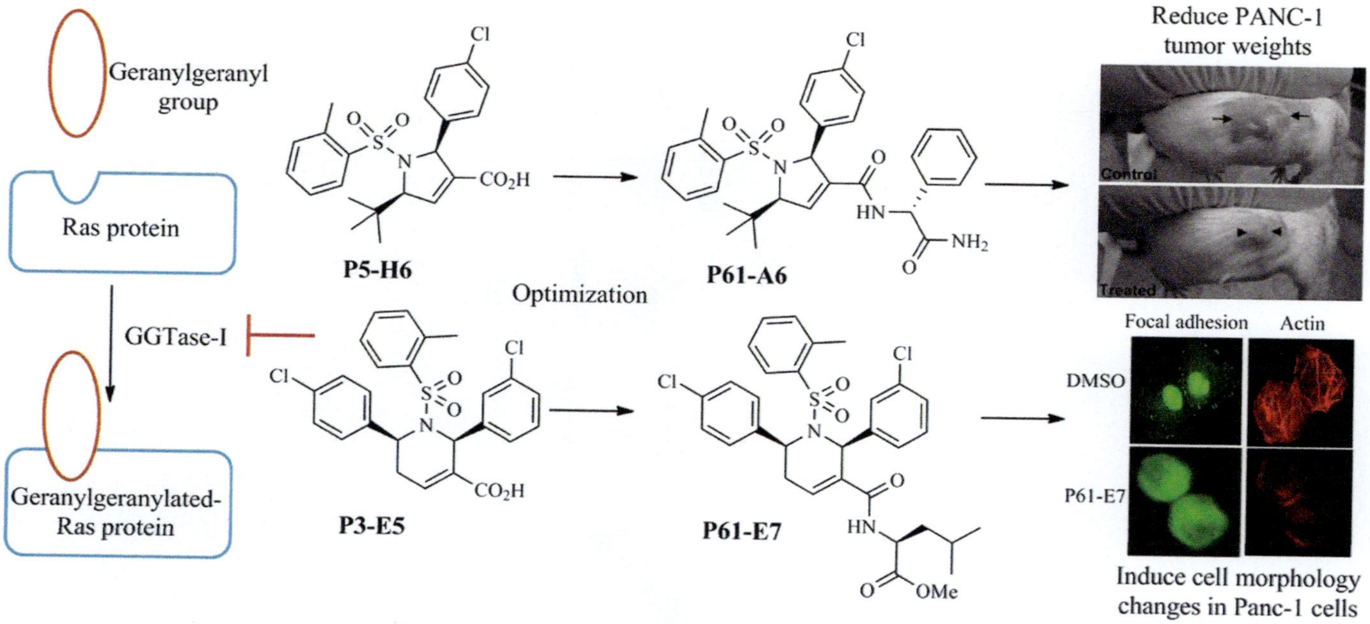

FIGURE 18.5 Small-molecule inhibitors of protein geranylgeranylation.

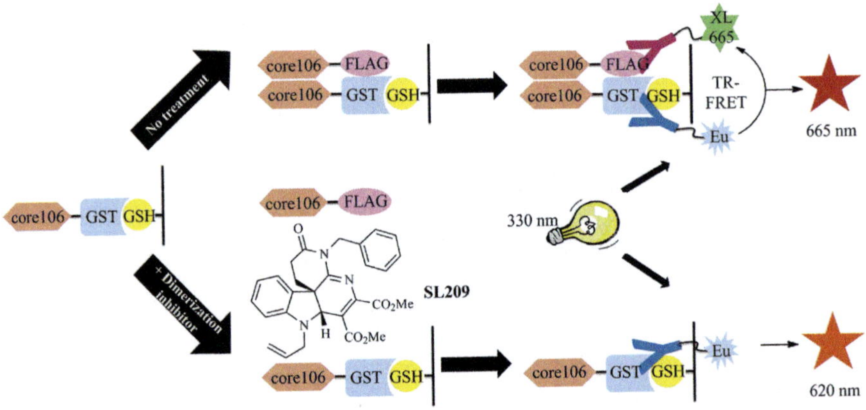

FIGURE 18.6 SL209 inhibit heterodimerization of GST-core106/FLAG-core106.

description of the chemical compounds using a set of characteristics such as molecular mass, lipophilicity, and topological features to name a few, Dobson conceptualized the chemical space as "the total descriptor space that encompasses all the small carbon-based molecules that could in principle be created" [30]. The chemical space is vast, as shown by the calculations of the total number of possible compounds that could be made (e.g., 10^{60}) [31]. As discussed in subsequent sections, one of the major applications of chemical space is the comparison of compound data sets to design novel collections and/or select existing compound libraries for experimental or computational screening. A second major application of the notion of chemical space is the classification of bioactive compounds according to their biological properties. The latter application is based on the hypothesis that similar molecules have similar bioactivity [32,33]. Therefore, it is expected that, at least in principle, the "biologically active space" [34] would be formed by separated clusters of compounds, each one associated with a different receptor [29]. The latter concept has led to the field of biology-oriented synthesis (BIOS), which aims to target "bioactivity islands" with compound data sets containing core structures of compound classes that are relevant to nature [12,35].

10.3 GENERAL ASPECTS OF CHEMOINFORMATIC METHODS TO ANALYZE THE CHEMICAL SPACE

Chemoinformatics, also referred in the literature as *cheminformatics* or *chemical information science* has been defined as "the application of informatic methods to solve chemical problems" [36] or as "a scientific field based on the representation of molecules as objects (graphs or vectors) in a chemical space" [37]. Additional definitions are reviewed by Varnek and Baskin [37] and Willett [38]. Chemoinformatics plays a key role in the diversity analysis of compound collections and the mining of chemical space. Major aspects of chemoinformatics include the representation of chemical compounds, storing and mining information in databases, and generating and analyzing data [36].

Based on the definition of chemical space by Dobson as discussed above, Le Guilloux et al. pointed out the need to define chemical spaces for chemoinformatic applications using a restricted number of compounds and number of descriptors [39]. To this end, a large number of molecular representations, such as physicochemical properties, fingerprint-based representations, and molecular scaffolds, have been used to describe chemical spaces and to compare compound databases. The use of a comprehensive description of the molecules often leads to multidimensional spaces that require dimension-reduction techniques to visualize such multidimensional arrays and extract useful information. In fact, molecular representation and data-reduction techniques are two major factors that have a strong influence on the analysis and visual representation of chemical spaces [40,41]. A main aspect of dimension-reduction techniques is to maintain a large percentage of the information from higher-dimensional space [34,42]. Different sets of descriptors and parameters used to define the space where the molecules will be located [39] may change the neighborhood relationships in chemical space [40]. One of the strategies proposed to minimize such dependence is to use multiple representations [43–45].

Several multidimensional data mining tools and visualization techniques are used to analyze SAR, perform structural analysis, and in other applications for drug discovery projects such as principal moments of inertia plots [46] and multifusion similarity maps [41], which have been used widely [27,31,47,48]. Additional approaches are hierarchical clustering, decision trees, multidimensional scaling, genetic algorithms, neural networks, and support vector machines. These and other techniques are reviewed elsewhere [27,42,49–51].

Two commonly used approaches to represent chemical spaces are self-organizing maps [52] and principal components analysis (PCA) [53]. Figure 10.1 illustrates an application of PCA to generate a visual representation of the ADME-related chemical space or "ADMET space" of six compound collections: namely, approved drugs, natural products from TCM and ZINC [54], a large collection of in-house combinatorial libraries, commercial vendor compounds, and a generally diverse collection obtained from the National Cancer Institute database [55]. To generate the plots in Figure 10.1, 24 ADME-related properties computed using the program QikProp (QikProp, version 3.4, Schrödinger LLC, New York, 2011) were subject to PCA. The first two principal components account for 74.9% of the variance. Figure 10.1a shows the ADME space of the combinatorial libraries; the other collections are in the background for reference. Figure 10.1b compares the ADME space of the combinatorial libraries and drugs showing that some of the combinatorial libraries occupy the same space as the drugs while other compounds cover neglected regions of the drug–ADME space. In contrast, diverse compounds in the National Cancer Institute database, commercial vendor molecules, and natural products from ZINC occupy the same area of drugs (Figure 10.1c to e). The latter observation is not surprising since compounds in these three last collections are typically selected to be "drug-like"—at least similar to currently known drugs. Figure 10.1f is a comparison of the ADME space of the combinatorial library compounds, drugs, and natural products from TCM. Interestingly, TCM covers a vast region of this property space, including unexplored areas of drugs and the combinatorial libraries. In general, the combinatorial libraries cover a large area shared with drugs and TCM.

10.4 CHEMOINFORMATIC-BASED ANALYSIS OF LIBRARIES USING DIFFERENT REPRESENTATIONS

Molecular representation is at the core of chemoinformatic applications. There are two major types of representation: graphs and descriptor vectors [37,56]. Graph methods are used to perform structural and substructural analyses. These approaches are straightforward to interpret and enable easy communication with medicinal chemists and biologists, as illustrated elsewhere in the chapter. Representation using descriptor vectors is commonly used in chemoinformatics for database processing, clustering, similarity searching, and developing descriptive and predictive models of SAR: for example, QSPR/QSAR models and activity landscape models [37]. Currently, more

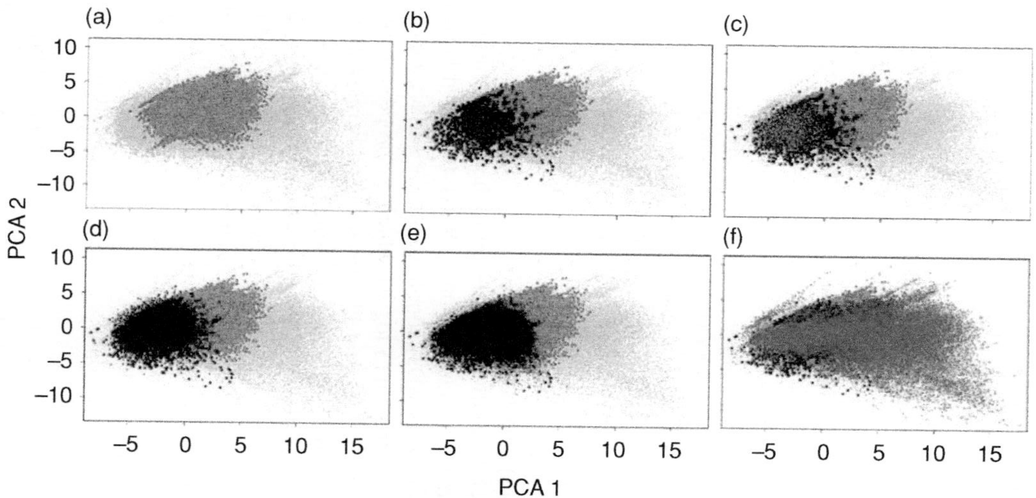

FIGURE 10.1 Application of principal components analysis (PCA) to generate a visual representation of the chemical space. The figure shows an ADME space representation of combinatorial libraries (green circles), approved drugs (red squares), diverse compounds from the National Cancer Institute (NCI) database (yellow triangles), commercial vendor compounds (blue diamonds), natural products from ZINC (black dots), and natural products from TCM (magenta dots). For better visualization, panels (a) to (f) show in the same graph one, two, or three libraries in color and the remaining libraries in gray for reference. In all panels, compounds are shown in the same coordinates. The first two principal components account for 74.8% of the variance. (*See insert for color representation of the figure.*)

than 5000 types of descriptors are available [57]. The choice of descriptors used to analyze compound data sets gives rise to different types of chemical spaces. Varnek and Baskin have pointed out that "unlike real physical space, a chemical space is not unique: each ensemble of graphs and descriptors defines its own chemical space" [37].

Sections 10.4.1 to 10.4.4 focus on a discussion of representative examples of the chemoinformatic characterization of the chemical space and molecular diversity of compound libraries using different structure representations.

10.4.1 Physicochemical Properties and Medicinally Relevant Chemical Spaces

Physicochemical properties have been used to develop the classical rules to define drug-like [58] and lead-like criteria [59] and Congreve's rule of 3 for fragment-based lead discovery [60]. It is worth noting that the seminal Lipinski's rule of 5 [58] has been revised over the past decade. For example, Walters et al. present a comprehensive review of studies showing the changes over the years in the physicochemical properties of compounds synthesized for drug discovery programs [61]. It is largely documented that "new molecular entities are moving away from the traditional drug space" [62,63] and that "as new targets emerge and optimization tools advance, the oral drug like space might expand" [64]. Dow et al. noted that the current "biologically relevant chemical space" can be considerable skewed by the uneven exploration of chemical space by chemical synthesis [35].

Physicochemical properties frequently used to describe chemical libraries include molecular weight (MW), number of rotatable bonds (RBs), hydrogen-bond acceptors (HBAs), hydrogen-bond donors (HBDs), topological polar surface area (TPSA), and the octanol/water partition coefficient ($S \log P$), which are properties commonly used as descriptors to represent lead-like, drug-like, or medicinally relevant chemical spaces. An advantage of using these properties to compare databases is that they are intuitive and straightforward to interpret. Characterization of compound libraries using additional properties related to ADME attributes are being explored increasingly (see Figure 10.1). This is exemplified by the work of Wager et al., who analyzed the ADME space of marketed central nervous system (CNS) drugs and an in-house collection [65] and the rich review of Ritchie et al. covering tools to visualize and represent ADME-related properties of data sets [51].

A number of comparisons of chemical data sets in the medicinally relevant chemical spaces have been published for different purposes. For example, the chemical space of 15 combinatorial libraries organized in three bis-diazacyclic "libraries from libraries" (LOLs) was recently compared to approved drugs using PCA and the distribution of six medicinally relevant properties: MW, RB, HBA, HBD, TPSA, and $S \log P$ (see above). The LOL strategy is a solid-phase DOS technique [66], where multiple scaffolds are generated from the same starting material (Figure 10.2a). Increasing skeletal diversity is known to be a very efficient way to increase structural diversity [67]. LOLs are designed to have different physicochemical and biological properties from those of the parent library [68]. This approach has been employed to synthesize

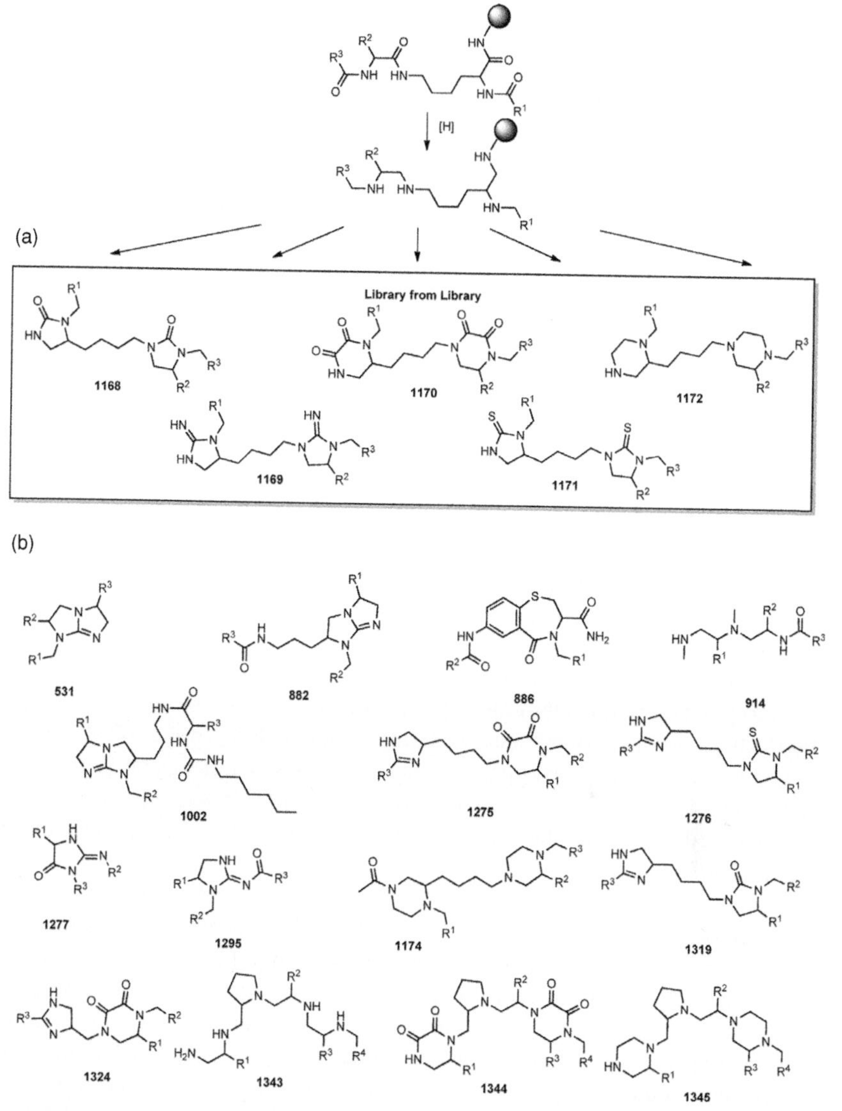

FIGURE 10.2 (a) Example of one LOL. Each library within an LOL shares the same number of diversity positions, identical building block side-chain functionalities, and the same number of compounds. (b) Core scaffolds of small-molecule synthetic combinatorial libraries included in the scaffold ranking plate of the Torrey Pines Institute for Molecular Studies.

numerous small-molecule libraries and discover novel active compounds for a wide range of therapeutic applications [68,69]. Comparison of the combinatorial libraries with drugs using molecular properties showed that combinatorial libraries have a different degree of overlap with the property space of drugs. In the same study, quantitative methods were used to demonstrate, the increased structural and

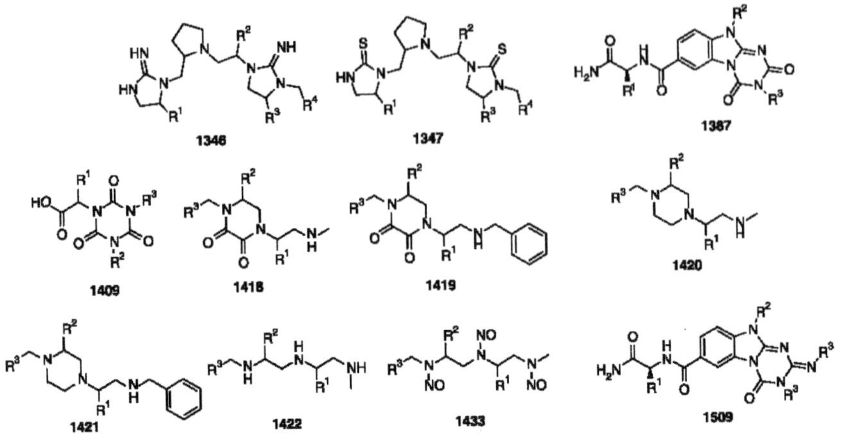

FIGURE 10.2 (*Continued*)

property diversity of the LOLs over individual libraries. That work represented the first comprehensive quantification of the diversity of LOLs, providing a solid approach to comparing and contrasting the diversity of DOS libraries with existing drugs or any other compound collection [48]. LOLs have also been analyzed using molecular fingerprints (see below) [70].

As part of an ongoing effort to identify novel inhibitors of natural origin for the emerging epigenetic target DNA methyltransferase (DNMT) [71], the chemical space of two natural product collections, including compounds from the TCM database, were compared to the property space of approved drugs, synthetic commercial compounds, and a focused library directed to identify DNMT. It was concluded that the DNMT-focused library and the two natural products collections have molecules with properties similar to those of approved drugs [72].

In a separate study, the chemical space of more than 28,000 compounds from the TCM database was compared to a large collection of small-molecule combinatorial libraries with 30 different scaffolds (Figure 10.2b) and a database of a commercial vendor library. The small-molecule libraries are part of the *scaffold ranking plate* that has been designed over the years to introduce diversity in structures and chemical properties. Experimental evidence has shown that this plate is a valuable tool in the rapid selection of the most promising scaffolds for further screening and identifying novel lead compounds [1,19,23]. Six medicinally relevant properties were used to define the chemical space: MW, RB, HBA, HBD, TPSA, and $S \log P$ (see above). That study revealed that, in general, compared to the other collections considered in that work, TCM has the largest values of HBD, HBA, $S \log P$, and TPSA. Concerning the size as measured by MW, TCM had the largest molecules overall. It was also concluded that a large proportion of the compounds in the combinatorial libraries have different degrees of overlap with the property space of currently approved drugs [55]. Of note, studies comparing drugs launched prior to 1983 ("old" drugs) and drugs launched between 1983 and 2002 ("new" drugs) reveal that the mean and median RB and MW values are larger in new drugs [62,63].

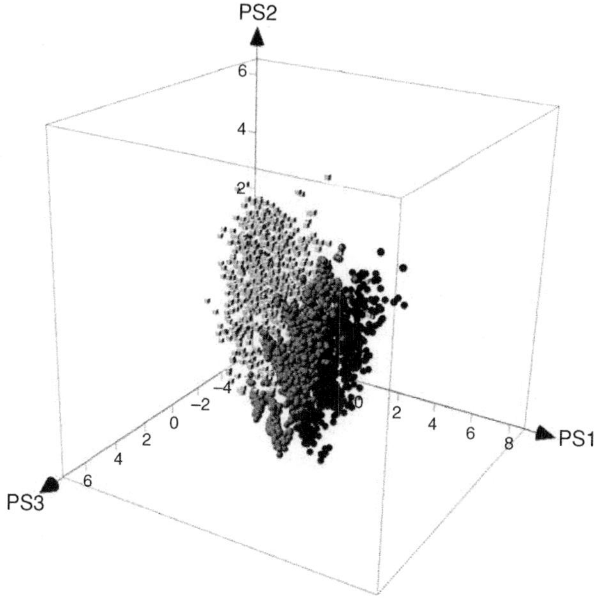

FIGURE 10.3 Visual representation of the chemical space of combinatorial libraries with the core scaffolds of **1170** (dark gray), **1171** (light gray) in Figure 10.2a, and a set of approved drugs (light gray). The plot was generated using ChemGPS-NP prediction scores calculated using the online tool ChemGPS-NPWeb. Of note, the combinatorial libraries obtained with the LOL approach occupy different regions of the chemical space.

Also using the six properties described above, the medicinally relevant chemical space of 2477 natural products from a commercial vendor was recently compared to 5963 synthetic compounds from academic sources using methods such as DOS, and 6152 synthetic compounds from a commercial vendor (as representative of common screening collections). The result of this comparison shows that the DOS compounds considered in that study are heavier and more lipophilic than the natural products or the synthetic commercial compounds [47].

Figure 10.3 shows a visual representation of the property-based chemical space of two combinatorial libraries obtained using an LOL approach and a collection of approved drugs. The visual representation was obtained using the recently developed Web-based public tool ChemGPS-NPWeb [73,74]. ChemGPS-NP [73,75] is a PCA-based global chemical positioning system [76] tuned for exploration of biologically relevant chemical space. The first four dimensions of the ChemGPS-NP map capture 77% of data variance. The first dimension (principal component 1, PC1) represents size, shape, and polarizability (the main contribution is size); PC2 is associated with aromatic and conjugation-related properties (the main influence is aromaticity); PC3 describes lipophilicity, polarity, and hydrogen-bond capacity (the major contribution is lipophilicity); and PC4 expresses flexibility and rigidity. Chemical compounds can be positioned onto this map using interpolation in terms of PCA score prediction. Further details of this method are provided elsewhere [74]. Figure 10.3 clearly illustrates that libraries obtained using the LOL method (see above) occupy different regions of chemical space. ChemGPS-NPWeb has recently been used to compare the chemical space of approved drugs with TCM and compounds derived from in-house combinatorial libraries available in PubChem [77]. Several other examples of the comparison of compound collections relevant in drug discovery are reviewed elsewhere [31,61].

10.4.2 Molecular Complexity

Molecular complexity has emerged as an attractive criterion to guide the selection of chemical libraries to explore the currently neglected chemical space [78,79]. Several measures of complexity have been proposed [80–83]. Using the fraction of saturated carbons as a simple and intuitive measure of complexity, Lovering et al. showed that more complex molecules have higher success rates in the drug discovery process [84]. The authors of that work also suggested that compounds with increased complexity, as captured by a measure of carbon bond saturation, might increase selectivity.

The hypothesis of Lovering et al. was supported experimentally by Clemons et al. [85], who screened three major types of compound collections—commercial compounds, natural products, and synthetic molecules obtained from academic groups—across 100 diverse proteins using techniques such as DOS. Figure 10.4 illustrates a comparison of the complexity of the three collections using two measures employed in that work: stereochemical complexity, calculated as the portion of carbon atoms that are stereogenic (Figure 10.4a), and shape complexity, calculated as the ratio of sp^3-hydridized carbon atoms to total sp^3- and sp^2-hybridized carbons (Figure 10.4b). The distributions were generated with the data reported by Clemons et al. [85]. The box plots, for which a rich statistical interpretation is presented by Ritchie and Macdonald [86], clearly show the increased complexity of the natural products collection over the commercial compounds and the set of diverse synthetic molecules. By contrast, the commercial compounds showed the lowest complexity. After screening these three compound databases, which differed in molecular complexity, the authors of that work concluded that increasing the content of sp^3-hydridized and stereogenic atoms relative to compounds from commercial sources enhanced the binding selectivity and frequency. The results of Clemons et al. are in agreement with the complexity model of Hann, Leach, and co-workers [87,88].

The pivotal work of Lovering et al. [84] using the fraction of saturated carbons to measure molecular complexity (or "aliphatic indicator of a molecule" as defined initially by Yan and Gasteiger [89]) encouraged the use of this measure to survey the complexity of different compound databases. Thus, analysis of the chemical complexity of the compounds in the Molecular Libraries Small Molecule Repository using the fraction of saturated carbons leads to the conclusion that natural products and collections from academic and other research institutes are more complex than libraries from commercial vendors [79]. Chen et al. demonstrated the increased complexity of natural products over drugs, clinical candidates, and bioactive molecules [90]. Walters et al. analyzed the distribution of the fraction of saturated carbons for

(a)

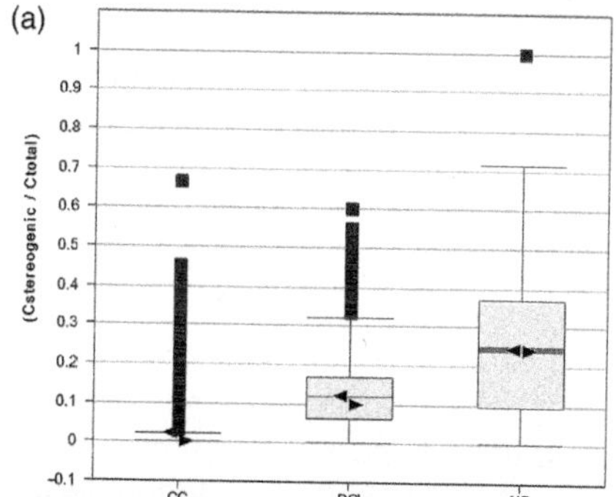

	CC	DC'	NP
Max	0.667	0.600	1.000
Q3	0.000	0.167	0.370
Median	0.000	0.100	0.240
Q1	0.000	0.063	0.095
Min	0.000	0.000	0.000
Mean	0.022	0.120	0.243
StdDev	0.053	0.078	0.171

(b)

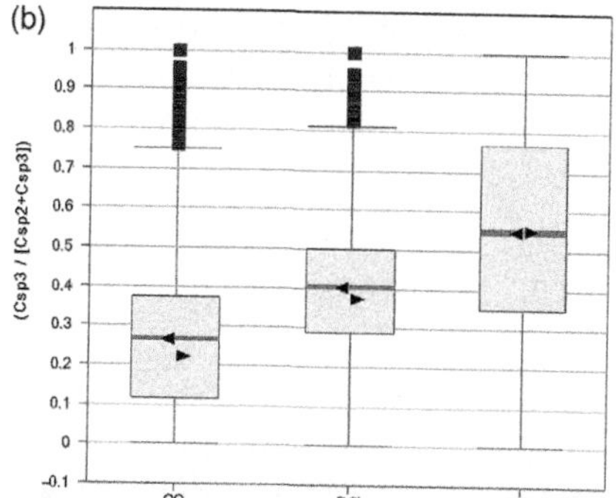

	CC	DC'	NP
Max	1.000	1.000	1.000
Q3	0.375	0.500	0.769
Median	0.222	0.375	0.552
Q1	0.118	0.286	0.350
Min	0.000	0.000	0.000
Mean	0.266	0.404	0.549
StdDev	0.203	0.166	0.261

FIGURE 10.4 Box plots of the molecular complexity measures of 6152 commercial compounds (CCs), 5963 diverse compounds from the academic synthetic chemistry community (DC′), and 2477 natural products (NP). The boxes enclose data points with values within the first (Q1) and third (Q3) quartiles (i.e., enclose 50% of the data); the arrowheads pointing to the right and left denote the median and mean distributions, respectively; the gray band inside the box indicates the 95% confidence interval of the mean value; the lines above and below indicate the upper and lower nonoutlier values [calculated as Q3 + 1.5 × (Q1 − Q3) and Q1 − 1.5 × (Q1 − Q3), respectively]. The squares indicate outliers. Selected statistics of each distribution are summarized in the table. (a) Measure of stereochemical complexity calculated as the portion of carbon atoms that are stereogenic. (b) Measure of shape complexity calculated as the ratio of sp^3-hydridized carbon atoms to total sp^3- and sp^2-hybridized carbons. The data sets were obtained from the work reported by Clemons et al. [85].

415,284 compounds published in the *Journal of Medicinal Chemistry* between 1959 and 2009 showing that since the mid-1970s, molecules published in this journal have consistently had less sp^3 character than that of marketed drugs [61].

The molecular complexity, physicochemical, and structural diversity profile of a large collection of 30 small-molecule combinatorial libraries was also surveyed using a complementary set of chemoinformatic approaches [55]. In the same study, the complexity and profile of physicochemical properties of the natural products database TCM was also analyzed. It was concluded that the high molecular complexity and structural uniqueness of the combinatorial libraries and natural product collections are indicative that these databases are suitable for interrogating novel regions of the chemical space. Results of this and other studies also supported the rationale to develop more complex, natural product–like libraries for improved drug discovery [91].

It should be emphasized that increasing molecular complexity is not the only criterion to consider when screening existing collections or designing new libraries; other properties have to be balanced. For example, a recent analysis reviewed by Leach and Hann [88] suggests that off-target promiscuity increases with molecular weight and, in particular, with lipophilicity. Since the evidence that supports the fact that molecular complexity is associated with improved drug discovery is growing, it has been anticipated that the computational and experimental characterization of DOS, other combinatorial libraries, natural products, and other screening collections will be conducted to explore their potential to expand the medicinally relevant chemical space [55].

10.4.3 Scaffold Analysis

The scaffold or molecular framework is used to describe the core structure of a molecule, and it is one of the concepts most widely used in medicinal chemistry and drug discovery [92]. Applications of the scaffold content in small data sets or large compound collections [4] include studying the SAR/SPR of a set of molecules with measured biological activity or other property; analyzing the structural diversity of compound collections, and evaluating the performance of virtual screening methodologies to retrieve novel scaffolds [93].

Several scaffold analyses of compound databases have been performed for different purposes [94–105]. For example, Bemis and Murcko analyzed a collection of drugs to identify common structural features and found that a rather small number of scaffolds accounts for a large fraction of molecules in the database [94]. This observation is in line with a more recent work of Langdon et al. that analyzed the scaffold diversity of approved drugs, compounds from commercial vendors, focused libraries targeted to kinases, compounds from the ChEMBL database [106], and molecules from in-house collections. It was found that most of the compounds in the databases analyzed are contained in a small number of well-represented scaffolds and that there are a large number of scaffolds with only one compound [107].

Scaffold content analysis of more than 24 million organic compounds in the CAS Registry showed that a small percentage of frameworks occur in a large percentage of compounds (e.g., the most common 5.0% of the hetero frameworks were found in 75.5% of the compounds) [108]. This observation supports the hypothesis that the currently known medicinally relevant chemical space is biased to the rather small fraction of compounds that have been made (see above) [35].

With the goal of identifying properties that are associated with biological activity, the properties and composition of heteroaromatic ring systems that are present in bioactive molecules has been compared to the properties of a large collection of virtual ring systems [109]. In that study, Ertl et al. concluded that the number of unique scaffolds is very small (ca. 0.5% in the databases analyzed) and that the bioactive aromatics scaffolds form bioactivity islands [109]. Authors suggested that the most important properties responsible for the separation of active and inactive areas of chemical space seem to be size of the scaffolds, their heteroatom composition, and their stability [109]. In an independent study, Pitt et al. showed that there is a large portion of synthetically accessible aromatic rings that can potentially be made that has not yet been explored [110].

Scaffold content analysis of compound libraries annotated with biological activity has given rise to the identification scaffolds frequently found in bioactive compounds. For example, the authors reported a chemotype-based hierarchical classification of the AIDS database at the National Cancer Institute in order to identify systematically the scaffolds associated with anti-AIDS activity [111]. Such selected scaffolds can serve as a basis to help chemists to identify optimal scaffold replacements. Ertl recently described a system, which was implemented at a pharmaceutical company, to identify bioisosteric scaffolds in a database with more than 7000 scaffolds extracted from bioactive compounds [112]. In this work, Ertl present an excellent discussion of the computational approaches that have been developed over the years in the field of automatic scaffold replacement [112].

According to a recent analysis of the public databases annotated with biological activity (BindingDB [113], ChEMBL [106], and PubChem [77]), the number of different scaffolds calculated for active molecules against human targets is more than 76,000 [114]. Several other scaffold analyses of different data sets are described in recent excellent works of scaffold analysis [107,114,115].

Scaffold content analyses of compound collections usually report the most frequent scaffolds found in the databases. To illustrate this point, Figure 10.5 shows the five most frequent scaffolds found in four types of compound collections: approved drugs,

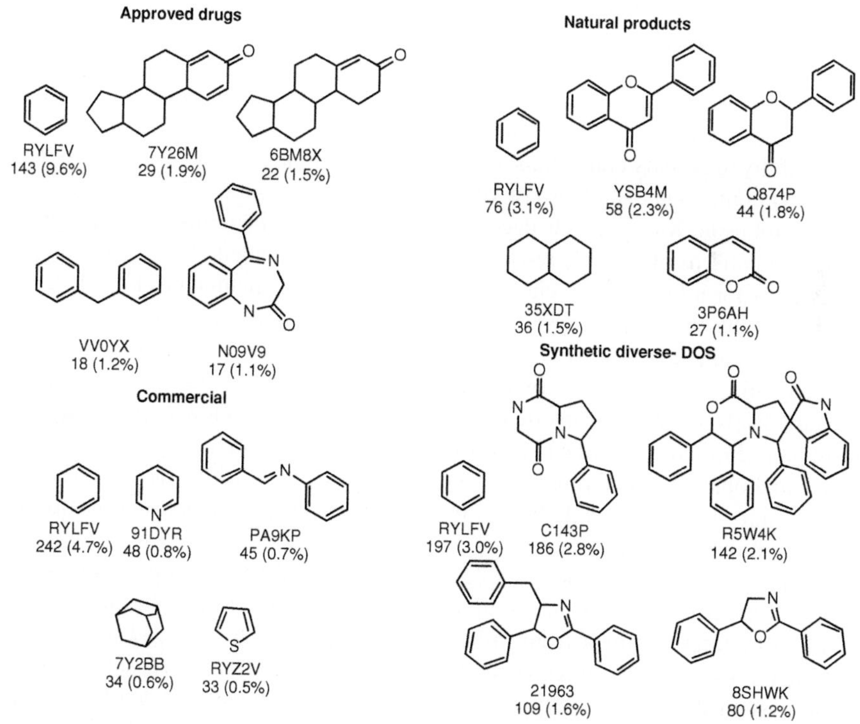

FIGURE 10.5 Most frequent scaffolds (cyclic systems or molecular frameworks) found in 1490 approved drugs, a collection of 2472 natural products, 6623 synthetic diverse compounds obtained from academic groups using techniques such as DOS, and 6152 compounds from a commercial vendor library. The three last data sets were obtained from the work reported by Clemons et al. [85]. The frequency and percentage are displayed.

natural products from a commercial vendor, diverse compounds from the academic synthetic chemistry community, and general screening compounds from a commercial source. Data sets were obtained from the work of Clemons et al. [85]. In all data sets, the most frequent scaffold is benzene, as reported previously for other collections of drugs and several other data sets [72,94,105,116]. The flavone and coumarine scaffolds that are frequent in this natural product collection (Figure 10.5) are also found frequently in the databases of natural products implemented in ZINC [72,105]. The structural complexity of some of the frequent scaffolds in the synthetic collection obtained from academic groups is remarkable.

Similar to the dependence of the chemical space with structure representation using descriptors, scaffold analysis depends on the definition of scaffold. It has been pointed out that a preferred scaffold representation is objective, invariant, and not dependent on the data set. There are various ways to derive the scaffold of a molecule computationally in a systematic and consistent manner that have been reviewed elsewhere [92,99]. One group of commonly used representations are the *atomic frameworks* of Bemis and Murcko, defined as the union of ring systems and linkers in a molecule [94]. This definition is similar to the *topological scaffolds* of Xu, defined as ring bonds and linker bonds but not chain bonds [96,117], or the *cyclic systems* of Xu and Johnson [118,119], which result by iteratively removing the side chains of the molecule (Figure 10.6). The cyclic systems are part of the chemotype methodology developed by Johnson and Xu and are calculated using the program Molecular Equivalent Indices (MEQI). The latter approach has been used successfully to classify collections of combinatorial libraries, drugs, natural products, and other compound databases [72,105,116,120].

The Johnson–Xu approach decomposes compounds in terms of characteristic structural patterns of variable resolution and complexity called *chemotypes* and provides tools for a hierarchical classification based on the chemotypes [18,119]. Figure 10.6 shows that a cyclic system skeleton is obtained if all atoms are set to a single atom type (e.g., carbon) and all bonds are set to single bonds in the cyclic system. Deleting all atoms attached to two other nonhydrogen atoms from the cyclic system skeleton leads to a reduced cyclic system skeleton, which is a low level of structural resolution. In Figure 10.6 the curved lines designate multibond links connecting the ellipsoidal objects, which represent general ring structures, while the corresponding straight lines represent single bonds. Note the hierarchical relationship at different levels of structural resolution.

Figure 10.7 shows a simple example of the hierarchical relationship among chemotypes at four different levels of resolution [111]. This figure depicts the classification of five structures obtained from in-house combinatorial libraries and deposited in PubChem. The five structures at the bottom of the figure share the same reduced cyclic system skeleton chemotype, which corresponds to the root node of the hierarchical tree. Moving from the root node to the leaf nodes corresponds to increasing structural resolution. At the level of the hierarchy lying just below the root node, the structures are classified into two cyclic system skeleton chemotypes. Three chemotypes are shown at the next lower level, which corresponds to cyclic systems. The lowest level of the hierarchy, which contains the leaf nodes, corresponds to the highest

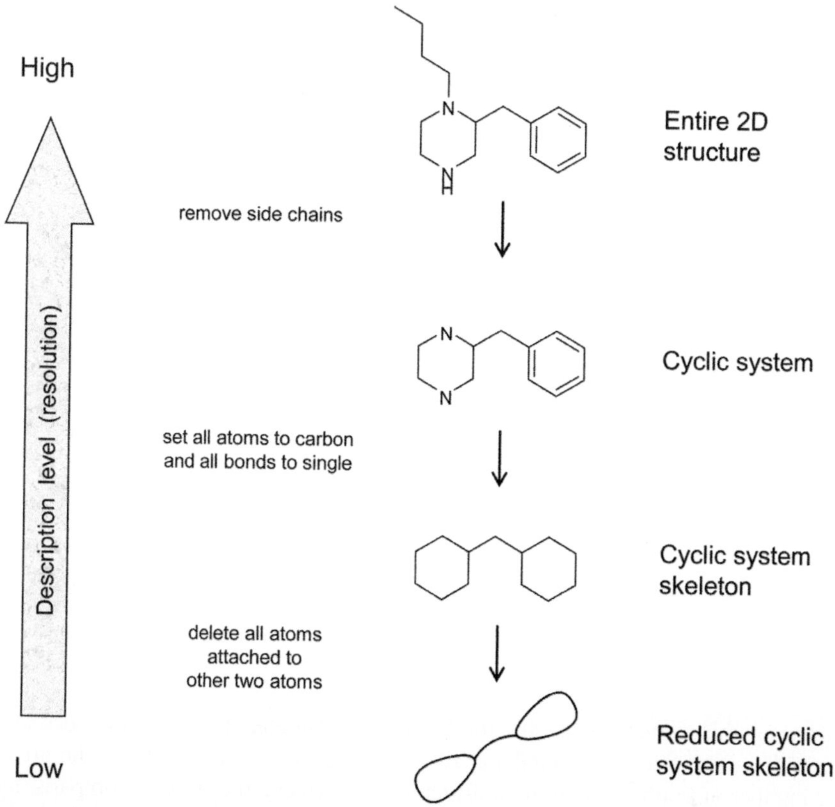

FIGURE 10.6 Example of scaffold definition (cyclic system) using MEQI. The cyclic system skeleton and reduced cyclic system skeleton of one chemical structure exemplifies different levels of structural resolution. The highest level corresponds to the entire two-dimensional structure. Removal of all side chains leads to the cyclic system level. Setting all nonhydrogen atoms to carbon and all bonds to single generates cyclic system skeletons. Finally, deleting all nonhydrogen atoms attached to two nonhydrogen atoms yields reduced the cyclic system skeletons, which is the lowest level of structural resolution. The curved lines designate multibond links connecting the ellipsoidal objects, which represent general ring structures, while the corresponding straight lines represent single bonds.

resolution and consists of entire molecules, including side chains. From this simple example it can be seen that as the level of structural resolution increases, moving down the tree, the number of chemotypes increases, and correspondingly, the number of structures belonging to each chemotype class decreases. Of note, each chemotype at a given level of the hierarchy represents an equivalence class. Equivalence classes partition the set of molecules so that molecules of a given chemotype do not lie in any other chemotype class at the same level of structural resolution [111].

Measuring and comparing the scaffold diversity of compound collections is not a trivial task. First, it depends on the specific approach to generating the scaffolds,

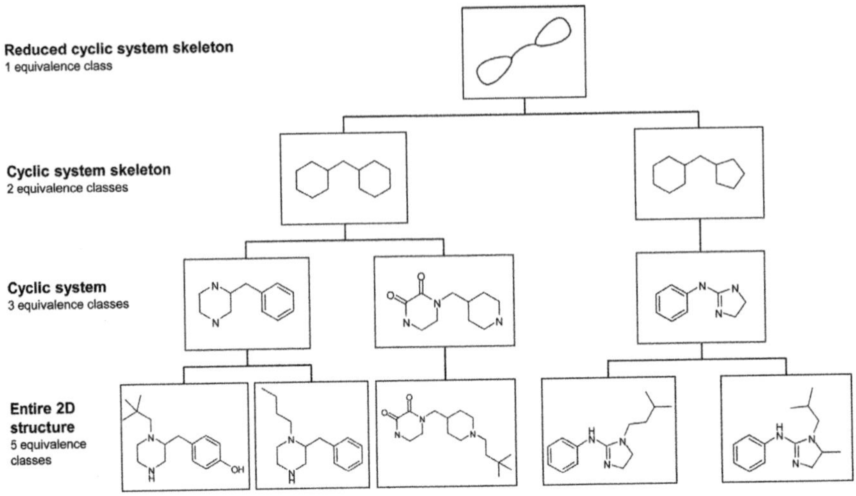

FIGURE 10.7 Depiction of the chemotype-based hierarchical classification tree for the five molecules located at the bottom of the figure. The root of the hierarchical tree corresponds to the lowest-resolution structure. Chemotypes at a given level of the hierarchy are equivalence classes. Note that exocyclic bonds of the diketopiperazine are considered part of the cyclic system, but they are not part of the cyclic system skeleton and the reduced cyclic system skeleton.

including the resolution or structural details considered (see above). Second, it depends on the size of the database. In general, the larger the database, the greater the number of scaffolds, making it difficult to construct a meaningful comparison of libraries of different size [67,100]. The scaffold diversity will depend not only on the number of scaffolds but also on the distribution of the molecules on those scaffolds (e.g., the scaffold representation) [107]. Frequently, scaffold diversity is measured based on frequency counts and fractions: for example, the fraction of cyclic systems relative to the data set size, and the fraction of singletons (cyclic systems containing only one compound) relative to the data set size and relative to the number of cyclic systems. Although these measures provide an idea of the scaffold diversity, they do not contain detailed information regarding the particular distribution of the scaffolds. To address this issue, scaffold recovery curves, also called cumulative scaffold frequency plots [107,108,116], have been used. These curves can be further characterized by obtaining the fraction of cyclic systems required to retrieve, for example, 50% of the corresponding database and the area under the curve [116].

In addition to the number of scaffolds and scaffold retrieval curves, the specific distribution of compounds in the n most populated cyclic systems can be quantified with an entropy-based information metric [116]. The *Shannon entropy* (SE) [121,122] of a population of P compounds contained in n cyclic systems is defined as

$$\mathrm{SE} = -\sum_{i=1}^{n} p_i \log_2 p_i \qquad p_i = \frac{c_i}{P} \tag{10.1}$$

where p_i is the relative frequency of the cyclic system i in a population of P compounds containing a total of n distinct cyclic systems; c_i corresponds to the absolute number of molecules containing a particular cyclic system i. The values of SE range between 0 and $\log_2 n$ and hence depend on n, but not explicitly on P. If SE = 0, all P compounds possess only a single cyclic system. If SE = $\log_2 n$, the P compounds are distributed uniformly among the n cyclic systems (i.e., maximum cyclic system diversity on the data set). To normalize the SE values for different values of n, the *scaled SE* (SSE) is defined as

$$\mathrm{SSE} = \frac{\mathrm{SE}}{\log_2 n} \tag{10.2}$$

The values of SSE range between 0, where all P compounds are contained in one cyclic system, and 1.0, where each cyclic system contains an equal number of compounds. Thus, SSE values closer to 1.0 indicate large scaffold diversity within the n most populated cyclic systems. This measure is discussed extensively elsewhere [116] and has been used recently to characterize the scaffold diversity of natural products in the TCM database implemented in ZINC, natural products from a commercial vendor, approved drugs, and other compound collections [72]. As an example, Figure 10.8 shows the scaffold diversity of four compound collections using the measure of Shannon entropy. The compound databases, analyzed above (see Figure 10.5), includes approved drugs, natural products from a commercial vendor, diverse compounds from academic groups synthesized by methods such as DOS, and commercial compounds. The figure shows the distribution of compounds in the 10 most populated scaffolds (cyclic systems), which were computed with MEQI. Each panel also shows the corresponding value of scaled Shannon entropy. As discussed above, SSE values closer to 1.0 indicate that the molecules are more equally distributed in the cyclic systems; hence, it is an indicator of large diversity. In turn, smaller SSE values indicate that most of the molecules are distributed in a few cyclic systems, denoting lower diversity. From the SSE values for the top 10 most populated cyclic systems in Figure 10.8, it can be concluded that natural products and the synthetic compounds from academic groups are the most diverse, with the largest SSE values (0.94 and 0.95, respectively). In contrast, the commercial database showed the lowest scaffold diversity (SSE = 0.74) according to this measure.

10.4.4 Structure Fingerprints and Multiple Representations

The chemical space of compound databases is usually compared using a single representation. However, as discussed above, there are no unique chemical spaces since they depend on the molecular descriptors. To address the dependence of chemical space with representation, conclusions obtained from multiple methods (e.g., multiple fingerprint representations) have been proposed. The combination or aggregation of methods is a common practice in consensus scoring [123] for modeling receptor–ligand interactions, data fusion for similarity searching [124], consensus activity landscape modeling in SAR analysis [125,126], and clustering [127]. The dependence of

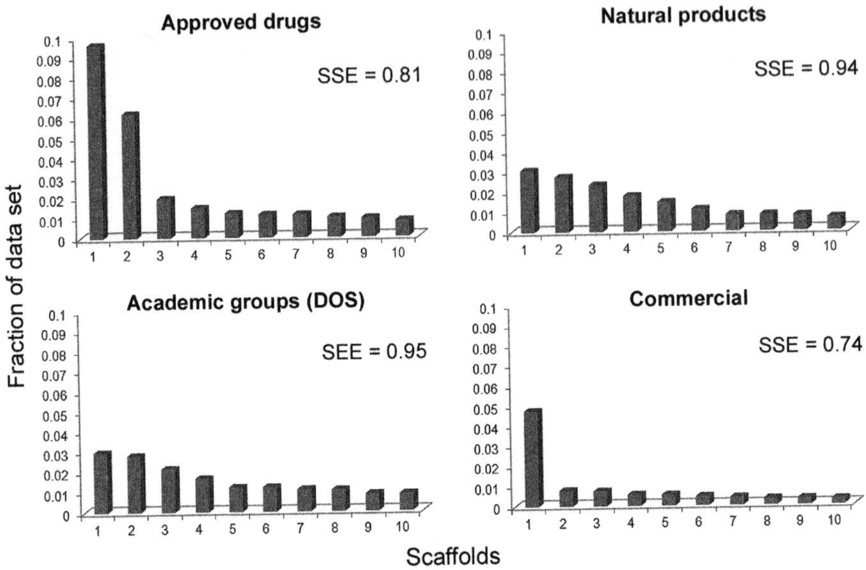

FIGURE 10.8 Distribution of compounds in the top 10 most populated scaffolds found in 1490 approved drugs, a collection of 2472 natural products, 6623 synthetic diverse compounds obtained from academic groups using techniques such as DOS, and 6152 compounds from a commercial vendor library. The last three data sets were obtained from the work reported by Clemons et al. [85] (see also Figure 10.5). Scaled Shannon entropy (SSE) values close to 1.0 indicate that the compounds are more evenly distributed in the scaffolds, suggesting large scaffold diversity. Smaller SSE values denote that most of the compounds are distributed in fewer scaffolds, indicating lower diversity.

chemical space with structure representation appears visually as in Figure 10.9, which shows a comparison of the chemical space of nine compound libraries with 250 compounds each. The compound databases are five in-house combinatorial libraries (red, yellow, gray, black, and cyan), natural products from a commercial provider (green), approved drugs (blue), synthetic commercial compounds (magenta), and compounds targeted to adenosine receptors (orange). Each visual representation of the chemical space was obtained by PCA of the similarity matrix computed using the Tanimoto similarity function and two different fingerprint representations: MACCS keys [128] and typed graph distance (TGD) [129] fingerprints. MACCS keys are a predefined set of structural keys; and TGD fingerprints are graph distances employing the following atom typing: donor, acceptor polar, anion, cation, and hydrophobe. The first three PCs account for 85.9% (MACCS) and 90.3% (TGD) of the variance, respectively. Figure 10.9 clearly illustrates the significant changes in the neighborhood relationships of the databases when different fingerprint representations are employed. Of course, this dependence may have profound effects not only on the visual representation of chemical spaces, as shown in Figure 10.9, but also on similarity searching and mining of the chemical space. Other examples of the visualization of chemical space with fingerprint representations are presented elsewhere [126].

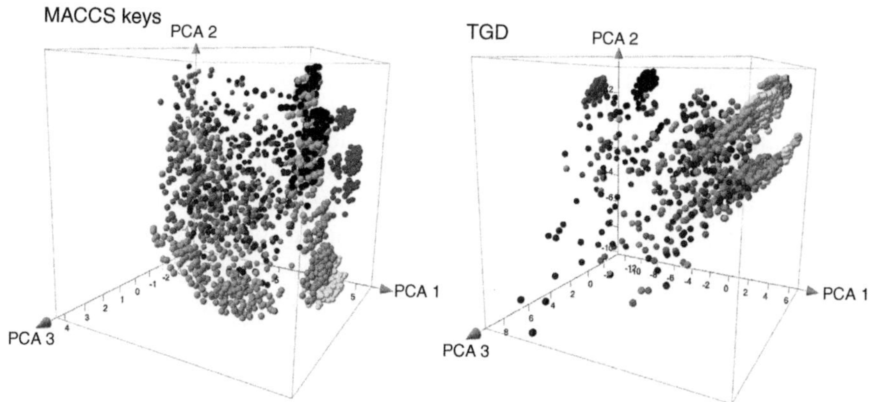

FIGURE 10.9 Dependence of chemical space with structure representation. Visual representation of the chemical space of a total of 2250 compounds selected from nine compound collections (250 compounds per data set). The data sets are natural products (green) and a general screening collection (magenta) from two commercial providers, five in-house combinatorial libraries (red, yellow, gray, black, and cyan), approved drugs (blue), and a set of compounds targeted to adenosine receptors (orange). The depiction of the chemical space was obtained by PCA of the similarity matrix computed by Tanimoto similarity and two different structure representations. The first three PCs account for 85.9% (MACCS) and 90.3% (TGD) of the variance, respectively. (*See insert for color representation of the figure.*)

To perform comprehensive comparisons of compound data sets, the use not only of different fingerprint representations but also of different criteria, such as physicochemical properties, structure fingerprints, and molecular scaffolds, has been proposed [105]. The basis of this approach is that each method has advantages and disadvantages. For example, the use of molecule properties has the advantage of being intuitive and straightforward to interpret. As discussed above, the drug-like [58] and lead-like criteria [59], and Congreve's rule of 3 [60] have been formulated using physicochemical properties. However, physicochemical properties do not provide information regarding the structural patterns, and molecules with different chemical structures can have the same or similar physicochemical properties. Also, chemotypes or scaffolds are straightforward to interpret by computational and medicinal chemists as well as biologists. For example, scaffold analysis has led to concepts such as scaffold hopping [130] and privileged structures [131,132]. However, one of the disadvantages of the scaffold analysis is the lack of information regarding structural similarity due to the side chains and the inherent similarity or dissimilarity of the scaffolds themselves. An obvious solution is the analysis not only of the molecular frameworks but also of the side chains, functional groups, and other substructural analysis strategies [133]. To consider the entire molecular structure in the comparisons, molecular fingerprints are widely used and have been applied to a number of chemoinformatic and computer-aided drug design applications [32,124,134]. Following the three criteria of physicochemical properties, structure fingerprints, and molecular scaffolds, a collection of natural products implemented in the ZINC database and four combinatorial libraries were compared [105]. Physicochemical properties included the six medicinally relevant properties discussed above and molecular fingerprints of different design [125,135]: namely, MACCS keys, TGD, and graph-based three-point pharmacophores (GpiDAPH3) [129]. The latter fingerprints are graph-based three-point pharmacophores employing any set of three possible atom types: pi system, donor, or acceptor. The use of multiple criteria allowed the authors to obtain a more comprehensive analysis of the density, coverage of chemical space scaffold, content, diversity, and structural similarity of combinatorial libraries and natural products compared to other collections, particularly known drugs. The use of different representations covering physicochemical properties, structural fingerprints, and molecular scaffolds has been used to compare other compound databases, such as natural products from TCM and DOS libraries [48,72].

10.5 RECENT TRENDS IN COMPUTATIONAL APPROACHES TO CHARACTERIZE COMPOUND LIBRARIES

As discussed in the preceding sections, the concept of the chemical space is widely employed in drug discovery and other research disciplines. However, there is not a unique definition of this concept. Similarly, the chemoinformatic characterization of compound libraries is a common practice in several research groups. Nevertheless, similar to the concept of chemical space, there is no unique, "best" protocol to use to conduct the analysis. Chemoinformatic characterization and mining of the chemical space is not a solved problem. The high dependence of chemical space with structure representation and the potential to have very high dimensional chemical spaces represent major challenges to mine, visualize, and extract useful information. Therefore, the research community is constantly developing and applying novel ways to address the major issues of characterization of compound libraries.

At the time of writing (April 2012), one of the latest advances in chemoinformatic approaches to mining and visualizing the chemical spaces of compound databases is the generation of delimited reference chemical spaces (DRCSs) [39,136]. In this work, Le Guilloux et al. analyzed 6.6 million unique screening compounds collected from 73 chemical providers and generated a representative contour that represents the overall shape of the chemical space of a very large proportion of the molecules analyzed in this work. Therefore, the DRCS method delimits the densest subspace spanned by a reference library in a reduced two-dimensional continuous space and has been used as a basis to develop subspace-specific diversity indices [136].

A recent development in the depiction of chemical space in low dimensions is represented by the latent trait model. This method was employed to reduce the dimensionality of a fingerprint array from 166-dimensional space to two-dimensional space, where it could be plotted and visualized. The chemical space of five combinatorial libraries and other reference collections was visualized using this method [70].

To generate visual representations of the chemical space using simple, yet meaningful descriptors, the research group of Reymond proposed a set of 42 integer value descriptors of molecular structure called molecular quantum numbers (MQNs)

[137]. These descriptors include atom, bond, polarity, and topology counts. The MQN space of PubChem, the entire chemical universe database GDB-13 (with 977 million molecules up to 13 atoms of C, N, O, S and Cl), and DrugBank have been developed and are freely available [138,139].

In an independent effort to represent the biologically relevant chemical space as a general reference, Rabal and Oyarzabal recently developed a novel descriptor called a ligand–receptor interaction fingerprint (LiRIf). This descriptor converts structural information into a one-dimensional string accounting for the plausible ligand–receptor interactions as well as for topological information, and it is useful to cluster, profile, and compare chemical libraries [140]. This work represents a significant advance toward the development of generally applicable representations of chemical space for which no reference space is required [140].

As discussed above, there is an increasing amount of biological information of compound data sets that have been screened across multiple biological endpoints. To analyze the SAR of complex chemogenomics data, the concept of chemotography (chemotype chromatography), has been proposed as an approach that combines visualization of chemical space with analysis of rich biological data [141]. The approach implemented by Lounkine et al. allows an easy comparison of different representations of chemical space [141].

Recent advances in scaffold analysis include the development of tree maps, which provide a novel way to visualize the distribution of molecules over scaffolds, and the molecular similarity of the scaffolds within a compound data set [107]. In that work, Langdon et al. used tree maps to visualize the scaffold space of drugs, in-house screening collections, focused libraries, compounds from ChEMBL, and other collections. Tree maps provide an easy way to display highly populated scaffolds and cluster similar scaffolds [107].

An activity landscape has been conceptualized as a chemical space with the addition of biological activity as another dimension [142]. This concept, which is at the interface between chemical space and biological space, has emerged as an attractive approach to analyze the SAR of compound collections systematically and detect *activity cliffs* (chemical compounds with highly similar structures but significantly different biological activities) [40] and *scaffold hops* (chemical compounds with highly dissimilar structures but significantly similar biological activities) [92]. Although a comprehensive review of activity landscapes is beyond the scope of this chapter, the interested reader is referred to Chapter 16 and to extensive reviews of progress in this field [126,142–144].

10.6 CONCLUDING REMARKS

Chemoinformatic characterization of the chemical space and molecular diversity of compound libraries is a central topic in drug discovery and other research areas. Unlike the cosmic universe, chemical space is not unique and it is highly influenced by the molecular representation. The criterion to describe the chemical space is a critical factor in deriving useful conclusions, and it is frequently based on the specific goals of the study and the resources available. Molecular properties, structure fingerprints, and molecular scaffolds are major types of structure representation that are used for quantitative study of the structure diversity and chemical space of compound databases such as LOLs and other collections obtained from DOS, natural products, compounds annotated with biological activity, marketed drugs, commercial vendor libraries, and virtual collections. These comparisons have provided basic information used to guide the selection, design, and synthesis of compound libraries for improved drug discovery. For example, frequent scaffolds identified in bioactive compounds and scaffolds present in natural products represent starting points to inspire the diversity-oriented synthesis of biologically relevant and natural product–like libraries that contain compounds with increased probability of having biological activity. Chemoinformatic analyses of various types of compound databases have shown that the medicinally relevant chemical space can be biased toward the currently known chemical organic universe (i.e., the organic compounds synthesized thus far). This observation is supported by several studies of scaffold content analyses of currently available compound collections, which indicate that just small portions of synthetically available scaffolds have actually been made. As such, the pharmaceutically relevant space may be greater than it is known currently, due to the still limited knowledge of pharmaceutically relevant targets and the complex relationships revealed by polypharmacology studies. To address this problem, increasing the molecular complexity of screening collections while balancing the physicochemical properties has been proposed as a feasible strategy for expanding the medicinally relevant chemical space. In this regard, DOS approaches such as LOLs represent promising strategies for exploring systematically neglected chemical and scaffold spaces.

Despite the fact that computational analysis of the chemical space and structure diversity of compound collections is conducted by industrial, academic, nonprofit, and other research groups, it is not a simple task. Characterizing chemical spaces poses challenges that the research community is pursuing actively. Innovative computational approaches are being developed in response to novel findings and emerging needs of the drug discovery community. For example, the large impact of ADME attributes in the clinical success of bioactive compounds has increased the awareness to employ ADME-related properties to visualize and mine ADME spaces. Although it has been conceptually attractive to dissect the biologically relevant chemical space in subspaces containing islands of compounds directed to specific therapeutic indications, the concept of polypharmacology challenges this idea. The constantly increasing chemogenomics data have stimulated the development of computational approaches to mine the multidimensional biological space of compound data sets and has led to novel concepts such as chemotography. The well-known dependence of chemical space with structure representation has boosted the development of approaches to make a consistent comparison of the chemical space across research groups. In this chapter we have illustrated recent efforts toward a consistent framework to represent the chemical space as demonstrated by the development of delimited reference chemical spaces, molecular quantum numbers, ChemGPS, and ligand-receptor interaction fingerprints.

Acknowledgments

The support of Fabian López-Vallejo in generating and interpreting the data in Figure 10.1, and the assistance of Jacob Waddell in developing a script to compute the data in Figure 10.8 and prepare the manuscript are greatly acknowledged. The assistance of Karen Gottwald in proofreading the manuscript is also much appreciated. The authors thank Mark Johnson for providing MEQI and the members of the research group and collaborators whose names are listed in various references for their substantial contributions to the research work reviewed in this chapter. This work was funded by the Multiple Sclerosis National Research Institute and by the State of Florida, Executive Office of the Governor's Office of Tourism, Trade, and Economic Development.

REFERENCES

1. R. A. Houghten, C. Pinilla, M. A. Giulianotti, J. R. Appel, C. T. Dooley, A. Nefzi, J. M. Ostresh, Y. P. Yu, G. M. Maggiora, J. L. Medina-Franco, D. Brunner, J. Schneider, *J. Comb. Chem.* **2008**, *10*, 3–19.
2. J. P. Kennedy, L. Williams, T. M. Bridges, R. N. Daniels, D. Weaver, C. W. Lindsley, *J. Comb. Chem.* **2008**, *10*, 345–354.
3. A. L. Harvey, *Drug Discov. Today* **2008**, *13*, 894–901.
4. T. Scior, P. Bernard, J. L. Medina-Franco, G. M. Maggiora, *Mini Rev. Med. Chem.* **2007**, *7*, 851–860.
5. A. L. Hopkins, *Nat. Chem. Biol.* **2008**, *4*, 682–690.
6. A. A. Shelat, R. K. Guy, *Curr. Opin. Chem. Biol.* **2007**, *11*, 244–251.
7. S. L. Schreiber, *Science* **2000**, *287*, 1964–1969.
8. R. Gozalbes, *Comb. Chem. High Throughput Screen.* **2011**, *14*, 428–428.
9. A. L. Harvey, R. L. Clark, S. P. Mackay, B. F. Johnston, *Expert Opin. Drug Discov.* **2010**, *5*, 559–568.
10. A. Ganesan, *Curr. Opin. Chem. Biol.* **2008**, *12*, 306–317.
11. A. T. Hauser, M. Jung, *Planta Med.* **2008**, *74*, 1593–1601.
12. A. Nören-Müller, I. Reis-Correa, H. Prinz, C. Rosenbaum, K. Saxena, H. J. Schwalbe, D. Vestweber, G. Cagna, S. Schunk, O. Schwarz, H. Schiewe, H. Waldmann, *Proc. Natl. Acad. Sci. U.S.A.* **2006**, *103*, 10606–10611.
13. C. Cordier, D. Morton, S. Murrison, A. Nelson, C. O'Leary-Steele, *Nat. Prod. Rep.* **2008**, *25*, 719–737.
14. J. Owens, C. Lipinski, *Drug Discov. Today* **2003**, *8*, 12–16.
15. CRC Dictionary of Natural Products, available at http://www.crcpress.com/; accessed Apr. 2012.
16. C. Y.-C. Chen, *PLoS ONE* **2011**, *6*, e15939.
17. Specs, available at http://www.specs.net/; accessed Apr. 2012.
18. R. L. Clark, B. F. Johnston, S. P. Mackay, C. J. Breslin, M. N. Robertson, A. L. Harvey, *Drug Discov. Today* **2010**, *15*, 679–683.

19. R. A. Houghten, C. Pinilla, J. R. Appel, S. E. Blondelle, C. T. Dooley, J. Eichler, A. Nefzi, J. M. Ostresh, *J. Med. Chem.* **1999**, *42*, 3743–3778.
20. R. E. Dolle, B. L. Bourdonnec, K. Worm, G. A. Morales, C. J. Thomas, W. Zhang, *J. Comb. Chem.* **2010**, *12*, 765–806.
21. R. Gozalbes, A. Pineda-Lucena, *Comb. Chem. High Throughput Screen.* **2011**, *14*, 548–558.
22. S. H. Fitzgerald, M. Sabat, H. M. Geysen, *J. Chem. Inf. Model.* **2006**, *46*, 1588–1597.
23. C. Pinilla, J. R. Appel, E. Borras, R. A. Houghten, *Nat. Med.* **2003**, *9*, 118–122.
24. R. Macarron, *Drug Discov. Today* **2006**, *11*, 277–279.
25. A. E. Carpenter, *Nat. Chem. Biol.* **2007**, *3*, 461–465.
26. S. H. Fitzgerald, M. Sabat, H. M. Geysen, *J. Comb. Chem.* **2007**, *9*, 724–734.
27. L. B. Akella, D. DeCaprio, *Curr. Opin. Chem. Biol.* **2010**, *14*, 325–330.
28. K. Martínez-Mayorga, T. L. Peppard, A. B. Yongye, R. Santos, M. Giulianotti, J. L. Medina-Franco, *J. Chemom.* **2011**, *25*, 550–560.
29. C. Lipinski, A. Hopkins, *Nature* **2004**, *432*, 855–861.
30. C. M. Dobson, *Nature* **2004**, *432*, 824–828.
31. J. L. Medina-Franco, K. Martínez-Mayorga, M. A. Giulianotti, R. A. Houghten, C. Pinilla, *Curr. Comput. Aided Drug Des.* **2008**, *4*, 322–333.
32. M. A. Johnson, G. M. Maggiora, *Concepts and Applications of Molecular Similarity*, Wiley, New York, **1990**.
33. Y. C. Martin, J. L. Kofron, L. M. Traphagen, *J. Med. Chem.* **2002**, *45*, 4350–4358.
34. M. vonKorff, K. Hilpert, *J. Chem. Inf. Model.* **2006**, *46*, 1580–1587.
35. M. Dow, M. Fisher, T. James, F. Marchetti, A. Nelson, *Org. Biomol. Chem.* **2012**, *10*, 17–28.
36. T. Engel, *J. Chem. Inf. Model.* **2006**, *46*, 2267–2277.
37. A. Varnek, Baskin II, *Mol. Inf.* **2011**, *30*, 20–32.
38. P. Willett, *WIREs Comput. Mol. Sci.* **2011**, *1*, 46–56.
39. V. Le Guilloux, L. Colliandre, S. Bourg, G. Guenegou, J. Dubois-Chevalier, L. Morin-Allory, *J. Chem. Inf. Model.* **2011**, *51*, 1762–1774.
40. G. M. Maggiora, *J. Chem. Inf. Model.* **2006**, *46*, 1535–1535.
41. J. L. Medina-Franco, G. M. Maggiora, M. A. Giulianotti, C. Pinilla, R. A. Houghten, *Chem. Biol. Drug Des.* **2007**, *70*, 393–412.
42. D. M. Maniyar, I. T. Nabney, B. S. Williams, A. Sewing, *J. Chem. Inf. Model.* **2006**, *46*, 1806–1818.
43. J. L. Medina-Franco, K. Martínez-Mayorga, A. Bender, R. M. Marín, M. A. Giulianotti, C. Pinilla, R. A. Houghten, *J. Chem. Inf. Model.* **2009**, *49*, 477–491.
44. J. Pérez-Villanueva, R. Santos, A. Hernández-Campos, M. A. Giulianotti, R. Castillo, J. L. Medina-Franco, *Med. Chem. Commun.* **2011**, *2*, 44–49.
45. J. L. Medina-Franco, A. B. Yongye, J. Pérez-Villanueva, R. A. Houghten, K. Martínez-Mayorga, *J. Chem. Inf. Model.* **2011**, *51*, 2427–2439.
46. W. H. B. Sauer, M. K. Schwarz, *J. Chem. Inf. Comput. Sci.* **2003**, *43*, 987–1003.
47. P. A. Clemons, J. A. Wilson, V. Dancik, S. Muller, H. A. Carrinski, B. K. Wagner, A. N. Koehler, S. L. Schreiber, *Proc. Natl. Acad. Sci. U.S.A.* **2011**, *108*, 6817–6822.

48. F. López-Vallejo, A. Nefzi, A. Bender, J. R. Owen, I. T. Nabney, R. A. Houghten, J. L. Medina-Franco, *Chem. Biol. Drug Des.* **2011**, *77*, 328–342.
49. D. K. Agrafiotis, M. Shemanarev, P. J. Connolly, M. Farnum, V. S. Lobanov, *J. Med. Chem.* **2007**, *50*, 5926–5937.
50. M. Wawer, E. Lounkine, A. M. Wassermann, J. Bajorath, *Drug Discov. Today* **2010**, *15*, 630–639.
51. T. J. Ritchie, P. Ertl, R. Lewis, *Drug Discov. Today* **2011**, *16*, 65–72.
52. D. Digles, G. F. Ecker, *Mol. Inf.* **2011**, *30*, 838–846.
53. I. T. Jolliffe, *Principal Component Analysis,* 2nd ed., Springer-Verlag, New York, **2002**.
54. J. J. Irwin, B. K. Shoichet, *J. Chem. Inf. Model.* **2005**, *45*, 177–182.
55. F. López-Vallejo, M. A. Giulianotti, R. A. Houghten, J. L. Medina-Franco, *Drug Discov. Today* **2012**, *17*, 718–726.
56. G. M. Maggiora, V. Shanmugasundaram, in: *Chemoinformatics and Computational Chemical Biology: Methods in Molecular Biology*, Vol. 672, J. Bajorath, Ed., Springer-Verlag, New York, **2011**, pp. 39–100.
57. R. Todeschini, V. Consonni, *Handbook of Molecular Descriptors*, Wiley-VCH, Weinheim, Germany, **2000**.
58. C. A. Lipinski, F. Lombardo, B. W. Dominy, P. J. Feeney, *Adv. Drug Deliv. Rev.* **1997**, *23*, 3–25.
59. S. J. Teague, A. M. Davis, P. D. Leeson, T. Oprea, *Angew. Chem. Int. Ed.* **1999**, *38*, 3743–3748.
60. M. Congreve, R. Carr, C. Murray, H. Jhoti, *Drug Discov. Today* **2003**, *8*, 876–877.
61. W. P. Walters, J. Green, J. R. Weiss, M. A. Murcko, *J. Med. Chem.* **2011**, *54*, 6405–6416.
62. B. Faller, G. Ottaviani, P. Ertl, G. Berellini, A. Collis, *Drug Discov. Today* **2011**, *16*, 976–984.
63. P. D. Leeson, A. M. Davis, *J. Med. Chem.* **2004**, *47*, 6338–6348.
64. H. Zhao, *Drug Discov. Today* **2011**, *16*, 158–163.
65. T. T. Wager, R. Y. Chandrasekaran, X. Hou, M. D. Troutman, P. R. Verhoest, A. Villalobos, Y. Will, *ACS Chem. Neurosci.* **2010**, *1*, 420–434.
66. M. D. Burke, E. M. Berger, S. L. Schreiber, *Science* **2003**, *302*, 613–618.
67. R. J. Spandl, A. Bender, D. R. Spring, *Org. Biomol. Chem.* **2008**, *6*, 1149–1158.
68. J. M. Ostresh, G. M. Husar, S. E. Blondelle, B. Dorner, P. A. Weber, R. A. Houghten, *Proc. Natl. Acad. Sci. U.S.A.* **1994**, *91*, 11138–11142.
69. A. Nefzi, J. M. Ostresh, J. Yu, R. A. Houghten, *J. Org. Chem.* **2004**, *69*, 3603–3609.
70. J. R. Owen, I. T. Nabney, J. L. Medina-Franco, F. López-Vallejo, *J. Chem. Inf. Model.* **2011**, *51*, 1552–1563.
71. J. L. Medina-Franco, T. Caulfield, *Drug Discov. Today* **2011**, *16*, 418–425.
72. J. Yoo, J. L. Medina-Franco, *Comput. Mol. Biosci.* **2011**, *1*, 7–16.
73. J. Larsson, J. Gottfries, S. Muresan, A. Backlund, *J. Nat. Prod.* **2007**, *70*, 789–794.
74. J. Rosen, A. Lovgren, T. Kogej, S. Muresan, J. Gottfries, A. Backlund, *J. Comput. Aided Mol. Des.* **2009**, *23*, 253–259.
75. J. Larsson, J. Gottfries, L. Bohlin, A. Backlund, *J. Nat. Prod.* **2005**, *68*, 985–991.
76. T. I. Oprea, J. Gottfries, *J. Comb. Chem.* **2001**, *3*, 157–166.

77. Y. L. Wang, J. W. Xiao, T. O. Suzek, J. Zhang, J. Y. Wang, Z. G. Zhou, L. Y. Han, K. Karapetyan, S. Dracheva, B. A. Shoemaker, E. Bolton, A. Gindulyte, S. H. Bryant, *Nucleic Acids Res.* **2012**, *40*, D400–D412.
78. P. Selzer, H.-J. Roth, P. Ertl, A. Schuffenhauer, *Curr. Opin. Chem. Biol.* **2005**, *9*, 310–316.
79. S. Dandapani, L. A. Marcaurelle, *Nat. Chem. Biol.* **2010**, *6*, 861–863.
80. S. H. Bertz, *J. Am. Chem. Soc.* **1981**, *103*, 3599–3601.
81. R. Barone, M. Chanon, *J. Chem. Inf. Comput. Sci.* **2001**, *41*, 269–272.
82. T. K. Allu, T. I. Oprea, *J. Chem. Inf. Model.* **2005**, *45*, 1237–1243.
83. A. Schuffenhauer, N. Brown, P. Selzer, P. Ertl, E. Jacoby, *J. Chem. Inf. Model.* **2006**, *46*, 525–535.
84. F. Lovering, J. Bikker, C. Humblet, *J. Med. Chem.* **2009**, *52*, 6752–6756.
85. P. A. Clemons, N. E. Bodycombe, H. A. Carrinski, J. A. Wilson, A. F. Shamji, B. K. Wagner, A. N. Koehler, S. L. Schreiber, *Proc. Natl. Acad. Sci. U.S.A.* **2010**, *107*, 18787–18792.
86. T. J. Ritchie, S. J. F. Macdonald, *Drug Discov. Today* **2009**, *14*, 1011–1020.
87. M. M. Hann, A. R. Leach, G. Harper, *J. Chem. Inf. Comput. Sci.* **2001**, *41*, 856–864.
88. A. R. Leach, M. M. Hann, *Curr. Opin. Chem. Biol.* **2011**, *15*, 489–496.
89. A. X. Yan, J. Gasteiger, *QSAR Comb. Sci.* **2003**, *22*, 821–829.
90. H. Chen, O. Engkvist, N. Blomberg, J. Li, *Med. Chem. Commun.* **2012**, *3*, 312–321.
91. G. L. Thomas, C. W. Johannes, *Curr. Opin. Chem. Biol.* **2011**, *15*, 516–522.
92. N. Brown, E. Jacoby, *Mini Rev. Med. Chem.* **2006**, *6*, 1217–1229.
93. M. D. Mackey, J. L. Melville, *J. Chem. Inf. Model.* **2009**, *49*, 1154–1162.
94. G. W. Bemis, M. A. Murcko, *J. Med. Chem.* **1996**, *39*, 2887–2893.
95. L. Xue, J. Bajorath, *J. Mol. Model.* **1999**, *5*, 97–102.
96. J. Xu, *J. Med. Chem.* **2002**, *45*, 5311–5320.
97. W. H. B. Sauer, M. K. Schwarz, *Chimia* **2003**, *57*, 276–283.
98. M. A. Koch, A. Schuffenhauer, M. Scheck, S. Wetzel, M. Casaulta, A. Odermatt, P. Ertl, H. Waldmann, *Proc. Natl. Acad. Sci. U.S.A.* **2005**, *102*, 17272–17277.
99. M. Krier, G. Bret, D. Rognan, *J. Chem. Inf. Model.* **2006**, *46*, 512–524.
100. A. Monge, A. Arrault, C. Marot, L. Morin-Allory, *Mol. Divers.* **2006**, *10*, 389–403.
101. A. Schuffenhauer, P. Ertl, S. Roggo, S. Wetzel, M. A. Koch, H. Waldmann, *J. Chem. Inf. Model.* **2007**, *47*, 47–58.
102. K. Grabowski, G. Schneider, *Curr. Chem. Biol.* **2007**, *1*, 115–127.
103. M. J. Wester, S. N. Pollock, E. A. Coutsias, T. K. Allu, S. Muresan, T. I. Oprea, *J. Chem. Inf. Model.* **2008**, *48*, 1311–1324.
104. K. Grabowski, K.-H. Baringhaus, G. Schneider, *Nat. Prod. Rep.* **2008**, *25*, 892–904.
105. N. Singh, R. Guha, M. A. Giulianotti, C. Pinilla, R. A. Houghten, J. L. Medina-Franco, *J. Chem. Inf. Model.* **2009**, *49*, 1010–1024.
106. A. Gaulton, L. J. Bellis, A. P. Bento, J. Chambers, M. Davies, A. Hersey, Y. Light, S. McGlinchey, D. Michalovich, B. Al-Lazikani, J. P. Overington, *Nucleic Acids Res.* **2012**, *40*, D1100–D1107.
107. S. R. Langdon, N. Brown, J. Blagg, *J. Chem. Inf. Model.* **2011**, *51*, 2174–2185.

108. A. H. Lipkus, Q. Yuan, K. A. Lucas, S. A. Funk, W. F. Bartelt, R. J. Schenck, A. J. Trippe, *J. Org. Chem.* **2008**, *73*, 4443–4451.
109. P. Ertl, S. Jelfs, J. Mühlbacher, A. Schuffenhauer, P. Selzer, *J. Med. Chem.* **2006**, *49*, 4568–4573.
110. W. R. Pitt, D. M. Parry, B. G. Perry, C. R. Groom, *J. Med. Chem.* **2009**, *52*, 2952–2963.
111. J. L. Medina-Franco, J. Petit, G. M. Maggiora, *Chem. Biol. Drug Des.* **2006**, *67*, 395–408.
112. P. Ertl, *Bioorg. Med. Chem.* **2012**, *20*, 5436–5442.
113. T. Q. Liu, Y. M. Lin, X. Wen, R. N. Jorissen, M. K. Gilson, *Nucleic Acids Res.* **2007**, *35*, D198–D201.
114. Y. Hu, D. Stumpfe, J. Bajorath, *J. Chem. Inf. Model.* **2011**, *51*, 1742–1753.
115. A. Schuffenhauer, T. Varin, *Mol. Inf.* **2011**, *30*, 646–664.
116. J. L. Medina-Franco, K. Martínez-Mayorga, A. Bender, T. Scior, *QSAR Comb. Sci.* **2009**, *28*, 1551–1560.
117. B. Liu, A. Lu, L. Zhang, H. Liu, Z. Liu, J. Zhou, *Internet Electron. J. Mol. Des.* **2004**, *3*, 143–149; available at http://www.biochempress.com.
118. Y. Xu, M. Johnson, *J. Chem. Inf. Comput. Sci.* **2001**, *41*, 181–185.
119. Y. J. Xu, M. Johnson, *J. Chem. Inf. Comput. Sci.* **2002**, *42*, 912–926.
120. F. López-Vallejo, R. Castillo, L. Yépez-Mulia, J. L. Medina-Franco, *J. Biomol. Screen.* **2011**, *16*, 862–868.
121. C. E. Shannon, W. Weaver, *The Mathematical Theory of Communication*, University of Illinois Press, Urbana, IL, **1963**.
122. J. W. Godden, J. Bajorath, in: *Reviews in Computational Chemistry*, Vol. 23, K. B. Lipkowitz, T. R. Cundari, Eds., Wiley, Hoboken, NJ, **2007**, pp. 263–289.
123. M. Feher, *Drug Discov. Today* **2006**, *11*, 421–428.
124. P. Willett, *Drug Discov. Today* **2006**, *11*, 1046–1053.
125. A. Yongye, K. Byler, R. Santos, K. Martínez-Mayorga, G. M. Maggiora, J. L. Medina-Franco, *J. Chem. Inf. Model.* **2011**, *51*, 1259–1270.
126. J. L. Medina-Franco, A. B. Yongye, F. López-Vallejo, in: *Statistical Modeling of Molecular Descriptors in QSAR/QSPR*, Vol. 2, D. Matthias, V. Kurt, B. Danail, Eds., Wiley-VCH, Weinheim, Germany, **2012**, pp. 307–326.
127. C.-W. Chu, J. D. Holliday, P. Willett, *Bioorg. Med. Chem.* **2012**, *20*, 5366–5371.
128. MACCS structural keys, Symyx Software, San Ramon, CA, 2002.
129. Molecular Operating Environment (MOE), version 2011.10, Chemical Computing Group Inc., Montreal, Quebec, Canada.
130. G. Schneider, W. Neidhart, T. Giller, G. Schmid, *Angew. Chem. Int. Ed.* **1999**, *38*, 2894–2896.
131. B. E. Evans, K. E. Rittle, M. G. Bock, R. M. DiPardo, R. M. Freidinger, W. L. Whitter, G. F. Lundell, D. F. Veber, P. S. Anderson, R. S. L. Chang, V. J. Lotti, D. J. Cerino, T. B. Chen, P. J. Kling, K. A. Kunkel, J. P. Springer, J. Hirshfield, *J. Med. Chem.* **1988**, *31*, 2235–2246.
132. J. S. Mason, I. Morize, P. R. Menard, D. L. Cheney, C. Hulme, R. F. Labaudiniere, *J. Med. Chem.* **1999**, *42*, 3251–3264.
133. H. O. Villar, M. R. Hansen, R. Kho, *Curr. Comput. Aided Drug Des.* **2007**, *3*, 59–67.
134. A. Bender, R. C. Glen, *Org. Biomol. Chem.* **2004**, *2*, 3204–3218.

135. J. Waddell, J. L. Medina-Franco, *Bioorg. Med. Chem.* **2012**, *20*, 5443–5452.
136. L. Colliandre, V. Le Guilloux, S. Bourg, L. Morin-Allory, *J. Chem. Inf. Model.* **2012**, *52*, 327–342.
137. K. T. Nguyen, L. C. Blum, R. van Deursen, J.-L. Reymond, *ChemMedChem* **2009**, *4*, 1803–1805.
138. R. V. Deursen, L. C. Blum, J.-L. Reymond, *J. Chem. Inf. Model.* **2010**, *50*, 1924–1934.
139. M. Awale, J.-L. Reymond, *Bioorg. Med. Chem.* **2012**, *20*, 5372–5378.
140. O. Rabal, J. Oyarzabal, *J. Chem. Inf. Model.* **2012**, *52*, 1086–1102.
141. E. Lounkine, P. Kutchukian, P. Petrone, J. W. Davies, M. Glick, *Bioorg. Med. Chem.* **2012**, *20*, 5416–5427.
142. J. Bajorath, L. Peltason, M. Wawer, R. Guha, M. S. Lajiness, J. H. Van Drie, *Drug Discov. Today* **2009**, *14*, 698–705.
143. A. M. Wassermann, M. Wawer, J. Bajorath, *J. Med. Chem.* **2010**, *53*, 8209–8223.
144. D. Stumpfe, J. Bajorath, *J. Med. Chem.* **2012**, *55*, 2932–2942.

11

DNA-ENCODED CHEMICAL LIBRARIES

Luca Mannocci

11.1 INTRODUCTION

11.1.1 Drug Discovery Today: A Formidable Challenge

The isolation of bioactive molecules is a formidable task in chemistry, biology, and pharmaceutical sciences. With the advent of the new millennium, sequencing of the human genome, as well as developments in proteomic research [1,2] and transcriptomic analysis [3,4], have massively aided our understanding of living systems and the discovery of a multitude of new biological targets associated with human diseases [5]. Biomedical scientists are rapidly gathering increased knowledge of the mechanisms of diseases; nonetheless, more and more often the elucidation of protein biological functions requires the availability of highly specific small organic ligands (chemical genetic approach) [6]. Therefore, the identification of agents capable of specific binding to validate target proteins remains a major challenge for both academic and industrial laboratories.

Moreover, with an aging population, biopharmaceutical research is constantly facing a briskly increasing demand for more and better drugs. Until now, only 10% of the totally known disease-associated genes have been used in drug discovery campaigns [7]. Following the quote by Nobel prize-winning J.W. Black, "the most fruitful basis for the discovery of a new drug is to start with an old drug" [8,9], major pharmaceutical industries appear to spend over U.S.$50 billion per year on R&D to pursue only a limited number of drug targets. Indeed, omitting biosimilars, salt forms, imaging agents, and vitamins and other supplements, merely 6% of the over 20,000 drug

Diversity-Oriented Synthesis: Basics and Applications in Organic Synthesis, Drug Discovery, and Chemical Biology, First Edition. Edited by Andrea Trabocchi.

products available on the market are "unique drugs" (of which about 90% are small organic molecules and 10% are biopharmaceutical, i.e., proteins and nucleic acid derivatives) [7]. Barely 324 different drug–target protein pathways (of which 266 identified in the human genome and 58 in the pathogen genome) and 130 privileged "druggable" protein domains (of the 16,000 protein families and 10,000 different folds) account for all the drug targets used in pharmaceutical developments so far [7].

Paradoxically, the gap between the number of drug molecules and drug targets available is *not* due to a lack of chemical compounds; rather it reflects a lack of adequate drug discovery tools for the efficient identification of optimal binding molecules among a vast, fast-growing chemical space [10].

In the era of molecular medicine, there is an inevitable need to develop more efficient and versatile drug discovery techniques, capable of rapidly generating and interrogating massive collection of chemical compounds in order to improve drastically the ability to tackle new target classes, difficult to treat with conventional drug discovery approaches.

11.1.2 Selecting Chemicals

Every pharmaceutical agent is, to begin with, a molecule that binds selectively to one or more protein targets. Thus, the identification of molecules capable of specific protein recognition is at the heart of modern drug discovery technologies.

Low-molecular-weight compounds (small organic molecules) are extremely powerful tools; such compounds are generally able to quickly penetrate cell membranes and initiate biomolecular modifications, eventually leading to the desired therapeutic effects. However, the identification of small-molecular probes capable of selectively binding to any target protein of choice is frequently a massively cumbersome task [11,12].

Conventional approaches for the identification of small molecular biological agents often require the discrete screening [high-throughput screening (HTS)] of very large chemical collections (compound libraries), comprising up to a few million compounds. Frequently, to expand the chemical space and explore underrepresented classes of chemical structures, rational design approaches [13,14], combinatorial chemistry [15,16], a fragment-based approach [11,17,18], and diverse-oriented synthesis [19–21] are employed.

Virtual screening [22,23] and structure–activity relationships (SARs) [24] may sensibly reduce the number of compounds that need to be evaluated. Nonetheless, high-throughput-screening campaigns routinely require the repetitive screening of hundreds of thousands to millions of chemical compounds against a single drug target in order to identify a handful of candidates with adequate biological properties (Figure 11.1a) [12].

Despite the practical burdens (tremendously demanding in terms of library synthesis, compound management, robotics, and logistics) and the intrinsic inefficiency, high-throughput screening is still the drug discovery technology of choice for the identification of small-molecule hits against a given biological target [25].

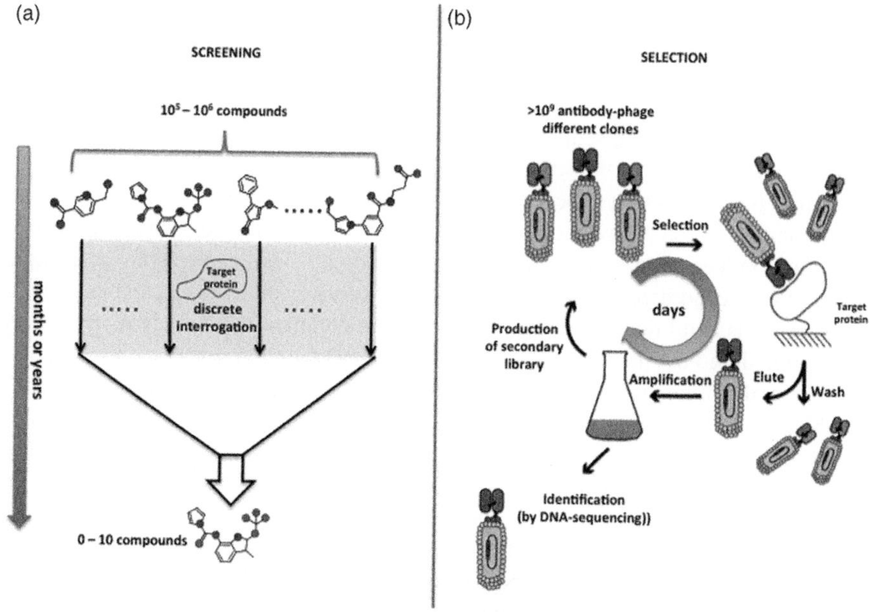

FIGURE 11.1 Screening vs. selection approaches. (a) Hundreds of thousands to millions of chemical compounds are discretely "screened" against a target protein. Only proteins for which a biochemical assay has been developed can be used in screening campaigns. Such an approach (i.e., high-throughput screening) is extremely demanding in terms of costs, time, and logistics. (b) Very large repertoires of biomacromolecules can be selected simultaneously by affinity capture against virtually any target protein of choice. After applying selection pressure, nonbinders are washed away, while binding compounds are amplified by bacterial infection and identified by DNA sequencing of the corresponding "genotype." Eventually, to improve the binding properties of the enriched candidates, a second-generation (affinity maturation) library can be reconstituted from the selected population and serve as an input for further rounds of selection. The entire process can be performed in just a few days.

In sharp contrast to the isolation of small organic molecules, the identification of specific biopolymer ligands to virtually any biological target antigen can be successfully and elegantly accomplished by applying display techniques such as phage display [26–29], yeast display [30], ribosome display [31,32], *m*RNA display [33–35], covalent display [36], and other conceptually similar methodologies [37,38]. Different from high-throughput screening, such technologies rely on the rapid assembling and selection of large repertoires of functional polypeptides (e.g., antibody libraries) containing billions of library members [39]. By means of native biosynthetic machineries, highly complex mixtures of recombinant DNA-genotype are individually translated into the corresponding molecules (i.e., proteins, RNA). Subsequently, after imposing a selection pressure on the entire library of candidates, compounds with optimal binding properties are preferentially enriched. The intimate linkage between the displayed phenotype (e.g., antibody) and the corresponding genotype (e.g., encoding gene) allows for PCR amplification of the genetic information, and the consequent translation into a succeeding population of enriched biomolecular compounds suitable for further rounds of selection (a process often termed *biomolecular evolution*) (Figure 11.1b) [27,39,40]. After multiple cycles of translation and selection, sequencing of the enriched genome information yields to the final identification of the selected binding molecules (Figure 11.1b).

In this light, *selection* procedures are intimately connected with the Darwinian evolutionary approach applied by nature to discover molecules with appropriate functions. Indeed, nature simultaneously imposes a selection pressure on a large population of candidate molecules. The cell that survives proliferates, and its genetic information is amplified and translated in the next population of bioactive molecules that will undergo the next selection process.

It is worth noting that in sharp contrast to the high-throughput *screening* approach, in which compound libraries are screened individually (i.e., one molecule at a time), based on a specific functional or binding assay, in selection approaches the display libraries are interrogated simultaneously, imposing the same selection conditions on all library members (Figure 11.1). Time and costs for selections are, in first-order approximation, independent of library size, since all library members are interrogated in a single experiment. Conversely, screening efforts tend to the increase linearly with library size, due to the discrete nature of the assays required.

Over the last 20 years, display technologies have shown themselves to be enormously effective for the identification of biopharmaceutical candidates [37,39,41–43]. However, such compounds (i.e., antibodies) are often unable to interfere with the majority of intracellular processes. Therefore, it would be conceivable to develop conceptually analog selection methodologies (e.g., similar to phage display) for the discovery of more target specific biologically effective low-molecular-weight compounds (Figure 11.2).

As small organic molecule libraries cannot be biosynthesized by native biosynthetic machineries, DNA-encoded chemical library technology emerged as the most promising and fascinating alternative to fill such a gap, allowing for the facile linkage of unnatural chemotypes to corresponding amplifiable genotypes (Figure 11.2).

11.1.3 Chapter Overview

In this chapter the concept of DNA encoding and the basic principles of DNA-encoded chemistry are described, including the fundaments of DNA conjugation and DNA-templated synthesis. Subsequently, with a focus on combinatorial synthesis, the history of DNA-encoded chemical libraries will be retraced, from the pioneering paper of Lerner and Brenner in 1992 to recent developments as an effective drug discovery tool.

In the second part of the chapter, alternative strategies for the construction of DNA-encoded chemical libraries, as well as the various selection and decoding approaches, are thoroughly surveyed. Moreover, the latest developments in DNA-encoded chemical libraries are reviewed, examining closely recent and future possible implications for drug discovery programs. Finally, looking globally at the technology

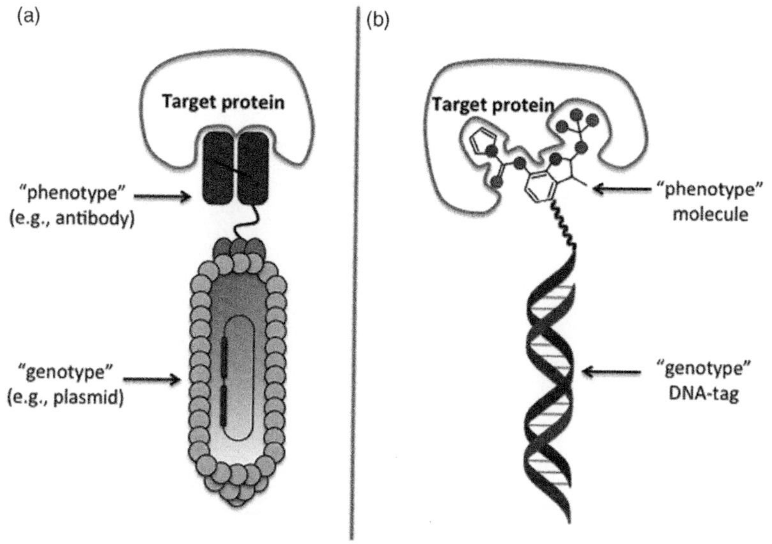

FIGURE 11.2 Analogy of a display technology (e.g., antibody phage display) and DNA-encoded chemical libraries. (a) Antibody fragment displayed in scFv format as a fusion protein to the pIII protein on the tip of the phage particle. The phage particle provides physical linkage between the protein-binding properties (phenotype) and the genetic information coding for the antibody (genotype). (b) DNA-encoded chemical libraries as a conceptual translation of the antibody phage display technology. Small organic molecules are covalently linked to unique DNA-oligonucleotides. The DNA tag (genotype) serves as an amplifiable identification bar code for the small organic molecule displayed (phenotype). (From [44], with permission of The Royal Society of Chemistry.)

as a whole, the achievements, current challenges, and outlooks of DNA-encoded chemical library technology are discussed.

11.2 DNA-ENCODED CHEMICAL LIBRARIES

11.2.1 DNA Encoding

Following the revolutionary discovery of the DNA structure by Watson and Crick in 1953 [45,46], scientists have been immediately fascinated by its amenable properties toward manipulation, hybridization, amplification, and sequencing. To date, researchers have pioneered the implementation of these principles from nanotechnology and material sciences, to chemical biology and drug discovery, far beyond their natural functions.

In 1992, inspired by the covalent linkage between the DNA genotype and antibody phenotype of display technologies such as phage display, Brenner and Lerner introduced the concept of associating PCR-amplifiable DNA-base identification tags with synthetic chemotypes (e.g., small organic molecules) to facilitate the identification of biologically active compounds after affinity-based selection experiments [47].

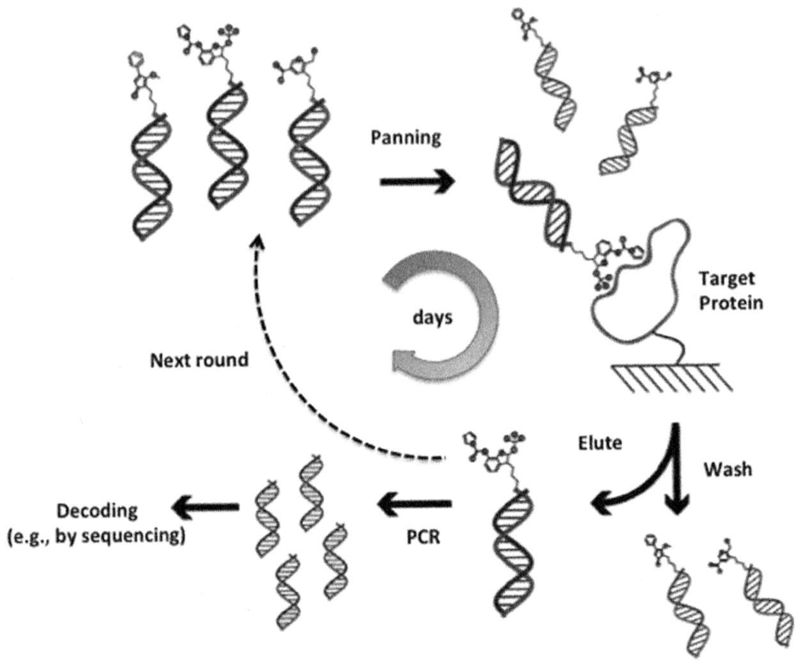

FIGURE 11.3 DNA-encoded chemical library panning. Initially, the DNA-encoded chemical library is incubated with a target protein of choice immobilized on a solid support. Following the removal of nonbinders by washing cycles of the solid support, DNA tags of the retained library members are PCR-amplified. Eventually, decoding (e.g., by DNA sequencing) and deconvoluting the enriched DNA codes reveals the identity of the compounds selected. In principle, the panning cycle can be iterated multiple times (dashed arrow), yielding further enrichment of binding library members over nonbinding compounds. The process can be performed in just a few days, and multiple targets can be selected in parallel.

Drawing an analogy between conceptually similar display technologies, the authors postulated that DNA-encoded small molecules could be selected in vitro against virtually any target protein of choice [47]. As already popularized by other polypeptide-based technologies and in resemblance to the gold-mining process of washing gravel in a pan to separate gold from contaminants, the selection procedure was termed *panning*.

In short, after imposing a selection pressure (e.g., incubating the DNA-encoded library with the target protein of interest immobilized on a solid support), washing cycles allow for the enrichment of specific binding moieties whose DNA codes are eventually PCR-amplified and decoded (e.g., by DNA sequencing, Figure 11.3). Relative quantification of library members before and after the selection process leads to the identification of enriched small-molecule structures. In principle, such a panning procedure can be repeated multiple times, yielding to further enrichment of binding library members over nonbinding compounds (Figure 11.3, dashed arrow).

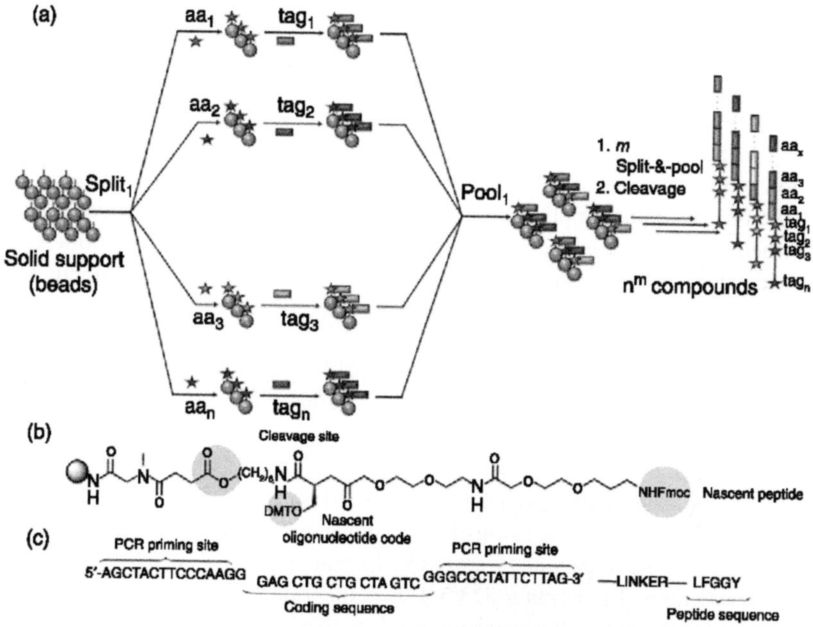

FIGURE 11.4 DNA-encoded chemical library by stepwise peptide/oligonucleotide synthesis on the same reaction bead. (a) Monomeric chemical compounds and coding-oligonucleotide tags are attached on the same solid support in an alternating fashion. The library can, in principle, be probed for binding to a selected target protein of interest. Here "aa" represents the different amino acids, and "tag" refers to a DNA sequence encoding the corresponding amino acid added in the stepwise procedure. (b) Schematic representation of the support by Brenner and Lerner and by Janda in the first practical implementation of the methodology. The cleavable linker enables easy detachment of the oligonucleotide-encoded peptide after synthesis, while O-DMT-protected serine and *N*-Fmoc-protected lysine allows for the bidirectional synthesis of oligonucleotide and peptide sequences. (c) Leucine–enkephalin pentapeptide (YGGFL) oligonucleotide conjugate after cleavage from beads. The conjugate was shown to bind to the anti-leucine–enkephalin antibody 3-E7 as efficiently as the reference peptide ($K_D = 7.1$ nM). Notably, the cognate oligonucleotide code could be amplified by standard PCR.

In the same theoretical article, Brenner and Lerner anticipated a split–pool-based combinatorial approach for the construction of DNA-encoded libraries, in which monomeric chemical compounds and coding-oligonucleotide tags are attached on the same solid support in an alternating fashion (Figure 11.4a) [47]. This strategy was demonstrated experimentally in collaboration with Janda's group in 1993 [48].

To test the feasibility of a stepwise peptide/oligonucleotide synthesis on the same reaction bead, a functionally active leucine–enkephalin pentapetide (YGGFL) was synthesized in five rounds of alternated synthesis [48]. Controlled-pore glass beads, serving as structural linkers between the nascent chemical molecules and the coding oligonucleotides, were used to facilitate oligonucleotide synthesis. The solid support was chemically modified with a trifunctional appendage (i.e., succinylaminohexanolsarcosine) comprising the following features: (1) an acid-labile cleavable site for ensuring easy detachment of the DNA-encoded molecule after library synthesis, (2) an *O*-DMT, and (3) an *N*-Fmoc moiety, to fulfill orthogonal chemistry requirements for the regiospecific elongation of the emerging coding oligonucleotide and peptide sequence, respectively (Figure 11.4b) [48]. After releasing the final compound from the solid support by acid cleavage, the correct peptide sequence was confirmed by Edman sequencing and radiolabeled using a T4-polynucleotide kinase. As anticipated, binding assays revealed the nanomolar affinity of the synthetic DNA-tagged pentapeptide (Figure 11.4c) toward antileucine–enkephalin antibody (3-E7) [49], thus indicating that the potential steric hindrance of the oligonucleotide tag does not significantly affect the binding interaction [48]. Furthermore, the oligonucleotide appendage could be efficaciously amplified and sequenced by means of conventional polymerase chain reactions (PCRs), confirming the specificity and stability of the coding sequence [48].

In the same year, Gallop and co-workers at Affymax, accomplished the synthesis of the first DNA-encoded collection of small-molecule polypeptides, using a synthetic strategy very similar to the one proposed by Brenner et al. (Figure 11.4) [47,50]. The library was designed to contain 823,540 heptapeptide sequences (7^7members). By means of seven cycles of alternate split–pool synthesis, different D- and L-amino acid building blocks were assembled in combinatorial fashion on small spherical beads (Figure 11.4, $n = m = 7$). To trace the process by which each peptide sequence had been assembled, dinucleotide DNA tags were introduced through a parallel combinatorial procedure [50]. However, with the aim of avoiding depurination of deoxyadenosine and deoxyguanosine during the trifluoroacetic acid cleavage of the compounds from the support, an acid-resistant deoxyadenosine was used (c^7dA) [51], whereas deoxyguanosine nucleotides were deliberately excluded from the coding oligonucleotides [50].

The final DNA-conjugated-heptapetide library was subjected to "on-bead" screening against a fluorescently labeled antipeptide antibody that selectively binds the RQFKVVT peptide sequence. A fluorescence-activated sorting instrument (FACS) allowed the isolation of those beads to which the antibody bound tightly. PCR amplification and DNA sequencing of the DNA tags attached to the sorted beads yielded the identification of the expected antibody-specific heptapeptide (RQFKVVT) and other analog sequences capable of binding the target with comparable affinities (K_D values ranging between 0.3 and 1400 nM) [50].

Even though these results marked the first milestone in the development of DNA-encoded chemistry technology, the intrinsic limitations in terms of library size (which could not be greater than the number of beads) and the need for a peculiar orthogonal chemistry, combined with a cumbersome solid-phase DNA synthesis, seriously hampered the evolution of this technology, which was never practically employed for drug discovery purposes.

Research in the field of DNA encoding remained essentially unknown until the end of the 1990s. Alternative chemical and physical strategies for the encoding of small compounds had been pursued extensively with the aim of generating very large libraries in a single test tube. Mass spectrometric–based tagging, peptide encoding,

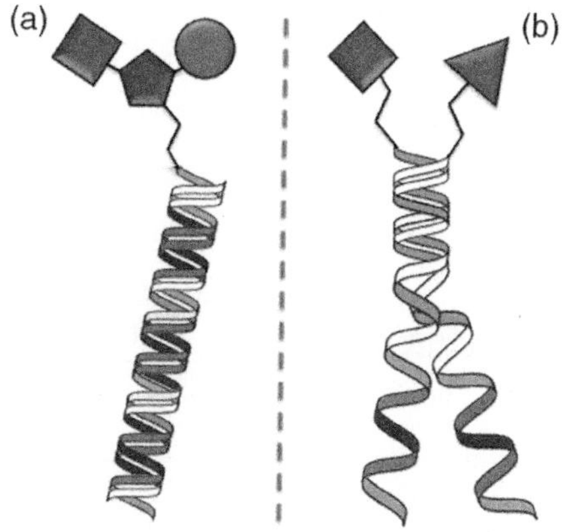

FIGURE 11.5 Second generation of DNA encoded chemical libraries DNA-encoded library formats displaying chemical compounds attached directly to the coding oligonucleotides. (a) Single-pharmacophore format: a single oligonucleotide is covalently linked to a putative binding molecule; (b) dual-pharmacophore format: multiple pairing oligonucleotides each displaying a covalently linked chemical compound.

haloaromatic tagging, and electronic encoding are a few of the numerous encoding strategies explored in that decade [52,53]. The lack of signal amplification upon binding to the target at extremely low concentrations (typically in the subnanomolar range) was an insurmountable limit for detection of the library members after the selection process. The stunning amplifiability properties of the DNA appeared to be the exceptional tool.

Therefore, at the beginning of the new millennium, DNA-encoded chemical library technology experienced a revival. The technology was pursued through exploration of completely new avenues. The bead as structural linkers between the chemical moiety and the coding oligonucleotide (see Figure 11.4) was omitted through attaching the candidate-binding molecule directly to the DNA-coding fragment. Today, this second generation of DNA encoded chemical libraries can be conveniently classified as 'single- and dual-pharmacophore libraries, depending on whether a single or multiple oligonucleotides display a covalently linked molecule (Figure 11.5).

11.2.2 Single-Pharmacophore DNA-Encoded Chemical Libraries

The covalent linkage of a unique chemical entity to a single extremity of an oligonucleotide tag represents the simplest format of a DNA-encoded chemical library. The major advantage of single-pharmacophore libraries is that following selection against a target antigen of choice, the selected binding compounds can be used straightforwardly as hits for further drug discovery developments (e.g., medicinal chemistry optimization), resynthesizing the compound without the oligonucleotide appendage.

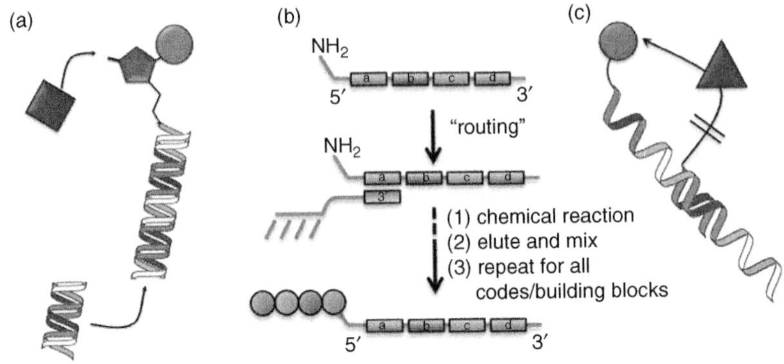

FIGURE 11.6 Single-pharmacophore DNA-encoded chemical libraries. DNA-encoded library formats: (a) library DNA conjugate: multiple sets of building blocks and corresponding DNA-coding tags are assembled stepwise by split–pool combinatorial reactions; (b) library DNA routed: DNA tags are used both for library encoding and for programming the library assembly; (c) library DNA directed: multiple building blocks are assembled on the coding DNA-oligonucleotide using DNA-directed reactions.

Numerous strategies have been explored for the assembly of single-pharmacophore DNA-encoded libraries, featuring the implementation of various combinatorial chemistry techniques. Nonetheless, such libraries can be mainly grouped as follows: (a) library DNA conjugate, essentially based on a stepwise combinatorial split–pool synthesis and encoding (Figure 11.6a; see Section 11.2.2.1) [54–58]; (b) library DNA-routed (Figure 11.6b, see Section 11.2.2.2) in which the DNA appendage is used both for encoding and for programming the library assembly [59]; and (c) library DNA directed, in which DNA-directed reactions are used for the synthesis and encoding of the compounds displayed (Figure 11.6c, see Section 11.2.2.3) [60,61].

As shown in Figure 11.6, independent of the strategy pursued for library construction, the molecule displayed is generally assembled by the combinatorial incorporation of multiple sets of building blocks in iterative fashion. Therefore, the use of highly efficient chemical reactions, which work well in a DNA-compatible solvent such as water, and preserve the integrity of the DNA tags, is often crucial for the synthesis of high-quality single-pharmacophore DNA-encoded chemical libraries.

11.2.2.1 DNA-Conjugate Libraries Single-pharmacophore DNA-encoded chemical libraries can be efficiently synthesized by straight DNA conjugation of building blocks and reagents in combinatorial fashion. Philochem (a spin-off company at ETH Zürich, Switzerland) and GlaxoSmithKline (former Praecis Pharmaceuticals) independently invented and industrially developed various formats of DNA-conjugate libraries for drug discovery purposes [54–56,58,62].

Philochem has pioneered the pool–split–pool approach to the construction of various DNA-conjugate libraries [54,55,57]. The strategy, as depicted in Figure 11.7, features five sequential steps: (i) linking of a first set of n chemical compounds (i.e., $n = 20$) to distinct 5′-reactive oligonucleotides (i.e., synthetic 5′-amino-oligonucleotide);

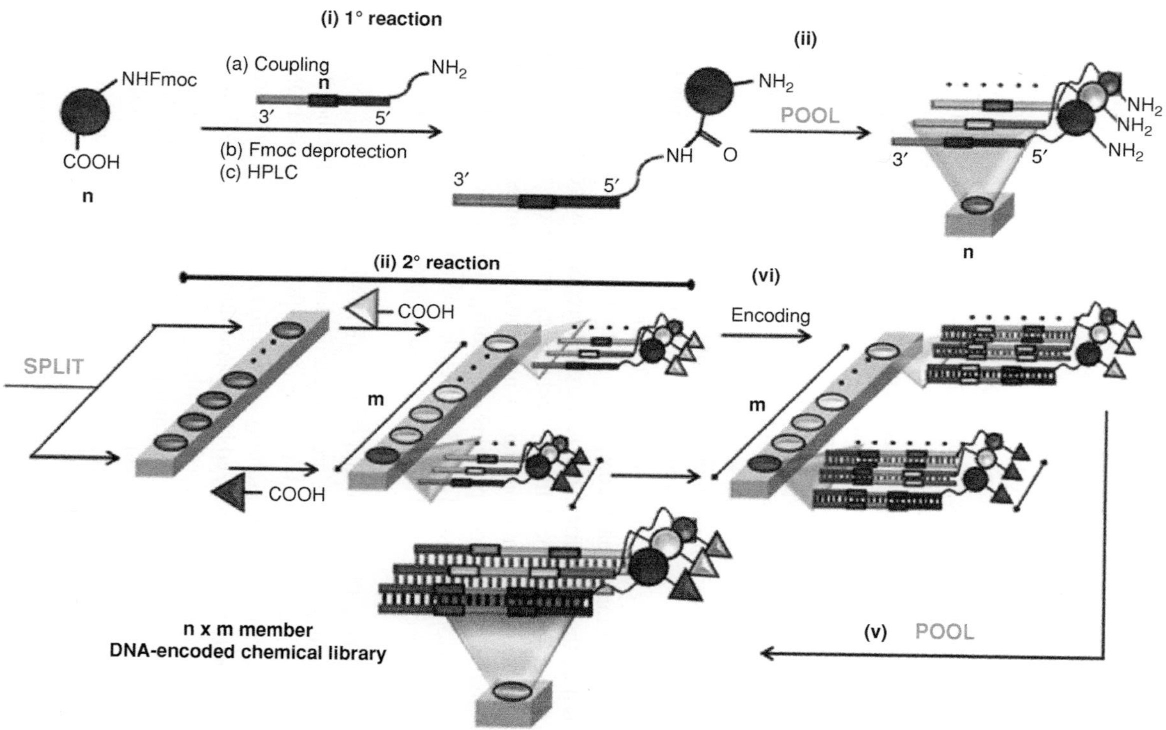

FIGURE 11.7 Schematic representation of the split–pool–split approach pioneered by Philochem for the construction of single-pharmacophore DNA-encoded chemical libraries based on two sets of building blocks. The strategies feature (i) conjugation of the first set of chemical compounds (e.g., *n* Fmoc amino acids) to distinct amino-modified oligonucleotides, followed by Fmoc removal and HPLC purification; (ii) pooling of the reactions and splitting into *m* separate reaction vessels, corresponding to the reactions for the insertion of the second building block; (iii) chemical incorporation of the second building block; (iv) encoding of the second building block (e.g., by hybridization of partially complementary oligonucleotides, followed by Klenow fragment-mediated DNA polymerization); (v) pooling of the encoded reactions to yield a DNA-encoded chemical library containing $n \times m$ molecules. (From [54]: copyright © 2008 National Academy of Sciences, U.S.A.) (*See insert for color representation of the figure.*)

(ii) HPLC purification, mass spectrometry characterization, and mixing of equimolar amounts of each DNA-conjugate compound to generate a first sublibrary pool; (iii) splitting of the pool into separate reaction vessels and incorporation of a second set of building blocks (i.e., $m = 200$); (iv) enzymatic DNA tagging (i.e., by staggered hybridization of a partially complementary oligonucleotides and subsequent Klenow-assisted DNA polymerization) to univocally encode the identity of the last chemical moiety incorporated; and (v) pooling of the encoded reactions to yield the final DNA-encoded chemical library.

In 2008, Mannocci et al. first demonstrated the methodology in the construction of a DNA-encoded chemical library containing 4000 compounds [54]. The library was spiked with a DNA-oligonucleotide conjugated to D-desthiobiotin, a biotin analog with nanomolar affinity to streptavidin, at the expected concentration of a single library compound [54,63]. High-throughput-sequencing analysis performed on the library before and after selection on streptavidin-coated Sepharose beads revealed the enrichment of the desthiobiotin positive control as well as of other classes of structurally related streptavidin-binding compounds (Figure 11.8) [54,63]. Notably, this work represents the first implementation of high-throughput-sequencing technology to decode DNA-encoded chemical library selections [54].

Extending the construction scheme depicted in Figure 11.7 to the use of alternative chemical reactions (e.g., Diels–Alder, Sonogashira, Suzuki, Mitsunobu) and suitable building blocks, Buller et al. efficiently synthesized various single-pharmacophore DNA-encoded chemical libraries (each comprising up to 10^6 DNA-conjugate compounds) [55,57]. Using these libraries, the researchers reported successful application of the DNA-encoded chemical library technology to the de novo identification of ligands possessing low-micromolar to nanomolar affinities against several pharmaceutically relevant target proteins, including human serum albumin [64], Bcl-xL [65], polyclonal human IgG [54], matrix metalloprotease 3 (MMP-3) [54], carbonic anhydrase IX [55], TNFα [57], and interleukin-2 [66] (see also Section 11.4).

Moreover, the scientists at Philochem and ETH Zürich demonstrated that single-pharmacophore DNA-conjugate libraries are also particularly suitable for the affinity optimization of previously discovered lead structures [62]. Such focused DNA-encoded chemical libraries (also termed *affinity maturation libraries*, as popularized by phage display) can be rapidly synthesized by incorporation of a known lead structure during the combinatorial assembly of the library. In 2010, ETH Zürich and Philochem described the synthesis of a benzamidine-based single-pharmacophore affinity maturation library, which allowed for the isolation of a >10,000-fold improved benzamidine-based trypsin inhibitor, with single-digit nanomolar potency and exquisite selectivity over closely related serine proteases (Figure 11.9) [62].

Iterating the split–pool–split process illustrated in Figure 11.7, it is possible to generate DNA-conjugate libraries comprising more than two sets of building blocks and over 10^8 DNA-encoded small molecules [54,55,57,58]. In 2009, Praecis/GSK accomplished the synthesis of a triazine-based DNA-conjugate library containing 802,160,640 compounds performing four rounds of stepwise pool–split–pool reactions and encoding by DNA ligation (Figure 11.10) [58].

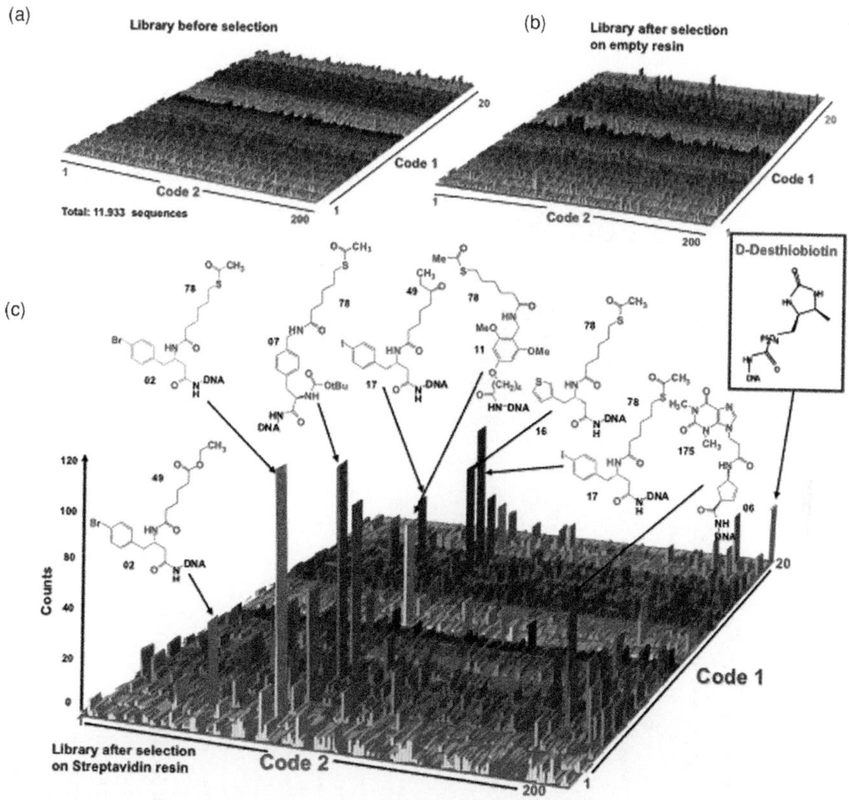

FIGURE 11.8 High-throughput-sequencing results before and after selection against streptavidin-coated Sepharose resin. (a–c) Histogram plots represent the 4000 library members in the x–y plane and the corresponding sequence counts on the z-axis. (a,b) Sequencing before selection and after selection on an empty resin showed equally distributed frequencies of library members; (c) streptavidin selection results exhibited a preferential enrichment of D-desthiobiotin (known streptavidin binder with nanomolar affinity spiked into the library prior to selection) and of additional structurally related compounds. The chemical structures of some of the most relevant straptavidin binders are indicated together with the library identification number. Building blocks used in the two synthetic steps are highlighted in green and red, respectively. Molecules identified with at least 30 counts exhibited after resynthesis preferential binding toward streptavidin, (K_D values ranging between 350 nM and 11 μM). In contrast, compounds tested with fewer than 10 counts did not exhibit appreciable binding to streptavidin (K_D > 50 μM). (From [54]; copyright © 2008 National Academy of Sciences, U.S.A. (*See insert for color representation of the figure.*)

Speculating that double-stranded oligonucleotide would be able to shelter the DNA from potential chemical lesions during library synthesis, a short covalently linked DNA duplex was chosen as a headpiece [58]. Following acylation and deprotection of 192 Fmoc amino acids onto the initial scaffold, a triazine ring was introduced using cyanuric chloride. Subsequently, the remaining chlorines of the triazine headpiece

(a)

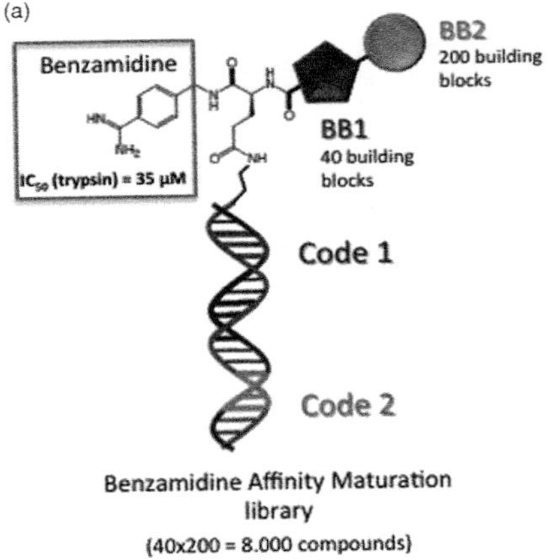

(b)

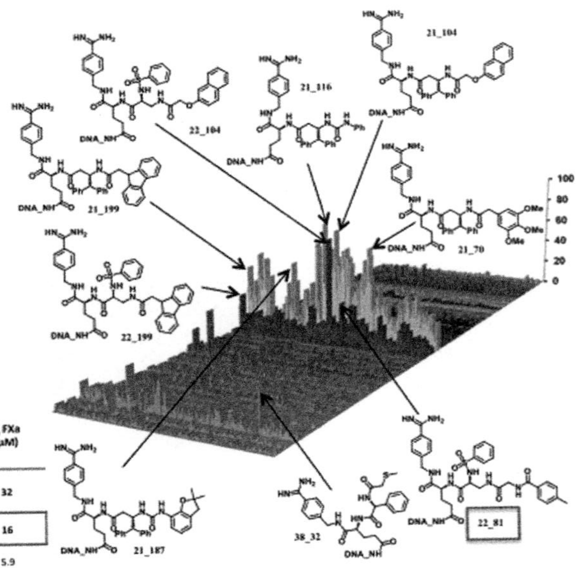

(c)

Entry	Compound	K_i Trypsin (μM)	K_i Thrombin (μM)	K_i uPA (μM)	K_i FXa (μM)
1	Benzamidine	35	over	35	32
2	Gln_22_81	0.0030	3.1	50	16
3	Ala_22_99	0.0052	0.052	7.9	5.9
4	Gln_22_199	0.019	1.9	14	over
5	Ala_21_116	0.023	8.2	over	over

FIGURE 11.9 Benzamidine-based affinity maturation DNA-encoded library. (a) Schematic representation of a library member: a library of 8000 (40 × 200) DNA-tagged benzamidine-based compounds was assembled by sequential stepwise addition of two different sets of building blocks onto a core structure consisting of (4-carbamimidoylbenzyl)glutamamide; (b) after selection on trypsin-coated resin, high-throughput-sequencing analysis revealed a strong enrichment of structurally related compounds; (c) table summarizing the inhibition constants of some hits, identified and resynthesized after trypsin selection. Twenty-one compounds were individually resynthesized as glutamine (Gln) or alanine (Ala) derivatives and incubated at different concentrations with four different serine proteases (trypsin, thrombin, uPA, and FXa). The reaction was started by addition of the fluorogenic substrate Z-GGR-AMC and the change of fluorescence signal was recorded. Gln_22_81 exhibited an IC_{50} value of 3.0 nM toward trypsin and excellent specificity toward closely related proteases. (Adapted from [62], with permission; copyright © 2010 American Chemical Society.)

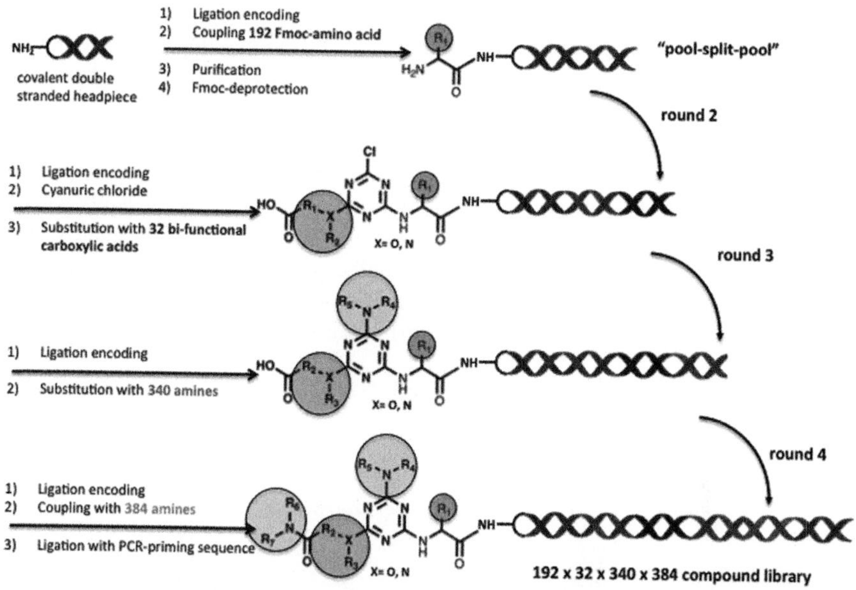

FIGURE 11.10 Praecis/GSK triazine-based DNA-conjugate library containing over 10^8 compounds. Pool–split–pool assembling of four different sets of building blocks and corresponding DNA tags on a triazine scaffold yielded a DNA-encoded library containing nearly 1 billion compounds (192 × 32 × 340 × 384 members). A covalent double-stranded oligonucleotide was used as a headpiece to shelter the DNA from chemical lesions during library synthesis.

were stepwise substituted with a small set of 32 bifunctional carboxylic acids and with 340 amines, respectively, in two consecutive rounds of split–pool–split synthesis. Eventually, the carboxylic function installed in round 2 was activated and reacted with additional 384 amines, to yield the final library containing nearly 1 billion compounds (Figure 11.10) [58].

Selection and high-throughput-sequencing decoding of the library against p38 MAPK (p38 mitogen-activated protein kinase) revealed a cluster of conserved trisynthon compounds to be preferentially enriched showing no preference for the identity of the amine introduced in cycle 3 [58]. Resynthesis and structure analysis of representative compounds from this family led to the identification of structurally related triazine compounds (as potential side products of the library synthesis), with EC_{50} in the low-nanomolar range [58]. Notably, the crystal structure of one compound bound to p38 MAPK displayed the carboxamide group, used initially to link the oligonucleotide tag, pointing toward the solvent, thus not actively contributing to binding with the target protein [58].

Despite the promising results reported by Praecis/GSK, the real efficacy of larger chemical libraries, containing more than 10^6 compounds, is currently under debate due to a number of technical limitations, including the limited purity of the library, sampling problems associated with high-throughput-sequencing decoding (e.g., the library results being much larger than the number of DNA sequences that can be efficaciously sequenced) and the drug-likeness of the final molecular candidates (see also Section 11.5).

It is worth noting that library size and chemical diversity of the compounds displayed depend exclusively on the number of split–pool cycles and building blocks introduced during library assembly. Therefore, size and diversity are typically gained at the expense of a growing molecular weight and decreasing library quality, due to incomplete reactions (Figure 11.11). Eventually, smaller libraries, comprising up to a few millions of compounds, are intrinsically of superior quality and typically contain compounds that may better fulfill drug-like criteria such as Lipinski's rule

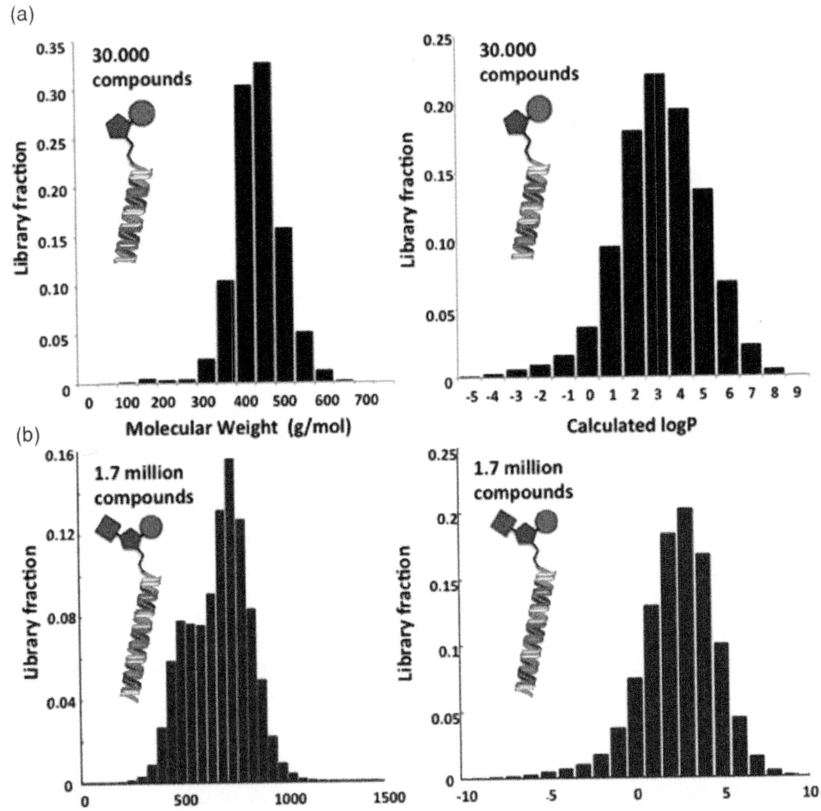

FIGURE 11.11 Molecular weight and log P distribution of two different single-pharmacophore DNA-encoded chemical libraries. (a) Two building blocks single-pharmacophore DNA-encoded library containing 30.000 compounds; (b) single-pharmacophore DNA-encoded library based on three building blocks comprising 1.7 million compounds. Small libraries are intrinsically of superior quality and contain compounds that may better fulfill drug-like criteria. (From [44], with permission of The Royal Society of Chemistry.)

of 5 [67,68] and yield drug candidates more suitable for later medicinal chemistry optimizations (Figure 11.11).

11.2.2.2 DNA-Routed Libraries In 2004, Halpin and Harbury established another milestone in DNA-encoded chemical library research [59,69]. For the first time they developed a variation of the split–pool library construction in which the DNA appendage served not only as a PCR-amplifiable identification tag but also as a code for programming the stepwise library assembly.

As illustrated in Figure 11.12, different oligonucleotide templates are physically separated ("routed") by flowing into sequential cartridges containing appropriate anticodon DNA sequences, and processed in discrete chemical reactions [69]. Subsequently, the subpopulations are combined and flown through a second set of

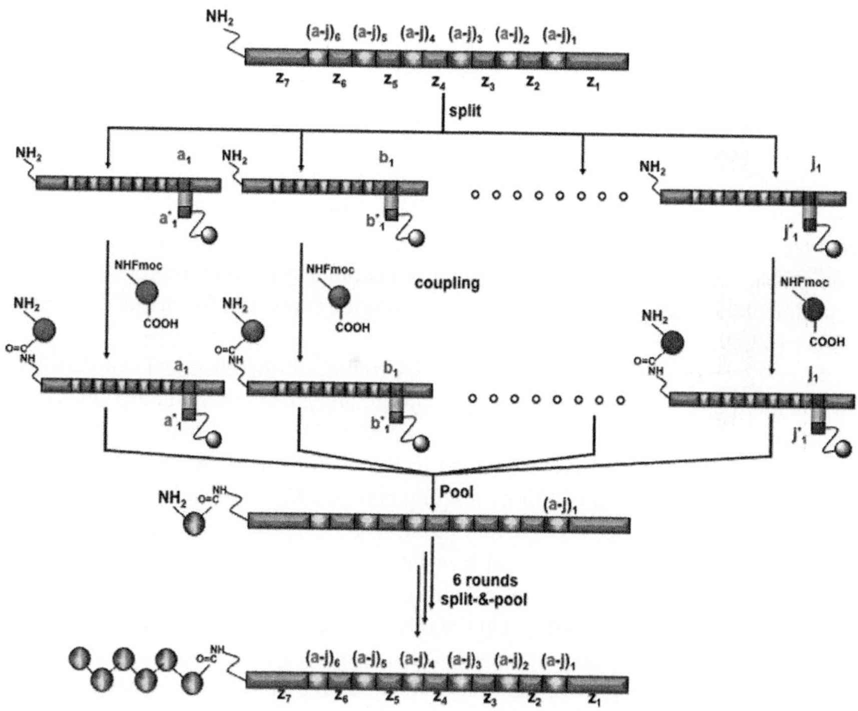

FIGURE 11.12 DNA-encoded chemical library by DNA routing. An initial combinatorial library comprising 10^6 different 340-mer-oligonucleotide templates was generated by PCR assembly of smaller oligonucleotides. Each oligonucleotide contains six coding regions, each encoding for 10 different amino acids [(a to j)$_{1\text{-}6}$], and seven constant domains ($z_{1\text{-}7}$). The template library was split using complementary oligonucleotides bound on affinity chromatography resins [(a* to j*)$_{1\text{-}6}$] (DNA routing). Following conjugation to the corresponding Fmoc-amino acid and deprotection, oligonucleotides were eluted and the initial template library reconstituted. Six rounds of DNA routing yielded to the final library of 10^6 different *N*-acylated pentapeptides. (*See insert for color representation of the figure.*)

anticodon columns for further chemical modifications. Such a split–pool process can be iterated for all codon regions until all the desired building blocks are incorporated on the initial DNA templates.

In a first practical implementation of the DNA-routed strategy (also termed *DNA display*), a double-stranded library containing 10^6 different 340-mer-oligonucleotide templates was generated by combinatorial PCR assembly of smaller oligonucleotides (Figure 11.12) [59]. Following the conversion of the DNA-duplex format into a single strand (by reverse transcription and base-mediated hydrolysis of the RNA strand), the oligonucleotide library was chemically translated into 10^6 *N*-acylated pentapeptides using 10 different Fmoc-protected amino acids and nine carboxylic acids for the terminal *N*-acylation (*N*-terminus unmodified derivatives were also included) (Figure 11.12). Five of the six coding regions (20 bases each) present on the initial oligonucleotide template allowed for unambiguous encoding of the pentapeptidic sequences, while the sixth coding region ensured the identity of the carboxylic acids used for the final *N*-acylation (Figure 11.12).

Acylated leucine–enkephalin pentapetides were intentionally included as positive controls during library synthesis [59]. The library sequencing before and after two rounds of selection against a single-digit nanomolar leucine–enkephalin antibody (the same as that used by Brenner et al., in model selection experiments) [48,49] revealed a remarkable round-to-round enrichment of the leucine–enkephalin consensus sequences [59]. Notably, the eluted DNA from each step was PCR-amplified and served as a evolutionary input for the subsequent round of chemical translation and selection. In this regard, the work of Halpin and Harbury represents the first experimental implementation of molecular evolution using a DNA-encoded chemical library [48,59,70].

More recently, Wrenn et al. applied DNA routing to the chemical translation of an octamer peptoid library, comprising approximately 100 million compounds [71]. Six rounds of evolutionary selection against *N*-CrkSH3 (*N*-terminal SH3 domain of the proto-oncogene Crk protein) yielded to the identification of peptoids that after resynthesis revealed binding affinity ranging between $K_D = 16$ and 97 μM [71].

Although at first glance DNA-routing strategies appear to be cumbersome and tedious, the possibility of chemical translation and diversification holds promises for the construction of large libraries whose sizes increase exponentially after each round of synthesis and encoding. On the other hand, from a drug discovery point of view, the simple polypeptidic structures which have been assembled so far using such technology may not represent the drug-like structures in which the pharmaceutical industry is traditionally interested in [68].

11.2.2.3 DNA-Directed Libraries The ability of a DNA double helix to direct region-specific chemical reactions by bringing DNA-linked reagents in close proximity through Watson–Crick base pairing has long been known [72,73]. But it wasn't until 2001 that Gartner and Liu introduced the concept of DNA-templated synthesis (DTS) to show that such a proximity effect is a general strategy to locally increase the effective molarity and thus the reaction rates [74].

In DTS, two reactants are covalently linked to two complementary DNA strands, each serving as either a template or a hybridization strand (reagent). Annealing of the two oligonucleotides confines the reagents to the same region in space. The local increase in reaction molarity allows the desired chemical reaction to occur, even at nominal concentrations that are several orders of magnitude lower than necessary for conventional solution-phase reaction. Interestingly, Gartner, Liu, and co-workers demonstrated that within a distance of 30 nucleotides, the proximity effect is nearly constant [74,75], thus enabling the use of various oligonucleotide architectures, featuring multiple reagent-coding sequences at different positions on the same DNA template [60,75,76].

Figure 11.13 shows three different DTS architectures developed for DNA-encoded chemical library construction [70]. The end-of-helix setup (Figure 11.13a) is currently the most widely used approach, thanks to its versatility in incorporating different chemical structures in multistep synthesis and its straightforward compatibility with current procedures of library selection and decoding (e.g., high-throughput-sequencing techniques) [60,70].

Using such architecture, the Harvard group described the first implementation of DTS for the construction of DNA-encoded small-molecule libraries [60]. A proof-of-concept library was synthesized using 48-mer 5′-lysine-modified oligonucleotide

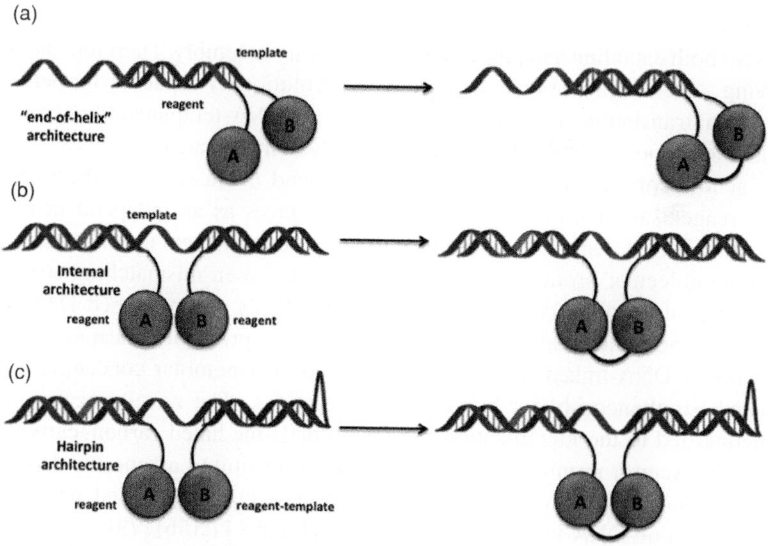

FIGURE 11.13 DTS architectures. (a) End-of-helix architecture: reagents are displayed on 5′ and 3′ extremities of two complementary oligonucleotide sequences (DTS-reagent and template). It is currently the most commonly used for DNA-encoded chemical library construction. (b) Internal architecture: reagents displayed on the top of two oligonucleotides are confined in a limited reaction space by means of a longer oligonucleotide template carrying both complementary sequences. (c) Similar to the internal embodiment, in hairpin configuration a longer oligonucleotide serves as both reagent and template oligonucleotide.

templates containing three adjacent DTS-coding regions. Three consecutive DNA-programmed acylations and one Wittig macrocyclization reaction allowed for the construction of a (64 + 1)-membered macrocycle library (4 × 4 × 4 plus a built-in phenyl sulfonamide derivative, which was known to bind carbonic anhydrase with nanomolar affinity) (Figure 11.11). Following panning experiments against carbonic anhydrase and PCR amplification of the library before and after selection, enzymatic digestion of the *NlaIII* domain (used as a coding sequence for sulfonamide positive control) revealed remarkable enrichment of the expected phenyl sulfonamide macrocycle, thus confirming the successful application of DTS principles to DNA-encoded chemical library construction [60].

More recently, using a strategy very similar to the one described in Figure 11.14, the same laboratory reported the construction of a larger macrocyclic fumaramide library containing 13,824 [(12 × 12 × 12) × 8] structures, starting from eight independent scaffolds and 36 different building blocks [61].

The library was used for in vitro selection against 36 different target proteins. Resynthesis of the selected macrocyclic structures resulted in the identification of various kinase inhibitors and activators. The two most potent compounds isolated specifically inhibited Src kinase with IC_{50} values of 680 and 960 nM, respectively (see Table 11.1) [77]. Starting from these compounds, Georghiou et al. developed and characterized in molecular detail a second generation of macrocycles with potencies as high as $IC_{50} \leq 4$ nM [78].

In analogy with DNA-routed libraries (see Section 11.2.2.2), the DNA appendage serves for both encoding and programming the library assembly. Therefore, templates surviving selection process can be used as an evolutionary input for further rounds of chemical translation and selection. However, for DNA-templated libraries, such a evolutionary process has not been experimentally demonstrated yet.

As at low concentrations DTS reactions depend exclusively on the hybridization of reagent and template, each DNA pair behaves as an independent molecular reactor [70,79]. Only DNA reactants linked to complementary oligonucleotides yield intramolecular products, whereas reactions between mismatching strands are substantially negligible. Following this principle, Liu's group applied DTS for the discovery of new chemical reactions [79,80]. In a first proof-of-concept experiment, two pools of DNA-linked substrates were mixed at nanomolar concentrations and evaluated simultaneously for bond-forming reactions. The novel reaction discovery method led to the identification of a new Pd(II)-mediated carbon–carbon bond reaction between a terminal alkyne and a monosubstituted alkene, as depicted in Figure 11.15a [79]. Notably, the newly discovered DNA-templated reaction also operated in a non-DNA-templated format setup (Figure 11.15b) [79].

11.2.2.3.1 DPC Platform On the groundwork by Liu's group on DNA-templated synthesis (DTS) and its implications in DNA-encoded chemical library construction, Ensemble Therapeutics developed DPC (DNA-programmed chemistry platform), an integrated system for the synthesis and selection of DNA-encoded libraries based on macrocyclic scaffolds (also known as *Ensemblins*) [81].

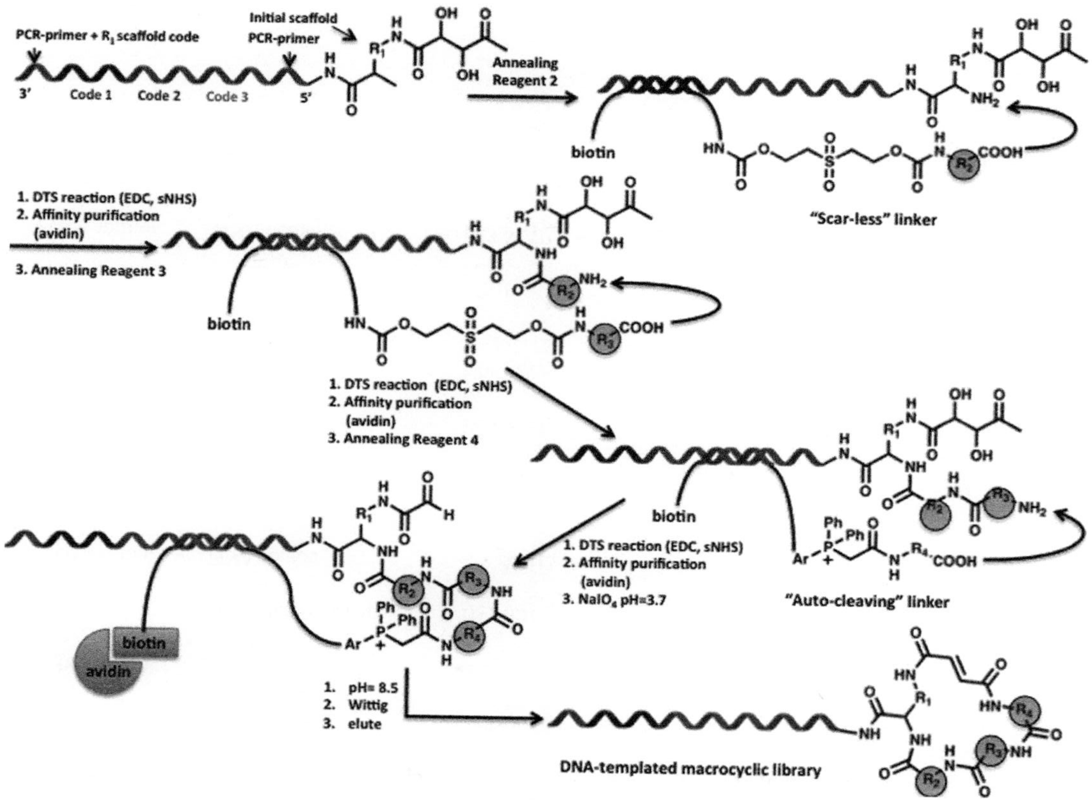

FIGURE 11.14 General scheme for DNA-templated synthesis of DNA-encoded macrocyclic libraries. A discrete number of DNA-linked scaffolds simultaneously undergo three DNA-programmed reaction steps (two acylations sNHS/EDC mediated and one Wittig macrocyclization) with different DNA-templated reagents, yielding the final single-stranded DNA-encoded macrocyclic library. Conjugate intermediates are purified by affinity capture on avidin (or streptavidin) resins and elution in denaturing conditions. Scarless linkers (i.e., base labile sulfone linkers) allow the introduction of suitable functional groups for the next reaction step, while autocleaving linkers are used to trigger macrocyclic ring-closure reaction at the end of library assembly.

TABLE 11.1 Applications Reported for the De Novo Discovery of Small Molecules

Reference	Size	Format[a]	Target Protein	Selection Results	Institution/Company	Section
Halpin & Harbury 2004 [59]	1 million	DNA routed	Anti-enkephalin mAb	Proof of principle	Stanford University	11.2.2.2
Gartner et al. 2004 [60]	65	DNA templated	Carbonic anhydrase	Proof of principle	Harvard University	11.2.2.3
Melkko et al. 2004 [92]	138	Dual pharmacophore (ESAC)	Streptavidin	Proof of principle	ETH Zürich	11.2.3
Melkko et al. 2007 [95]	550	Dual pharmacophore (ESAC)	Trypsin	98 nM (based on lead structure)	ETH Zürich	11.2.3
Wrenn and Harbury 2007 [115]	100 million	DNA routed	N-CrkSH3	16 μM (focused library)	Stanford University	11.2.2.2
Mannocci et al. 2008 [54]	4,000	DNA conjugate (2)	Streptavidin	0.35 μM (de novo)[b]	ETH Zürich/Philochem AG	11.2.2.1
Mannocci et al. 2008 [54]	4,000	DNA conjugate (2)	Poly-IgG	Chromatography resin	ETH Zürich/Philochem AG	11.2.2.1
Tse et al. 2008 [61]	13,824	DNA templated	Not reported		Harvard University	11.2.2.3
Dumelin et al. 2008 [64]	550	Dual pharmacophore (ESAC)	HSA	3.2 μM (de novo)[d]	ETH Zürich/Philochem AG	11.2.3
Scheuermann et al. 2008 [94]	550	Dual pharmacophore (ESAC)	MMP-3	9.9 μM (de novo)[d]	ETH Zürich/Philochem AG	11.2.3
Clark et al. 2009 [58]	ca. 7 million	DNA conjugate (3)	p38 MAP kinase	≤2 nM (based on lead structure)[c]	GSK/Praecis Pharmaceuticals	11.2.2.1
Clark et al. 2009 [58]	ca. 7 million	DNA conjugate (3)	Aurora A kinase	0.27–6.3 μM (de novo)[c]	GSK/Praecis Pharmaceuticals	11.2.2.1
Clark et al. 2009 [58]	ca. 800 million	DNA conjugate (4)	p38 MAP kinase	0.25 μM (de novo)[c,d]	GSK/Praecis Pharmaceuticals	11.2.2.1

Buller et al. 2009 [56]	4,000	DNA conjugate (2)	TNF	20 μM (de novo)[b]	ETH Zürich/Philochem AG	11.2.2.1
Buller et al. 2009 [56]	4,000	DNA conjugate (2)	Bcl-xL	10 μM (de novo)[b]	ETH Zürich/Philochem AG	11.2.2.1
Hansen et al. 2009 [90]	100	DNA templated (YoctoReactor)	Anti-enkephalin mAb	Proof of principle	Vipergen	11.2.2.3
Melkko et al. 2010 [65]	4,000	DNA conjugate (2)	Bcl-xL	0.93 μM (de novo)[b]	Philochem AG	11.2.2.1
Kleiner et al. 2010 [77]	13,824	DNA templated	Src kinase	0.68 μM (de novo)[c]	Harvard University	11.2.2.3
Mannocci et al. 2010 [62]	8,000	DNA conjugate (2)	Trypsin	3.0 nM (based on lead structure)[c]	Philochem AG	11.2.2.1
Buller et al. 2011 [116]	1 million	DNA conjugate (3)	CAIX	0.5 μM (based on lead structure)[c]	ETH Zürich/Philochem AG	11.2.2.1
Daguer et al. 2011 [100]	62,500	Dual pharmacophore	Carbonic anhydrase	0.3 μM (based on lead structure)	Université de Strasbourg-CNRS (ISIS)	11.2.3
Ciobanu et al. 2011 [99]	37,485	Dual pharmacophore	DC-SIGN	0.127 μM (based on lead structure)	Université de Strasbourg-CNRS (ISIS)	11.2.3
Leimbacher et al. 2012 [66]	30,000	DNA conjugate (2)	IL2	2.5 μM (de novo)[b]	ETH Zürich/Philochem AG	11.2.2.1

Source: From [44], with permission of The Royal Society of Chemistry.

[a] A number in parentheses indicates the number of different building blocks and coding regions used for library construction.

[b] K_D measurements by fluorescence polarization.

[c] IC_{50} measurements.

[d] Possible hydrolysis or alcoholysis products with about 7 nM potency have been reported [57].

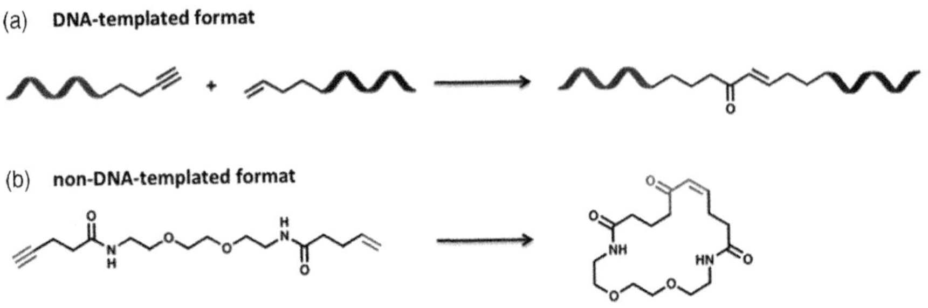

FIGURE 11.15 Characterization of a new alkyne–alkene macrocyclization reaction. (a) In DNA-templated format, the reaction was performed incubating the reagents with 500 mM Na_2PdCl_4 at 37°C for 1 h or at 25°C for 20 min. The reaction was analyzed by PAGE and MALDI–TOF mass spectrometric analysis. (b) In non-DNA-templated format, the macrocyclic enone product was characterized by ^{1}H NMR, ^{13}C NMR, COSY, UV–visible spectrometry, and high-resolution mass electrospray. The product formation probably proceeds through the following steps: formation of Pd(II)–alkynyl intermediate; insertion of the alkene into the Pd–alkyne bond; β-hydride elimination to form a conjugated enyne; Pd(II)-mediated hydration of the alkyne to form an enol π-allyl Pd complex; tautomerization and π-allyl Pd protonation to generate the trans enone.

Macrocyclic compounds represent a niche class of chemical scaffolds. Typically, they are constituted by 12 or more ring atoms and display several appendages around the cyclic structures [82] (Figure 11.16). Despite the relatively high molecular weight (500 to 2000 Da), such compounds are often capable of establishing very effective surface interactions with their target protein, modulating its biological function [82–85]. Additionally, several naturally occurring substances, which have been so far demonstrated to have highly relevant biological properties, often exhibit cyclic conformations very similar to synthetic macrocycles [82–85].

Macrocyclic compounds with sufficiently diverse structures and substituents are not easily accessible through traditional organic synthesis. Indeed, most of the preparations of large ring-compounds are plagued by slow rates of intramolecular

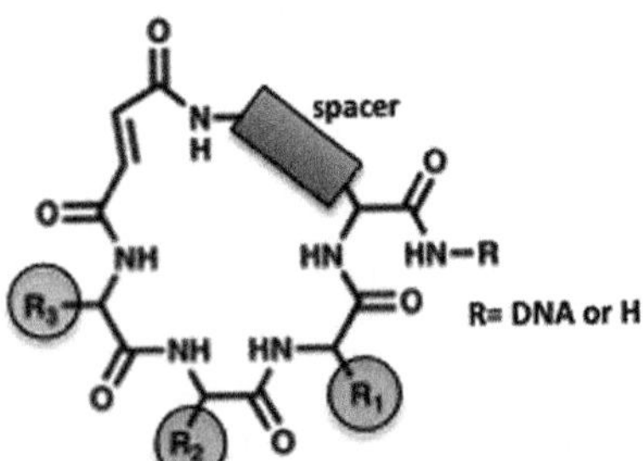

FIGURE 11.16 General structure of an ensemblins library compound. According to this structure, Ensemble has generated various libraries, each containing up to 6×10^4 members.

cyclization. Reactions between precursors (i.e., oligomerization and polymerization) become competitive and difficult to avoid [86]. Therefore, the principal method of favoring the intramolecular cyclization consists in slowing the rate of intermolecular reactions by performing the macrocyclization at very low reagent concentration (high dilution conditions) [86]. By contrast, in DNA-templated synthesis, the proximity effect (see above) is concentration independent and enables the simultaneous and specific assemble of thousands of macrocyclic variants in one pot. According to such a principle, Ensemble Thera-peutics has generated several peptidic macrocycle libraries, each containing up to 6×10^4 structures [81].

11.2.2.3.2 Chemetics Nuevolution is a drug discovery company headquartered in Copenhagen. Since 2001 it has developed several DNA-encoded chemical libraries employing conventional split–pool techniques (see Section 11.2.2.1) as well as its proprietary DNA-template-based technology (called *chemetic*). Although Nuevolution reported the use of chemetics in numerous patent applications on DNA-encoded chemical library assembly, to date there is no peer-reviewed publication exhaustively describing the methodology and how it is practically pursued for the construction of DNA-encoded chemical libraries.

Such an approach seems to rely primarily on the control of the association/dissociation equilibrium between the template and the reagent oligonucleotides to modulate the DNA-templated reactivity, and thus library construction.

In a 2004 U.S. patent application, the Danish company presented the first example of the chemetic strategy for library synthesis. In virtue of the proximity effect (see above), random DNA sequences mediated the enzymatic incorporation (e.g., by means of a polymerase) of unnatural DNA-nucleotide building blocks (each displaying two functional moieties and a cleavable linker) by a reaction between neighboring reactive groups (Figure 11.17a) [87]. Eventually, the linker cleavage yielded the final double-stranded DNA-encoded library (Figure 11.17a).

According to the authors, after library selection, the enriched DNA codons can be PCR-amplified and used as molecular evolution templates (see Section 11.2.2.2) into an additional round of library resynthesis and selection [87].

In a later patent, the construction strategy described before has been extended to the use of partially complementary DNA-hairpin templates to prime the annealing of short complementary DNA-conjugate reagents (Figure 11.17b) [88]. Following ligation of DNA-linked reagent to the hairpin coding sequence, denaturation of the duplex structure allows the DNA-templated reaction to occur (Figure 11.17b) [88].

In a more recent patent, Nuevolution described the latest example of chemetic strategy for the assembly of DNA-encoded chemical libraries (called *zipper-box approach*), as depicted schematically in Figure 11.17c [89]. Multiple DNA-conjugate reagents, each containing two hybridization domains, are annealed consecutively on an independent single-stranded DNA template ("zipper box"). According to the patent application, it is possible to modulate the corresponding DNA-templated reaction controlling the association/dissociation of two adjacent DNA-conjugate reagents (e.g., performing PCR-like thermal cycles or using appropriate high/low salt concentration buffers) [89]. If a bifunctional DNA-linked moiety is included, iteration of the

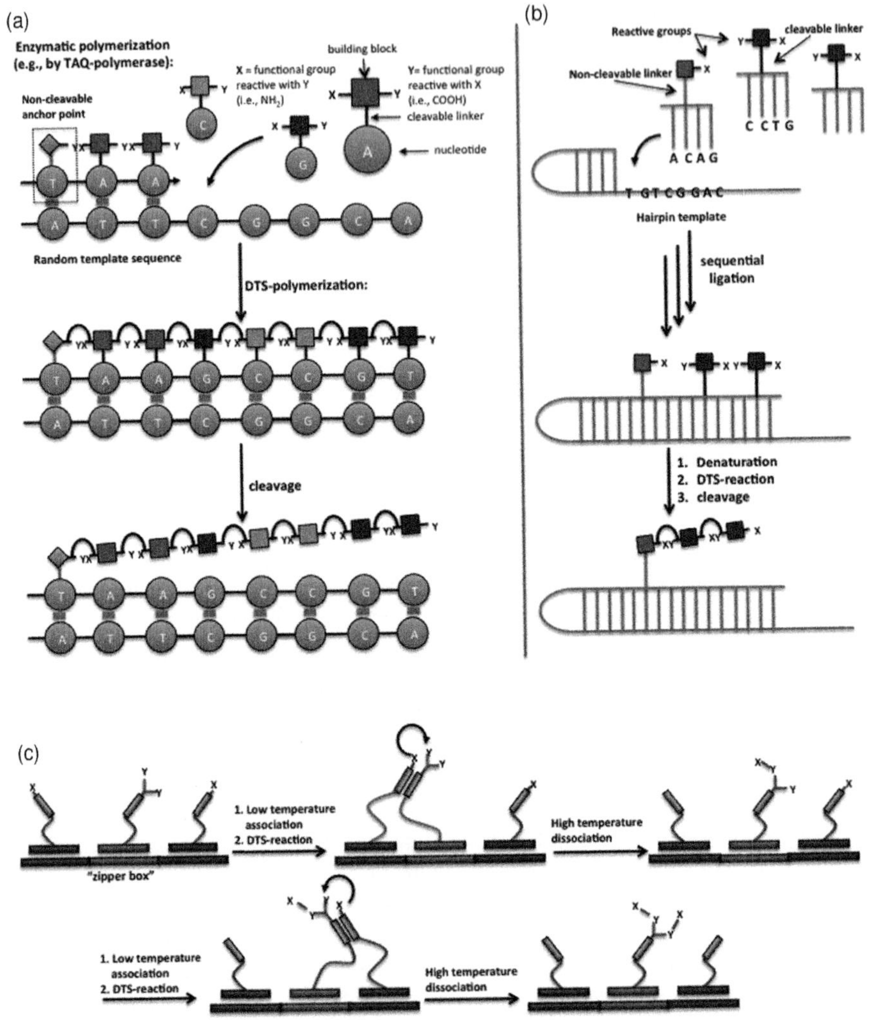

FIGURE 11.17 Nuevolution chemetic strategies for DNA-encoded chemical library assembly. (a) Random DNA sequences serve as a template for the enzymatic incorporation of unnatural DNA-nucleotide building blocks (each displaying two functional moieties and a cleavable linker). By virtue of the proximity effect, adjacent reactive groups polymerize. Linker cleavage yields to the final double-stranded DNA-encoded library. (b) DNA hairpin primes the annealing of short complementary DNA-conjugate reagents. Ligation and denaturation of the hairpin sequence allow DNA-templated reactions to occur. Cleavage of the linkers generates the final DNA-conjugate products. (c) Zipper-box approach: Multiple DNA-conjugate reagents, each containing two hybridization domains, are consecutively annealed on a complementary single-stranded DNA template. Association/dissociation of two adjacent DNA-conjugate reagents enables DTS reactions to take place. If a bifunctional DNA-linked moiety is included, the iteration of the association/dissociation cycle allows multiple one-pot DNA-templated reactions to occur on a single DNA-linked core structure.

association/dissociation cycle may allow multiple one-pot DNA-templated reactions to take place on a single DNA-linked core structure (Figure 11.17c) [89].

Nuevolution claims to be able to build a 100-million member library in two to three days, and today their libraries comprise over 1 billion chemical structures (see http://www.nuevolution.com/). Although Nuevolution has disclosed several collaborations with major pharmaceutical companies and to have assembled over 1 billion chemical structures with its proprietary technology, the role of chemetic in a DNA-encoded chemical library field and in drug discovery has yet to be demonstrated.

11.2.2.3.3 YoctoReactor Vipergen AS is a Danish company that has recently developed a unique DNA-directed approach for the construction of DNA-encoded chemical libraries, termed YoctoReactor technology (yR). The key feature of yR is the formation of a highly stable three-dimensional DNA complex called DNA junction (Figure 11.18a) [90,91]. Such architecture allows for confining suitable DNA-conjugate reagents within a volume on the order of a yoctoliter (10^{-24} liter, hence the name *YoctoReactor*) [90,91].

In virtue of the proximity effect (see Section 11.2.2.3), DNA-directed reactions take place at the center of the reactor, delivering the corresponding building blocks on the initial reaction site (Figure 11.18a). It is worth noting that in contrast to the DPC approach (see above) [70], in yR technology there is no structural distinction between a reagent and a template oligonucleotide. Each DNA-conjugate compound equally contributes to the three-dimensional structural framework of the junction, serving simultaneously as both template and reagent for another DNA-linked reactive moiety (Figure 11.18a).

Figure 11.18b depicts the use of Vipergen's approach to the synthesis of a three-way DNA-encoded chemical library. Before library assembly, individual building blocks are linked via cleavable or noncleavable linkers to distinct hairpin-shaped DNA oligonucleotides, each containing (1) an appropriate annealing domain for three-dimensional DNA-junction formation, as well as (2) a compound-specific coding sequence, at the distal end of the oligonucleotide [90,91].

Reactions for library assembling are programmed to occur in stepwise fashion following DNA-junction formation. Initially, DNA-building block conjugates $a_{1\text{-n}}$ and $b_{1\text{-m}}$ undergo the first DNA-directed reaction (Figure 11.18b). Following encoding of building block $b_{1\text{-m}}$ by ligation of the appropriate DNA extremities, cleavage of the linker enables the irreversible delivery of the compound to the initial reaction site (Figure 11.18b). Iteration of the process enables incorporation of additional building blocks (i.e., $c_{1\text{-p}}$) and corresponding DNA-coding tags. After completion of the programmed reactions, the molecular reactor is dismantled (performing a single round of PCR) and the final small-molecule library is displayed on linear double-stranded DNA templates (Figure 11.18b) [90,91].

Like other DNA-directed approach, yR is compatible with multiple rounds of selection and chemical translation (termed *rolling translation* by Vipergen) [48,70, 90]: PCR-amplified DNA-coding tags selected in the first round of panning serve as yR templates for synthesis of the next-generation library in the subsequent round of selection (Figure 11.18c).

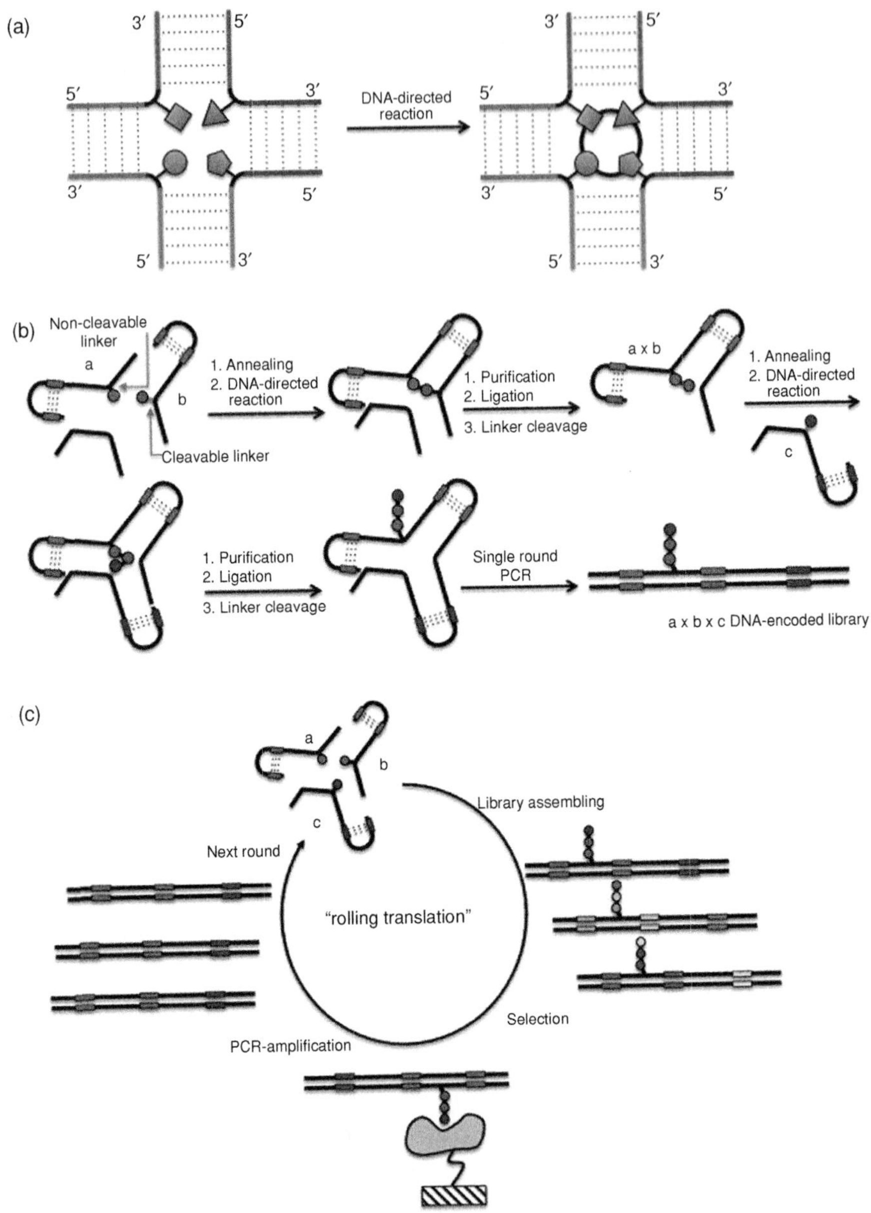

FIGURE 11.18 Vipergen's YoctoReactor technology. (a) A highly stable three-dimensional DNA complex (DNA junction) confines DNA-conjugate reagents within a volume on the order of 10^{-24} liter (yoctoliter). By means of the proximity effect, DNA-directed reactions take place at the center of the molecular reactor, delivering the corresponding building blocks on the initial reaction site. (b) Synthesis of a three-way DNA-encoded chemical library. Individual building blocks are linked via cleavable or noncleavable linkers to distinct hairpin-shaped DNA

In 2009, as a proof-of-concept demonstration, Vipergen applied the principles of yR technology to generate a 100-member DNA-encoded pentapeptide library. In a model selection conducted over two rounds of panning and library resynthesis, >150,000-fold enrichment of known ligand (ENK, [Leu]-enkephalin) toward the anti-ENK antibody 3-E was reported [49,90].

In principle, yR supports DNA frameworks with higher dimensions (e.g., four- and five-way DNA junctions) and the construction of DNA-encoded small molecule collections comprising up to billions of different compounds [90,91]. However, to date, there is no peer-reviewed report on the efficacy of the Vipergen's YoctoReactor approach and of its rolling translation for the de novo discovery of small-molecule ligands against a given biological target.

11.2.3 Self-Assembled DNA-Encoded Chemical Libraries (Dual-Pharmacophore Libraries)

Philochem AG, in collaboration with Neri's group at ETH Zürich, has pioneered a conceptually different approach for the construction of very large high-quality DNA-encoded chemical libraries. This strategy, also known as *encoded self-assembling chemical* (ESAC) *libraries*, features the combinatorial assembly of complementary sublibraries by means of Watson–Crick base pairing to form stable heterodimers (Figure 11.19a) [92]. The methodology can be extended to a triple-pharmacophore library by applying Hoogsteen base pairing (Figure 11.19a) [92]. Each sublibrary consists of an oligonucleotide containing both a tagging domain, univocally encoding the covalently linked chemical moiety (at either the 5′- or 3′-etremity), and a hybridization domain featuring self-assembling with a further sublibrary pool (Figure 11.19a) [63,92].

In sharp contrast with single-pharmacophore methodologies, the ESAC approach does not involve chemical reactions between building blocks to generate complex

FIGURE 11.18 (*Continued*) oligonucleotides. *a*, building block 1, and *b*, building block 2, undergo the first DNA-directed reaction. Following encoding of building block 2 by enzymatic ligation, cleavage of the linker enables irreversible delivery of the compound to the initial reaction site. Iteration of the process enables incorporation of the third set of building blocks (*c*) and corresponding DNA-coding tags. After the reactions are completed, the hairpin-DNA reactor is converted into a linear oligonucleotide, yielding to the final library of small molecules displayed on double-stranded DNA templates. (c) Rolling translation: PCR-amplified DNA-coding tags selected in the first round of panning serve as yR templates for chemical translation of the next-generation library in a subsequent round of selection. Before entering the next round of translation and selection, the double-stranded DNA templates are site-specifically digested to expose single-stranded DNA codons, which effectively allow hybridization with the corresponding building block conjugates. (*See insert for color representation of the figure.*)

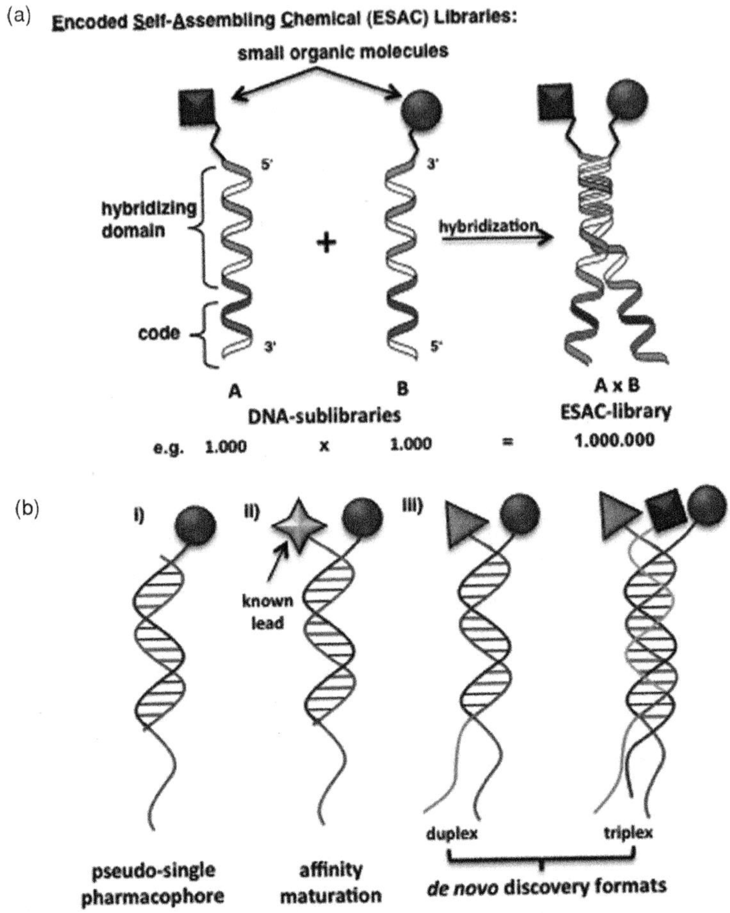

FIGURE 11.19 Encoded self-assembling chemical (ESAC) library technology. (a) Combinatorial self-assembly of relatively small partially complementary sublibraries (i.e., each containing 1000 members) by DNA hybridization allows the generation of large combinatorial libraries (i.e., 1 million combinations). Typically, the first sublibrary displays chemical entities attached at the 5′-extremity of single-stranded oligonucleotides, whereas in the second sublibrary the chemical moieties are connected to the 3′-extremity of the encoding oligonucleotides. The display of two neighboring chemical entities may enable simultaneous binding of two compounds to adjacent nonoverlapping binding sites on a target protein, in full analogy to fragment-based drug discovery approaches. (b) ESAC technology can be used in at least three different embodiments: (i) pseudo-single-pharmacophore format; (ii) affinity maturation setting; (iii) heteroduplex (or heterotriplex) library for de novo discovery of binding molecules.

libraries with multiple dimensions of diversities. Library members are individually synthesized (e.g., by coupling of a reactive moiety to amino-modified DNA fragment) and HPLC purified, thus enabling highly reliable quality controls (e.g., MS analysis, spectrophotometric quantification) and featuring almost 100% display for all library conjugates [63,92]. Moreover, by means of self-assembly, relatively small ESAC sublibraries (e.g., a comprising 1000 members) easily yield very large DNA-encoded repertoires (e.g., a 1,000,000-member ESAC library), retaining the initial high quality and purity (Figure 11.19a) [92].

ESAC technology can be used in at least three different embodiments (Figure 11.19b): (i) each sublibrary can be used independently for affinity capture and selection experiments (pseudo-single-pharmacophore format); (ii) a sublibrary can be annealed with an oligonucleotide displaying a known binder to the target and used for lead optimization selections (affinity maturation setup); and (iii) two (or three) independent sublibraries can be assembled to form a combinatorial heteroduplex (or heterotriplex) library in order to perform panning experiment for the de novo discovery of bi- or tridentate binding molecules [92].

It is worth noting that in analogy to traditional fragment-based methodologies, the compounds displayed in ESAC architecture are not covalently linked to each other. Therefore, the relative flexibility between the adjacent binding moieties may enable the simultaneous engagement of two distinct nonoverlapping binding sites on the same target protein surface, yielding to a substantial increase in binding affinity (chelate effect, Figure 11.20) [92,93].

Initially, microarray-based methodologies have been used for decoding in ESAC library selection experiments [63,64,94,95]. However, to avoid the loss of connectivity information between fragments during the readout, a novel deconvolution strategy compatible with high-throughput-sequencing decoding, in which essentially both coding regions of the DNA heteroduplexes are transferred on the same oligonucleotide strand, has recently been developed [96].

In analogy with traditional fragment-based drug discovery methodologies, preferential binding pairs have to be connected into a single drug-like high-affinity compound. Although determination of the optimal linkage (length, flexibility, and position) between the selected binding fragments selected by combinatorial medicinal chemistry can be relatively cumbersome, Melkko and co-workers successfully employed ESAC technology in various discovery campaigns, including the de novo identification of binders [94] as well as the improvement of suboptimal ligands [92,95].

In particular, ESAC technology has been applied to the identification of a general class of portable albumin binders based on 4-(*p*-iodophenyl)butyric acid structures [64]. Such a compound has been used to improve the pharmacokinetic properties of small organic molecules [97] as well as the tumor uptake of tumor-targeting antibodies [98]. Additionally, the newly discovered compound, in conjugation with various contrast agents (e.g., fluorescein or gadolinium complexes for MRI applications), was shown to considerably enhance their in vivo imaging performance [64].

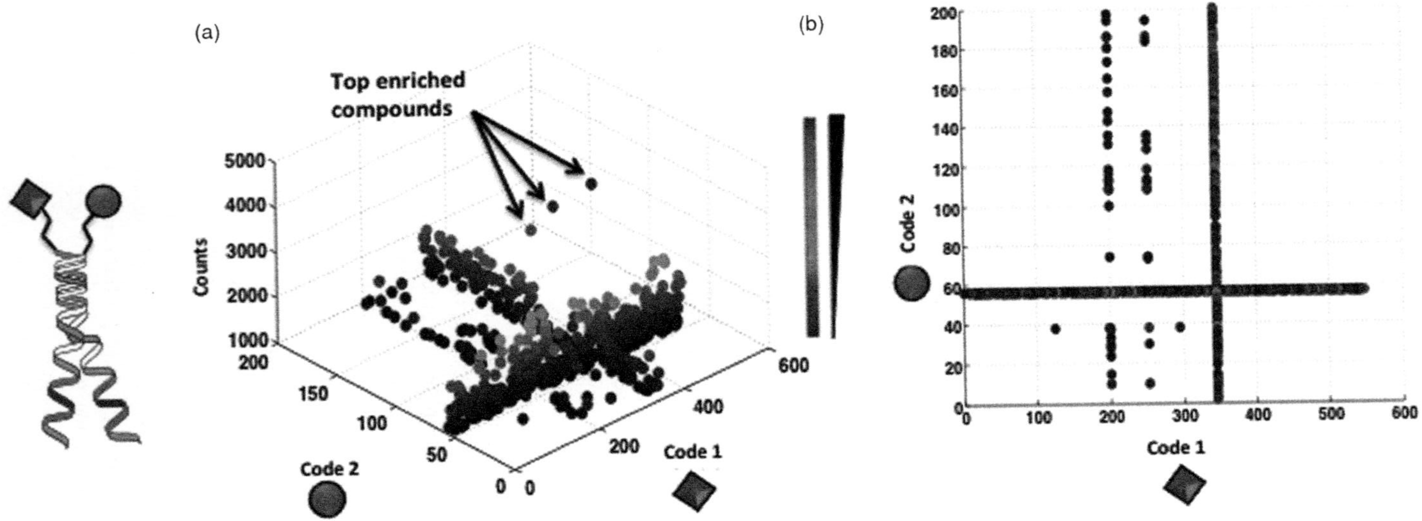

FIGURE 11.20 Example of the chelate effect. High-throughput-sequencing decoding of selection using an ESAC library, containing 120,000 different combinations (i.e., 200 × 600 compounds). Fingerprints revealed a preferential enrichment of compounds (indicated dots, in part a), at the cross points of three orthogonal lines (part b), thus indicating a simultaneous and cooperative engagement of two adjacent nonoverlapping binding sites on the same target protein by the two neighboring ESAC pharmacophores.

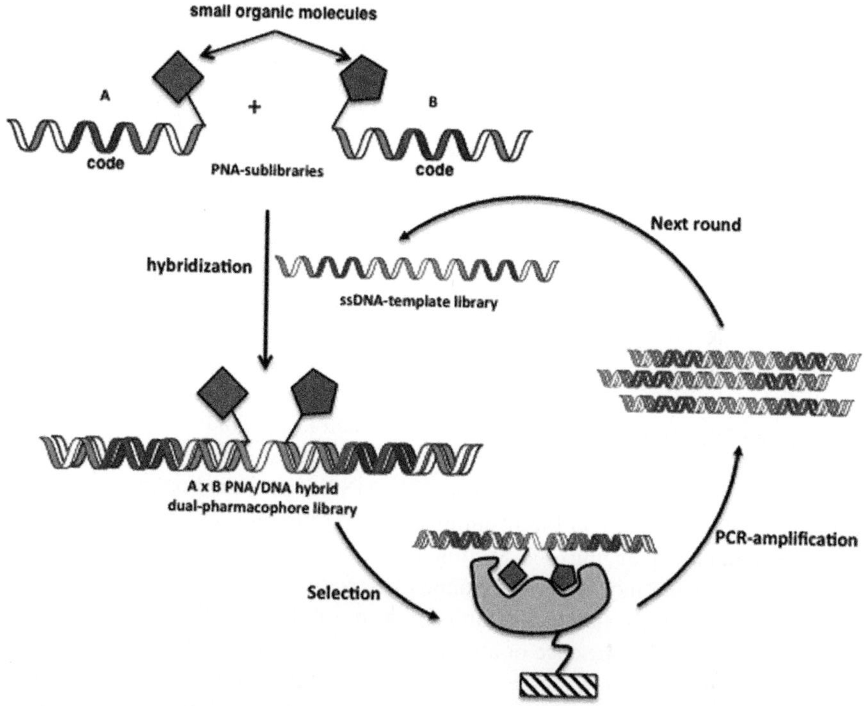

FIGURE 11.21 PNA/DNA combinatorial self-assembly encoded library. Two independent PNA sublibraries, each displaying a chemical moiety and a compound-specific coding region, are hybridized on a complementary DNA-template library. Following selection and PCR amplification of the DNA tags selected, dsDNA is converted into ssDNA. The second-generation library is reassembled, exposing the templates to an excess of the fragment PNA sublibraries.

More recently, Winssinger's group has developed a dual-pharmacophore strategy conceptually similar to ESAC technology, based on the combinatorial self-assembly of two PNA libraries on single-stranded DNA templates (Figure 11.21) [99,100]. This procedure allows for the formation of a hybrid PNA/DNA structure which encodes the chemical moieties in a manner compatible with PCR amplification (Figure 11.21).

In principle, all the chemistry can be performed in an automated fashion (using a common peptide synthesizer for the synthesis of the PNA conjugates and an oligonucleotide synthesizer for the DNA templates) [100]. However, for the hypothetical construction of a 1-million item compound library (1000 × 1000 library member), it would require the laborious and expensive synthesis of 2000 PNA conjugates as well as a tedious combinatorial enzymatic assembling of 1 million different DNA templates (e.g., by mix and split of 2000 preconstituted DNA templates).

Nonetheless, the strategy presented by Daguer et al. lends itself to the molecular evolution of its library members by the reannealing of PNA fragments after selection onto the initial DNA templates, a feature not yet anticipated by any other dual-pharmacophore technologies (Figure 11.21) [100].

11.3 SELECTION AND DECODING

11.3.1 In Vitro Selection Strategies

As described previously (see Sections 11.1.2 and 11.2.1), DNA-encoded chemical libraries can be selected ("panned") by affinity capture against virtually any target protein of interest (Figure 11.3) [101,102], in analogy with other display technologies (i.e., antibody-phage display) (Figure 11.1b) [40,103,104]. The libraries are typically incubated with the target protein of choice immobilized on a solid support at very low concentration (picomolar to femtomolar concentration of individual library members). Subsequently, repeated washing steps allow for separating the "bound" fraction from the "unbound" portion of the library. The DNA tags of the bound library members are PCR-amplified and the resulting amplicon mixture deconvoluted (e.g., by high-throughput sequencing, see Section 11.3.2) in order to determine the relative concentration of the individual compounds before and after applying the selection pressure [54,56,58]. Eventually, identified hits are validated in biological and/or biochemical assays in the presence or absence of the DNA tag [54,56,58].

It is worth noting that in sharp contrast with traditional screening methodologies that discretely assess the biological properties of the individual molecules, selection strategies are capable of interrogating the entire displayed library simultaneously, imposing the same selection pressure to all library members (see also Section 11.1.2).

As for antibody phage display, in vitro selections with DNA-encoded chemical libraries can be performed employing various methodologies and conditions. The choice of the most suitable strategy depends primarily on the nature of the target protein and the scope of the selection [104]. So far, the use of antigens immobilized on CNBr-activated sepharose resin and His-tagged proteins on metal affinity chromatography resin (IMAC) has proved to be the most versatile and efficient method of processing DNA-encoded chemical libraries [54–56,65,94].

Very recently, a novel technology termed *interaction-dependent PCR* (IDPCR) has been developed to perform one-pot multiple selections in the homogeneous phase without the need to immobilize the target protein on a solid support [105]. In essence, binding between DNA-conjugated targets and ligands assists the formation of double-stranded DNA hairpins otherwise too short to hybridize. The hairpin duplex is then used to prime the final PCR amplification [105].

Selection stringency can be enhanced by increasing the number of washing steps or varying the concentration/coating density of the target protein [62,95]. Blocking agents (e.g., hearing sperm DNA, bovine serum albumin) and detergents (e.g., Tween-20) have been demonstrated to prevent unspecific binding and improve the accuracy of the final fingerprints [54,56,58,65,94].

Last but not least, as for other display technologies, selections using DNA-encoded chemical libraries can be directed toward a specific epitope or pocket of the target protein of choice, either by displacement of a known ligand or by performing multiple panning experiments on a set of related proteins [62,103].

11.3.2 Decoding of DNA-Encoded Chemical Libraries

As mentioned elsewhere, after panning the DNA tags of the compounds selected are amplified by PCR. Binding hits are identified by comparing the relative frequency of the individual library member before and after imposing selection pressure [54–56,62].

During the initial development of DNA-encoded chemical libraries, subcloning and sequencing of the individual colonies have been used to decode the amplicon mixtures [60,71,92]. Although Sanger sequencing offers a relatively long and accurate sequencing read length, it is cost-effective up to 1000 sequences and only useful when the amplicon population converges rapidly to a small number of active sequences (e.g., decoding of library mixtures containing up to 10^2 compounds).

To get high-resolution maps of large DNA-encoded chemical libraries, microarray-based techniques have been developed. Hybridization of fluorescent DNA amplicons to complementary DNA microarrays proved to be a versatile strategy, capable of giving a global profiling of the DNA distribution without being subject to statistical effects as when sampling individual population members [63,80,92,106]. However, due to physical limitations on the number of probes that can be arrayed over the support and the low discrimination rate of closely related DNA sequences, this technology could not be implemented conveniently for the deconvolution of DNA mixtures containing more than a few thousand different DNA codes.

Recent advances in ultrahigh-throughput DNA sequencing, which enable the inexpensive analysis of more than 10 million sequences per sequencing run [107–111], provided an invaluable tool for the decoding of last-generation DNA-encoded chemical libraries. Today, using state-of-the-art high-throughput-sequencing platforms, it is possible to acquire selection information about libraries containing millions of compounds in less than a few days and at a cost of about 1000 euros per selection.

In 2008, Mannocci et al. described the first implementation of Roche's 454 high-throughput-sequencing technology for the decoding of DNA-encoded chemical libraries [54]. The technology, based on emulsion PCR and subsequent pyrosequencing of DNA-coated beads on ultrahigh-density picoliter plate, was initially able to provide approximately 1 million sequence tags per sequencing experiment [107]. To date, the latest generation of the 454 GS FLX titanium sequencer yields up to 3 million sequences with a 400- to 500-bp read length in a run of about 10 h [108–111].

More recently, an Illumina/Solexa platform [112] has been used in the decoding of DNA encoded chemical libraries [55,113]. Such technology relies on the attachment of single-stranded DNA fragments to a flow cell surface. Bridge amplification of the individual DNA template hybridized on the surface enables sequencing by synthesis by means of reversible terminators. Although compared to Roche's 454 technology, Illumina yields a shorter read length (up to 100 bp) and requires a longer run time (several days vs. 10 to 24 h for Roche's 454), the Illumina platform generates approximately 40 times more sequencing data (about 20 Gb) per sequencing run at a similar cost [108–111]. For reasons such as these, Illumina technology quickly became the industry standard for DNA-encoded chemical library decoding.

In general, high-throughput-sequencing decoding simply requires the PCR insertion of a platform-specific DNA adapter into the selection amplicon sequences. Therefore, PCR bar coding of individual selection experiments is often used to maximize the capabilities of the sequencing throughput, facilitating the parallel analysis of multiple selections in a single sequencing run.

Depending on the library format, sequence counts obtained after deconvolution of sequencing results can conveniently be plotted as depicted in Figure 11.22a to c. However, statistical analyses are required to accurately describe the distribution of library sequences before and after the selection process [56]. For example, quantile–quantile plots and fitted negative binomial density functions allow the determination of p values and of other significant statistical parameters, crucial for the reliable identification of the compounds enriched (Figure 11.22d) [56].

As DNA-encoded chemical libraries are growing steadily in size [58], alternative high-throughput-sequencing technologies may also be considered (e.g., SOLiD sequencing [108]) for efficient library decoding. Perhaps advances in technologies based on single-molecule sequencing [114] may further facilitate the processing of a very large DNA-encoded chemical library of up to 1 billion chemical entities in just a few days.

11.4 DRUG DISCOVERY BY DNA-ENCODED CHEMICAL LIBRARIES

Over the past decade, several pharmaceutical companies and academic laboratories have integrated DNA-encoded strategies in the de novo discovery of small molecules capable of selective binding to specific biological targets. Table 11.1 summarizes a number of such applications as they have been reported by some of the major companies and academic groups currently active in this field.

Remarkable efforts in the implementation of DNA-encoded chemical library technology for drug discovery purposes have been reported by GlaxoSmithKline (formerly Praecis Pharmaceuticals) and by Philochem AG in collaboration with ETH Zürich.

In 2008, Dumelin et al. at Philochem described the use of DNA-encoded chemical libraries for the identification and characterization of a novel class of highly specific portable albumin binders based on 4-(p-iodophenyl)butyric acid structure, with affinities to human serum albumin ranging from 3.2 to 55 μM (entry 9, Table 11.1) [64]. Conjugates of the highest-affinity ligand to a general carbonic anhydrase inhibitor (acetazolamide) or to antibody fragments (specific for tumor-associated antigens) have been demostrated to improve the pharmacokinetic properties of the small organic molecules [96] as well as the tumor uptake of tumor-targeting antibodies [98]. Additionally, conjugation of the newly discovered albumin binder to blood pool contrast agents (fluorescein and Gd-DTPA) showed a significantly increase in their

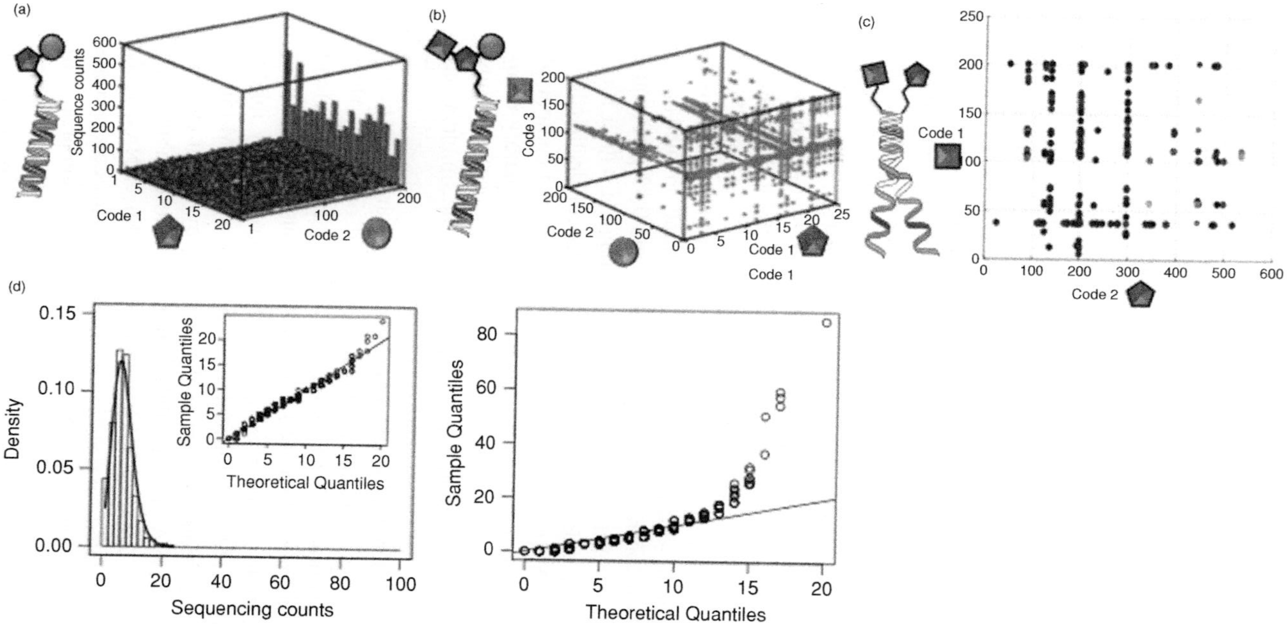

FIGURE 11.22 Decoding of DNA-encoded chemical libraries. (a) High-throughput sequencing fingerprint resulting from a two-building-block single-pharmacophore DNA-encoded chemical library comprising 200 × 20 compounds. Bars at the right end of the fingerprint indicate a series of structurally related compounds preferentially enriched over the background. (b) Pseudo-four-dimensional plot of the high-throughput-sequencing results obtained after selection of a three-building-block single-pharmacophore library comprising 1 million compounds (200 × 200 × 25). The three codes are presented on the axis (x, y, z), while the dot size correlates with the sequence counts. In the oval a family of structurally related compounds with the same building blocks 1 and 2 but a different building block 3 is highlighted. (c) Two-dimentional representation of high-throughput-sequencing selection results using a dual-pharmacophore (ESAC) library. The dot shades correlate with the sequence counts. (d) Statistical analysis of high-throughput-sequencing decoding results. Fitting of the sequence counts following high-throughput sequencing of the library prior to selection allows determination of the negative binomial density function. Subsequently, for each experimental sequence count, the corresponding quantile plot is compared to the same quantile derived from the fitted negative binomial density function (inset). Before applying selection pressure, no significant deviations from the diagonal are observed, indicating good model fit (inset). After selection, substantial deviations from the diagonal reveal enriched candidate compounds.

circulatory half-life and superior imaging performance with respect to the parental compounds [64].

ESAC dual-pharmacophore libraries (see Section 11.2.3) have led to the identification of novel inhibitors of stromelysin-1 (MMP-3), a matrix metalloproteinase involved in both physiological and pathological tissue remodeling processes (entry 10, Table 11.1) [94]. Initially, a 550-memebred ESAC sublibrary was panned on human MMP-3 covalently coupled to CNBr-activated Sepharose resin. Microarray-based decoding enabled the identification of a lead binding structure, which was subsequently conjugated to the amino-modified 3′-extremity of a 24-mer oligonucleotide capable of pairing with the previous ESAC sublibrary and used as a lead in a further affinity maturation selection experiment. Eventually, decoding and chemical conjugation of synergistically binding molecules in the absence of the DNA appendage yielded a MMP3 inhibitor with an IC_{50} value as high as 9.9 μM [94].

The advent of high-throughput-sequencing technologies for decoding of DNA-encoded chemical libraries enormously facilitates the identification of bioactive molecules from libraries of unprecedented size (see Section 11.3.2) [54]. Indeed, shortly afterward, Philochem in collaboration with MedImmune (formerly Cambridge Antibody Technology) described the implementation of high-throughput sequencing for the discovery of protein–protein interaction inhibitors. Selections against Bcl-xL, an antiapoptotic target protein for cancer therapy, using a 4000-member DNA-encoded library (two building blocks single pharmacophore format; see Section 11.2.2.1) allowed for the identification of several structurally related binding compounds with dissociation constants ranging between 60 and 0.93 μM (entry 17, Table 11.1) [65]. The highest-affinity binder was able to compete with a Bak-derived BH3 peptide (an antagonist of Bcl-xL function) and to induce cell death in Raji cells with an EC_{50} value of 77 μM [65]. Notably, the newly discovered compound displayed an indomethacin moiety, a previously reported anti-inflammatory compound capable of inducing apoptosis in various cancer cell lines [117–119], probably by suppression of the STATs/Bcl-xL signaling pathway [120] and the simultaneous inhibition of multidrug-resistance protein 1 function [121].

In 2009, together with ETH Zürich, Philochem described a selection against tumor necrosis factor (TNF) deploying a 4000-compound library based on Diels–Alder cycloaddition reactions (see Section 11.2.2.1; entry 14, Table 11.1) [56]. Among the selected structures, the best ligand displayed a dissociation constant of 20 μM. Medicinal chemistry optimization yielded a compound able to completely inhibit TNF-mediated killing of L-M fibroblast at concentrations as high as 300 μM [56].

Diels–Alder cycloaddition reaction has been explored further by Buller et al. in the construction of a 1-million DNA-encoded library based on the combinatorial assembly of three different sets of building blocks and corresponding DNA-coding fragments [55]. This library facilitated the isolation of several inhibitors against the tumor-associated antigen carbonic anhydrase IX [97], with IC_{50} as potent as 240 nM (entry 20, Table 11.1) [55]. A fluorescently labeled derivative of one of the most potent compounds revealed an in vitro specific accumulation in hypoxic tumor tissues overexpressing carbonic anhydrase IX and in vivo tumor targeting of human colorectal adenocarcinoma and human renal cell carcinoma xenograft models [55].

Very recently, Philochem and ETH Zürich used a high-quality DNA-encoded chemical library comprising 30,000 drug-like compounds to identify specific ligands to the proinflammatory cytokine interleukin-2 (IL-2) (entry 23, Table 11.1) [66]. After panning and decoding of the library, the isolated compounds exhibited affinity to the target in the low-micromolar range ($K_D = 2.5$ μM) and a complete and selective inhibition of IL-2 activity in a T-cell proliferation assay ($IC_{50} = 32$ μM), without any cytotoxicity to primary fibroblast cell cultures up to 128 μM concentrations [66]. Additionally, the binding mode of the newly discovered compound class was confirmed by computational docking based on the crystal structure of IL-2 [66,121]. The docking revealed the 2-methyl-1*H*-indole moiety of 67 of 100 structurally related compounds positioned in the same hydrophobic groove of the cytokine with a strong propensity for binding aromatic structures [66,121].

In 2009, Praecis/GSK reported the use of high-throughput-sequencing techniques for the decoding of very large DNA-encoded chemical libraries containing 7 million and 800 million DNA-encoded compounds, respectively (see Section 11.2.2.1; entries 11 to 13, Table 11.1) [58]. Selection experiments yielded to the identification of inhibitors against Aurora A and p38 MAP kinases with submicromolar potencies [58].

In 2010, Kleiner et al. described the first application of a DNA-templated library for the de novo discovery of bioactive molecules (entry 18, Table 11.1) [77]. Selections using a 13,824-member macrocycle library against 36 different target proteins enabled the isolation of novel classes of kinases modulators [77]. Compounds selected included ATP-competitive Src inhibitors (with IC_{50} as potent as 680 and 960 nM), p38α-MAPKAP2 cascade inhibitor (11 μM) as well as VEGFR2 activators (up to 300% activity enhancement upon treatment with a 100μM compound) [77].

DNA-encoded chemical libraries have also been successfully used to convert previously discovered lead structures into new derivatives with enhanced potency and specificity. Indeed, benzamidine-based dual- and single-pharmacophore lead optimization (affinity maturation) libraries allowed the isolation of up to over 10,000-fold improved trypsin inhibitor, with IC_{50} in the single-digit nanomolar range and excellent selectivity with respect to other related serine proteases (entries 4 and 19, Table 11.1) [62,95]. Eventually, rapid and efficient DNA-encoded affinity maturation strategies may be alternative (or complementary) to conventional medicinal chemistry approaches to the optimization of compounds hits.

11.5 DNA-ENCODED CHEMICAL LIBRARIES: PROSPECTS AND OUTLOOK

The appeal of DNA-encoded chemical library technology is indubitably due to the stunning sensitivity of DNA encoding/decoding and the unrivalled opportunity to generate ready-to-screen chemical libraries comprising thousands to millions of small organic molecules in a single test tube [54,55,58,90,96].

A DNA-encoded library expands the realm of selection approaches, so far used exclusively for the efficient isolation of high-affinity-binding polypeptides, to the world of small organic molecules. En masse interrogation of chemical libraries by means of affinity-based selection strategies offers enormous technical benefits over conventional screening approaches. First, the cost, time, robotics, and storage needs are radically minimized. Second, the simple affinity capture of small molecule on an immobilized target protein does not require the development of complicated biological assays. As a result, artifacts commonly observed in traditional high-throughput-screening campaigns are circumvented (e.g., compound aggregation, insolubility, abnormal fluorescence absorption/quenching), and the deploying of target proteins, such as components involved in protein-protein interactions, particularly challenging to tackle with conventional screening approaches, is allowed, [56,65]. Third, selection procedures compel low amounts of target protein (i.e., a few milligrams), and multiple panning experiments can be performed rapidly in parallel, applying various experimental conditions (e.g., different protein supports, coating density, number of washing steps, re-panning) or adopting alternative selection strategies (e.g., presence of related proteins, competition with ligands previously discovered, addition or absence of cofactors and substrates).

In principle, other naturally (i.e., RNA) and nonnaturally (i.e., PNA) occurring nucleic acid residues have been considered for the encoding of chemical libraries. However, ribonucleic acids are very susceptible to RNAse cleavage, thus making the handling and storage of RNA-encoded libraries very inconvenient. Conversely, more stable and chemically versatile macromolecules, such as peptide nucleic acids [122], cannot be amplified by PCR, restraining their use to libraries of smaller size (up to about 10^4 compounds), in combination with low-throughput decoding strategies such as microarray-based technologies [123–126].

DNA-encoded chemical library technology has benefited profoundly from the latest high-throughput-sequencing developments. Cutting-edge deep-sequencing platforms, capable of gathering millions of DNA-sequence information per sequencing run in just a few days [107,111,127,128], allowed simultaneous evaluation of thousands to millions of structurally related compounds (including stereoisomers and enantiomers). This approach straightforwardly provided instant SAR databases after each selection experiment, invaluable sets of information for the design and improvement of lead structures by medicinal chemistry optimization and/or further DNA-encoded affinity maturation cycles (see Section 11.2.2.1) [54,55].

To date, sequencing throughput defines itself the natural limit for the largest library, which can be consistently interrogated. As shown in Figure 11.22d (Section 11.3.2), rigorous statistical analysis is indispensable for establishing an accurate correlation between sequencing counts and the relative abundance of the individual library compounds before and after imposing the selection pressure. Therefore, it is crucial to obtain a high degree of sequencing coverage after decoding with respect to library size. For the appropriate assessment of a selection experiment using a 1 million compound library, approximately 10 million raw sequencing "reads" are typically required. Considering that sequencing power is currently settled to about 50 million DNA tags per run (at a cost of about 5000 euros), it is not yet conceivable to screen DNA-encoded libraries comprising more than a few million member compounds. Although several strategies have been implemented for the construction of single-pharmacophore DNA-encoded chemical libraries, the performance of a library ultimately depends on its design and purity. If libraries can be constructed including up to billions of chemical entities, the gain in library size and chemical diversity is often paralleled by an undesirable increase in the average molecular weight beyond the generally accepted drug-like criteria (following Lipinski's rule of 5) [67,68]. Furthermore, the number of reaction steps performed for library assembly is inversely proportional to library quality, due to incomplete reactivity of the reactions used in the split–pool strategy [58]. Eventually, single-pharmacophore libraries generated by the combinatorial assembly of two or three different sets of building blocks and containing up to a few million small-molecules typically display compounds that better fulfill drug-like requirements and are more suitable for further medicinal chemistry optimization steps (Figure 11.20).

In contrast to single-pharmacophore libraries, large dual-pharmacophore collections can easily be generated with unrivaled purity by self-assembly of relatively small DNA-encoded sublibraries (i.e., 10^3 to 10^4 member) [63]. The dual display of two independent chemical moieties on the distinct extremities of the double-stranded oligonucleotide (see Figure 11.16, Section 11.2.3) may enable the simultaneous engagement of adjacent nonoverlapping binding pockets (or clefts) on the same target protein [93]. In analogy to fragment-based drug discovery approaches, the individual moieties must be reformatted into a single organic molecule by means of appropriate linkers and linking chemistry [63,64,94,95].

Last but not least, DNA-encoded chemical libraries are particularly appealing for the optimization (affinity maturation) of previously discovered lead binding molecules. Recently, lead-based DNA-encoded libraries have been used for the identification of compounds with >10,000-fold improved potency with respect to the parental compound and exquisite selectivity toward the selected target protein [62]. Application of such libraries to drug discovery is part of the most recent research in the DNA-encoded chemical library field (see Table 11.1, Section 11.4).

11.6 CONCLUSIONS

In conclusion, DNA-encoded chemical libraries are coming of age. Remarkable developments have been gained rapidly over the last decade. Today, DNA-encoded chemical libraries hold the promise to revolutionize the way to discover novel ligands, pointing to a future where scientists can design and routinely en masse interrogate millions of chemical compounds, ideally using fully integrated platforms on which synthetic compounds such as like chemical genes truly evolve through the selection steps as components of an artificial immune system, which quickly raises high-specific-binding compounds against profoundly different biomacromolecular antigens.

While we wait for the first drug candidate in clinical trials stemming from a DNA-encoded chemical library, several pharmaceutical industrial and academic laboratories have already implemented DEL technologies in various drug discovery research programs (see Table 11.1). Expanding the palette of DNA-compatible chemical reactions for ad hoc library synthesis, as well as performing selections on protein targets in native environment (e.g., using living cells or cell lysate), remain the most challenging tasks for future advances of this technology in the drug discovery field. Only time will tell whether DNA-encoded chemical library technology will play an important role in drug discovery and be able to cope with third-millennium pharmaceutical challenges [8,129].

Acknowledgments

I would like to thank all the people, research groups, and companies mentioned above who have provided their exceptional work and experience for the preparation of this chapter. Particularly, I heartily thank Yixin Zhang for critical reading of the manuscript and invaluable and inspiring contributions.

REFERENCES

1. S. D. Patterson, R. H. Aebersold, *Nat. Genet.* **2003**, *33*, 311–323.
2. R. B. Stoughton, *Annu. Rev. Biochem.* **2005**, *74*, 53–82.
3. O. Morozova, M. Hirst, M. A. Marra, *Annu. Rev. Genom. Hum. Genet.* **2009**, *10*, 135–151.
4. B. Domon, R. Aebersold, *Science* **2006**, *312*, 212–217.
5. R. Kramer, D. Cohen, *Nat. Rev. Drug Discov.* **2004**, *3*, 965–972.
6. R. L. Strausberg, S. L. Schreiber, *Science* **2003**, *300*, 294–295.
7. J. P. Overington, B. Al-Lazikani, A. L. Hopkins, *Nat. Rev. Drug Discov.* **2006**, *5*, 993–996.
8. V. J. Haupt, M. Schroeder, *Brief. Bioinf.* **2011**, *12*, 312–326.
9. T. N. Raju, *Lancet* **2000**, *355*, 1022.
10. B. Munos, *Drug Discov.* **2009**, *8*, 959–968.
11. P. J. Hajduk, J. Greer, *Nat. Rev. Drug Discov.* **2007**, *6*, 211–219.
12. J. Drews, *Science* **2000**, *287*, 1960–1964.
13. S. Ghosh, A. Nie, J. An, Z. Huang, *Curr. Opin. Chem. Biol.* **2006**, *10*, 194–202.
14. P. Schneider, Y. Tanrikulu, G. Schneider, *Curr. Med. Chem.* **2009**, *16*, 258–266.
15. R. A. Houghten, C. Pinilla, S. E. Blondelle, J. R. Appel, C. T. Dooley, J. H. Cuervo, *Nature* **1991**, *354*, 84–86.
16. K. S. Lam, S. E. Salmon, E. M. Hersh, V. J. Hruby, W. M. Kazmierski, R. J. Knapp, *Nature* **1991**, *354*, 82–84.
17. M. Pellecchia, I. Bertini, D. Cowburn, C. Dalvit, E. Giralt, W. Jahnke, T. L. James, S. W. Homans, H. Kessler, C. Luchinat, B. Meyer, H. Oschkinat, J. Peng, H. Schwalbe, G. Siegal, *Nat. Rev. Drug Discov.* **2008**, *7*, 738–745.
18. D. C. Rees, M. Congreve, C. W. Murray, R. Carr, *Nat. Rev. Drug Discov.* **2004**, *3*, 660–672.
19. D. S. Tan, *Nat. Chem. Biol.* **2005**, *1*, 74–84.
20. D. Morton, S. Leach, C. Cordier, S. Warriner, A. Nelson, *Angew. Chem. Int. Ed.* **2009**, *48*, 104–109.

21. S. L. Schreiber, *Science* **2000**, *287*, 1964–1969.
22. D. B. Kitchen, H. Decornez, J. R. Furr, J. Bajorath, *Nat. Rev. Drug Discov.* **2004**, *3*, 935–949.
23. J. Bajorath, *Nat. Rev. Drug Discov.* **2002**, *1*, 882–894.
24. T. Scior, P. Bernard, J. L. Medina-Franco, G. M. Maggiora, *Mini Rev. Med. Chem.* **2007**, *7*, 851–860.
25. L. M. Mayr, D. Bojanic, *Curr. Opin. Pharmacol.* **2009**, *9*, 580–588.
26. J. McCafferty, A. D. Griffiths, G. Winter, D. J. Chiswell, *Nature* **1990**, *348*, 552–554.
27. G. Winter, A. D. Griffiths, R. E. Hawkins, H. R. Hoogenboom, *Annu. Rev. Immunol.* **1994**, *12*, 433–455.
28. G. P. Smith, *Science* **1985**, *228*, 1315–1317.
29. T. Clackson, H. R. Hoogenboom, A. D. Griffiths, G. Winter, *Nature* **1991**, *352*, 624–628.
30. E. T. Boder, K. D. Wittrup, *Nat. Biotechnol.* **1997**, *15*, 553–557.
31. J. Hanes A. Pluckthun, *Proc. Natl. Acad. Sci. U.S.A.* **1997**, *94*, 4937–4942.
32. J. M. Kim, H. J. Shin, K. Kim, M. S. Lee, *Mol. Biotechnol.* **2007**, *36*, 32–37.
33. R. W. Roberts, J. W. Szostak, *Proc. Natl. Acad. Sci. U.S.A.* **1997**, *94*, 12297–12302.
34. A. D. Keefe, J. W. Szostak, *Nature* **2001**, *410*, 715–718.
35. D. S. Wilson, A. D. Keefe, J. W. Szostak, *Proc. Natl. Acad. Sci. U.S.A.* **2001**, *98*, 3750–3755.
36. J. Bertschinger, D. Grabulovski, D. Neri, *Protein Eng. Des. Sel.* **2007**, *20*, 57–68.
37. C. Heinis, T. Rutherford, S. Freund, G. Winter, *Nat. Chem. Biol.* **2009**, *5*, 502–507.
38. M. G. Cull, J. F. Miller, P. J. Schatz, *Proc. Natl. Acad. Sci. U.S.A.* **1992**, *89*, 1865–1869.
39. M. Silacci, S. Brack, G. Schirru, J. Mårlind, A. Ettorre, A. Merlo, F. Viti, D. Neri, *Proteomics* **2005**, *5*, 2340–2350.
40. K. M. Esvelt, J. C. Carlson, D. R. Liu, *Nature* **2011**, *472*, 499–503.
41. H. K. Binz, A. Pluckthun, *Curr. Opin. Biotechnol.* **2005**, *16*, 459–469.
42. B. Liu, L. Huang, C. Sihlbom, A. Burlingame, J. D. Marks, *J. Mol. Biol.* **2002**, *315*, 1063–1073.
43. G. Walsh, *Nat. Biotechnol.* **2010**, *28*, 917–924.
44. L. Mannocci, M. Leimbacher, M. Wichert, J. Scheuermann, D. Neri, *Chem. Commun.* **2011**, 12747–12753.
45. J. D. Watson, F. H. Crick, *Nature* **1953**, *171*, 737–738.
46. J. D. Watson, F. H. Crick, *Nature* **1953**, *171*, 964–967.
47. S. Brenner, R. A. Lerner, *Proc. Natl. Acad. Sci. U.S.A.* **1992**, *89*, 5381–5383.
48. J. Nielsen, S. Brenner, K. D. Janda, *J. Am. Chem. Soc.* **1993**, *115*, 9812–9813.
49. T. Meo, C. Gramsch, R. Inan, V. Höllt, E. Weber, A. Herz, G. Riethmüller, *Proc. Natl. Acad. Sci. U.S.A.* **1983**, *80*, 4084–4088.
50. M. C. Needels, D. G. Jones, E. H. Tate, G. L. Heinkel, L. M. Kochersperger, W. J. Dower, R. W. Barrett, M. A. Gallop, *Proc. Natl. Acad. Sci. U.S.A.* **1993**, *90*, 10700–10704.
51. L. McConlogue, M. A. Brow, M. A. Innis, *Nucleic Acids Res.* **1988**, *16*, 9869.
52. A. W. Czarnik, *Proc. Natl. Acad. Sci. U.S.A.* **1997**, *94*, 12738–12739.
53. I. Hughes, in: *Drug Discovery and Development*, Vol. 1, M. S. Chorghade, Ed., Wiley, Hoboken, NJ, **2006**, pp. 129–167.

54. L. Mannocci, Y. Zhang, J. Scheuermann, M. Leimbacher, G. De Bellis, E. Rizzi, C. Dumelin, S. Melkko, D. Neri, *Proc. Natl. Acad. Sci. U.S.A.* **2008**, *105*, 17670–17675.
55. F. Buller, M. Steiner, K. Frey, D. Mircsof, J. Scheuermann, M. Kalisch, P. Buhlmann, C. T. Supuran, D. Neri, *ACS Chem. Biol.* **2011**, *6*, 336–344.
56. F. Buller, Y. Zhang, J. Scheuermann, J. Schäfer, P. Bühlmann, D. Neri, *Chem. Biol.* **2009**, *16*, 1075–1086.
57. F. Buller, L. Mannocci, Y. Zhang, C. E. Dumelin, J. Scheuermann, D. Neri, *Bioorg. Med. Chem. Lett.* **2008**, *18*, 5926–5931.
58. M. A. Clark, R. A. Acharya, C. C. Arico-Muendel, S. L. Belyanskaya, D. R. Benjamin, N. R. Carlson, P. A. Centrella, C. H. Chiu, S. P. Creaser, J. W. Cuozzo, C. P. Davie, Y. Ding, G. J. Franklin, K. D. Franzen, M. L. Gefter, S. P. Hale, N. J. Hansen, D. I. Israel, J. Jiang, M. J. Kavarana, M. S. Kelley, C. S. Kollmann, F. Li, K. Lind, S. Mataruse, P. F. Medeiros, J. A. Messer, P. Myers, H. O'Keefe, M. C. Oliff, C. E. Rise, A. L. Satz, S. R. Skinner, J. L. Svendsen, L. Tang, K. van Vloten, R. W. Wagner, G. Yao, B. Zhao, B. A. Morgan, *Nat. Chem. Biol.* **2009**, *5*, 647–654.
59. D. R. Halpin, P. B. Harbury, *PLoS Biol.* **2004**, *2*, E174.
60. Z. J. Gartner, B. N. Tse, R. Grubina, J. B. Doyon, T. M. Snyder, D. R. Liu, *Science* **2004**, *305*, 1601–1605.
61. B. N. Tse, T. M. Snyder, Y. Shen, D. R. Liu, *J. Am. Chem. Soc.* **2008**, *130*, 15611–15626.
62. L. Mannocci, S. Melkko, F. Buller, I. Molnàr, J. P. Bianké, C. E. Dumelin, J. Scheuermann, D. Neri, *Bioconjug. Chem.* **2010**, *21*, 1836–1841.
63. C. E. Dumelin, J. Scheuermann, S. Melkko, D. Neri, *Bioconjug. Chem.* **2006**, *17*, 366–370.
64. C. E. Dumelin, S. Trüssel, F. Buller, E. Trachsel, F. Bootz, Y. Zhang, L. Mannocci, S. C. Beck, M. Drumea-Mirancea, M. W. Seeliger, C. Baltes, T. Müggler, F. Kranz, M. Rudin, S. Melkko, J. Scheuermann, D. Neri, *Angew. Chem. Int. Ed.* **2008**, *47*, 3196–3201.
65. S. Melkko, L. Mannocci, C. E. Dumelin, A. Villa, R. Sommavilla, Y. Zhang, M. G. Grütter, N. Keller, L. Jermutus, R. H. Jackson, J. Scheuermann, D. Neri, *ChemMedChem* **2010**, *5*, 584–590.
66. M. Leimbacher, Y. Zhang, L. Mannocci, M. Stravs, T. Geppert, J. Scheuermann, G. Schneider, D. Neri, *Chem. Eur. J.* **2012**, *18*, 7729–7737.
67. C. A. Lipinski, *J. Pharmacol. Toxicol. Methods* **2000**, *44*, 235–249.
68. C. A. Lipinski, F. Lombardo, B. W. Dominy, P. J. Feeney, *Adv. Drug Deliv. Rev.* **2001**, *46*, 3–26.
69. D. R. Halpin, P. B. Harbury, *PLoS Biol.* **2004**, *2*, E173.
70. X. Li, D. R. Liu, *Angew. Chem. Int. Ed.* **2004**, *43*, 4848–4870.
71. S. J. Wrenn, R. M. Weisinger, D. R. Halpin, P. B. Harbury, *J. Am. Chem. Soc.* **2007**, *129*, 13137–13143.
72. R. Naylor, P. T. Gilham, *Biochemistry* **1966**, *5*, 2722–2728.
73. L. E. Orgel, *Cold Spring Harbor Symp.* **1987**, *52*, 9–16.
74. Z. J. Gartner, D. R. Liu, *J. Am. Chem. Soc.* **2001**, *123*, 6961–6963.
75. Z. J. Gartner, R. Grubina, C. T. Calderone, D. R. Liu, *Angew. Chem. Int. Ed.* **2003**, *42*, 1370–1375.

76. L. H. Eckardt, K. Naumann, W. M. Pankau, M. Rein, M. Schweitzer, N. Windhab, G. von Kiedrowski, *Nature* **2002**, *420*, 286.

77. R. E. Kleiner, C. E. Dumelin, G. C. Tiu, K. Sakurai, D. R. Liu, *J. Am. Chem. Soc.* **2010**, *132*, 11779–11791.

78. G. Georghiou, R. E. Kleiner, M. Pulkoski-Gross, D. R. Liu, M. A. Seeliger, *Nat. Chem. Biol.* **2012**, *8*, 366–374.

79. M. W. Kanan, M. M. Rozenman, K. Sakurai, T. M. Snyder, D. R. Liu, *Nature* **2004**, *431*, 545–549.

80. Y. Chen, A. S. Kamlet, J. B. Steinman, D. R. Liu, *Nat. Chem.* **2011**, *3*, 146–153.

81. N. K. Terrett, *Drug Discov. Today. Technol.* **2010**, *7*, e97–e104.

82. E. M. Driggers, S. P. Hale, J. Lee, N. K. Terrett, *Drug Discov.* **2008**, *7*, 608–624.

83. F. Feyen, F. Cachoux, J. Gertsch, M. Wartmann, K. H. Altmann, *Acc. Chem. Res.* **2008**, *41*, 21–31.

84. S. L. Schreiber, *Cell* **1992**, *70*, 365–368.

85. S. L. Schreiber, G. R. Crabtree, *Immunol. Today* **1992**, *13*, 136–142.

86. A. Parenty, X. Moreau, J. M. Campagne, *Chem. Rev.* **2006**, *106*, 911–939.

87. H. Pedersen, A. H. Gouliaev, T. Franch, C. Klarner Sams, E. Kampmann Olsen, F. Abilgaard Slok, G. Nystrup Husemoen, J. Fielding, L. Hyldtoft, M. Norregaard-Madsen, M. Anders Godskesen, S. Schroder Glad, T. Thisted, P.-O. Freskgard, A. Holtmann, US 2010/0016177 A1, Jan. 21, **2010.**

88. T. Franch, S. Nyboe Jacobsen, T. Ravin Rasmussen, S. Neve, H. Pedersen, A. Haahr Gouliaev, US 2006/0246450 A1, Nov. 2, **2006.**

89. H. Pedersen, A. Holtmann, T. Franch, A. Haahr, J. Felding, US 2005/0221316 A1, Oct. 6, **2005.**

90. M. H. Hansen, P. Blakskjaer, L. K. Petersen, T. H. Hansen, J. W. Højfeldt, K. V. Gothelf, N. J. Hansen, *J. Am. Chem. Soc.* **2009**, *131*, 1322–1327.

91. T. Heitner, N. V. Hansen, *Expert Opin. Drug Discov.* **2009**, *4*, 1201–1213.

92. S. Melkko, J. Scheuermann, C. E. Dumelin, D. Neri, *Nat. Biotechnol.* **2004**, *22*, 568–574.

93. S. Melkko, C. E. Dumelin, J. Scheuermann, D. Neri, *Chem. Biol.* **2006**, *13*, 225–231.

94. J. Scheuermann, C. E. Dumelin, S. Melkko, Y. Zhang, L. Mannocci, M. Jaggi, J. Sobek, D. Neri, *Bioconjug. Chem.* **2008**, *19*, 778–785.

95. S. Melkko, Y. Zhang, C. E. Dumelin, J. Scheuermann, D. Neri, *Angew. Chem. Int. Ed.* **2007**, *46*, 4671–4674.

96. J. Scheuermann, D. Neri, *ChemBioChem* **2010**, *11*, 931–937.

97. J. K. Ahlskog, C. E. Dumelin, S. Trüssel, J. Marlind, D. Neri, *Bioorg. Med. Chem. Lett.* **2009**, *19*, 4851–4856.

98. S. Trüssel, C. Dumelin, K. Frey, A. Villa, F. Buller, D. Neri, *Bioconjug. Chem.* **2009**, *20*, 2286–2292.

99. M. Ciobanu, K. T. Huang, J. P. Daguer, S. Barluenga, O. Chaloin, E. Schaeffer, C. G. Mueller, D. A. Mitchell, N. Winssinger, *Chem. Commun.* **2011**, *47*, 9321–9323.

100. J. P. Daguer, M. Ciobanu, S. Alvarez, S. Barluenga, N. Winssinger, *Chem. Sci.* **2011**, 625–632.

101. J. Scheuermann, C. E. Dumelin, S. Melkko, D. Neri, *J. Biotechnol.* **2006**, *126*, 568–581.

102. S. Melkko, C. E. Dumelin, J. Scheuermann, D. Neri, *Drug Discov. Today* **2007**, *12*, 465–471.

103. H. R. Hoogenboom, *Nat. Biotechnol.* **2005**, *23*, 1105–1116.

104. G. M. Makara, J. Athanasopoulos, *Curr. Opin. Biotechnol.* **2005**, *16*, 666–673.

105. L. M. McGregor, D. J. Gorin, C. E. Dumelin, D. R. Liu, *J. Am. Chem. Soc.* **2010**, *132*, 15522–15524.

106. M. M. Rozenman, M. W. Kanan, D. R. Liu, *J. Am. Chem. Soc.* **2007**, *129*, 14933–14938.

107. M. Margulies, M. Egholm, W. E. Altman, S. Attiya, J. S. Bader, L. A. Bemben, J. Berka, M. S. Braverman, Y. J. Chen, Z. Chen, S. B. Dewell, L. Du, J. M. Fierro, X. V. Gomes, B. C. Godwin, W. He, S. Helgesen, C. H. Ho, G. P. Irzyk, S. C. Jando, M. L. Alenquer, T. P. Jarvie, K. B. Jirage, J. B. Kim, J. R. Knight, J. R. Lanza, J. H. Leamon, S. M. Lefkowitz, M. Lei, J. Li, K. L. Lohman, H. Lu, V. B. Makhijani, K. E. McDade, M. P. McKenna, E. W. Myers, E. Nickerson, J. R. Nobile, R. Plant, B. P. Puc, M. T. Ronan, G. T. Roth, G. J. Sarkis, J. F. Simons, J. W. Simpson, M. Srinivasan, K. R. Tartaro, A. Tomasz, K. A. Vogt, G. A. Volkmer, S. H. Wang, Y. Wang, M. P. Weiner, P. Yu, R. F. Begley, J. M. Rothberg, *Nature* **2005**, *437*, 376–380.

108. E. Pettersson, J. Lundeberg, A. Ahmadian, *Genomics* **2009**, *93*, 105–111.

109. S. Fox, S. Filichkin, T. C. Mockler, *Methods Mol. Biol.* **2009**, *553*, 79–108.

110. O. Morozova, M. A. Marra, *Genomics* **2008**, *92*, 255–264.

111. S. C. Schuster, *Nat. Methods* **2008**, *5*, 16–18.

112. D. R. Bentley, S. Balasubramanian, H. P. Swerdlow, G. P. Smith, J. Milton, C. G. Brown, et al., *Nature* **2008**, *456*, 53–59.

113. F. Buller, M. Steiner, J. Scheuermann, L. Mannocci, I. Nissen, M. Kohler, C. Beisel, D. Neri, *Bioorg. Med. Chem. Lett.* **2010**, *20*, 4188–4192.

114. M. Xu, D. Fujita, N. Hanagata, *Small* **2009**, *5*, 2638–2649.

115. S. J. Wrenn, P. B. Harbury, *Annu. Rev. Biochem.* **2007**, *76*, 331–349.

116. F. Buller, M. Steiner, K. Frey, D. Mircsof, J. Scheuermann, M. Kalisch, P. Buhlmann, C. T. Supuran, D. Neri, *ACS Chem. Biol.* **2011**, *6*, 336–344.

117. S. H. Ferreira, S. Moncada, J. R. Vane, *Nat. New Biol.* **1971**, *231*, 237–239.

118. S. Aggarwal, N. Taneja, L. Lin, M. B Orringer, A. Rehemtulla, D. G. Beer, *Neoplasia* **2000**, *2*, 346–356.

119. Y. C. Ou, C. R. Yang, C. L. Cheng, S. L. Raung, Y. Y. Hung, C. J. Chen, *Eur. J. Pharmacol.* **2007**, *563*, 49–60.

120. D. J. de Groot, M. van der Deen, T. K. Le, A. Regeling, S. de Jong, E. G. de Vries, *Br. J. Cancer* **2007**, *97*, 1077–1083.

121. M. R. Arkin, M. Randal, W. L. DeLano, J. Hyde, T. N. Luong, J. D. Oslob, D. R. Raphael, L. Taylor, J. Wang, R. S. McDowell, J. A. Wells, A. C. Braisted, *Proc. Natl. Acad. Sci. U.S.A.* **2003**, *100*, 1603–1608.

122. M. Egholm, O. Buchardt, L. Christensen, C. Behrens, S. M. Freier, D. A. Driver, R. H. Berg, S. K. Kim, B. Norden, P. E. Nielsen, *Nature* **1993**, *365*, 566–568.

123. F. Debaene, N. Winssinger, *Methods Mol. Biol.* **2009**, *570*, 299–307.

124. J. Harris, D. E. Mason, J. Li, K. W. Burdick, B. J. Backes, T. Chen, A. Shipway, G. Van Heeke, L. Gough, A. Ghaemmaghami, F. Shakib, F. Debaene, N. Winssinger, *Chem. Biol.* **2004**, *11*, 1361–1372.

125. N. Svensen, J. J. Diaz-Mochon, M. Bradley, *Chem. Commun.* **2011**, *47*, 7638–7640.
126. H. D. Urbina, F. Debaene, B. Jost, C. Bole-Feysot, D. E. Mason, P. Kuzmic, J. L. Harris, N. Winssinger, *ChemBioChem* **2006**, *7*, 1790–1797.
127. M. Kircher, J. Kelso, *BioEssays* **2010**, *32*, 524–536.
128. T. D. Otto, *Nat. Rev. Microbiol.* **2011**, *9*, 633.
129. S. M. Paul, D. S. Mytelka, C. T. Dunwiddie, C. C. Persinger, B. H. Munos, S. R. Lindborg, A. L. Schacht, *Nat. Rev. Drug Discov.* **2010**, *9*, 203–214.

PART III

SCREENING METHODS AND LEAD IDENTIFICATION

12

EXPERIMENTAL APPROACHES TO RAPID IDENTIFICATION, PROFILING, AND CHARACTERIZATION OF SPECIFIC BIOLOGICAL EFFECTS OF DOS COMPOUNDS

Eduard A. Sergienko and Susanne Heynen-Genel

12.1 INTRODUCTION

Historically, a small number of protein classes provided a disproportionally large pool of current medications. Arguably, this suggests that some target classes are more amenable to regulation with small molecules. Alternatively, this prevalence may also have emerged stochastically through a predominant reliance on specific approaches for identification of initial drugs and their targets. Completion of sequencing and mapping the human genome promises to infuse the drug discovery field with new representatives and classes of therapeutically relevant targets. Gene microarrays and RNA*i*-based approaches for detection and quantification of the transcriptional products have already benefited immensely from identification of novel genes. These techniques are now used routinely in research labs and provide invaluable information on overall correlations between pathophysiological conditions and certain genes. Changes in the level of gene transcriptional activity in the cells, bearing disease phenotypes, are taken to indicate a potential association for protein products of the genes affected with the disease. Unfortunately, ascribing a link between a disease and a specific protein function is not always possible: many proteins are known to take part in multiple, frequently unrelated cellular activities.

Diversity-Oriented Synthesis: Basics and Applications in Organic Synthesis, Drug Discovery, and Chemical Biology, First Edition. Edited by Andrea Trabocchi.

As a rule of thumb, the better studied a protein is, the larger the scope of its potential cellular activities is reported in the literature, suggesting that some degree of multifunctionality [1] might be a common attribute of all biological macromolecules. For example, it is not uncommon for metabolic enzymes to form complexes with other cellular components, affecting each other's spatial distribution and functions. Each protein activity is mediated through binding to a different partner, either a small-molecule metabolite or a cellular macromolecule, and is likely to rely on a distinct binding area within the protein molecule. Unfortunately, variations observed in the protein expression levels associated with diseases are not capable of singling out which of the known (or yet-unknown) activities of the protein might be involved in the pathophysiological manifestation. Thus, selective small-molecule compounds, targeting a specific binding event, are necessary for zooming in on specific protein functions involved with a disease. These compounds, commonly referred to as *chemical probes*, provide invaluable chemical biology tools.

Chemogenomics is a research discipline responsible for identification of these selective molecules, with the overall goal of finding chemical probes for each potential binding site of macromolecules [2]. Understandably, discrimination of binding sites through selectivity of the site recognition is a prerequisite for the biological utility of chemical probes. In a broad sense, chemogenomics represents "aligning" the chemical universe with a universe of biological macromolecules, or more specifically, their binding sites. This is clearly an unattainable task if one is to rely on a pairwise compound-target matching approach. Recent advances in the synthetic chemistry and biological screening disciplines will certainly make this task less daunting, even if not readily attainable in the new future.

One of the major synthetic chemistry advances over the past decade is an emergence and maturation of diversity-oriented synthesis (DOS) methodologies, described in detail in other chapters. This approach enables the production of libraries of thousands to tens of thousands of structurally diverse compounds. Theoretically, DOS compound collections represent a renaissance of combinatorial library approaches fashionable a decade and a half ago. In reality, DOS methodologies surpass the limits inherent in combinatorial libraries of the past by augmenting the appendage and functional group variability nascent in combinatorial libraries with unprecedented scaffold and stereochemical diversity embedded in the design of DOS libraries and enabled by the aforementioned advances in synthetic chemistry.

DOS compounds are carefully designed and created through a series of intricate synthetic steps. Despite the complex chemistry leading to the generation of DOS collections, the resulting compound structures are foreseeable and could be confirmed easily, or if necessary, established through straightforward and conclusive tests providing an objective picture of a compound's degree of diversity. On the other hand, the spectrum of biological activity of any given DOS compound is an unknown and a priori unforeseeable property; it has to be established through extensive experimental testing against a diverse panel of biological targets. The methodology elected for testing compounds is normally defined by the size of compound collections, which are on the order of thousands to tens of thousands of compounds in a typical DOS collection. Testing sizable compound collections is usually performed in the confines of *high-throughput screening* (HTS), a discipline of the biomolecular sciences. HTS relies on a specialized, constantly expanding group of biological tests that are amenable to miniaturization and screening in high-density formats. The design and optimization of these tests, known as *screening assays*, is usually the domain of *assay development*, a sister discipline within biomolecular sciences.

12.2 BASIC PRINCIPLES OF HTS

The main goal of a high throughput screening project is the identification of lead compounds with a desired specific set of properties. These properties are defined by the requirements of the future use of the lead compounds. Nevertheless, to be considered useful, all lead compounds have to possess some general properties. First, the compounds identified have to be genuine modulators of the target of interest, not of the assay utilized for screening. Second, the lead compounds need to be sufficiently potent and efficacious in modulating the target of interest. Finally, they need to be sufficiently specific and selective, so their modulation of the target of interest could be distinguished from any potential off-target effects. These expectations define the screening process and its components.

12.2.1 Specifics of HTS Assays

HTS provides an environment in which biological processes are queried through screening assays that establish a link between cellular and macromolecular phenomena/functions and a quantifiable signal (Figure 12.1). HTS became the central methodology within drug discovery for testing small-molecule compound collections and for the generation and optimization of chemical leads. It emerged and matured in response to a demand of the drug discovery industry in screening large-scale compound libraries and was supported by a concomitant development of screening instrumentation and methodologies.

Most HTS assays are performed in microplates, in which the wells represent separate compartments of nanoliter to microliter volume. Diversity of plate formats exist; nonetheless, the great majority of HTS plates adhere to Society for Biomolecular Screening plate format, established to ensure interchangeability and compatibility of instrumentation utilized for screening. Most frequently used plates come in 96-, 384- and 1536-well densities (Figure 12.2). Variations in plate materials, color additives, and surface treatment make them suitable for different screening applications.

Screening plates could be seen as a unit of an HTS-based experiment. As such, they contain internal controls that allow for judging the success of the experiment. These control wells are used to assess the overall data quality and to serve as the reference markers of efficacy for screened compounds. Working reagent solutions are dispensed in the wells with the help of microchannel pipetters or automated liquid handlers.

There are three main types of samples within HTS plates. First, wells containing all assay components but lacking the tested compounds are called *negative control wells*; they represent unaffected biological process corresponding to a zero-efficacy

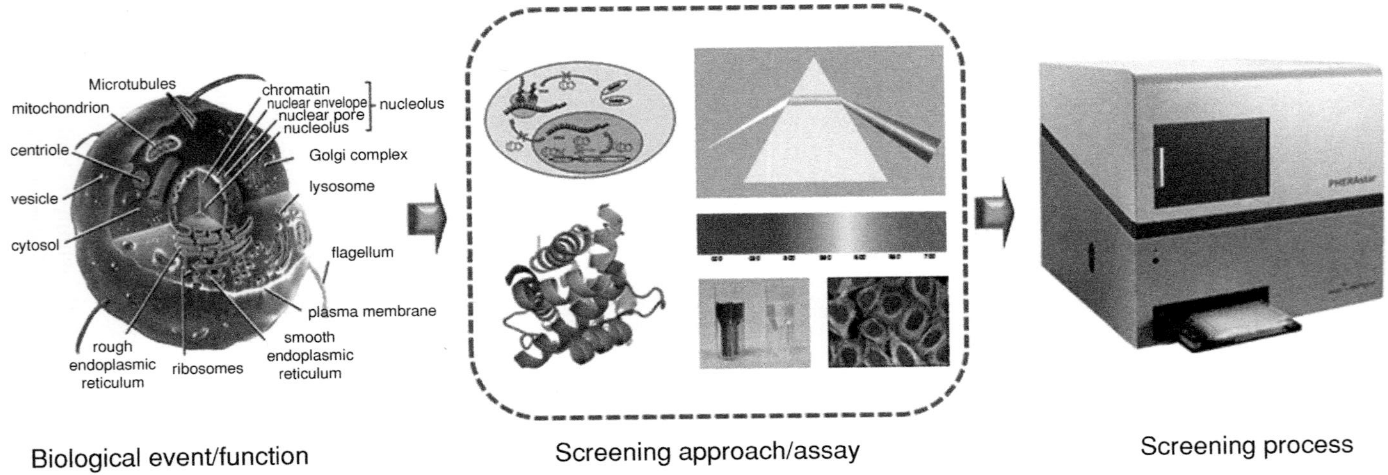

FIGURE 12.1 HTS assays provide a link between biological phenomena and a signal measured using special HTS instrumentation. (Adapted from Image of the 3D Cell is from http://www.ebi.ac.uk/microarray/biology_intro.html.) (*See insert for color representation of the figure.*)

FIGURE 12.2 Most frequently used plates come in 96-, 384- and 1536-well densities. Courtesy of Maggie Bartlett, NHGRI/NIH

marker. Second, wells exemplifying the desired inducible effect sought after with the screened compounds provide *positive control wells*; they are usually mimicked by the addition of a known modulator establishing a maximal-efficacy marker. Finally, the majority of the wells contain the complete assay mixture and the tested chemical compounds at a desired concentration. Control well compositions are selected such that the compound wells are expected to fall in the range between the values of the positive and negative controls.

One of the major challenges in HTS is selection of an compound appropriate concentration for primary screening. On the one hand, too high a concentration would lead to the assay being plagued by undesirable effects of the compounds on the assay. For example, the number of compounds precipitating out of solution increases dramatically with increasing compound concentration. On the other hand, setting the concentration too low is also detrimental, leading to a decreased rate of finding active compounds. To establish the optimal concentration, a limited set of compound plates is usually tested at different compound concentrations during HTS assay implementation stage.

12.2.2 Assay Performance Measures

The cost of HTS is directly proportional to the number of screened plates. Therefore, plates that have failed and have to be retested incur an additional and avoidable cost to the overall process. Clearly, a decrease in the number of retested plates or, preferably, their complete elimination is desired. Thus, the screening process needs to be extremely reliable and robust. Certain parameters obtained from the screening data, termed *assay performance measures*, are utilized to gauge the precision and reliability of the screening data. In HTS, these parameters are usually calculated from the data of positive and negative control wells. The same parameters are also critical during the development of the screening assays; they provide a basis for objective comparison of various assay conditions and assessment of their effect on assay performance.

One of these parameters is the signal-to-background ratio (S/B) that evaluates the strength of the signal in comparison with background measurements, such as signal in the presence of enzyme as compared to a background of enzyme-independent reaction. Although most assays require S/B values above 10 to achieve high-quality HTS data, some assay modalities perform well even with S/B ≤ 5. Understandably, the level of dispersal in the experimental data has a significant impact on an acceptable range of S/B values. For example, protein thermal shift assays (discussed in a later section) usually have S/B < 1.1; nevertheless, the approach is very robust and works extremely well, due to the tremendous precision of the measurements. Therefore, the use of S/B usually goes hand in hand with another assay performance parameter, the signal-to-noise ratio (S/N), which represents the strength of the signal in relation to the noise (signal scatter). To take into account the noise associated with both signal and background measurements, the S/N of an assay is calculated as a ratio of distance between positive and negative controls and the Pythagorean sum of their standard deviations. In practical terms, the higher the S/N values, the easier it is to distinguish genuine compound effects from inherent experimental noise of the assay. For most assay types, S/N ≥ 10 is desired, specifically when S/B is low.

S/B and S/N values are most useful and provide excellent guidance during the assay development stage, helping to optimize the assays to a desired level. However, in HTS these parameters are insufficient for evaluating the screening data. The most commonly utilized parameter for evaluating the quality of HTS data is the Z′-factor, which merges the information provided by S/B and S/N [3]:

$$Z'\text{-factor} = 1 - \frac{3\sigma_+ + 3\sigma_-}{|\mu_+ - \mu_-|} \tag{12.1}$$

The Z′-factor method relies on experimental value probability distributions and translates them into an estimate of statistical separation of the assay controls. The distribution of measurements in HTS experiments is described using normal (Gaussian) probability distribution curves characterized with mean values (μ) and standard deviation (σ) of measurements. According to the distribution theory, the range of values outlined by 3 standard deviations on either side of the mean value is expected to contain 99.8% of replicate tests. Therefore, most values of each control are expected to fall within these boundaries. The Z′-factor parameter takes this into account and defines the distance between the distribution curves of the two controls relative to the distance between the controls.

Assays advancing to HTS are expected to have a Z′-factor between 0.5 and 1, allowing us to clearly distinguish controls, inactive compounds, and compounds active in the assay. On the other hand, even assays with a Z′-factor between 0 and 0.5 could provide invaluable information, especially if the biological system studied cannot be reduced to a more robust assay. Usually, the lower the number of compounds screened, the lower the Z′-factor that could be tolerated, because the larger proportion of compounds could be accommodated through the reconfirmation stage. Lower Z'-factor values simply increase the chance of missing the active compounds and/or enriching the pool of active compounds with those that appear active for stochastic reasons, the situation correctable through replicate samples.

12.2.3 Primary Hit Selection Criteria

Compounds with a desired activity in primary HTS assay are called *primary hits* or *primary positives*. The primary screening process is usually designed as a process for decreasing the number of compounds for the follow-up work; it is aimed at discarding out unwanted compounds while retaining those that appear to genuinely affect the target. This hit identification process relies on selection criteria defined by several factors: the size of the screened library, the overall level of efficacy demonstrated by the compounds within the library, and the throughput of the most critical assay at the follow-up stage. In majority of HTS campaigns, compounds are selected using either relative efficacy or statistical significance as the main selection criterion of their effect in the assay.

Primary HTS is usually designed to establish apparent compound efficacy at a set concentration utilized in screening. Relative compound efficacy is usually calculated using positive and negative control values and expressed as a percent response:

$$\text{relative efficacy} = \frac{X - \text{NC}}{\text{PC} - \text{NC}} \cdot 100\% \tag{12.2}$$

where X is the experimental value of activity corresponding to the compound well within a plate; NC and PC are mean values of negative and positive control wells of the plate, respectively. Compound efficacy is expected to correlate with its potency, which is usually designated as EC_{50}, in an inverse nonlinear fashion, such that a 50% response corresponds to the EC_{50} value equal to the screening concentration. Therefore, the primary hits could be selected based on their relative efficacy being above a certain preset cutoff value. For example, 50% cutoff would select compounds with EC_{50} values below the concentration used in primary HTS.

The second approach relies on the statistical significance of the compound effect, measured through the standard score, also known as the Z-score, parameter values [4]. The value of the Z-score parameter correlates the compound differential effect expressed in units of the assay dispersion:

$$\text{Z-score} = \frac{X - \mu}{\sigma} \tag{12.3}$$

where X is the experimental value of activity corresponding to the compound well within a plate; μ and σ are the mean and standard deviation of the population of inactive samples. Since the great majority of compounds are expected to be inactive in the primary HTS, routinely, μ and σ are calculated using compound wells. Utilization of negative control wells for calculation of μ and σ values is more appropriate for libraries with a high hit rate. As a useful reference, Z-score = 3 corresponds to the activity value expected from less than 0.1% of samples in the inactive population.

Small-molecule compounds have diverse physical and chemical properties. For any assay, some percentage of compounds interfere with an assay and appear as hits while having no true effect on the biological target. This phenomenon is independent of the screening method employed. Although the number and identity of interfering compounds will vary between the assays, the fact of their existence for any assay is unaltered. These compounds are called *false positives*. The inverse, when a genuinely active compound appears inactive in the assay (*false negative*), is rarer, but still possible. Although diverse in nature and copious in numbers, false-positive compounds are generally innocuous and are easily identified in the follow-up hit confirmation studies. Conversely, false negatives are generally unidentifiable in screening data and therefore usually represent lost opportunities.

Arguably, the major reason for false negatives is compound precipitation in the assays. Great majority of compounds in traditional screening collections have preferential solubility in organic solvents; they have limited solubility in aqueous solutions and precipitate at high enough concentration. For some compounds, precipitation occurs as early as in the compound storage plates, since hygroscopic DMSO tends to develop a high water content, decreasing the compound solubility. On the other hand, the major compound insolubility is observed when compound solutions are diluted into the intermediate dilution or the assay plates. The precipitation is especially pronounced in the intermediate dilution plates, where the concentration of compounds is too high for an aqueous solution containing less than 10% DMSO. Utilization of acoustic dispensers and low-volume pin-tools, commonly used at specialized screening facilities for compound dispensing, helps to avoid the intermediate dilution step, significantly reducing compound precipitation.

12.2.4 Quality Control of HTS Data

HTS data quality control (QC) relies on the performance of the assay as judged by assay performance measures. All plates within an HTS run are expected to have roughly the same S/B, S/N, and Z′-factor values. Systemic errors resulting from failures of dispensing instruments or employed assays are easily identified through QC of plate statistics. The data from failed plates are usually discarded requiring retesting of these plates. As most compounds in small-molecule libraries, with the exception of a focused or cherry-picked collection, are expected to have no effect on the target of interest, compound wells could be used for additional data QC. Severe differences in the hit rate between plates, or a repeated pattern of hits within multiple plates, serve as warning signs of a potential problem. Luckily, plates with detectable minor patterns of signal distribution could be recovered through HTS data analysis using appropriate software packages.

Unfortunately, randomly failed wells are not discernible through plate data analysis. Nevertheless, advanced analysis of the screening results may be used to help recover these false negatives. Indeed, most traditional compound libraries, composed of proprietary scaffolds and commercially available collections, typically contain a number of close analogs within each scaffold. This property of chemical collections is greatly beneficial for nascent identification of structure–activity relationship (SAR) in scaffolds. It could also serve for cross-confirmation of screening data and recovery of false negatives. Since most representatives of any given scaffold share significant similarity, they are expected to demonstrate a similar sign of activity within the screen. Therefore, false negatives would appear as outliers in comparison with their analogs; these compounds would be included for a retest in a hit-confirmation stage and thus could potentially be recovered.

Lead identification and optimization is a highly collaborative process. Aforementioned HTS data safeguards exemplify close interdependence of disciplines adjacent to HTS. The primary function of biologists in this process is to design and optimize the sensitive and robust assays, providing a reliable platform for screening. Automation/HTS specialists ensure that the instruments and robotics systems are in working condition for the reliable execution of assays in HTS. In addition, they perform global HTS data analysis aimed at identification and correction of plate signal patterns. Cheminformatics specialists perform HTS scaffold analysis using data generated and feed back into the assembling of a list for hit confirmation, helping to eliminate frequent hitters and recover potential false negatives.

12.2.5 Stages of Lead Identification Projects

HTS is an integral part of most lead identification and optimization projects (Figure 12.3). HTS could be viewed as an initial enrichment process, aimed at sieving out inactive compounds, preferably without discarding the active compounds. Other project stages that follow the HTS stage are generally responsible for confirmation and characterization of initial hits as well as further optimization of the most attractive scaffolds.

Typically, a screening assay is designed and optimized during the assay development stage. The design and optimization of the assay usually takes into account all aspects of the project, especially the properties of the screening target, the properties of the chemical library, and the desired properties of the probe compounds. Primary HTS is commonly performed using the single-concentration single-replicate mode. Active compounds are identified using a distribution curve of all compounds or, alternatively, could be selected using the percent response with respect to positive and negative controls, as was described above.

The primary hit list is compiled from HTS data through hit analysis. At this stage, the hit list is compared to all other assays previously utilized to screen the same chemical library. Promiscuous compounds are normally removed from the list, while the rest of the hits proceed into reconfirmation. Initial reconfirmation is performed using single-concentration replicate wells for each hit. Confirmed hits undergo further reconfirmation in dose–response mode using replicate serial-dilution curves for each compound. The same hits are normally also tested in additional (secondary) assays aimed either at confirmation of the hits (orthogonal assays) or identification of compounds interfering with the assay (interference assays).

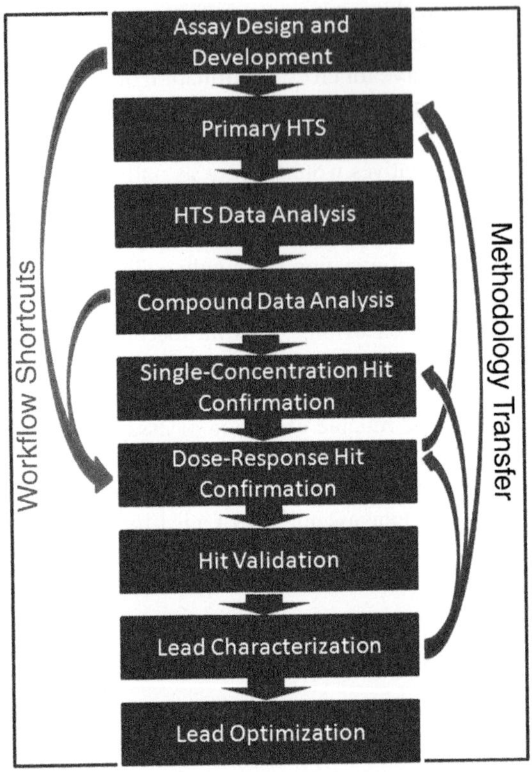

FIGURE 12.3 Stages of HTS-based lead identification and optimization projects.

Confirmed compounds are normally resupplied in dry powder; their purity, chemical identity, and biological activity are confirmed in hit validation stage. In the lead characterization stage, the compounds are confirmed and profiled using secondary biological assays as well as the assays for appropriate counter-targets to establish specificity. Structure–activity relationship (SAR) and mechanism of action (MOA) studies, also part of lead characterization, allow identification of the scaffolds possessing the desired properties. In the lead optimization stage, several rounds of SAR are performed to optimize the desired attributes of the lead compounds. For example, lead compounds within drug discovery projects are usually tested in a standard in vitro pharmacology panel to establish and optimize compounds' solubility and stability and eliminate undesired reactivity.

Depending on the number of compounds passing through the various stages, several shortcuts could be utilized to speed up the process (depicted with arrows on the left side in Figure 12.3). A large number of primary hits could be profiled directly at the single-concentration stage with respect to their specificity, selectivity, and MOA attributes. This shortcut allows an enrichment of the hit set with compounds possessing the desired properties. On the other hand, a hit list containing only a small number of primary hits could proceed directly to dose–response confirmation in the primary and secondary assays. The arrows on the right side in Figure 12.3 represent some advanced methodologies utilized for screening (see below).

12.2.6 Special HTS Modalities

An interesting approach was proposed by the NIH Chemical Genomics Center. The approach, *quantitative high-throughput screening* (qHTS), is based on performing primary HTS in a dose–response mode through, screening the same compound plate at seven or more concentrations [5]. Understandably, the approach is costlier than a single-concentration HTS, but it offers some great benefits. First, the approach helps eliminating false negatives and false positives, as it relies on data curve fitting and translates into embedded data confirmation. For example, the qHTS approach allows identification of active compounds that would otherwise be missed due to their precipitation at high concentrations. Second, qHTS allows determination of compound's potency and selectivity values, if also involves counter screen assays. Thus, hits could be selected and prioritized based on their potency and relative selectivity.

Another approach is based on testing compounds in the presence of varied concentrations of protein ligand, allowing to establish the MOA of primary hits, as described in the next section. This approach, which could be called mechanistic HTS (mHTS, by analogy with qHTS), has been utilized successfully by us [6,7] and by others [8]. An obvious drawback of the mHTS approach is the additional cost. On the other hand, the benefit of this approach is that it allows prioritization of hits based on MOA and could be cost-effective for screening campaigns with a high number of primary hits. Performing the primary HTS in a single assay and using mHTS at the hit confirmation stage may provide an acceptable compromise.

12.2.7 Principles of Assay Design

The goal of the assay development is to establish the most appropriate conditions for HTS. The assays are optimized through varying different assay parameters and aimed at improving assay performance. Precision and accuracy are two major parameters that evaluate the assay performance and sensitivity. *Precision* affects the reproducibility of repeated measurements, while *accuracy* defines the degree of closeness between the true values and those determined experimentally. Interestingly, conditions that favor precision in HTS assays do not normally provide acceptable assay accuracy. Thus, optimization of an HTS assay finds an acceptable balance between precision and accuracy.

Each HTS assay employs numerous components that could be classified into two main groups. Some of these components, such as enzyme, substrate, or cells, are necessary in the assay and are called the *principal assay components*. Other components, such as secondary buffer or cell media additives, could be omitted or substituted and represent the supporting assay components of the assay.

Most HTS assays have more than one principal component. A principal component that is present at a concentration lower than that of all the other ones is called a *limiting component*; other principal components represent nonlimiting components. For example, in an HTS assay employing an enzymatic reaction, the limiting and nonlimiting components are the enzyme and its substrates, respectively. The signal in the assay is directly proportional to the concentration of the limiting component; thus, its concentration defines the S/B value of the assay. The lowest concentration of the limiting component that is discernible from the assay background is called the *limit of detection* (LOD) [9]. Numerically, the LOD is equal to 3 standard deviations of the background measures, divided by the slope of the linear dependence between the assay signal and the concentration of the limiting component. For most screening applications, maintaining the concentration of the limiting component above $10 \times$ LOD is required.

At a high enough concentration of the limiting component, the linear dependence between the measured signal and its concentration breaks, indicating either a change in the limiting component or inability of the instrumentation to handle high signal. The limit of linearity defines the concentration corresponding to an upper limit of quantitation. The parameter may not be important for the identification of inhibitors, as long as the concentration of the limiting compound in the assay is below the upper limit of quantitation. On the other hand, this parameter can play a critical role in assays aimed at the identification of activators, as it will define the upper limit of an apparent activation observed in the assay. A project performed at the Conrad Prebys Center for Chemical Genomics (CPCCG) at the Sanford–Burnham Medical Research Institute (SBMRI) provides an example of a successful assay optimization for concurrent identification of both inhibitors and activators in a single assay [10].

Interestingly, the time component of a steady-state enzymatic reaction possesses the properties of a limiting component and requires similar optimization. The concentration of the limiting component translates directly into signal strength and is critical for assay precision. On the other hand, concentrations of the nonlimiting components define differential sensitivity of the assay in identifying modulators with various MOAs. It is a common knowledge that uncompetitive inhibitors of enzymes are more potent in the presence of a high concentration of substrates, while competitive inhibitors are more potent at a low concentration of substrates. The concentration of a nonlimiting component corresponding to its apparent K_m (or K_d in the case of a binding ligand) provides the best overall sensitivity for identification of modulator classes with various MOAs.

At a concentration of the nonlimiting component equal to K_m (K_d), the EC_{50} value of competitive and uncompetitive compounds is $2 \times K_i$, whereas the EC_{50} value of noncompetitive inhibitors is independent of the substrate concentration and equal to K_i. Thus, the nonlimiting component concentration defines the accuracy of the EC_{50} values determined in the screening. Utilization of a nonlimiting component concentration equal to K_m (K_d) at the primary HTS stage is desired especially in screening DOS compounds, since it permits identification of compounds with diverse MOAs in a single assay. A compound's MOA and the dissociation constant (K_i) are usually established during the lead characterization stage.

Interestingly, the concentration of the limiting component in the assay also affects the accuracy of the HTS results, since it sets a lower boundary for the EC_{50} values observed in a dose–response assay [9]. Therefore, achieving the lowest possible concentration of the limiting component is beneficial for SAR and MOA studies. The supporting components of the assay could frequently help decrease the LOD for the limiting component, allowing for utilization of a lower enzyme concentration and thus decreasing the lower boundary of the EC_{50} values.

12.3 COMMON ASSAY METHODS AND TECHNIQUES

HTS assays link cellular phenomena or macromolecular functions with quantifiable physical parameters of biological systems (as shown in Figure 12.1). Diverse biological processes such as cell death, translational activity of genes, enzyme activity, and protein–protein interactions can be characterized in HTS assays reliably and quantitatively. In their turn, these biological processes could be coupled with a variety of detection signals, resulting in a wide diversity of HTS assays. The major groups of HTS assays, together with their detection techniques and biological phenomena, are discussed in this section.

12.3.1 HTS Detection Approaches

HTS assays are commonly grouped according to their detection approaches. Detection based on spectrophotometric properties is by far the most common in screening, in major part due to technical and technological simplicity of instrumental implementation. A variety of monochromator-, optical filter-, or diode array–based spectrophotometers provide a basis for the flourishing of spectroscopic HTS assays. Absorbance, fluorescence, and luminescence phenomena are utilized in building HTS assays. Colorimetric assays, based on the absorbance of light, dominated the field of spectrophotometric detection in the past; however, fluorescence and especially luminescence approaches, due to their unparalleled sensitivity, have gained significantly in popularity over the past decade.

Fluorescence approaches utilize reemission of the light energy absorbed and stored temporarily by a fluorophore. Three major HTS fluorescence approaches are fluorescence intensity (FI), fluorescence polarization (FP), and fluorescence (Förster) resonance energy transfer (FRET). While FI makes use of a routine excitation-emission phenomenon, FP assays utilize depolarization of light through reemission to gauge the rate of molecular rotation of fluorophore molecules, providing a basis for binding studies. Given that a significant mass change is required to generate a reliable FP assay, these assays usually employ low-molecular-weight ligands binding to large macromolecules. The third type of fluorescence approach, FRET, relies on nonradiative energy transfer between fluorescence donor and acceptor moieties to study biomolecular interactions. This energy transfer requires dipole–dipole coupling; importantly, the efficiency of energy transfer is inversely proportional to the sixth degree of the distance separating the donor and acceptor dipoles, allowing for a sensitive gauge of intermolecular distances in the range 10 to 100 Å.

Utilization of fluorophores with long lifetimes allows for time-gated fluorescence data acquisition, employed in time-resolved fluorescence (TRF) and FRET (TR-FRET). The underlying phenomenon, sometimes called *photoluminescence*, bridges the fluorescence and luminescence domains. The most frequently utilized slow-decaying fluorophores are two lanthanides, europium and terbium, whose caged forms could be attached to one of two binding partners, directly or through a labeled antibody. Long-lived fluorescence of lanthanides provides sustained light output spanning several hundreds microseconds; for comparison, the lifespan of common fluorophores is measured on the pico- and nanosecond scale requiring simultaneous excitation and emission for their detection. Therefore, time-gated data acquisition allows for a significant increase in assay sensitivity through selective elimination of short-lived background light sources, such as nonspecific fluorescence and light reflection and scattering.

The sensitivity provided by the third main spectrophotometric detection technique, luminescence, could easily surpass that of the TRF approaches. This is due to the extremely low (near-zero) background inherent in luminescence. This advantage of luminescence has long been appreciated and made use of in the blotting protein detection techniques. Luminescent approaches could be grouped into bioluminescent and chemiluminescent techniques, depending on whether a participation and catalytic reaction of a special enzyme from a class of luciferases is critical for photon generation.

Two additional approaches blend some of the features of fluorescence and luminescence phenomena. Similar to FRET, bioluminescence resonance energy transfer (BRET) relies on nonradiative energy transfer though dipole–dipole coupling; however, it employs a luminescent entity as the energy donor, as opposed to the fluorescence moiety employed in FRET approaches. Strong inverse correlation dependence of the efficiency of energy transfer on the distance between donor and acceptor molecules, inherent in all resonance energy transfer applications, makes it suitable for studying biomolecular interactions in vitro and within live cells. Variations in luciferases, their substrates, and fluorescent acceptor moieties resulted in several BRET applications developed over the years.

The second approach is called amplified luminescent proximity homogeneous assay (AlphaScreen). This assay approach is aimed at detecting biomolecular interactions through specialized bead chemistry. In these assays, two types of beads, donor and acceptor, are present. The donor bead, which contains a photosensitizer, phthalocyanine, generates singlet oxygen molecules storing extra energy in a single excited electron upon excitation at 680 nm. Singlet oxygen is capable of transferring its energy to thioxene moieties of the acceptor beads, resulting in light emission in the range 520 to 620 nm. The singlet oxygen's 4-μs half-life defines the limited range of its diffusion equal to about 200 nm away from the donor bead. Therefore, emission light is generated only when donor and acceptor beads are present in close proximity.

Each detection technique has inherent strengths and weaknesses. The main weakness of any particular detection approach is a potential compound interference. Most of detection approaches have specific sets of compound properties that interfere with the assay; moreover, no approach is immune to compound interference artifacts. This detection-aligned specificity of artifacts provides a basis for their identification and elimination, either through screening of an unrelated target using the same detection technique or through retesting hits against the target of interest using an orthogonal assay with a different detection mode. Both of these approaches are heavily utilized in follow-up hit confirmation and validation studies.

12.3.2 The Great (Biological) Divide

The HTS assays are designed to query and provide information on the status of biological targets in the presence of varied "perturbagens" such as interference RNA or small molecules. The assays could employ intact cells or purified biological materials and measure various functional characteristics of this biological matter, ranging from enzymatic activity to phenotypic changes in the appearance of cells. All screening assays could be divided into two major groups: cell-based and cell-free assays. As the names imply, *cell-based assays* employ intact native or engineered cells. In contrast, *cell-free assays* employ cellular lysates and purified macromolecules: most commonly, proteins. The advantages and disadvantages of both of these approaches are outlined below.

One of the main advantages of the cell-based assays is that the target of interest is studied as an integral part of interconnected cellular pathways of other macromolecules. This integration is greatly beneficial for identification of modulators of multiple targets in a single assay, yet could also give rise to off-target compound effects. In the light of the integral approach, selection of an appropriate cellular background is probably the most important step in the development of a cell-based assay. Incorrect cellular background compounded with overexpression of the target of interest would lead to misbalance of natural interactions and signaling, and would result in generation of uninterpretable screening data. If available, primary patient-derived cells offer a very attractive and relevant biological model. Continued progress in the field of induced pluripotent stem (iPS) cells promises to overcome the current limitations of the primary cells, their limited supply and inability to proliferate, making iPS cells extremely attractive for future HTS.

Another advantage of cell-based assays is that they help with identification of physiologically relevant molecules. Indeed, cellular membranes provide a set of natural barriers, driving off compounds with suboptimal membrane permeability. On the other hand, the barrier properties of cellular membranes could also be disadvantageous, especially during early stages of a lead identification project. Certainly, compound membrane permeability is easily altered later in the project, providing that good starting-point compounds were not omitted simply on the basis of their polarity. Another "inevitable evil" of cell-based assays is the presence of serum in cell culture media that is, on the one hand, required to keep the cells alive, yet, on the other hand, is characterized by a high capacity for nonspecific binding of compounds. To address this problem, an increasing number of low-serum or serum-free media formulations have been developed for varying cell types in recent years.

Similar to cell-based assays, cell-free assays offer their own significant advantages, which have made them very popular. Although some cell-free assays employ cell lysates, a great majority of them are biochemical assays that are based on purified macromolecules present in a buffer of optimal composition, greatly reducing nonspecific binding of compounds. Biochemical assays are extremely attractive, as they provide easily interpretable quantitative results, unhindered by compound membrane permeability or nonspecific protein binding. On the other hand, since biochemical assays employ highly diluted solutions of pure protein preparations, they frequently face a challenge of keeping the intracellular proteins in their native state and combating undesired protein unfolding, oxidation, or nonspecific binding to organic polymer surfaces.

12.3.3 Common Biochemical Screening Methods

In addition to being classified into cell-based and cell-free assay groups, most assays could be further affiliated with either binding or functional assays. Both of these approaches are generally applicable to either cell-based or biochemical assays. Quite frequently, the name *binding* refers to assays that employ displacement of a known ligand or a protein partner from a binding complex with the target of interest. To distinguish them from assays that actually detect compound binding, the term *displacement assays* is more appropriate for describing this type of assays.

Functional biochemical assays most frequently involve a single catalytic principal enzyme in reaction mixtures. However, some assays are built to reconstitute a multistep pathway. In some cases, selection of this approach is dictated by an inability to generate a substrate for a protein of interest; in this case, an enzyme preceding the target of interest in the native pathway is utilized to generate the substrate. An additional benefit of employing a pathway assay is that it provides an opportunity of targeting multiple potential targets in a single assay. Examples of successful pathway assays are provided by ubiquitination, SUMOylation, and neddylation assays performed at the CPCCG [11].

In a functional biochemical assay, a compound's binding is gauged though changes in substrate binding or catalytic efficiency of the active site. The enzyme catalytic activity is usually monitored directly through generation of a product or consumption of a substrate. In some cases, a product of the reaction is converted further in a coupled assay, leading to the generation of a detectable moiety. For example, many metabolic reactions could be coupled with those consuming or generating NAD(P)H or ATP. NAD(P)H is detectable directly through absorbance at 340 nm or fluorescence at 460 nm; alternatively, its concentration could be assessed by using bacterial luciferase to provide luminescent signal or a resazurin/diaphorase reaction, leading to the generation of the fluorescent product resorufin. The concentration of ATP can be assessed using firefly luciferase to generate luminescence. Other reaction types dealing with protein modification and not easily coupled with a detectable chemistry (e.g., dephosphorylation or proteolysis) are commonly monitored with the help of antibodies that recognize the product or via a mock reaction with an artificial fluorogenic or luminogenic substrate.

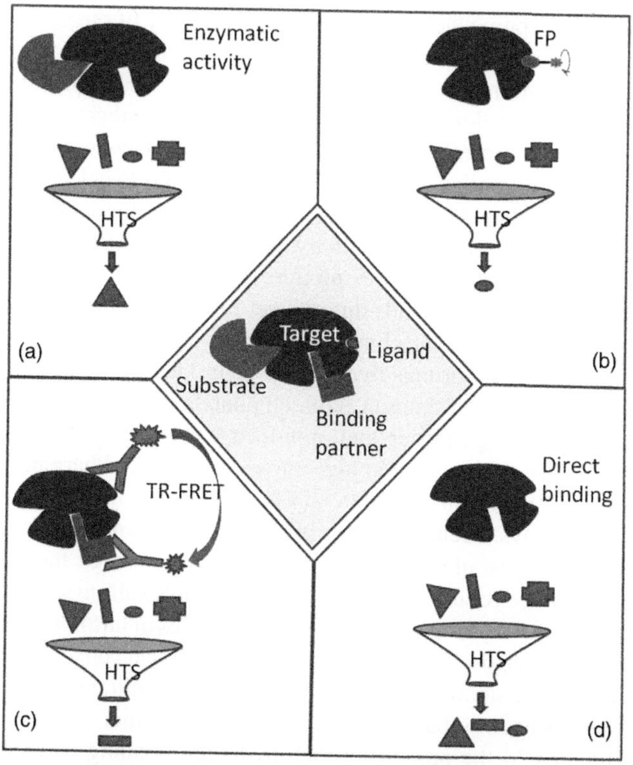

FIGURE 12.4 Diverse biochemical screening approaches.

It is quite common for an enzyme or protein to have more than one binding site or partner. For example, p38 MAP kinase has three ligand-binding sites, corresponding to Mg^{2+}ATP, protein substrate, and MKP1 dual-specificity protein phosphatase, as well as an allosteric site generated by movement of the enzyme's activation loop. The center panel of Figure 12.4 illustrates a protein similar to p38. This hypothetical protein performs catalytic conversion of a substrate and binds to a small-molecule ligand as well as to a protein partner, thus providing multiple venues for designing functional and displacement assays (Figure 12.4a to c).

One main commonality between all displacement and enzymatic assays is that they are sensitive to a binding of small molecules in a specific ("orthosteric") site and frequently are insensitive to the binding in allosteric sites. Thus, screening with any single ligand of choice will result in a narrow set of compounds targeting the specific ligand-binding site, as illustrated through the outcomes of the "screening funnels" shown in Figure 12.4a to c. Therefore, compounds against different binding sites of a single target could be identified but would require a panel of several displacement assays. This paradigm is not only costly, in general, but clearly impossible for a protein with no known ligands. Thus, detection of direct compound binding (Figure 12.4d) may be a preferred approach, especially for novel, poorly characterized targets.

Several detection approaches are available for the study of direct ligand binding. For example, surface plasmon resonance (SPR) detects binding to an immobilized analyte and provides information on the kinetics of binding and dissociation. Complementarily to SPR, microcalorimetry detects ligand binding either through isothermal heat release (isothermal titration calorimetry, ITC) or changes in the thermodynamics of the protein (differential scanning calorimetry, DSC). Nuclear magnetic resonance could detect binding through changes in spectral data for either labeled protein residues or a ligand itself. Mass spectrometry analysis could potentially identify binding of multiple ligands from a complex mixture of compounds; however, the need to separate bound and unbound ligands through gel-filtration steps makes it insensitive to weak binders. The mentioned technologies provide a powerful arsenal for studying protein–ligand binding; unfortunately, none of them have sufficient throughput for screening tens to hundreds of thousands of compounds against numerous targets.

An example of a high-throughput non-mechanism-based binding screening approach is the ThermoFluor technology pioneered by 3D Pharmaceuticals [12]. This approach is also known as differential scanning fluorimetry (DSF) and protein thermal shift assays. Technically, the protein thermal denaturation is performed on a real-time PCR type of instrument and monitored through the binding of an environment-sensitive dye such as 1-anilinonaphthalene-8-sulfonate or Sypro Orange [13]. These dyes bind to hydrophobic patches and surfaces of unfolding protein giving rise to a significant increase in fluorescence; their fluorescence is high in hydrophobic environment while significantly quenched in aqueous milieu (Figure 12.5). The screening of compounds is based on their effect on thermal stability of proteins. Compounds binding to the protein through hydrogen bonding and van der Waals interactions are expected to stabilize the protein molecules and result in the increased temperature of protein thermal denaturation, as visualized by the shift in transition temperature T_m (Figure 12.5b and c). This approach is normally based on 384-well generic PCR plates and could provide a throughput of five to seven plates per hour per instrument.

Another example of the target-based, mechanism-agnostic binding HTS approach is provided by resonant waveguide grating (RWG) technology. This technology utilizes evanescent waves generated at the boundary between a waveguide high-refractive-index layer deposited on the glass plate bottom and the low-refractive-index solution in the well. These waves extend into the solution above the glass bottom surface and are sensitive to any change in refractive index in the local region. This approach can detect direct binding of small-molecule ligands to protein molecules immobilized at the surface of the well. Understandably, this approach requires utilization of special plates with a high-quality grating layer that are much more expensive than generic screening plates, yet it provides an effective and rapid method for HTS using direct binding assay in high-density formats.

Noteworthy, a mechanism-agnostic binding approach not only provides an environment for unbiased identification of compounds targeting all potential binding sites on a protein molecule, but importantly, serves as an inherent enrichment process for finding sites on the macromolecule surface that are most suitable to binding with a particular set of compounds. This represents a great advantage for

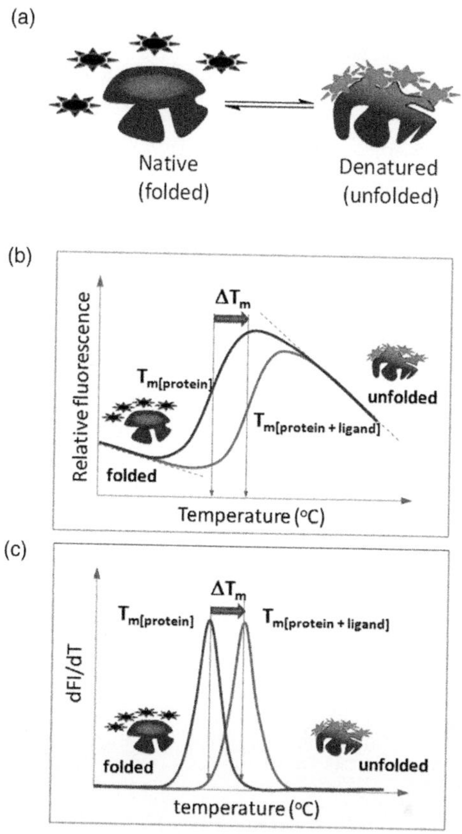

FIGURE 12.5 Basics of DSF approach for screening proteins. Environment-sensitive dye binds to the hydrophobic patches of unfolded protein: (a) resulting in fluorescence increase visualized in melting curves; (b) and their first derivatives; (c) for a protein alone and together with its ligand.

screening libraries that have an extreme diversity of compounds, such as those present in DOS collections.

12.3.4 Common Cell-Based Assays

Cell-based assays could be divided primarily into plate reader-based assays and imaging (high-content screening, HCS) assays; the latter are discussed in detail in the next section. The main difference between these assay types is the level of the information resolution: measured signal is generated by the cells and media present in the wells of plate-reader assays, while HCS assays are capable to subcellular localization tracking and background subtraction through capturing microscopy images and employing population statistics analysis. Similarly to biochemical assays, all cell-based assays could be subdivided into functional and binding groups. Assignment of assays to

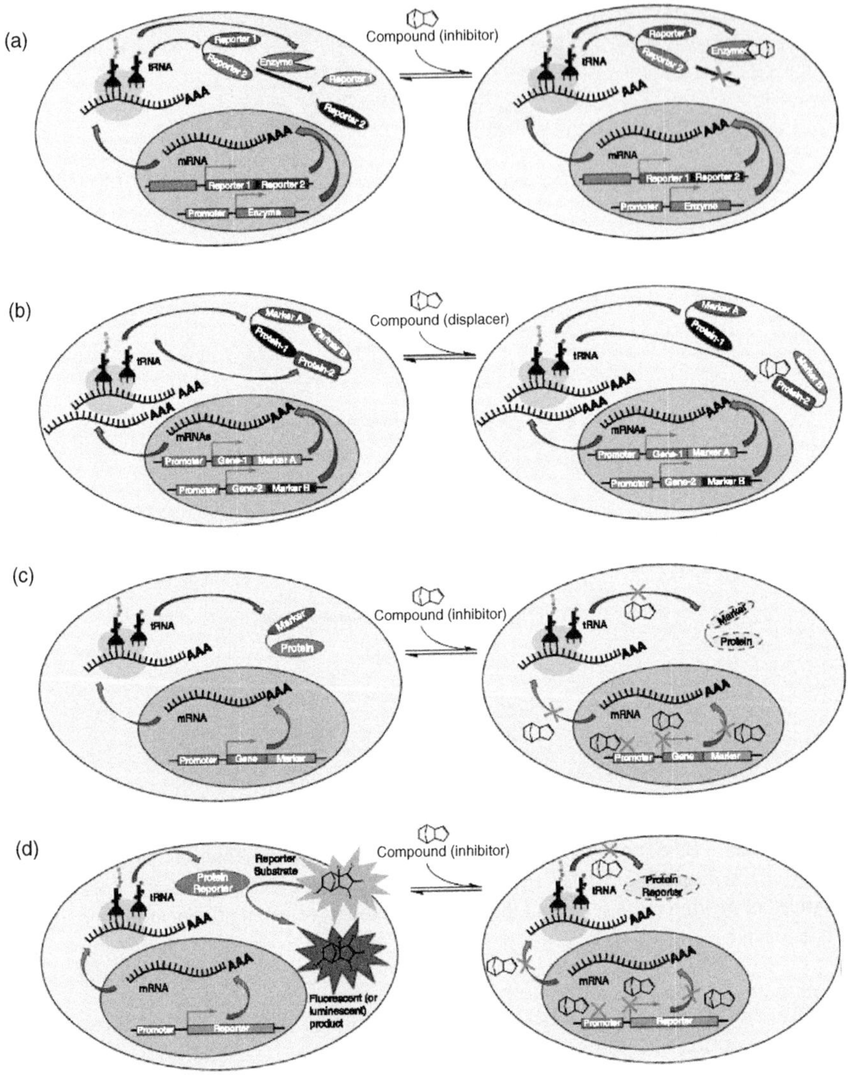

FIGURE 12.6 Cell-based assays: (a) functional cell-based assay; (b) binding cell-based assay; (c) reporter gene assay using a reporter with a directly measurable optical property; (d) reporter gene assay using a reporter with catalytic activity measured through fluorogenic or luminogenic substrates. (*See insert for color representation of the figure.*)

these groups is not always unambiguous, as multiple intracellular events could lead to a similar outcome that is being observed, especially in the plate-reader assays.

Functional cell-based assays are illustrated by an assay diagram in Figure 12.6a. In this assay, a protein with catalytic activity (e.g., a protease, as shown in the figure) is expressed in the cell along with its substrate. Active enzyme would cleave the substrate, and inhibiting activity of the enzyme should lead to accumulation of uncleaved substrate molecule. The substrate could employ a pair of variants of green fluorescent protein (GFP) with fluorescence properties consistent with FRET detection. Alternatively, utilization of a pair consisting of a luciferase and an acceptor fluorophore protein would provide a substrate with BRET properties. Utilization of a cell-penetrable luciferase substrate would make it possible to perform live cell measurements and determination of a time course of the reaction. On the other hand, conversion of a nonfluorescent reporter substrate could also be assessed using TR-FRET or AlphaScreen approaches concomitant with lysing the cells.

A binding cell-based assay is usually limited to study of protein-protein interactions (PPIs) (Figure 12.6b); achieving consistent intracellular concentration of a fluorophore-tagged small molecule may be challenging due to membrane permeability. Proteins of interest could be tagged with a GFP variant or a luciferase/fluorophore pair for FRET or BRET detection, respectively. These assays would be able to measure protein–protein interaction and its disruption with a compound within live cells. Two-hybrid systems are also popular for screening PPIs visualizing the interaction though reporter gene activity.

Reporter gene assays are perhaps the most popular modality among cell-based assays. A wide variety of gene reporters is currently available. Two major groups of reporters are illustrated in Figure 12.6c and d. One of them supplies a reporter with a directly measurable optical property, such as the fluorescence of GFP-tagged protein (Figure 12.6c). The second group provides a reporter with a catalytic activity that is measured through the use of fluorogenic or luminogenic substrates (Figure 12.6d).

Similar to biochemical assays, the cell-based assays are also prone to interference from compound groups consistent with the detection method utilized. In addition, all cell-based assays rely on the network of cellular proteins and pathways that may be sensitive to compound inhibition. One major effect of compounds resulting in assay-specific artifacts of cell-based assays is cytotoxicity of compounds. It can be especially pronounced when the cells are exposed to a high concentration of compounds for an extended time. On the other hand, cytotoxicity might be a sought-after event, as in HTS for an anticancer project. In these assays, the number of live cells is monitored through a variety of cellular parameters, such as intracellular ATP concentration or the respiratory activity of mitochondria. Alternatively, parameters associated with cell death such as activity of caspases or cellular membrane leakage could be utilized to measure the number of damaged or dead cells.

Cytotoxicity and reporter gene, along with a number of other assays, such as proliferation and cell motility assays, are part of a large group of "black-box" phenotypic assays. In these assays, the same outcome could be achieved through affecting numerous proteins and pathways. Some of these assays could be handled by traditional plate reader assays; however, many of them are more robust when special technologies such as label-free and image-based detection technologies are utilized.

The aforementioned RWG detection approach is applicable not only for biochemical assays but also for cell-based assays. This approach represents a relatively new but mature field called *label-free biosensor technology*. Another representative of this technological field is an impedance-based approach that requires the use of special screening plates and is only applicable to cell-based assays. In cell-based applications, label-free approaches epitomize phenotypic assays, making it possible to detect major structural reorganizations unaltered primary mammalian cells. Although many phenotypic assays could be performed in a plate reader format, the HCS approach described in the next section represents a much more robust and elegant platform for their execution.

The great advantage of phenotypic assays with respect to characterization of the biological effects of DOS compounds should not be underestimated. They allow the ultimate multiplex approach by targeting hundreds (if not thousands) of proteins in a single test. Although the identification of compounds' intracellular targets may prove to be challenging, methodological developments within the proteomics and functional genomics fields should help to streamline the task [14].

12.3.5 Image-Based Screening

A specialized subset of functional cell-based assays relies on quantification of cellular phenotypic changes made possible by advances in the throughput of fully automated microscopy systems and image analysis [15]. These imaging assays are often referred to as *image-based high-content screening assays*, due to their information-rich nature. The majority of these assays image a defined subregion of each well of a microwell plate and quantify the morphologic [shape, size, and number of (sub-)cellular structures], granulometric (texture or spatial distribution of biological markers), and fluorometric (fluorophore intensities) changes at the individual cell (or another object of interest) level [16]. These cell level data are then analyzed using population statistics approaches to summarize the response of each well's cell population or assay-specific cell subpopulation (Table 12.1). Commonly, the cell population data extracted from image-based assays are multiparametric, since the images acquired provide not just fluorescent intensity information but also spatial information and an object count per well. These multiparametric data allow for simultaneous evaluation of multiple phenotypic parameters at the single cell level and are often used to extract compound toxicity or cell cycle effects in addition to the specific phenotypic target under investigation [17,18]. Table 12.1 summarizes phenotypic changes typically investigated using image-based assays. Most of these make use of the two-dimensional nature of microscopy images, which allows for the interrogation of spatial changes of fluorescent markers such as (sub-)cellular marker location, spatial distribution, morphology of (sub-)cellular structures, and fluorescent biomarker colocalization.

Another advantage of imaging assays is the ability to multiplex several (routinely, up to five different fluorescent channels) endogenous markers or fluorescent reporters in one screening assay. Multiplexing several markers, which investigate different parallel pathways or related biological effects, for example, can provide insights into the potential mechanism of action of compounds or, at a minimum, help validate the biological response. Although dedicated large-scale image storage (typically, tens of terabytes) is required, saving the acquired images of the screens for manual review or future reanalysis can prove advantageous. Visual evaluation of images of the primary screening hits allows for manual rejection of assay artifacts not identified automatically by the image quantification algorithm. In some cases, changes deviating from the

TABLE 12.1 General Phenotypic Changes Typically Investigated Using Image-Based High Content Screening Assays

Imaging Assays Investigating . . .	Descriptive Examples	Literature Examples
(Sub-)cellular location	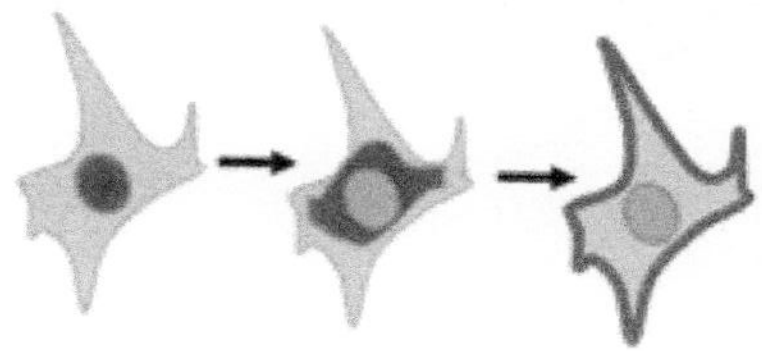 Fluorescent marker located in nucleus, cytoplasm, perinuclear region, or cell membrane	Estrogen receptor nuclear translocation [17]
Spatial distribution	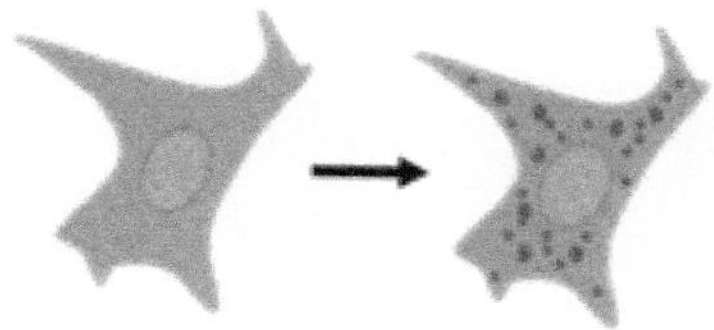 Homogeneous vs. punctuate distribution; long vs. short fibers; large vs. small vesicles	β-arrestin-based GPCR redistribution assay (TransFluor) [19]; POD nuclear foci assay [20]
Co-localization	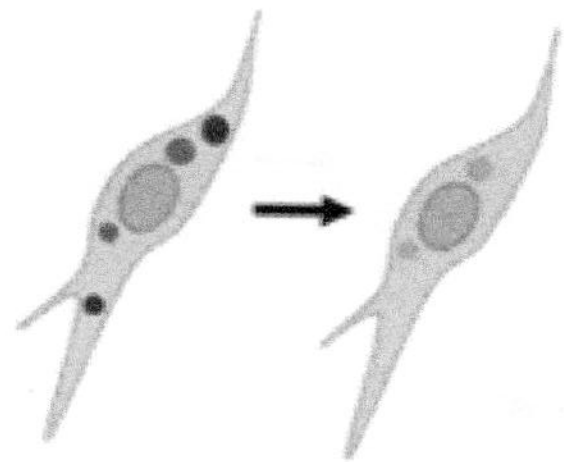 Fluor 1–labeled protein co-localized with fluor 2–labeled cell organelle; two proteins labeled with different fluorophores co-localized with each other	Lipid storage vesicles [21]

(continued)

TABLE 12.1 (Continued)

Imaging Assays Investigating . . .	Descriptive Examples		Literature Examples
Shape/size/number of cells or cellular structures	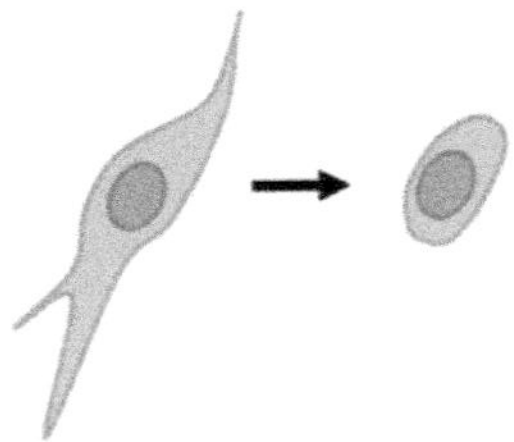	Spread-out vs. rounded cell; size or number of intracellular vesicles; length, area, or number of cytoskeletal structures	Invadopodia assay [22]; neurite outgrowth assay [23]
Cell subpopulation response	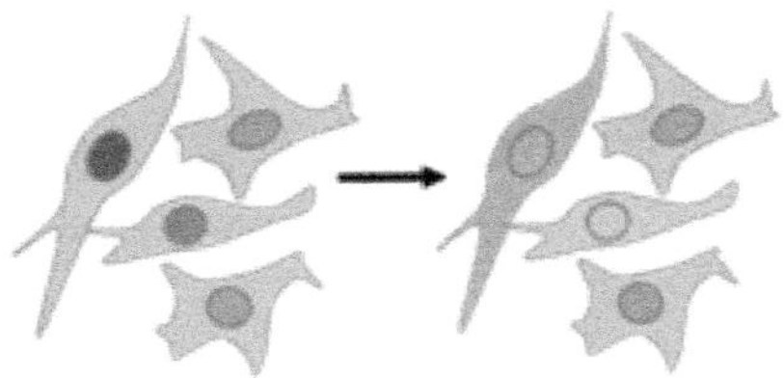	Response of transinfected cells only; response of one cell type in a mixed population	Neuronal assay using cell co-cultures [24]
Multiplexed markers	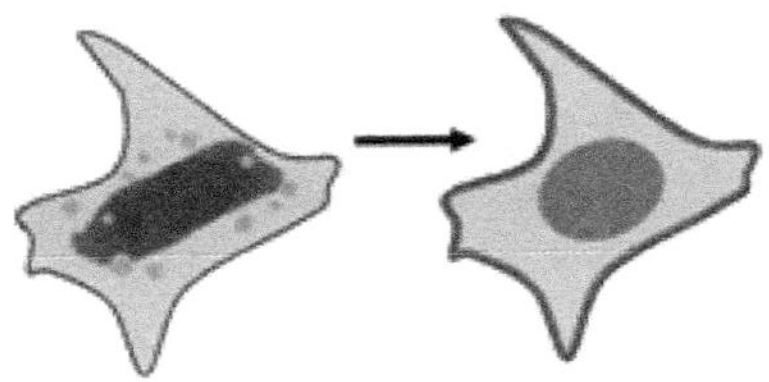	Multiple cytoskeletal markers; apoptosis markers probing different signaling pathways; phophorylated vs. ubiquitinated vs. acetylated vs. total protein; differentiation markers of stem cells	Differentiation of stem cells into different lineages [25]

expected phenotype can be identified during manual image review, and the analysis algorithm can be adjusted to measure additional parameters characterizing the unexpected effect. For example, in an image-based screen aiming to identify perturbagens involved in ciliogenesis by quantifying the percentage of ciliated cells, manual image review identified a phenotype with increased cilia length [26]. Subsequently, the size of the cilia detected was added to the analysis algorithm, and the project succeeded in identifying perturbagens for a decrease in number of ciliated cells as well as a stabilized cilia phenotype. This exemplifies the potential of image-based screening assays to identify unexpected phenotypes or to distinguish subtle differences in phenotypes that may provide additional insight into cellular profiling of DOS compound libraries.

Extraction and accurate quantification of multiple biologically relevant effects allows for rapid characterization of compounds, making image-based high-content assays an ideal choice for screening of DOS-based libraries. However, image-based assays have inherently lower throughput than that of cellular plate reader–based assays, due to longer image acquisition times (typically, 15 min to 2 h per 384-well plate, depending on the assay design). Specialized (and expensive) high-content screening equipment and infrastructure are required to prepare, image, and analyze image-based assay plates [15]. Additionally, advanced knowledge in image analysis, population statistics, and data mining is paramount to being able to mine and correctly interpret the rich biological information resulting from these imaging assays.

As the image-based high-content screening field matured over the last decade, new image-based screening modalities emerged to better characterize the biological systems. The recent increase in image acquisition speeds and improvement in image handling, analysis, and storage allowed for observation of live cell phenotypic changes over time. Temporal phenotypic changes can be evaluated for short periods (several seconds) at fast imaging rates (10 to 100 Hz) in kinetic image cytometry, where each well is imaged for the entire duration before moving to the next well. Examples of kinetic image cytometry assays are assays measuring action potentials (calcium waves) of cardiomyocytes or ion channel activity. Time-lapse imaging performs multiple imaging runs of an entire well plate over long periods of time (hours to days): for example, to evaluate subtle changes in cell cycle and mitosis [27,28]. Live cell imaging requires high-content screening systems with environmental control to maintain incubation conditions during imaging and an image-data infrastructure to deal with terabytes of data per experiment. The same technological advances that allowed for live cell imaging also enabled investigating more complex biological systems by evaluating the phenotypic changes in three-dimensional biological model systems, such as tumor spheroids, or small organisms such as *Caenorhabditis elegans*. Specimen preparation for such assays required sophisticated assay design and is often the limiting factor as to throughput for these types of assays. However, they mimic the true biological system or disease better than do two-dimensional cell monolayers, and an increase in the use of these assays should be expected as plate preparation methodologies and imaging techniques improve.

The advances in technologies described above have also helped to push some functional cell assays to the medium throughput required for screening compound libraries of several thousand compounds. Migration, invasion, cell motility, and chemotaxis assays have been used routinely in basic research for decades. Special assay plates and screening approaches have been developed and are now commercially available to enable use of these assays in a medium-throughput format. Most of these assays are more robust when used with image-based screening, but some can also be optimized successfully for use with plate readers. While such complex functional assays are very useful for evaluation of DOS compound effects, it is often very difficult to decipher their mechanism of action, due to the large number of cellular components, signaling events, and pathways involved in complex biological behavior such as cell movement. For such projects to succeed, a good strategy for target or MOA deconvolution is critical.

12.4 FUTURE PERSPECTIVES

DOS compound collections provide an unprecedented level of chemical diversity. In addition to traditional target- and mechanism-based screening approaches, they could benefit significantly from utilization of mechanism-agnostic and phenotypic assays. Holding their promise, DOS collections are likely to lead to the identification of new target classes or the discovery of novel modulation mechanisms for difficult targets. Utilization of qHTS and mHTS screening modalities may be particularly attractive, as they allow rapid profiling and characterization of compounds and are especially beneficial for small compound collections.

REFERENCES

1. For definition, see K. Kirschner, H. Bisswanger, *Annu. Rev. Biochem.* **1976**, *45*, 143–166.
2. (a) E. Jacoby, *Mol. Biosyst.* **2006**, *2*, 218–220; (b) H. Kubinyi, *Ernst Schering Res. Found. Workshop* **2006**, *58*, 1–19.
3. J. H. Zhang, T. D. Chung, K. R. Oldenburg, *J. Biomol. Screen.* **1999**, *4*, 67–73.
4. C. Brideau, B. Gunter, B. Pikounis, A. Liaw, *J. Biomol. Screen.* **2003**, *8*, 634–647.
5. J. Inglese, D. S. Auld, A. Jadhav, R. L. Johnson, A. Simeonov, A. Yasgar, W. Zheng, C. P. Austin, *Proc. Natl. Acad. Sci. U.S.A.* **2006**, *103*, 11473–11478.
6. TNAP project: (1) PubChem Assay ID 1012, http://pubchem.ncbi.nlm.nih.gov/assay/assay.cgi?aid=1012&loc=ea_ras; (2) PurChem AID 1135, http://pubchem.ncbi.nlm.nih.gov/assay/assay.cgi?aid=1135&loc=ea_ras.
7. PMI project: (1) PubChem AID 1209, http://pubchem.ncbi.nlm.nih.gov/assay/assay.cgi?aid=1209&loc=ea_ras; (2) PubChem AID 1220, http://pubchem.ncbi.nlm.nih.gov/assay/assay.cgi?aid=1220&loc=ea_ras.
8. M. Wei, R. Wynn, G. Hollis, B. Liao, A. Margulis, B. G. Reid, R. Klabe, P. C. Liu, M. Becker-Pasha, M. Rupar, T. C. Burn, D. E. McCall, Y. Li, *J. Biomol. Screen.* **2007**, *12*, 220–228.
9. E. A. Sergienko, in: *Chemical Genomics*, H. Fu, Ed., Cambridge University Press, New York, **2012**, pp. 159–172.

10. (a) E. A. Sergienko, Y. Su, X. Chan, B. Brown, A. Hurder, S. Narisawa, J. L. Millan, *J. Biomol. Screen.* **2009**, *14*, 824–837; (b) E. A. Sergienko, J. L. Millan, *Nat. Protoc.* **2010**, *5*, 1431–1439.

11. PubChem AID 485343, http://pubchem.ncbi.nlm.nih.gov/assay/assay.cgi?aid=485343&loc=ea_ras; PubChem AID 2011, http://pubchem.ncbi.nlm.nih.gov/assay/assay.cgi?aid=2011&loc=ea_ras.

12. D. Matulis, J. K. Kranz, F. R. Salemme, M. J. Todd, *Biochemistry* **2005**, *44*, 5258–5266.

13. F. H. Niesen, H. Berglund, M. Vedadi, *Nat. Protoc.* **2007**, *2*, 2212–2221.

14. S. Heynen-Genel, L. Pache, S. K. Chanda, J. Rosen, *Expert Opin. Drug Discov.* **2012**, *7*, 955–968.

15. S. Haney, *High Content Screening: Science, Techniques, and Applications*, Wiley, Hoboken, NJ, **2008**.

16. S. Heynen-Genel, J. H. Price, in: *Handbook of Medical Image Processing and Analysis*, Vol. 1, 2nd ed., Isaac Bankman, Ed., Academic Press, San Diego, CA, **2008**, pp. 453–463.

17. A. Dull, E. Goncharova, G. Hager, J. B. McMahon, *J. Steroid Biochem. Mol. Biol.* **2010**, *122*, 341–351.

18. D. W. Young, A. Bender, J. Hoyt, E. McWhinnie, G. W. Chirn, C. Y. Tao, J. A. Tallarico, M. Labow, J. L. Jenkins, T. J. Mitchison, Y. Feng, *Nat. Chem. Biol.* **2008**, *4*, 59–68.

19. P. Zhao, H. Sharir, A. Kapur, A. Cowan, E. B. Geller, M. W. Adler, H. H. Seltzman, P. H. Reggoi, S. Heynen-Genel, M. Sauer, T. D. Chung, Y. Bai, W. Chen, M. G. Caron, L. S. Barak, M. E. Abood, *Mol. Pharmacol.* **2010**, *78*, 560–568.

20. K. W. Yip, M. Cuddy, C. Pinilla, M. Giulanotti, S. Heynen-Genel, S. Matsuzawa, J. C. Reed, *J. Biomol. Screen.* **2011**, *16*, 251–258.

21. P. M. McDonough, R. M. Agustin, R. S. Ingermanson, P. A. Loy, B. M. Buehrer, J. B. Nicoll, N. L. Prigozhina, I. Mikic, J. H. Price, *Assay Drug Dev. Technol.* **2009**, *7*, 440–460.

22. M. Quintavalle, L. Elia, J. H. Price, S. Heynen-Genel, S. A. Courtneidge, *Sci. Signal.* **2011**, *4*, ra49.

23. N. M. Radio, J. M. Breier, T. J. Shafer, W. R. Mundy, *Toxicol. Sci.* **2008**, *105*, 106–118.

24. L. S. Kaltenbach, M. M. Bolton, B. Shah, P. M. Kanju, G. M. Lewis, G. J. Turmel, J. C. Whaley, O. J. Trask, Jr., D. C. Lo, *J. Biomol. Screen.* **2010**, *15*, 806–819.

25. M. D. Treiser, E. H. Yang, S. Gordonov, D. M. Cohen, I. P. Androulakis, J. Kohn, C. S. Chen, P. V. Moghe, *Proc. Natl. Acad. Sci. U.S.A.* **2010**, *107*, 610–615.

26. J. Kim, J. E. Lee, S. Heynen, E. Suyama, K. Ono, K. Y. Lee, T. Ideker, P. Aza-Blac, J. G. Gleeson, *Nature* **2010**, *464*, 1048–1051.

27. N. Harder, F. Mora-Bermúdez, W. J. Godinez, A. Wünsche, R. Eils, J. Ellenberg, K. Rohr, *Genome Res.* **2009**, *19*, 2113–2124.

28. B. Neumann, T. Walter, J.-K. Hériché, J. Bulkescher, H. Erfle, C. Conrad, P. Rogers, I. Poser, M. Held, U. Liebel, C. Cetin, F. Sieckmann, G. Pau, R. Kabbe, A. Wünsche, V. Satagopam, M. H. A. Schmitz, C. Chapuis, D. W. Gerlich, R. Schneider, R. Eils, W. Huber, J.-M. Peters, A. A. Hyman, R. Durbin, R. Pepperkok, J. Ellenberg, *Nature* **2010**, *464*, 721–727.

13

SMALL-MOLECULE MICROARRAYS

HONGYAN SUN

13.1 INTRODUCTION

The microarray concept traces its origin to 1991, when Fodor et al. introduced the high-density peptide microarray in their seminal work [1]. Over the past two decades, significant development in this field has seen microarray emerging as a powerful and robust platform for both basic science and biotechnology. As a miniaturized analytical tool, microarray offers numerous advantages over traditional screening platforms [2,3]. For example, the miniaturization property of microarray significantly reduces the need for expensive and precious bioreagents. It also provides a flat screening surface and notably simplifies the tedious washing and incubation process. Automation and parallelization are two important properties and essential features of microarrays. Both functions contribute to the superiority of microarrays over traditional screening platforms. They help to facilitate the rapid screening of thousands of spatially addressed molecules in a high-throughput manner, reducing the screening time significantly. In recent years we have seen tremendous development in microarray technology. This approach has been applied to screen virtually any type of biomolecule, including DNAs, small molecules, proteins, live cells, and even tissues [4–8].

Biologically active small molecules are progressively gaining importance as biomedical research tools. Compared with the traditional genetic approach, small molecules offer a temporal and reversible route to modulate protein functions [9]. Small molecules can play an important role in elucidating various cellular mechanisms. This is especially notable for small molecules that are capable of selectively inhibiting or activating a given protein or regulating a specific signaling pathway. In high-throughput studies involving small molecules, researchers are often confronted

Diversity-Oriented Synthesis: Basics and Applications in Organic Synthesis, Drug Discovery, and Chemical Biology, First Edition. Edited by Andrea Trabocchi.

with two obstacles: how to efficiently create diverse small molecule collections with intriguing biological activity, and how to rapidly and effectively interrogate their functions with different biological assays. To overcome the hindrance, small-molecule microarray (SMM) rises to the call.

SMM is generally considered a state-of-the-art platform. In conjunction with solid-phase combinatorial chemistry, it serves as a potent and preeminent technology to interrogate the biological activity of small-molecules in a high-throughput manner. In 1999, MacBeath et al. demonstrated site-specific immobilization of small-molecules onto an array, making it the first scientific description and documented evidence of the early work on SMM. The team proved that the technology could lead to successful identification of target protein/small molecule pairs [10]. In 2002, the same group successfully identified a small-molecule modulator for yeast transcription inhibitor Ure2p by screening a 3780-member small-molecule library on the microarray [11]. This remarkable breakthrough offers uncontested proof of the hypothesis that it is possible to use small molecules to modulate selectively a target protein's activity. In addition, the result established unambiguous demonstration of the distinctive potential of a microarray platform for small-molecule screening research.

Over the past decade, SMM technology has experienced flourishing development in many aspects, including small-molecule library synthesis, microarray fabrication, and microarray application (Figure 13.1). In the following sections, the recent development and evolution of SMM are elaborated and illustrated. Although it is not possible to present exhaustive information and cover every detail of SMM within the scope of this chapter, we focus on some of the most important highlights in the development of this technology.

13.2 CHEMICAL LIBRARY DESIGN AND SYNTHESIS

A fundamental and crucial step for SMM fabrication is to create a diverse library of biologically active compound collections. The libraries are derived mainly from the following sources: commercially available libraries, isolated natural products, and synthetic combinatorial libraries. Synthetic combinatorial libraries, for example, have been shown to be a robust strategy to use to produce huge libraries of small molecules. With advancements in solid-phase synthesis and combinatorial chemistry, the synthesis of small-molecule libraries has been further accelerated.

Solid-phase chemistry was introduced by Merrifield in the 1960s for peptide synthesis [12,13]. It provides a convenient and facile method to separate excess reagents and soluble by-products from resin-bound products. With solid-phase chemistry, researchers can avoid the tediousness of intermediate purification at each step of the synthesis. It should be noted, however, that the chemistry utilized for synthesis throughout the entire process must be highly efficient and selective. Otherwise, the impurities accumulated will have a considerably negative impact on the overall purity of the final product.

Split-and-pool synthesis is another technology that significantly accelerates the process of generating huge libraries of compounds. The technology was introduced by

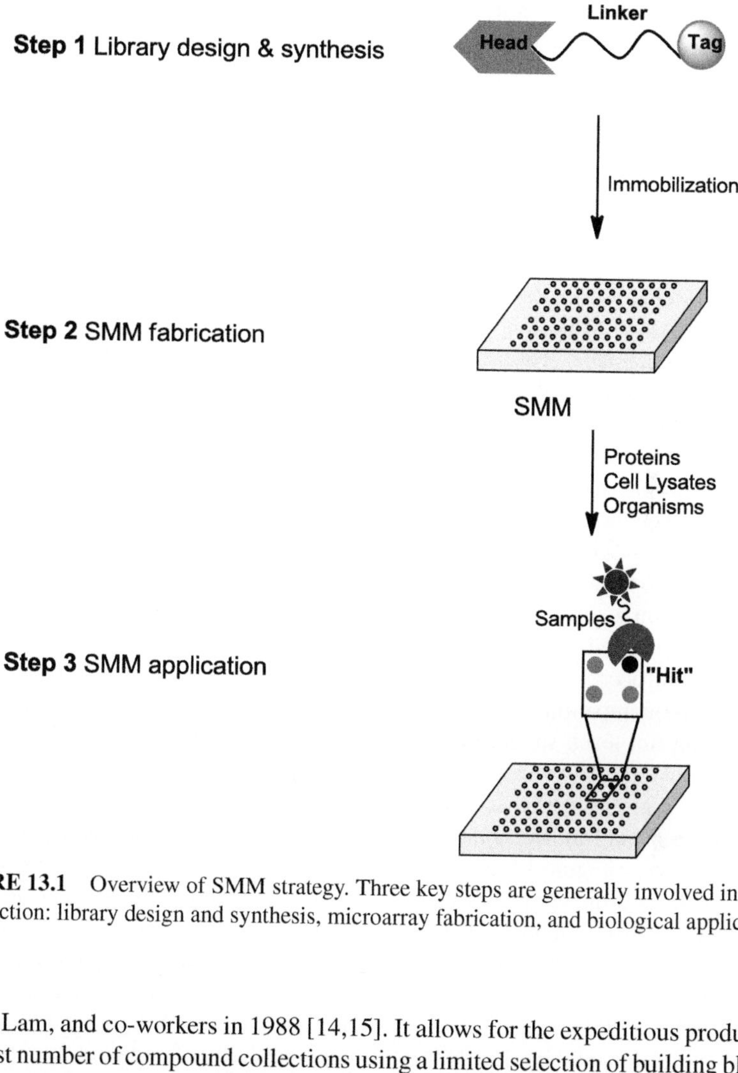

FIGURE 13.1 Overview of SMM strategy. Three key steps are generally involved in SMM construction: library design and synthesis, microarray fabrication, and biological application.

Furka, Lam, and co-workers in 1988 [14,15]. It allows for the expeditious production of a vast number of compound collections using a limited selection of building blocks. Taking advantage of the evolution of robotic technology, pharmaceutical companies can now easily produce more than 100,000 new compounds each year.

To design a small-molecule library compatible with microarray screening, the following features are generally considered as a requisite: (1) a tag that allows for the small molecules to be stably immobilized onto the array, (2) a linker to minimize the steric hindrance between proteins and the solid support surface, and (3) a protein recognition head or element that endows the binding affinity toward the protein target [16]. One key consideration during library design is that the small molecules should have the potential to possess relevant biological activity. The design of the molecules could be based on a natural product scaffold, which exhibits biological activity, or known synthetic drugs. Once a scaffold with interesting biological activity is targeted, a proper synthetic route can be designed to facilitate achieving the synthesis. Over the past 20 years, many efficient library synthetic approaches, such as diversity-oriented synthesis and fragment-based methods, have been developed to facilitate the production of small-molecule libraries. In the following sections we explore these approaches in greater detail.

13.2.1 Diversity-Oriented Synthesis

Diversity-oriented synthesis (DOS) was devised by Schreiber in 2000 [17]. It established a milestone as an innovative strategy to create diverse libraries for probing biological mechanisms. Through this ingenious method, synthetic chemists are able to generate compound collections with rapidly both skeletal and stereochemical diversity. These molecules can be screened to unravel novel lead compounds for multiple protein targets by exploiting a variety of biological assays, including protein-binding assay and cell-based phenotypic assay. On the other hand, the DOS approach has also posed several challenges to synthetic chemists. For example, DOS synthetic libraries require products with skeletal diversity and high purity. To implement the DOS approach appropriately, the chemistry employed must be highly efficient because it is performed on a solid support and involves multiple steps of synthesis. The synthetic route designed must also meet the requirement to provide complex and diverse structures with accessible building blocks. Furthermore, the planned reaction at each step must be compatible with various intermediates generated during the synthesis. Last but not least, stereoselectivity is also a key parameter for a DOS library. Stereoselective reactions should be chosen carefully in the design.

In 2002, a group led by Schreiber applied the DOS methodology to construct a 3780-member small-molecule library using a "one bead–one stock solution" approach [11]. As shown in Figure 13.2a, the overall design of this small-molecule library was based on the 1,3-dioxane scaffold because of its rigidity, which is an important parameter of a drug candidate. The scaffold could be synthesized in a stereoselective manner. After stepwise solid-phase synthesis, the 1,3-dioxanes were released from solid support and immobilized onto a chlorinated glass slide through a hydroxyl group handle. The fabricated array was then incubated with fluorescently labeled yeast transcription inhibitor Ure2p. Eight compounds were identified as Ure2p binder, and one of the compounds, uretupamine A, was found capable of inhibiting Ure2p activity with excellent selectivity in vivo. Subsequent SAR analysis revealed a more potent analog, uretupamine B ($K_d = 7.5\ \mu M$). The result offers indisputable evidence that it is possible to combine DOS strategy and microarray platform to discover new small molecules, which are capable of selectively modulating protein functions.

In the following years, the Schreiber group led this research further by synthesizing a library containing 12,396 compounds using three different DOS pathways [18]. The libraries were generated through 1,3-dioxane, biaryl, and Diels–Alder synthetic pathways, respectively. The group printed these compounds onto chlorinated glass slides as before, and screened them by sandwich assay with Hap3p, a subunit of yeast transcription factor complex (Figure 13.2b). The three microarrays were probed with Hap3p-GST fusion protein and visualized using a fluorescently labeled anti-GST antibody. A true Hap3p binder called haptamide A was revealed in the array screening. It was later verified by biophysical SPR experiment ($K_d = 5.03\ \mu M$). Further SAR analysis revealed a more potent ligand, haptamide B of Hap3p with a K_d value of 0.33 μM. Subsequent cell experiments also showed that haptamide A is a reversible inhibitor for Hap3p-mediated transcription in vivo.

In both aforementioned experiments, chlorinated glass slides were utilized to capture DOS-derived compounds containing a hydroxyl group. Literature reports, however, showed that there are more synthetic libraries with carboxylic acids and phenols than those with a hydroxyl group. Because of this, Schreiber et al. employed an alternative immobilization strategy to anchor small molecules with a phenol group. They synthesized a library of 6336-member phenol-containing fused bicycles and tetracycles with up to six stereogenic centers [19]. The compounds were then printed

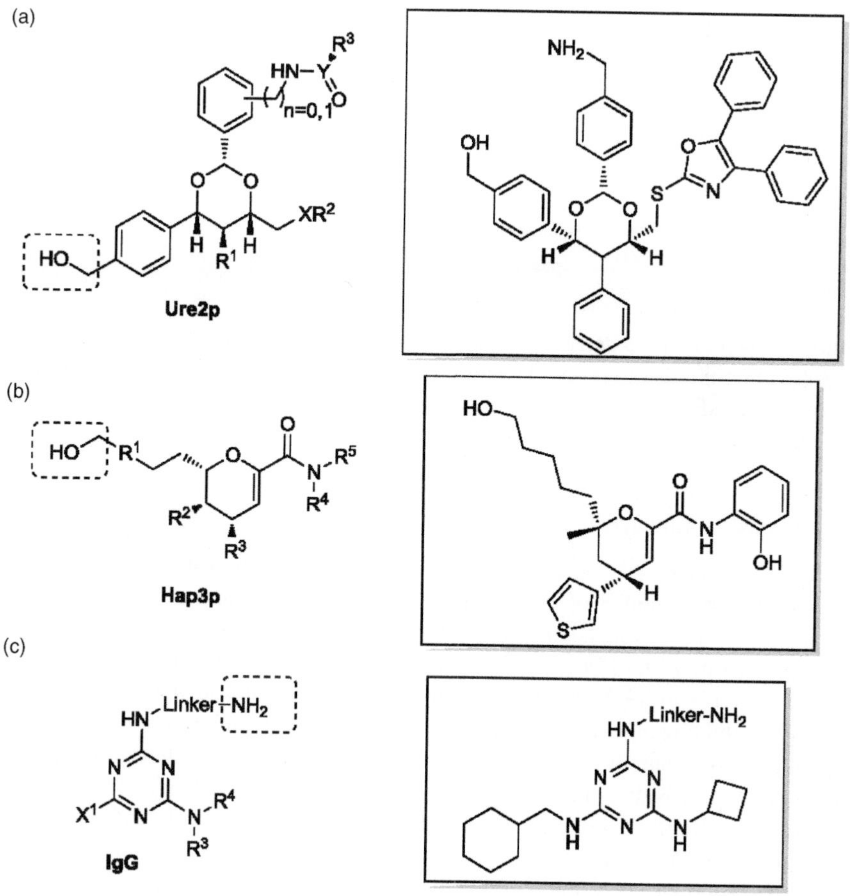

FIGURE 13.2 Selected examples of libraries used in SMM and hits identified in the array experiments.

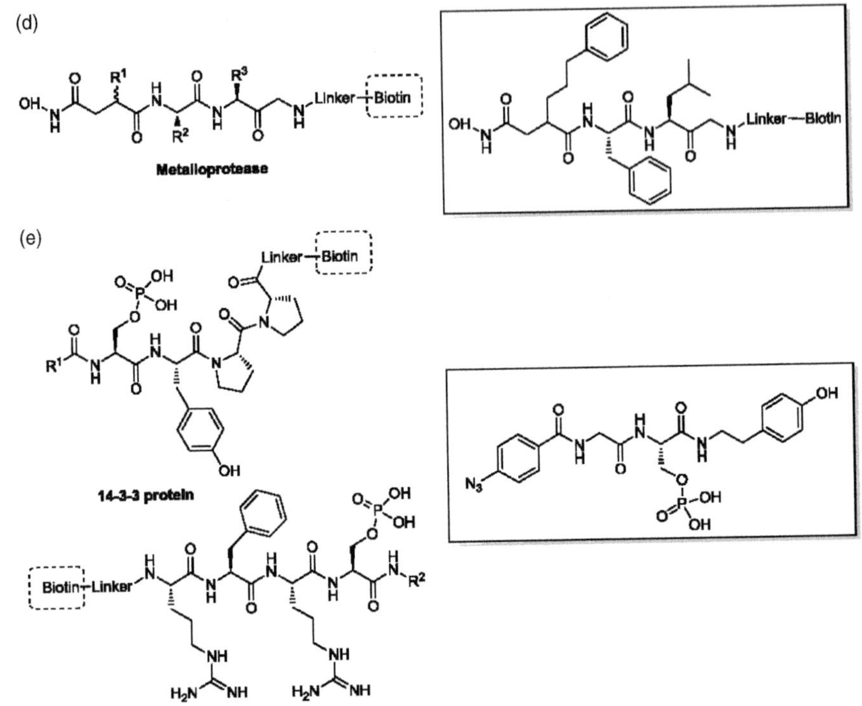

FIGURE 13.2 (*Continued*)

onto diazobenzylidene-derivatized glass slides, which were able to catch phenol and other acidic functional groups. The high-density microarrays were probed with Cy5-labeled calmodulin. 16 putative hits were identified from the array experiments, and the most potent ligand showed a K_d value of around 0.12 μM.

13.2.2 Other Libraries

As we have discussed, the DOS approach has clearly demonstrated its capability to produce large number of compounds with skeletal and stereochemical diversity. Uttamchandani et al., on the other hand, formulated another efficient approach to generating small-molecule libraries with biological activity [20]. In their library design, as illustrated in Figure 13.2c, the 1,3,5-triazine scaffold was chosen. It exhibits threefold symmetry, thereby allowing for versatile modifications and minimum regiochemical complications during the synthesis. Apart from that, it has been indicated in the literature that 1,3,5-triazine presents some interesting biological activity [21,22]. At the start of the synthesis, a linker was incorporated to facilitate immobilization of small molecules onto the array. A library of 2688 compounds with an amine group was synthesized and anchored onto an *N*-hydroxysuccinimide-functionalized glass slide. By probing the array with fluorescently labeled IgG, these workers discovered four novel IgG binders. The best hits identified showed a K_d value of around 2 μM in biophysical SPR validation experiments. The small molecule discovered may serve as an alternative to protein A and protein G for industrial antibody production.

Peptide-based libraries have frequently been used in microarray technology because of their excellent bioactivities and availability, due to well-established peptide chemistry. Peptide arrays can be employed to decipher substrate specificity and the inhibition patterns of various classes of enzymes. In 2007, a 1400-member peptide hydroxamate library was synthesized and immobilized onto avidin-functionalized glass slides by Uttamchandani et al. (Figure 13.2d) [23]. The fabricated array was used to acquire the inhibitor fingerprints of four metalloproteases: thermolysin, collagenase, carboxypeptidase, and anthrax lethal factor (LF). The design of a peptide-based library featured a universal hydroxamate warhead that could bind to a zinc ion in the active-site pocket of the enzymes, and a distinct peptide fragment that determines the binding specificity between the enzymes and the ligands. With this strategy, the group was able to obtain unique inhibitor fingerprints with all four enzymes tested. Several potent inhibitors against anthrax LF were identified and could be used for further research development.

Carbohydrates represent another rich and important source for small-molecule libraries. Carbohydrate arrays have proved to be a valuable tool to use to investigate interactions between sugars and other biomolecules [24]. Microarray applications of carbohydrates could be valuable for disease diagnosis and vaccine development. In the past there was immense difficulty in creating libraries with structurally diverse carbohydrates. This hindrance has been alleviated with recent developments in oligosaccharide synthesis. It is now even possible to synthesize defined carbohydrates automatically on the solid support [25], and the duration of the synthesis is reduced from months to days. In 2004, Blixt et al. constructed a carbohydrate array with 200 synthetic and natural glycan sequences [26]. The team investigated a number of fluorescently labeled glycan-binding proteins (GBPs), such as C-type lectins, siglecs and galectins. The glycan array not only confirmed that the GBP ligand specificity was consistent with previous reports, but also revealed several new interactions.

The fragment-based approach is another method that has found wide use in designing potent enzyme inhibitors. This strategy has recently been adopted by Wu et al. for SMM application [27]. The group designed two hybrid libraries based on the original 14-3-3 peptide substrate RFRpSYPP. As depicted in Figure 13.2e, the pS residue was retained, and the two flanking peptide fragments were replaced by acid and amine building blocks, respectively. In total, 243-member N-terminal and 50-member C-terminal hybrid libraries were synthesized and immobilized onto array. By screening with the 14-3-3 protein, potent binders from the two sublibraries were identified and reconstituted into new pS-containing small molecules. Experiments showed that one of the reconstituted compounds exhibited strong binding affinity to 14-3-3 and could compete for 14-3-3/ligand binding in vitro. Further experiments showed that it could induce apoptosis and cell cycle arrest in cell-based assays.

Multiple component reactions (MCRs) have also found use in SMM. Lin et al. constructed an SMM via Ugi four-component reactions on cellulose supports [28]. First, a photolabile linker was introduced onto a solid support for immobilization purposes. After UV illumination, the photolabile linker could be cleaved to allow postsynthetic biological screening such as agar overlay assay. Different building blocks containing carboxylic acid, aldehyde, and isocyanide were applied onto a cellulose support. The Ugi MCR could occur efficiently with the addition of water. The compounds produced through in situ MCR displayed high purity, with most of them showing a purity level above 90%. This approach may hold substantial promise for creating small-molecule libraries in the future.

13.3 FABRICATION OF SMMs

Following the completion of library synthesis, small molecules from 384-well plates can be deposited automatically onto the glass slides by high-precision robotic printers. Depending on the type of microarray spotter used, small molecules could be deposited onto glass slides by contact printing or piezoelectric mode. The typical spot size for microarray is around 100 μm in diameter, and the typical spot volume is around 1 nL. This miniaturized assay saves a substantial amount of expensive bioreagent compared with other assay platforms, such as microplate-based assays.

Immobilization technology sits at the heart of array fabrication. It is of paramount importance to immobilize biomolecules stably onto a solid support during microarray fabrication. Three approaches are generally used for the fabrication process: (1) a noncovalent immobilization approach, (2) a covalent immobilization approach, and (3) an in situ synthesis approach.

13.3.1 Noncovalent Immobilization Approach

Noncovalent interactions have commonly been employed for SMM immobilization. The strong interaction between two complementary DNA stands is frequently utilized for decoding purposes in combinatorial chemistry. In 2007, Schroeder et al. introduced an elegant DNA-directed strategy to immobilize cell-specific peptide ligands onto an array [29]. In this approach, various biotinylated peptide ligands were first coupled to streptavidin–DNA conjugates. The resulting peptide–DNA conjugates were then hybridized to a DNA microarray containing complementary DNA strands. The peptide array obtained could be used for the cell adhesion study of NIH-3T3 fibroblasts. Experiments showed that cells could grow properly on the spot in the presence of suitable ligands. This method not only proved the feasibility of fabricating a live cell array in a DNA-directed manner, but also created additional functions, such as controlled release of cells on an array. Melkko et al. realized another creative design [30]. The team generated self-assembling chemical libraries by linking two different small molecules together through a specially designed DNA tag. The tag consists of two parts: a unique oligonucleotide sequence for decoding purpose that can be hybridized onto a DNA microarray, and an assembly oligonucleotide sequence for mediating the self-assembly of two molecules. Using this approach they were able to identify tight binders against proteins with multiple pockets, such as human serum albumin and bovine carbonic anhydrase (see also Chapter 11).

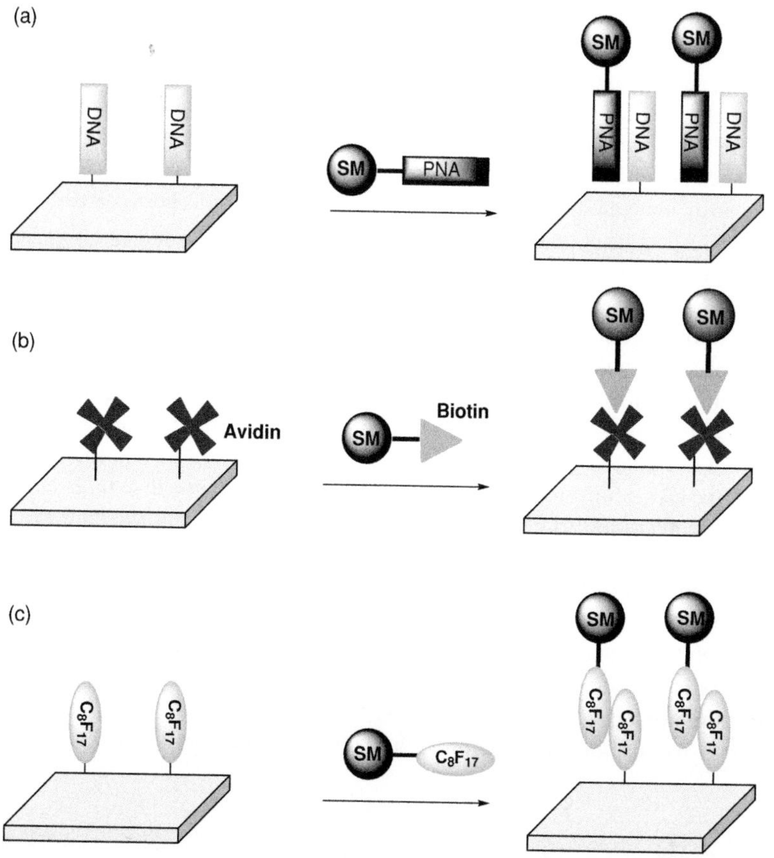

FIGURE 13.3 Selected examples of noncovalent immobilization approaches in SMM: (a) small molecules with a PNA tag hybridized onto a DNA microarray; (b) small molecules with a biotin tag immobilized onto an avidin-coated slide surface; (c) small molecules with a fluorous tag absorbed onto a fluorinated glass slide.

By exploiting the strong interaction between DNA and peptide nucleic acid (PNA), Winssinger et al. developed another new method for the preparation of SMMs (Figure 13.3a) [31]. In an early experiment, the group linked cathepsin inhibitors covalently with a PNA tag whose sequence encodes the identity of the inhibitors. The advantage of a PNA tag lies in its synthetic flexibility and resistance to degradation. These small molecule–PNA conjugates were then allowed to incubate with cathepsin. After passing through size-exclusion chromatography, the small molecule–PNA conjugates that bind to cathepsin were isolated and hybridized onto a DNA array to reveal their identity. The method has proved to be highly sensitive. It could detect as little as 0.4 pmol of cathepsin. The approach was subsequently extended to profile protease activity in crude cell lysates and dust mite extracts [32,33].

Another well-known noncovalent interaction pair is streptavidin and biotin. The affinity between streptavidin and biotin is very strong. It can resist various harsh conditions such as extreme pH, organic solvent, and heat. It has been used widely for biochemistry studies, including protein enrichment and protein–protein interaction assays. In 2002, Lesaicherre et al. applied this technique successfully to immobilize biotinylated peptides onto avidin-functionalized glass slides (Figure 13.3b) [34]. The peptide array generated was used to study specific interactions between peptides and antibodies. Apart from providing immobilization functions, avidin can also serve as a molecular layer. It effectively blocks nonspecific binding of proteins with the slide surface. The approach was subsequently applied by Uttamchandani and Yao in a series of substrate fingerprinting studies of various proteins [35].

Fluorous–fluorous interaction has been exploited by several groups to probe non-covalent immobilization of small molecules onto an array (Figure 13.3c). Ko et al. developed a fluorous-based approach by immobilizing carbohydrates with a C_8F_{17} tail onto a fluorinated glass slide [36]. The group demonstrated that the interaction between the C_8F_{17} tail and the fluorinated surface is robust enough to withstand stringent wash conditions. After immobilization onto an array, the carbohydrates could recognize a fluorescently labeled lectin specifically. Recently, Vegas et al. constructed a fluorous-based SMM to screen histone deacetylase (HDAC) inhibitors [37]. Twenty fluorous-tagged small molecules were printed onto an array and screened against HDAC2, HDAC8, and HDAC3/NCoR2. The method produced an excellent signal-to-noise ratio and proved its potential as an effective tool in identifying selective HDAC inhibitors. In a similar approach, Nicholson et al. proved that a fluorous-based microarray was capable of detecting protein–ligand binding events [38].

13.3.2 Covalent Immobilization Approach

Unlike noncovalent interactions, covalent immobilization generates a firm and stable linkage between small molecules and an array surface. For example, small molecules containing primary amines could be firmly immobilized onto epoxy-, aldehyde-, and NHS-coated slides. Depending on whether the immobilization tag is unique, covalent immobilization methods can be further divided into random immobilization and site-specific immobilization. If there is more than one immobilization tag present in the of molecules, it will lead to random immobilization. As a result, the immobilized molecules could possess multiple orientations in the same spot. This will lead further to inhomogeneous molecular interactions. Introducing a unique tag during library synthesis is often tedious. Nevertheless, the time and effort spent pay off, as they ensure that the immobilized molecules display uniform orientation, thereby enhancing molecular interaction.

The Schreiber group has developed several array immobilization approaches to anchor small molecules derived from a DOS library. The group first fabricated a chlorinated glass slide to catch DOS-derived compounds containing a hydroxyl group. Following that, they developed a strategy to coat the glass surface with a diazobenzyldiene group, which could catch phenol and other acidic functional groups. To immobilize small molecules with various functional groups, the team went further and developed an isocynanate-mediated approach, which could capture small molecules containing alcohols, amines, carboxylic acids, thiols, phenols, and so on. In a recent report, the group has adopted this method successfully to immobilize more than 15,000 compounds onto glass slides and screen their binding affinity toward 100 different proteins. The compounds come from different sources, such as commercial compounds, natural products, and synthetic libraries from academic laboratories [39].

In recent years, many research groups have developed various site-specific approaches to immobilize small molecules onto an array. Houseman et al., for example, utilized Diels–Alder chemistry to immobilize peptide–cyclopentadiene conjugates onto a self-assembled monolayer (SAM) displaying a benzoquinone handle [40]. The process is demonstrated in Figure 13.3.2a [38]. SAMs can minimize nonspecific absorption of proteins onto a solid surface and eliminate the blocking procedure in the subsequent array screening process. The reaction between benzoquinone and cyclopentadiene is highly specific and extremely fast. Through this immobilization strategy, the group was able to study peptide phosphorylation with Src kinase by various compatible techniques, such as surface plasmon resonance (SPR) and a fluorescence method. The group subsequently extended their work to adopt a photochemical approach to generating patterning substrates [40] which could be used for biospecific interaction studies. In this approach, the gold surface was modified by SAMs with nitroveratryloxycarbonyl (NVOC)-protected hydroquinone. Under ultraviolet irradiation, specific regions of the gold surface could be activated to release a free hydroquinone group. This allowed for subsequent oxidation to yield benzoquinone, which reacted further with diene-containing molecules. Through this technique, fluoresceine and peptides with RGD sequence were immobilized successfully onto a gold surface. The same group also developed another complementary strategy using maleimide-functionalized SAMs to anchor thiol-terminating molecules [41,42]. They immobilized carbohydrate–thiol conjugates onto a gold surface and studied the lectin-binding profiles with SPR and fluorescent techniques. The results of the research demonstrated that lectin could recognize the corresponding carbohydrate ligands on the chip.

Native chemical reaction is well-known for its chemoselectivity and bioorthogonal activity. It has been used by Yao's group to investigate the phosphorylation of peptides by Src kinase on microarray (Figure 13.4b) [34,43]. As a proof-of-concept experiment, the group first immobilized peptides with N-terminal cysteine onto thioester-functionalized glass slides and probed the binding specificity with their corresponding fluorescently labeled antibodies. The method was found highly specific and efficient. Only the terminal cysteine of the peptides would react with the thioester group on the glass surface when other internal cysteines and nucleophiles were present. Driven by these positive results, the group delved further to study the phosphorylation pattern of a peptide library against a kinase. Using putative p60c-src substrate as the template, various peptides were generated via different peptide library synthetic strategies, including alanine-, deletion-, and positional-scanning peptide libraries. These N-terminal cysteine-containing peptides were then spotted onto PEGylated thioester slides and used to profile the phosphorylation pattern of

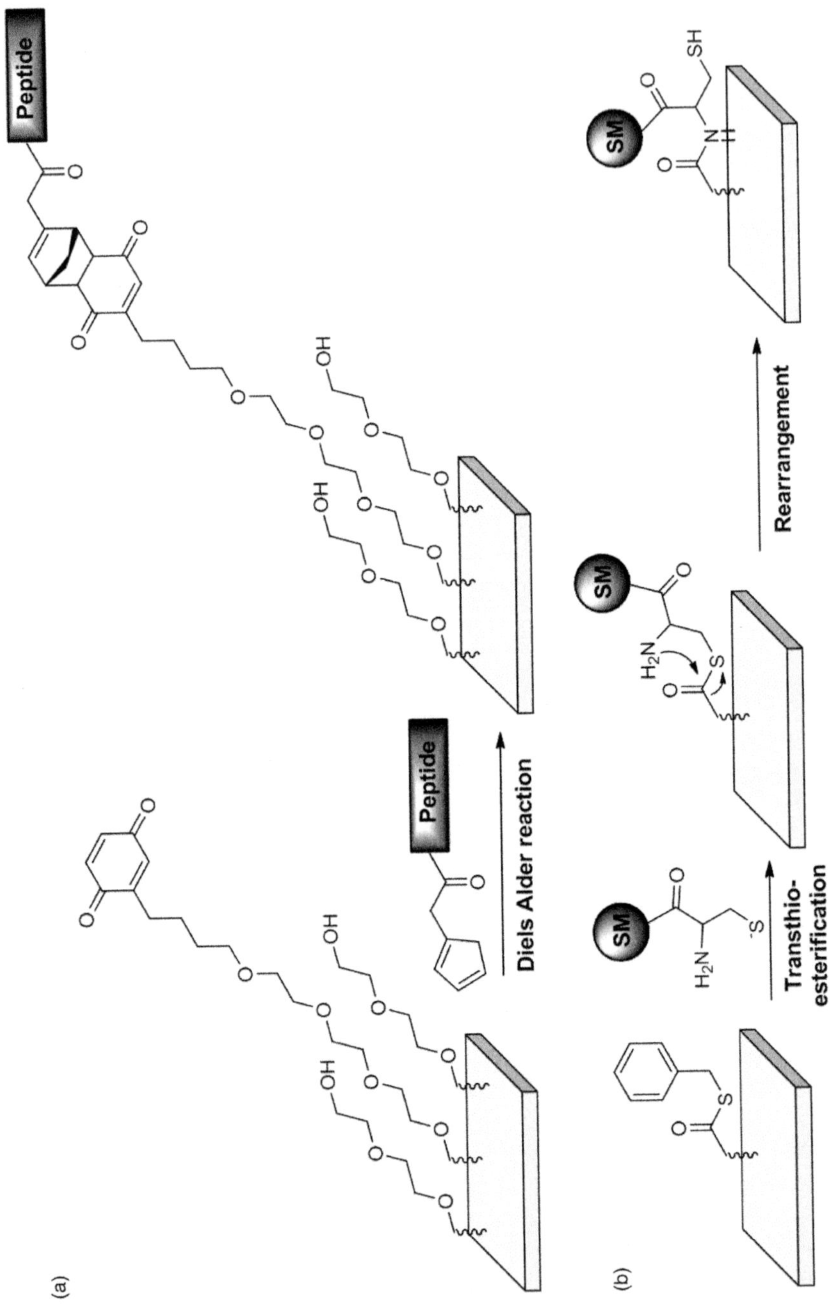
(a)
Peptide
Diels Alder reaction
(b)
SM
Transthio-
esterification
Rearrangement

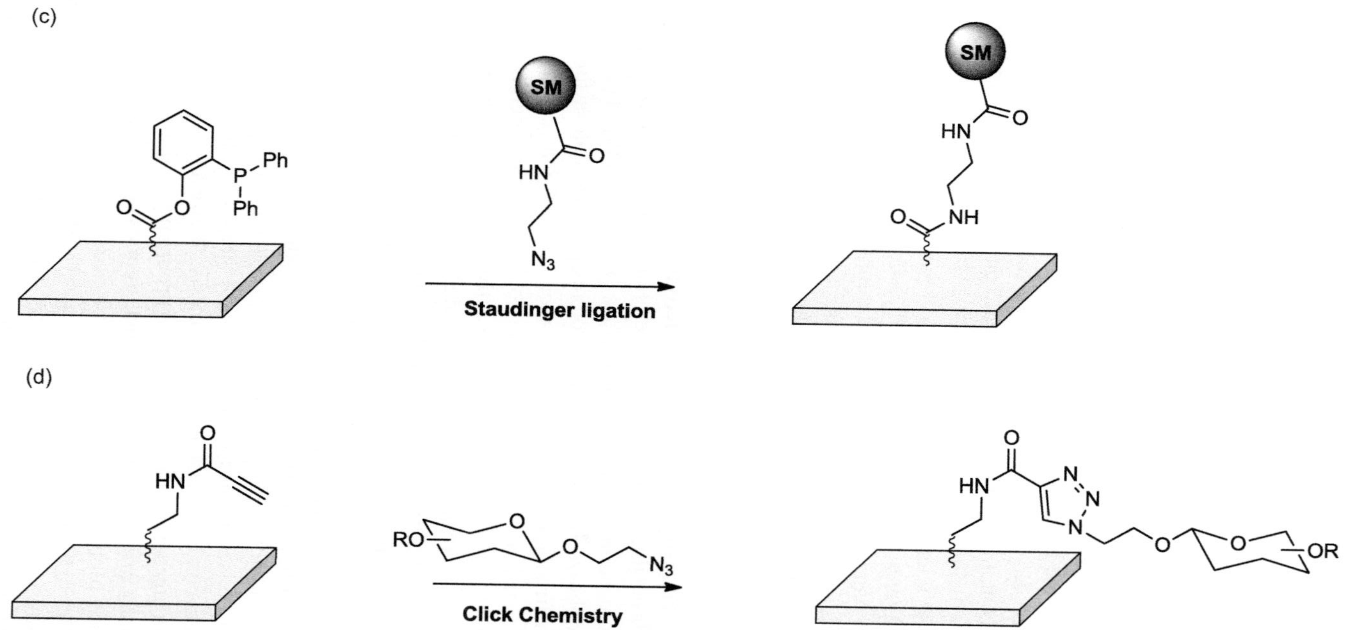

FIGURE 13.4 Selected examples of covalent immobilization approaches through various chemoselective reactions: (a) Diels-Alder reaction; (b) native chemical ligation; (c) Staudinger ligation; (d) click chemistry.

a tyrosine kinase, p60c-src.The results of the experiment indicated clearly that a peptide array can serve as a competent approach to assaying kinase activity.

Besides native chemical reactions, many other chemoselective reactions have been explored for small molecule immobilization. Falsey et al. prepared a new type of microchip by installing a glyoxylyl group on the slide surface [44]. The authors successfully anchored aminooxy- and cysteine-containing small molecules onto glyoxylyl-derived glass slides with two different chemical reactions. They employed this method in a protein-binding study, a kinase assay, and a cell adhesion study. Using a similar process, Melnyk et al. synthesized glyoxylyl-linked peptides and immobilized these peptides onto semicarbazide-derivatized glass slides [45]. The group showed that the peptide array constructed was highly specific and sensitive to the antibody tested. Kohn et al. and Soellner et al., independently, developed methods to anchor azide-containing small molecules onto phosphine-derivatized slides using Staudinger ligation (Figure 13.4c) [46,47]. It is noteworthy that Staudinger ligation is very mild and efficient even in the presence of water and oxygen.

In 2004, the Wong group functionalized the surface of microtiter plates with a cleavable alkyne handle and introduced azide-containing oligosaccharides onto the solid surface (Figure 13.4d) [48]. Through the addition of Cu(I), click chemistry could occur efficiently between an alkyne and an azide in aqueous solution. The group demonstrated that this approach could be used for lectin/antibody binding studies. Additionally, a cleavable linker allows easy characterization of carbohydrates and quantitative analysis with MS analysis after DTT cleavage. Although this immobilization chemistry was done with microtiter plates, the method can readily be applied to glass slides. Through thiol-ene click chemistry, Jonkheijm et al. also used a new photochemical method for surface patterning [49]. The approach can be used to immobilize olefin-containing proteins/peptides onto a thiol-derivatized surface through irradiation. The immobilization reaction was complete within 10 min, and the dimension of surface patterning could reach the submicrometer range. In 2005, Lee and Shin employed a chemoselective immobilization approach to link hydrazide substrates covalently to epoxide-coated slides [50]. No cross-reaction was observed in the presence of amines and hydroxyl groups, and the method was shown to be highly selective. It was also noted that strong nucleophiles such as thiols would undergo a slow reaction with epoxide-derivatized surface at low pH values, whereas hydrazide could react with epoxide efficiently at pH 3 to 5.

The aforementioned approaches share the fact that they all require a unique tag in the presence of molecules. This is to allow for the site-specific attachment of molecules onto the solid surface. Introducing a special tag into molecules can be realized through combinatorial library synthesis. However, some small-molecule libraries without any tag or reactive functional groups pose difficulty to SMM screening. It is especially evident for natural product libraries derived from natural sources. To overcome this problem, Kanoh et al. developed a random covalent immobilization approach through a photoaffinity reaction [51]. They modified the glass surface with a photoaffinity linker containing diazirine. The surface underwent irradiation to activate the diazirine group and release the carbine species. The carbine species were then inserted in the neighboring molecules to form strong covalent bonds. Various natural products, including rapamycin, cyclosporine A, digioxin, and FK506, have successfully been immobilized onto a diazirine-functionalized glass surface with their affinity toward the binding proteins retained.

13.3.3 In Situ Synthesis Approach

An alternative method for SMM fabrication is the in situ synthesis of small-molecule libraries. The entire synthetic process is executed directly on the array surface. With this approach, synthetic chemists can eliminate the tedious cleavage and spotting process. It should be noted that the chemistry used for the in situ synthesis of a small-molecule library has a direct impact on the quality of the array. Therefore, it must be extremely efficient and clean.

In 2004, Shaginian et al. described a light-directed approach for the in situ synthesis of SMMs [52]. Three orthogonal safety-catch photolabile protecting groups were used for the selective deprotection of multiple reactive sites (Figure 13.5a). With this method in hand, the authors successfully constructed a 2×2 SMM without using a photomask. Bowerman et al., on the other hand, developed a four-step cycle for the in situ synthesis of peptoids with MeNPOC group, which is depicted in Figure 13.5b [53]. In this design, MeNPOC glycolic acid was first linked to the amine-functionalized surface. UV irradiation was then applied to remove the MeNPOC group and revealed the hydroxyl group. The hydroxyl group could be activated by tosyl chloride and reacted with a primary amine to complete a monomer unit synthesis. In this approach, the light-sensitive MeNPOC group can be deprotected selectively at specific positions with a digital photolithograph technique. In theory, up to 100,000 peptoids can be synthesized in situ using this method. In the next stage of the research, the group extended their strategy to construct a cyclic peptide array, using different chemically orthogonal groups [54]. A cyclic peptide, Glu(Fmoc)–Dpr(ivDde)–Dpr(NPPOC)–Dpr)–Gly–OH, was covalently anchored onto amine-coated slides. NPPOC group was first removed for the introduction of the first diversity. The Fmoc and ivDde groups were selectively removed, and free amines were released accordingly. The amine group could be protected by the NPPOC group and further modified by other carboxylic acids in a light-controlled manner.

An in situ synthesis approach has also been adopted for the construction of an oligosaccharide array. As illustrated in Figure 13.5c, Ban and Mrksich prepared a disaccharide array by directly assembling the carbohydrates on a phenol-functionalized surface [55]. The phenol group served as a nucleophile which could link the first carbohydrate building block covalently onto a slide surface. The protecting group of the first carbohydrate was removed selectively to reveal a hydroxyl group. The hydroxyl group could react with the second carbohydrate building block to yield the desired disaccharide. With this approach, the authors constructed an array of 24 disaccharides and used the array to profile the substrate specificity of β-1,4-galactosyltransferase in a label-free manner.

Yet another creative design was made by Anderson et al. The team created a polymer microarray by spotting different combinations of 25 unique monomers onto pHEMA-coated slides [56]. With long-wavelength UV irradiation, 576 different polymers were obtained in situ. The polymer array generated was used to screen against 1700 human embryo stem cells to detect new types of biomaterial, which could help to modulate stem cell proliferation and differentiation.

13.4 APPLICATIONS OF SMM

One of the most fundamental purposes of biomedical research is to disclose specific interactions between small molecules and proteins with therapeutic value. After the identification of potent and selective small molecules for a given therapeutic protein, the molecule can be further designed and optimized into an inhibitor and

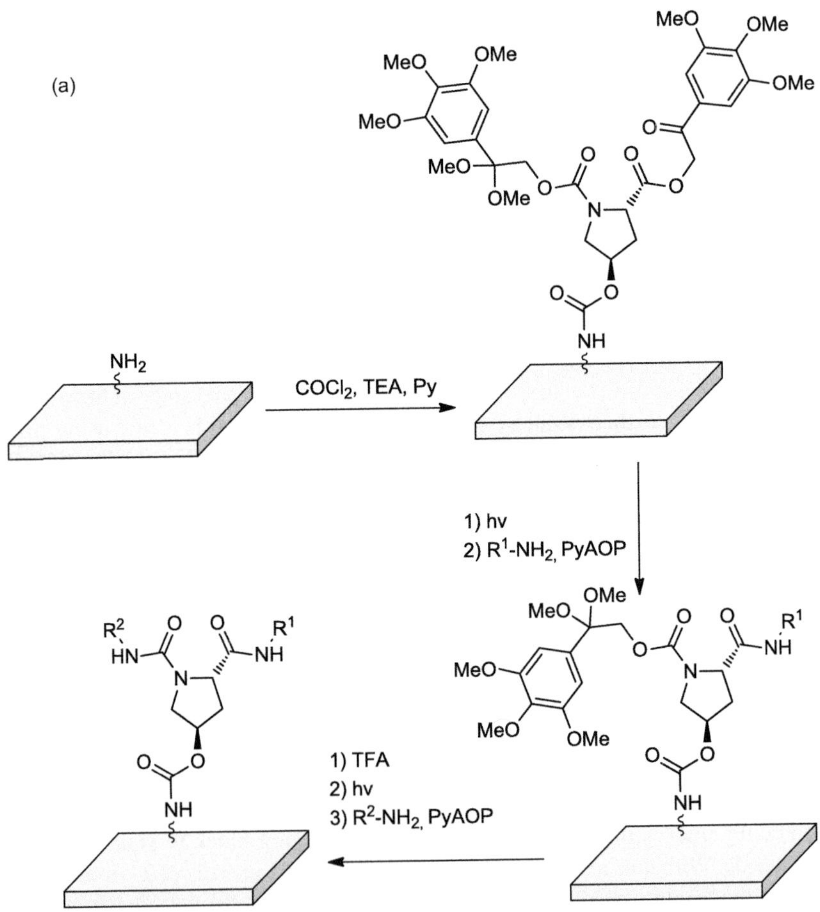

FIGURE 13.5 Selected examples of SMM construction via in situ synthesis: (a) fabrication of SMM using orthogonal safety-catch photolabile protective groups; (b) in situ synthesis of peptoid library via a four-step cycle; (c) in situ synthesis of a disaccharide array.

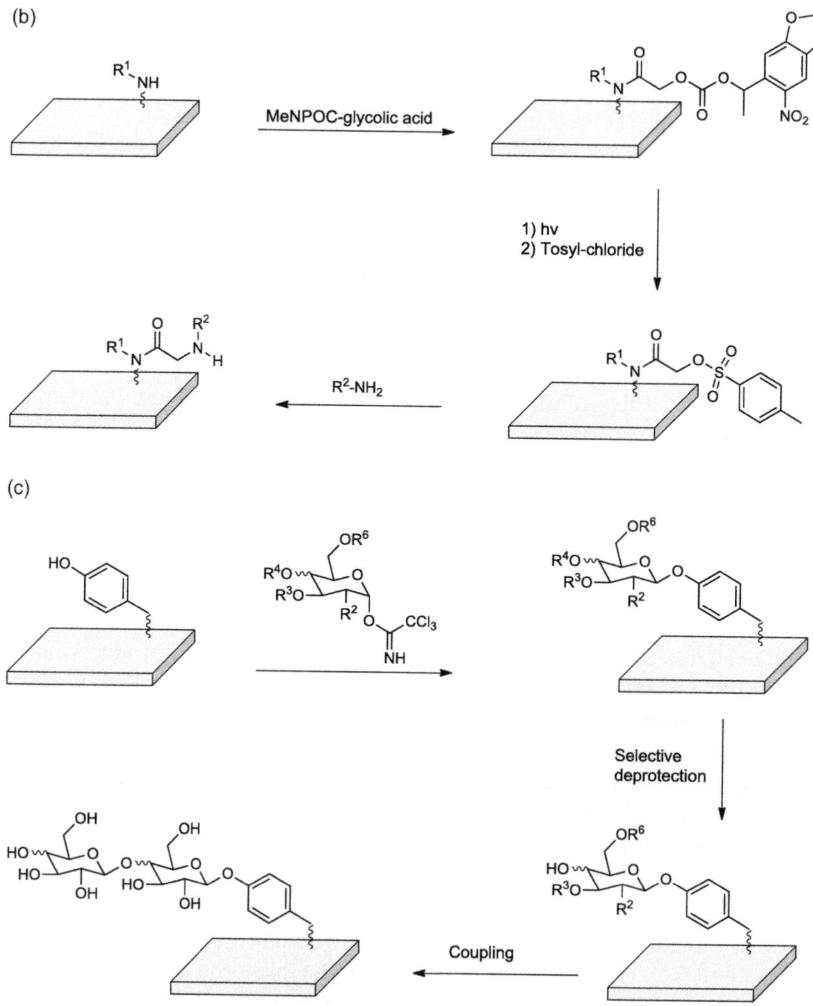

FIGURE 13.5 (*Continued*)

ultimately "tuned" into a potential drug. SMMs were originally established as a dynamic tool to discover potent and selective protein ligands but were then developed into a widely used platform technique. In recent years, many different research groups have explored new biological applications with strong enthusiasm. These uses include enzyme substrate profiling, nanodroplet microarray, and metabolizing enzyme toxicity assay among many others.

13.4.1 Protein Ligand Discovery

Small-molecule arrays are well-established, convenient tools for protein ligand discovery. Under standard procedures, proteins of interest can be labeled directly with fluorescent dyes and applied onto SMMs. The fluorescent reading will provide qualitative information on the interaction of small molecules and proteins. As an alternative approach, researchers can use sandwich-based assays, which provide high specificity and sensitivity. In this process the protein is first incubated with SMM. After multiple washing cycles, fluorescently labeled antibodies can be applied to identify the hits on the array. Corresponding negative control should be included in the experiment design because some antibodies might bind to microarray nonspecifically. If quantitative data such as K_d are required, concentration-dependent experiments can be performed on the array directly by applying fluorescent-labeled proteins of various concentrations. Using quantitative data has several advantages. For example, it gives information on how strong the small molecule binds to proteins. It also helps to significantly reduce the probability of obtaining false positives.

In 2004, Wong et al. synthesized 18,000 enatioenriched 1,3-dioxanes and screened their activity on a microarray [57]. This is the first report that stereochemically diverse sets of small molecules had been screened on the array. The library was synthesized by one bead–one stock solution technology. By probing the array with a fluorescently labeled calmodulin, the workers showed clear evidence that a heptamethyleneimine moiety is important for calmodulin-binding activity. In a zebrafish phenotypic assay, the authors further identified an inhibitor of cardiovascular function in an enantiomer-dependent manner. Later, Bradner et al. extended the strategy to immobilize around 10,000 small molecules from a DOS library and from natural products onto an isocyanate-functionalized glass slide (Figure 13.6a) [58]. Several bioactive compounds were detected for their ability to bind to their corresponding antibodies on the array. The authors demonstrated further that with the strategy they were able to directly screen small molecules with cellular lysates from human cells without prior purification of the protein. This technique offers a valuable and workable approach to identifying small-molecule ligands of the proteins that are inaccessible through standard protein purification process.

In 2005, Reddy and Kodadek synthesized a 7680-member peptoid library using a submonomer approach, and printed the peptoid library onto a maleimide-functionalized surface [59]. Unique and reproducible protein fingerprints were observed after applying three different fluorescently labeled proteins: MBP, GST, and ubiquitin. This approach can serve as a useful tool for protein identification without knowing the identity of the small molecules on the array. It can also be applied to other complicated fingerprinting assay, such as the diagnosis of cancers.

13.4.2 Enzyme Substrate/Inhibitor Profiling

To acquire the substrate specificity of enzymes, several groups have developed different strategies with SMMs. Enzymes such as kinases, phosphatases, and proteases play important roles in the regulation of many cellular functions. The characterization of an enzyme activity pattern could help researchers to gain a more comprehensive view of the catalytic mechanism and the properties of the enzyme. The Yao group synthesized and immobilized a variety of fluorogenic coumarin derivatives onto an array and used it to screen against four different classes of enzymes [60]. Three enzymes

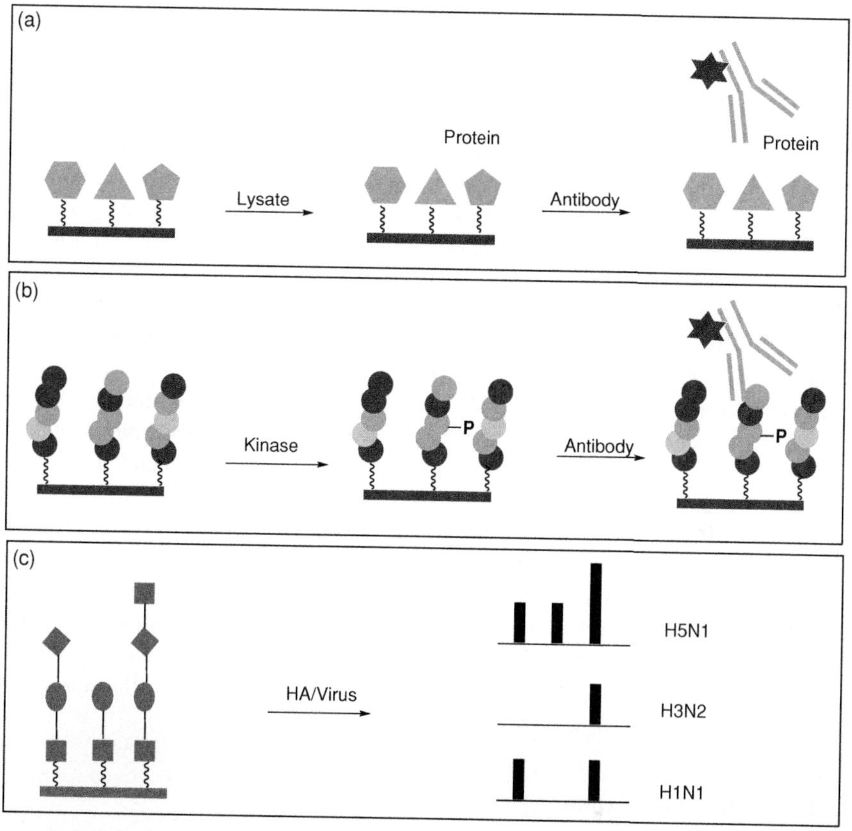

FIGURE 13.6 Biological applications of SMMs: (a) the process of screening cellular lysates without prior protein purification; (b) peptide microarray used for kinase substrate identification; (c) carbohydrate microarray used for virus detection.

were observed capable of cleaving their corresponding substrates selectively. In a similar setup, Salisbury et al. synthesized a combinatorial library with 361 peptide coumarin conjugates and used it to study the substrate fingerprints of proteases on the array [61]. The group was able to obtain a comprehensive proteolytic fingerprint with a thrombin enzyme in a single experiment. The approach also provided important quantitative data to determine the preference of the substrates, in this case kcat/km.

Gosalia and Diamond developed a novel nanodroplet SMM approach to circumvent the tedious immobilization process. In their study, a library of 352 compounds without any tag was first dispensed in glycerol droplets onto a plain glass slide [62]. Following that, the enzyme and its corresponding substrate were added in aerosol form to the discrete nanodroplets on the glass slide. The printed nanodroplets can last for a considerably long period during the screening process because of the high boiling point of glycerol. With this method, the group was able to obtain distinct inhibition profiles of these small molecules. The Yao group made a similar use of a protease-sensitive surface to obtain the inhibitor fingerprints of metalloproteases [63]. The plain glass surface was first functionalized with a layer of fluorogenic substrate. Nanodroplets with 400 inhibitors and enzymes were then applied to this protease-sensitive glass surface. The inhibitor fingerprinting profiles could be derived according to the fluorescent readout. The best inhibitor identified in the experiment showed a K_i value of 2.4 nM against thermolysin. The aforementioned approaches could present a promising tool for the discovery of new and selective inhibitors in a high-throughput manner.

Another research group from Kyushu University, Han et al., undertook an extensive investigation on the substrate specificity of various kinases by chemoselectively immobilizing 290 Tyr peptides and 1100 Ser/Thr peptides onto a glass surface (Figure 13.6b) [64]. Many new, potent, and selective peptides were identified for each kinase. The peptide array not only identified a previously reported consensus motif, but also disclosed some novel motifs. Waldmann's and Yao's groups, independently, developed phosphopeptide arrays to decipher substrate specificity against various protein phosphatases [46,65,66]. Several groups have also generated carbohydrate arrays to test the activity of glycosyltransfreases. Bryan et al. screened potential inhibitors of fucosyltransferase using a carbohydrate array [67]. Park and Shin immobilized 20 different carbohydrate probes on the array and investigated the substrate specificity of galactosyltransferase [68]. These novel strategies are helpful for understanding the substrate specificity of enzymes and provided invaluable insights for inhibitor design and discovery.

13.4.3 Other Applications

Apart from the aforementioned applications, SMMs have found many other novel applications in light of recent technological advancements. Bailey et al. developed an array-based system to screen the toxicity of small molecules against mammalian cells [69]. In this design, small molecules were embedded in the biodegradable poly(lactic acid)–poly(glycolic acid) copolymer (PLGA) and assembled in a microarray. Cells were then seeded over the array. As the polymer degraded, the compounds could be released into neighboring cells in a controlled environment. The strategy was demonstrated by applying toxic compounds against 1549 cells, as well as by synthetic lethal screening experiments. In 2005, Lee et al. developed a metabolizing enzyme toxicity assay on the array to investigate the cytotoxicity of these prodrugs after P450 activation [70]. The authors successfully assembled mixtures of sol–gel-encapsulated P450 and prodrugs in a microarray. A cell monolayer was then applied to this sol–gel array. After incubation, the cells can be removed and stained with fluorescence reagents for conventional microarray analysis. This technology may present an inexpensive and alternative approach for human metabolism and toxicology screening with liver slices.

During their recent research, Liang, Liao, and co-workers have constructed a sialosides array to profile the binding pattern against various influenza hemagglutinin (HA) subtypes (Figure 13.6c) [71,72]. 27 Sialosides were synthesized and screened with HA receptors and even complete viruses. Results showed that the HA receptor has binding patterns highly similar to those of viruses, especially avian viruses. A minimum set of five oligosaccharides can be used to differentiate various HA subtypes (i.e., H1, H3, H5, H7, and H9). The Yao group, on the other hand, constructed a peptide aldehyde array to study the substrate specificity against various cysteine proteases [73]. Distinct substrate fingerprinting profiles were observed with four cysteine proteases: caspase-3, caspase-7, cruzain, and rhodesain. The peptide array was shown capable of differentiating the various stages of red blood cells after parasitic infection.

In a recent experiment, a total of 10,800 small molecules have been immobilized covalently onto an array by Chen et al. to identify compounds that bind to fluorescently labeled Aβ peptide [74]. The hits identified from SMM were then subject to a secondary assay, an Aβ-induced toxicity assay, with PC12 cells. One compound was found to reduce the cytotoxicity of Aβ42 in a dose-dependent manner. It could provide a feasible approach to exploring Aβ aggregation pathways and contribute to the treatment of Alzheimer's disease. Lately, Luo et al. has developed a unique quantitative electroactive microarray to study stem cell differentiation [75]. In this approach, the hydroquinone molecule serves as a quantitative marker for surface intensity. After oxidation, it is converted to a quinone, which can be used to anchor small molecules with various functional groups. A variety of small-molecule ligands with precisely controlled ligand density were immobilized on the transparent surface to study human mesenchymal stem cell (hMSC) differentiation. Both the composition and the density of the ligands were found to have a direct influence on the rate of adepogenic differentiation.

13.5 SUMMARY AND OUTLOOK

In the past decade we have witnessed many innovative and exciting developments in SMM technology. Numerous library synthetic strategies have been designed to create large libraries of small molecules to probe biological functions. Among these evolutionary strategies, DOS has unequivocally underscored its potential as a robust tool for generating compound collections that resemble the properties of natural products. With advancements in organic chemistry, in particular with the development of novel stereoselective reactions and synthetic planning strategies, we can expect DOS to undergo further development and unfold more possibilities. The horizon of SMM research will undoubtedly be broadened henceforth.

In the past decade we have also seen various exciting surface chemistry techniques with excellent efficiency developed for microarray immobilization. SMM has already matured into a remarkably competent and expedient technology. Through SMM, not only purified proteins, but also crude cellular lysates, can be used to discover novel ligands. An entire cell and even an entire organism can be used in a microarray. The future of SMM research holds great promise by moving toward clinical applications. Through synergy between research and clinical institutes, SMM will unleash its potential as a powerful platform for disease diagnosis in the years to come. With continued research and development, it can be envisaged that additional innovations and creative transformations will take root in this field.

Acknowledgment

The author would like to acknowledge a City University of Hong Kong Strategic Research Grant (project 7002721) for financial support.

REFERENCES

1. S. P. Fodor, J. L. Read, M. C. Pirrung, L. Stryer, A. T. Lu, D. Solas, *Science* **1991**, *251*, 767–773.
2. M. Uttamchandani, J. Wang, S. Q. Yao, *Mol. Biosyst.* **2006**, *2*, 58–68.
3. H. Sun, S. Chattopadhaya, J. Wang, S. Q. Yao, *Anal. Bioanal. Chem.* **2006**, *386*, 416–426.
4. Y. Fang, A. G. Frutos, J. Lahiri, *J. Am. Chem. Soc.* **2002**, *124*, 2394–2395.
5. H. Zhu, M. Bilgin, R. Bangham, D. Hall, A. Casamayor, P. Bertone, N. Lan, R. Jansen, S. Bidlingmaier, T. Houfek, T. Mitchell, P. Miller, R. A. Dean, M. Gerstein, M. Snyder, *Science* **2001**, *293*, 2101–2105.
6. J. Ziauddin, D. M. Sabatini, *Nature* **2001**, *411*, 107–110.
7. D. L. Rimm, R. L. Camp, L. A. Charette, J. Costa, D. A. Olsen, M. Reiss, *Cancer J.* **2001**, *7*, 24–31.
8. L. Horvath, S. Henshall, *Pathology* **2001**, *33*, 125–129.
9. A. Tavassoli, A. D. Hamilton, D. R. Spring, *Chem. Soc. Rev.* **2011**, *40*, 4269–4270.
10. G. MacBeath, A. N. Koehler, S. L. Schreiber, *J. Am. Chem. Soc.* **1999**, *121*, 7967–7968.
11. F. G. Kuruvilla, A. F. Shamji, S. M. Sternson, P. J. Hergenrother, S. L. Schreiber, *Nature* **2002**, *416*, 653–657.
12. R. B. Merrifield, *J. Am. Chem. Soc.* **1963**, *85*, 2149–2154.
13. R. B. Merrifield, *Angew. Chem. Int. Ed. Engl.* **1985**, *24*, 799–810.
14. A. Furka, F. Sebestyen, M. Asgedom, G. Dibo, *Int. J. Pept. Protein Res.* **1991**, *37*, 487–493.
15. K. S. Lam, S. E. Salmon, E. M. Hersh, V. J. Hruby, W. M. Kazmierski, R. J. Knapp, *Nature* **1991**, *354*, 82–84.
16. H. Wu, J. Ge, M. Uttamchandani, S. Q. Yao, *Chem. Commun.* **2011**, *47*, 5664–5670.
17. S. L. Schreiber, *Science* **2000**, *287*, 1964–1969.
18. A. N. Koehler, A. F. Shamji, S. L. Schreiber, *J. Am. Chem. Soc.* **2003**, *125*, 8420–8421.
19. D. Barnes-Seeman, S. B. Park, A. N. Koehler, S. L. Schreiber, *Angew. Chem. Int. Ed.* **2003**, *42*, 2376–2379.
20. M. Uttamchandani, D. P. Walsh, S. M. Khersonsky, X. Huang, S. Q. Yao, Y. T. Chang, *J. Comb. Chem.* **2004**, *6*, 862–868.
21. H. E. Park, S. Y. Lee, H. Y. Ahn, J. C. Shin, Y. T. Chang, Y. A. Joe, *J. Appl. Pharmacol.* **2003**, *11*, 85–90.
22. D. W. P. M. Lowik, C. R. Lowe, *Eur. J. Org. Chem.* **2001**, 2825–2839.
23. M. Uttamchandani, W. L. Lee, J. Wang, S. Q. Yao, *J. Am. Chem. Soc.* **2007**, *129*, 13110–13117.
24. T. Horlacher, P. H. Seeberger, *OMICS*, **2006**, *10*, 490–498.
25. P. H. Seeberger, *Chem. Soc. Rev.* **2008**, *37*, 19–28.

26. O. Blixt, S. Head, T. Mondala, C. Scanlan, M. E. Huflejt, R. Alvarez, M. C. Bryan, F. Fazio, D. Calarese, J. Stevens, N. Razi, D. J. Stevens, J. J. Skehel, I. van Die, D. R. Burton, I. A. Wilson, R. Cummings, N. Bovin, C. H. Wong, J. C. Paulson, *Proc. Natl. Acad. Sci. U.S.A.* **2004**, *101*, 17033–17038.
27. H. Wu, J. Y. Ge, S. Q. Yao, *Angew. Chem. Int. Ed.* **2010**, *49*, 6528–6532.
28. Q. Lin, J. C. O'Neill, H. E. Blackwell, *Org. Lett.* **2005**, *7*, 4455–4458.
29. H. Schroeder, B. Ellinger, C. F. Becker, H. Waldmann, C. M. Niemeyer, *Angew. Chem. Int. Ed.* **2007**, *46*, 4180–4183.
30. S. Melkko, J. Scheuermann, C. E. Dumelin, D. Neri, *Nat. Biotechnol.* **2004**, *22*, 568–574.
31. N. Winssinger, J. L. Harris, B. J. Backes, P. G. Schultz, *Angew. Chem. Int. Ed.* **2001**, *40*, 3152–3155.
32. N. Winssinger, S. Ficarro, P. G. Schultz, J. L. Harris, *Proc. Natl. Acad. Sci. U.S.A.* **2002**, *99*, 11139–11144.
33. J. Harris, D. E. Mason, J. Li, K. W. Burdick, B. J. Backes, T. Chen, A. Shipway, G. Van Heeke, L. Gough, A. Ghaemmaghami, F. Shakib, F. Debaene, N. Winssinger, *Chem. Biol.* **2004**, *11*, 1361–1372.
34. M. L. Lesaicherre, M. Uttamchandani, G. Y. Chen, S. Q.Yao, *Bioorg. Med. Chem. Lett.* **2002**, *12*, 2079–2083.
35. M. Uttamchandani, S. Q. Yao, *Curr. Pharm. Des.* **2008**, *14*, 2428–2438.
36. K. S. Ko, F. A. Jaipuri, N. L. Pohl, *J. Am. Chem. Soc.* **2005**, *127*, 13162–13163.
37. A. J. Vegas, J. E. Bradner, W. P. Tang, O. M. McPherson, E. F. Greenberg, A. N. Koehler, S. L. Schreiber, *Angew. Chem. Int. Ed.* **2007**, *46*, 7960–7964.
38. R. L. Nicholson, M. L. Ladlow, D. R. Spring, *Chem. Commun.* **2007**, 3906–3908.
39. P. A. Clemons, N. E. Bodycombe, H. A. Carrinski, J. A. Wilson, A. F. Shamji, B. K. Wagner, A. N. Koehler, S. L. Schreiber, *Proc. Natl. Acad. Sci. U.S.A.* **2010**, *107*, 18787–18792.
40. B. T. Houseman, J. H. Huh, S. J. Kron, M. Mrksich, *Nat. Biotechnol.* **2002**, *20*, 270–274.
41. W. S. Dillmore, M. N. Yousaf, M. Mrksich, *Langmuir* **2004**, *20*, 7223–7231.
42. B. T. Houseman, E. S. Gawalt, M. Mrksich, *Langmuir* **2003**, *19*, 1522–1531.
43. M. Uttamchandani, E. W. Chan, G. Y. Chen, S. Q. Yao, *Bioorg. Med. Chem. Lett.* **2003**, *13*, 2997–3000.
44. J. R. Falsey, M. Renil, S. Park, S. J. Li, K. S. Lam, *Bioconjug. Chem.* **2001**, *12*, 346–353.
45. O. Melnyk, X. Duburcq, C. Olivier, F. Urbès, C. Auriault, H. Gras-Masse, *Bioconjug. Chem.* **2002**, *13*, 713–720.
46. M. Kohn, M. Gutierrez-Rodriguez, P. Jonkheijm, S. Wetzel, R. Wacker, H. Schroeder, H. Prinz, C. M. Niemeyer, R. Breinbauer, S. E. Szedlacsek, H. Waldmann, *Angew. Chem. Int. Ed.* **2007**, *46*, 7700–7703.
47. M. B. Soellner, K. A. Dickson, B. L. Nilsson, R. T. Raines, *J. Am. Chem. Soc.* **2003**, *125*, 11790–11791.
48. M. C. Bryan, F. Fazio, H. K. Lee, C. Y. Huang, A. Chang, M. D. Best, D. A. Calarese, O. Blixt, J. C. Paulson, D. Burton, I. A. Wilson, C. H. Wong, *J. Am. Chem. Soc.* **2004**, *126*, 8640–8641.
49. P. Jonkheijm, D. Weinrich, M. Köhn, H. Engelkamp, P. C. Christianen, J. Kuhlmann, J. C. Maan, D. Nüsse, H. Schroeder, R. Wacker, R. Breinbauer, C. M. Niemeyer, H. Waldmann, *Angew. Chem. Int. Ed.* **2008**, *47*, 4421–4424.

50. M. Lee, I. Shin, *Angew. Chem. Int. Ed.* **2005**, *44*, 2881–2884.
51. N. Kanoh, S. Kumashiro, S. Simizu, Y. Kondoh, S. Hatakeyama, H. Tashiro, H. Osada, *Angew. Chem. Int. Ed.* **2003**, *42*, 5584–5587.
52. A. Shaginian, M. Patel, M. H. Li, S. T. Flickinger, C. Kim, F. Cerrina, P. J. Belshaw, *J. Am. Chem. Soc.* **2004**,*126*, 16704–16705.
53. S. Li, D. Bowerman, N. Marthandan, S. Klyza, K. J. Luebke, H. R. Garner, T. Kodadek, *J. Am. Chem. Soc.* **2004**, *126*, 4088–4089.
54. S. Li, N. Marthandan, D. Bowerman, H. R. Garner, T. Kodadek, *Chem. Commun.* **2005**, *5*, 581–583.
55. L. Ban, M. Mrksich, *Angew. Chem. Int. Ed.* **2008**, *47*, 3396–3399.
56. D. G. Anderson, S. Levenberg, R. Langer, *Nat. Biotechnol.* **2004**, *22*, 863–866.
57. J. C. Wong, S. M. Sternson, J. B. Louca, R. Hong, S. L. Schreiber, *Chem. Biol.* **2004**, *11*, 1279–1291.
58. J. E. Bradner, O. M. McPherson, R. Mazitschek, D. Barnes-Seeman, J. P. Shen, J. Dhaliwal, K. E. Stevenson, J. L. Duffner, S. B. Park, D. S. Neuberg, P. Nghiem, S. L. Schreiber, A. N. Koehler, *Chem. Biol.* **2006**, *13*, 493–504.
59. M. M. Reddy, T. Kodadek, *Proc. Natl. Acad. Sci. U.S.A.* **2005**, *102*, 12672–12677.
60. Q. Zhu, M. Uttamchandani, D. B. Li, M. L. Lesaicherre, S. Q. Yao, *Org. Lett.* **2003**, *5*, 1257–1260.
61. C. M. Salisbury, D. J. Maly, J. A. Ellman, *J. Am. Chem. Soc.* **2002**, *124*, 14868–14870.
62. D. N. Gosalia, S. L. Diamond, *Proc. Natl. Acad. Sci. U.S.A.* **2003**, *100*, 8721–8726.
63. J. Wang, M. Uttamchandani, L. P. Sun, S. Q. Yao, *Chem. Commun.* **2006**, 717–719.
64. X. M. Han, T. Sonoda, T. Mori, G. Yamanouchi, T. Yamaji, S. Shigaki, T. Niidome, Y. Katayama, *Comb. Chem. High Throughput Screen.* **2010**, *13*, 777–789.
65. H. Y. Sun, L. P. Tan, L. Q. Gao, S. Q. Yao, *Chem. Commun.* **2009**, 677–679.
66. H. Y. Sun, C. H. S. Lu, M. Uttamchandani, Y. Xia, Y. C. Liou, S. Q. Yao, *Angew. Chem. Int. Ed.* **2008**, *47*, 1698–1702.
67. M. C. Bryan, L. V. Lee, C. H. Wong, *Bioorg. Med. Chem. Lett.* **2004**, *14*, 3185–3188.
68. S. Park, I. Shin, *Org. Lett.* **2007**, *9*, 1675–1678.
69. S. N. Bailey, D. M. Sabatini, B. R. Stockwell, *Proc. Natl. Acad. Sci. U.S.A.* **2004**, *101*, 16144–16149.
70. M. Y. Lee, C. B. Park, J. S. Dordick, D. S. Clark, *Proc. Natl. Acad. Sci. U.S.A.* **2005**, *102*, 983–987.
71. C.-H. Liang, S.-K. Wang, C.-W. Lin, C.-C. Wang, C.-H. Wong, C.-Y. Wu, *Angew. Chem. Int. Ed.* **2011**, *50*, 1608–1612.
72. H. Y. Liao, C. H. Hsu, S. C. Wang, C. H. Liang, H. Y. Yen, C. Y. Su, C. H. Chen, J. T. Jan, C. T. Ren, C. H. Chen, T. J. R. Cheng, C.-Y. Wu, C.-H. Wong, *J. Am. Chem. Soc.* **2010**, *132*, 14849–14856.
73. H. Wu, J. Ge, P.-Y. Yang, J. Wang, M. Uttamchandani, S. Q. Yao, *J. Am. Chem. Soc.* **2011**, *133*, 1946–1954.
74. J. Chen, A. H. Armstrong, A. N. Koehler, M. H. Hecht, *J. Am. Chem. Soc.* **2010**, *132*, 17015–17022.
75. W. Luo, E. W. L. Chan, M. N. Yousaf, *J. Am. Chem. Soc.* **2010**, *132*, 2614–2621.

14

YEAST AS A MODEL IN HIGH-THROUGHPUT SCREENING OF SMALL-MOLECULE LIBRARIES

Irene Stefanini, Carlotta De Filippo, and Duccio Cavalieri

14.1 INTRODUCTION

14.1.1 The Quest for Rapid and Smart Biological Assays

The pharmaceutical industry is facing a difficult period. The coincidence of pending patent expirations for major drug classes and a series of falling success rates in the development of innovative therapeutics has affected the research and development sectors significantly [1,2]. By chance or necessity, the gaining social impact of specific diseases such as cancer and Alzheimer's, has obliged most companies to investigate the same or similar targets and approaches [3]. Actually, the PubChem [4] and ChemBank databases encompass nearly 57 and 2.5 million compounds, respectively. Furthermore, synthetic chemistry has reached an unprecedented level of sophistication and execution, allowing for the rapid, economically affordable, and often parallelizable synthesis of a large plethora of molecules [5,6]. When studying a molecule selected either for its role in the onset of a disease or as a suitable therapeutic target, the availability of widespread chemical libraries represents a precious launch pad for research on molecules able to modify the biological functions of the target. During the past two decades, the "targetphilia" [7]—the target-based drug discovery approach—allowed for the identification of novel lead compounds and therapeutic candidates. Despite these promising results, the overall approval rate for new chemical entities has maintained a flat trend [8]. Yet the targeted approach forbids the

Diversity-Oriented Synthesis: Basics and Applications in Organic Synthesis, Drug Discovery, and Chemical Biology, First Edition. Edited by Andrea Trabocchi.

identification of biological modulations theoretically exploited by the wide variability of available natural and synthetic compounds acting on entities different from the target selected. Given the availability of this wide number and variety of compounds, neglecting their potential biological modulation activity would be an unquantifiable waste of effort and opportunity. Obviously, the disposal of this vast plethora of compounds is not related strictly to the ability to determine their biological potential. When the number of variables increases, the use of rapid biological screenings, providing easily handled, exhaustive, extensive, and understandable results, becomes a central quest.

14.1.1.1 The State of the Art: From Isolated Targets to Systems Over the years, the pharmaceutical industry has relied on several different technologies to assess the biological activities of screened compounds. For a long time, the initial biological screenings aimed at drug discovery purposes were carried out by means of in vitro assays. Mainly, isolated entities (i.e., enzymes) were used to identify compounds having functional effects; otherwise, conventional high-throughput LC-MS-MS and newer MS approaches were used [9]. Although ligand–receptor binding assays were considered the "gold standard," many drawbacks are associated with these approaches, and their fallouts are easily quantifiable when considering the high rate of drug development abortions at the preclinical stage. The application of *in silico* approaches, based on the modeling of novel compounds able to bind and modulate targetable entities, only partially prevented the failure of new bioactive compounds. Although the use of a molecule isolated from its natural environment clears the presence of eventual background noises, it forbids the prediction of characteristics that make a selected compound a "druggable" molecule. As an example, the ability of the molecule selected to permeate the cell membrane and its metabolism are often unpredictable a priori. In addition, eventual side effects are hardly predictable. Moreover, the utilization of purified target molecules relies on the one target–one disease concept [10]. On the contrary, a great many disorders, such as cancer and neurodegenerative diseases, revealed a higher complexity, often being associated with several molecules or with complete pathway dysfunction. Cell-based assays thus became a useful tool in the selection of bioactive compounds not biased by the problems described above. A cell-based approach does not require the selection of a specific target. A desirable phenotype such as fungal cell death when searching for novel antifungals can be selected, and then the molecules are screened for their ability to induce such a phenotype. At the same time, this approach allows for the identification of compounds active in inducing the selected phenotype and the characterization of otherwise hardly predictable mechanisms lying at the basis of this phenotype. The other side of the coin is that, once a bioactive compound is selected, its cellular target has to be identified.

The more is known about the cell system used for the assay, the less complicated is the identification of molecular targets lying at the basis of the phenotype modulation induced by a selected compound or perturbation. Aiming at this, the use of a model system is preferable to utilization of the final (clinical) target, often represented by the still poorly characterized *Homo sapiens*. Although there are obvious limitations to using a microorganism to identify potential human drug targets, use of the well-characterized *Saccharomyces cerevisiae* yeast showed striking advantages in several applications.

14.1.2 *Saccharomyces cerevisiae* as a Model

14.1.2.1 S. cerevisiae Biology *Saccharomyces cerevisiae* is the world's most popular yeast, being instrumental in baking and brewing since ancient times. During at least the past 9000 years, humans have taken advantage of the properties of *S. cerevisiae* for wine, bread, and beer production. The first wine fermentation observed dates back 9000 years, based on the discovery of residues of ancient *S. cerevisiae* DNA in Chinese pots [11] and in jars at the King Scorpion tomb in Abydos (3150 B.C.) [12]. After a thousand years of appreciation of the products of its fermentations, *S. cerevisiae* has gained increasing interest among scientists. Thanks to several features that make it an easy-to-handle organism, this budding yeast has made possible the discovery and deepening of fundamental biological processes often shared with higher eukaryotic cells.

The *S. cerevisiae* cell has a mean diameter of 8 to 10 μm and it is covered by a cell wall composed by a small fraction of chitin, a polymer of the glucose derivative *N*-acetyl glucosamine. In addition to chitin, the wall is composed of other molecules that constitute a cementing matrix: polysaccharides (80 to 90%), proteins, lipids, polyphosphates, and inorganic ions. *S. cerevisiae* cells exist in two forms: haploid (a single copy of each one of the 16 chromosomes) or diploid (two copies of each chromosome). Haploid cells in turn exist in two forms, called *mating types*: a or α. The budding yeast can multiply by both sexual and asexual reproduction (Figure 14.1). The mating of two haploid cells generates a diploid cell, which in adverse conditions can sporulate. Sporulation, driven by meiosis, is a process through which the yeast

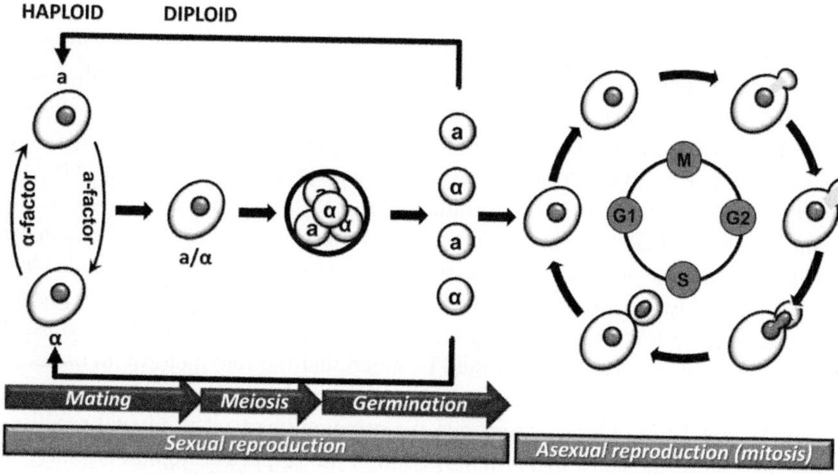

FIGURE 14.1 Cell cycle of *Saccharomyces cerevisiae*.

diploid cell assumes a quiescent but viable form, the sporal form. When surrounding conditions get better, each spore germinates, giving origin to a vital cell. The *S. cerevisiae* haploid genome is composed by 16 linear chromosomes, for a total of nearly 1206 Mb, with 5570 genes coding proteins (nearly 70% of the genomic DNA). The *S. cerevisiae* genome was the first to be sequenced completely [13]. Moreover, the majority of the ORF (open reading frame) has been annotated, allowing for the identification and functional characterization of the respective gene products. The availability of the complete yeast genome sequence allowed for the construction of mutants knocked out for every gene.

14.1.2.2 Model Characteristics One of the most relevant advantages rising from the use of *S. cerevisiae* as a model in high-throughput screenings is given by the availability of its completely sequenced and almost entirely annotated genome. The deep knowledge of the yeast genome and the relatively high efficiency of homolog recombination facilitated the microorganism genetic manipulation, thus providing many techniques for dissection, manipulation, and transformation of the genome. Taking advantage of these characteristics, a wide collection of mutants has been generated. A central role in the study of the effects of molecules and perturbations is held by the collection composed of strains each deleted for a single gene. Such a collection has been developed both in homozygote (encompassing deletion of only nonessential genes) and heterozygote (composed of strains deleted for every gene annotated) genetic backgrounds. Knowledge of *S. cerevisiae* is not restricted to its genome but is also supported by a deep knowledge of the metabolism (both aerobic and anaerobic). Moreover, an encouraging advantage of using *S. cerevisiae* as a model in high-throughput screenings rises from the availability and economical affordability of the necessary growing media and handling procedures.

S. cerevisiae cells show high similarity with higher eukaryotic cells such as human cells, sharing a substantial evolutionary conservation of the principal biological processes (i.e., cell cycle, DNA reparation, several metabolic pathways). By virtue of these characteristics, the budding yeast has been defined as an "honorary mammal" [14]. In a recent study [15], the excellence of *S. cerevisiae* as a model organism has been reassessed at the molecular level by full proteome comparison with several organisms. The advent of Next-Generation Sequencing (NGS) techniques allowed for the more rapid and accurate sequencing of complete genomes of every organism, thus paving the way to disclosure of an almost unexplored universe of genetic and molecular features. In their work, Karathia et al. [15] compared protein sequences obtained by an *in silico* translation of newly sequenced genomes of several organisms with the *S. cerevisiae* protein sequence, which is almost completely annotated. Thus, they defined the level of conservation of the primary structure of proteins, protein functions, and the protein pathway, giving a measure of the reliability of yeast to model the organisms analyzed. It should be noted that the vast majority of essential genes in *S. cerevisiae* are conserved in many other organisms and in *Homo sapiens* as well (Figure 14.2). These genes code for proteins involved in pivotal reactions for the survival of the cells, and a malfunction in the respective human orthologs is often associated with severe diseases [16,17]. The "honorary mammal" *S. cerevisiae*

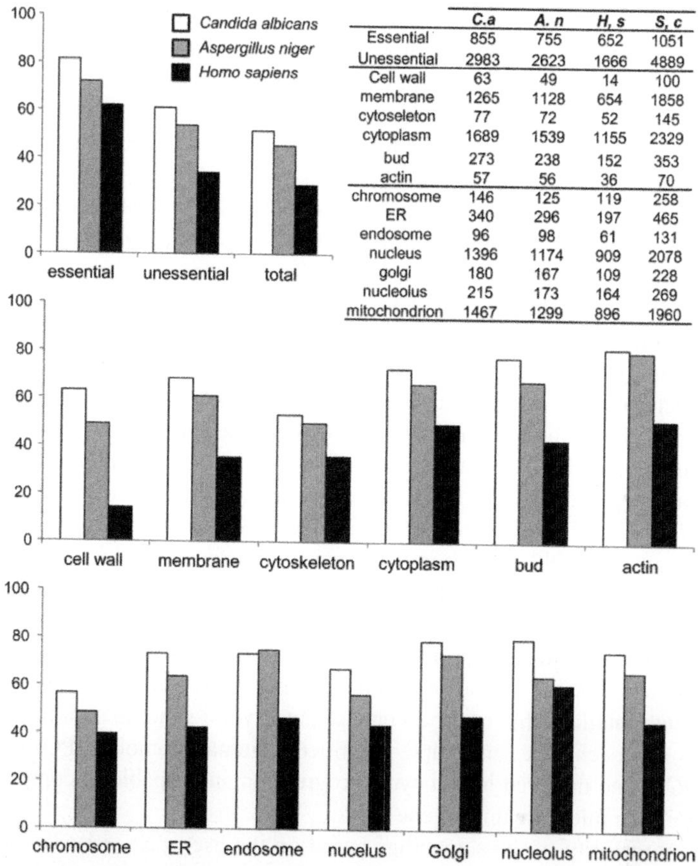

	C.a	A. n	H, s	S, c
Essential	855	755	652	1051
Unessential	2983	2623	1666	4889
Cell wall	63	49	14	100
membrane	1265	1128	654	1858
cytoseleton	77	72	52	145
cytoplasm	1689	1539	1155	2329
bud	273	238	152	353
actin	57	56	36	70
chromosome	146	125	119	258
ER	340	296	197	465
endosome	96	98	61	131
nucleus	1396	1174	909	2078
golgi	180	167	109	228
nucleolus	215	173	164	269
mitochondrion	1467	1299	896	1960

FIGURE 14.2 Analysis of proteins conserved among *S. cerevisiae*, *Candida albicans*, *Aspergillus niger*, and *Homo sapiens* cells. Histogram bars represent the percentage of *S. cerevisiae* proteins conserved in the organism. Proteins are divided by functional class membership. The number of proteins for every class in every organism is given in the table.

represents a middle ground between fungi and higher eukaryotic cells. Obviously, this microorganism is genetically closer to other fungi with respect to other model organisms, and at the same time it bears several cellular functions conserved with respect to mammalian cells. Unique among model organisms, *S. cerevisiae* can be used for modeling both fungal pathogens and higher eukaryotic cells.

Over the past 20 years, the number of invasive fungal infections has risen, primarily because of the increase in the number and survival of severely immunocompromised patients [18]. Despite the development of more active and selective antifungals and the standard use of prophylaxis, fungal infections still represent an important clinical emergency, provoking high rates of morbidity and mortality. Thus, because of the insurgence of resistance and the ineffectiveness of classical antifungals against mold infections, the quest for novel antifungal drugs has elicited renewed interest. The pathogens mainly associated with fungal infections belong to the *Candida* (yeast) and *Aspergillus* (mold) species. Given their pathogenicity and our poor knowledge of their genomes, metabolisms, and phenotypes, the modeling of such fungi with *S. cerevisiae* provided a fundamental contribution to the identification of novel antifungal drugs on several occasions. As for pathogenic fungi, knowledge of the genome, metabolism, and phenotype for *Homo sapiens* is too poor to exploit human cells directly for drug discovery purposes. Moreover, human cells are hardly cultivable, and the tools for their genetic and proteic manipulation are limited. Once again, *S. cerevisiae* was found to be a good model. The most renowned examples of the insights on mammalian cells obtained by means of the yeast model concerns studies on cancer. Given its genetic similarity with higher eukaryotic cells, this microorganism has been used to deduce functional and mechanistic aspects of eukaryotic proteins and protein systems. In addition, heterologous expression of human proteins in *S. cerevisiae* is possible, and this approach allows for both the identification of mutations associated with diseases and the functional characterization of aberrant human enzyme functions. In 1993, Strand et al. [19] predicted that mutations in the human orthologs of three yeast genes (*MSH2*, *MLH1*, and *PML1*) might be the main factors causing hereditary nonpolyposis colorectal cancer (HPNCC). This indication was confirmed later by Fishel et al. [20], who proposed that mutations in either hMSH2 (the ortholog of the yeast gene *MSH2*) or hMLH1 underlie the vast majority of cases of HPNCC. Clonal complementation has also been used to identify human protein functions. Expression of human cDNA in *S. cerevisiae* allowed for the selection of clones expressing human genes able to complement mutations that occurred on a selected yeast gene. In a pioneering study using this approach, Lee and Nurse identified a human homolog (POLD1) of the yeast *CDC2* gene involved in cell cycle progression and capable of complementing mutations in the microorganism gene [21].

The vast majority of the screenings based on the use of *S. cerevisiae* used only one genetic background, that of the laboratory strain S288c. Nevertheless, the use of this strain is limited by the accumulation of genetic defects in this strain, such as the insertion of a transposon in the gene *HAP1*, making the strain unable to sustain proper respiratory metabolism [22]. The effects on the yeast metabolic asset given by these genetic defects could mask the effect of a given perturbation (i.e., forbidding the identification of compounds modulating respiration). To prevent such a limitation, parallel use of the widely used S288c yeast strain and of a second deeply characterized laboratory strain, W303, was proposed, having 98% genetic identity with S288c but able to sustain proper respiratory metabolism since bearing a functional *HAP1* gene [23].

All the aforementioned qualities make yeast the best model for studies on fundamental cellular processes of relevance also in higher eukaryotic cells, and for the identification of determinants of stress and drug response [24]. Gene expression in yeast cells (wild-type or mutant), measured using DNA microarrays, helps to identify both the primary target of stimuli/compounds and any secondary consequences. The primary output is a quantitative assessment of the changes in gene expression for almost every gene in the yeast genome. The effects on gene expression resulting from chemical (compound/stimulus) and genetic (gene mutation or deletion) perturbations can be compared to identify the primary target of the stimulus. *S. cerevisiae* has become a model organism to study how eukaryotic cells respond to stress and define the specific role of stress-induced proteins [25–27]. Moreover, the high degree of evolutionary conservation of the stress pathways between yeast and higher eukaryotes indicates that yeasts are a valuable model system for characterization of the stress response in more complex organisms. When challenged with different stressors, such as heat, ethanol, metal ions, high osmolarity, and oxidants, yeast cells display common molecular mechanisms of defense. The common events have been identified as the general stress response. Different stressors can induce reactive oxygen species production that triggers the oxidative stress response, leading to the acquisition of resistance to subsequent oxidative stress. The connection between the response elicited by different stress agents and the oxidative stress response is given primarily by common induction of the transcription of *CTT1* gene encoding the cytosolic catalase. Indeed, many stress-related genes harbor on their promoter a STRE (stress response element) sequence. By means of DNA microarrays, Gasch et al. [25] explored the genomic expression patterns in *S. cerevisiae* responding to diverse environmental changes. They studied the processes by which yeast cells respond to temperature, hyper- and hypo-osmotic and hydrogen peroxide shocks, superoxide-generating drugs (i.e., menadione), sulfhydryl-oxidizing agents (i.e., diamide), disulfide-reducing agents (i.e., dithiothreitol), amino acid starvation, nitrogen source depletion, and progression into a stationary phase. This approach makes it possible to identify a large set of genes sharing similar drastic responses to almost all these perturbations and to recognize later confirmed involvement of the transcription factors Yap1p [28], as well as Msn2p [29] and Msn4p [30], in the response induction. The use of DNA microarrays, combined with yeast mutants resistant to a certain drug, or yeasts carrying deletions in genes important for the response to a given drug, allows not only for an understanding of the mechanisms of drug action, but also for elucidating the mechanisms of resistance and adaptation to the toxic agent. Assays of the expression response in yeast cells to treatment with the immunosuppressant rapamycin [31] showed significant similarity in the metabolic response profile to the diauxic shift [32].

14.2 CHEMICAL GENETICS AND *S. cerevisiae*

Since its beginning concurrent with Mendal seminal studies, classical genetics has played a central role in the dissection of gene and protein functions. Genetic approaches can be divided into two main branches: forward and reverse genetics (Figure 14.3). In forward genetics, random mutations are induced in the genome of a model organism; then a desirable phenotype is selected (i.e., growth rate decrease, morphology or phenotype alterations), and the mutation inducing such a phenotype is characterized (Figure 14.3a) [33]. As suggested by its name, reverse genetics acts in the opposite direction: the role of a gene in the cell is investigated by studying the phenotypic effects of its specific mutation (Figure 14.3b) [34]. The principal problem arising from the use of genetic approaches is that the mutations in essential

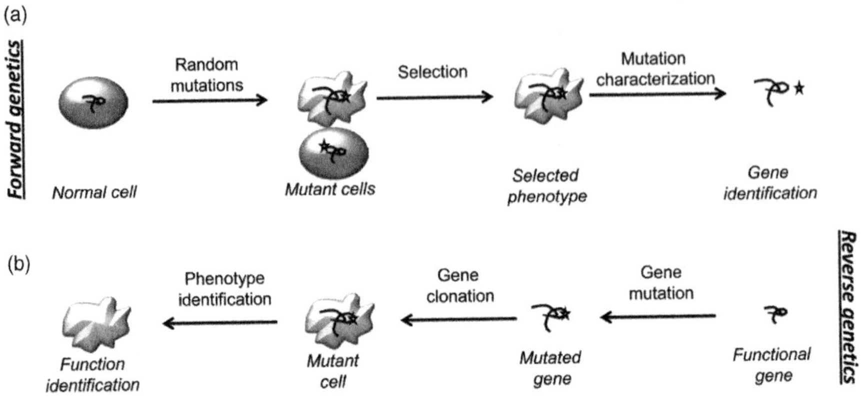

FIGURE 14.3 Schematic representation of the classical genetic approaches. (a) Forward genetics select the gene whose mutation induces a selected phenotype. (b) Reverse genetics investigate the function of a gene by observing the effects of its mutation.

genes often induce cell death, thus limiting the functional and mutational study to nonessential genes. Actually, the effects induced by nonlethal mutations could also be overlooked by classical genetic approaches, because of the ability of the mutant organism to compensate for the loss of the gene function prompted by the mutation. A further limitation of the genetic approach is that most mutations are not conditional; thus, they cannot be turned on or off at will. Even if the mutation is designed to be conditional (i.e., when the mutated allele is under an inducible promoter), the expression of the allele entails a stimulus that itself induces a cell stress, thus concealing the effect of the mutation.

A relatively new approach to the dissection of biological functions is based on RNA interference (RNA*i*). By using small RNA molecules able to bind specific *m*RNAs selectively, expression of the protein coded by the targeted *m*RNA can be inhibited. In this way, several disadvantages of the classical genetic approaches are bypassed. Nevertheless, the target inhibition cannot be modulated, as the effects are indeed "off-target effects." Although there are several limitations in using classical genetics approaches, their application to the model organism *S. cerevisiae* made it possible to learn fundamental lessons in cell biology that could be generalized to mammalian cells. One of the most representative and fruitful efforts in their direction was that of Leland H. Hartwell. Thanks to his studies based on *S. cerevisiae*, which rewarded him with the Nobel Prize in Medicine in 2001, the cell cycle and the principal actors involved in its control and progression were deeply understood. More than 100 genes were identified as being involved in cell cycle control and determining the pathway of cell cycle regulation events [35]. The pivotal role of the class of CDC (cell division control) genes in *S. cerevisiae* cell cycle was then confirmed in mammalian cells by two others scientists, who shared the Nobel prize in 2001 with L. H. Hartwell. One of them, Paul Nurse, discovered the first human equivalent of yeast's CDC genes: the cyclin-dependent kinase 1 (the *CDK1* gene) [36].

Thanks to the synthesis of large chemical libraries, the ideation of a new approach (chemical genetics) free of the limitations associated with classical genetics became possible. Chemical genetics was first proposed by S.L. Schreiber in 1998 as "the study of genes by modulation of protein function with small molecules" [37]. Small molecules, easily able to permeate cell membranes, can perturb cellular targets rapidly and often reversibly. Moreover, the perturbation can affect several different features of the target. Indeed, a small molecule can perturb a selected protein, or one function of a multifunctional protein, or even disrupt protein–protein interactions. Nevertheless, chemical genetics also has disadvantages: (1) it is limited by the availability of characterized compounds, and (2) the study of the effects, when a poorly characterized cell system is used, is confusing and often unsuccessful, like looking for a needle in a haystack. With regard to the availability of the molecules characterized, the synthesis of wide chemical libraries, encompassing either compounds assembled by modifying a shared scaffold or those based on a diversity-oriented synthesis (DOS) mindset, gives a vast scenario of possibilities to be tested. Thus, the availability of a well-characterized biological system becomes pivotal, making the yeast *S. cerevisiae* a powerful instrument for developing insights as to both the biological functions of targeted entities and the activity of the molecules tested. The logic lying at the base of chemical genetics is similar to that on which classical genetic approaches are based. In a chemical genetics approach, exogenous ligands are used to alter the function of a single gene product within an entire cellular context (Figure 14.4). Generally, the forward genetic approaches are exploited when the aim of the investigation is the identification of molecules able to induce a specific phenotypic effect. On the other hand, reverse chemical genetic approaches are fundamental when trying to assess the effects of targeting an entity known to be involved in interesting processes

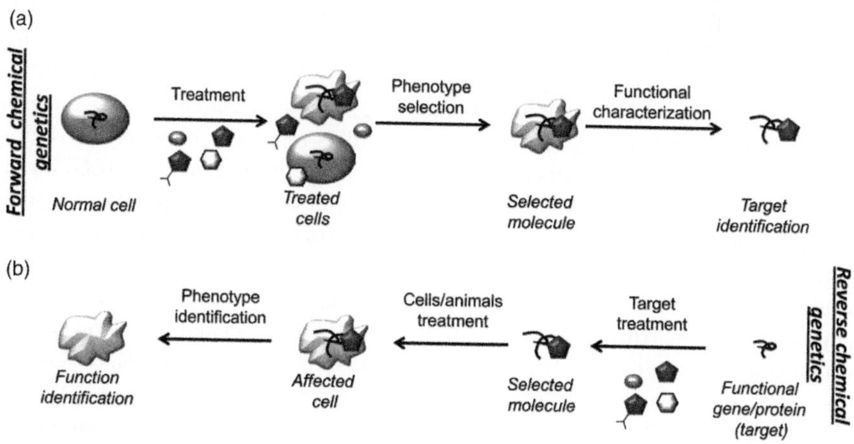

FIGURE 14.4 Schematic representation of chemical genetics approaches. (a) Forward chemical genetics select the molecule able to induce a desired phenotype. (b) Reverse chemical genetics investigate the effects induced by a ligand of the gene selected by observing the effects of their binding.

(i.e., genes involved in cancer insurgence). In the latter case, background knowledge of the gene function is required. The use of small molecules in a chemical genetic mindset gives another striking advantage to the common advantages arising from the approach itself. Indeed, these molecules usually act rapidly at the intracellular level, thanks to their low chemical hindrance, which makes them more prone to permeate the cell membrane with respect to bigger molecules possessing similar polarity.

14.2.1 Forward Chemical Genetics

As in forward genetics, forward chemical genetics entails the introduction of a perturbation on a cell. In the first case, the perturbation is represented by a genetic mutation; in the latter, this effect is given by a small molecule. A molecule able to induce a chosen phenotype is then selected, and the entity (i.e., protein or DNA) whose targeting by the molecule induces the phenotypic effect is identified (Figure 14.4a). The use of *S. cerevisiae* with a forward chemical genetics mindset to discover novel compounds able to induce a desired phenotype is favored by the possibility of easily treating a large number of different cell cultures simultaneously and of monitoring the phenotype selected. By using multiple-well parallelized cultures, it is possible to monitor simultaneously the effects of large numbers of compounds in large numbers of yeast strains, thus briefly obtaining a wide amount of input–effect combinations. To that aim, all forward and reverse chemical genetics approaches can be scaled to the high-throughput level.

14.2.1.1 Haploinsufficiency Profiling The principle behind drug-induced haploinsufficiency is that lowering the dosage of a single gene targeted by a chemical compound could make the cell sensitized to the molecule [38]. As a preliminary test, Giaever et al. constructed six heterozygous *S. cerevisiae* strains, each one deleted in a gene encoding a known drug target [38]. Whereas the wild-type yeast strain growth is affected by treatment with the compound, the heterozygous strain deleted in one copy of the gene coding the targeted protein will be more susceptible to the treatment with the same compound (Figure 14.5). On the basis of previous investigations, indicating that *ALG7* overexpression confers resistance to the glycosylation inhibitor tunicamycin, Giaever et al. [38] generated a heterozygous *alg7/ALG7* strain and evaluated the effects of treatment with tunicamycin on its growth rate with respect to the wild-type strain. As expected, the treatment decreased the growth rate of the deletant strain, thus giving the proof of principle for this assay. The same results were achieved when treating with other compounds (fluconazole, benomyl, 3-aminotriazole, and hydroxyurea), the strains deleted for the relative molecule targets, further confirming use of the method with a broad range of molecules. Later, the test was expanded to a larger number of deletant strains [39], again confirming the utility of this approach for drug target identification. In their work, Baez et al. monitored by O.D. measurement the growth rate of a set of more than 5000 heterozygous *S. cerevisiae* deletant strains treated with dihydromotuporamine C (dhMotC, an inhibitor of angiogenesis and metastasis) or with DMSO (the drug diluent) as the control [39]. Thanks to this method, it was possible to identify the sphingolipid metabolism as

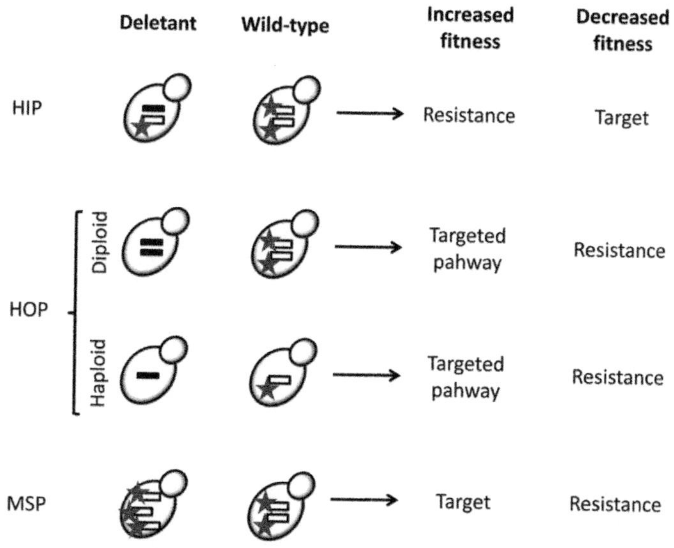

FIGURE 14.5 Schematic representations of three chemical genetics approaches. White rectangles represent functional genes; black rectangles represent deleted genes. "Fitness" refers to the growth rate of the deletant strain with respect to the wild-type strain. The black star represents the active chemical compound.

dhMotC target in yeast, whose inhibition induces a decrease in ceramide levels. The dhMotC effect on the ceramide cell content was also confirmed on human cells, thus demonstrating that the drug way of action hypothesized in the yeast system can be translated to higher eukaryotic cells. The use of the growth rate as a measure of the effects of a compound is based on a considerable assumption: The phenotypic effect of the compound tested has to be the variation in wild-type cells' growth rate. Thus, this quantification is useful when the mode of action of anticancer or antifungal drugs must be investigated. Nevertheless, the cell-based screening procedure proposed by Baetz et al. can be adapted to screen numerous drug-induced phenotypes other than growth rate variation, such as the activation of signaling pathways and of transcription when an optically measurable reporter effect is modulated.

14.2.1.2 Homozygous Profiling In contrast to haploinsufficiency profiling, homozygous profiling (HOP) is based on deletion of the complete set of the gene. In the HOP approach, yeast strains are used in a homozygote (diploid) or haploid context (Figure 14.5). In the first case, both gene copies are deleted; in the latter, the unique gene copy present is deleted. This approach is based on the idea that when the target of a cytotoxic molecule is lacking (because of complete deletion of the gene coding it), the yeast cell is not affected by the treatment. As is easily predictable, only the effects on genes that are not essential for cell survival can be investigated using this method. This limitation forbids the identification of entities whose targeting by a molecule induces cell death, thus resulting in a considerable restriction in the discovery of the mode of action of novel compounds having interesting pharmacological features (i.e., antifungal or anticancer drugs). In any case, useful insights on the mode of action of novel compounds can be achieved using this approach. Indeed, the observation of increased fitness of a homozygous deletant strain treated with a cytotoxic compound can be ascribed to the lack of an entity downstream in the pathway encompassing the direct target of the molecule. On the other hand, the HOP approach can also be exploited for the identification of genes involved in pathways buffering the compound or drug effects. Indeed, if a compound is lethal for the cell, the presence of a gene coding for a protein involved in its elimination (i.e., efflux or metabolism) grants to the cell the possibility of surviving the treatment. The deletion of this response-associated gene reduces the cell defenses, making the deletant yeast cells more susceptible to the compound with respect to wild-type cells. Using this approach, the mechanisms driving a cell's response to nickel compound toxicity were defined by the identification of functional categories overrepresented in the group of hypersensible deletant strains [40]. The use of *S. cerevisiae* as a model for drug discovery can be affected by the presence of several multidrug resistance (MDR) mechanisms, which can prevent the identification of biologically active compounds just because of the abnormal (with respect to other eukaryotic cells) extrusion of chemical compounds from the cell. The use of HOP, and in particular of homozygous strains deleted in genes involved in MDR, is therefore useful as an initial screen for the selection of novel bioactive compounds [41].

Taking advantage of the HOP approach, Dudley et al. were able to generate a profile of pleiotropy genes [42]. Pleiotropy occurs when a single gene influences several phenotypic traits. As a consequence, a mutation or inactivation (i.e., caused by a chemical compound) of this gene may have an effect on some or all traits simultaneously. This is a considerable disadvantage for forward chemical genetics approaches, since the noisy background of the effect of the targeting forbids identification of the drug target. In addition, a major challenge in the analysis of pleiotropic genes is determining whether every phenotype associated with a mutation results from the loss of a single function or of multiple functions encoded by the same gene. Dudley et al. succeeded in identifying genes whose deletion made the cell resistant to the perturbation (both chemical and physical perturbations) by observing both dimension and density of the colonies generated by the deletant cells separately undergoing several stresses. The definition of a response pattern to the broad and diversified collection of conditions studied for each gene deletion made it possible to propose a functional gene grouping criterion; if treatment with a specific perturbation induces the same phenotype on two or more different deletant strains, the genes for which these strains are deleted are probably involved similarly in the process affected by the treatment.

14.2.1.3 Multicopy Suppression Profiling: Dosage Suppression The multicopy suppression profiling (MSP) approach (Figure 14.5) is based on the principle that cells bearing an increased copy number of a specific gene express an increased level of the gene product. If the gene encodes for the target of a specific cytotoxic compound, the overexpressing cell should tolerate higher drug levels. For this application, a new strain library was generated; a yeast chromosomal library harbored on a multicopy vector was introduced into a wild-type yeast strain. The individual transformants can be tested for resistance to the inhibitory compound, and resistant strains can be selected as able to survive a lethal concentration of the compound. As observed for approaches described previously, in this case also the main limitation to drug target identification is that only a subset of potentially interesting compounds has a growth-inhibitory phenotype in yeast. Nevertheless, this approach was useful for the identification of the molecular targets of several known antifungals, such as ketoconazole [43] and soraphen [44]. A yeast library consisting of strains transformed with plasmids of over 80% (5800 ORFs) of all yeast ORFs, whose expression is controlled by a galactose-inducible promoter, was created [45]. The supplementation of galactose in the culture medium induces the high-level expression of the ORF included in the plasmid. Although the considerable advantage of using an inducible promoter is that the expression can be induced at specific times during the course of the experiment, galactose induction often exceeds the endogenous gene expression level, thus resulting in toxicity to the host cell [46]. It is worth noting that 15% of yeast genes are toxic to the cell when overexpressed. Nevertheless, it is possible to take advantage of these toxic effects, despite their apparent disadvantage. Indeed, the toxic genes can give useful insights into how regulation of the gene products can alter the cell physiology, and they can be used as a starting point for chemical suppressor screens to identify compounds that suppress the genic toxicity. This approach, called *chemical suppression*, has been employed successfully to identify human genes that, when expressed in yeast, result in a growth defect [47]. Tugendreich et al. suggested that an overexpression of the human *p38*, a MAP kinase suggested to play an important role in the development of AIDS and HIV-associated neurocognitive disorders [48], induces yeast cell death. Thus, the ability of novel molecules to nullify the toxic gene effect can be exploited as potential drugs for HIV treatment when looking for p38 inhibitors.

14.2.2 Reverse Chemical Genetics

Reverse chemical genetics is based on a process that starts with the selection of a specific protein. Chemical libraries are then screened for a ligand of the protein chosen, and the ligand is finally used to determine the phenotypic consequences of altering the function of the protein in a cellular context (Figure 14.3b).

14.2.2.1 Yeast Two-Hybrid Assay The yeast two-hybrid (Y2H) method was first described in 1989 by Fields and Song [49], and later developed to the high-throughput level [50]. This method allows for the assessment of functional protein–protein or protein–DNA interactions. The premise behind the assay is the activation of a reporter gene through the binding of a transcription factor to a reporter-specific promoter (a DNA sequence located upstream and regulating the expression of the coding sequence). In the two-hybrid method, the transcription factor is divided into two fragments: the binding domain (BD, responsible for the DNA binding) and the activating domain (AD, responsible for the activation of transcription) (Figure 14.6).

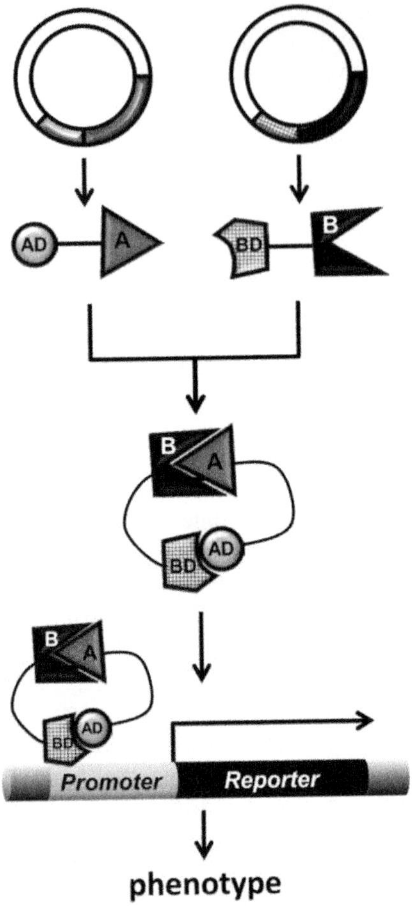

FIGURE 14.6 Schematic representation of the Y2H assay. The genes coding two investigated proteins are inserted into different plasmids in frame with genes coding for two parts of the transcription factor. The proteins are expressed and, if they bind, the transcription factor becomes functional, thus expressing the measurable reporter gene.

These two fragments, able to reconstruct a functional transcription factor when interacting physically, are fused with two proteins (A and B) whose interaction has to be assessed. Separate plasmids are generated, one containing protein A in a frame with the BD, and the other containing protein B in a frame with the AD (Figure 14.6). The expression of both genes leads to the synthesis of two chimeric proteins composed of the proteins studied associated with the two fragments of the transcription factor. If the two proteins do interact, the two transcription factor fragments can bind, thus forming a functional transcription factor. Then, the functional transcription factor induces a measurable expression of the reporter gene. The classical transcription factor–reporter system used in the two-hybrid assay encompasses the exploiting

GAL4 as a transcription factor and *lacZ* as a gene reporter. Expression of the *lacZ* gene is observable thanks to the enzymatic ability of the gene product to metabolize the substrate X-gal (which can be added to the culture medium) to a blue compound [51]. The two-hybrid assay, hypothesized in a yeast cellular background, has been extended to bacterial cells to discover protein–protein interactions [52].

Nevertheless, the use of such methodology in the *S. cerevisiae* system guarantees several advantages. Indeed, the two-hybrid assay can be used in the budding yeast to assess the interactions occurring between both yeast-specific and heterologous proteins (such as mammalian proteins). Whereas the bacterial cells do not bear the mechanisms to posttranscriptionally modify the expression product (i.e., to eliminate eventual introns), yeast cells do. One of the main limitations to use of the yeast two-hybrid assay is that such a method is useless when one of the proteins whose interactions is to be studied is a receptor or, in general, a membrane-anchored protein. Indeed, a protein bound to the membrane is stationary, enabling translocation of the complex studied and of the newly assembled transcription factor to the nucleus and to the promoter of the reporter gene.

Aiming at the reverse chemical genetics scope, Y2H can be used when the target protein is known and has been selected. As an example, when the protein inducing a disease is known and new molecules able to inhibit its function have to be screened, Y2H acquires a fundamental role in high-throughput screenings [53,54]. The Y2H assay can be applied with a forward chemical approach as well. Indeed, thanks to advancements in genomic sequencing and genetic handling techniques, it is now possible to screen the effects of sets of molecules on all the combinations of protein couples interactions. With an evolved version of this approach, Chidley et al. revealed that the tetrahydrobiopterin biosynthesis was the target of the known anti-inflammatory drug sulfasalazine [55]. In another study, the Y2H assay was used to verify the antiviral activity of a novel peptide selected previously by an in vitro assay [56]. In this work, the NDFRSKT heptapeptide was selected for its ability to bind purified proteins of the AIV subtype H9N2 (avian influenza virus H9N2). Thanks to the Y2H assay, it was possible to assume that the antiviral activity of the peptide selected was given by the inhibition of the viral replication achieved by binding of the peptide to the hemagglutinin (HA) protein, which is involved in viral linking to host cells before infection. An elegant evolution of the Y2H methodology has been developed to detect interactions between proteins and small molecules, the yeast three-hybrid assay (Y3H, Figure 14.7) [57]. In this case, three chimeric entities are generated, the first composed of the BD of the transcription factor and a protein (A), the second composed of the AD of the transcription factor and another protein (B). The third chimera is composed of two bounded small molecules, one ligand for the A protein (sm) and a second ligand for the B protein (lig). The binding of the two small molecules to their corresponding proteins allows for generation of the functional transcription factor, thus resulting in the induction of the reporter phenotypic effect. The utility of the Y3H assay as a tool for identifying the binding proteins of a compound was demonstrated by Henthorn et al. by screening a mouse cDNA library for clones encoding unknown proteins able to bind methotrexate [58]. Furthermore, the suitability of the Y3H method for the de novo identification of small-molecule

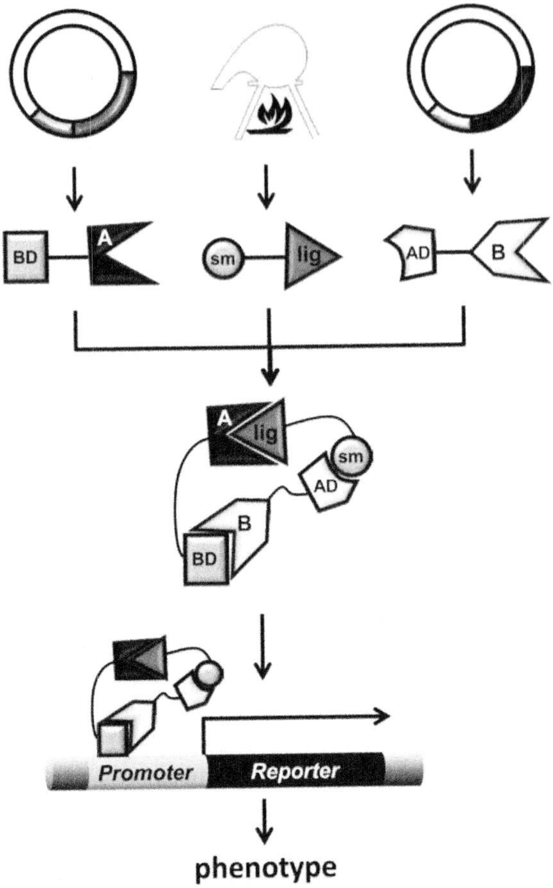

FIGURE 14.7 Y3H assay. Three chimeric entities are needed. The chemical chimera consists of the compound investigated covalently bound to a known ligand of protein B. The second chimera consists of protein B and one part of the transcription factor. The third chimera consists of the second part of the transcription factor and protein A, whose targeting by the chemical compound under investigation is assayed. If the entire complex gets assembled, the transcription factor is functional, thus expressing the measurable reporter gene.

binding proteins was demonstrated in a study by Becker et al. [59] designed to explore the targets of small-molecule-inhibiting kinases. In this study they showed that the Y3H assay gave useful insights into the characterization of the interaction of ATP-competitive active-site kinase inhibitors with both serine/threonine and tyrosine kinases. The findings that interactions with affinities in a range of pharmacological interest as kinases can be detected with Y3H suggest that this assay can be used in the context of a broader range of synthetic small molecules. Since the chemical chimera can be substituted for by a bifunctional small molecule, this approach is useful to assess the biological activity of such compounds [60].

14.3 CHEMICAL GENOMICS AND *S. cerevisiae*

All the aforementioned chemical genetics approaches are based on single-strain assays. Indeed, in every procedure, identification of the genotype associated with a selected phenotype is the fundamental request for the success of the investigation. Actually, in their work setting the foundations for the HIP approach, Giaever et al. also proposed a way to specifically recognize different deletant strains from a pool of deletant strains [38]. The ability to identify the deletion of a specific strain is possible because of the bar coding of the deletion. In more detail, every functional gene is substituted for by the KanMX cassette, conferring resistance on the antifungal kanamycin, flanked by two unique 20-bp sequences (bar codes) specific for the deletion. The presence of these specific and identifiable tags makes possible the culturing of pooled deletant strains, thus reducing the experimental variability of the treatment while reducing the culturing efforts otherwise necessary for parallel treatments. An additional advantage in using a competitive culture assay is the request for less of the compound being studied (one to two orders of magnitude). A yeast bar-coded knockout collection has been generated consisting of a complete set of deletion strains, including haploid strains of both yeast mating types and heterozygous and homozygous diploid deletions (Figure 14.8).

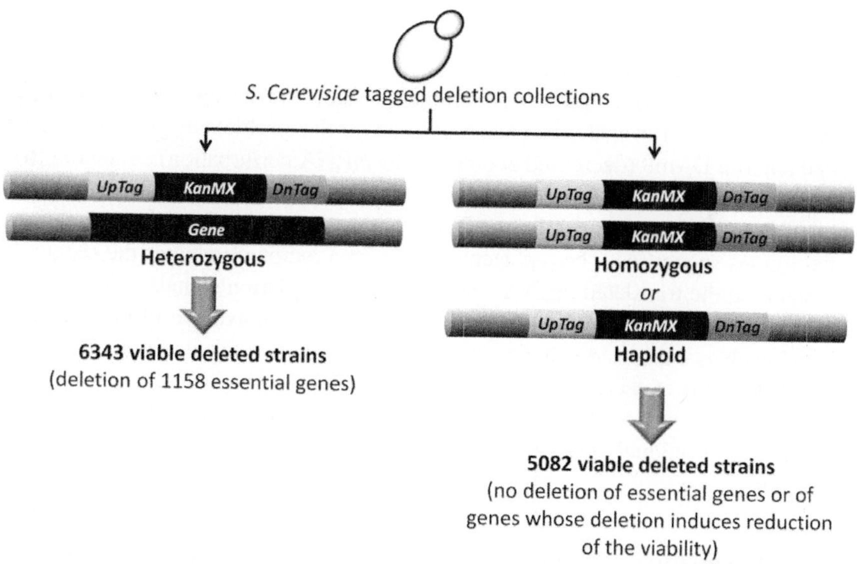

FIGURE 14.8 Available yeast bar-coded deletion collections. A functional gene is deleted through substitution with a genetic cassette conferring the deleted strain kanamycin resistance (*KanMX*). The presence of two deletion-specific 20-mer sequences flanking the *KanMX* cassette allows for identification of the deletion. The deletion of a single gene copy in diploid cells allows for the study of essential and nonessential genes; the deletion of both gene copies in diploid cells, or of the unique copy in haploid cells, allows for the study of nonessential genes only.

Since early applications of this approach, initially restricted to a few pooled deletant strains, bar-code identification was carried out by their hybridization to specific microarrays [38,61]. Nowadays, the use of NGS techniques is applicable as a more quantitatively and qualitatively reliable identification of bar codes [62].

14.3.1 Competitive Growth Assay Based on Heterozygote Strains

HIP was one of the first assays taking advantage of the competitive growth strategy. Thanks to this approach, using a unique treatment it is possible to assay the response of 6343 deletant strains, corresponding to a strain for every single yeast identified ORF. Indeed, the heterozygote background makes the generation of yeast strains deleted for essential genes possible and still viable thanks to the presence of at least one copy of the gene. Bar codes that decrease in abundance after treatment with respect to the control allow for identification of the strains deleted for genes whose product is targeted by the condition tested. The feasibility and strength of the competitive assay has been demonstrated by screening novel compounds [61,63]; this approach has then been applied to a dissection of the mode of action of novel synthetic and natural compounds [64–66].

When treating the heterozygote bar-coded collection with a molecule whose molecular target is under investigation, it is necessary to use a sublethal concentration of the compound studied. Indeed, such a concentration will maintain constant the growth rate of strains bearing both copies of the target gene and at the same time will lower the growth rate of the strain heterozygote for that gene. Nevertheless, such worldly wisdom is sometimes not sufficient, and the gene dosage targeted should be decreased further. Yan et al. proposed a solution for that problem by generating a bar-coded haploid DAmP (decreased abundance by *m*RNA perturbation) collection [67]. This new approach is not based on gene deletion but, rather, makes use of the replacement of the downstream noncoding sequence of the gene (3′-UTR) with nonfunctional heterologous sequences. The replacement of such a region eliminates the sequences that stabilize the translated *m*RNA—such as polyadenylation signals—thus causing a reduced *m*RNA accumulation [68]. Only about 10% on average of the wild-type protein levels is expressed, whereas expression of the gene selected in heterozygote bar-coded strains is decreased to 50% with DAmP. By using different heterologous nonfunctional sequences for replacement of different genes' 3′-UTR regions, every gene targeted is uniquely identifiable, thus allowing for pooled treatment of the entire collection.

14.3.2 Competitive Growth Assay Based on Haploid or Homozygous Strains

The competitive growth assay based on haploid or homozygous strains is analogous to the assay based on heterozygote strains, with the exception that the strains are completely deleted for nonessential genes in either haploid or diploid strains. The idea driving this approach is the same as that of the HOP assay, but extended to a genomic scenario. In the heterozygote strain-based competitive growth assay, the genes targeted by the molecule are identifiable by the bar codes overrepresented after the treatment. On the contrary, in the homozygote background the overrepresented bar-coded strains are deleted in genes able to recover, when functional, the toxic effect of the compound. Furthermore, this assay discloses fundamental information on the mode of action of compounds lacking a direct protein target, such as in the case of DNA-damaging agents. A great advancement in disclosing the genetic requirements for the resistance to DNA-damaging agents was achieved thanks to this approach [69]. Treatment with 12 known DNA-damaging compounds of the competitively grown homozygous bar-coded collection allowed for the confirmation of the involvement of genes known to have a role in DNA metabolism, but also the discovery of new genes responsible for that response. Moreover, comparison of the profile of deletant strains affected by treatment with the DNA-damaging agent disclosed a fingerprint ("signature") of the resistance to every given compound. Compounds with similar biological effects are supposed to lead to similar chemical genetics profiles [70]. A great deal of effort in that direction is shown in the work of Parsons et al. [71], endowing the scientific community with a compendium of parallel fitness profiles obtained by treatment of the yeast haploid deletion collection with 82 chemical perturbations. In addition to the outstanding data set, Parsons et al. also proposed a method for the dissection of pathways and proteins affected by the chemicals tested, based on clustering analysis and probabilistic sparse matrix factorization.

14.3.3 Comparative Expression Profiling

The transcriptional profile of a cell is the most representative portrait of cell status. The definition of gene expression levels gives a detailed snapshot of the arrangement of all the structures, molecular components, and reactions present in the cell at a given moment of its life or in a given condition. When the condition studied is treatment with a new or known molecule, the molecular effects affecting the complete cell system can be observed by transcriptional analysis. The action of the molecule on the cell can be observed both if the compound affects a specific target (observing the domino effect induced by the perturbation of this entity on the cell organization) and in the presence of a nonselective disturbance. Several efforts in recent years have made possible the development of a method able to capture, analyze, and demonstrate the entire expression profile of a cell. To that end, DNA microarray technology represented a pivotal step forward [72]. The laser scanning of labeled mRNA hybridized to spotted cDNA gives a quantitative measurement of the expression level of every annotated ORF of genomically sequenced organisms. In *S. cerevisiae*, microarray approaches have been used widely to measure gene expression upon cell exposure to a variety of conditions and stresses [73–76]. A number of studies analyzed the yeast expression patterns of several different physiological conditions using differently designed arrays based on probes composed by either DNA fragments [72,77] or oligonucleotides [78,79].

cDNA microarrays have been used to study yeast gene expression during fundamental events such as growth [80], sporulation [81], and in the cell cycle [82]. A brilliant example of the application of transcriptional analysis to the dissection of the drug mode of action is represented by a study of Hardwick et al. [31].

Here, the researchers observed the effects induced at the transcriptional level by rapamycin, a known immunosuppressant drug used to prevent rejection in organ transplantation. Thanks to a comparison of the transcriptional profile induced by the rapamycin treatment with that representing the metabolic response profile to diauxic shift [80], the mechanism of resistance to the toxic agent was identified. In particular, they showed how rapamycin inhibits Tor1p and Tor2p, ultimately resulting in cellular responses characteristic of nutrient deprivation through a mechanism involving translational arrest. This suggested that Tor proteins modulate directly the glucose activation and nitrogen discrimination pathways and the pathways that respond to the diauxic shift (glycolysis and the citric acid cycle). Similarly important was the contribution given by this approach to the assessment of the mode of action of alkylating agents (e.g., methyl methanesulfonate) [83]. Treatment with an alkylating agent yields dramatic changes in the expression of genes involved in DNA repair and identified many genes not previously associated with DNA repair processes. Responsive genes fall into several expected classes, such as stress response/detoxification, DNA repair/replication, cell cycle, signal transduction, cell wall biogenesis, and membrane transport, and many unexpected categories, such as nitrogen and sulfur metabolism, carbohydrate metabolism/fermentation, and *m*RNA processing. The same authors produced evidence for a cell program designed to eliminate and replace alkylated proteins. Actually, proteins are known substrates for methyl methanesulfonate alkylation. The fact that genes involved in protein degradation are up-regulated in response to methyl methanesulfonate suggested that alkylated proteins may be targeted for degradation and that their elimination may be important for cellular recovery.

As for the identification of patterns responsive to chemical perturbation, genome-wide gene expression patterns have been used to validate drug targets and to identify secondary drug targets [32,84]. Marton et al. [84] used DNA microarrays to compare genome-wide gene expression patterns in wild-type and mutant yeast cells to identify primary drug targets as well as the secondary cellular consequences of drug exposure. The logic behind the Marton et al. study is that the cellular consequences of genetic and pharmacological inhibition should be very similar; that is, removing the gene from the genome and chemically inhibiting the activity of the protein product of that gene should be functionally equivalent. Thus, by comparing chemically induced gene expression profiles of wild-type yeast with the profiles of many genetic deletion strains should allow for identification of the primary target of the drug. The method has been validated by treating yeast mutant strains defective in calcineurin, immunophilins, or other genes with the immunosuppressants cyclosporin A or FK506. The presence or absence of the characteristic drug "signature" pattern of altered gene expression in drug-treated cells with a mutation in the gene encoding a putative target established whether or not that target was required to generate the drug signature. By analyzing multidrug resistance yeast mutants (*PDR1-3* and/or *PDR3-7*) with DNA microarrays, De Risi et al. identified all target genes that appeared to be involved in transport or in membrane lipids and cell wall biosynthesis [32]. Several targets seem to contribute to protect the cell from a variety of stresses. This work illustrated the coordinated use of genome-based and biochemical approaches to delineate a cellular pathway modulated by the protein target of a small molecule. In conclusion, yeast will continue to be a valuable tool for studying the mechanisms of action and resistance to drugs and in designing new drugs. The value of a genomic approach to an understanding of fundamental biological processes is connected to the simultaneous examination of expression patterns of all genes for branch-point enzymes. Similarly, the definition of the expression profile induced by the treatment of yeast cells with a selected drug or molecule allows for the identification of genes mediating the drug response. Nevertheless, developing the complete transcriptional profile of a cell might be as complicated as looking for a needle in a haystack. Indeed, a huge number of genes can change in their expression level, thus making difficult the disclosure of a biologically meaningful portrait. Deep knowledge of the central metabolism and the use of bioinformatics tools for data mining, hierarchical organization, and clustering analysis is thus required [85,86]. Aiming at a simplification of the results and at other tasks, many different analyses have been proposed since the high-throughput microarray technique begin to gain interest. The expression levels associated with each gene can be analyzed using a supervised or an unsupervised approach. The first is applicable when further information on the profiles is available. As an example, by attributing a characteristic (e.g., diseased/normal, treated/untreated) to a sample it is possible to build a classifier able to predict the sample status from the expression profile, through the identification of specific signatures. Unsupervised data analysis consists of grouping together objects (be they genes or samples) by clustering the expression profiles and allowing for the identification of common features [87]. Clustering is not a new technique, as it was developed initially for phylogenetic analyses. Many different ways to calculate distances (and thus similarities/differences) describing transcriptional profile differences are available [86,88,89], and several tools have been developed with this aim. The combination of sample and gene clustering allows for the identification of groups of genes whose expression can distinguish one group of samples (a subcluster) from another sample subcluster. Other approaches have been developed aiming at the disclosure of a molecular profile or biological effect induced by the perturbation studied. In the early days, procedures were based on the utilization of patterns of genes grouped as belonging to the same domain or pathway. The principal domains defined by the Gene Ontology Project [90] are: *cellular component*, the parts of a cell or its extracellular environment; *molecular function*, the elemental activities of a gene product at the molecular level, such as binding or catalysis; and *biological process*, operations or sets of molecular events pertinent to the functioning of integrated living units (i.e., cells, tissues, organs, and organisms). The Gene Ontology Consortium, in a collaboration among three model organism databases (FlyBase for *Drosophila* [91], SGD for *S. cerevisiae* [92], and MGD for mouse [93]), developed controlled vocabularies (ontologies) that describe gene products in terms of their associated domain. Also, pathway databases have been annotated as a result of information gained from the different research domains. The combination of genetic, functional, and proteomic data allowed for the generation of lists of functionally or metabolically related genes. The main pathways data sets curated for *S. cerevisiae* are (1) KEGG [94], a collection of manually drawn pathway maps representing the molecular interaction and reaction networks for metabolism, genetic information processing, cellular processes, and organismal systems; (2) Reactome [95], a free online open-source curated pathway database encompassing many areas of human biology; and (3) YOUNG [96], encompassing lists of target genes for known transcription factors. The possibility of obtaining an overview of the effects of a compound on a cell through transcriptional and pathway analyses allows not only for a restriction on the number of possible targets (molecules or complexes) that are affected directly or indirectly by the treatment, but also for predicting other features, such as the insurgence of resistance to the candidate drug. Pathway-based microarray analysis methods look for patterns of gene expression variation in any predefined set of genes. Several tools aiming at this analysis have been developed. As an example, EuGene Analyser [97], by calculating a *p*-value using the Fisher exact test (a measure of the significance of a pathway's enrichment in transcriptionally altered genes), also associates with every output *p*-value a sign reflecting the relative number of up-regulated genes with respect to down-regulated genes for each pathway. The use of pathways and gene ontology analyses discloses important effects of the treatment on the cell. Nevertheless, the identification of a specific entity targeted by the treatment can be confused with downstream effects. The use of external data might result in a driving tool for the disclosure of otherwise unidentifiable information. The procedure lying at the basis of such a comparison, *meta-analysis*, refers to "the quantitative review and synthesis of the results of related but independent studies" [98], and its application to microarray technologies outputs does not represent an exception. Previously, meta-analysis was applied to estimate an average effect of different studies when their results had conflicting conclusions or to evaluate the degree of agreement of the biological results obtained by various studies [99,100]. In addition, such an approach was exploited for the identification of common traits of effects induced by different perturbations. Meta-analysis approaches have been used widely in different fields [99,101] and key rules have been proposed [102,103]. With a drug discovery mindset, meta-analysis can take advantage of a brilliant tool proposed by Hughes et al. in 2000, the Rosetta compendium, a reference database of expression profiles corresponding to 300 diverse mutations and chemical treatments in *S. cerevisiae* [85]. The idea driving such a comparison is that if a newly assayed molecule induces a transcriptional profile similar to the one induced by the deletion of a nonfundamental gene, the functional gene might be targeted directly or indirectly by the compound being studied. Beltrame et al. proposed moving the comparison of the transcriptional profile of different treatments from the gene level to the pathway level [104]. For every condition investigated a pathway signature is generated, which is a set of descriptors that recapitulate the biologically meaningful pathways related to the variable (the perturbation) of interest. Whereas gene-based approaches are generally biased by the analytical procedures employed, the method based on pathway signatures groups similar samples together successfully irrespective of the experimental design. By clustering the pathway signatures calculated for every condition investigated, the similarities among different treatments can be identified. Use of this approach gave interesting insights for completion of the functional characterization of PPARalpha, a ligand-activated transcription factor using a meta-analysis approach on data obtained from several different organisms

[105] and for a dissection of the way of action of novel chemical compounds on *S. cerevisiae* [23,84].

14.4 CONCLUSIONS: THE ROUTE OF DRUG DISCOVERY WITH THE BUDDING YEAST

The drug discovery process is a complex path consisting of several sequential steps and requiring both chemical and biological expertise. From the time a decision to start is made, be it selecting a target to be explored or exploring the biological effects induced by a novel compound, the process encounters frequent and crucial crossroads. No matter what the starting decision is, utilization of the proper tool to use in choosing the direction to take in the steps that follow gains pivotal importance. High-throughput screenings based on the use of chemical genetic approaches to *S. cerevisiae* do allow for the selection of chemical compounds able to modulate a phenotype selected from the yeast model. Parallel treatment with different compounds of yeast strains selected

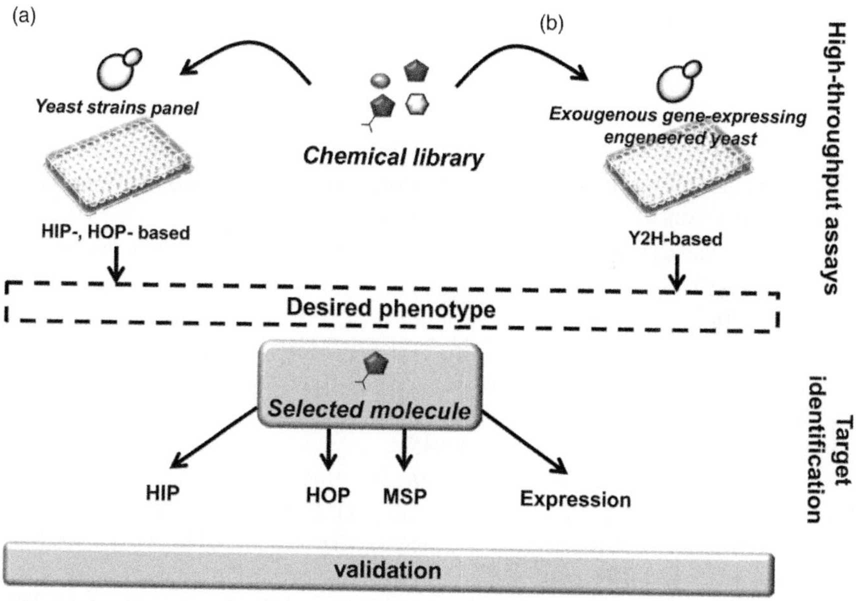

FIGURE 14.9 Route of drug discovery with *S. cerevisiae cells*. Parallelized cultures of selected strains of *S. cerevisiae* allow for the selection of chemical compounds able to modify biological activities. (a) The treatment of selected *S. cerevisiae* deletant strains with a chemical library allows for the identification of yeast genes whose product is affected by the molecules, and the selection of biologically active compounds. (b) By expressing exogenous proteins in *S. cerevisiae* (belonging to a different organism), the effect of a chemical library can be assessed on a set of targetable entities associated with human diseases. Once a small number of compounds has been selected by high-throughput assays as inducing a selected phenotype, the way of action of each molecule can be explored and targets can be identified, then validated.

because of a specific phenotype were revealed to be a powerful tool. Once one or a few compounds are selected, their mode of action has to be investigated. From a pharmaceutical point of view, identification of the molecular target both makes it possible to assess selectivity for the specific target and represents an initial indication of the possibility of transferring the observed effect to different organisms. In this case, *S. cerevisiae* was also shown to be a useful model organism. In conclusion, the simple yeast model *S. cerevisiae* represents a good choice both for the selection of molecules when screening complex compound libraries and for dissection of the way of action of a selected molecule, thus acting as a beacon along the dark path of drug discovery (Figure 14.9).

The continuously increasing number, precision and reliability of novel techniques and the parallel growth of exploitable chemical compound libraries will grant the possibility to explore previously unpredictable universes with pharmaceutical and medical interests.

REFERENCES

1. Y. L. Bennani, *Drug Discov. Today* **2012**, *17* Suppl., S31–S44.
2. S. M. Paul, D. S. Mytelka, C. T. Dunwiddie, C. C. Persinger, B. H. Munos, S. R. Lindborg, A. L. Schacht, *Nat. Rev. Drug Discov.* **2010**, *9*, 203–214.
3. D. Cavalla, R. Minhas, *Drug Discov. Today* **2010**, *15*, 230–234.
4. E. Bolton, Y. Wang, P. A. Thiessen, S. H. Bryant, in: *Annual Reports in Computational Chemistry*, American Chemical Society, Washington, DC, **2008**.
5. K. C. Nicolaou, D. Vourloumis, N. Winssinger, P. S. Baran, *Angew. Chem. Int. Ed. Engl.* **2000**, *39*, 44–122.
6. R. Noyori, *Nat. Chem.* **2009**, *1*, 5–6.
7. M. Williams, *Biochem. Pharmacol.* **2005**, *70*, 1707–1716.
8. J. Dorren Corbett, Drug approvals slipped in 2010, *Wall Street Journal* December 31 **2010**.
9. J. Weigelt, *EMBO Rep.* **2009**, *10*, 941–945.
10. A. Schrattenholz, K. Groebe, V. Soskic, *Methods Mol. Biol.* **2010**, *662*, 29–58.
11. P. E. McGovern, J. Zhang, J. Tang, Z. Zhang, G. R. Hall, R. A. Moreau, A. Nunez, E. D. Butrym, M. P. Richards, C. S. Wang, G. Cheng, Z. Zhao, C. Wang, *Proc. Natl. Acad. Sci. U.S.A.* **2004**, *101*, 17593–17598.
12. D. Cavalieri, P. E. McGovern, D. L. Hartl, R. Mortimer, M. Polsinelli, *J. Mol. Evol.* **2003**, *57*, S226–S232.
13. A. Goffeau, B. G. Barrell, H. Bussey, R. W. Davis, B. Dujon, H. Feldmann, F. Galibert, J. D. Hoheisel, C. Jacq, M. Johnston, E. J. Louis, H. W. Mewes, Y. Murakami, P. Philippsen, H. Tettelin, S. G. Oliver, *Science* **1996**, *274*, 546–567.
14. M. A. Resnick, B. S. Cox, *Mutat. Res.* **2000**, *451*, 1–11.
15. H. Karathia, E. Vilaprinyo, A. Sorribas, R. Alves, *PLoS ONE* **2011**, *6*, e16015.
16. D. Petranovic, J. Nielsen, *Trends Biotechnol.* **2008**, *26*, 584–590.
17. M. G. Smith, M. Snyder, *Curr. Protoc. Hum. Genet.* **2006**, Chap. 15(Unit 15), 6.

18. J. R. Perfect, *Am. J. Med.* **2012**, *125*, S39–S51.
19. M. Strand, T. A. Prolla, R. M. Liskay, T. D. Petes, *Nature* **1993**, *365*, 274–276.
20. R. Fishel, M. K. Lescoe, M. R. Rao, N. G. Copeland, N. A. Jenkins, J. Garber, M. Kane, R. Kolodner, *Cell* **1993**, *75*, 1027–1038.
21. M. G. Lee, P. Nurse, *Nature* **1987**, *327*, 31–35.
22. M. Gaisne, A. M. Becam, J. Verdiere, C. J. Herbert, *Curr. Genet.* **1999**, *36*, 195–200.
23. I. Stefanini, A. Trabocchi, E. Marchi, A. Guarna, D. Cavalieri, *J. Biol. Chem.* **2010**, *285*, 23477–23485.
24. P. Perego, G. S. Jimenez, L. Gatti, S. B. Howell, F. Zunino, *Pharmacol. Rev.* **2000**, *52*, 477–492.
25. A. P. Gasch, P. T. Spellman, C. M. Kao, O. Carmel-Harel, M. B. Eisen, G. Storz, D. Botstein, P. O. Brown, *Mol. Biol. Cell* **2000**, *11*, 4241–4257.
26. J. Verghese, J. Abrams, Y. Wang, K. A. Morano, *Microbiol. Mol. Biol. Rev.* **2012**, *76*, 115–158.
27. M. Zdralevic, N. Guaragnella, L. Antonacci, E. Marra, S. Giannattasio, *Sci. World J.* **2012**, 912147.
28. L. A. Rowe, N. Degtyareva, P. W. Doetsch, *Mech. Ageing Dev.* **2012**, *133*, 147–156.
29. A. Sadeh, D. Baran, M. Volokh, A. Aharoni, *J. Cell. Sci.* **2012**, *125*, 3333–3342.
30. A. Sadeh, N. Movshovich, M. Volokh, L. Gheber, A. Aharoni, *Mol. Biol. Cell* **2011**, *22*, 3127–3138.
31. J. S. Hardwick, F. G. Kuruvilla, J. K. Tong, A. F. Shamji, S. L. Schreiber, *Proc. Natl. Acad. Sci. U.S.A.* **1999**, *96*, 14866–14870.
32. J. DeRisi, B. Van Den Hazel, P. Marc, E. Balzi, P. Brown, C. Jacq, A. Goffeau, *FEBS Lett.* **2000**, *470*, 156–160.
33. N. Nguyen, L. M. Judd, A. Kalantzis, B. Whittle, A. S. Giraud, I. R. Van Driel, *Am. J. Physiol. Gastrointest. Liver Physiol.* **2011**, *300*, G1–G11.
34. F. Radtke, A. Wilson, G. Stark, M. Bauer, J. Van Meerwijk, H. R. MacDonald, M. Aguet, *Immunity* **1999**, *10*, 547–558.
35. L. H. Hartwell, *Biosci. Rep.* **2002**, *22*, 373–394.
36. P. M. Nurse, *Biosci. Rep.* **2002**, *22*, 487–499.
37. S. L. Schreiber, *Bioorg. Med. Chem.* **1998**, *6*, 1127–1152.
38. G. Giaever, D. D. Shoemaker, T. W. Jones, H. Liang, E. A. Winzeler, A. Astromoff, R. W. Davis, *Nat. Genet.* **1999**, *21*, 278–283.
39. K. Baetz, L. McHardy, K. Gable, T. Tarling, D. Reberioux, J. Bryan, R. J. Andersen, T. Dunn, P. Hieter, M. Roberge, *Proc. Natl. Acad. Sci. U.S.A.* **2004**, *101*, 4525–4530.
40. A. Arita, X. Zhou, T. P. Ellen, X. Liu, J. Bai, J. P. Rooney, A. Kurtz, C. B. Klein, W. Dai, T. J. Begley, M. Costa, *BMC Genomics* **2009**, *10*, 524.
41. A. Trabocchi, I. Stefanini, M. Morvillo, L. Ciofi, D. Cavalieri, A. Guarna, *Org. Biomol. Chem.* **2010**, *8*, 5552–5557.
42. A. M. Dudley, D. M. Janse, A. Tanay, R. Shamir, G. M. Church, *Mol. Syst. Biol.* **2005**, *1*, 2005.0001.
43. H. Launhardt, A. Hinnen, T. Munder, *Yeast* **1998**, *14*, 935–942.
44. H. F. Vahlensieck, L. Pridzun, H. Reichenbach, A. Hinnen, *Curr. Genet.* **1994**, *25*, 95–100.

45. H. Zhu, M. Bilgin, R. Bangham, D. Hall, A. Casamayor, P. Bertone, N. Lan, R. Jansen, S. Bidlingmaier, T. Houfek, T. Mitchell, P. Miller, R. A. Dean, M. Gerstein, M. Snyder, *Science* **2001**, *293*, 2101–2105.

46. R. Sopko, D. Huang, N. Preston, G. Chua, B. Papp, K. Kafadar, M. Snyder, S. G. Oliver, M. Cyert, T. R. Hughes, C. Boone, B. Andrews, *Mol. Cell* **2006**, *21*, 319–330.

47. S. Tugendreich, E. Perkins, J. Couto, P. Barthmaier, D. Sun, S. Tang, S. Tulac, A. Nguyen, E. Yeh, A. Mays, E. Wallace, T. Lila, D. Shivak, M. Prichard, L. Andrejka, R. Kim, T. Melese, *Genome Res.* **2001**, *11*, 1899–1912.

48. K. E. Medders, M. Kaul, *J. Neuroimmune Pharmacol.* **2011**, *6*, 202–215.

49. S. Fields, O. Song, *Nature* **1989**, *340*, 245–246.

50. P. O. Vidalain, M. Boxem, H. Ge, S. Li, M. Vidal, *Methods* **2004**, *32*, 363–370.

51. M. J. Casadaban, J. Chou, S. N. Cohen, *J. Bacteriol.* **1980**, *143*, 971–980.

52. J. C. Hu, *Trends Microbiol.* **2001**, *9*, 219–222.

53. N. Lentze, D. Auerbach, *Expert Opin. Ther. Targets* **2008**, *12*, 505–515.

54. C. Chidley, H. Haruki, M. G. Pedersen, C. Fellay, S. Moser, K. Johnsson, *Chimia* **2011**, *65*, 720–724.

55. C. Chidley, H. Haruki, M. G. Pedersen, E. Muller, K. Johnsson, *Nat. Chem. Biol.* **2011**, *7*, 375–383.

56. M. Rajik, F. Jahanshiri, A. R. Omar, A. Ideris, S. S. Hassan, K. Yusoff, *Virol. J.* **2009**, *6*, 74.

57. E. J. Licitra, J. O. Liu, *Proc. Natl. Acad. Sci. U.S.A.* **1996**, *93*, 12817–12821.

58. D. C. Henthorn, A. A. Jaxa-Chamiec, E. Meldrum, *Biochem. Pharmacol.* **2002**, *63*, 1619–1628.

59. F. Becker, K. Murthi, C. Smith, J. Come, N. Costa-Roldan, C. Kaufmann, U. Hanke, C. Degenhart, S. Baumann, W. Wallner, A. Huber, S. Dedier, S. Dill, D. Kinsman, M. Hediger, N. Bockovich, S. Meier-Ewert, A. F. Kluge, N. Kley, *Chem. Biol.* **2004**, *11*, 211–223.

60. T. W. Corson, N. Aberle, C. M. Crews, *ACS Chem. Biol.* **2008**, *3*, 677–692.

61. P. Y. Lum, C. D. Armour, S. B. Stepaniants, G. Cavet, M. K. Wolf, J. S. Butler, J. C. Hinshaw, P. Garnier, G. D. Prestwich, A. Leonardson, P. Garret-Engele, C. M. Rush, M. Bard, G. Schimmack, J. W. Phillips, C. J. Roberts, D. D. Shoemaker, *Cell* **2004**, *116*, 121–137.

62. A. M. Smith, L. E. Heisler, J. Mellor, F. Kaper, M. J. Thompson, M. Chee, F. P. Roth, G. Giaever, C. Nislow, *Genome Res.* **2009**, *19*, 1836–1842.

63. G. Giaever, P. Flaherty, J. Kumm, M. Proctor, C. Nislow, D. F. Jaramillo, A. M. Chu, M. I. Jordan, A. P. Arkin, R. W. Davis, *Proc. Natl. Acad. Sci. U.S.A.* **2004**, *101*, 793–798.

64. N. Bivi, M. Romanello, R. Harrison, I. Clarke, D. C. Hoyle, L. Moro, F. Ortolani, A. Bonetti, F. Quadrifoglio, G. Tell, D. Delneri, *Genome Biol.* **2009**, *10*, R93.

65. A. Lopez, A. B. Parsons, C. Nislow, G. Giaever, C. Boone, *Prog. Drug Res.* **2008**, *66*, 239–271.

66. J. Doostzadeh, R. W. Davis, G. N. Giaever, C. Nislow, J. W. Langston, *Toxicol. Sci.* **2007**, *95*, 182–187.

67. Z. Yan, M. Costanzo, L. E. Heisler, J. Paw, F. Kaper, B. J. Andrews, C. Boone, G. Giaever, C. Nislow, *Nat. Methods* **2008**, *5*, 719–725.

68. M. Schuldiner, S. R. Collins, N. J. Thompson, V. Denic, A. Bhamidipati, T. Punna, J. Ihmels, B. Andrews, C. Boone, J. F. Greenblatt, J. S. Weissman, N. J. Krogan, *Cell* **2005**, *123*, 507–519.

69. W. Lee, R. P. St. Onge, M. Proctor, P. Flaherty, M. I. Jordan, A. P. Arkin, R. W. Davis, C. Nislow, G. Giaever, *PLoS Genet.* **2005**, *1*, e24.

70. J. A. Brown, G. Sherlock, C. L. Myers, N. M. Burrows, C. Deng, H. I. Wu, K. E. McCann, O. G. Troyanskaya, J. M. Brown, *Mol. Syst. Biol.* **2006**, *2*, 2006.0001.

71. A. B. Parsons, A. Lopez, I. E. Givoni, D. E. Williams, C. A. Gray, J. Porter, G. Chua, R. Sopko, R. L. Brost, C. H. Ho, J. Wang, T. Ketela, C. Brenner, J. A. Brill, G. E. Fernandez, T. C. Lorenz, G. S. Payne, S. Ishihara, Y. Ohya, B. Andrews, T. R. Hughes, B. J. Frey, T. R. Graham, R. J. Andersen, C. Boone, *Cell* **2006**, *126*, 611–625.

72. M. Schena, D. Shalon, R. W. Davis, P. O. Brown, *Science* **1995**, *270*, 467–470.

73. D. Cavalieri, J. P. Townsend, D. L. Hartl, *Proc. Natl. Acad. Sci. U.S.A.* **2000**, *97*, 12369–12374.

74. C. Varela, J. Cardenas, F. Melo, E. Agosin, *Yeast* **2005**, *22*, 369–383.

75. J. P. Townsend, D. Cavalieri, D. L. Hartl, *Mol. Biol. Evol.* **2003**, *20*, 955–963.

76. R. C. Fry, M. S. Demott, J. P. Cosgrove, T. J. Begley, L. D. Samson, P. C. Dedon, *BMC Genomics* **2006**, *7*, 313.

77. D. Shalon, S. J. Smith, P. O. Brown, *Genome Res.* **1996**, *6*, 639–645.

78. L. Wodicka, H. Dong, M. Mittmann, M. H. Ho, D. J. Lockhart, *Nat. Biotechnol.* **1997**, *15*, 1359–1567.

79. D. J. Lockhart, H. Dong, M. C. Byrne, M. T. Follettie, M. V. Gallo, M. S. Chee, M. Mittmann, C. Wang, M. Kobayashi, H. Horton, E. L. Brown, *Nat. Biotechnol.* **1996**, *14*, 1675–1680.

80. J. L. DeRisi, V. R. Iyer, P. O. Brown, *Science* **1997**, *278*, 680–686.

81. S. Chu, J. L. DeRisi, M. Eisen, J. Mulholland, D. Botstein, P. O. Brown, I. Herskowitz, *Science* **1998**, *282*, 699–705.

82. P. T. Spellman, G. Sherlock, M. Q. Zhang, V. R. Iyer, K. Anders, M. B. Eisen, P. O. Brown, D. Botstein, B. Futcher, *Mol. Biol. Cell* **1998**, *9*, 3273–3297.

83. S. A. Jelinsky, P. Estep, G. M. Church, L. D. Samson, *Mol. Cell. Biol.* **2000**, *20*, 8157–8167.

84. M. J. Marton, J. L. DeRisi, H. A. Bennett, V. R. Iyer, M. R. Meyer, C. J. Roberts, R. Stoughton, J. Burchard, D. Slade, H. Dai, D. E. Bassett, Jr., L. H. Hartwell, P. O. Brown, S. H. Friend, *Nat. Med.* **1998**, *4*, 1293–1301.

85. T. R. Hughes, M. J. Marton, A. R. Jones, C. J. Roberts, R. Stoughton, C. D. Armour, H. A. Bennett, E. Coffey, H. Dai, Y. D. He, M. J. Kidd, A. M. King, M. R. Meyer, D. Slade, P. Y. Lum, S. B. Stepaniants, D. D. Shoemaker, D. Gachotte, K. Chakraburtty, J. Simon, M. Bard, S. H. Friend, *Cell* **2000**, *102*, 109–126.

86. M. B. Eisen, P. T. Spellman, P. O. Brown, D. Botstein, *Proc. Natl. Acad. Sci. U.S.A.* **1998**, *95*, 14863–14868.

87. A. Brazma, J. Vilo, *FEBS Lett.* **2000**, *480*, 17–24.

88. P. Tamayo, D. Slonim, J. Mesirov, Q. Zhu, S. Kitareewan, E. Dmitrovsky, E. S. Lander, T. R. Golub, *Proc. Natl. Acad. Sci. U.S.A.* **1999**, *96*, 2907–2912.

89. S. Tavazoie, J. D. Hughes, M. J. Campbell, R. J. Cho, G. M. Church, *Nat. Genet.* **1999**, *22*, 281–285.

90. M. Ashburner, C. A. Ball, J. A. Blake, D. Botstein, H. Butler, J. M. Cherry, A. P. Davis, K. Dolinski, S. S. Dwight, J. T. Eppig, M. A. Harris, D. P. Hill, L. Issel-Tarver, A. Kasarskis, S. Lewis, J. C. Matese, J. E. Richardson, M. Ringwald, G. M. Rubin, G. Sherlock, *Nat. Genet.* **2000**, *25*, 25–29.

91. FLYBASE. FlyBase: A Database of Drosophila Genes & Genomes, **2011**.

92. SGD_PROJECT. Saccharomyces Genome Database [online], **2011**; available at www.SGD.com; accessed Apr. 22, 2012.

93. MGD_PROJECT. Mouse Genome Database [online], **2011**; available at http://www.informatics.jax.org/; accessed Apr. 22, 2012.

94. M. Kanehisa, S. Goto, M. Hattori, K. F. Aoki-Kinoshita, M. Itoh, S. Kawashima, T. Katayama, M. Araki, M. Hirakawa, *Nucleic Acids Res.* **2006**, *34*, D354–D357.

95. G. Joshi-Tope, M. Gillespie, I. Vastrik, P. D'Eustachio, E. Schmidt, B. De Bono, B. Jassal, G. R. Gopinath, G. R. Wu, L. Matthews, S. Lewis, E. Birney, L. Stein, *Nucleic Acids Res.* **2005**, *33*, D428–D432.

96. T. I. Lee, N. J. Rinaldi, F. Robert, D. T. Odom, Z. Bar-Joseph, G. K. Gerber, N. M. Hannett, C. T. Harbison, C. M. Thompson, I. Simon, J. Zeitlinger, E. G. Jennings, H. L. Murray, D. B. Gordon, B. Ren, J. J. Wyrick, J. B. Tagne, T. L. Volkert, E. Fraenkel, D. K. Gifford, R. A. Young, *Science* **2002**, *298*, 799–804.

97. D. Cavalieri, C. Castagnini, S. Toti, K. Maciag, T. Kelder, L. Gambineri, S. Angioli, P. Dolara, *Bioinformatics* **2007**, *23*, 2631–2632.

98. S. L. Normand, *Stat. Med.* **1999**, *18*, 321–359.

99. C. Romualdi, C. De Pitta, L. Tombolan, S. Bortoluzzi, F. Sartori, A. Rosolen, G. Lanfranchi, *BMC Genomics* **2006**, *7*, 287.

100. R. G. Jenner, R. A. Young, *Nat. Rev. Microbiol.* **2005**, *3*, 281–294.

101. K. C. Roet, K. Bossers, E. H. Franssen, M. J. Ruitenberg, J. Verhaagen, *Exp. Neurol.* **2011**, *229*, 10–45.

102. A. Ramasamy, A. Mondry, C. C. Holmes, D. G. Altman, *PLoS Med.* **2008**, *5*, e184.

103. P. Cahan, F. Rovegno, D. Mooney, J. C. Newman, G. St. Laurent 3rd, T. A. McCaffrey, *Gene* **2007**, *401*, 12–18.

104. L. Beltrame, L. Rizzetto, R. Paola, P. Rocca-Serra, L. Gambineri, C. Battaglia, D. Cavalieri, *PLoS ONE* **2009**, *4*, e4128.

105. D. Cavalieri, E. Calura, C. Romualdi, E. Marchi, M. Radonjic, B. Van Ommen, M. Müller, *BMC Genomics* **2009**, *10*, 596.

15

VIRTUAL SCREENING METHODS

Jürgen Bajorath

15.1 INTRODUCTION

This chapter is an overview of current *computational compound screening* methods [1]. Introductory sections are followed by a discussion of relevant methodologies. Computational compound screening is commonly also referred to as *in silico screening* or *virtual screening* (VS). The latter term first appeared in the literature in 1997 [2] and is currently probably the most popular name. In VS, computational methods are applied to screen large compound databases for novel active compounds (i.e., with specific activity against a target of interest). As such, VS is often rationalized as a complement to experimental high-throughput screening (HTS) [3–5]. In its essence, VS is a hit identification approach, analogous to HTS. One might argue that VS approaches might not really be required for practical purposes given the very large HTS capacities that are available at present in the pharmaceutical industry. However, this is clearly not the case. As has recently been pointed out [6], VS has produced a considerable number of success stories, including the identification of first-in-class inhibitors and novel active compounds in cases where HTS has not been successful. Furthermore, VS frequently is a method of choice when assays are complex, time consuming, and/or difficult to format for high throughput. Moreover, in academic settings, where large HTS facilities are rarely available, VS is often the only feasible approach to hit identification. However, VS can certainly not be expected to produce novel active compounds en masse. The majority of successful VS applications currently yield limited numbers of hits that are predominantly weakly or moderately potent, although there are exceptions [1,6–9]. The potency of HTS and VS hits is often comparable, but HTS typically produces large numbers of hits because

Diversity-Oriented Synthesis: Basics and Applications in Organic Synthesis, Drug Discovery, and Chemical Biology, First Edition. Edited by Andrea Trabocchi.

large numbers of compounds are tested experimentally. By contrast, VS attempts to rationalize candidate selection by reducing the number of test compounds as much as possible, and hence yields in successful instances only small numbers of novel active compounds. In fact, a major attraction of the VS approach is its ability to significantly reduce the number of candidate compounds for experimental evaluation. For example, current screening libraries often contain several million compounds whose chemical information content can be processed computationally in a target-directed manner to identify small subsets of compounds, on the order of hundreds of thousands, which have a high likelihood to be active. If required assay systems are complicated, this reductionist approach is highly attractive, even if HTS capacity is available. Although active compounds are usually enriched in database selection sets, the specificity of VS calculations is generally limited regardless of the methods that are employed. Consequently, the majority of candidate compounds in database selections are usually false positives. Thus, if VS is successful, it can realistically be expected to find a few novel active compounds in a database selection set: for example, one to five among 100 to 200 database candidates [1,7]. On the other hand, HTS is often plagued with large numbers of apparent (weakly potent) hits, many of which turn out to be false positives in secondary assays. Hence, HTS hit triaging and assay follow-up usually require much time and effort. Clearly, any experimental or computational screening technology has its limitations. While hit discovery in pharmaceutical research and development is clearly dominated by HTS, VS approaches are established in drug discovery settings and their use is expected to further increase as even larger numbers of compounds and screening targets become available in the future. Over the past approximately 15 years, VS has become increasingly popular in both pharmaceutical research and academia [10–13], and there is no end in sight. In addition to practical relevance, the compound reductionist concept underlying VS approaches, the search for "active needles" in large "compound haystacks," and the prediction of structure–activity relationships (SARs) present intellectually stimulating tasks and challenging aspects for computational method development. These challenges catalyze many exploratory activities in academia and in the pharmaceutical industry to further improve current VS methodologies or develop conceptually new methods. Hence, it is fair to say that the VS field is in constant flux, in terms of theoretical developments and practical applications.

15.2 BASIC VIRTUAL SCREENING CONCEPTS

VS efforts require specific input information. In principle, this can be the three-dimensional structure of the target protein or knowledge of already known active compounds. Accordingly, one generally distinguishes between structure-based VS (SBVS) [14,15] and ligand-based VS (LBVS) [16,17].

15.2.1 Structure- and Ligand-Based Virtual Screening

SBVS makes explicit use of target structures as screening templates, whereas LBVS extrapolates from known active compounds to identify structurally diverse molecules that have similar activity. Hence, in LBVS, known active compounds are used as reference molecules. In principle, such molecules are not required for SBVS, although the knowledge of active compounds also provides additional information for structure-based approaches (e.g., known actives can be modeled into binding sites to guide compound selection). It is important to note that SBVS and LBVS approaches are not mutually exclusive. In many cases, especially for popular pharmaceutical targets, both target structure and ligand information is available and SBVS and LBVS can be applied in parallel or in a coordinated manner. Furthermore, SBVS efforts greatly benefit from the availability of x-ray structures of ligand–target complexes, which makes it possible to study compound binding modes directly and use this information to guide docking studies. This knowledge usually greatly improves the likelihood of identifying novel active compounds. However, SBVS is in principle a de novo approach to hit identification, whereas LBVS is always comparative in nature: that is, it attempts to relate chemical and structural features of known active compounds and candidate molecules to each other as a basis for predictions. Accordingly, in LBVS, the question of ligand novelty plays a central role.

15.2.2 Scaffold Analysis

One is always interested in identifying hits that depart structurally from already known actives: those that contain previously unobserved core structures (also called *scaffolds*) [18]. Therefore, the primary LBVS goal is generally referred to as *scaffold hopping* [19,20]. By contrast, one is usually not interested in identifying molecules that are closely related to or are structural analogs of known active compounds, because such compounds can be selected "by eye" or by substructure searching, without the need to apply VS methods. However, the assessment of scaffold hopping potential in LBVS is generally complicated and often biased by several factors. For example, chemically distinct scaffolds in compounds having similar activity might span a wide structural range from very similar ones (which might be distinguished only by individual heteroatom positions and/or bond orders) to essentially unrelated ones. Thus, scaffold hops often represent very different degrees of chemical distance. Identifying active compounds with similar scaffolds is not very challenging; however, finding virtually unrelated scaffolds that represent compounds with similar activity is rather challenging. It has generally been difficult to compare the scaffold hopping potential of different LBVS methods (i.e., to assess the chemical distances that distinguish newly identified hits from reference molecules). Recently, however, a first distance function has been introduced to quantify the chemical distance between scaffolds [21]. By applying this function, it is possible to evaluate and compare numerically the degree of difficulty involved in scaffold hopping exercises. Figure 15.1 shows exemplary scaffolds hops and reports the interscaffold distances.

15.2.3 Methodological Complexity

It is often assumed that methodological complexity in LBVS scales with search performance and scaffold hopping potential. However, this is an unsubstantiated and

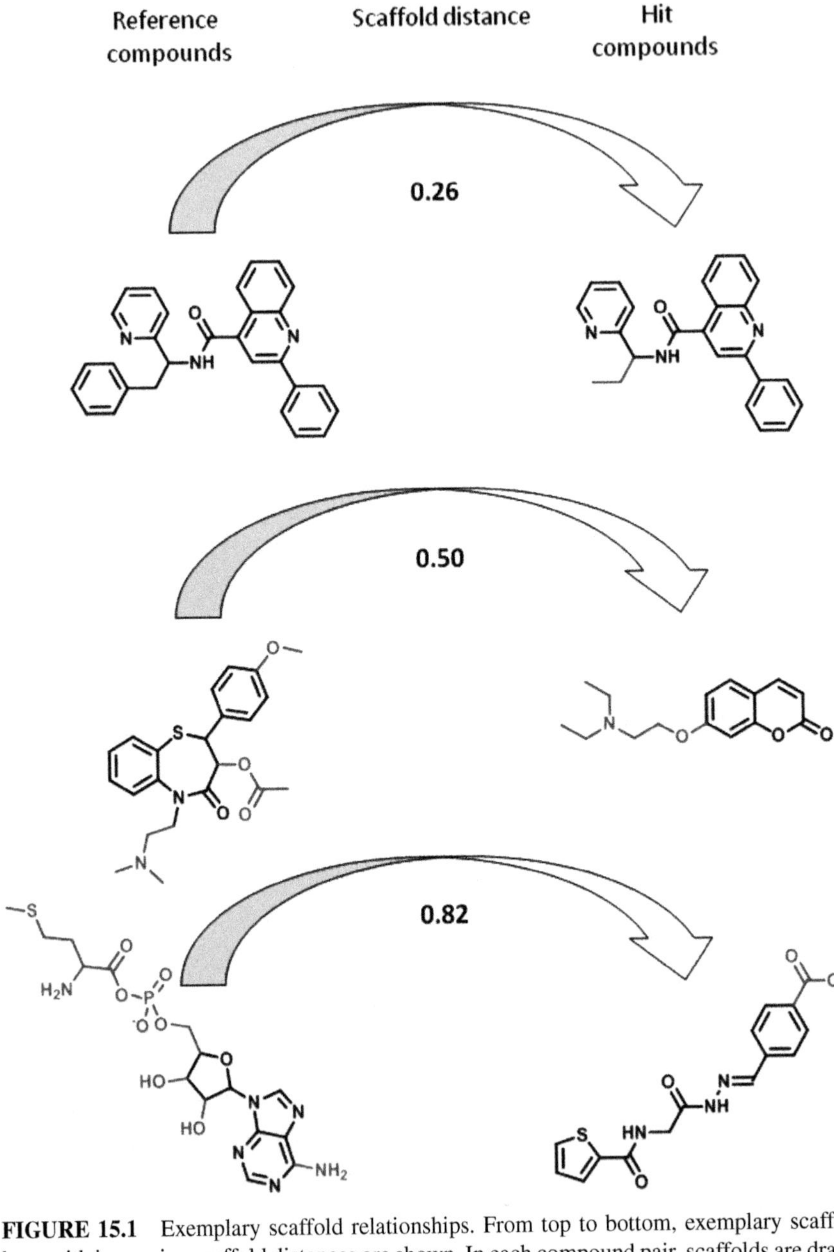

FIGURE 15.1 Exemplary scaffold relationships. From top to bottom, exemplary scaffold hops with increasing scaffold distances are shown. In each compound pair, scaffolds are drawn with bold lines. Larger distances correspond to structurally increasingly dissimilar scaffolds. Examples are taken from REPROVIS database [103] including sets 13, 2, and 25 (from top to bottom).

largely incorrect assumption [17]. Rather, LBVS methods of very different sophistication and computational complexity have been shown to be capable of identifying structurally diverse active compounds, depending on the targets and compound classes under investigation. In some instances, relatively simple molecular representations and methods might be successful, whereas complex approaches fail; in others, methods of varying complexity consistently identify hits or, alternatively, no method succeeds [7,17]. The general compound class and target dependence of VS approaches is currently not well understood. However, in LBVS, it is evident that scaffold hopping ability does not correlate with methodological complexity. For example, relatively simple similarity search tools, such as molecular fingerprints (i.e., bit-string representations of molecular structure and properties, as discussed further below), have demonstrated scaffold hopping potential [22,23]. A recent investigation has revealed mechanisms by which fingerprints recognize structurally diverse active compounds [24]. In general, care must be taken to judge method performance. This applies particularly to VS method evaluation in benchmark calculations. In this case, known active compounds are added to screening databases as potential hits and their retrieval (or recall) rates are calculated as a measure of VS performance [17]. In such theoretical investigations, VS performance is typically overestimated and does not scale with success rates in practical applications [1,8,9]. This is a direct consequence of artificial system setups, where optimized active compounds are taken from the literature as hits and added to background databases. Optimized active compounds are generally larger and chemically or topologically more complex than screening database compounds and thus easier to distinguish from them than are typical screening hits [25,26]. Thus, artificially high recall rates are often observed in benchmark calculations in both LBVS and SBVS.

15.3 MOLECULAR SIMILARITY IN VIRTUAL SCREENING

All LBVS methods explore the similarity between active reference molecules and candidate compounds. This is a fundamental aspect of the LBVS concept [27]. However, molecular similarity can be evaluated from different points of view.

15.3.1 Local vs. Global Similarity

Any similarity assessment is dependent on the molecular representations chosen, such as molecular fingerprints or numerical property descriptors [28]. Importantly, in LBVS, one needs to distinguish between local and global molecular similarity comparisons. *Local similarity* refers to the presence of specific functional groups or molecular features in a given topological environment or a specific geometric arrangement. Here, only active parts (known or putative) of molecules are considered, and this local view of similarity is central to the pharmacophore concept [29], as discussed below, and also to quantitative structure–activity relationship (QSAR) modeling [30]. QSAR analysis is applied primarily to congeneric series of compounds to predict chemical modifications that lead to improvements in potency. Accordingly,

QSAR modeling is in principle not designed for LBVS and scaffold hopping. Although QSAR models have occasionally also been adapted for LBVS [30,31], their use is much less frequent than other LBVS approaches. Therefore, QSAR modeling is not discussed in detail herein.

In contrast to local similarity assessment, *global similarity* methods rely on whole-molecule comparisons, without considering substructures known or thought to be important for activity. Conceptually, the assessment of global similarity is based on the *similarity-property principle*, which has been introduced in the context of a milestone molecular similarity publication [32]. This principle simply states that chemically/structurally similar molecules have a tendency to display similar biological activities. Global similarity assessment is at the core of many LBVS methods. Because global molecular comparisons do not require knowledge of activity-determining features, they are often employed when little or no pharmacophore information is available.

15.3.2 Molecular Representations

As stated earlier, similarity evaluation always depends on the molecular representations chosen. In addition, metrics or functions applied to quantify the degree of similarity between compounds also influence similarity assessment. Importantly, a molecular representation can also be rationalized as a chemical reference space [28], which provides the framework for molecular comparisons and similarity analysis. Distance relationships in chemical reference spaces are discussed further below. Regardless of the nature of chemical reference spaces, the most important paradigm in LBVS is that molecular similarity must correlate with activity similarity, following the similarity-property principle. However, defined relationships between molecular similarity and activity similarity are difficult to establish and essentially unavailable at present [17,27]. This is already implicit in the little understood compound class dependence of LBVS methods mentioned above. Moreover, it is also well known that small modifications of active compounds can either dramatically improve or reduce their potency and that active and inactive compounds might often be very similar and distinguishable only by minute chemical differences [33]. The presence of such local similarity determinants falls outside the applicability domain of global similarity approaches in LBVS [27]. Small chemical changes with large-magnitude effects are also inconsistent with the basic ideas underlying scaffolds hopping, which are conceptually more related to the similarity-property principle. Regardless, the use of chemical reference spaces is of central importance to similarity analysis in LBVS, and different types of reference spaces can be generated. Literally thousands of numerical property descriptors are available for space generation [34], which are calculated primarily from molecular graphs (two-dimensional descriptors) or compound conformations (three-dimensional descriptors). Navigating numerical descriptor spaces is the basic operation underlying many compound classification techniques used in LBVS. Chemical space navigation involves the determination of intercompound distances. If fingerprints are used as a molecular representation, the overlap between fingerprints of different compounds is quantified in a pairwise manner as a measure of molecular similarity using metrics such as the well-known Tanimoto coefficient [35]. Importantly, chemical reference spaces are not always designed for global similarity analysis. Rather, they can also be local in nature. For example, all possible three- or four-point pharmacophore arrangements for predefined pharmacophore feature sets can be calculated to constitute a reference space and stored in fingerprint format [36]. The comparison of such pharmacophore fingerprints of test compounds then represents a local similarity assessment that is confined to possible pharmacophore patterns.

15.4 SPECTRUM OF VIRTUAL SCREENING APPROACHES

Having discussed some of the principles and foundations of VS, the spectrum of current VS approaches is described as illustrated in Figure 15.2. SBVS approaches include primarily docking techniques and structure-based pharmacophore models

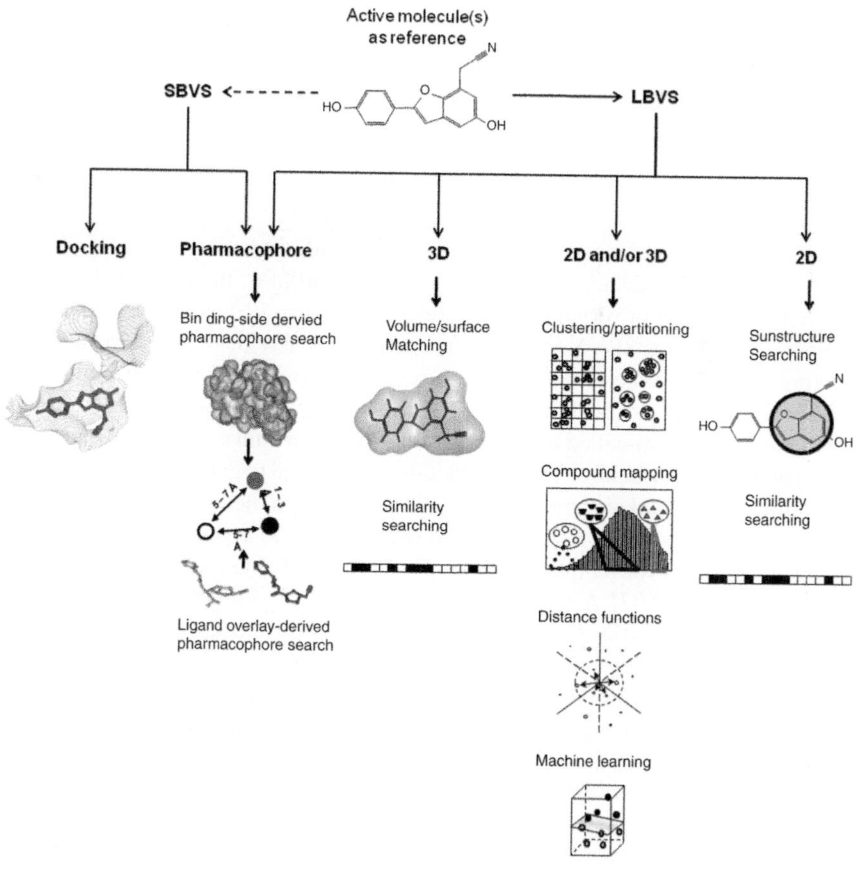

FIGURE 15.2 Virtual screening approaches. Major categories of SBVS and LBVS methodologies are illustrated.

utilized for compound database searching. The LBVS field is methodologically more heterogeneous, and alternative classifications of LBVS methods are possible. Herein, LBVS approaches are divided broadly into similarity search and compound classification methods. As discussed below, both areas include many different methodological concepts and algorithms. For example, the similarity search category covers two-dimensional fingerprints, pharmacophore models, three-dimensional pharmacophore fingerprints, and shape queries. Regardless of their algorithmic differences, most similarity search methods produce rankings of database compounds with respect to reference molecules. Herein, compound classification approaches are divided further into basic classification methods such as clustering and partitioning (for which many different algorithms exist) and machine learning approaches that are becoming increasingly popular in LBVS. Compound classification methods typically produce a binary readout by labeling test compounds as "active" or "inactive." However, database rankings can also be produced by some machine learning approaches. Essentially all SBVS and LBVS approaches assign the likelihoods of activity to test compounds (on the basis of rankings) or yield binary active/inactive class label predictions. However, none of these approaches predicts actual potency values. Exceptions to this rule are QSAR models adapted for VS. Apart from QSAR, very few approaches have considered compound potency as a VS search parameter.

15.5 DOCKING

The discussion of specific VS approaches begins with a brief account of molecular docking. For further details concerning docking algorithms, the interested reader is referred to a number of very informative reviews that are cited in this section. Ligand docking is the major SBVS approach. In fact, if we view the VS field as a whole, docking is the single most popular VS technique on the basis of published applications [1]. In general, ligand docking aims at predicting the binding conformation and orientation, i.e., the so-called pose, of a test compound within a predefined active-site region of the target structure. Scoring functions are applied to guide the posing process and, ultimately, to rank database compounds according to their likelihood of activity. Currently, the state of the art in this field is fully flexible ligand docking into rigid template structures (or template structures with limited side-chain flexibility in the active-site region) [9,37]. A variety of different docking algorithms and programs have been introduced [38–42]. Despite many algorithmic differences in conformational search and pose finding, and also differences in force field–based or empirical scoring functions, all methods have in common that the docking process is divided into energy calculation-assisted posing and final scoring and ranking. Force field–based scoring functions treat intra- and intermolecular interactions using molecular mechanics–type potentials, whereas empirical scoring functions contain energy terms that are specifically adjusted to reproduce experimental structures of ligand–target complexes or fit binding data [43]. In addition, potentials of mean force are derived via statistical analysis of distributions of atomic ligand–protein interactions and favor types of interactions that are frequently observed in complex x-ray structures [44].

Currently, regardless of the specifics of docking algorithms and scoring functions, posing is more reliable than scoring. This means that binding modes of active ligands can often be modeled with reasonable to very good accuracy but that it is rather difficult to distinguish between active and inactive compounds on the basis of calculated scores. The major reason for this complication is that it is not possible to compute free energies of binding accurately for ligand–target complexes in a consistent manner (although successful case studies are occasionally reported in the literature). In this context, it must also be taken into account that the vast majority of database compounds are inactive but that many of these inactives might yield a "science fiction" pose that displays reasonable chemical and shape complementarity to the active site. In fact, even based on visual inspection, taking chemical knowledge and intuition into account, it is usually difficult to distinguish between "true" active compounds and false positives. Hence, compound selection from database rankings is far from being trivial. Given the well-known limitations of current scoring schemes, essentially any successful docking application reported in the literature involves visual inspection and subjective analysis of docked complexes in order to select candidate compounds for experimental testing, a situation that has not changed over the past decade [9,45].

Because the computational requirements of flexible ligand docking are generally high, it is often attempted to reduce the size of the screening database prior to docking. For this purpose, LBVS and SBVS are often carried out in a sequential manner (although truly integrated LBVS/SBVS approaches are yet to be introduced). Typically, if at least a few active compounds are known for a docking target, similarity search calculations are carried out to preselect database compounds for docking that display at least some remote chemical similarity to the known actives. This preselection scheme might work against the identification of structurally completely novel hits but reduces the size of a source database by at least an order of magnitude, or even more. Reducing 1 million database compounds to a subset containing about 100,000 or fewer compounds enables fully flexible ligand docking within a reasonable time frame.

15.6 SIMILARITY SEARCHING

Similarity searching is a traditional LBVS approach. Regardless of the specifics of similarity search methods, they generally produce rankings of database compounds relative to reference molecules in the order of decreasing similarity and thus decreasing probability of activity. It is also common practice in LBVS to prefilter screening databases and omit compounds with potential toxicity and other liabilities or non-compliance with preformulated property range rules such as Lipinski's rule-of-5 [46]. However, such filtering techniques are not considered a part of the spectrum of similarity search approaches. This also applies to substructure searching that is an "all or nothing" matching technique.

Pharmacophore models [47,48] and fingerprints represents the major similarity search approaches. Pharmacophore searching currently is the most popular LBVS approach; the majority of prospective LBVS applications reported in the literature employ pharmacophore queries [8]. In addition, there are also other types of methods for which fewer molecular representations and algorithms have been introduced, such as shape similarity searching [49]. Pharmacophore and shape searching are primarily three-dimensional methods (i.e., they are based three-dimensional molecular representations. However, two-dimensional pharmacophore fingerprints are also available, and molecular shape representations can be approximated from molecular graphs. Fingerprints can be derived from two-dimensional or three-dimensional molecular representations, with pharmacophore fingerprints being the most popular category of three-dimensional fingerprints [50].

15.6.1 Pharmacophores

Currently, four-point pharmacophore models represent the standard for three-dimensional database searching [29,50]. In this case, pharmacophore queries are used to search conformational databases of test compounds for hits that match the four features and six interfeature distances of the model. Importantly, precise matches of interfeature distances are rarely possible, and hence distance ranges are used to relax the query and identify compounds that match it closely. Pharmacophore features (or functions) that are assigned to feature points typically include hydrophobic groups, aromatic centers, negatively charged groups, positively charged groups, hydrogen-bond donors, and hydrogen-bond acceptors. Pharmacophores are derived from spatial alignments of active compounds in known or, more frequently, putative bioactive conformations, which is often referred to as *pharmacophore mapping*. This process has been automated in many modeling programs [51]. A key assumption underlying pharmacophore mapping is that the bioactive compounds share the same mechanism of action and binding mode. Pharmacophore models cannot only be derived from ligand superpositions but also based on active-site features of targets with known x-ray structures. In this case, pharmacophores are modeled to be complementary to important interaction sites in proteins. In pharmacophore searching, subsets of database compounds are identified that match the query and/or that can be ranked because of three-dimensional similarity criteria. In addition to individual pharmacophore models, pharmacophore fingerprints record large ensembles of predefined pharmacophore patterns (i.e., combinations of geometric arrangements and specific features). Each pattern is assigned to a single bit position in the fingerprint. Based on systematic conformational search, patterns are recorded that are matched by test compounds. Pharmacophore fingerprint overlap between active reference and database compounds is then calculated as a measure of similarity. Pharmacophore models and pharmacophore fingerprints are illustrated schematically in Figure 15.3.

Pharmacophore searching and fingerprinting has an important distinction. In pharmacophore fingerprinting, no match to a specific pharmacophore model is evaluated. Rather, overlap in generally defined pharmacophore space (i.e., the union of all predefined patterns) is quantified based on systematic conformational search. The underlying idea is that the greater the overlap of two molecules in pharmacophore space is, the more likely they are to have similar activity. Thus, in contrast to

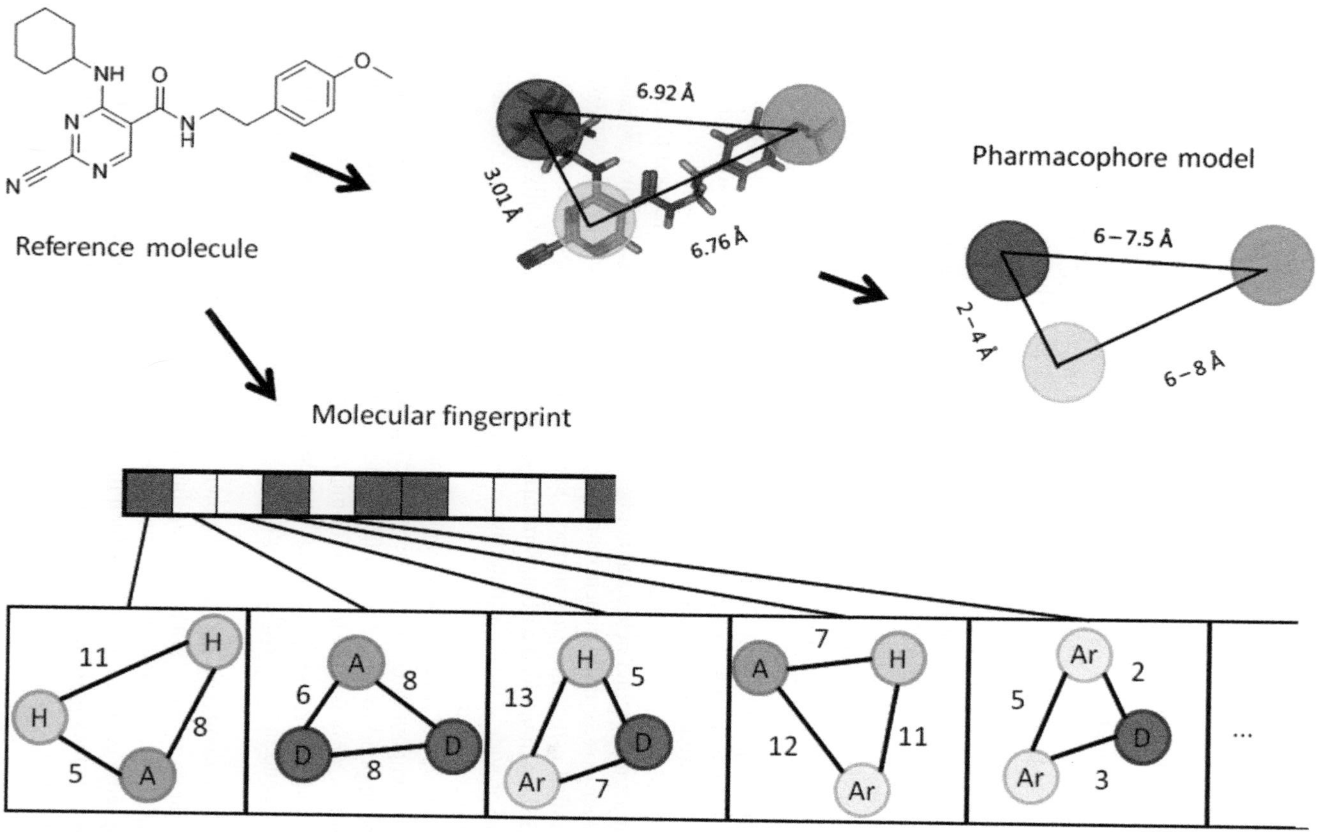

FIGURE 15.3 Pharmacophore representations. A three-point pharmacophore query derived from a single reference molecule and a three-point pharmacophore fingerprint of the same molecule are schematically represented. (*See insert for color representation of the figure.*)

three-dimensional database searching, pharmacophore fingerprinting does not rely on bioactive compound conformations, the prediction of which represents a major caveat in ligand-based drug design and LBVS.

15.6.2 Two-Dimensional Fingerprints

15.6.2.1 Fingerprint Design and Comparison Two-dimensional fingerprints are also popular descriptors and search tools for LBVS. Fingerprints of very different design, complexity, and length have been introduced over the years, including, among others, atom pairs [52] and fragment sets [53], topological pathways through molecules [54], and combinatorial fingerprints [55,56]. The latter capture layered atom environments in a compound-specific manner. In addition, other fingerprints monitor two-dimensional pharmacophore features such as atom triplets or quadruplets [22], in analogy to three-dimensional pharmacophore fingerprints. Combinatorial fingerprints, in particular extended connectivity fingerprints [56], have become increasingly popular in recent years and are currently among the best-performing fingerprint descriptors [22,23]. In addition, new types of fingerprints continue to be introduced (albeit not very frequently), such as bonded atom pairs accounting for short-range atom environments [57] or topology fingerprints that systematically enumerate subgraphs up to a predefined size [58]. Fingerprints are either keyed or hashed in their design. In keyed fingerprints, each bit position is associated with a specific feature and monitors its presence or absence (or sometimes also its count) in a molecule. By contrast, in hashed fingerprints, features are mapped to overlapping bit segments (and hence cannot be interpreted in chemical terms). Regardless of whether two- or three-dimensional fingerprints are keyed or hashed, their overlap is quantified using similarity metrics, the most popular being the Tanimoto coefficient [35], as mentioned earlier. Similarity coefficient values typically range from 0 to 1 and provide a basis for database ranking according to decreasing similarity to reference compounds.

For the use of multiple reference compounds, data fusion methods that further increase compound recall have been investigated intensely. Data fusion can principally be applied to different similarity metrics (which is not often done), different fingerprints, or rankings obtained for multiple reference compounds (using the same metric and fingerprint). The latter approach is most widely applied in similarity searching and involves rank averaging or maximum rank selection [59,60].

15.6.2.2 Complexity Effect A known conundrum of fingerprint searching is the molecular complexity effect [25,26]. Increasingly large and topologically complex molecules have a tendency to increase the bit density of fingerprints, regardless of their specific chemical features, which in turn causes a statistical tendency for larger similarity values in database searching, as illustrated in Figure 15.4. This statistical tendency favors false-positive detections and, consequently, reduces the recall of active compounds in benchmark calculations. In practical applications, it leads to similarity values of many database compounds that are comparable in magnitude to those of potential hits, which increases the background noise of the calculations and

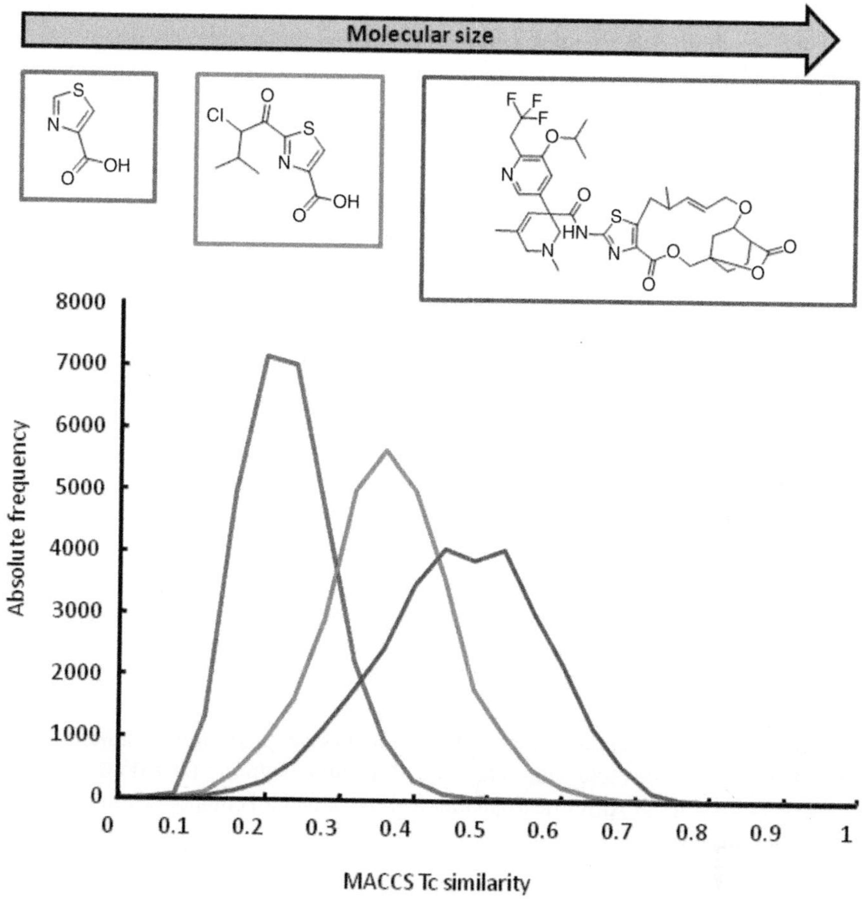

FIGURE 15.4 Biased similarity calculations. The influence of the molecular complexity effect on fingerprint calculations is illustrated. Three molecules of increasing size are shown (red, green, and blue, respectively). Each molecule is searched against the ZINC compound database [104] and the resulting distributions of MACCS T_c values are compared. As can be seen, apparent similarity to database compounds increases with molecular complexity and size. (*See insert for color representation of the figure.*)

hence makes it more difficult to identify new hits. It has been shown that the use of large and complex reference compounds generally leads to low hit rates [26], which is a direct consequence of the complexity effect. In principle, the complexity effect can be eliminated from similarity searching by using similarity metrics that weight equally fingerprint bits that are set on and off [61] or by using fingerprints that yield a constant bit density for all compounds, regardless of their topological complexity and size [62]. The former approach might be controversial because in this case, the absence of features contributes indirectly to similarity assessment.

15.6.2.3 Fingerprint Engineering In fingerprint engineering, which represents an emerging approach, it is attempted to modify fingerprints having a constant format in a rational manner to further increase their search performance. For example, complexity effects can be eliminated by merging a fingerprint with its complement [63], which generates a version that has twice the length of the original fingerprint but always a constant bit density of 50%. This type of fingerprint modification falls into the spectrum of fingerprint engineering approaches. Moreover, through engineering, segments from two or more different keyed fingerprints can be combined into new fingerprint designs with further improved search performance compared to their parental fingerprints [64]. For this purpose, feature selection methods are used to identify subsets of bit positions in different fingerprints that are most important for recognizing compounds belonging to a given activity class. Bit segments isolated from different fingerprints are then recombined. This approach transforms generally applicable fingerprints into compound class-specific search tools.

15.7 COMPOUND CLASSIFICATION

In addition to similarity searching, compound classification is another principal approach to LBVS for which many conceptually diverse methods have been introduced. Compound classification approaches typically do not produce compound rankings but classify compounds as active or inactive (a process often referred to as *class label prediction* in machine learning). However, there are exceptions, as discussed below. Many popular compound classification methods are machine learning approaches. Given the increasing relevance of machine learning for LBVS, machine learning approaches are discussed separately. First, standard compound classification techniques are reviewed.

15.7.1 Chemical Reference Spaces

Compound classification methods operate in chemical reference spaces, typically molecular descriptor spaces (where the selection of n numerical descriptors constitutes an n-dimensional space). A principal challenge of chemical space design for LBVS is that the descriptors selected must be activity-relevant: that is, capable of differentiating between compounds having a desired activity and inactive molecules. The position of compounds in chemical reference space is determined by their descriptor vectors (i.e., their coordinates). In general terms, the smaller the distance between two compounds in chemical reference space, the more similar the compounds are. If a compound is active, another one close to it should also have a high probability to be active (provided that the space representation is indeed activity-relevant). Thus, a general approach to identifying novel active compounds would be adding known active reference molecules to chemical space representations into which databases have been projected and selecting candidate compounds that closely map to these references. Therefore, distance functions have been introduced to mine the neighborhood of active reference compounds in more or less complex chemical spaces

[65]. Such distance functions can be applied to produce (distance-based) compound rankings. Binary fingerprints consisting of m features can be rationalized as a special case of an m-dimensional reference space, with each dimension having either unit length or a length of zero.

Chemical space navigation is usually more complicated in high- than in low-dimensional descriptor spaces and, accordingly, the paradigm of low dimensionality has been a cornerstone of activity-oriented compound classification. Low dimensionality of chemical space representations can either be achieved by a priori design of low-dimensional spaces through the use of orthogonal multicomponent descriptors [66] or, alternatively, by dimension reduction of original high-dimensional descriptor spaces. For dimension reduction, statistical methods such as principal component analysis [67] or multidimensional scaling [68] are available. However, it has also been shown that low-dimensional spaces are not essential for effective compound classification and LBVS. For example, mapping algorithms have been introduced to assign database compounds to combinations of descriptor value ranges that are defined by sets of reference molecules in chemical spaces of iteratively increasing dimensionality [69]. This increase is facilitated by adding new descriptors per iteration. Following this approach, database compounds that are most similar to references are identified by increasing the dimensionality and hence chemical resolution of descriptor spaces. Dimension extension is terminated when only a small database selection set remains.

15.7.2 Clustering and Partitioning

Clustering is a classical approach to compound classification for which many different algorithms have been introduced [70]. In LBVS, cluster analysis is carried out by adding active reference compounds to a database and inspecting resulting clusters into which references fall. It is then assumed that compounds in these clusters might be good candidates for hit identification. As a form of unsupervised learning, clustering depends on distance relationships in chemical reference space, as discussed above. Regardless of the specifics of clustering algorithms (e.g., whether they might be hierarchical or nonhierarchical) [70], the clustering process involves the pairwise comparison of compound distances, which generally precludes the applicability of clustering methods to very large compound data sets. As a consequence, as screening databases have substantially grown in size, partitioning methods have become popular for LBVS [71], especially cell-based partitioning, which is based on the low-dimensionality paradigm for chemical space representations [66,71]. The principal difference between clustering and partitioning is that the latter approach does not depend on pairwise compound distance comparisons. Rather, in cell-based partitioning, compounds are assigned to subsections of orthogonal low-dimensional reference space (e.g., descriptor spaces with six to eight dimensions). These sections, called *cells*, are obtained by binning of descriptor axes. Assignment of compounds to such cells is done based on their descriptor coordinates. Candidate compounds are then selected that populate the same cell(s) as reference molecules.

Despite their long history in compound classification, the development of advanced cluster algorithms continues to be an area of active research. Recently, emphasis has been put on combining different clustering approaches or integrating them with other computational methods. For example, clustering has been combined with principal component analysis to add structural diversity criteria to compound selection [72]. Furthermore, cluster analysis has recently been integrated with probabilistic models of activity to incorporate biological activity information in the clustering process [73]. Moreover, clustering has been combined with molecular network analysis to classify active compounds [74]. This approach is also applicable to LBVS.

15.8 MACHINE LEARNING

Machine learning approaches for supervised learning have become increasingly popular for LBVS. In general, learning sets of known active compounds (positive training examples) and inactive compounds (negative training examples) are used to build models of activity for class label prediction. Confirmed inactive compounds might be taken from HTS data, but randomly selected database compounds assumed to be inactive are often used as negative training examples. For LBVS, the most popular machine learning methods include neural networks and self-organizing maps (SOMs) [75–77], kernel methods [78], especially support vector machines (SVMs) [79,80], decision trees [81], and Bayesian methods [82], the latter being probabilistic approaches for activity prediction. The computational architectures of SOMs, decision trees, and SVMs are illustrated in Figure 15.5.

15.8.1 Self-Organizing Maps vs. Decision Trees

Neural networks and decision trees are paradigmatic examples for machine learning approaches that have "black box" character or are interpretable in chemical terms, respectively. This distinction is frequently considered in machine learning. Neural networks for LBVS are derived based on learning sets of compounds and chosen

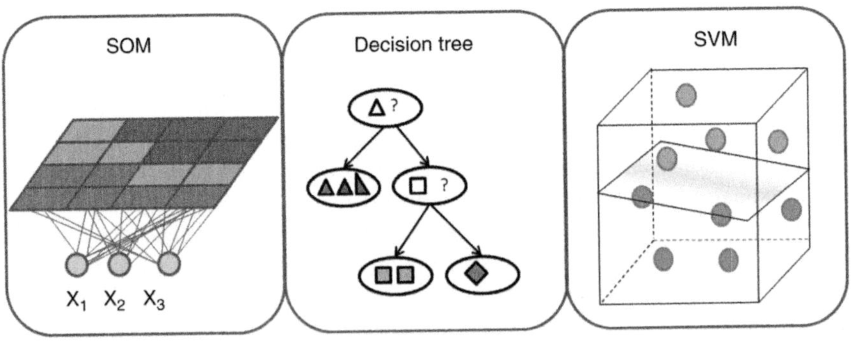

FIGURE 15.5 Representative machine learning approaches. From left to right, SOMs, decision trees, and SVMs are illustrated schematically. (*See insert for color representation of the figure.*)

descriptors. This setting is common for machine learning approaches. Neural network simulations build descriptor-based models for class label prediction by deriving pathways through arrays of computational neurons that best distinguish between positive and negative training examples. Once the model is built, it is used to predict the class label (active versus inactive) of screening database compounds. However, the model does not reveal why a compound might be predicted as active or inactive. This remains hidden, which also applies to SOMs, a special neural network architecture designed to map compounds from descriptor reference spaces onto a two-dimensional neuron grid. The SOM is trained to group positive and negative training examples on distinct regions of the map and separate them from each other. Then, test compounds are projected onto the SOM. Because SOMs start from higher-dimensional descriptor spaces, this approach is also a dimension reduction method. A trained SOM does not reveal why a compound was assigned to an active region in the neuron grid, analogously to other neural nets. By contrast, decision trees separate training compounds along descriptor pathways. Each descriptor represents a decision point to divide a learning set along the tree. Typically, a yes/no decision is made if the presence or absence of a feature is detected, or, alternatively, it is determined whether or not compounds fall into a specific value range of a chosen numerical descriptor. During training, trees are constructed that best separate active and inactive compounds in terminal leaf nodes. A model is derived by recursively partitioning compounds along the tree that yields a meaningful separation. If this is the case, combinations of selected descriptors form pathways that are signatures of a given biological activity, whereas other pathways enrich inactive compounds along the tree. Importantly, pathways in a decision tree are directly interpretable as feature/value range sequences that establish classification rules, different from methods with black box character. The tree structure so derived is then used to screen a database for active compounds. Ensembles of independently derived decision trees, capturing different descriptors and pathways, are often combined to yield random forest models where predictions made independently are subjected to consensus scoring schemes [83,84]. Such random forest models typically further increase the LBVS performance of individual decision trees.

15.8.2 Support Vector Machines

The currently most popular machine learning approaches in LBVS are SVMs and Bayesian classifiers. SVMs represent a class of algorithms to project training sets that are not linearly separable into chemical reference spaces of higher dimensionality where a separating hyperplane can be derived. Thus, SVMs are designed to depart into opposite direction from the paradigm of low dimensionality, which provides the basis for cell-based partitioning, as discussed. In high-dimensional space representations, SVMs construct a maximum-margin hyperplane that yields the largest possible distance to the nearest positive and negative training examples. A key aspect of SVM modeling is that high-dimensional descriptor spaces are not explicitly constituted and mapped. Rather, a kernel function is applied to determine the degree of similarity between compounds in a higher-dimensional representation. For example,

Gaussian or other polynomial kernel functions are often used in LBVS in combination with numerical property descriptors or two-dimensional fingerprints, but simple linear kernels have also been employed successfully [85–87]. The kernel approach is flexible and makes it possible to evaluate the similarity of objects using different features and similarity measures. A variety of kernel functions have been introduced for SVM modeling, including both ligand and target kernels that capture rather different information for similarity assessment [88,89], such as graph or descriptor similarity (compounds) and sequence or binding-site similarity (target proteins). Ligand and target kernel can be combined for model building. For the assessment of compound similarity, a Tanimoto kernel has been introduced [90] that is widely applied in LBVS, given the popularity of the Tanimoto coefficient. A special feature of the SVM approach is that it cannot only be used for class label predictions but also for database ranking [87,91,92]. This is accomplished by calculating the signed distance of test compounds from the separating hyperplane. The underlying idea is that the farther test compounds on the active side are away from the hyperplane (and negative training examples), the greater the likelihood of activity becomes. By contrast, the farther compounds are away from the hyperplane (and positive training examples) on the inactive side, the more likely they are to be inactive. Similarity searching has also been compared directly to SVM classification and ranking. A noteworthy finding has been that SVM calculations using two-dimensional fingerprints as descriptors typically produce higher compound recall and hit rates than similarity searching using these fingerprints [87]. Currently, SVM learning is a preferred approach for LBVS using two-dimensional molecular representations, which often yields better results than other methods. Another kernel-based approach adapted for LBVS is binary kernel discrimination [93,94], which also uses fingerprints as descriptors and combines the kernel function methodology with likelihood estimates of compound feature probability distributions.

15.8.3 Bayesian Methods

Statistical modeling approaches based on Bayes' theorem [82] have become very popular in LBVS. Bayesian modeling methods for activity prediction generally derive numerical probabilities of compound activity based on feature (descriptor value) distributions in training sets. Therefore, Bayesian methods also produce data set rankings. In LBVS, naive Bayesian classifiers using fingerprints as descriptors are widely used [95,96]. Following this approach, it is assumed that all features are independent of each other (representing the naive assumption) and that numerical descriptor values are normally distributed. These assumptions are approximations for chemical descriptor data. Feature probabilities are derived from the frequencies observed for positive and negative training examples. Based on calculated likelihood estimates, a probability of activity is assigned to each test compound based on its descriptor values. Despite the underlying approximations, naive Bayesian classifiers have been proven to be versatile and robust machine learning approaches for compound classification and hit identification. They have been derived for many different compound activity classes and have also been used for *in silico* profiling of compounds against arrays of targets. Extensions of Bayesian classifiers have been introduced for LBVS such as Bayesian (inference) networks [97,98]. In these networks, nodes are random Bayesian variables that are connected by edges if they are involved in conditional dependency relationships. Such networks represent a more complex form of Bayesian classification than naive classifiers, but follow essentially the same principles. However, networks might not always offer advantages compared to simpler similarity methods in LBVS (as also observed for other VS approaches of increasing complexity, as discussed above). Furthermore, Bayesian screening has been combined with concepts from information theory to predict recall rates of LBVS calculations for a given system setup, that is, a chosen molecular representation (such as a fingerprint), a set of active reference compounds, and a particular screening database [99,100]. This type of analysis can help to identify preferred descriptors for practical LBVS applications targeting a specific biological activity, which are most likely to distinguish active compounds from other database molecules. Finally, attempts have recently also been made to combine outcomes of different machine learning approaches, such as neural nets, decision forests, SVMs, and naive Bayesian classifiers into meta-classifiers for LBVS [101,102].

15.9 CONCLUSIONS

In this chapter, an overview over contemporary VS methods has been provided. Initially, VS has been positioned relative to HTS in the context of drug discovery research. Then, fundamental VS concepts have been introduced that should be taken into consideration, regardless of the methods that might be used. The methodological spectrum of VS approaches has been described. Furthermore, different requirements for SBVS and LBVS have been discussed, and a brief account of ligand docking as the primary SBVS approach has been provided. Moreover, the critically important role of chemical reference space design and molecular similarity assessment for LBVS has been rationalized and intrinsic limitations have been discussed. Emphasis has been put on highlighting the conceptual diversity of LBVS that is methodologically much more heterogeneous than SBVS and covers approaches of greatly varying design and complexity. The LBVS field can essentially be divided into similarity search and compound classification methods. It has been pointed out that machine learning plays an increasingly important role for LBVS. Taken together, pharmacophore searching, SVMs, and Bayesian methods currently dominate the LBVS field. However, there is no single approach that is superior to others, and LBVS/SBVS success strongly depends on the compound classes, biological activities, and targets that are under investigation. LBVS and SBVS calculations are often carried out in a sequential (and sometimes parallel) manner, especially to limit the number of test compounds for computationally demanding docking calculations. However, truly integrated LBVS and SBVS approaches that incorporate molecular similarity assessment into the computational analysis of ligand–target interactions are yet to be reported, which provides interesting opportunities for future research and method development activities.

Acknowledgment

The author is grateful to Dagmar Stumpfe for review of the chapter and help with illustrations.

REFERENCES

1. P. Ripphausen, B. Nisius, L. Peltason, J. Bajorath, *J. Med. Chem.* **2010**, *53*, 8461–8467.
2. D. Horvath, *J. Med. Chem.* **1997**, *40*, 2412–2423.
3. M. F. M. Engels, P. Venkatarangan, *Curr. Opin. Drug Discov. Dev.* **2001**, *4*, 275–283.
4. J. Bajorath, *Nat. Rev. Drug Discov.* **2002**, *1*, 882–894.
5. T. Polgar, G. M. Keseru, *Comb. Chem. High Throughput Screen.* **2011**, *14*, 889–897.
6. D. Stumpfe, P. Ripphausen, J. Bajorath, *Future Med. Chem.* **2012**, *4*, 593–602.
7. D. Stumpfe, J. Bajorath, in: *Methods and Principles in Medicinal Chemistry. Virtual Screening. Principles, Challenges, and Practical Guidelines*, C. Sotriffer, Ed., Wiley-VCH, Weinheim, Germany, **2011**, pp. 73–103.
8. P. Ripphausen, B. Nisius, J. Bajorath, *Drug Discov. Today* **2011**, *16*, 372–376.
9. P. Ripphausen, D. Stumpfe, J. Bajorath, *Future Med. Chem.* **2012**, *4*, 603–613.
10. W. P. Walters, M. T. Stahl, M. A. Murcko, *Drug Discov. Today* **1998**, *3*, 160–178.
11. T. Langer, R. D. Hoffmann, *Curr. Pharm. Des.* **2001**, *7*, 509–527.
12. H. Xu, D. K. Agrafiotis, *Curr. Top. Med. Chem.* **2002**, *2*, 1305–1320.
13. I. Muegge, *Mini Rev. Med. Chem.* **2008**, *8*, 927–933.
14. B. K. Shoichet, *Nature* **2004**, *432*, 862–865.
15. D. B. Kitchen, H. Decornez, J. R. Furr, J. Bajorath, *Nat. Rev. Drug Discov.* **2004**, *3*, 935–949.
16. F. L. Stahura, J. Bajorath, *Curr. Pharm. Design* **2005**, *11*, 1189–1202.
17. H. Geppert, M. Vogt, J. Bajorath, *J. Chem. Inf. Model.* **2010**, *50*, 205–216.
18. Y. Hu, D. Stumpfe, J. Bajorath, *J. Chem. Inf. Model.* **2011**, *51*, 1742–1753.
19. G. Schneider, W. Neidhart, T. Giller, G. Schmid, *Angew. Chem. Int. Ed.* **1999**, *38*, 2894–2896.
20. N. Brown, E. Jacoby, *Mini Rev. Med. Chem.* **2006**, *6*, 1217–1229.
21. R. Li, D. Stumpfe, M. Vogt, H. Geppert, J. Bajorath, *J. Chem. Inf. Model.* **2011**, *51*, 2507–2514.
22. M. Vogt, D. Stumpfe, H. Geppert, J. Bajorath, *J. Med. Chem.* **2010**, *53*, 5707–5715.
23. E. J. Gardiner, J. D. Holliday, C. O'Dowd, P. Willett, *Future Med. Chem.* **2011**, *3*, 405–414.
24. K. Heikamp, J. Bajorath, *J. Chem. Inf. Model.* **2011**, *51*, 2254–2265.
25. Y. Wang, H. Eckert, J. Bajorath, *ChemMedChem* **2007**, *2*, 1037–1042.
26. Y. Wang, J. Bajorath, *J. Chem. Inf. Model.* **2008**, *48*, 75–84.
27. H. Eckert, J. Bajorath, *Drug Discov. Today* **2007**, *12*, 225–233.
28. J. Bajorath, *J. Chem. Inf. Comput. Sci.* **2001**, *41*, 233–245.
29. J. S. Mason, A. C. Good, E. J. Martin, *Curr. Pharm. Design* **2001**, *7*, 567–597.

30. E. X. Esposito, A. J. Hopfinger, J. D. Madura, *Methods Mol. Biol.* **2004**, *275*, 131–214.
31. A. Tropsha, A. Golbraikh, *Curr. Pharm. Des.* **2007**, *13*, 3494–3504.
32. M. A. Johnson, G. M. Maggiora, Eds., *Concepts and Applications of Molecular Similarity*, Wiley, New York, **1990**.
33. H. Kubinyi, *Perspect. Drug Discov. Des.* **1998**, *9–11*, 225–252.
34. R. Todeschini, V. Consonni, in: *Methods and Principles in Medicinal Chemistry*, Vol. 41, R. Mannhold, H. Kubinyi, G. Folkers, Eds., Wiley, New York, **2009**.
35. J. D. Holliday, C. Y. Hu, P. Willett, *Comb. Chem. High Throughput Screen.* **2002**, *5*, 155–166.
36. O. F. Guner, *Curr. Top. Med. Chem.* **2002**, *2*, 1321–1332.
37. A. R. Leach, B. K. Shoichet, C. E. Peishoff, *J. Med. Chem.* **2006**, *49*, 5851–5855.
38. N. Brooijmans, I. D. Kuntz, *Annu. Rev. Biophys. Biomol. Struct.* **2003**, *32*, 335–373.
39. D. S. Goodsell, A. J. Olson, *Proteins* **1990**, *8*, 195–202.
40. G. Jones, P. Willet, R. C. Glen, *J. Mol. Biol.* **1995**, *245*, 43–53.
41. M. Rarey, B. Kramer, T. Lengauer, G. Klebe, *J. Mol. Biol.* **1996**, *261*, 470–489.
42. D. T. Moustakas, P. T. Lang, S. Pegg, E. T. Pettersen, I. D. Kuntz, N. Brooijmans, R. C. Rizzo, *J. Comput. Aided Mol. Des.* **2006**, *20*, 601–609.
43. G. Klebe, *Drug Discov. Today* **2006**, *11*, 580–594.
44. I. Nobeli, J. B. O. Mitchell, A. Alex, J. M. Thornton, *J. Comput. Chem.* **2001**, *22*, 673–688.
45. D. B. Kitchen, H. Decornez, J. R. Furr, J. Bajorath, *Nat. Rev. Drug Discov.* **2004**, *3*, 935–949.
46. C. A. Lipinski, F. Lombardo, B. W. Dominy, P. J. Feeney, *Adv. Drug Deliv. Rev.* **1997**, *23*, 3–25.
47. S.-Y. Yang, *Drug Discov. Today.* **2010**, *15*, 444–450.
48. K.-H. Kim, N. D. Kim, B.-L. Seong, *Expert. Opin. Drug Discov.* **2010**, *5*, 205–222.
49. T. S. Rush, J. A. Grant, L. Mosyak, A. Nicholls, *J. Med. Chem.* **2005**, *48*, 1489–1495.
50. J. S. Mason, I. Morize, P. R. Menard, D. L. Cheney, C. Hulme, R. F. Labaudiniere, *J. Med. Chem.* **1999**, *42*, 3251–3264.
51. T. Langer, E. M. Krovat, *Curr. Opin. Drug Discov. Dev.* **2003**, *6*, 370–376.
52. R. E. Carhart, D. H. Smith, R. Venkataraghavan, *J. Chem. Inf. Comput. Sci.* **1985**, *25*, 64–73.
53. J. M. Barnard, G. M. Downs, *J. Chem. Inf. Comput. Sci.* **1997**, *37*, 141–142.
54. C. A. James, D. Weininger, *Daylight Theory Manual*, Daylight Chemical Information Systems, Inc., Irvine, CA, **2009**.
55. A. Bender, H. Y. Mussa, R. C. Glen, S. Reiling, *J. Chem. Inf. Comput. Sci.* **2004**, *44*, 1708–1718.
56. D. Rogers, M. Hahn, *J. Chem. Inf. Model.* **2010**, *50*, 742–754.
57. H. E. A. Ahmed, M. Vogt, J. Bajorath, *J. Chem. Inf. Model.* **2010**, *50*, 487–499.
58. P. Liu, D. K. Agrafiotis, D. N. Rassokhin, *J. Chem. Inf. Model.* **2011**, *51*, 2843–2851.
59. M. Whittle, V. J. Gillet, P. Willett, J. Loesel, *J. Chem. Inf. Model.* **2006**, *46*, 2193–2205.
60. M. Whittle, V. J. Gillet, P. Willett, J. Loesel, *J. Chem. Inf. Model.* **2006**, *46*, 2206–2219.
61. M. Fligner, J. Verducci, P. Blower, *Technometrics* **2002**, *44*, 110–119.

62. H. Eckert, J. Bajorath, *J. Chem. Inf. Model.* **2006**, *46*, 2515–2526.
63. B. Nisius, J. Bajorath, *Chem. Biol. Drug Des.* **2010**, *75*, 152–160.
64. B. Nisius, J. Bajorath, *ChemMedChem* **2010**, *5*, 859–868.
65. J. W. Godden, J. Bajorath, *J. Chem. Inf. Model.* **2006**, *46*, 1094–1097.
66. R. S. Pearlman, K. M. Smith, *Perspect. Drug Discov. Des.* **1998**, *9*, 339–353.
67. L. Xue, J. Bajorath, *J. Chem. Inf. Comput. Sci.* **2000**, *40*, 801–809.
68. D. K. Agrafiotis, D. N. Rassokhin, V. S. Lobanov, *J. Comput. Chem.* **2001**, *22*, 488–500.
69. H. Eckert, I. Vogt, J. Bajorath, *J. Chem. Inf. Model.* **2006**, *46*, 1623–1634.
70. P. Willett, *Similarity and Clustering in Chemical Information Systems*, Research Studies Press, Letchworth, UK, **1987**.
71. F. L. Stahura, J. Bajorath, *Curr. Med. Chem.* **2003**, *8*, 707–715.
72. S. Rännar, P. L. Andersson, *J. Chem. Inf. Model.* **2010**, *50*, 30–36.
73. E. Lounkine, F. Nigsch, J. L. Jenkins, M. Glick, *J. Chem. Inf. Model.* **2011**, *51*, 3158–3168.
74. S. Saito, T. Hirokawa, K. Horimoto, *J. Chem. Inf. Model.* **2010**, *51*, 61–68.
75. J. Gasteiger, A. Teckentrup, L. Terfloth, S. Spycher, *J. Phys. Org. Chem.* **2003**, *16*, 232–245.
76. H. Bauknecht, A. Zell, H. Bayer, P. Levi, M. Wagener, J. Sadowski, J. Gasteiger, *J. Chem. Inf. Comput. Sci.* **1996**, *36*, 1205–1213.
77. A. Teckentrup, H. Briem, J. Gasteiger, *J. Chem. Inf. Model.* **2004**, *44*, 626–634.
78. N. Cristianini, J. Shawe-Taylor, *An Introduction to Support Vector Machines and Other Kernel-Based Learning Methods*, Cambridge University Press, Cambridge, UK, **2000**.
79. C. Cortes, V. Vapnik, *Mach. Learn.* **1995**, *20*, 273–297.
80. R. N. Jorissen, M. K. Gilson, *J. Chem. Inf. Model.* **2005**, *45*, 549–561.
81. A. I. I. I. Rusinko, M. W. Farmen, C. G. Lambert, P. L. Brown, S. S. Young, *J. Chem. Inf. Comput. Sci.* **1999**, *39*, 1017–1026.
82. R. O. Duda, P. E. Hart, D. G. Stork, *Pattern Classification*, 2nd ed., Wiley-Interscience, New York, **2000**, pp. 20–83.
83. D. S. Palmer, N. M. O'Boyle, R. C. Glen, J. B. O. Mitchell, *J. Chem. Inf. Model.* **2007**, *47*, 150–158.
84. T. M. Ehrman, D. J. Barlow, P. J. Hylands, *J. Chem. Inf. Model.* **2007**, *47*, 264–278.
85. Q. Li, A. Bender, J. Pei, L. Lai, *J. Chem. Inf. Model.* **2007**, *47*, 1776–1786.
86. C. Y. Liew, X. H. Ma, X. Liu, C. W. Yap, *J. Chem. Inf. Model.* **2009**, *49*, 877–885.
87. H. Geppert, T. Horváth, T. Gärtner, S. Wrobel, J. Bajorath, *J. Chem. Inf. Model.* **2008**, *48*, 742–746.
88. L. Jacob, J.-P. Vert, *Bioinformatics* **2008**, *24*, 2149–2156.
89. A. M. Wassermann, H. Geppert, J. Bajorath, *J. Chem. Inf. Model.* **2009**, *49*, 2155–2167.
90. L. Ralaivola, S. J. Swamidass, H. Saigo, P. Baldi, *Neural. Netw.* **2005**, *18*, 1093–1110.
91. A. Wassermann, H. Geppert, J. Bajorath, *J. Chem. Inf. Model.* **2009**, *49*, 582–592.
92. S. Agarwal, D. Dugar, S. Sengupta, *J. Chem. Inf. Model.* **2010**, *50*, 716–731.
93. G. Harper, J. Bradshaw, J. C. Gittins, D. V. S. Green, A. R. Leach, *J. Chem. Inf. Comput. Sci.* **2001**, *41*, 1295–1300.

94. D. J. Wilton, R. F. Harrison, P. Willett, J. Delaney, K. Lawson, G. Mullier, *J. Chem. Inf. Model.* **2006**, *46*, 471–477.

95. T. Ewing, J. C. Baber, M. Feher, *J. Chem. Inf. Model.* **2006**, *46*, 2423–2431.

96. A. E. Klon, D. J. Diller, *J. Chem. Inf. Model.* **2007**, *47*, 1354–1365.

97. A. Abdo, B. Chen, C. Mueller, N. Salim, P. Willett, *J. Chem. Inf. Model.* **2010**, *50*, 1012–1020.

98. A. Abdo, N. Salim, *J. Chem. Inf. Model.* **2010**, *51*, 25–32.

99. M. Vogt, J. Bajorath, *J. Chem. Inf. Model.* **2007**, *47*, 337–341.

100. M. Vogt, J. Bajorath, *ChemMedChem* **2007**, *2*, 1311–1320.

101. C. Kramer, B. Beck, T. Clark, *J. Chem. Inf. Model.* **2010**, *50*, 404–414.

102. F. Cheng, Y. Yu, J. Shen, L. Yang, W. Li, G. Liu, P. W. Lee, Y. Tang, *J. Chem. Inf. Model.* **2011**, *51*, 996–1011.

103. P. Ripphausen, A. M. Wassermann, J. Bajorath, *J. Chem. Inf. Model.* **2011**, *51*, 2467–2473.

104. J. J. Irwin, B. K. Shoichet, *J. Chem. Inf. Model.* **2005**, *45*, 177–182.

16

STRUCTURE–ACTIVITY RELATIONSHIP DATA ANALYSIS: ACTIVITY LANDSCAPES AND ACTIVITY CLIFFS

JÜRGEN BAJORATH

16.1 INTRODUCTION

In this chapter, the concepts of activity landscapes [1] and activity cliffs [1,2] are discussed. An *activity landscape* is generally defined as "any graphical representation that integrates similarity and potency relationships between compounds having a specific biological activity" [1]. Furthermore, *activity cliffs* are "formed by pairs (or groups) of structurally similar or analogous compounds having large differences in potency" and thought to be the most prominent features of activity landscapes [2]. In chemoinformatics and medicinal chemistry, the study of activity landscapes and cliffs has become increasingly popular in recent years, as highlighted, for example, by a symposium at a national meeting of the American Chemical Society in 2010 [3]. This popularity is due primarily to the ability of the activity landscape approach to analyze structure–activity relationships (SARs) systematically on a large scale and extract available SAR information from large and heterogeneous compound data sets [1]. In addition, the activity cliff concept is chemically intuitive and has become a trendy term in the literature (occasionally used in a rather loose manner and without clear scientific specifications; see below).

The computational study of activity landscapes is a still evolving area of research, but a number of alternative landscape representations have already been reported. For

Diversity-Oriented Synthesis: Basics and Applications in Organic Synthesis, Drug Discovery, and Chemical Biology, First Edition. Edited by Andrea Trabocchi.

reasons mentioned above (and others discussed in the following), the analysis of different landscape views is attractive for medicinal chemistry applications. However, the choice of molecular representations and the assessment of compound similarity, which are important factors for activity landscape quality and interpretability (as discussed further below), present significant challenges for computational modeling. The generation of activity landscapes is a topic of research at the interface of chemoinformatics and medicinal chemistry, and so is the large-scale analysis of SARs, which is highly relevant for medicinal chemistry but would not be possible without specialized computational methods. Accordingly, the following discussion is divided into three sections. In the first section, numerical SAR analysis functions are introduced that depart from classical QSAR formalisms [4], are developed specifically to capture quantitative SAR information contained in large compound data sets and are also of central relevance for activity landscape modeling. In the second section, alternative activity landscape representations are discussed that are based on different design philosophies and have different information content. In the third section, the discussion focuses on activity cliffs, and potential applications of the activity cliff concept are critically evaluated.

16.2 NUMERICAL SAR ANALYSIS FUNCTIONS

Different types of SAR analysis function have been introduced [5]. However, these numerical functions generally have in common that they systematically compare chemical similarity and potency of specifically active compounds in a pairwise manner. It is important to note that the assessment of compound similarity plays a role here that differs from other applications of similarity analysis in the chemoinformatics field [6]. In many of these applications, molecular similarity is quantified as a measure of activity similarity, although no well-defined quantitative relationship exists between calculated similarity values and biological activity [6]. However, molecular similarity calculated as a putative measure of activity similarity follows the similarity property principle stating that "similar compounds should have similar bioactivity" [7]. This principle is a central paradigm in chemoinformatics. For numerical SAR analysis functions, similarity evaluation plays a different role because these functions also explicitly involve the comparison of potency values. Thus, for this purpose, utilizing reliable and robust representations and measures of structural similarity/relatedness (see below) is much more important than finding representations that are particularly relevant for a given bioactivity. In fact, consistent use of chosen molecular representations and similarity measures is an essential condition for the comparability of quantitative SAR assessment across different compound classes.

16.2.1 Structural Similarity vs. Activity Similarity

A prototypic SAR analysis function provided the basis for the generation of structure–activity similarity (SAS) maps [8], which represent the first activity landscape representation as defined herein, (discussed) further below. Structural similarity and activity similarity of data set compounds are compared in a pairwise manner, and activity similarity is defined for a pair of compounds i and j as follows:

$$\mathrm{sim}_{\mathrm{act}}(i,j) = 1 - \frac{|P_i - P_j|}{P_{\max} - P_{\min}} \tag{16.1}$$

In this equation, P_i and P_j are experimental potency values of compounds i and j, respectively, and $P_{\max} - P_{\min}$ is the difference between the maximum and minimum potency occurring in the compound data set. Hence, for each possible compound pair, a normalized potency difference is obtained as well as a calculated pairwise similarity value, typically Tanimoto similarity based on molecular fingerprints, one of the standard similarity assessments in chemoinformatics [6]. Albeit not formally integrated in an equation, the combined evaluation of molecular and activity similarity underlying SAS maps represents the essence of an SAR analysis function.

16.2.2 SAR Index

The SAR index (SARI) [9] consists of separately calculated continuity and discontinuity scores. The "raw" continuity score is derived as the potency weighted mean of pairwise compound dissimilarity within a data set A. This continuity score highly weights structurally diverse compounds with high potency and small potency differences:

$$\mathrm{cont}_{\mathrm{raw}}(A) = \underset{\{(i,j)\in A|i\neq j\}}{\text{weighted mean}} \left(\frac{1}{1+\mathrm{sim}(i,j)}\right) = \frac{\sum_{\{(i,j)\in A|i\neq j\}} \left(\mathrm{weight}(i,j)\cdot \frac{1}{1+\mathrm{sim}(i,j)}\right)}{\sum_{\{(i,j)\in A|i\neq j\}} \mathrm{weight}(i,j)}$$

$$\mathrm{weight}(i,j) = \frac{P_i \cdot P_j}{1+|P_i - P_j|} \tag{16.2}$$

P stands for potency and sim(i,j) for the similarity of compounds i and j (also calculated as Tanimoto fingerprint similarity).

In addition, the raw discontinuity score is derived as the average pairwise potency difference between compounds multiplied by pairwise similarity:

$$\mathrm{disc}_{\mathrm{raw}}(A) = \underset{\left\{\substack{(i,j)\in A|\mathrm{sim}(i,j)>T,\\ |P_i-P_j|>1}\right\}}{\text{mean}} (|P_i - P_j| \cdot \mathrm{sim}(i,j)) \tag{16.3}$$

This discontinuity score emphasizes structurally similar compounds with large potency differences. Because the discontinuity score should account for the presence of activity cliffs, only pairs of similar compounds (above a predefined similarity threshold T) with at least one order of magnitude difference in potency are taken into account for dissimilarity score calculation.

The raw continuity and discontinuity scores are converted into Z-scores, which are then mapped onto the value range from 0 to 1 by calculating their cumulative probabilities with respect to an assumed normal value distribution. The resulting normalized continuity and discontinuity scores are combined to yield the SARI score that balances their contributions:

$$\mathrm{SARI}(A) = \frac{1}{2}\left\{\mathrm{cont}_{\mathrm{norm}}(A) + [1 - \mathrm{disc}_{\mathrm{norm}}(A)]\right\} \tag{16.4}$$

This scoring scheme was originally designed to globally classify SAR phenotypes of compound data sets into three categories: continuous (high SARI scores), discontinuous (low scores), or heterogeneous SARs (intermediate scores around 0.5). Global SAR heterogeneity might result from combinations of locally continuous and discontinuous SARs (represented by different compound subsets/series) or from the presence of SAR continuity near activity cliffs. The latter situation occurs when structural modifications of active compounds are possible without significant changes in potency as long as critically important R-groups are conserved that satisfy essential binding interactions (e.g., a functional group in a series of inhibitors that complexes an ion in the active site of an enzyme).

16.2.3 Per-Compound Discontinuity Score

A variant of the SARI discontinuity score has been introduced to capture the contributions of individual compounds to SAR discontinuity [10]. A compound introduces significant SAR discontinuity if its potency differs dramatically from that of its immediate structural neighbors. The per-compound discontinuity score is defined as follows:

$$\mathrm{disc}(i) = \frac{\sum\limits_{\{j \mid \mathrm{sim}(i,j)>0.65,\, i\neq j\}} \mathrm{potdiff}(i,j) \times \mathrm{sim}(i,j)}{|\{j \mid \mathrm{sim}(i,j) > 0.65,\, i \neq j\}|} \tag{16.5}$$

Here potdiff is the potency difference between a pair of compounds and sim is their calculated similarity. The raw score is obtained by comparing a compound to all other molecules that are more similar to it than the predefined threshold T (here a fingerprint Tanimoto similarity greater than 0.65). These scores are normalized by using the individual scores of all compounds in the data set for Z-score calculation [10]. According to this formalism, pairs of structurally very similar compounds with large differences in potency obtain a per-compound score close to 1 and mark prominent activity cliffs.

16.2.4 Structure–Activity Landscape Index

The structure–activity landscape index (SALI) score [11] is defined as

$$\mathrm{SALI}(i,j) = \frac{P_i - P_j}{1 - \mathrm{sim}(i,j)} \tag{16.6}$$

In this formula, P means potency and sim represents similarity. Thus, SALI and the per-compound discontinuity score are analogous formulations. However, the SALI score is not normalized and therefore has an infinite value range. It is designed to identify activity cliffs in a data set and provides the basis for the generation of SALI graphs [11], an activity cliff-centric activity landscape model, as discussed below.

Considering the SAR analysis functions above, an important distinction to be made is that the formalism underlying SAS maps and SARI are global in nature; that is, these functions capture the SAR information content of an entire data set, whereas the per-compound discontinuity score and SALI are local functions that focus primarily on the identification of local SAR discontinuity and activity cliffs (see below). It should also be noted that local functions cannot account for SAR heterogeneity in a compound set because the detection of SAR heterogeneity requires consideration of multiple compound subsets and hence the application of a global scoring scheme.

16.3 PRINCIPLES AND INTRINSIC LIMITATIONS OF ACTIVITY LANDSCAPE DESIGN

General characteristics of activity landscapes include that their design is data oriented (especially tailored toward large compound data sets) and descriptive in nature rather than predictive. Thus, activity landscape views should reveal SAR patterns in compound data and provide an intuitive access to available SAR information. Accordingly, landscape views should graphically represent and complement the results of numerical SAR analysis. In addition, landscape models have also been introduced that do not incorporate numerical SAR analysis schemes. Regardless, an important aspect of activity landscapes is that they are not designed to facilitate QSAR-like or binary activity predictions; rather, their focus is on data representation and knowledge extraction. Accordingly, most activity landscape representations do not impose a predefined SAR model on compound activity data.

If we adhere to the general definition of an activity landscape, as presented above, we need to be concerned about three major aspects. These include the derivation of similarity and activity/potency relationships and the integration of these relationships within a graphical framework. This leads to a number of considerations concerning chemical references spaces, the way molecular similarity is evaluated, and the assessment and selection of experimental measurements. The latter aspect will also be of particular importance for the evaluation of activity cliff distributions, as discussed further below.

16.3.1 Chemical Reference Space

The design of chemical references spaces is critical for many chemoinformatics applications, including compound activity predictions [6], and this also applies to activity landscape modeling. However, as mentioned above, here the requirements differ from space design for virtual screening and activity prediction. This is the case because a prototypic activity landscape is formed by compounds sharing the same specific activity but having different potencies. Hence, chemical reference space is primarily not required to account for bioactivity-specific features, but must reflect different degrees of molecular similarity (structural relatedness). However designed, a chemical space representation forms a foundation of activity landscape generation, as distances between compounds in chemical space must reflect pairwise similarity relationships in a meaningful manner. Chemical space representations of different complexity might be chosen. For example, if we consider the calculation of fingerprint distances, which are utilized, for example, in the context of the SAS map, SARI, and SALI, "coordinate-free" reference spaces of relatively low complexity are obtained that are defined solely by all pairwise compound distances in a given data set. In principle, any chemical reference space is transformed into an activity landscape by adding a biological "hypersurface" to it that accounts for compound potency information. Depending on the nature of the landscape design, this can be accomplished in different ways: for example, by calculating activity similarity from potency values. However, the well-known variance of many chemical space representations represents the most critical variable in activity landscape design.

16.3.2 Similarity Assessment

Activity landscape methods typically rely on calculated whole-molecule similarity, which is greatly influenced by chosen molecular (chemical space) representations. When different descriptors such as binary fingerprints or numerical molecular property descriptors are used, pairwise compound similarity or dissimilarity (distance) relationships usually change, often in a significant manner. Because activity landscape modeling involves exhaustive enumeration of pairwise compound similarity/dissimilarity relationships, even moderate changes in pairwise relationships might distort landscapes. Ensuing changes in activity landscape topology inevitably alter the SAR information content of alternative landscape representations of a given compound data set. For example, the higher the chemical "resolution" of chosen descriptors, the more dissimilar compounds will on average appear to be, which influences landscape topology in a systematic manner. The representation dependence of similarity/dissimilarity value distributions must be taken into account when analyzing and comparing activity landscape models (see below). The choice of molecular representations typically has a greater influence on the information content and interpretability of activity landscape models than alternative similarity metrics. As is the case for chemical reference spaces in chemoinformatics, there are no generally preferred solutions. However, because in medicinal chemistry SAR information is ultimately evaluated on the basis of two-dimensional molecular graphs, graph-based measures of structural similarity such as fragment or topological fingerprints [6] often yield robust landscape views that are interpretable from a chemical perspective [1]. For any chosen molecular descriptors, it is essential for activity landscape analysis that calculated similarity values are chemically intuitive and readily interpretable, as discussed in the following.

16.4 ACTIVITY LANDSCAPE REPRESENTATIONS

In the following, activity landscape models of different design are introduced. Currently available activity landscapes essentially can be divided into "normal" graphs, molecular networks, and three-dimensional models. Different types of activity landscapes are displayed in Figure 16.1.

16.4.1 Three-Dimensional Models

Probably the most intuitive way to rationalize an activity landscape is to envision the addition of the biological hypersurface as a third dimension to a two-dimensional projection of chemical space. Then, activity landscape views become rather similar to geographical landscapes, and their topology reminds us of plains, mountains, valleys, and so on. The attractiveness of such idealized (theoretical) three-dimensional landscapes for the discussion and rationalization of SAR features had been recognized long before comparable modes were derived for actual data sets [12]. In an idealized three-dimensional model, hypothetical potency information is represented as a contiguous surface without indicating the positions of individual compounds (which are located in the two-dimensional projection of chemical space). In such three-dimensional models, gently sloped areas represent regions of local SAR continuity where gradual changes in chemical structure are accompanied by small or moderate changes in potency. In regions of SAR continuity, scaffold hopping [13] can occur; that is, a transition can be facilitated from one chemotype to another without a dramatic change in potency. It should be noted that for compound subsets forming regions of local SAR continuity, it might be possible to build linear models of activity as a function of structural modifications, consistent with the classical QSAR paradigm.

By contrast, rugged areas represent regions of local SAR discontinuity where small changes in chemical structure lead to large changes in potency. In these regions, large peaks represent activity cliffs, the extreme form of SAR discontinuity. In Figure 16.1, an idealized "variable" activity landscape is shown that combines smooth and rugged regions and hence corresponds to global SAR heterogeneity.

Equivalent three-dimensional models can also be built on the basis of "real data," also shown in Figure 16.1. Despite the intuitive nature and attractiveness of theoretical three-dimensional landscape views, the first real data models have only recently been introduced [14]. For various compound data sets, coordinate-free chemical reference spaces formed by pairwise fingerprint distances were generated and projected onto the x–y plane of a coordinate system through multidimensional scaling [15]. From sparsely distributed potency values of test compounds, a contiguous surface was generated along the z-axis through the application of interpolation functions. The potency surface is color-coded according to surface elevation, applying a continuous color spectrum (from green to red). Thus, prominent activity cliffs appear as red peaks (Figure 16.1). White regions correspond to interpolated surface area where compound data are lacking. These regions can be regarded as "SAR holes" for a given data set. Similar to idealized three-dimensional landscape views, compound data-based

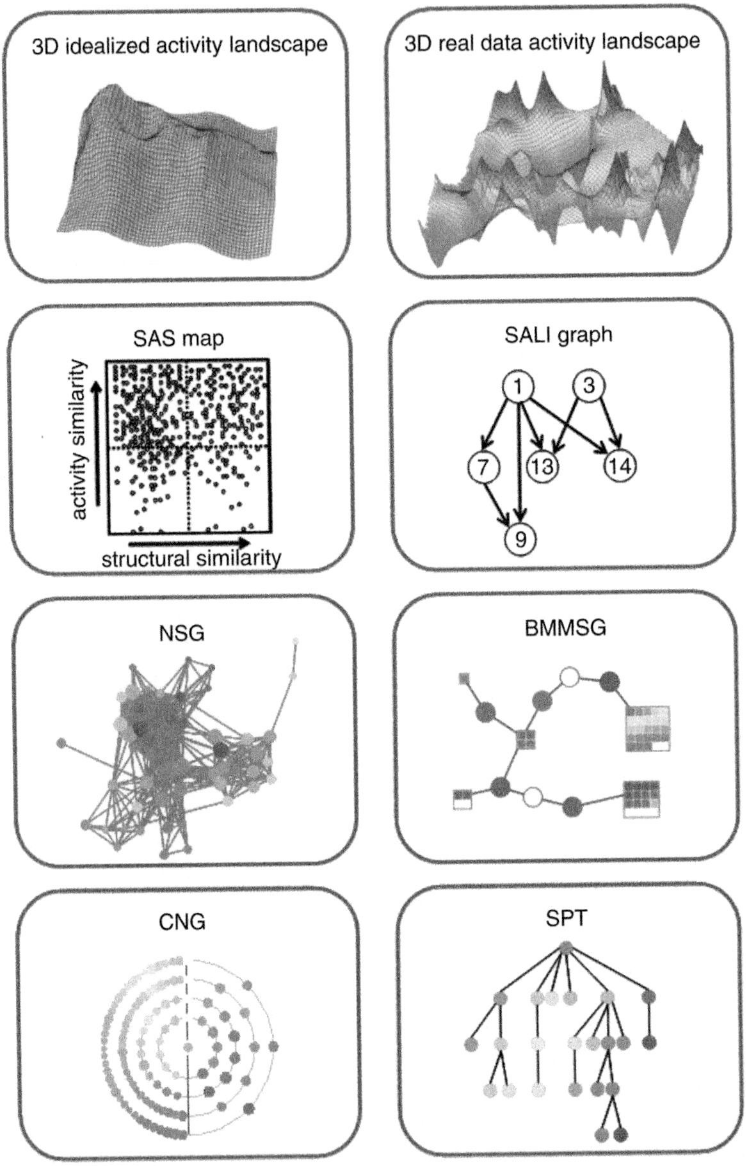

FIGURE 16.1 Activity landscape representations. Alternative activity landscape views are shown. Details are discussed in the text. For the SALI graph, NSG, and BMMSG, small representative sub-graphs are displayed that are taken from larger graphs of compound data sets. (*See insert for color representation of the figure.*)

three-dimensional models also reveal characteristic topologies that correspond to specific SAR features of compound data sets, in analogy to idealized models. Three-dimensional models are designed primarily to provide global views of SAR features. In addition, these three-dimensional models are particularly useful as a diagnostic tool to illustrate the in-part dramatic dependence of landscape topology on chosen molecular representations [14]. Of course, three-dimensional landscapes might also be rationalized as nonlinear QSAR-type model and hence could be considered for activity prediction. However, in our experience, interpolation of larger surfaces from sparsely distributed potency data is in general not sufficiently accurate to predict compound potencies, in particular, in activity cliff regions.

16.4.2 SAS Maps

SAS maps, first reported by Shanmugasundaram and Maggiora in 2001 [8], are a prototype of graph-based activity landscape models. In SAS maps, systematically computed pairwise structural and activity similarity (see above) are plotted against each other such that each data point reports a pairwise compound comparison (Figure 16.1). These data points are then color-coded according to the potency value of the more active compound of each pair. This data representation can be modified in different ways. For example, compound potency differences or the sum of compound potency values might be utilized instead of calculated activity similarity [9]. A SAS map is best rationalized to consist of four sections that capture different SAR characteristics (Figure 16.1). The upper-left section contains compound pairs with high activity similarity and low structural similarity and hence corresponds to a scaffold hopping region. The upper-right section contains compounds with high structural and high activity similarity such as analogs with comparable potency. The lower-left section is characterized by the presence of low structural and low activity similarity and is thus not very informative from an SAR perspective. By contrast, compound pairs falling into the lower-right section have high structural similarity and low activity similarity. Accordingly, this section represents the activity cliff region of an SAS map.

Yongye et al. have introduced a number of modifications and extensions of the original SAS map. For example, consensus SAS maps have been generated by calculating mean similarity values for different two- and three-dimensional fingerprints [16] in order to balance the dependence of landscape topology on individual molecular representations. Another variant utilizes logarithmic potency differences against two or three targets for compound pairs instead of single-target activity similarity, thereby producing two-dimensional dual activity difference (DAD) [17] or three-dimensional triple activity difference (TAD) maps [18]. In these cases, dots representing compound pairs are color-coded by fingerprint similarity. DAD maps follow the ideas underlying the design of selectivity landscapes [19] and TAD maps the principle of multitarget activity landscapes [20], another recent extension of the activity landscape concept (see below), which is also addressed through the introduction of structure multiple activity similarity (SmAS) maps and the structure multiple activity landscape index (SmALI), a variant of the SALI formalism [21]. In this case, activity similarity is obtained through calculations of Tanimoto similarity of real-valued vectors containing compound potency values against three targets. In analogy, for the computation of SALI scores, this multitarget activity similarity replaces single-target potency differences for each compound pair.

16.4.3 Molecular Networks

Recent activity landscape design efforts have also focused on molecular network representations. In general, in these types of landscapes, nodes represent compounds and edges pairwise compound similarity relationships or, alternatively, activity cliffs.

16.4.3.1 SALI Graph The SALI graph is based on the SALI formalism and is an activity cliff-centric representation of an activity landscape (Figure 16.1) [11]. Here, edges indicate activity cliffs and are drawn as an arrow from the less to the more potent compound of a pair. Two compounds are connected by a directed edge if their SALI score exceeds a given threshold value, for example, a score greater than 60, 70, or 80% of all pairwise scores in a data set. By applying different threshold values, networks of activity cliffs of increasing magnitude are obtained. The SALI formalism can also be applied for the assessment of computational SAR models [22]. For this purpose, it is determined how many edges in a SALI graph are predicted correctly by a given computational model. A SALI curve is calculated that reports the percentage of correctly directed edges as a function of the SALI threshold values applied to generate the SALI graphs.

16.4.3.2 Network-like Similarity Graph The network-like similarity graph (NSG) design represents a prototype of a similarity-based compound network as an activity landscape (Figure 16.1) [10]. The major characteristic of this landscape representation is its ability to reveal relationships between global and local SAR features in large data sets. In this network, nodes represent compounds that are color-coded according to their potency values and scaled in size according to the per-compound discontinuity score (as discussed above) that quantifies the individual contribution of each molecule to local SAR discontinuity. In addition, edges represent pairwise Tanimoto similarity relationships. An edge is drawn between two nodes if their similarity value exceeds a predefined threshold. The NSG can be annotated with additional information, for example, results of cluster analysis and compound cluster discontinuity scores to characterize compound subset SARs [10]. In an NSG, combinations of large red (highest potency in a data set) and green (lowest potency) nodes connected by edges mark the most significant activity cliffs in the set. In addition, NSGs identify different local continuous or discontinuous SAR environments. Their combination provides a view of the global SAR character of a compound data set, which can also be compared with global SARI scoring. NSG analysis is also applicable to SAR mining of high-throughput screening (HTS) data [23].

Several extensions of the NSG concept have been introduced. For example, compound potency measurements have been combined with molecular mechanism-of-action information for receptor ligands by modifying the node color code [24]. In this case, primary node colors indicate a specific mechanism of action (e.g., agonists, partial agonists, inverse agonists, or antagonists) and different potency values are reflected by color transparency levels. This NSG variant makes it possible to select subsets of similar compounds from regions of apparent mechanistic heterogeneity and study structural modifications that lead to mechanism hopping [24].

Furthermore, the NSG formalism has been extended to derive selectivity landscapes for compounds active against pairs of targets [19]. Here, potency ratios (logarithmic potency differences) are utilized instead of compound potency values. The topology of the single-target potency-based NSGs and a dual-target selectivity-based NSG is conserved because it is determined only by compound similarity relationships. For the representation of selectivity landscapes, per-compound discontinuity scores are also calculated on the basis of potency ratios. Hence, potency-and selectivity-based NSGs can be compared for a given compound data set, and different types of relationships between single-target activity cliffs and selectivity cliffs can be studied.

A logical extension of selectivity landscapes for target pairs is the design of multitarget activity landscapes for compound sets with activity against three or more targets. Conceptually, this represents a more difficult problem than transforming activity landscapes into selectivity landscapes because "vectors" of potency information for multiple targets must be accounted for and compared in a consistent manner. A first multitarget landscape representation has recently been introduced [20]. For this purpose, an NSG variant was designed in which a ternary potency code was assigned to each compound node on the basis of a potency binning scheme, that is, by classifying potency values against a given target as highly ("2"), moderately ("1"), or weakly ("0") potent. For node scaling, a compound discontinuity score was calculated over all targets to account for the introduction of SAR discontinuity in multitarget space. Furthermore, a modified color code was applied to nodes indicating selected potency profiles. On the basis of this landscape representation, subsets of compounds can be prioritized for further study that are characterized by a high degree of SAR discontinuity in multitarget space and that form multitarget activity cliffs. Subsequently, an alternative multitarget activity landscape design was reported where compounds were encoded as arrays of pairwise target potency relationships and classified in self-organizing maps [25]. Furthermore, extensions of SAS maps have also been introduced to represent multitarget SARs (see above) [18,21]. For example, pairwise compound activity similarity over multiple targets has been calculated as Tanimoto similarity of real-valued potency vectors [21].

16.4.3.3 Bipartite Matching Molecular Series Graph A general caveat for the interpretability of activity landscape representations in medicinal chemistry is the use of calculated (Tanimoto) similarities. This is the case because structural relationships that are based on calculated whole-molecule similarity values are often not easy to interpret in chemical terms. Accordingly, a molecular network-based activity landscape design has recently been introduced in which calculated similarities are replaced with well-defined and immediately interpretable substructure relationships between active compounds [26]. This activity landscape is termed a bipartite matching molecular series graph (BMMSG; Figure 16.1). Substructure relationships are determined systematically on the basis of the matched molecular pair (MMP) formalism [27]. An MMP is defined as a pair of compounds that are distinguished only by the exchange of a single substructure (which might, for example, be an R-group or a ring). Following this new landscape idea, all compounds in a data set that only differ by a single substructure at a specific site are classified as a matching molecular series (MMS). The BMMSG network is bipartite, that is, it contains two different types of nodes. There are molecule nodes representing individual compounds (colored by potency in analogy to NSGs) and set nodes (white) representing the invariant substructure of an MMS. Molecule nodes are connected to a set node by an edge if they belong to the MMS represented by the set node. Arrays of molecule nodes connected to a set node can be represented as "supernodes" such that individual compounds are indicated as colored squares within a large set node square (Figure 16.1), which reduces the complexity of the graph. Characteristic node patterns emerging from a BMMSG include, for example, SAR hotspots, which are supernodes displaying a clear potency progression. Molecule and set nodes in BMMSGs are associated with compound and invariant fragment structures, respectively, and edges with exchanged substructures, which further aids in the chemical interpretation of the network. A characteristic feature of BMMSGs is that compound data sets typically yield series of disjoint subgraphs containing compound subsets with substructure relationships that are distinct from others. Given its focus on structural interpretability, the BMMSG data structure has also been adopted to rationalize the structural basis of mechanism hopping in analogy to molecular mechanism-based NSGs [28].

16.4.4 Compound-Centric Activity Landscape Views

The activity landscapes discussed thus far represent complete data sets and provide global or global vs. local views of SAR features. However, this is not a prerequisite for landscape representation. It is also possible—and meaningful for many applications—to generate local landscape views that are focused on individual compounds and their chemical neighborhood or, alternatively, on series of analogs.

16.4.4.1 Similarity-Potency Tree The similarity-potency tree (SPT) data structure [29] (Figure 16.1) captures the immediate SAR environment of a chosen reference compound, which is used as a root node of a tree structures in which edges connect nearest neighbors. This means that an edge is drawn between a compound and the compound to which it is most similar in the data set. Compounds are represented as nodes and colored by potency. The radius of the structural environment around the reference compound is determined by a predefined Tanimoto similarity threshold value and the structural similarity of compounds relative to the root deceases along the tree. Horizontal and vertical node patterns with well-defined potency progression indicate the presence of interpretable SAR information. SPTs are generated systematically for all data set compounds (i.e., a data set comprising n compounds yields n trees) and ranked by quantifying available interpretable patterns [29]. If a data set contains SAR information, it can usually be appreciated by inspecting a limited number of highly ranked trees, which often capture overlapping local SAR environments that emphasize SAR hotspots. Thus, this compound-centric activity landscape design represents a versatile data structure for local SAR analysis and SAR data mining.

Global and local SAR analysis using activity landscape models can also be combined in an effective manner. This has been accomplished, for example, through NSG-SPT analysis [30]. Here, the basic idea is to initially monitor a data set globally using an NSG representation, select from it the most interesting local SAR environments (compound subsets), and then inspect them at high resolution in series of SPTs where each subset compound is used once as a reference. NSG-SPT analysis has been applied successfully to extract SAR information from a large and noisy phenotypic screening data set with more than 13,000 potential antimalarial hits [30].

16.4.4.2 Chemical Neighborhood Graph Another compound-centric activity landscape representation is provided by a chemical neighborhood graph (CNG) [31], shown in Figure 16.1. In this case, the neighborhood of a reference compound is also defined as the subset of compounds that exceed a predefined similarity threshold relative to the reference. However, the graphical analysis scheme is distinct from the NSG or SPT designs. Different from molecular network-based activity landscapes, the CNG does not contain edges and basically represents a set of potency-colored nodes (rather than a formal graph). In CNGs, similarity relationships are captured by arranging all compound nodes falling into a neighborhood on concentric circles around the reference. Each circle represents a range of similarity values, and increasing radii of circles reflect decreasing similarity to the central node. CNGs provide a detailed view of the potency distribution within organized (and also often overlapping) chemical neighborhoods and are ranked by quantifying similarity and potency value distributions that are indicative of interpretable SAR information [31].

16.4.4.3 Representation of Analog Series Analog series can also be analyzed in detail in "mini-landscape" formats, which might be regarded as a special case of compound-centric activity landscapes. In medicinal chemistry, the standard way to represent analog series is the use of R-group tables. These tables contain the invariant structural core of a series with labeled substitution sites and report R-groups available at each site together with the potency of the corresponding analogs. As an extension of R-group tables, different versions of SAR maps have been introduced [32,33] that report analog series after standard R-group decomposition in a matrix format where each cell represents a unique combination of R-groups applying a potency-based color code.

Going beyond the conventional R-group table format and its extensions, hierarchical tree-like local activity landscape representations have been introduced specifically for analyzing analog series [34,35]. These combinatorial analog graphs (CAGs) systematically organize analog series on the basis of substitution site combinations and calculate discontinuity scores for (overlapping) compound subsets sharing the same substitution site or combinations of sites. Because the data structure is applied exclusively to analyze series of analogs, whole-molecular similarity values for discontinuity score calculations are best replaced by local similarity measures such as R-group-based pharmacophore edit distances [35]. CAGs reveal substitution sites or combinations that are SAR hotspots. Furthermore, given their hierarchical organization, CAGs also identify SAR holes (different from those in three-dimensional landscapes, as discussed above), that is, substitution site combinations that have not yet been explored [34]. Moreover, side-by-side comparison of CAGs for series of analogs active against several targets enables the study of multitarget SARs [35].

16.5 DEFINING AND IDENTIFYING ACTIVITY CLIFFS

Having reviewed different approaches to activity landscape design, the following discussion focuses on activity cliffs, the most prominent features of landscape representations. In the literature, the term *activity cliff* probably first appeared in 1991 [36]. If we consider the general definition of an activity cliff (see above), it is not surprising that the activity cliff concept has become rather popular in chemoinformatics and medicinal chemistry given its chemically intuitive nature and immediate link to SAR information.

16.5.1 Similarity and Potency Criteria

However, the definition of activity cliffs also raises two important questions. First, how similar must compounds be to qualify as cliff partners? Figure 16.2 illustrates potential ambiguities of activity cliffs designated on the basis of calculated similarity values. Two pairs of compounds forming activity cliffs are shown. In both instances, Tanimoto similarity is high, but there is a clear difference between these compound pairs. One cliff is formed between two very closely related analogs, whereas the compounds forming the other cliff display a large chemical change in their core structures. Thus, from a chemistry perspective, the structural/similarity relationships characterizing these cliffs are certainly not equivalent. Second, what are acceptable potency value ranges and significant potency differences? A cliff formed between two compounds with 10 nM and 10 μM potency would not be considered comparable to a cliff formed by two compounds with 1 μM and 100 μM potency, although in both cases there is a 100-fold difference in potency and the potency value ranges overlap. From an SAR point of view, the first cliff would be more interesting than the other because it involves a highly potent compound rather than two weakly potent compounds.

These examples illustrate potential complications associated with the activity cliff concept. There currently are no generally accepted similarity and potency criteria to define activity cliffs. However, for any activity cliff analysis, such criteria must be clearly specified (which is often not the case in the literature). Of course, the dependence of activity landscape topology on chosen molecular representations and similarity measures also applies to the formation of activity cliffs, and their distribution is expected to change when alternative molecular representations are used. This has been well illustrated for activity landscape models of a given data set calculated with different fingerprints or other descriptors [14,16]. Hence, if comparisons between different data sets are made, it is essential to use the same representation

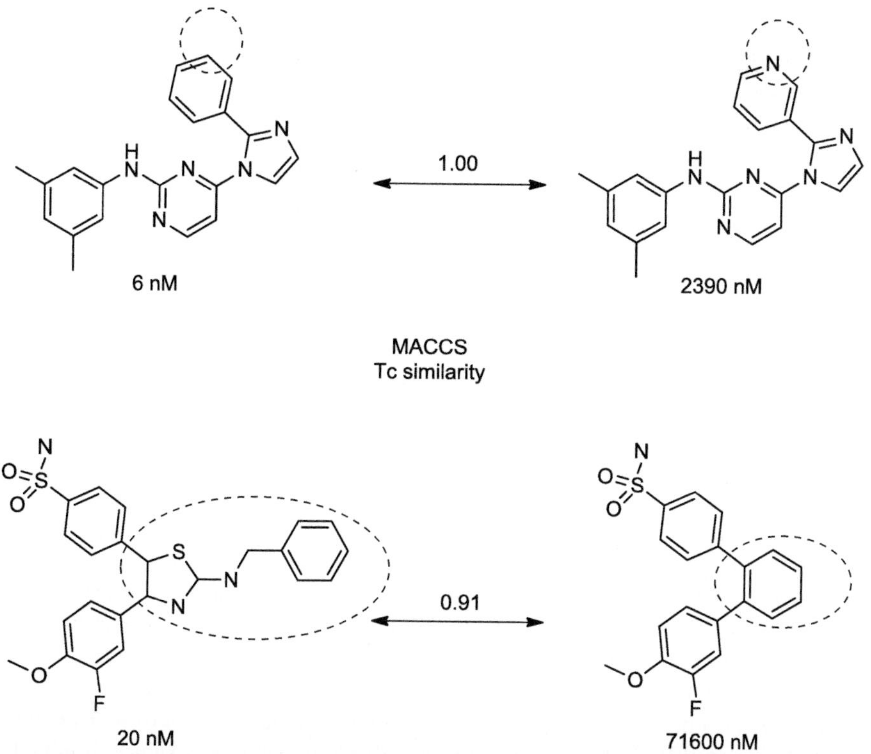

FIGURE 16.2 Exemplary activity cliffs involving structural relationships of different nature. The activity cliff at the top is formed by two vascular endothelial growth factor receptor-2 antagonists that are only distinguished by a single nitrogen substituent (yielding a MACCS Tanimoto similarity of 1.0). The cliff at the bottom is formed by two cyclooxygenase-1 inhibitors. Here, a central ring system is replaced. These two inhibitors also yield a high MACCS Tanimoto similarity value of 0.91, but the structural similarity of the cliff-forming compounds is much more remote than in the other case.

throughout. In addition, it is also important to avoid the use of inappropriate molecular descriptors. For example, multiproperty molecular representations that increasingly abstract from molecular structure are not suitable for the analysis of activity cliffs. As a general guideline, any calculated similarity that cannot be immediately appreciated on the basis of two-dimensional molecular graphs should be avoided to characterize activity cliffs (and activity landscapes). The use of such descriptors tends to "flatten" activity landscape and separate partners of chemically reasonable cliffs.

As an alternative to Tanimoto similarity, hierarchical structural relationships between active compounds, Bemis and Murcko scaffolds [37], and cyclic skeletons [38] might be utilized as similarity criteria for cliff formation. Starting from complete molecules, Bemis and Murcko scaffolds abstract from R-groups but retain linker fragments between rings. Cyclic skeletons further abstract from Bemis and Murcko scaffolds by converting all heteroatoms to carbon and setting all bond orders to 1. Thus, a unique cyclic skeleton covers multiple topologically equivalent scaffolds. Following this structural hierarchy, Bemis and Murcko scaffolds and analog series they represent might be regarded as structurally similar if they yield the same cyclic skeleton (i.e., if they are topologically equivalent). This structural similarity criterion has been applied, for example, in the exploration of coordinated activity cliffs [39]. The application of this criterion can also lead to ambiguities, for example, when the heteroatom content of topologically equivalent scaffolds substantially differs. However, in many instances, this simple structural organization scheme has proven to be a robust indicator of closely related analogs and otherwise similar compounds, leading to a meaningful representation of activity cliffs [39].

16.5.2 Continuum of Activity Cliffs vs. Discrete States

If cliff partners are analogs or structurally related to the extent that their similarity is evident on the basis of visual inspection, activity cliffs are defined in a meaningful way. From molecular graph- or network-based activity landscape representations presented herein, such cliffs can readily be selected. Importantly, the application of local SAR analysis functions such as SALI or per-compound discontinuity scores will consistently identify the most prominent cliffs that are present in a compound data set for a chosen molecular representation, regardless of their potential significance for medicinal chemistry. A characteristic feature of SALI is that it is designed to detect a continuum of activity cliffs, which is appropriate for the characterization of the activity cliff distribution in a data set. However, through continuous scoring, only the ordering of cliffs is determined, not their absolute magnitude. Hence, it must be investigated separately at which score threshold level activity cliffs become significant with respect to the potency distribution within a compound set. Alternatively, chemical significance criteria can be predefined in order to focus the analysis exclusively on activity cliffs meeting these requirements. Hence, activity cliffs might be defined as discrete states, rather than a continuum, by specifying permitted potency ranges, required potency differences, and similarity threshold values. A discrete definition of activity cliffs is meaningful for applications such as data set comparisons or frequency-of-occurrence analysis. For example, the following activity cliff definition has been proposed [40]: An activity cliff is formed between two compounds if at least one of them is potent in the nanomolar range and if there is an at least a 100-fold difference in potency between them. Furthermore, cliff-forming compounds must yield a Tanimoto similarity of 0.55 calculated with the extended connectivity fingerprint with bond diameter 4 (ECFP4) [41], which corresponds approximately to a value of 0.85 calculated with MACCS structural keys [42]. For many (but not all) data-mining applications this is a meaningful definition of activity cliffs.

16.5.3 Experimental Data

In addition to the molecular representation dependence of activity landscapes and cliffs, their characteristics are also influenced by experimental data variability. However, this potential complication has thus far been investigated much less than the similarity conundrum. In the context of activity landscapes, experimental data variability refers to potency measurements. A systematic analysis of activity cliff formation using alternative compound potency measurements has recently been carried out for BindingDB compounds [43,44]. Applying the definition of discrete activity cliffs stated above, it was found that activity cliff distributions in many different data sets changed significantly when means of multiple available IC_{50} or K_i values were used or minimum and maximum values instead. On average, only about 50% of activity cliffs were conserved when multiple measurements were utilized in different ways. In cases where all available potency values fell within an order of magnitude, fewer activity cliffs were detected than when the data were more variable. Furthermore, K_i measurements yielded consistently fewer activity cliffs than IC_{50} measurements [44]. Different from IC_{50}, K_i values represent equilibrium constants and are thus not dependent on assay conditions and are less error-prone. Taken together, these findings show that approximate potency measurements and increasing data variability generally produce larger numbers of activity cliffs than more accurate potency values. Thus, in addition to the choice of molecular representations, the types of experimental potency values and their confidence level must be considered carefully in the analysis of activity cliffs.

16.5.4 Different Types of Activity Cliffs

In addition to "standard" activity cliffs formed by pairs of similar compounds, other types and variants of activity cliffs have been introduced. For example, a special category of activity cliffs are "R-cliffs" in analog series that result from R-group replacements at specific substitution sites [33]. R-cliffs have been defined on the basis of matrix extensions of R-group tables (see above). Furthermore, going beyond conventional single-target activity cliffs, selectivity cliffs [19] and multitarget activity cliffs [20] have been introduced. *Selectivity cliffs* are formed by pairs of compounds with significantly different potencies against one or both targets of a target pair and are the cardinal feature of selectivity landscapes. In addition, *multitarget cliffs* are formed by pairs of compounds with different potencies against series of targets. In Figure 16.3, the formation of single- and multitarget activity cliffs and selectivity cliffs is illustrated. Multitarget cliffs are "directed" if one compound has consistently high potency against its targets, or "undirected" if the cliff partners display differential (inverse) potency against one or more targets [20].

Considering the molecular representation dependence, activity cliffs have been specified that are consistently formed in different chemical reference spaces called *consensus activity cliffs* [45]. Such consensus cliffs are hence thought to be the least descriptor-dependent. Another type of consensus activity cliff might be conceptualized considering the influence of experimental measurements and data variability. These experimental consensus cliffs would be conserved when alternative potency measurements and multiple potency values are utilized. It should be interesting to compare representation-independent activity cliffs with experimentally invariant cliffs and explore potential overlap between these populations in different compound data sets.

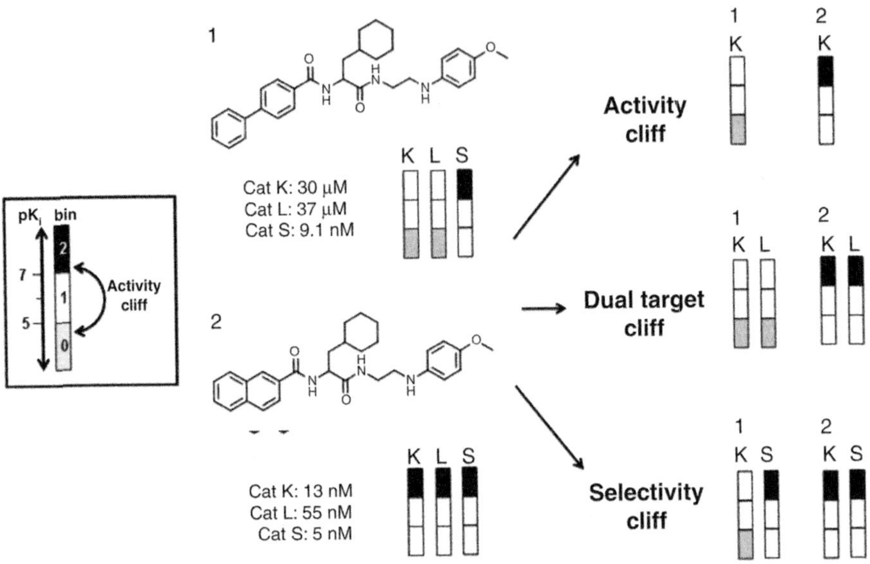

FIGURE 16.3 Dual-target activity and selectivity cliffs. For a pair of inhibitors active against three cathepsin thiol proteases (K, L, and S) potency values are binned (as reported on the left) to illustrate the formation of special types of cliffs. Based on their relative potencies against the three cathepsins, the two inhibitors form single-target cliffs, a directed dual-target cliff, and a selectivity cliff. Further details are discussed in the text.

Although activity landscapes and activity cliffs are typically derived from small-molecule data, activity cliffs might also be rationalized at the target structure level. SAR discontinuity detected in compound data sets results at least in part from the modulation of critical ligand–receptor interactions through chemical substitutions. It has been shown that the formation of activity cliffs can often be reconciled on the basis of interactions seen in x-ray structures of ligand–target complexes [46]. However, this is clearly not always the case because short-range interactions in x-ray structures reflect only a part of the binding process, in the course of which entropic effects and solvation/desolvation energies might also make large contributions. Nevertheless, it has also been attempted to define activity cliffs computationally solely on the basis of ligand–target interactions. This approach has been termed *identification of structure-based activity cliffs (ISAC)* [47]. Here, interaction energies between ligand and protein atoms are calculated using force field–based scoring functions to derive an interaction fingerprint for a given active compound. Then, SALI scores are derived using the interaction fingerprint instead of conventional fingerprints for the calculation of Tanimoto similarity of different ligands. If activity cliffs are formed by compounds with similar interaction fingerprints, individual contacts that differ between the cliff partners can be identified, and protein atoms (residues) involved in these discriminatory interactions are regarded as hotspots for activity cliff formation. Compared to other approaches defining activity cliffs, the distinguishing feature of the ISAC methodology is that chemical similarity of active compounds is replaced by ligand–target interaction similarity. In other words, although Tanimoto similarity is also calculated for ISAC-based SALI scoring, a unique molecular representation is utilized: a ligand–protein interaction space, which is distinct from conventional chemical reference spaces.

16.6 ACTIVITY CLIFF SURVEY

One of the most interesting questions concerning activity cliffs includes how often they actually occur in sets of bioactive compound and how they might be distributed over ligands of different target families. For assessing the relevance of activity cliff analysis in medicinal chemistry, these are important points to consider. In addition, it is attractive to investigate whether there might be molecular building blocks (structural motifs) that would be recurrent in activity cliffs.

16.6.1 Frequency of Cliff Formation

To determine how frequently activity cliffs occur, an analysis of BindingDB [43] and ChEMBL [48] compound data sets has recently been carried out (adhering to the definition of discrete activity cliffs referred to above) [40]. Based on this analysis, about 12% of all active compounds were found to be involved in the formation of at least one or two activity cliffs of considerable magnitude. Only about 4% of all activity cliffs were multitarget cliffs, and almost all of them were directed. Thus, although activity cliffs are formed with notable frequency, the majority of currently available cliffs are single-target cliffs, and cliff-forming compounds with different selectivity for multiple targets are extremely rare. In this study it was also determined that cliffs were distributed similarly over different target families, perhaps unexpectedly so. However, this observation suggests that activity cliffs will with high likelihood be identified in compound data sets with diverse bioactivities.

16.6.2 From Isolated to Coordinated Activity Cliffs

When analyzing activity cliffs systematically, one can often observe that they are not formed in isolation (i.e., by pairs of compound without structural neighbors) but that there are clusters of cliffs involving multiple compounds. To put these observations on a quantitative basis, a hypothetical data structure termed an *activity ridge* was introduced as a query for data mining [39]. The ridge structure was proposed to consist of a set of five or more compounds having nanomolar potency within an order of magnitude and another set of at least five compounds with 100-fold higher or lower potency, also within an order of magnitude. The compounds in these two sets would then consistently form pairwise ("combinatorial") activity cliffs. In a systematic search of 242 different compound data sets, a total of 125 activity ridges were indeed identified in 71 sets [39]. These ridge structures contained between 10 (minimally required) and 70 cliff-forming compounds. For comparison, in addition to 125 activity ridges, about 1500 isolated cliffs were identified across all data sets. The activity ridge structure is a model for the formation of coordinated activity cliffs. Given the considerable number of ridges that were detected, compound subsets forming coordinated activity cliffs are expected to be widely distributed. Such compound arrays represent a rich source of SAR information, more so than isolated cliffs.

16.6.3 Preferred R-Groups and Scaffolds

In another survey, about 200 chemical substitutions were identified to form activity cliffs with high frequency across different compound classes and biological targets [49]. This analysis was based on systematic extraction and comparison of MMPs from active compounds. Hence, a fairly large number of defined R-group replacements preferentially formed activity cliffs in different structural environments. Surprisingly, similar trends were detected for molecular scaffolds in an independent study. On the basis of currently available data, a total of about 100 Bemis and Murcko scaffolds of different size and composition were identified that displayed a statistically significant tendency to participate in activity cliffs across different target families [50]. The high propensity of selected molecular scaffolds of varying complexity to form activity cliffs is less intuitive than the involvement of specific chemical substitutions because core structures are usually involved in critical ligand–target interactions less frequently than R-groups. However, considerable numbers of chemical substitutions and molecular frameworks have been identified that should merit special consideration in the exploration of activity cliffs and in compound design.

16.7 ACTIVITY CLIFFS AND SAR INFORMATION

Because activity cliffs are regarded as the most prominent features of activity landscapes, they are usually the initial focal point of SAR analysis. However, this also raises theoretical and practical questions. For example, are there relationships between activity cliffs and other SAR components? Also, how should one best interpret activity cliff information in chemical terms in order to deduce SAR determinants?

16.7.1 SAR Discontinuity vs. Continuity

Continuous and discontinuous local SARs are often found to coexist in compound data sets, giving rise to global SAR heterogeneity. However, this does not necessarily mean that individual compound series are exclusively continuous or discontinuous in their SAR behavior. For example, careful analysis of SARI score combinations has revealed that SAR continuity might also be observed in the vicinity of activity cliffs, a SAR phenotype referred to as *heterogeneous-constrained* [9]. This means that modifications of active compounds might often be tolerated as long as one or more groups involved in critical ligand–receptor interactions are present. A textbook example of this SAR phenotype would be sulfonamide-containing inhibitors of carbonic anhydrases.

However, other links between activity cliffs and SAR continuity can also be observed. This has been revealed by searching for SAR pathways, which represent a predefined SAR model designed to mine NSG representations systematically [51,52]. Such pathways are formed by series of pairwise structurally similar compounds that follow a potency gradient. The underlying SAR model prioritizes pathways that consist of many overlapping pairs of similar compounds, span a large potency interval between the start and end compounds, and ideally follow linear potency gradient with small potency increases between neighboring compounds. Because these pathways consist of pairwise similar compounds, "scaffold hops" might occur along a path. Accordingly, SAR pathways are designed to navigate regions of local SAR continuity in activity landscapes. In systematic NSG mining, SAR pathways have often been identified that directly connect regions of SAR continuity to prominent activity cliffs [51,52], thereby establishing a close connection between SAR continuity and discontinuity. These insights have been further corroborated and extended by results of SAR-oriented particle swarm optimization [53,54]. An optimization strategy was devised to identify activity cliffs formed by compounds proximal to initialized particle positions and to select other compounds from the immediate structural neighborhood of prominent activity cliffs that form continuous local SARs [54]. In seven of 32 diverse compound data sets, subsets of at least five compounds were identified that represented SAR continuity and had structural neighbors forming large-magnitude activity cliffs. These findings revealed the presence of highly heterogeneous SAR microenvironments and provided further evidence for close links between SAR continuity and discontinuity.

16.7.2 Information Extraction from Activity Cliffs

The inspection of isolated activity cliffs in compound data sets reveals individual chemical changes leading to large potency differences between pairs of compounds. This information alone might not be sufficient to aid in the design of better compounds. The interpretation of activity cliffs becomes more informative if they are viewed in the context of other SAR trends and if the structural neighborhood of cliffs is analyzed. Moreover, because activity cliffs often do not occur in isolation but in a coordinated manner involving multiple compounds (see above), a search for cliffs might initially be better directed toward ridge-like structures from which more SAR information can be deduced than from isolated cliffs.

Because most compound data sets analyzed thus far were found to contain multiple activity cliffs, often in different local arrangements, the investigation of new compound series for established targets should involve mining of already known compound activity data. Together with prior knowledge of activity cliff–forming molecular building blocks, this might help to prioritize substitution patterns for chemical exploration in different structural environments. Finally, for the chemical interpretation of activity cliff information, alternative activity landscape representations should be taken into account, including those that complement or replace calculated similarity values with defined substructure relationships.

16.8 CONCLUDING REMARKS

In this chapter, activity landscape representations and activity cliffs, their most prominent features, have been discussed in the context of molecular similarity and large-scale SAR analysis. The organization and contents of this contribution partly follow two recent perspective articles that have presented thorough reviews of the activity landscape and activity cliff concepts [1,55].

In combination with numerical SAR analysis functions, activity landscape models have become popular because they enable the extraction of SAR information from large and heterogeneous compound data sets and the visualization of global and local SAR features. A variety of activity landscape designs has already been introduced that are often complementary in their use, such as global and compound-centric landscape views. Moreover, the activity landscape concept is being extended in different ways: for example, by exploring multitarget activity space.

A generally important feature of activity landscapes is their graphical nature. The role of SAR visualization techniques for making computational results accessible to medicinal chemists should not be underestimated. Graphical SAR exploration is supported in different ways. For example, in network-based activity landscapes, nodes (and sometimes edges) are often associated directly with corresponding compound (or fragment) structures, which provides a basis for interactive SAR analysis and inductive chemical interpretation of SAR networks. A characteristic feature of activity landscape design is that it forms an interface between chemoinformatics and medicinal chemistry. For further progress in this research area, including the development of novel computational approaches, and for a broader acceptance of activity landscape modeling among practicing medicinal chemists, it is also important that a number of activity landscape methods have already been made publicly available, including, for example, SALI graphs and various landscape tools implemented in the SARANEA software environment [56].

The design of activity landscapes and consistent analysis of SARs critically depend on a systematic assessment and integration of compound similarity and potency relationships. Thus, the concept of molecular similarity naturally plays a central role in activity landscape modeling. As discussed, the dependence of activity landscape topologies and SAR features on a chosen molecular representation and (to a lesser extent) similarity measure is the most significant caveat for landscape design and interpretation. Therefore, the exploration of alternative and intuitive ways to account systematically for molecular similarity, including, for example, the application of hierarchical structural and substructure relationships, should merit further investigation. In addition, activity landscape analysis is also affected by the choice of experimental measurements and their variability, which also requires careful consideration.

Despite these inherent limitations, modeling of activity landscapes provides significant opportunities for medicinal chemistry. The data-oriented nature of activity landscape views usually provides immediate access to SAR information available in large data sets and makes it possible to focus the analysis on compound subsets representing interesting local SARs. It is also straightforward to identify areas that have been explored thoroughly without revealing promising SAR trends (and to de-prioritize them for further analysis). Clearly, the identification of key compounds in large data sets (including HTS data) that reveal SAR determinants is a major attraction of activity landscape exploration. Although activity landscapes are first and foremost descriptive in nature and do not represent predictive models, compound design suggestions can be deduced from their analysis, which meets the primary interest of practicing medicinal chemists.

Different landscape designs convey SAR information in different ways, but activity cliffs are generally regarded as their most interesting features. However, this assumption also requires consideration. The study of activity cliffs reveals substitution sites that are critical for compound potency but does not necessarily suggest which compounds to make next. To understand how structural modifications alter compound potency in a defined and predictable manner, it is also required to explore SAR continuity in the vicinity of cliffs. Thus, although high SAR information content is associated with small structural changes of active compounds leading to large biological effects, a thorough rationalization of SARs requires consideration of this information within its chemical context. This is often (but not always) straightforward. For example, inspecting the environment of prominent activity cliffs might reveal the formation of coordinated cliffs involving multiple compounds, which provides more information about SAR determinants than do series of isolated cliffs. Furthermore, this might reveal patterns of SAR continuity and discontinuity within the same or similar series of compounds from which SAR determinants might also be deduced.

Activity cliffs are often discussed in a rather loose manner, which makes it difficult to judge a given analysis or compare results from different studies. Currently, there are no generally accepted criteria for the description and rationalization of activity cliffs. However, regardless of whether one considers activity cliffs as a continuum or as discrete states, it is essential to define them clearly for any computational or chemical investigation.

On a final note, it is well appreciated that lead optimization requires the consideration of multiple properties beyond compound potency. Hence, if the study of activity landscapes and cliffs should indeed become an integral component of chemical optimization efforts, it would ultimately be required to further extend the activity landscape concept and include multiple chemical and/or biological properties. First attempts in this direction are being made, and there is considerable potential for future developments.

Acknowledgment

The author would like to thank Dagmar Stumpfe for reviewing the manuscript and helping with illustrations.

REFERENCES

1. A. M. Wassermann, M. Wawer, J. Bajorath, *J. Med. Chem.* **2010**, *53*, 8209–8223.
2. G. M. Maggiora, *J. Chem. Inf. Model.* **2006**, *46*, 1535–1535.

3. J. Bajorath, G. M. Maggiora, M. Lajiness (Organizers), The emerging concepts of activity landscapes and activity cliffs and their role in drug research, *240th National Meeting of the American Chemical Society, Divisions of Chemical Information and Computers in Chemistry*, Boston, Aug. 22–26, **2010**.
4. E. X. Esposito, A. J. Hopfinger, J. D. Madura, *Methods Mol. Biol.* **2004**, *275*, 131–214.
5. J. Bajorath, L. Peltason, M. Wawer, R. Guha, M. J. Lajiness, J. H. Van Drie, *Drug Discov. Today* **2009**, *14*, 698–705.
6. H. Geppert, M. Vogt, J. Bajorath, *J. Chem. Inf. Model.* **2010**, *50*, 205–216.
7. M. A. Johnson, G. M. Maggiora, Eds., *Concepts and Applications of Molecular Similarity*, Wiley, New York, **1990**.
8. V. Shanmugasundaram, G. M. Maggiora, *Proceedings of the 222nd American Chemical Society National Meeting, Division of Chemical Information*, American Chemical Society, Washington, DC, Chicago, Aug. 26–30, **2001**, abstr. *77*.
9. L. Peltason, J. Bajorath, *J. Med. Chem.* **2007**, *50*, 5571–5578.
10. M. Wawer, L. Peltason, N. Weskamp, A. Teckentrup, J. Bajorath, *J. Med. Chem.* **2008**, *51*, 6075–6084.
11. R. Guha, J. H. Van Drie, *J. Chem. Inf. Model.* **2008**, *48*, 646–658.
12. G. M. Maggiora, V. Shanmugasundaram, M. S. Lajiness, T. N. Doman, M. W. Schulz, in: *Chemoinformatics in Drug Discovery*, T. I. Oprea, Ed., Wiley-VCH, Weinheim, Germany, **2005**, pp. 317–332.
13. J. Brown, E. Jacoby, *Mini Rev. Med. Chem.* **2006**, *6*, 1217–1229.
14. L. Peltason, P. Iyer, J. Bajorath, *J. Chem. Inf. Model.* **2010**, *50*, 1021–1033.
15. I. Borg, P. J. F. Groenen, *Modern Multidimensional Scaling: Theory and Applications*, 2nd ed., Springer-Verlag, New York, **2005**.
16. A. B. Yongye, K. Byler, R. Santos, K. Martínez-Mayorga, G. M. Maggiora, J. L. Medina-Franco, *J. Chem. Inf. Model.* **2011**, *51*, 1259–1270.
17. J. Pérez-Villanueva, R. Santos, A. Hernández-Campos, M. A. Giulianotti, R. Castillo, J. L. Medina-Franco, *Med. Chem. Commun.* **2011**, *2*, 44–49.
18. J. L. Medina-Franco, A. B. Yongye, J. Pérez-Villanueva, R. A. Houghten, K. Martínez-Mayorga, *J. Chem. Inf. Model.* **2011**, *51*, 2427–2439.
19. L. Peltason, Y. Hu, J. Bajorath, *ChemMedChem* **2009**, *4*, 1864–1873.
20. D. Dimova, M. Wawer, A. M. Wassermann, J. Bajorath, *J. Chem. Inf. Model.* **2011**, *51*, 256–288.
21. J. Waddell, J. L. Medina-Franco, *Bioorg. Med. Chem.*, **2012**, *20*, 5443–5452.
22. R. Guha, J. H. Van Drie, *J. Chem. Inf. Model.* **2008**, *48*, 1716–1728.
23. M. Wawer, L. Peltason, J. Bajorath, *J. Med. Chem.* **2009**, *52*, 1075–1080.
24. P. Iyer, D. Stumpfe, J. Bajorath, *J. Chem. Inf. Model.* **2011**, *51*, 1281–1286.
25. P. Iyer, J. Bajorath, *Chem. Biol. Drug Desi.* **2011**, *78*, 778–786.
26. M. Wawer, J. Bajorath, *J. Med. Chem.* **2011**, *54*, 2944–2951.
27. P. W. Kenny, J. Sadowski, in: *Chemoinformatics in Drug Discovery*, T. I. Oprea, Ed., Wiley-VCH, Weinheim, Germany, **2005**, pp. 271–285.
28. P. Iyer, J. Bajorath, *Med. Chem. Commun.* **2012**, *3*, 441–448.
29. M. Wawer, J. Bajorath, *J. Chem. Inf. Model.* **2010**, *50*, 1395–1409.
30. M. Wawer, J. Bajorath, *ACS Med. Chem. Lett.* **2011**, *2*, 201–206.

31. M. Wawer, S. Sun, J. Bajorath, *Int. J. High Throughput Screen.* **2010**, *1*, 15–27.
32. D. Agrafiotis, M. Shemanarev, P. Connolly, M. Farnum, V. Lobanov, *J. Med. Chem.* **2007**, *50*, 5926–5937.
33. D. Agrafiotis, J. J. M. Wiener, A. Skalkin, J. Kolpak, *J. Chem. Inf. Model.* **2011**, *51*, 1122–1132.
34. L. Peltason, N. Weskamp, A. Teckentrup, J. Bajorath, *J. Med. Chem.* **2009**, *52*, 3212–3224.
35. A. M. Wassermann, L. Peltason, J. Bajorath, *ChemMedChem* **2010**, *5*, 847–858.
36. M. J. Lajiness, in: *QSAR: Rational Approaches to the Design of Bioactive Compounds*, C. Silipo, A. Vittoria, Eds., Elsevier, Amsterdam, **1991**, pp. 201–204.
37. G. W. Bemis, M. A. Murcko, *J. Med. Chem.* **1996**, *39*, 2887–2893.
38. Y.-L. Xu, M. Johnson, *J. Chem. Inf. Comput. Sci.* **2001**, *41*, 181–185.
39. M. Vogt, Y. Huang, J. Bajorath, *J. Chem. Inf. Model.* **2011**, *51*, 1848–1856.
40. A. M. Wassermann, D. Dimova, J. Bajorath, *Chem. Biol. Drug Desi.* **2011**, *78*, 224–228.
41. D. Rogers, M. Hahn, *J. Chem. Inf. Model.* **2010**, *50*, 742–754.
42. *MACCS Structural Keys*, Symyx Software, San Ramon, CA, **2005**.
43. T. Liu, Y. Lin, X. Wen, R. N. Jorissen, M. K. Gilson, *Nucleic Acids Res.* **2007**, *35*, D198–D201.
44. D. Stumpfe, J. Bajorath, *J. Chem. Inf. Model.* **2011**, 51, 3131–3137.
45. J. L. Medina-Franco, K. Martínez-Mayorga, A. Bender, R. M. Marín, M. A. Giulianotti, C. Pinilla, R. A. Houghten, *J. Chem. Inf. Model.* **2009**, *49*, 477–491.
46. M. T. Sisay, L. Peltason, J. Bajorath, *J. Chem. Inf. Model.* **2009**, *49*, 2179–2189.
47. B. Seebeck, M. Wagener, M. Rarey, *ChemMedChem* **2011**, *6*, 1630–1639.
48. A. Gaulton, L. J. Bellis, A. P. Bento, J. Chambers, M. Davies, A. Hersey, Y. Light, S. McGlinchey, D. Michalovich, B. Al-Lazikani, J. P. Overington, *Nucleic Acids Res.* **2012**, *40*, D1100–D1107.
49. A. M. Wassermann, J. Bajorath, *J. Chem. Inf. Model.* **2010**, *50*, 1248–1256.
50. Y. Hu, J. Bajorath, *J. Chem. Inf. Model.* **2010**, *50*, 500–510.
51. M. Wawer, L. Peltason, J. Bajorath, *J. Med. Chem.* **2009**, *52*, 1075–1080.
52. M. Wawer, J. Bajorath, *ChemMedChem*, **2009**, *4*, 1431–1438.
53. J. Kennedy, R. C. Eberhart, *Proceedings of the IEEE International Conference on Neural Networks IV (ICN95)*, **1995**, 1942–1948.
54. V. Namasivayam, P. Iyer, J. Bajorath, *Chem. Biol. Drug Desi.* **2012**, *79*, 22–29.
55. D. Stumpfe, J. Bajorath, *J. Med. Chem.* **2012**, *55*, 2932–2942.
56. E. Lounkine, M. Wawer, A. M. Wassermann, J. Bajorath, *J. Chem. Inf. Model.* **2010**, *50*, 68–78.

PART IV

APPLICATIONS IN CHEMICAL BIOLOGY AND DRUG DISCOVERY

17

DIVERSITY-ORIENTED SYNTHESIS AND DRUG DEVELOPMENT: FACILITATING THE DISCOVERY OF NOVEL PROBES AND THERAPEUTICS

JEREMY R. DUVALL, EAMON COMER, AND SIVARAMAN DANDAPANI

17.1 INTRODUCTION

As our basic understanding of the underlying biology in disease areas has evolved, a fundamental challenge has emerged. New targets have been found that could hold the key to a cure long sought. Despite the uncovering of a wealth of biological targets in the genomic revolution [1], initial difficulties have instead led to the emergence of a new term: *undruggable*. This generally refers to challenging targets, such as protein–protein interactions (PPIs) and transcription factors, but ultimately is defined as any target linked to a disease that we as an industry failed to disrupt by the use of small molecules. As more challenging biological targets continue to emerge, small molecules will have to evolve to deliver novel therapeutics.

A compelling case for this was presented by GlaxoSmithKline (GSK) in their efforts to discover novel antibiotics [2]. As resistance emerges in different bacterial strains, the drug discovery industry has had limited success countering, with 20 antibiotics coming onto the market since 2000, most of which are derivatives of existing antibiotics and have known liabilities such as toxicity or resistance [3]. This is not due to a lack of trying, although with such limited success, many pharmaceutical companies are now leaving the field. GSK's effort focused on high-throughput-screening (HTS) campaigns against targets that had emerged from the sequencing

Diversity-Oriented Synthesis: Basics and Applications in Organic Synthesis, Drug Discovery, and Chemical Biology, First Edition. Edited by Andrea Trabocchi.

of the bacterial genome. Sixty-seven campaigns were run against specific targets, of which only 15 resulted in hits, and five of those resulted in leads. Three additional phenotypic cell–based assays were also run, but no viable leads were identified. Such poor screening results lead to at least two possible conclusions being formed. The first is that the biological targets screened are not ideal for modulation by a small molecule and are therefore undruggable. The second is that compounds traditionally used in HTS campaigns are poorly suited for the discovery of novel leads against these new types of biological targets. It is often a combination of both parameters, although scientists at GSK point to the lack of chemical diversity and compounds containing desirable chemical properties within their screening collection.

One result to come from the recent failures in drug discovery has been next-generation synthesis, a shift away from traditional small-molecule collections in an attempt to access novel chemical space. Several approaches have been introduced over the past decade in an attempt to define an optimal screening collection in which more tractable hits are identified for medicinal chemistry efforts. These include fragment-based drug discovery (FBDD) [4], diversity-oriented synthesis (DOS) [5], biology-oriented synthesis (BIOS) [6], function-oriented synthesis (FOS) [7], natural product–derived/target-oriented synthesis, and recently, lead-oriented synthesis [8]. Fragment-based drug discovery has served as a useful approach in the discovery of a variety of difficult targets, including protein–protein interactions. FBDD focuses on screening a small number of low-molecular-weight compounds against a known target using biophysical methods. Most often, FBDD hits are weak binders but have high ligand efficiencies. Progression of the hit compound to a lead with a strong binding affinity usually requires a crystal structure of the protein with the ligand. Biology-oriented synthesis utilizes the underlying framework of natural products for the synthesis of compound collections. Lead-oriented synthesis incorporates more drug-like properties into the screening collection so that compounds identified from a primary high-throughput screen can easily be advanced to in vivo proof-of-concept studies. Diversity-oriented synthesis strategies provide new types of small molecules that are currently lacking in most screening collections, so that chances of finding hits against new types of biological targets can be enhanced. Chemical complexity in the final product and the modular assembly of the core scaffolds are both hallmarks of DOS. Although these approaches are defined, hybrids of two or more of the concepts have been published. In the case of DOS, it has been utilized in FBDD [9] and used in a logical design that incorporates drug-like properties, similar to "lead-oriented synthesis" [10].

DOS allows access to novel chemical space and structural complexity. Although DOS is defined by the approach and not the compounds it produces, it is associated primarily with complex compounds that have been inspired by natural products. Screening collections have generally been dominated by flat, achiral compounds available from commercial sources, with complexity being introduced through screening small numbers of natural products (examples are shown in Figure 17.1). Increasing the content of structurally complex compounds to include DOS compounds provides novel chemical space compared to that of traditional small-molecule collections similar to GSK's, which failed in their antibacterial campaign. Complexity should not be

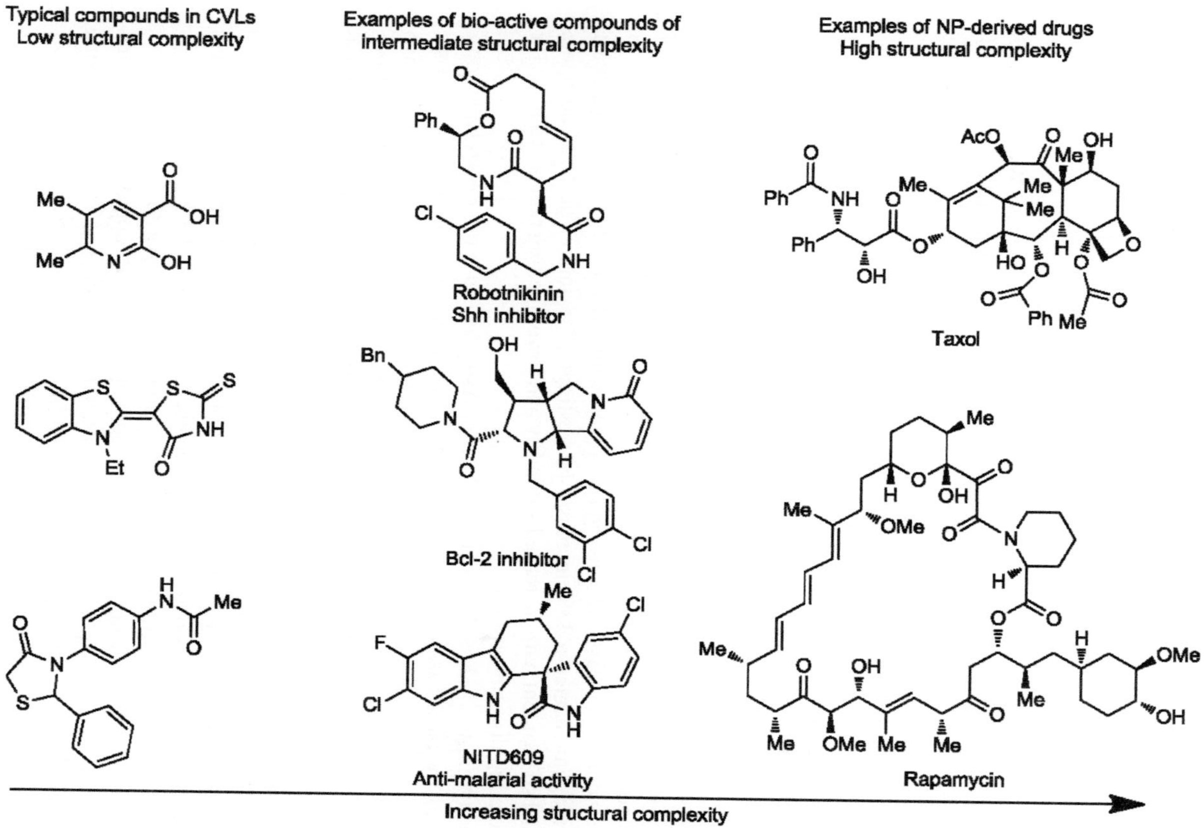

FIGURE 17.1 Example of complexity represented in various classes of compounds, including natural products, DOS-derived compounds, and members of commercially available libraries. (Adapted from [11], with permission of Macmillan Publishers Ltd.; copyright © 2010.)

overlooked, since an increasing amount of attention has focused on this feature and its correlation to drug-likeness.

Although complexity is desirable in screening collections, it is difficult to compute compared to more straightforward physicochemical properties which have played an important role in the design of small molecules for drug discovery. Lipinski's rule of 5 focused on molecular weight, log *P*, hydrogen-bond donors and acceptors, and their correlation to bioavailability [12]. Additional analyses have expanded these filters to other standard physicochemical properties, but only recently have complexity descriptors been calculated and used to correlate a small molecule's chances of success through clinical trials. Work done by Lovering et al. from Wyeth looked at both sp^3 content and stereogenic centers in this regard [13]. In this analysis, all compounds from 1980 onward were retrieved from the GVK BIO database representing discovery through phases I, II, and III and approved drugs. Complexity was measured by the fraction of sp^3 content (Fsp3), which is the number of sp^3 carbons divided by the total number of all carbons. As Fsp3 increases, the compound contains more three-dimensional character, which can correlate to increased complexity. The analysis shows not only the general increase in the average sp^3 content for every stage going from discovery to drugs, but that drugs have a 31% increase in the Fsp3 content compared to that of compounds in discovery. A similar analysis was done on compounds that contain at least one stereogenic center, another descriptor for complexity. The same enrichment is observed as the average is calculated for each class of compounds, ranging from discovery to drugs. In this case, 46% of the compounds in discovery contain at least one stereogenic center, while 61% of drugs contain at least one stereogenic center, a 33% increase. Interestingly, the authors go on to correlate increased sp^3 content with increased solubility and lower melting points, a descriptor for predicting drug absorption.

In addition, work done by Luker et al. from AstraZeneca further correlates complexity as measured by flatness and toxicity [14]. Their analysis focused on a set of internal small-molecule candidates that were investigated in rat and dog in vivo toxicology studies over the past 10 years. The authors were attempting to develop a predictive model for preclinical toxicology based off structural descriptors. Although many properties are described as being a contributing factor in a compound's toxicity, flatness, as measured by the ratio of aromatic atoms to total heavy atoms [15], was specifically cited as having a statistically relevant impact. Compounds that reached humans had an average of 49% of the atoms aromatic, while compounds that were stopped due to toxicity had an average aromatic atom count of 64%.

DOS is a synthetic strategy using short, modular pathways to access complex scaffolds. One such strategy is the build/couple/pair (B/C/P) approach, which has been utilized in several published library productions [16]. In the first step, highly functionalized, stereochemically enriched building blocks are built (Figure 17.2). Accessing all stereochemical permutations is highly beneficial for the generation of stereostructure–activity relationships (SSAR) [17] in future HTS campaigns. The building blocks are then coupled together with all possible stereochemical combinations to access all stereoisomers of a common intermediate. Pairing reactions combine complementary functional groups to rigidify the scaffold and accentuate the stereochemistry built into the scaffold.

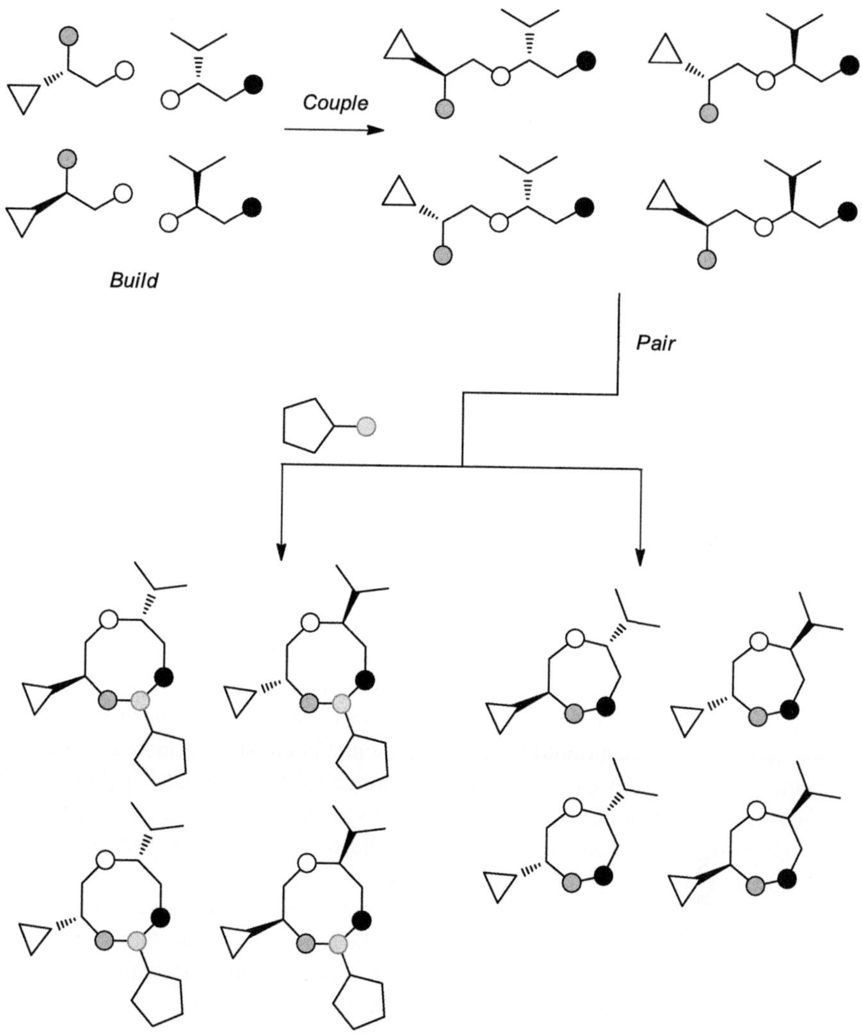

FIGURE 17.2 Overview of a build/couple/pair strategy.

Due to its modular approach, DOS strategies allow systematic access to diverse compounds. Already shown was the utilization of a build/couple/pair strategy allowing access to stereochemically diverse scaffolds in an efficient manner. Incorporation of desirable or more drug-like properties is also possible by upfront evaluation of the building blocks utilized, either in the synthesis of the scaffold or appendage diversification through library production.

The latter concepts was the focus of work presented by Akella and Marcaurelle [10]. Library design focused on a logical sparse matrix design, making compounds with appropriate building block combinations rather than a full matrix approach where all building block combinations are synthesized. The building block combinations are selected through a stepwise approach that first involves enumeration of a virtual library and application of a series of physicochemical property filters to eliminate any compounds that violate preset limits. The filters applied included molecular weight, clog*P*, rotatable bonds, and hydrogen-bond donors and acceptors. Once violators are filtered out, the final library is selected to cover the greatest amount of chemical space based on the maximal dissimilarity method, while retaining closely related analogs for built-in structure–activity relationships upon screening in biological assays.

High-throughput screening of DOS libraries is often enriched in structure–activity relationships (SAR) and/or SSAR to facilitate prioritization of hit clusters. Tractable SAR is an important concept that can be used to validate different hit clusters as well as to give confidence in developing hit compounds to lead compounds. Rather than synthesis of additional analogs after the original HTS to achieve confidence, upfront investment allows for this information to already be present in the screening collection. This principle is illustrated with examples later in the chapter.

Once a hit cluster is identified, medicinal chemistry is most often required for developing a lead candidate. Taking advantage of the initial DOS pathway allows a chemist to systematically access most if not every atom in the compound. Such an approach allows fundamental changes in the core skeleton rather than limiting the development of SAR to appendage sites only. This principle can be illustrated with a set of macrocycles recently published using a fragment-based domain shuffling approach [18]. The synthetic route utilizes a glycal amino alcohol, a γ-amino acid, and an *o*-fluorobenzoic acid to give 16-membered lactams (Figure 17.3a). If these compounds were found to be hits in an HTS campaign, one can easily imagine how a modular approach could facilitate medicinal chemistry. In just one example, the authors use a variety of amino acids to change substitution pattern, substitution groups, and linker length yet continue to utilize the synthetic route developed originally (Figure 17.3b). The same approach could be applied to the benzoic acid or amino alcohol motifs to probe additional SAR.

Despite the concept of DOS being relatively new, there is remarkable progress that gives reason to continue efforts in this area. In the remainder of the chapter we focus on examples of biologically active compounds derived from DOS pathways. When possible, the principles described above will be presented. Case Study 1 focuses on the identification of a small-molecule probe for the inhibition of β-cell apoptosis from a DOS screening collection after a more traditional set of small molecules failed. Case Study 2 looks at two campaigns focusing on antimalarial agents from DOS pathways. Case Study 3 looks at the successes in targeting protein–protein interactions and transcription factors using DOS compounds.

17.2 CASE STUDY 1: INHIBITION OF CYTOKINE-INDUCED β-CELL APOPTOSIS

Type 1 diabetes is primarily an autoimmune disease in which the insulin-secreting β-cells in the pancreas are destroyed by cytokine-induced apoptosis [19]. Hence small molecules that can protect β-cells from cytokine-induced apoptosis can serve

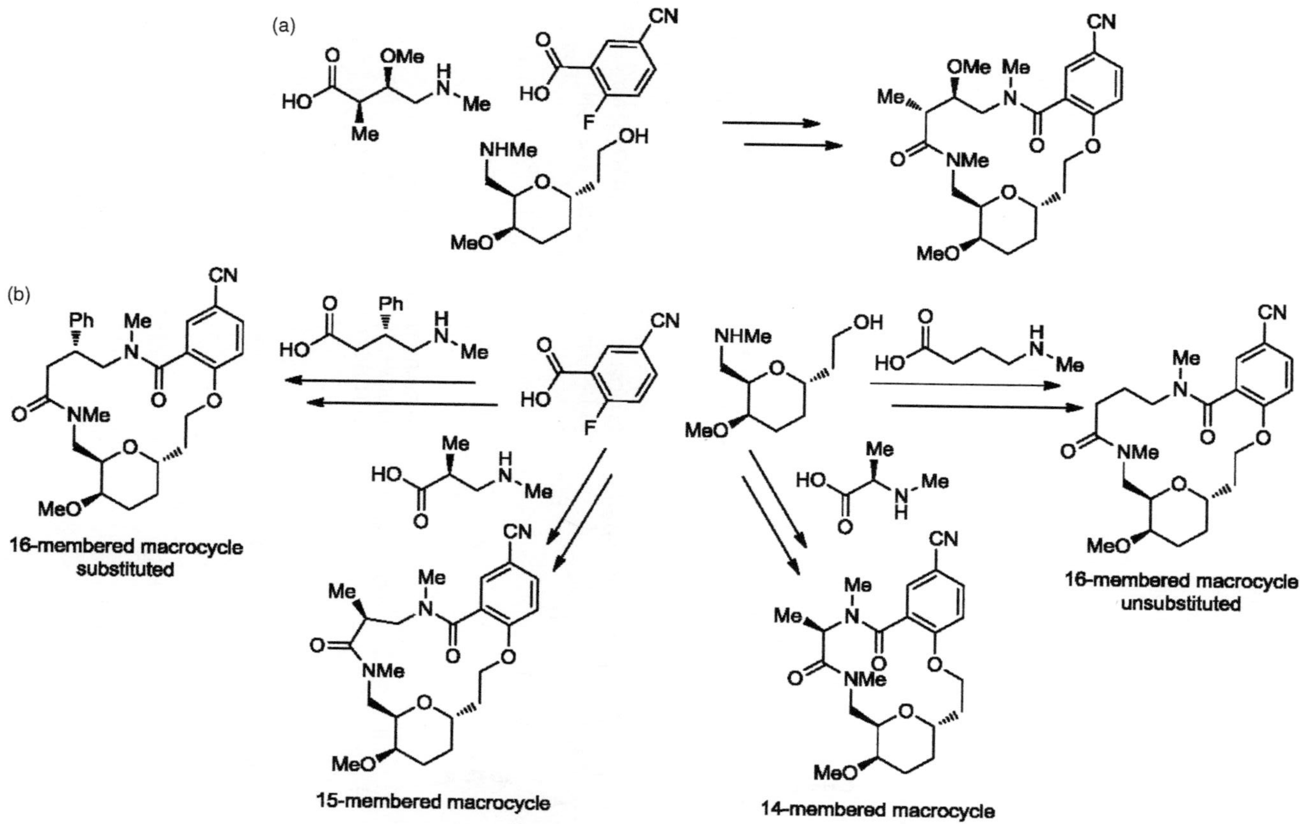

FIGURE 17.3 (a) Overview of a fragment-based domain shuffling approach to building macrocycles. (b) An example of substituting an amino acid motif to prove the structure–activity relationships.

as chemical tools to further explicate the complex biology of β-cells and can also lead to novel medicines against diabetes. Chou et al. developed a phenotypic assay to identify small molecules that can prevent cytokine-induced β-cell apoptosis [20]. They also developed a number of secondary assays to further characterize the small molecules identified through the phenotypic screen.

When the primary assay was conducted in the HTS mode, two different types of small-molecule collections were screened: (1) the Molecular Libraries Small Molecule Repository (MLSMR) collection and (2) a DOS-derived library at the Broad Institute. The MLSMR collection consists of over 300,000 small molecules derived primarily from commercially available small molecules. This collection was rich in heterocycles, had lower Fsp^3 content, and in general had scaffolds that are less complex in terms of both chemical architecture and stereochemical elements [11]. In contrast, the DOS collection screened in this assay was rich in medium-sized rings, had higher Fsp^3 content, and in general had more complex scaffolds, each with several stereocenters.

The synthesis of a representative DOS library included in this assay is described in Scheme 17.1 and was carried out in the Chemical Biology Platform at the Broad Institute using a build/couple/pair strategy [16c]. The synthesis began with the build phase and utilized the anti-selective Abiko asymmetric aldol reaction [21] between **1** and the protected amino acetaldehyde. After hydrolysis and protection of the secondary alcohol, the anti γ-amino acid derivative **RS-3** was obtained. The corresponding syn

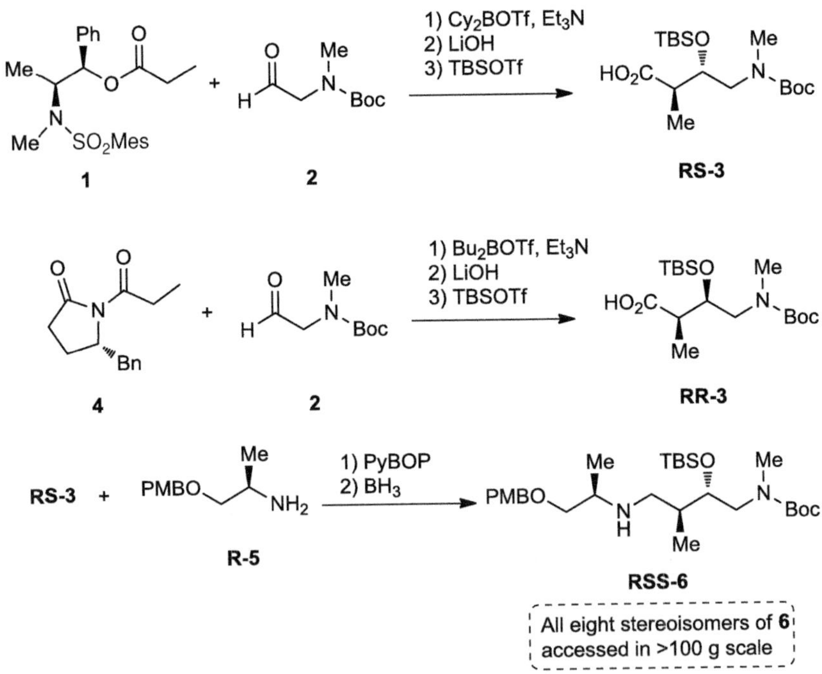

SCHEME 17.1 Build and couple to a linear template **6**.

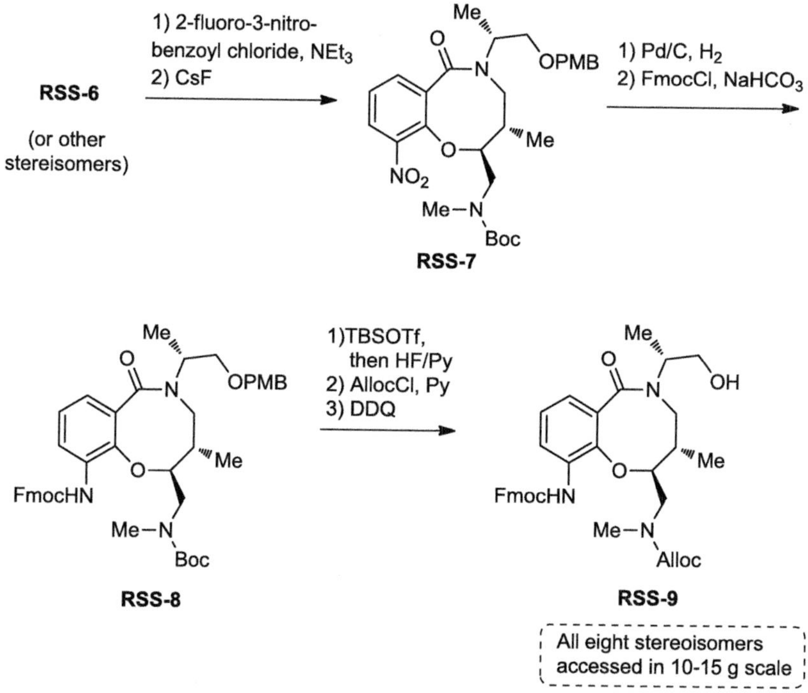

SCHEME 17.2 Pair using S_NAr.

γ-amino acid derivative **RR-3** was obtained in three steps from a syn-selective Evan's asymmetric aldol reaction [22] between **1** and the protected amino acetaldehyde **2**. The corresponding enantiomers (**SR-3** and **SS-3**, not shown) were also synthesized starting with the chiral antipodes of **1** and **4**. In the couple phase, **RS-3** was coupled with the protected alaninol **R-5** and reduced to provide the linear amine **RSS-6**.

All eight stereoisomers of **6** were accessed by a similar sequence of steps, and a robust synthetic pathway was established to access more than 100 g of each stereoisomer of **6**. The linear template **6** has two amino groups and two hydroxy groups that are orthogonally protected for a range of chemical modifications leading to a number of different DOS pathways. More than a dozen cyclic scaffolds have been accessed from **6**, and a representative library synthesis using an S_NAr pairing reaction is described in Schemes 17.2 and 17.3.

The linear amines **6** were acylated with 2-fluoro-3-nitrobenzoyl chloride, and the resulting amide was subjected to a S_NAr cycloetherification on treatment with CsF [23]. All four diastereomers underwent this cyclization smoothly, and hence all eight cyclic scaffolds **7** were obtained in high yields. The nitro group was reduced with Pd/C and the resulting aniline was protected with FmocCl. The Boc group on the secondary amine was then replaced with an Alloc group to be compatible with the subsequent solid-phase diversifications. Finally, the primary alcohol was unmasked

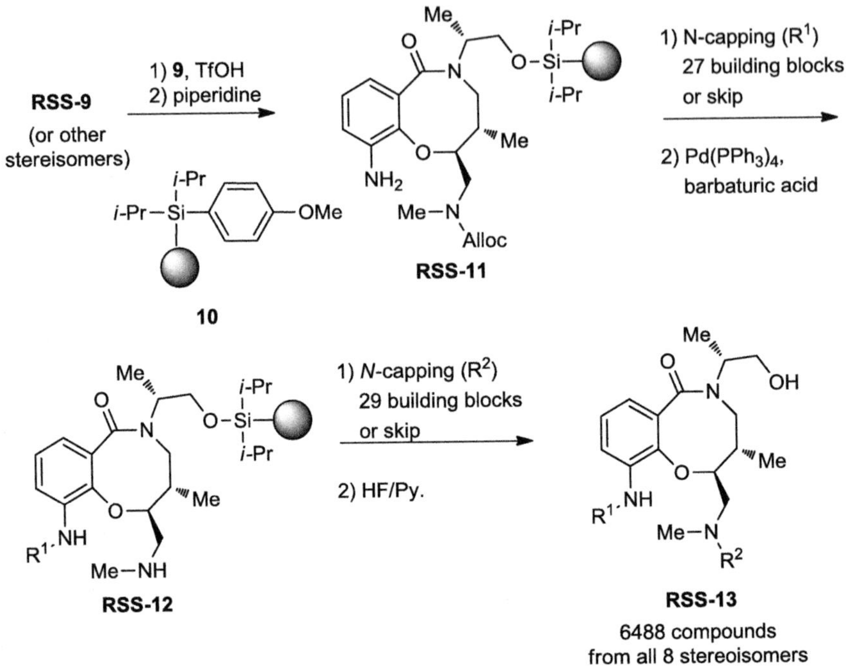

SCHEME 17.3 Library production using lanterns.

with the deprotection of the PMB group with DDQ. Using this seven-step sequence, all eight stereoisomers of **9** were obtained on a scale of 10 to 15 g.

Having assembled all eight stereoisomers of the cyclic core **9**, solid-phase diversifications of both the aniline and secondary amine were undertaken with SynPhase Lanterns [24] using a split–pool [25] approach. The solid-phase lanterns **10** carrying the silicon linker were first activated with triflic acid, and then the core **9** was added. The loading levels for the immobilization step were approximately 15 to 18 μmol per lantern. Washing the lanterns with 20% piperidine in DMF removed the Fmoc-protecting group. The lanterns were split and the free aniline was functionalized with 27 building blocks, including sulfonyl chlorides, isocyanates, acid chlorides, or formaldehyde. Some lanterns also skipped functionalization at this step. The lanterns were pooled and subjected to alloc deprotection using $Pd(PPh_3)_4$ and barbaturic acid. The lanterns were split again and the secondary amine was functionalized with 29 building blocks, including sulfonyl chlorides, isocyanates, or aldehydes. The lanterns that skipped the first diversification event also skipped the second diversification step. Finally, the compound was released from the lantern with HF/pyridine. Using this split–pool approach, a total of 6448 compounds were obtained from the eight stereoisomeric cores **9**.

A collection of nearly 20,000 DOS compounds, including the S_NAr library described above, was screened in the phenotypic assay to identify small molecules that can prevent cytokine-induced β-cell apoptosis. In parallel, the MLSMR collection, which contained over 300,000 compounds, was also screened in the same

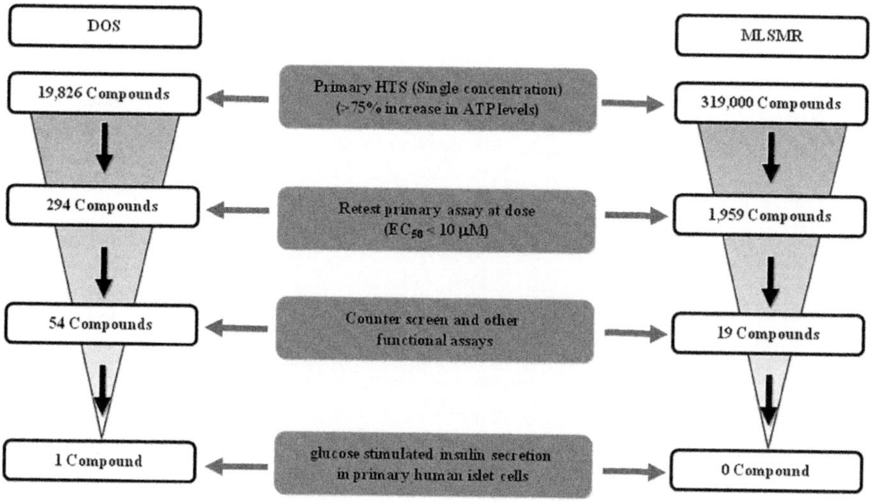

FIGURE 17.4 Primary HTS and hit triaging.

assay. Results from the primary assay and the subsequent hit triaging through a range of secondary assays are described in Figure 17.4. The primary HTS was conducted using the rat β-cell line INS-1E [26] in the presence of a cocktail of cytokines. ATP levels were determined as a surrogate for cell viability, and compounds increasing the normalized ATP levels by greater than 75% in this assay were considered hits. A total of 294 compounds (a hit rate of 1.5%) were identified as hits from the DOS collection, while 1959 compounds (a hit rate of 0.6%) scored as hits from the MLSMR collection. The stereochemical structure–activity relationship from the primary screen of the S_NAr library is shown in Figure 17.5. The 5*R*,6*R* configuration was required for activity at the internal stereocenters. However, the external (2*R* or 2*S*) configuration was less sensitive for activity. Analogs with urea substitutents on the aromatic ring were favored over sulfonamides and amides. At the secondary amine position, sulfonamides were preferred over ureas and amines. This SAR and SSAR coming directly out of the primary HTS was attributed to the upfront design of the library.

All of the hit compounds were retested in the primary assay at dose, and compounds with EC_{50} < 10 μM were advanced to a total of five secondary assays (counter screen and functional assays). Fifty-four compounds from the DOS collection and 19 compounds from the MLSMR collection were advanced to these secondary assays. Unfortunately, all the compounds from the MLSMR collection dropped out during the secondary assays. However, one DOS compound had a favorable outcome from this panel of secondary assays and was prioritized for medicinal chemistry efforts before evaluation of glucose-stimulated insulin secretion in primary human islet cells.

Having established a clean SSAR, more traditional structure–activity relationship (SAR) studies were undertaken to optimize potency. Three different regions of **SRR-14** (Table 17.1) were explored, and ultimately a more potent analog **SRR-15** (Figure 17.6) was identified when the *p*-methoxyphenylsulfonamide was switched to a benzodioxin based sulfonamide. The optimized **SRR-15** was characterized further

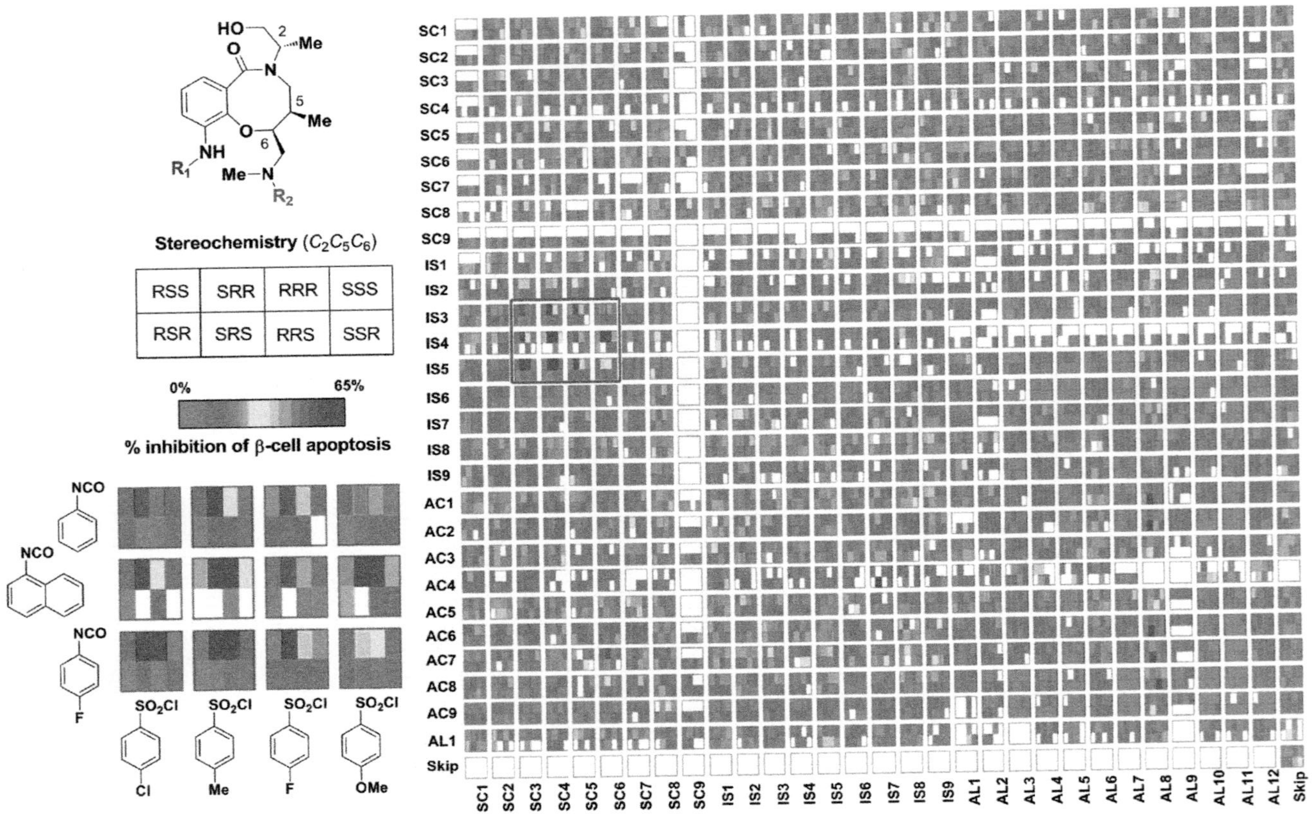

FIGURE 17.5 Primary screening data displayed as percent inhibition of β-cell apoptosis. The full matrix of building blocks is shown for R^1 (*y*-axis) and R^2 (*x*-axis). (From [23], with permission; copyright © 2011 American Chemical Society.) (*See insert for color representation of the figure.*)

TABLE 17.1 Stereochemical Structure–Activity Relationship

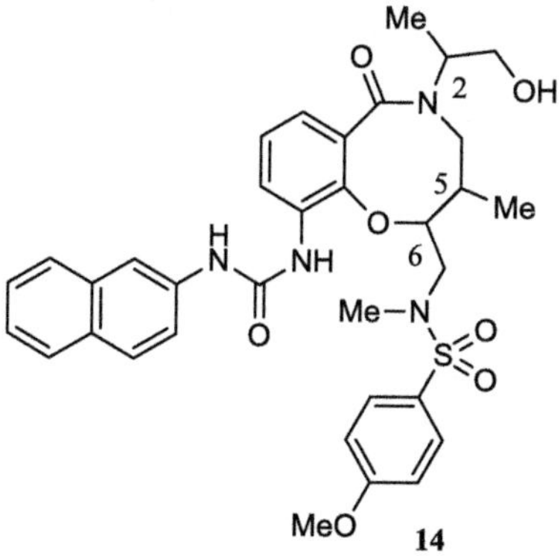

14

Configuration $C_2C_5C_5$	INS-1E Cell Viability[a] (EC_{50}, μM)	
	With Cytokines	Without Cytokines
SRR	2.71	IA
RRR	21	IA
SRS	NA	NA
SSR	IA	IA
RRS	26	IA
RSR	IA	IA
SSS	26	IA
RSS	IA	IA

[a]IA, inactive; NA, not available.

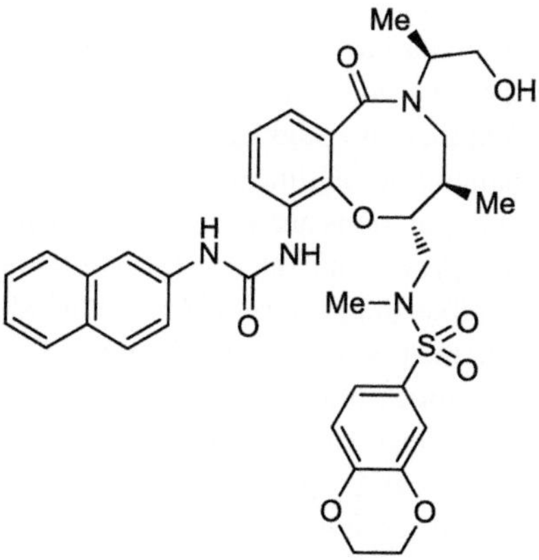

SRR-15

FIGURE 17.6 Structure of optimized compound **SRR-15**.

to explain its effects on β-cell biology. When β-cells undergo apoptosis, caspase-3 activity is high. However, **SRR-15** reduced the caspase-3 activity in a dose-dependent manner in apoptotic β-cells. Cytokine-induced β-cell apoptosis has been shown to decrease mitochondrial membrane potential, and **SRR-15** restored mitochondrial membrane potential, suggesting that β-cells are protected from cytokine-induced apoptosis. The most important physiological function of β-cells is to secrete insulin on stimulation with glucose. However, when β-cells were treated with cytokines, glucose-stimulated insulin secretion (GSIS) was abolished after two days. It is remarkable that when 5 μM of **SRR-15** was used, GSIS returned to normal levels even in the presence of cytokines. These data provide convincing evidence that **SRR-15** protects β-cells from apoptotic cell death induced by cytokines.

17.3 CASE STUDY 2: IDENTIFICATION OF ANTIMALARIALS

Malaria is a devastating worldwide parasitic disease that affects approximately 225 million people annually, killing 781,000 mostly young children in sub-Saharan Africa [27]. The disease is caused by five different species of malaria parasite, *Plasmodium falciparum* being the most virulent and deadly [28]. In terms of chemotherapy, a major challenge with this disease is resistance [29]. Drugs effective against the parasite lose their potency over time, with resistance to some drugs being documented almost as soon as it has been introduced [30]. This has led us to our current situation, where the parasite has gained resistance to almost every antimalarial developed. The only exceptions are artemisinin and its derivatives, but even here the signs are worrying, as populations in certain areas have experienced reduced clinical response to this drug, although detectable resistance has not been found [31]. Given the lack of an effective vaccine, it is imperative that chemotherapy remains a frontline solution to this problem.

While the search for antimalarials is centuries old, no new class of drug has been introduced into clinical practice since 1996 (Figure 17.7) [28]. The high risk of parasite resistance to current therapies highlights the need for additions to our current arsenal. An additional challenge in antimalarial development is safety, as the most vulnerable malaria patient populations are young children and pregnant women with limited access to medical supervision [32]. For the same reason, there is a need for antimalarials to be orally bioavailable for ease of administration, and for the cost of the drug to be low. A chemist approaching the antimalarial drug discovery effort is faced with these harsh realities plus the fact that these parasites have overcome almost every chemotype presented to them. The question then becomes: What is a starting point for a drug discovery program? Many of the drugs in development are based on existing pharmacophores (e.g., aminoquinolines and peroxides), chemically modified to improve upon their predecessors. It has been estimated that the number of distinct small-molecule scaffolds with genuine in vivo activity against *P. falciparum* is fewer than 30, with most of these not suitable for the clinic [33]. Although these compounds may represent an important strategy in the treatment of malaria, it has become increasingly difficult to manipulate these molecules to overcome resistance. It would be preferable to discover chemotypes with novel mechanisms of actions [34],

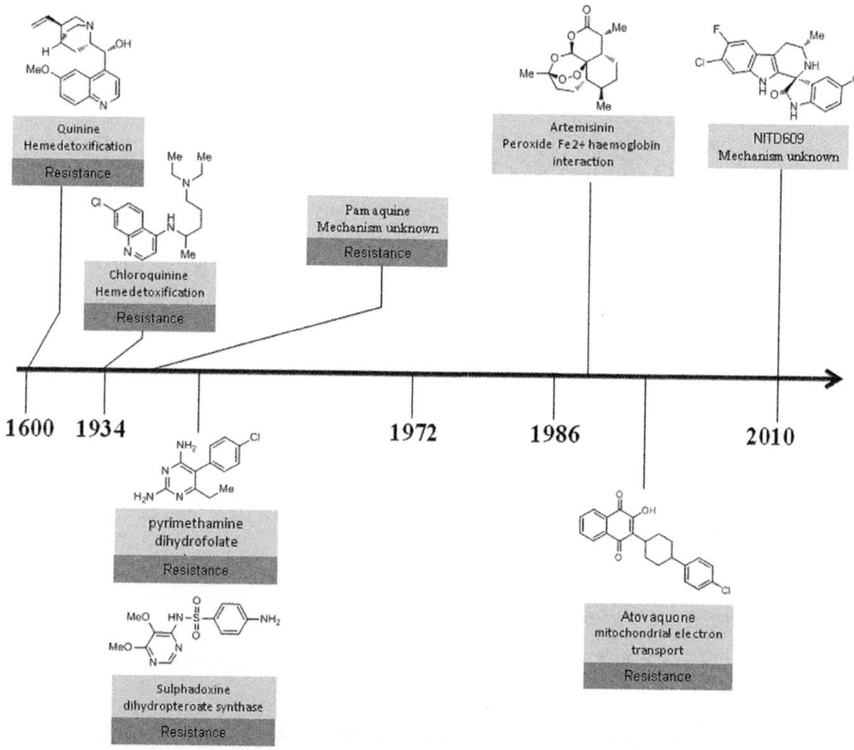

FIGURE 17.7 Time line for antimalarials with a new mechanism of action [40–43].

particularly as combination therapies with different targets are being used increasingly to lessen the rate of resistance. Recent efforts to uncover new antimalarial chemotypes [35] include large screening efforts involving up to 2 million compounds from GSK's screening collection [36]. This has produced some promising leads that have been developed [37,38], but none has yet progressed to the clinic. Given a generalized assumption that compounds with similar structures share a similar mode of action (the similarity principle) [39], if our goal is to produce antimalarials with new modes of action, diversity-oriented synthesis (DOS) can be an important strategy in the fight against this disease. In this section we highlight two examples of ways in which DOS is addressing needs in this area.

In 2010, Novartis reported a screening campaign of just 10,000 diverse compounds together with 2000 natural products, which found promising new antimalarial agents [44]. This study was a collaboration of the Novartis Institute for Tropical Diseases, the Genomics Institute of the Novartis Research Foundation (GNF), the Biomedical Primate Research Centre, and the Swiss Tropical Institute. The goal of this program was to develop antimalarials with new mechanisms of action, and their strategy incorporated diverse compounds, with new chemotypes, tested in phenotypic assays where no assumptions on the most relevant protein targets were made. Specifically, the GNF developed a high-throughput *P. falciparum* assay which enabled the screening of

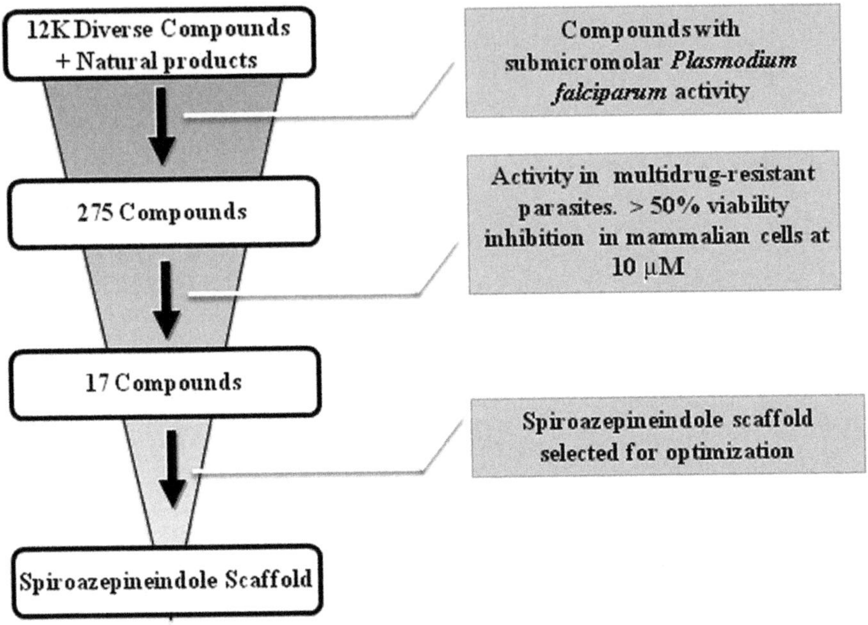

FIGURE 17.8 Summary of Novartis HTS screening campaign.

12,000 compounds in less than a week, a job that previously would have taken months to complete [45]. This led to the identification of 275 compounds with submicromolar activity (Figure 17.8). Primary hits that showed toxicity or which had unfavorable pharmacokinetic and physical properties were eliminated, and the spiroazepineindole **19** was selected as a starting point for further development.

This lead compound was prepared using Pictet–Spengler cyclization [46] and follows a build/couple/pair strategy for developing complex diverse libraries. Thus, synthesis of spiroazepineindole **19** involved an initial build phase (Figure 17.9) in

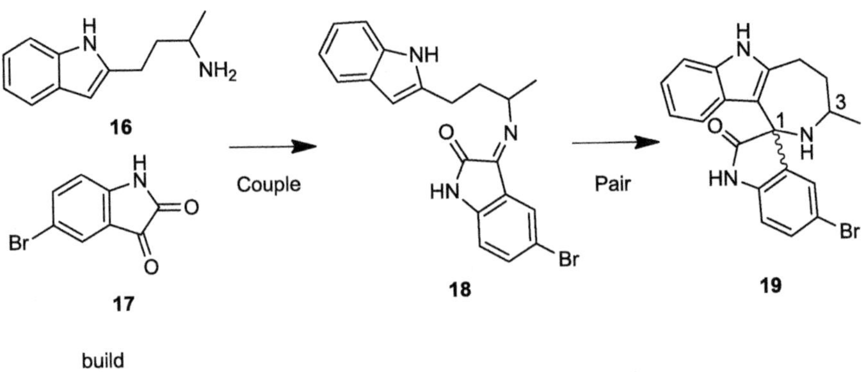

FIGURE 17.9 Preparation of spiroazepineindole **19** using a build/couple/pair strategy.

which compound **16** was prepared. This fragment contained orthogonal functionality that could initially partake in a condensation coupling reaction with 5-bromoisatin (**17**), followed by in situ diastereoselective pairing to afford the spiroazepineindole **19**. The importance of having access to every possible stereoisomer in a diversity library was seen upon chiral resolution of the reaction products. Testing the 1*R*,3*S* and 1*S*,3*R* stereoisomers of **19** in vitro against the NF54 parasite strain showed that the former was 250-fold more potent than its enantiomer (Table 17.2).

Libraries prepared using the build/couple/pair strategy facilitates subsequent medicinal chemistry by modularizing the synthetic route. For example, Novartis chemists could readily vary the size of the seven-membered ring primary hit (spiroazepineindole **19**) to six-membered spirotetrahydro β-carboline **25** (termed *spiroindolone*) simply by employing an indoleamine with one less methylene unit (Scheme 17.4). Thus, starting from D-tryptophanol, amine **23** could be obtained through overall reduction to a methyl group of the hydroxymethyl group in **20**. A diastereoselective Pictet–Spengler reaction provided the trans isomer (1*R*,3*S*)-**25** as the major component. Access to L-tryptophanol allowed for preparation of the enantiomer, and both substrates provided the separable cis isomer, making all possible stereoisomers obtainable. In practice, due to difficulty obtaining various starting materials, the Novartis chemists preferred a racemic route using chiral resolution to obtain pure compounds.

DOS libraries designed to access the full complement of stereoisomers allow for powerful SSAR and are invaluable in terms of both investigating potency (as seen above) and pharmacokinetic properties. As shown in Table 17.2, the in vitro potencies of the spirotetrahydro β-carboline series varied dramatically among the four possible stereoisomers. Additionally, significant differences between enantiomers were observed with respect to metabolic stability. For example, the highly potent 1*R*,3*S*-**25** had inferior metabolic stability and high clearance relative to its enantiomer 1*S*,3*R*-**25**, which displayed excellent pharmacokinetic properties but weak potency. Novartis chemists were able to overcome these liabilities by identifying parts of the molecule susceptible to CYP enzyme metabolism and blocking these sites with halogens to extend the compound's half-life. These standard medicinal chemistry techniques to improve clearance were no doubt aided by the ability to interchange fragments or portions of the molecule readily without having to develop new chemistry. Attempts to remove stereochemistry, such as the C3 methyl group (compound **26**) led to a decrease in activity. Compound **RS-27**, known as NITD609, has potency in the low-nanomolar range and shows no evidence of diminished potency against a number of drug-resistant strains, suggesting a new mechanism of action. This compound showed potency at early and late *P. falciparum* development, in contrast to the late (trophozoite)-stage specific activity observed with chloroquine, although NITD609 did not inhibit parasite growth as quickly as the artemisinin derivatives.

NITD609 showed no significant cytotoxicity against a panel of in vitro mamalian cell lines and was not mutagenic. Low binding to hERG and a series of human G-protein-coupled receptors, enzymes, and ion channels was also observed. In oral and intravenous PK studies in mice and rats, NITD609 had a good half-life ($T_{1/2}$ = 10 and 27.7 h, respectively), moderate volume of distribution (V_{SS} = 2.11 and 3.04

TABLE 17.2 Pharmacokinetic Profile and In Vitro Antimalarial Activity of Spiroindolones

Compound	Hepatic CL_{int}		In Vivo CL	NF54	Compound	Hepatic CL_{int}		In Vivo CL	NF54
	Mouse	Human	(mL/min/kg)	IC_{50} (nM)		Mouse	Human	(mL/min/kg)	IC_{50} (nM)
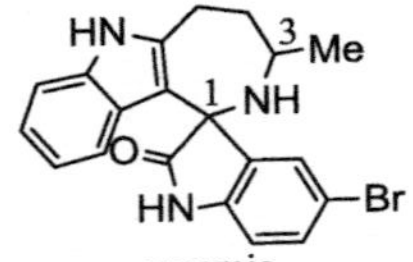 racemic **19**	—	—	—	90	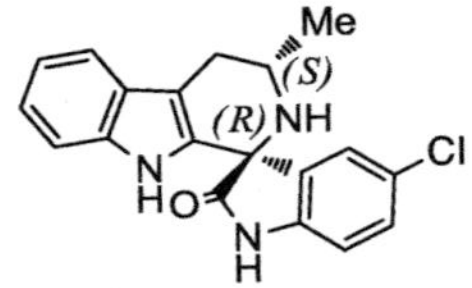 **RS-25**	High	High	49.66	9
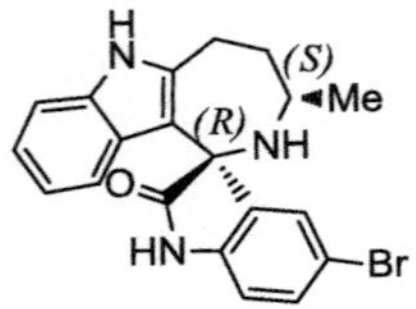 **RS-19**	—	—	—	20	**SR-25**	Low	Low	—	>5000
SR-19	—	—	—	>5000	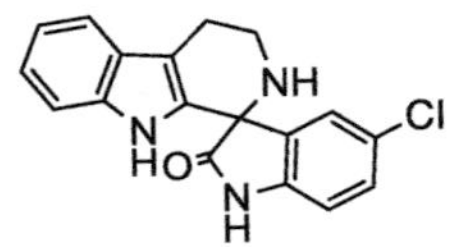 racemic **26**	—	—	—	126
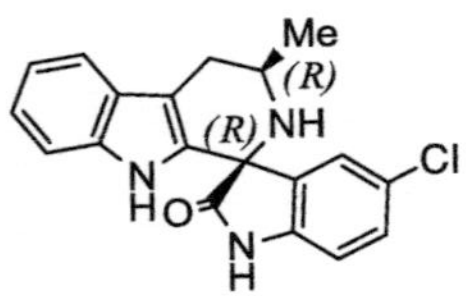 **RR-25**	—	—	—	444	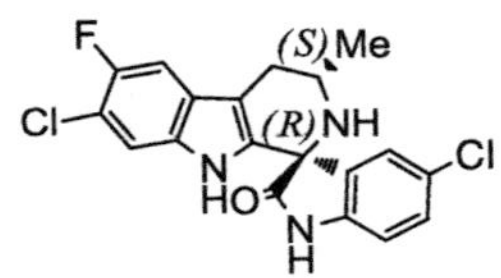 **RS-27**	Low	Low	9.75	0.9
SS-25	—	—	—	1808	**SR-27**	Low	Low	2.58	77

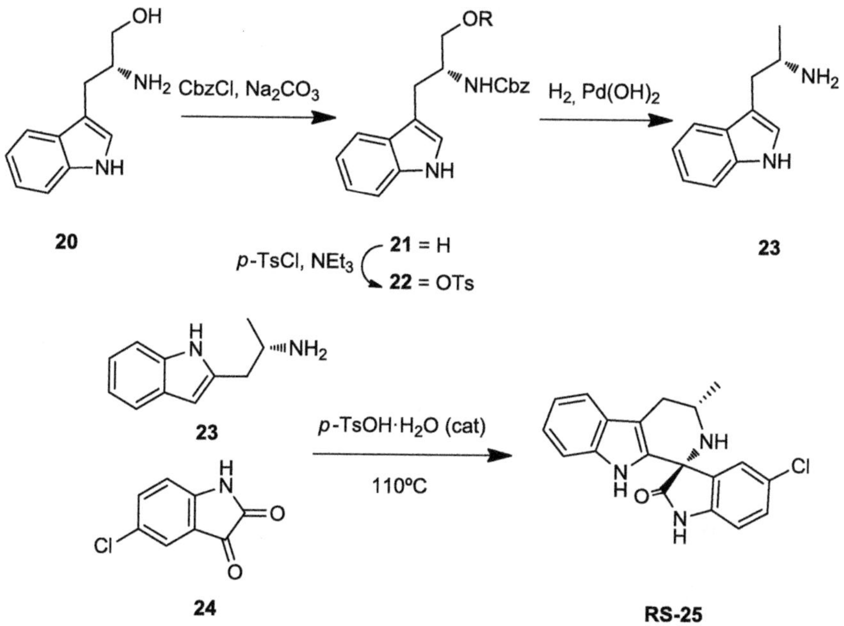

SCHEME 17.4 Preparation of spirotetrahydro β-carboline **25** (spiroindolone **25**).

L/kg), a low total systemic clearance (CL = 9.75 and 3.48 ml/min/kg) and excellent bioavailability ($F = 100\%$). Taken together, these properties are compatible with a once daily oral dosing regimen. In a *P. berghei* malaria mouse model, infection was cleared completely at a single oral dose of 100 mg/kg. Three daily oral doses of 50 mg/kg also afforded a complete cure. This is significant, as current antimalarial treatments require drugs to be taken between one and four times a day for up to seven days; thus the possibility of developing NITD609 into a once-daily oral drug could make treatment significantly easier in resource-deficient regions.

The goal of this study was to uncover new mechanisms of action in a cell-based assay. While cell-based screening has many advantages, such as presenting many targets for the compound to interact with in a biologically relevant setting, a major hurdle then becomes identification of the target of these small molecules. An excellent approach to target identification in antimalarials is to apply resistance selection coupled with next-generation sequencing [47]. In this approach, sublethal drug pressure is applied to a parasite strain until resistance to the drug is developed. Analysis of the genome of the drug-resistant strain can reveal the basis for resistance and the target of the compound. When a similar strategy was applied by Novartis researchers using NITD609 and the Dd2 parasite strain, it was discovered that three to four months of constant drug pressure was required to observe a 7- to 24-fold decrease in IC_{50} values. This high number of passages to achieve modest resistance could be interpreted as a positive attribute of the drug and suggests that the parasites have difficulty developing resistance. Importantly, none of the resistant parasites showed cross-resistance to a panel of known antimalarial agents, further suggesting a new mechanism of action. Analysis of the resistant line genome relative to the nonresistant parental reference line identified mutations in *pfatp4*, which is a cation transporting P-type adenosine triphosphatase. While further work is necessary to establish whether PfATP4 is the true target of NITD609 or part of a mechanism to reduce its effectiveness, its identification suggests a departure from the known mechanisms of action of the existing antimalarials. NITD609 is currently in phase IIa clinical trials and would represent the first antimalarial with a novel mechanism of action to progress to this stage in the last 20 years.

With the goals of creating a transformative screening collection, scientists at the Broad Institute created a diverse library of >100,000 compounds using a DOS strategy. Compounds in the Broad screening collection contain zero to three diversity sites and up to five stereogenic centers. A full matrix of stereoisomers was synthesized, giving rise to rich SSAR as well as SAR. In an attempt to discover novel antimalarials, a representative 8000-membered DOS "informer" subset of the larger screening collection was screened using a whole-cell assay [48]. For each compound selected to be in the screen, all possible stereoisomers were included. The assayed compounds were selected computationally for maximum diversity, representing all libraries and chemotypes in the overall collection. It was reasoned that compound selection in this informer set would enable the rapid development of SSAR and SAR for the prioritization and optimization of hit compounds. High-throughput screening of the compounds at 5 μM was conducted using a growth inhibition assay with multidrug-resistant Dd2 *P. falciparum* parasites (Figure 17.10). Greater than 90% growth inhibition was seen

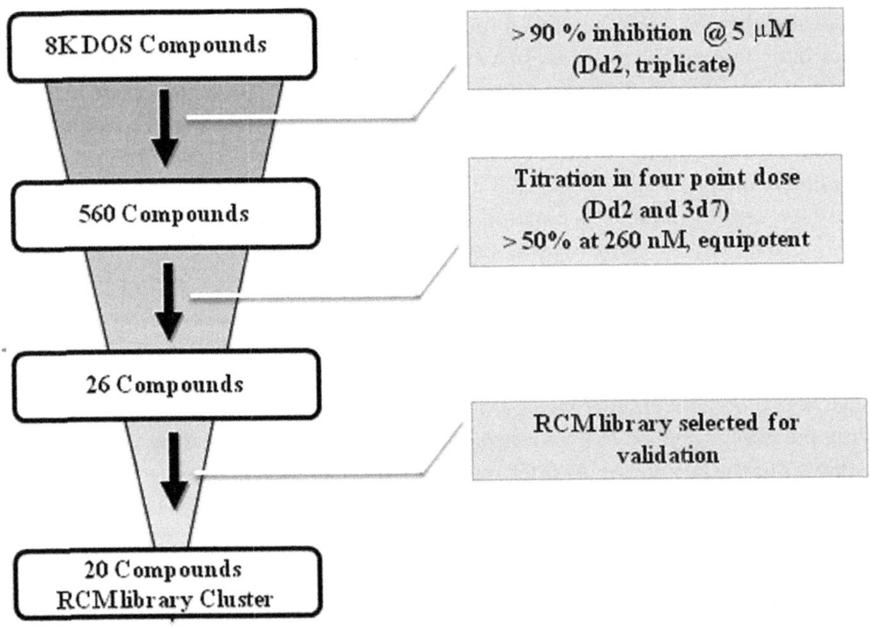

FIGURE 17.10 Summary of Broad Institute's HTS screening campaign.

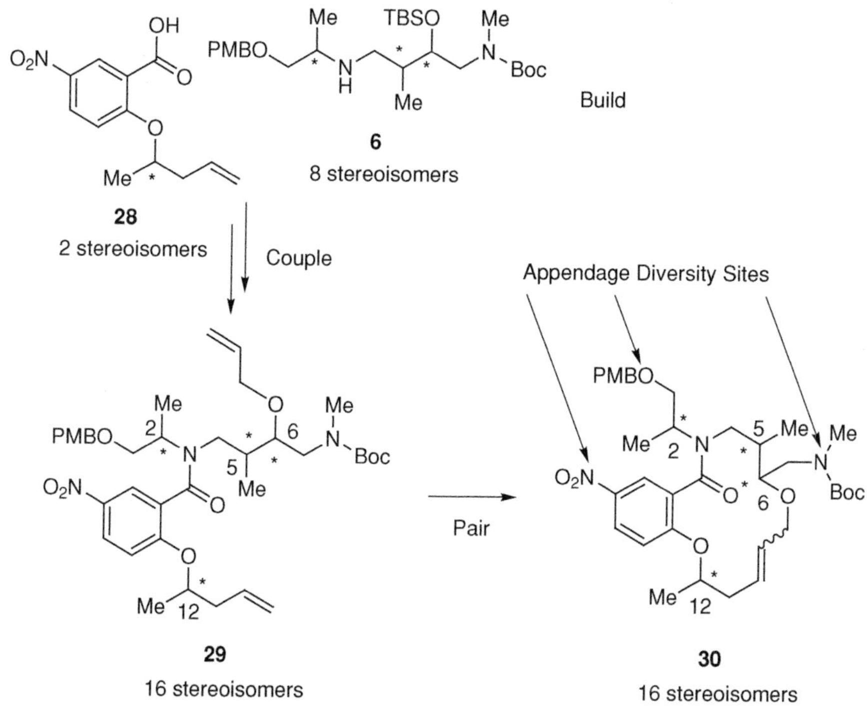

FIGURE 17.11 Preparation of macrocycles using the build/couple/pair strategy.

with 560 compounds, and titration at a four-point dose further narrowed the list to 26 candidates. Of these compounds, 20 compounds were from the ring-closing metathesis (RCM) macrocyclic library, and this scaffold was selected for further studies. The primary hit compound was titrated in a 12-point assay to give a potency (EC_{50}) of 120 nM against Dd2 erythrocytic parasites. The RCM library is a series of 14-membered macrocycles with four stereogenic centers. This chemotype has no parallels in malaria therapy, and indeed, macrocyclic rings are often underrepresented in traditional screening collections, probably because their syntheses are often considered difficult to develop.

The RCM library was prepared using a B/C/P strategy and involved an initial build phase of chiral building blocks using an asymmetric aldol reaction and chiral reagents (Figure 17.11). The carboxylic acid (**28**) and amine (**6**) building blocks were then coupled to provide the corresponding amide (**29**) in all possible stereochemical combinations. Intramolecular cyclizations (the pair phase) utilized the RCM macrocyclization reaction. Using this strategy, the complete stereochemical matrix of RCM compounds was created in a modular fashion, lending itself to the generation of powerful SSAR. The original RCM library, a subset of which was included in the informer set, used solid-phase chemistry to create a library of 14,440 macrocycles. Compound **30** contains a masked aniline, a protected secondary amine moiety, and a protected alcohol, all excellent functionality for medicinal chemistry investigation.

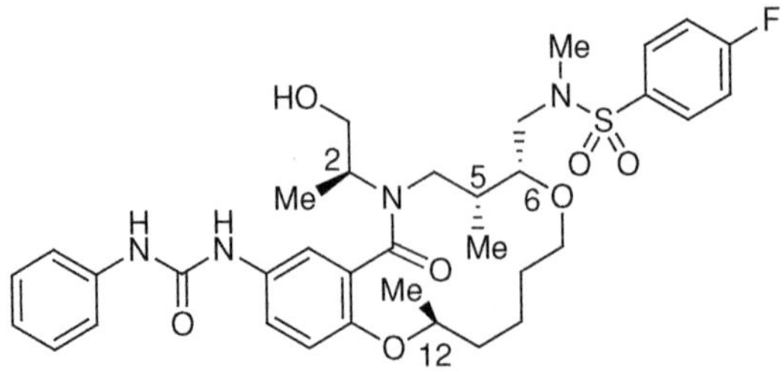

HTS Hit Compound **31**

Stereochemistry: SRRS
Dd2 (EC_{50}): 120 nM
3d7 (EC_{50}): 120 nM
Hemolysis (EC_{50}): >40 μM
Solubility (water): <500 nM

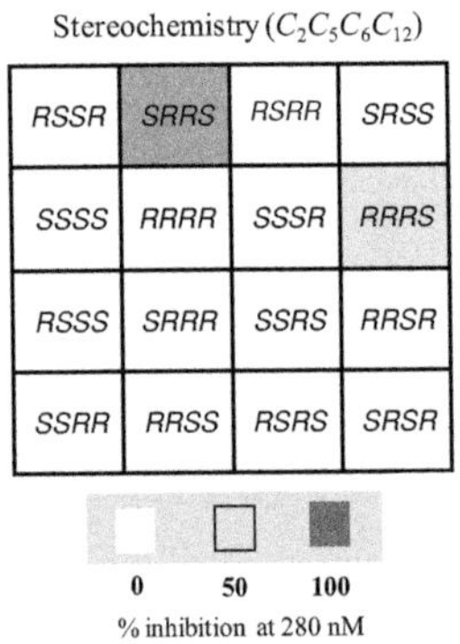

FIGURE 17.12 SSAR of hit compound **31**.

As was the case with NITD609, biological activity was dependent on the compound's stereochemistry. SSAR analysis of all 16 isomers of macrocycle **30** showed that biological activity is observed in just two isomers having the same ring configuration and differing only in the exocyclic C2 stereochemistry (Figure 17.12). The most potent compound, **SRRS-31**, was not toxic to erythrocytes but was insoluble in aqueous solution (<0.5 μM in water), which hindered analysis in further assays. SAR studies were initiated on this compound focusing on substituents at three areas of the molecule: (1) the aniline site, (2) the amine side chain, and (3) the lactam side chain, with a goal of improving potency and aqueous solubility.

The synthesis of analogs was performed in solution and followed the strategy reported by Marcaurelle et al. [16c]. In general, the analogs were prepared through an initial acylation of aromatic acid **28** with linear amine **6**, which was derived from an anti-selective aldol reaction adapting the ephedrine-based protocols reported by Inoue, Abiko, and co-workers [21]. The resulting amides were subjected to TBS deprotection and allylation to give RCM precursors **29** in three steps. The outcome of the RCM macrocyclization was found to be highly dependent on the stereochemistry at C5 and C6, with those derived from the anti-selective asymmetric aldol reaction being much more efficient and facile then those derived from syn-selective aldol reaction. Thus, compound **29** readily provided the 14-membered macrocycle as a mixture of *E*/*Z* isomers using the second-generation Hoveyda–Grubbs catalyst (Scheme 17.5). Reduction of a mixture of olefins and a nitro group provided the first opportunity for

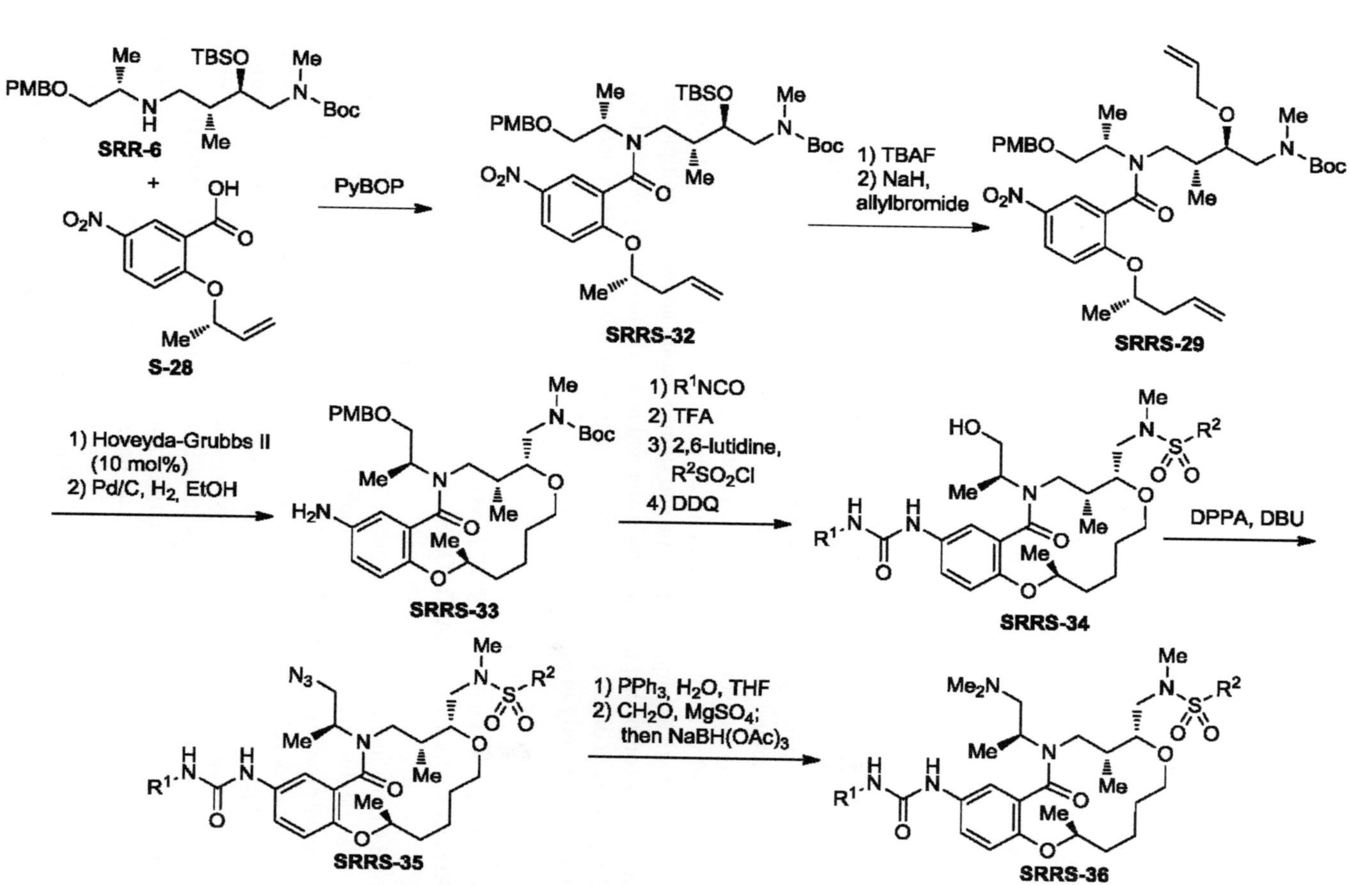

SCHEME 17.5 Solution-phase preparation of macrocycles.

derivatization, which in the case of **31** was achieved using an isocyanate. Subsequent removal of the Boc group allowed for derivatization at the amine side chain. Finally, the alcohol at C1 could be deprotected using DDQ, and nitrogen-based functional groups could be accessed with DPPA under Mitsunobu conditions.

Broad Institute scientists initially explored the aniline site (Table 17.3), and while all analogs saw a loss of potency, preparation of the 4-amino-3,5-dimethylisoxazoyl urea analog **40** provided valuable aqueous solubility. Attempted optimization of the amine side chain resulted in a loss of biological activity for all compounds, even with the regioisomeric 2-fluorophenylsulfonamide **42**, suggesting that this portion of the molecule contains an essential pharmacophore.

Substituents of the amide side chain were then investigated (Table 17.4). Interestingly, the *para*-methoxybenzyl ether **47**, which is accessed en route to alcohol **31**, had subnanomolar potency ($EC_{50} = 0.18$ nM). A range of groups were appended to the alcohol and showed similar improvements, but lacked solubility in water. The solution to the compound's poor solubility came from the introduction of nitrogen-based functional groups. As shown in Table 17.3, introduction of an azide produced a highly potent analog (**51**) but one still lacking aqueous solubility. Staudinger reduction, followed by bis-methylation of the resulting amine, affords compound **54** with both subnanomolar potency ($EC_{50} = 0.54$ nM) and solubility in water (water solubility = 124 μM), achieving the initial objective of discovering a highly potent antimalarial with good aqueous solubility. This compound is more potent than chloroquine or artesunate and is similar in potency to atovoquone. Compound **54** is a promising antimalarial with a unique chemotype and subnanomolar activity in multiple malaria strains. This compound is soluble in water and nontoxic to erythrocytes and HepG2 cells. Studies to discover the mechanism of action for this new class of antimalarial agents are currently in progress.

17.4 CASE STUDY 3: TARGETING PROTEIN–PROTEIN AND PROTEIN–DNA INTERACTIONS

Considerable attention has been cast of late on protein–protein interactions (PPIs) and their ability to be perturbed by small molecules. PPIs play a critical role in many biological processes, and the ability to stabilize [49] or disrupt these interactions is an approach to developing new therapeutics [50]. Recent literature has estimated about 130,000 PPIs in the human interactome, of which only about 8% have been identified [51]. In the past, these targets were largely avoided, due to the thought that the protein surfaces were too large and featureless to be influenced by a small molecule. However, evidence suggests that defined areas termed *hot spots*, on the protein surface are ideal for small-molecule interactions and can have a large influence on the function of the protein [52]. Another preconceived notion of these targets being "undruggable" could also have arisen on past failures of HTS campaigns. However, recent successes in disrupting PPIs either with structurally more complex compounds or those beyond the rule of 5 (BRo5) has indicated that the existing screening collections may have been a contributing factor [53–55].

TABLE 17.3 Comparison of Antimalarial Activity: SARs of the Aniline and Amine Substituents

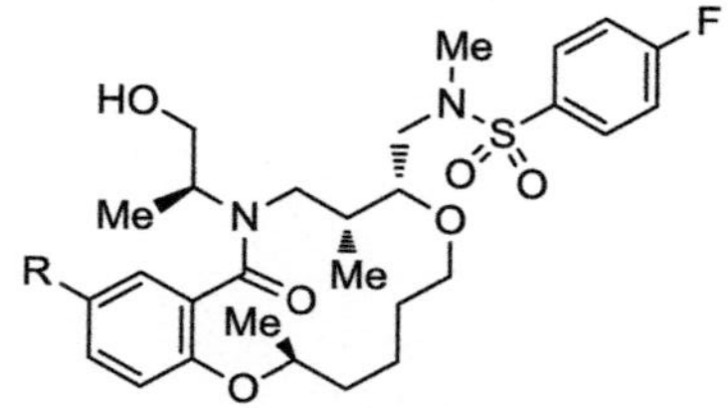			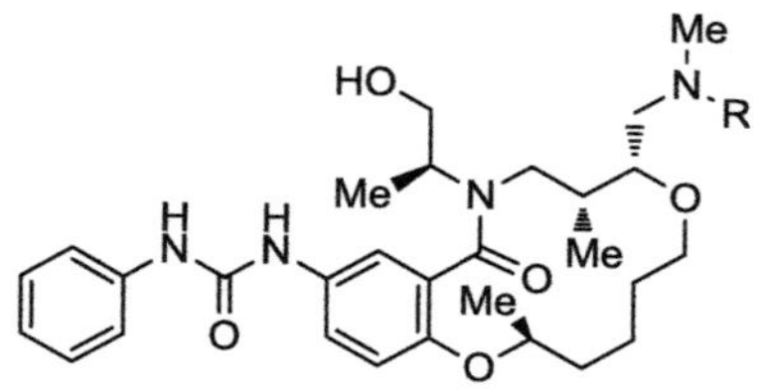		
R	Growth Inhibition (Dd2, EC_{50}, nM)	Solubility (Water, μM)	R	Growth Inhibition (Dd2, EC_{50}, nM)	Solubility (Water, μM)
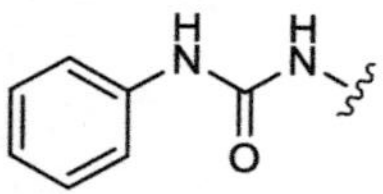 **31**	120	<0.5	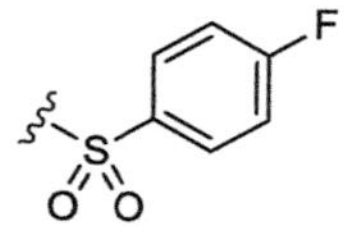 **31**	120	<0.5
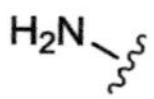 **37**	990	—	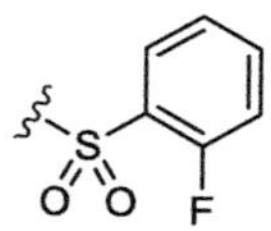 **42**	1100	<0.5
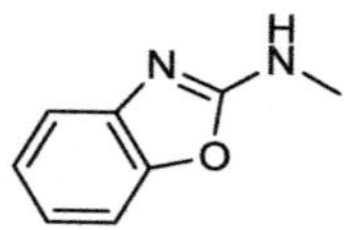 **38**	580	<0.5	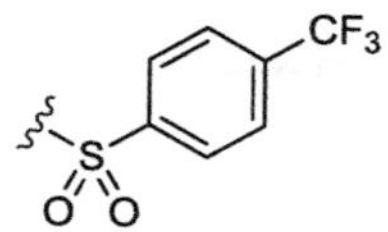 **43**	877	<0.5

(continued)

TABLE 17.3 ***(Continued)***

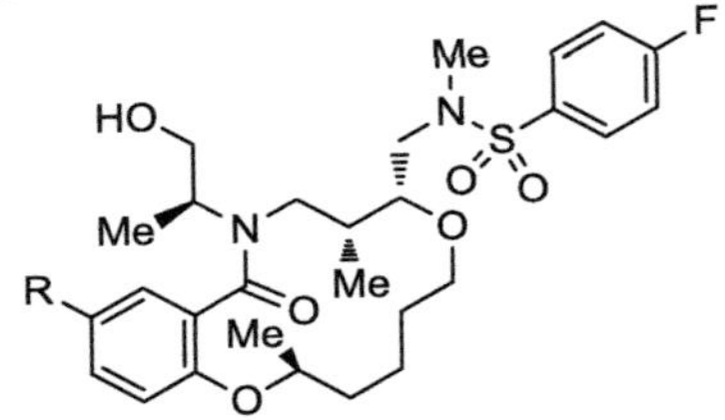

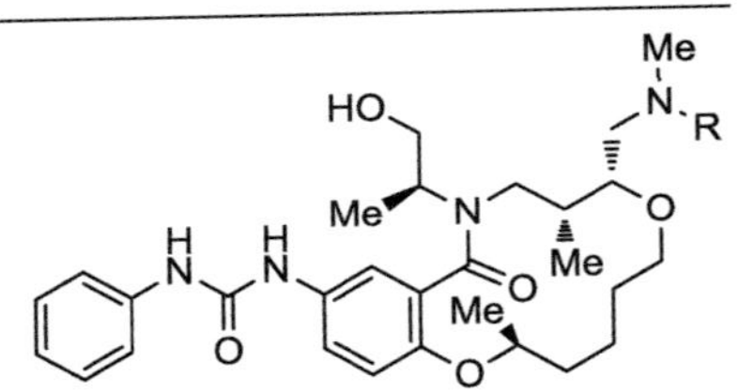

R	Growth Inhibition (Dd2, EC_{50}, nM)	Solubility (Water, μM)	R	Growth Inhibition (Dd2, EC_{50}, nM)	Solubility (Water, μM)
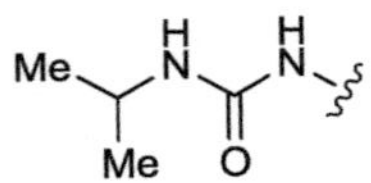 **39**	400	2.2	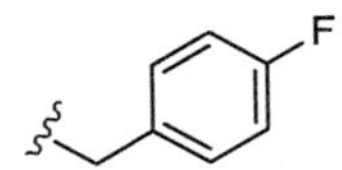 **44**	4300	<0.5
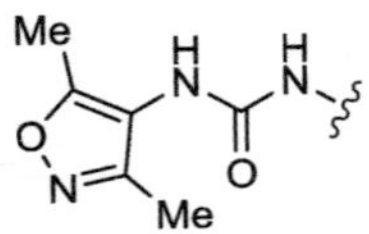 **40**	240	45	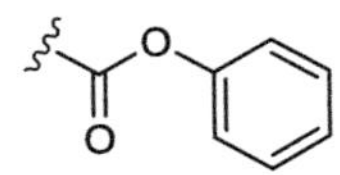 **45**	>5000	1.2
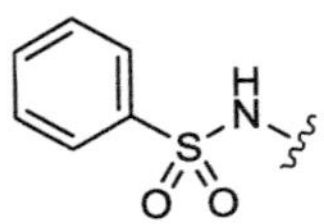 **41**	2400	—	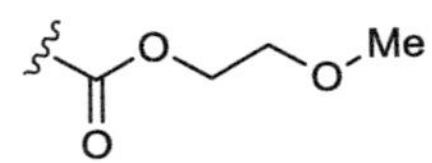 **46**	>5000	280

TABLE 17.4 Comparison of Antimalarial Activity: SARs of the Amide Substituent

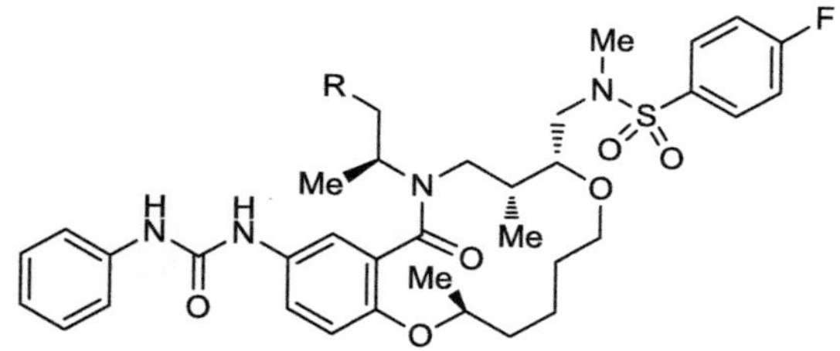

R	Growth Inhibition (Dd2, EC_{50}, nM)	Solubility (Water, μM)	R	Growth Inhibition (Dd2, EC_{50}, nM)	Solubility (Water, μM)
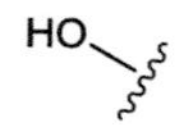 **31**	120	<0.5	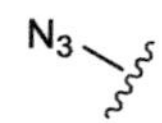 **51**	3.2	<0.5
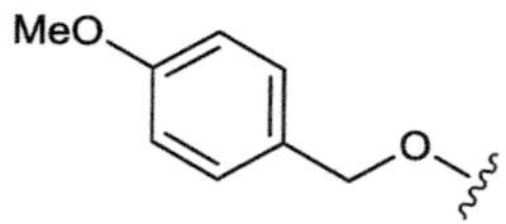 **47**	0.18	<0.5	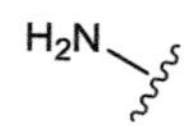 **52**	370	270

(*continued*)

TABLE 17.4 (*Continued*)

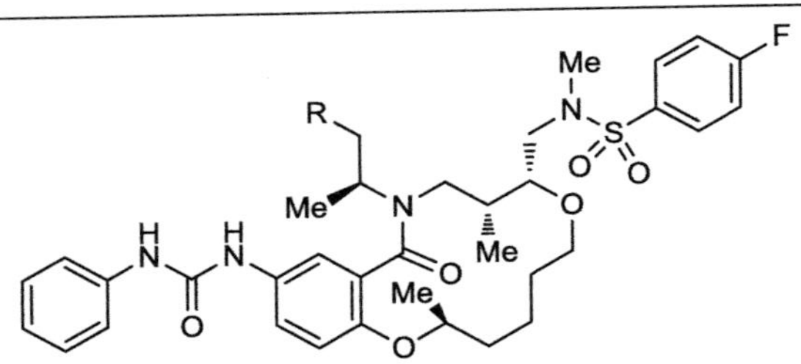

R	Growth Inhibition (Dd2, EC_{50}, nM)	Solubility (Water, μM)	R	Growth Inhibition (Dd2, EC_{50}, nM)	Solubility (Water, μM)
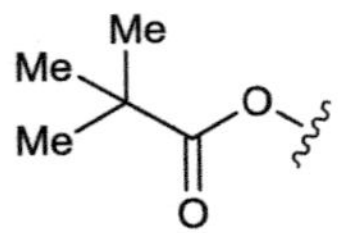 **48**	7.3	<0.5	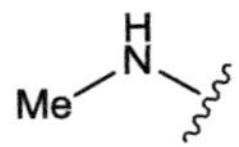 **53**	79	400
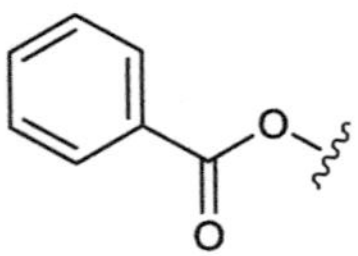 **49**	15	<0.5	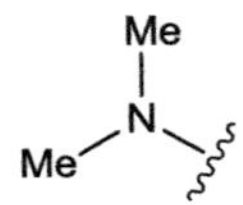 **54**	0.54	120
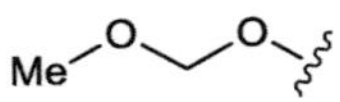 **50**	8.2	<0.5			

Although not receiving as much attention as protein–protein interactions, transcription factors also play an important role in biological systems, regulating gene expression based on environmental signals [56]. Like protein–protein interactions, disruption of protein–DNA interactions could have a tremendous therapeutic benefit, especially in scenarios of known increased expression having been tied to a disease pathway. However, for the same reasons as PPIs, transcription factors have largely not been targeted. Additionally, transcription factor inhibitors must have selective binding for normal gene expression to occur [57,58]. In this section we describe recent success in inhibiting both protein–protein and protein–DNA interactions with compounds derived from DOS libraries. First, we review work done at Infinity Pharmaceuticals targeting protein–protein interactions involving BCL-2. Second, we look at a small-molecule library synthesized at the Broad Institute, leading to a selective inhibitor of the transcription factor HOXA13.

β-Cell lymphoma 2 (BCL-2) is a pro-survival protein that is overexpressed in many types of cancer cells. This increase in protein levels shows that cell survival, as well as cell proliferation, could be an important role in tumor growth [59,60]. BCL-2 is one of five closely related pro-survival proteins in the mitochondrial pathway for programmed cell death; BCL-X_L, BCL-*w*, MCL1, and A1 are the other four proteins that promote cell life. These pro-survival proteins antagonize two pro-apoptotic proteins, BAX and BAK, which are localized on the mitochondrial outer membrane and can activate the caspase cascade, and ultimately apoptotic cell death, through release of cytochrome *c* into the cytosol. Programmed cell death is tightly regulated through these proteins and is essential for the removal of aged, damaged, or unnecessary cells. In normal cells, the function of BAX or BAK is suppressed by a pro-survival protein until an apoptotic signal is received. The conserved BCL-2 homology 3 (BH3) domain of pro-apoptotic proteins is the primary interaction with pro-survival proteins responsible for cell survival. In an environment where pro-survival proteins are overexpressed, disruption of these protein–protein interactions, either by an antagonist of a pro-survival protein or an agonist of a pro-apoptotic protein, is required for apoptotic cell death to occur. A limited set of small molecules have shown antagonist activity against one of the pro-survival proteins in the extended BCL-2 family and are currently in various stages of clinical trials [59,61].

The first such inhibitors of BCL-2 derived from a DOS strategy were developed from a diverse library of mono- and bicyclic lactones or the corresponding amide [62]. The synthetic sequence started with a tandem esterification/1,3-dipolar cycloaddition between allylic alcohol **55** and a nitrono ester using $Ti(O\textit{i}\text{-}Pr)_4$ to give bicyclic lactone **57** (Scheme 17.6). Appendage diversity is built into the library using a variety of nitrono ester building blocks (ortho, meta, and para substitution on the aryl ring) and through orthogonal protecting groups on the allylic alcohol, which allows for either substitution on the lactone or the generation of an exocyclic secondary alcohol. This approach allowed for the generation of 12 different scaffolds, including enantiomeric pairs for generation of SSAR. Library production on solid-support lanterns allowed for further appendage diversity through Sonogashira coupling.

Interestingly, further solid-phase chemistry generated additional skeletal diversity (Scheme 17.7). N–O bond cleavage resulted in monocyclic lactone **59**, while

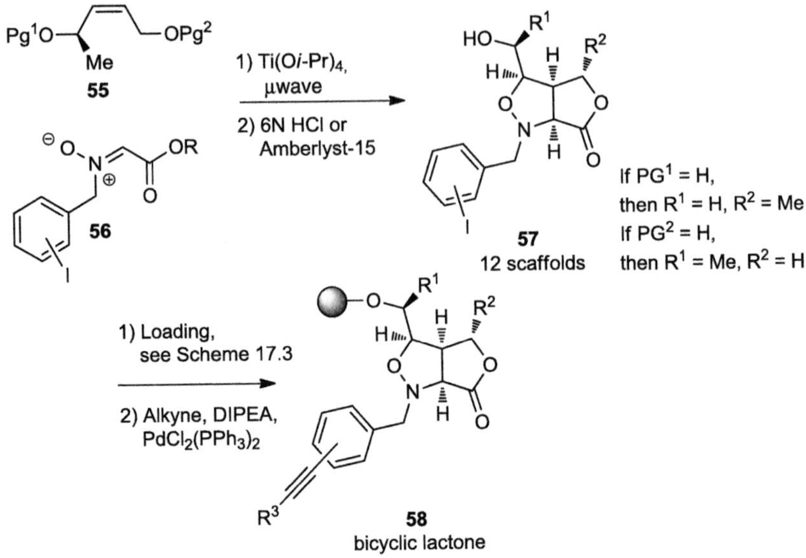

SCHEME 17.6 Synthesis of a bicyclic lactone scaffold used in the isoxazolidine library synthesis at Infinity Pharmaceuticals.

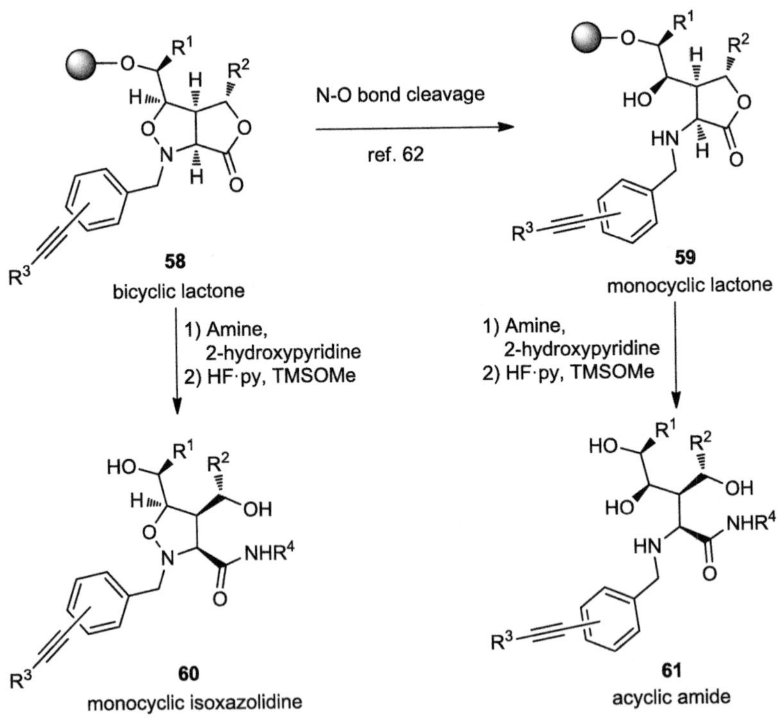

SCHEME 17.7 Access to bi-, mono-, and acyclic scaffolds in the isoxazolidine library, providing rich skeletal diversity.

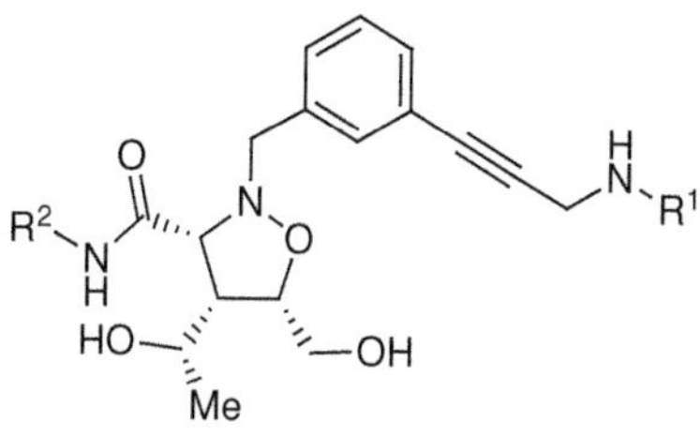

62
micromolar binding affinity to Bcl-2

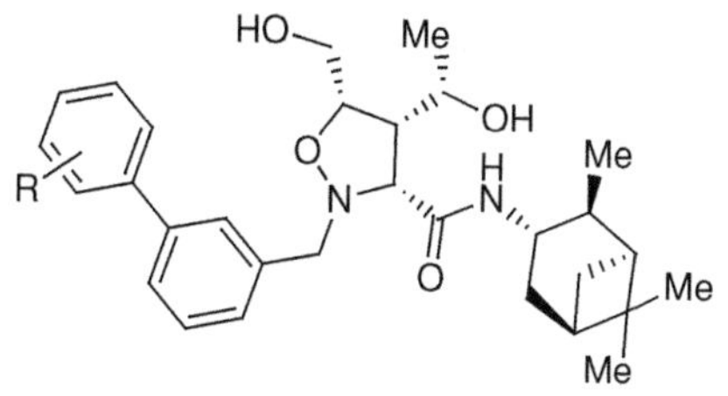

63
nanomolar binding affinity to Bcl-2

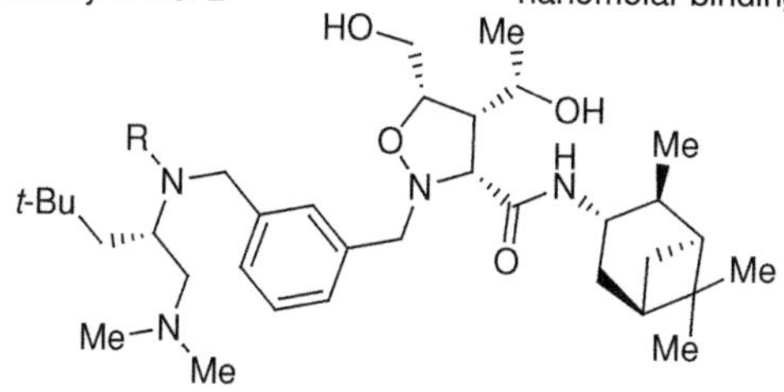

64
picomolar binding affinity to Bcl-2

FIGURE 17.13 Discovery of a novel inhibitor of BCL-2 with micromolar binding affinity from the primary screen using members of the isoxazolidine library. Subsequent optimization provided a compound with picomolar binding affinity to BCL-2.

amidation of the bi- or monocyclic lactone gave the corresponding mono- or acyclic amides, respectively. After cleavage from lanterns using HF/pyridine, a total of 20,000 unique compounds were generated for screening in biological assays.

This library was screened for activity against BCL-2 and BCL-xl at Infinity Pharmaceuticals. Based on a review of the patent [63], the monocyclic isoxazolidine scaffold showed micromolar binding affinity directly out of the primary screening deck, a remarkable feat given the limited success with targeting protein–protein interactions. Hits showed that a variety of alkyl alkynes could be used in Sonogashira cross-coupling, but meta-substitution proved to be critical for activity, showing the importance of including regioisomers in the original library design (**62**, Figure 17.13). The amide also showed that a variety of substituents resulted in activity. However, the bicyclic lactone showed no activity, illustrating that the amide was necessary for BCL-2 inhibition. The acyclic amide was also not mentioned, showing the necessity for the isoxazolidine ring. Another important aspect in the original screen was the generation of strong SSAR, as the enantiomer of the lead compounds showed no binding affinity for BCL-2.

Subsequent optimization of the original BCL-2 inhibitor from the isoxazolidine series has resulted in a picomolar binder for BCL-2. Although a variety of amines can be used to open the lactone and still show activity, the use of (+)-isopinocampheylamine resulted in the strongest activity. Continued optimization of the aryl substituent resulted in replacing the meta-substituted alkyne with a variety

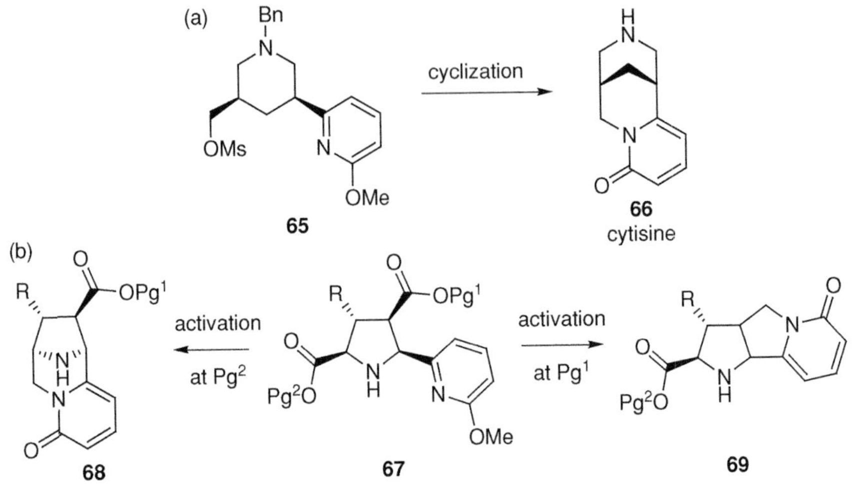

FIGURE 17.14 (a) Key cyclization in the total synthesis of cytisine; (b) key cyclization in a novel DOS library to access a bridged bicyclic scaffold and a tricyclic scaffold.

of substituted aryl ring systems, still at the meta-position, increasing activity to nanomolar binding affinity (**63**). Replacing the biaryl motif for a benzyl amine led to a number of single-digit nanomolar binders, as well as two picomolar binders for BCL-2 (**64**). Preclinical trials for these lead compounds are still ongoing.

The second such example from Infinity involves a natural product–inspired DOS library, which also provided novel inhibitors of BCL-2 through an HTS campaign [64]. Cytisine (**66**), a bridged-bicyclic pyridone natural product, contains a therapeutically relevant motif, as analogs have been used both as antismoking and antiphospatase therapeutics. A DOS library was designed utilizing a key cyclization described in the total synthesis of cytosine [65] to form the pyridone motif (Figure 17.14a). By starting with highly functionalized pyrrolidine (**67**), along with suitable ester functional groups, access to bridged bicyclic **68** and tricyclic **69** was achieved (Figure 17.14b).

The synthesis for both scaffolds starts with a [3 + 2] cyclization between common pyridinium imine **71** and the appropriate α,β-unsaturated ester using AgOAc (Scheme 17.8). The resulting pyrrolidine is then subjected to a series of protecting group manipulations and the key cyclization to afford the scaffolds for library synthesis. Two important aspects should be considered: The chemistry was scalable, having achieved the synthesis on >75 g of material, and the scaffold synthesis was carried out on both enantiomers of each scaffold, providing valuable SSAR data in future HTS campaigns. Subsequent capping of the secondary amine with a variety of different functional groups (R^1), as well as incorporating different esters, amides, and carboxylic acids (R^2) at the carbonyl position, provided closely related analogs in the library to generate SAR in biological assays. In all, some 15,000 compounds were prepared with acceptable purities for screening.

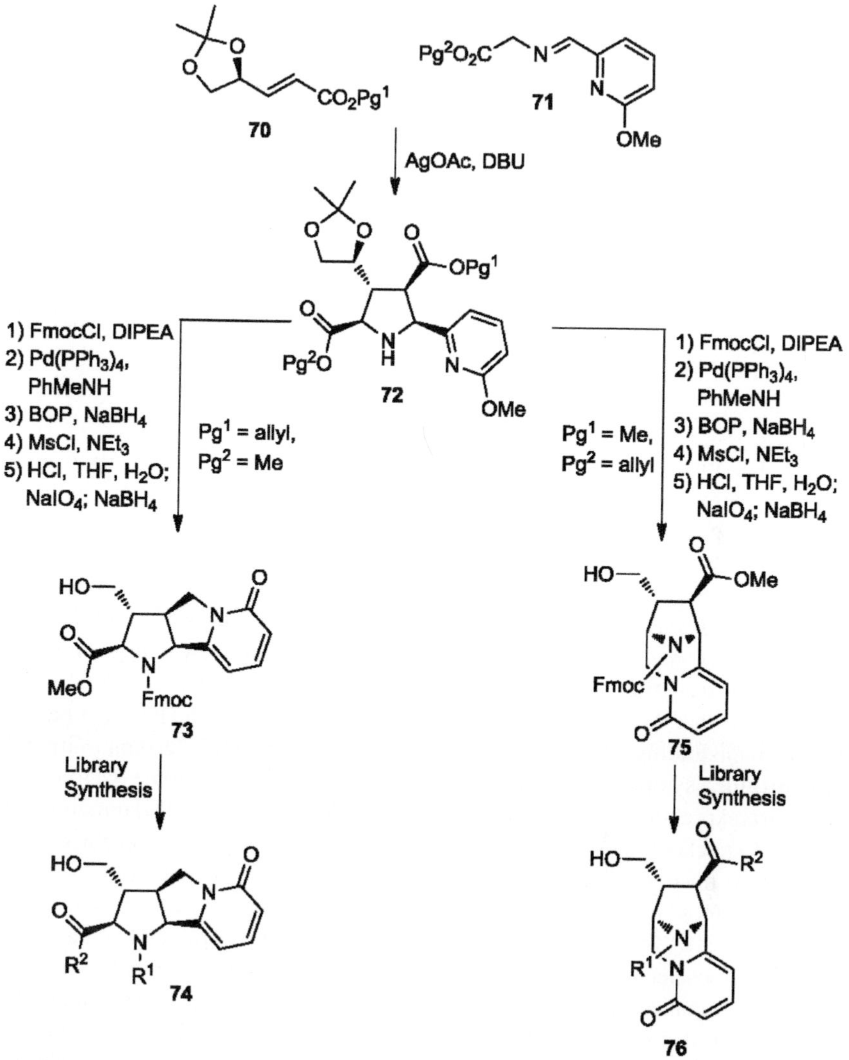

SCHEME 17.8 Synthesis of bridged bicyclic and tricyclic scaffolds and corresponding library synthesis, resulting in about 15,000 cytisine-inspired compounds.

The DOS library was screened against BCL-2 in a traditional competition binding assay against the BH3 peptide in the presence of an inhibitor. From the HTS, three different hit clusters emerged showing micromolar inhibition against BCL-2 (Figure 17.15). In all cases, structurally related analogs showed similar activity, giving further validation of the hits due to tractable SAR. Interestingly, the bridged-bicyclic pyridone hits had similar potency for both enantiomers, lending to a hypothesis of possible nonspecific binding. For both hit clusters from the tricyclic pyridone, stereochemistry was critical for activity, as only one enantiomer shows any activity against

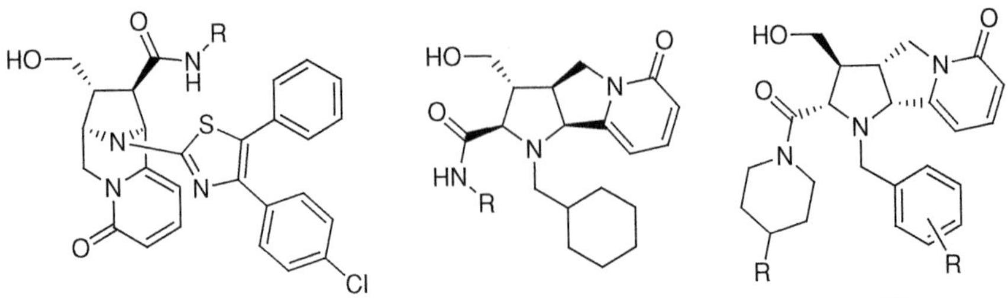

77
8 examples
Ki (μM) = 1.3 - 14.0

78
4 examples
Ki (μM) = 4.1 - 8.4

79
4 examples
Ki (μM) = 1.2 - 6.8

FIGURE 17.15 Three hit clusters that show inhibition against BCL-2.

BCL-2. The finding of three novel, micromolar BCL-2 inhibitors directly from the HTS campaign once again is remarkable considering the difficulty of identifying inhibitors of protein–protein interactions.

The transcription factor HOXA13 has been shown to regulate both tissue and tumor development. The HOX family of genes, which contain a DNA sequence known as the *homeobox*, is regulated through interaction with HOX proteins, transcription factors that can modulate specific target genes positively or negatively [66]. Two extreme examples of this regulation connected to a single gene, HOXA13, illustrate the role that these genes play in development. Mutations in HOXA13 have led to protein misfolding, cytoplasmic aggregation, and degradation leading to irregular limb growth, a symptom observed in the hand–foot–genital syndrome [67]. Alternatively, overexpression of the HOXA13 protein can lead to uncontrolled tumor growth. HOXA13 knockout mice showed drastically reduced tumor growth compared to that of wild-type mice, making it an attractable target for treating cancer [68].

Ng et al. took advantage of two different pathways starting from the same imine to give access to skeletally distinct scaffolds (Scheme 17.9) [69]. They had established a stereoselective cycloaddition between imine **80** and an anhydride to yield highly functionalized pyrrolidinones **83**. Chemistry established by Cushman et al. [70] used similar imines with the appropriate anhydride to give access to dihydroisoquinolinones. Interestingly, complete stereocontrol is observed to give the syn-product **81**, but when able, complete epimerization will take place, yielding the anti-product **82**.

Ng et al. used these two synthetic pathways to access skeletal, appendage, and stereochemical diversity. Library synthesis began with the generation of solid-supported imine **84**, which then reacted with the appropriate anhydride to give two different skeletons (Scheme 17.10). Appendage diversity was generated by coupling the resulting acid with a variety of amines. An additional layer of skeletal diversity was introduced with select building block combinations to give rise to fused tricyclic (**86**) and spirocyclic (**89**) ring systems. After cleavage from solid support, greater than 500 compounds were generated for screening in biological assays.

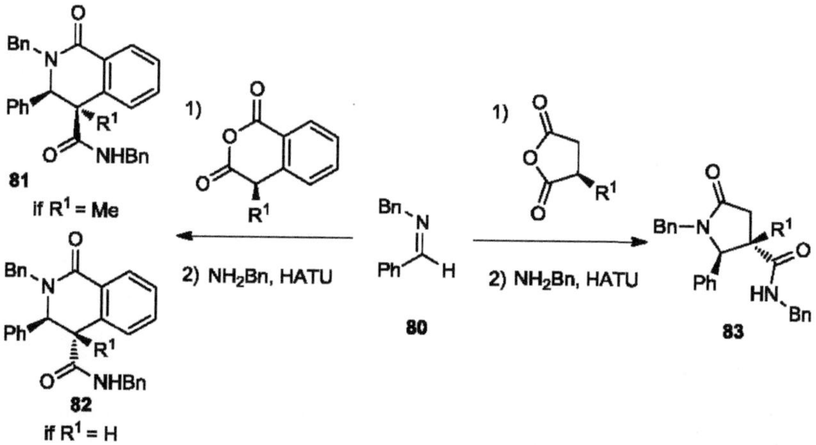

SCHEME 17.9 Synthetic strategy to access skeletally diverse lactams by Ng et al. [69].

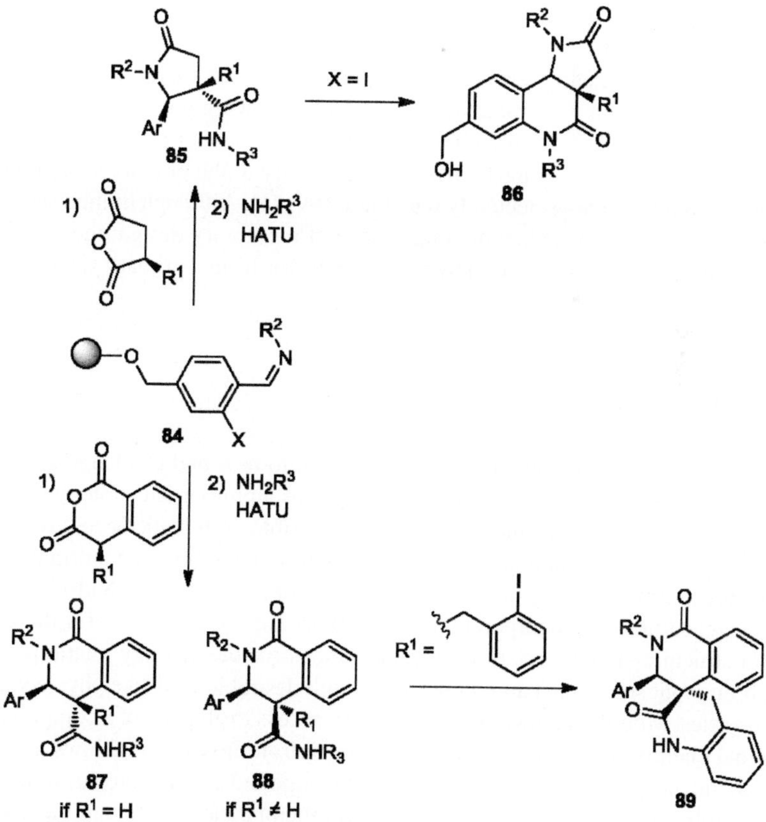

SCHEME 17.10 Library production on a solid phase to access appendage diversity across stereochemically and skeletally diverse scaffolds.

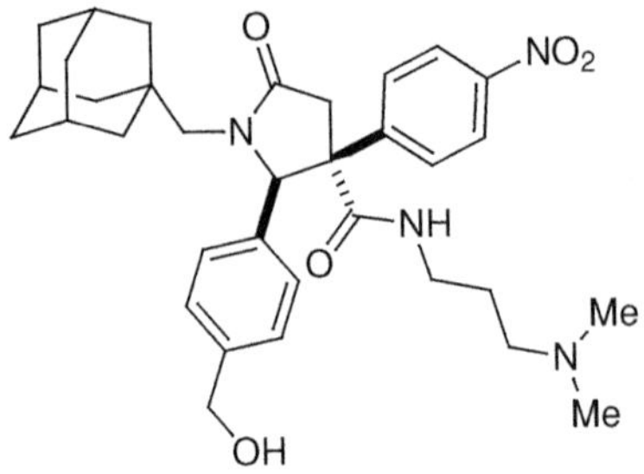

90
Selective inhibitor of
HOXA13 DNA interaction

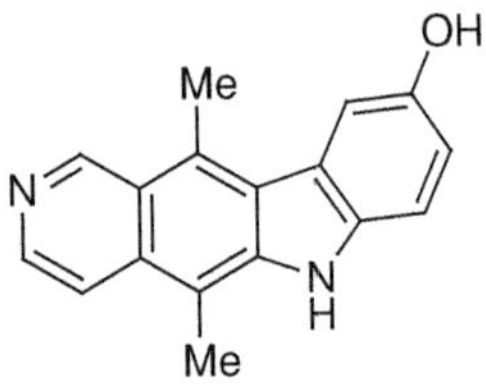

9-hydroxyelipticine

91
Unselective inhibitor of
HOXA13 DNA interaction

FIGURE 17.16 Selective and nonselective inhibitor of HOXA13 DNA interaction from the HTS of a pyrrolidinone library and known bioactives.

Screening of this library in a variety of assays revealed that two compounds inhibit the binding of HOXA13 with DNA. Two pyrollidinones (**90**, Figure 17.16, and a closely related analog not shown) were identified from the primary screen. The most potent showed an IC_{50} value of 6.5 μM while showing no nonspecific binding to DNA. The lack of nonspecific binding is important, as other compounds (such as 9-hydroxyelipticine, **91**, Figure 17.16) showed activity in the primary screen but are also known to interact nonselectively with the DNA [71]. Although limited screening data were available to establish the importance of the library design, the novelty of finding a micromolar inhibitor of a transcription factor from a primary screen cannot be ignored.

17.5 CONCLUSIONS

As probe and drug development continue to adapt to new and challenging biological targets, diversity-oriented synthesis will serve as an important strategy to access next-generation small molecules. DOS makes possible systematic access to natural product–like inspired compounds which incorporate complexity, an important feature that relates to lower toxicity and a greater chance of success through clinical trials. DOS-designed libraries incorporate near neighbors and stereoisomers for the generation of structure–activity relationships and stereostructure–activity relationships in high-throughput screening campaigns, along with desirable lead-like physicochemical properties. Modular pathways allow for easy access to all parts of a compound for medicinal chemistry purposes. Combined, this upfront investment allows access to novel chemical space, the prioritization of hit compounds, and strategies to develop lead or probe compounds, and, possibly, novel therapeutics. Although the concept is still relatively new, progress has shown that DOS will play a valuable role in the future of drug discovery.

REFERENCES

1. J. Drews, *Nat. Biotechnol.* **1996**, *14*, 1516–1518.
2. D. J. Payne, M. N. Gwynn, D. J. Holmes, D. L. Pompliano, *Nat. Rev. Drug Discov.* **2007**, *6*, 29–40.
3. (a) M. S. Butler, M. A. Cooper, *J. Antibiot.* **2011**, *64*, 413–425; (b) J. Bérdy, *J. Antibiot.* **2012**, *65*, 385–395.
4. (a) D. A. Erlanson, R. S. McDowell, T. O'Brien, *J. Med. Chem.* **2004**, *47*, 3463–3482; (b) M. Congreve, G. Chessari, D. Tisi, A. J. Woodhead, *J. Med. Chem.* **2008**, *51*, 3661–3680; (c) C. W. Murray, D. C. Rees, *Nat. Chem.* **2009**, *1*, 187–192.
5. (a) S. L. Schreiber, *Science* **2000**, *287*, 1964–1969; (b) M. D. Burke, S. L. Schreiber, *Angew. Chem. Int. Ed.* **2007**, *74*, 48–56; (c) T. E. Neilson, S. L. Schreiber, *Angew. Chem. Int. Ed.* **2007**, *74*, 48–56; (d) W. R. J. D. Galloway, A. Isidro-Llobet, D. R. Spring, *Nat. Commun.* **2010**, *1*, 1–13.
6. A. Nören-Müller, I. Reis-Corrêa, Jr., H. Prinz, C. Rosenbaum, K. Saxena, H. J. Schwalbe, D. Vestweber, G. Cagna, S. Schunk, O. Schwarz, H. Schiewe, H. Waldmann, *Proc. Natl. Acad. Sci. U.S.A.* **2006**, *103*, 10606–10611.
7. P. A. Wender, V. A. Verma, T. J. Paxton, T. H. Pillow, *Acc. Chem. Res.* **2008**, *41*, 40–49.
8. A. Nadin, C. Hattotuwagama, I. Churcher, *Angew. Chem. Int. Ed.* **2012**, *51*, 1114–1122.
9. A. W. Hung, A. Ramek, Y. Wang, T. Kaya, J. A. Wilson, P. A. Clemons, D. W. Young, *Proc. Nat. Acad. Sci. U.S.A.* **2011**, *108*, 6799–6804.
10. L. B. Akella, L. A. Marcaurelle, *ACS Comb. Sci.* **2011**, *13*, 357–364.
11. S. Dandapani, L. A. Marcaurelle, *Nat. Chem. Biol.* **2010**, *6*, 861–863.
12. C. A. Lipinski, F. Lombardo, B. W. Dominy, P. J. Feeney, *Adv. Drug Deliv. Rev.* **2001**, *46*, 3–26.
13. F. Lovering, J. Bikker, C. Humblet, *J. Med. Chem.* **2009**, *52*, 6752–6756.
14. T. Luker, L. Alcaraz, K. K. Chohan, N. Blomberg, D. S. Brown, R. J. Butlin, T. Elebring, A. M. Griffin, S. Guile, S. St.-Gallay, B.-M. Swahn, S. Swallow, M. J. Waring, M. C. Wenlock, P. D. Leeson, *Bioorg. Med. Chem. Lett.* **2011**, *21*, 5673–5679.
15. P. D. Leeson, S. A. St.-Gallay, M. C. Wenlock, *Med. Chem. Commun.* **2011**, *2*, 91–105.
16. See Reference 5. Also see (a) T. Uchida, M. Rodriquez, S. L. Schreiber, *Org. Lett.* **2009**, *11*, 1559–1562; (b) S. L. Schreiber, *Nature* **2009**, *457*, 153–154; (c) L. A. Marcaurelle, E. Comer, S. Dandapani, J. R. Duvall, B. Gerard, S. Kesavan, M. D. Lee IV, H. Liu, J. T. Lowe, J.-C. Marie, C. A. Mulrooney, B. A. Pandya, A. Rowley, T. D. Ryba, B.-C. Suh, J. Wei, D. W. Young, L. B. Akella, N. T. Ross, Y. L. Zhang, D. M. Fass, S. A. Reis, W.-N. Zhao, S. J. Haggarty, M. Palmer, M. A. Foley, *J. Am. Chem. Soc.* **2010**, *132*, 16962–16976.
17. (a) Q. S. Zhang, H. J. Lu, C. Richard, D. P. Curran, *J. Am. Chem. Soc.* **2004**, *126*, 36–37; (b) D. P. Curran, Q. Zhang, C. Richard, H. Lu, V. Gudipathi, C. S. Wilcox, *J. Am. Chem. Soc.* **2006**, *128*, 9561; (c) S. Dandapani, M. Jeske, D. P. Curran, *J. Org. Chem.* **2005**, *70*, 9447; (d) I. E. Wrona, J. T. Lowe, T. J. Turbyville, T. R. Johnson, J. Beignet, J. A. Beutler, J. S. Panek, *J. Org. Chem.* **2008**, *74*, 1897–1916.
18. E. Comer, H. Liu, A. Joliton, A. Clabaut, C. Johnson, L. B. Akella, L. A. Marcaurelle, *Proc. Nat. Acad. Sci. U.S.A.* **2011**, *108*, 6751–6756.
19. M. Cnop, N. Welsh, J. C. Jonas, A. Jorns, S. Lenzen, D. L. Eizirik, *Diabetes* **2005**, *54*, S97–S107.

20. D. H. Chou, N. E. Bodycombe, H. A. Carrinski, T. A. Lewis, P. A. Clemons, S. L. Schreiber, B. K. Wagner, *ACS Chem. Biol.* **2010**, *5*, 729–734.
21. (a) T. Inoue, J.-F. Liu, D. Buske, A. Abiko, *J. Org. Chem.* **2002**, *67*, 5250–5256; (b) A. Abiko, *Org. Synth. Coll. Vol. 10*, **2004**, 343–349; (c) A. Abiko, *Acc. Chem. Res.* **2004**, *37*, 387–395; (d) A. Abiko, J.-F. Liu, S. Masamune, *J. Am. Chem. Soc.* **1997**, *119*, 2586–2587.
22. D. A. Evans, *Aldrichim. Acta* **1982**, *15*, 23–32, and references therein.
23. D. H. Chou, J. R. Duvall, B. Gerard, H. Liu, B. A. Pandya, B.-C. Suh, E. M. Forbeck, P. Faloon, B. K. Wagner, L. A, Marcaurelle, *ACS Med. Chem. Lett.* **2011**, *2*, 698–702.
24. T. D. Ryba, K. M. Depew, L. A. Marcaurelle, *J. Comb. Chem.* **2009**, *11*, 110–116.
25. N. J. Ede, Z. Wu, *Curr. Opin. Chem. Biol.* **2003**, *7*, 374–379.
26. A. Merglen, S. Theander, B. Rubi, G. Chaffard, C. B. Wollheim, P. Maechler, *Endocrinology* **2004**, *145*, 667–678.
27. *World Malaria Report*, World Health Organization, Geneva, Switzerland, **2010**.
28. E. H. Ekland, D. A. Fidock, *Int. J. Parasitol.* **2008**, *38*, 743–747.
29. M. J. Mackinnon, K. Marsh, *Science* **2010**, *328*, 866–871.
30. J. E. Hyde, *Trends Parasitol.* **2005**, *21*, 494–498.
31. (a) S. Saralamba, W. Pan-Ngum, R. J. Maude, S. J. Lee, J. Tarning, N. Lindegardh, K. Chotivanich, F. Nosten, N. P. Day, D. Socheat, N. J. White, A. M. Dondorp, L. J. White, *Proc. Natl. Acad. Sci. U.S.A.* **2011**, *108*, 397–402; (b) M. I. Veiga, P. E. Ferreira, L. Jörnhagen, M. Malmberg, A. Kone, B. A. Schmidt, M. Petzold, A. Björkman, F. Nosten, J. P. Gil, *PLoS ONE* **2011**, *6*, e20212.
32. R. W. Snow, E. A. Okiro, P. W. Gething, R. Atun, S. I. Hay, *Lancet*, **2010**, *376*, 1409–1416.
33. D. C. Smithson, W. Armand Guiguemde, R. Kiplin Guy, in: *Burger's Medicinal Chemistry: Drug Discovery and Development*, Wiley, Hoboken, NJ, **2010**, pp. 1–110.
34. T. N. C. Wells, P. L. Alonso, W. E. Gutteridge, *Nat. Rev. Drug Discov.* **2009**, *8*, 879–891.
35. W. A. Guiguemde, A. A. Shelat, D. Bouck, S. Duffy, G. J. Crowther, P. H. Davis, D. C. Smithson, M. Connelly, J. Clark, F. Zhu, M. B. Jiménez-Díaz, M. S. Martinez, E. B. Wilson, A. K. Tripathi, J. Gut, E. R. Sharlow, I. Bathurst, F. E. Mazouni, J. W. Fowble, I. Forquer, P. L. McGinley, S. Castro, I. Angulo-Barturen, S. Ferrer, P. J. Rosenthal, J. L. DeRisi, D. J. Sullivan, J. S. Lazo, D. S. Roos, M. K. Riscoe, M. A. Phillips, P. K. Rathod, W. C. Van Voorhis, V. M. Avery, R. K. Guy, *Nature* **2010**, *465*, 311–315.
36. F. J. Gamo, L. M. Sanz, J. Vidal, C. de Cozar, E. Alvarez, J. L. Lavandera, D. E. Vanderwall, D. V. S. Green, V. Kumar, S. Hasan, J. R. Brown, C. E. Peishoff, L. R. Cardon, J. F. Garcia-Bustos, *Nature* **2010**, *465*, 305–310.
37. F. Calderón, D. Barros, J. M. Bueno, J. M. Coterón, E. Fernández, F. J. Gamo, J. L. Lavandera, M. L. León, S. J. F. Macdonald, A. Mallo, P. Manzano, E. Porras, J. M. Finador, J. Castro, *ACS Med. Chem. Lett.* **2011**, *2*, 741–746.
38. F. Calderón, J. Vidal-Mas, J. Burrows, J. Carlos de la Rosa, M.B. Jiménez-Díaz, M. Mulet, S. Prats, J. Solana, M. Witty, F. J. Gamo, E. A. Fernández, *ACS Med. Chem. Lett.* **2012**, *3*, 373−377.
39. M. Johnson, M. Lajiness, G. M. Maggiora, in: *QSAR: Quantitative Structure–Activity Relationships in Drug Design*, J. L. Fauchere, Ed., Alan R. Liss, New York, **1989**, pp. 167–171.
40. T. N. Wells, E. M. Poll, *Discov. Med.* **2010**, *9*, 389–398.
41. A. L. Baggish, D. R. Hill, *Antimicrob. Agents Chemother.* **2002**, *46*, 1163–1173.

42. A. W. Sweeney, C. R. B. Blackburn, K. H. Rieckmann, *Am. J. Trop. Med. Hyg.* **2004**, *71*, 187–189.

43. A. Nzila, *J. Antimicrob. Chemother.* **2006**, *57*, 1043–1054.

44. (a) M. Rottmann, C. McNamara, B. K. S. Yeung, M. C. S. Lee, B. Zou, B. Russell, P. Seitz, D. M. Plouffe, N. V. Dharia, J. Tan, S. B. Cohen, K. R. Spencer, G. E. González-Páez, S. B. Lakshminarayana, A. Goh, R. Suwanarusk, T. Jegla, E. K. Schmitt, H.-P. Beck, R. Brun, F. Nosten, L. Renia, V. Dartois, T. H. Keller, D. A. Fidock, E. A. Winzeler, T. T. Diagana, *Science* **2010**, *329*, 1175–1180; (b) B. K. S. Yeung, B. Zou, M. Rottmann, S. B. Lakshminarayana, S. H. Ang, S. Y. Leong, J. Tan, J. Wong, S. Keller-Maerki, C. Fischli, A. Goh, E. K. Schmitt, P. Krastel, E. Francotte, K. Kuhen, D. Plouffe, K. Henson, T. Wagner, E. A. Winzeler, F. Petersen, R. Brun, V. Dartois, T. T. Diagana, T. H. Keller, *J. Med. Chem.* **2010**, *53*, 5155–5164.

45. http://www.mmv.org/sites/default/files/uploads/docs/press_releases/MMV%20Project%20of%20the%20Year%202009.pdf; accessed June 25, 2012.

46. A. Pictet, T. Spengler, *Berichte* **1911**, *44*, 2030–2036.

47. K. Hayton, X. Z. Su, *Curr. Genet.* **2008**, *54*, 223–239.

48. R. W. Heidebrecht, Jr., C. Mulrooney, C. P. Austin, R. H. Barker, Jr., J. A. Beaudoin, K. C.-C. Cheng, E. Comer, S. Dandapani, J. Dick, J. R. Duvall, E. H. Ekland, D. A. Fidock, M. E. Fitzgerald, M. Foley, R. Guha, P. Hinkson, M. Kramer, A. K. Lukens, D. Masi, L. A. Marcaurelle, X.-Z. Su, C. J. Thomas, M. Weïwer, R. C. Wiegand, D. Wirth, M. Xia, J. Yuan, J. Zhao, M. Palmer, B. Munoz, S. Schreiber, *ACS Med. Chem. Lett.* **2012**, *3*, 112–117.

49. P. Thiel, M. Kaiser, C. Ottmann, *Angew. Chem. Int. Ed.* **2012**, *51*, 2–9.

50. J. Wanner, D. C. Fry, Z. Peng, J. Roberts, *Future Med. Chem.* **2011**, *3*, 2021–2038.

51. P. Braun, M. Tasan, M. Dreze, M. Barrios-Rodiles, I. Lemmens, H. Yu, J. M. Sahalie, R. R. Murray, L. Roncari, A.-S. de Smet, K. Venkatesan, J.-F. Rual, J. Vandenhaute, M. E. Cusick, T. Pawson, D. E. Hill, J. Tavernier, J. L. Wrana, F. P. Roth, M. Vidal, *Nat. Methods* **2009**, *6*, 83–90.

52. D. Kozakov, D. R. Hall, G.-Y. Chuang, R. Cencic, R. Brenke, L. E. Grove, D. Beglov, J. Pelletier, A. Whitty, S. Vajda, *Proc. Natl. Acad. Sci. U.S.A.* **2011**, *108*, 13528–13533.

53. J. A. Wells, C. L. McClendon, *Nature* **2007**, *450*, 1001–1009.

54. D. L. Sackett, D. Sept, *Nat. Chem.* **2009**, *1*, 596–597.

55. A. G. Tzakos, D. Fokas, C. Johannes, V. Moussis, E. Hatzimichael, E. Briasoulis, *Molecules* **2011**, *16*, 4408–4427.

56. A. N. Koehler, *Curr. Opin. Chem. Biol.* **2010**, *14*, 331–340.

57. N. S. Hegde, D. A. Sanders, R. Rodriguez, S. Balasubramanian, *Nat. Chem.* **2011**, *3*, 725–731.

58. K. Narasimhan, S. Pillay, N. R. Bin Ahmad, Z. Bikadi, E. Hazai, L. Yan, P. R. Kolatkar, K. Pervushin, R. Jauch, *ACS Chem. Biol.* **2011**, *6*, 573–581.

59. G. Lessene, P. E. Czabotar, P. M. Colman, *Nat. Rev. Drug Discov.* **2008**, *7*, 989–1000.

60. A. G. Letai, *Nat. Rev. Cancer* **2008**, *8*, 121–132.

61. For an additional Bcl antagonist involving DOS, see S. Di Micco, R. Vitale, M. Pellecchia, M. F. Rega, R. Riva, A. Basso, G. Bifulco, *J. Med. Chem.* **2009**, *52*, 7856–7867.

62. L. A. Marcaurelle, C. W. Johannes, *Prog. Drug Res.* **2008**, *66*, 187–216.

63. A. C. Castro, W. Deng, K. M. Depew, M. A. Foley, C. C. Fritz, A. T. Georges Evangelinos, M. J. Grogan, N. Hafeez, E. B. Holson, B. T. Hopkins, N. O. Koney, T. Liu, D. A. Mann, L. A. Marcaurelle, D. A. Snyder, D. J. Underwood, A. A. Wylie, L.-C. Yu, L. Zhang, WO 2006/009869, Jan. 26, **2006**.

64. L. A. Marcaurelle, C. Johannes, D. Yohannes, B. P. Tillotson, D. Mann, *Bioorg. Med. Chem. Lett.* **2009**, *19*, 2500–2503.

65. (a) B. T. O'Neill, D. Yohannes, M. W. Bundesmann, E. P. Arnold, *Org. Lett.* **2000**, *2*, 4201–4204; (b) D. Yohannes, K. Procko, L. A. Lebel, C. B. Fox, B. T. O'Neill, *Bioorg. Med. Chem. Lett.* **2008**, *18*, 2316–2319.

66. A. Falaschi, G. Abdurashidova, G. Biamonti, *Crit. Rev. Biochem. Mol.* **2010**, *45*, 14–22.

67. B. Utsch, C. D. McCabe, K. Galbraith, R. Gonzalez, M. Born, J. Dotsch, M. Ludwig, H. Reutter, J. W. Innis, *Am. J. Med. Genet. A* **2007**, *143A*, 3161–3168.

68. Z.-D. Gu, L.-Y. Shen, H. Wang, X.-M. Chen, Y. Li, T. Ning, K.-N. Chen, *Cancer Res.* **2009**, *69*, 4969–4973.

69. P. Y. Ng, Y. Tang, W. M. Knosp, H. S. Standler, J. T. Shaw, *Angew. Chem. Int. Ed.* **2007**, *46*, 5352–5355.

70. (a) M. Cushman, J. Gentry, F. W. Dekow, *J. Org. Chem.* **1977**, *42*, 1111–1116; (b) M. Cushman, E. J. Madaj, *J. Org. Chem.* **1987**, *52*, 907–915.

71. J. B. Le Pecq, D. Nguyen, C. Gosse, C. Paoletti, *Proc. Natl. Acad. Sci. U.S.A.* **1974**, *71*, 5078–5082.

18

DOS-DERIVED SMALL-MOLECULE PROBES IN CHEMICAL BIOLOGY

Nicholas Hill, Lingyan Du, and Qiu Wang

18.1 INTRODUCTION

Small molecules are essential both as instruments of modern medicine and as probes for studying biological processes [1]. By interacting with macromolecules, small molecules can perturb polypeptide and nucleic acid function and modulate a wide range of processes.

Traditionally, natural products have been a valuable source of structural diversity and have enabled the development of compounds for both fundamental biological studies and clinical applications [2]. However, successes have generally come about on a case-by-case basis, and their use has been limited by a number of challenges [1a,3]. In past decades, chemical genetics has emerged as a systematic approach to exploring biological processes with small-molecule probes by identifying small molecules that target a desired phenotype (forward chemical genetics) or the function of a specific protein (reverse chemical genetics) [4].

Diversity-oriented synthesis (DOS), by design, aims to generate small-molecule collections having a high degree of structural and functional diversity spanning large regions of biologically relevant chemical space [5]. In many cases the chemical space represented by a DOS library may include regions of known bioactivity (which, by definition, are regions where biologically active agents have already been discovered) and "untapped" regions of chemical space that may contain molecules with biologically relevant properties that have thus far escaped the attention of researchers and, perhaps, even nature. Furthermore, a DOS strategy can facilitate the optimization of

Diversity-Oriented Synthesis: Basics and Applications in Organic Synthesis, Drug Discovery, and Chemical Biology, First Edition. Edited by Andrea Trabocchi.

an initial hit and can streamline an efficient scaled synthesis of an optimized probe. The logic underlying a variety of DOS strategies was reviewed extensively in earlier chapters.

Integration of the DOS approach into chemical genetics has emerged as a powerful tool for systematic probing of the chemical space relevant to the biological processes inherent in a wide range of human diseases [6]. Efforts in this regard have produced new and important probes for dissecting complex biological processes and lead compounds for developing novel treatments of diseases. In this chapter we present several recent studies aimed at identifying novel small-molecule probes from DOS-based libraries and DOS-based optimization strategies.

18.2 DOS-DERIVED SMALL-MOLECULE PROBES

Growing efforts directed to discovering bioactive molecules from DOS libraries have led to many important new biological probes. Tables 18.1 and 18.2 provide a summary of recent DOS-derived bioactive small molecules identified using a forward and reverse chemical genetics approach. A set of examples are discussed below, with a particular focus on representing the extensive utilities of a DOS strategy in probe development, ranging from targeting challenging biological pathways and expanding the collections of important biological probes to achieving therapeutically desirable phenotypes.

18.3 DEVELOPING SMALL-MOLECULE PROBES OF COMPLEX BIOLOGICAL PATHWAYS

18.3.1 Inhibitors of Sonic Hedgehog Signaling

The Hedgehog (Hh) signaling pathway plays an important role in embryonic development and cellular differentiation [52]. In mammals, the major components in the Hh signaling pathway include the Hh ligands (Sonic, Desert, and Indian), the 12-transmembrane receptor Patched (Ptc1), the pseudo-G-protein-coupled signal transducer Smoothened (Smo), and the glial (Gli) family of transcription factors [52,53]. The Hh ligands are secreted glycoproteins, among which the Sonic Hedgehog protein (Shh) is best characterized. In the absence of ligands, Ptc1 binds to and inhibits the signal transducer Smo on the cell membrane. The inhibitory effect of Ptc1 on Smo is reversed upon binding of ShhN, the active 20-kDa N-terminal fragment of Shh, to the extracellular domain of Ptc1, which initiates downstream signaling by enhancing the expression and nuclear translocation of the Gli transcription factors (Gli1, Gli2, and Gli3) and subsequent up-regulation of the Shh target genes (Figure 18.1) [43,54]. Aberrant activity of the Shh signaling pathway is associated with the development of various cancers [53]. For example, mutations in the *Ptch1* gene cause nevoid basal cell carcinoma syndrome, in which excessive ectopic activation of Shh signaling is observed. Individuals having the mutations exhibit an increased risk of developing

TABLE 18.1 DOS-Derived Bioactive Small Molecules by Forward Chemical Genetics

Structure	Name	Activity	Assay	Ref.
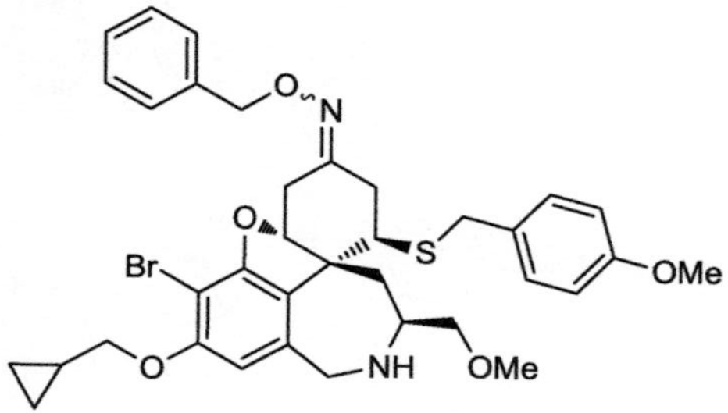	Galanthamine analog secramine	Secretory pathway inhibitor; blocks protein trafficking at 2 μM	Cell-based phenotypic assay with VSVG-EGFP	7
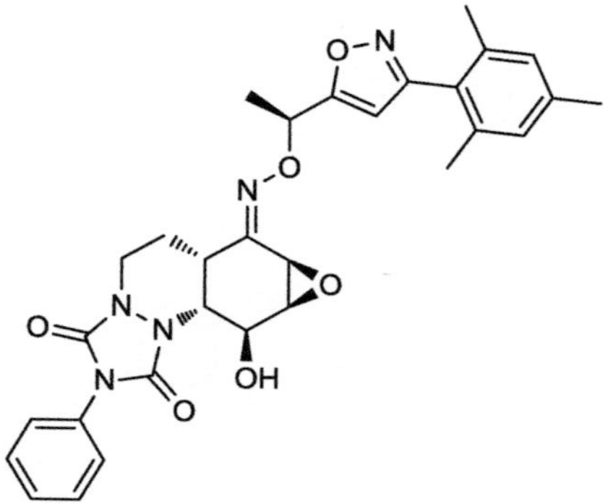	Epoxyquinol BUCMLD-B10A11	Hepatitis C replication inhibitor; $EC_{50} < 0.5$ μM	HCV replicon system where the HCV structural genes are replaced by a luciferase reporter	8

(*continued*)

TABLE 18.1 (*Continued*)

Structure	Name	Activity	Assay	Ref.
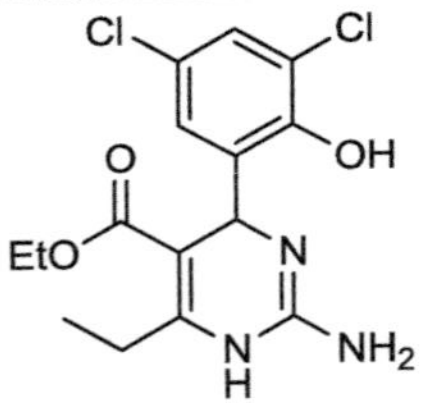	Dihydropyrimidine emmacin	Prokaryote-selective dihydrofolate reductase (DHFR) inhibitor; $IC_{50} = 5.4\ \mu M$	EMRSA bacterial proliferation phenotypic assay	9
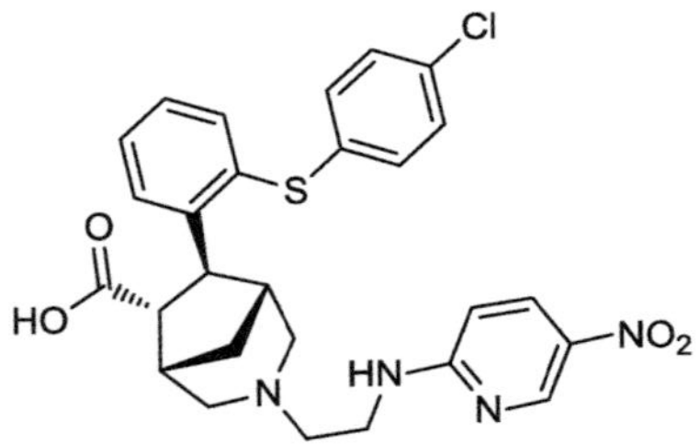	Homopiperidine gemmacin	Cell membrane–disrupting antibiotic; $MIC_{50} = 8.0\ \mu g/mL$ against EMRSA-15	Minimum inhibitory concentration (MIC) assay	10
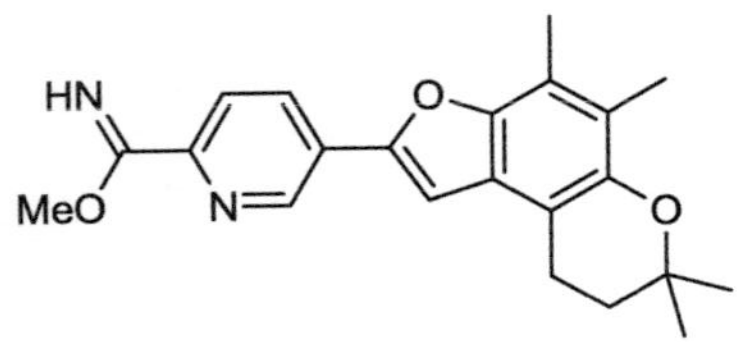	Furochromane analog	*Mycobacterium tuberculosis* (MTB) inhibitor; $MIC_{95} = 0.6\ \mu g/mL$	Microdilution resazurin assay	11

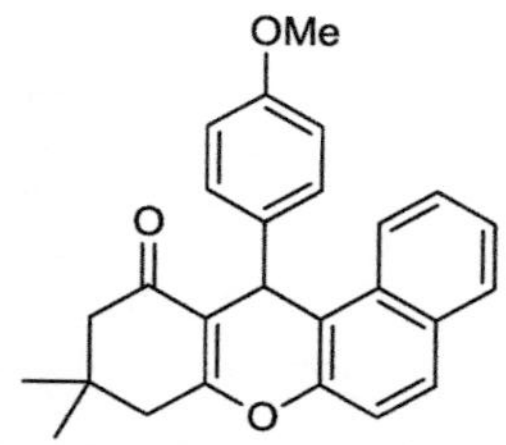	Benzochromene analog	Antiproliferative; IC_{50} = 4.0 μg/mL against KB carcinoma cells	Cell proliferation inhibition	12
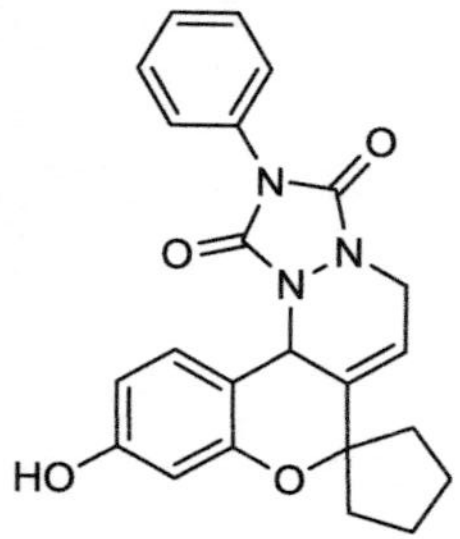	Benzopyran–triazoledione analog	Osteoclastogenesis inhibitor; IC_{50} = 5.09 μM	Cell-based tartrate-resistant acid phosphatase (TRAP) activity assay	13
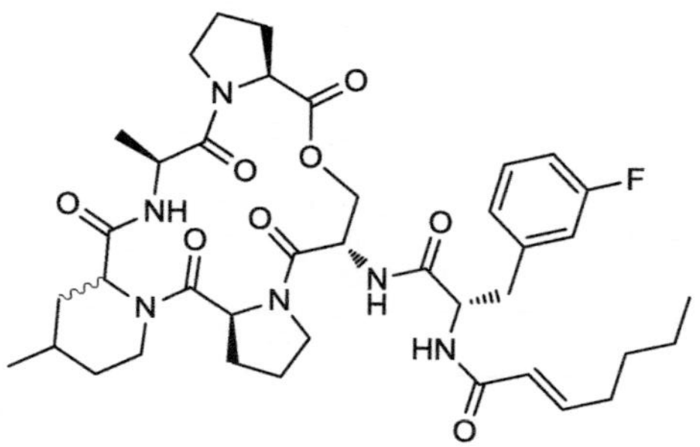	Acyldepsipeptide analog	Antibacterial activity; MIC = 0.6 μg/mL against MRSA	Minimum inhibitory concentration assay	14

(*continued*)

TABLE 18.1 (*Continued*)

Structure	Name	Activity	Assay	Ref.
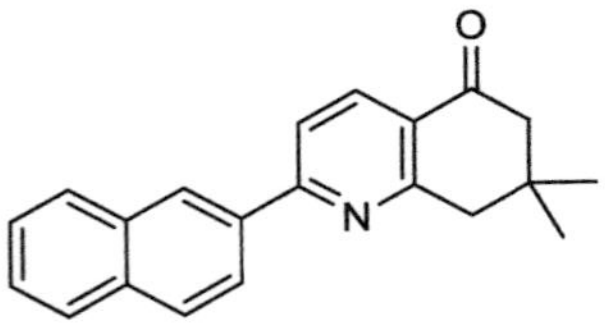	Dihydroquinolin-5-one analog	*Mycobacterium tuberculosis* (MTB) inhibitor; MIC = 1.56 μg/mL	Minimum inhibitory concentration assay	15
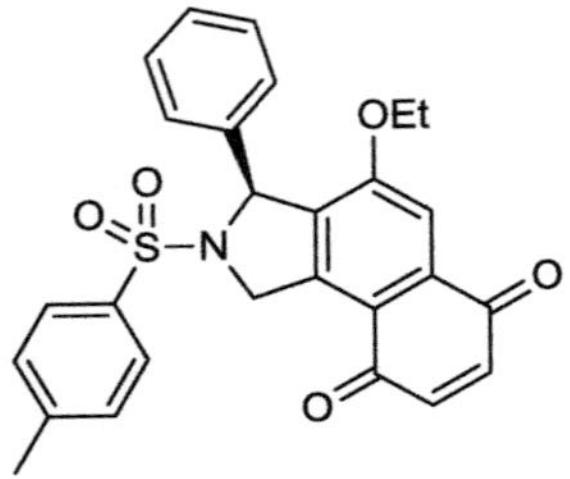	Naphthoquinone analog	Cell migration inhibitor; IC_{50} = 14.8 μM	Wound closure assay with breast cancer cells (MDA-MB-231)	16
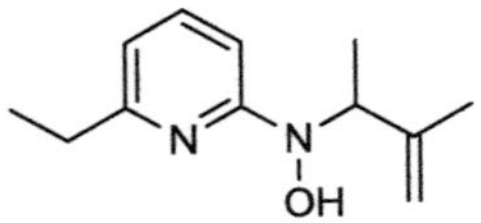	*N*-Pyridin-2-yl hydroxylamine analog	Antibacterial activity; MIC_{90} = 2 μM against *Micrococcus luteus*	MIC assay using the broth microdilution method	17
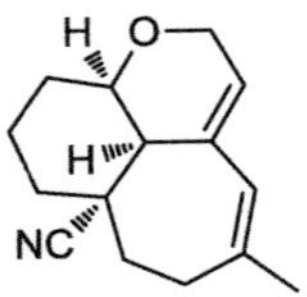	Synthetic sesquiterpene analog	Antitrypanosomal activity; IC_{50} = 0.55 μg/mL	Activity against *Trypanosoma brucei*	18

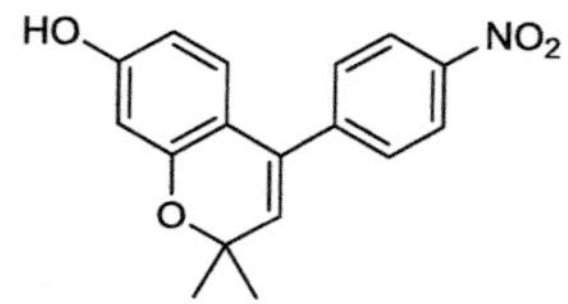	Benzopyran analog	Osteogenic activity; 6.8-fold increase in activity of ALP at 2 μM	Cell-based alkaline phosphatase assay	19
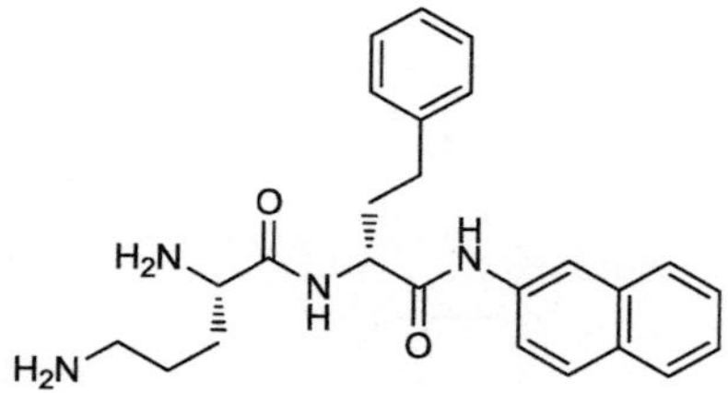	C-capped dipeptide BU-005	7.2-fold reduction in MIC of chloramphenicol at 25 μg/mL in *S. coelicolor*	Minimum inhibitory concentration assay	20
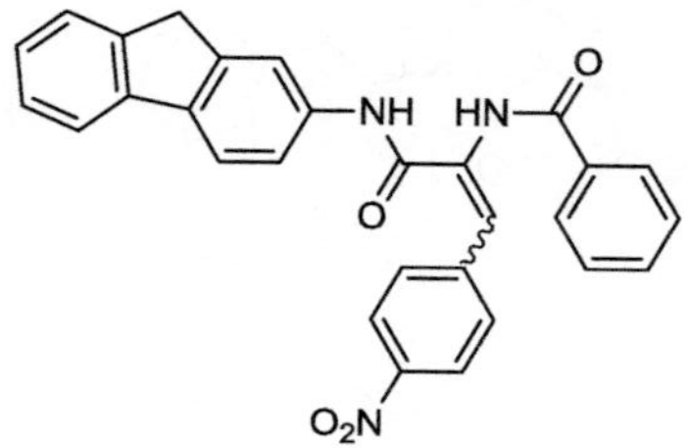	Dipeptide mimetic	Cytotoxic activity; Inhibition rate of 57.2% at 0.1 μg/mL	Cytotoxicity against colon carcinoma SW1116 cells	21
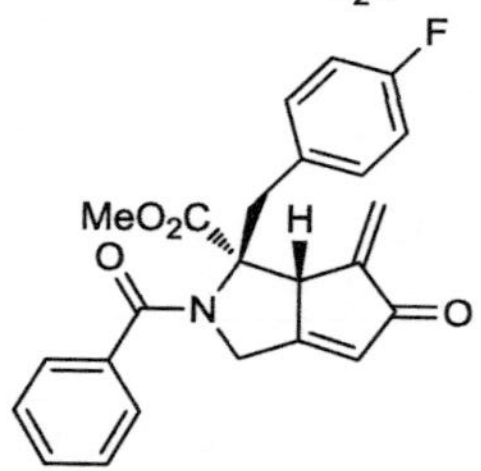	α-Methylene cyclopentenone MARPIN	DNA damage checkpoint inhibitor; $IC_{50} = 7.7$ μM for phorphorylation of Chk1	Cell-based assay detection of phospho-Chk1	22

(*continued*)

TABLE 18.1 (*Continued*)

Structure	Name	Activity	Assay	Ref.
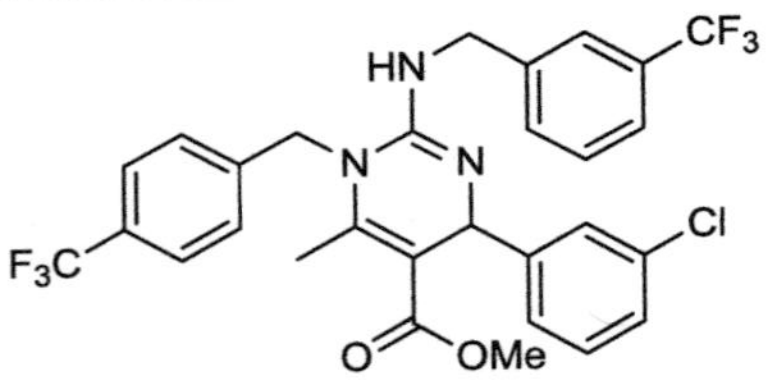	Polycyclic ketal, indoline alkaloid, and guanidine chemotypes	*Plasmodium falciparum* inhibitor (antimalaria); $IC_{50} = 0.02$ μM for a guanidine analog	SYBR green dye DNA staining assay	23
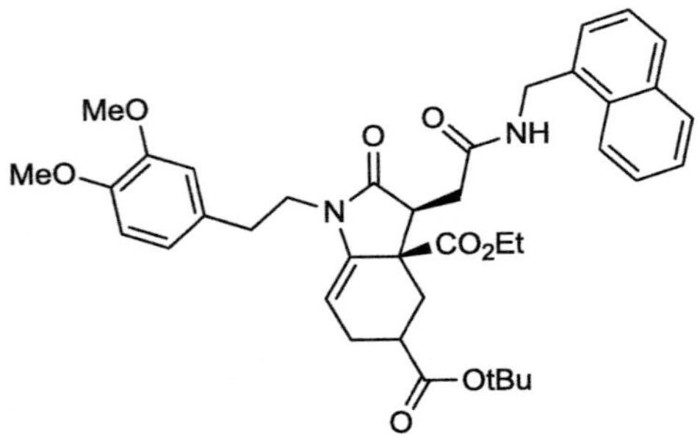	Fused-bicyclic lactam analog	Glucose transport inhibitor; Depletion of ATP, $IC_{50} = 1$ μM	Cell-based assay determining ATP level using antimycin A–sensitized A549 or CHO cells	24
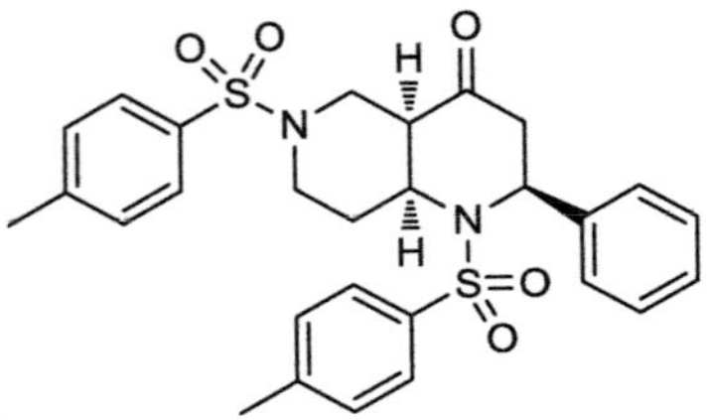	Naphthyridinone analog	Endothelial cell activator; activation of inflammatory pathways at 10 μM	Cellular production of the chemokine MIP1β	25

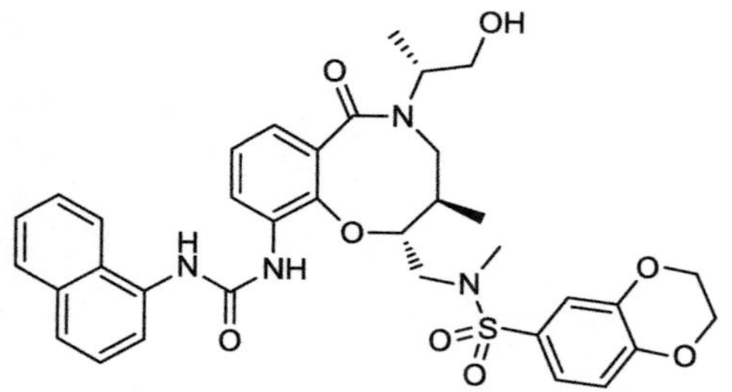	Macrocycle BRD-0476	Suppressor of β-cell apoptosis; $EC_{50} = 0.78$ μM	Cytokine-induced β-cell death	26
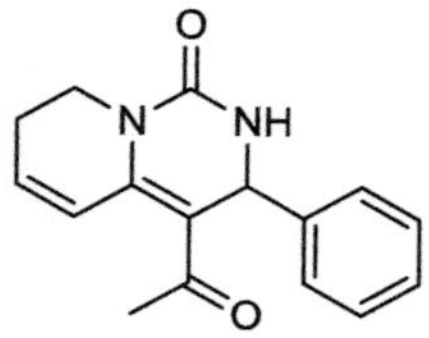	Pyridopyrimidinone CMLDBU6128	Orthopoxvirus inhibitor; $IC_{50} = 5.3$ μM at 12 h postinfection	Fluorescent detection of viral fusion proteins	27
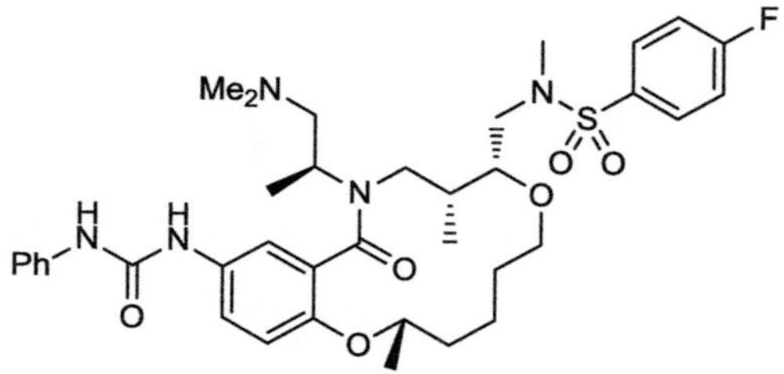	Macrocycle ML238	Anti-*Plasmodium falciparum* (antimalarial) activity; $GI_{50} = 0.54$ nM	Blood-stage malaria assay with a DAPI reporter	28
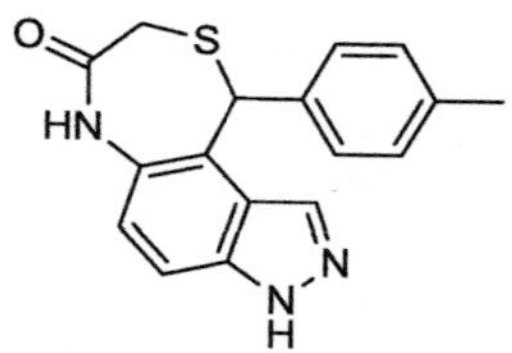	1,4-Thiazepine analog	Cytotoxic activity inhibition rate of 66.82% at 1 mg/mL	Cytotoxicity against carcinoma HCT-116 cells	29

TABLE 18.2 DOS-Derived Bioactive Small Molecules by Reverse Chemical Genetics

Structure	Name	Activity	Assay	Ref.
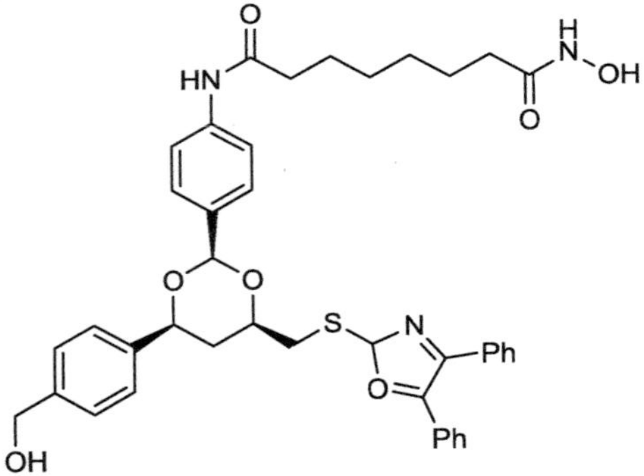	Hydroxamic acid tubacin	Tubulin and histone deacetylase inhibitors; tubacin increases α-tubulin acylation levels at 125 nM	Cell-based cytoblot assays using acylation-specific antibodies	30
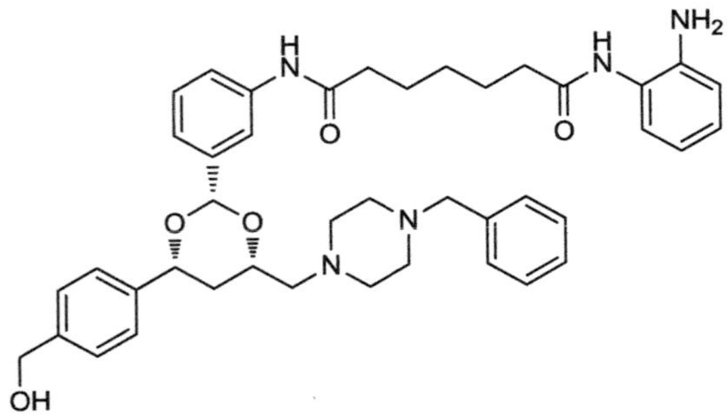	*o*-Aminoanilide histacin	Histacin increases histone acylation levels at 20 μM	Cell-based cytoblot assays using acylation-specific antibodies	31

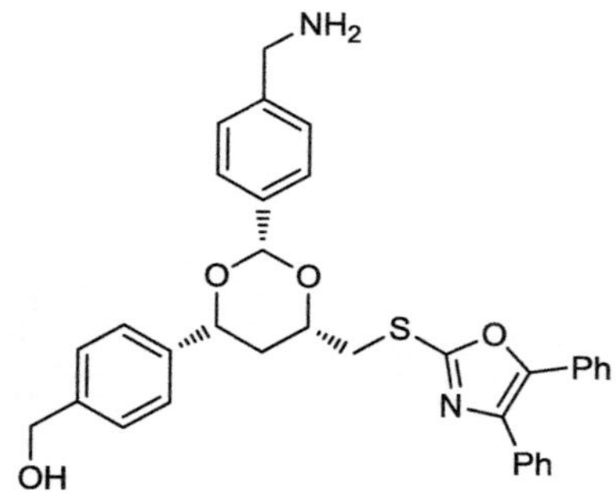	1,3-Dioxane uretupamine B	Transcription factor repressor Ure2p binder; $K_d = 7.5$ μM	Small-molecule microarray binding screen	32
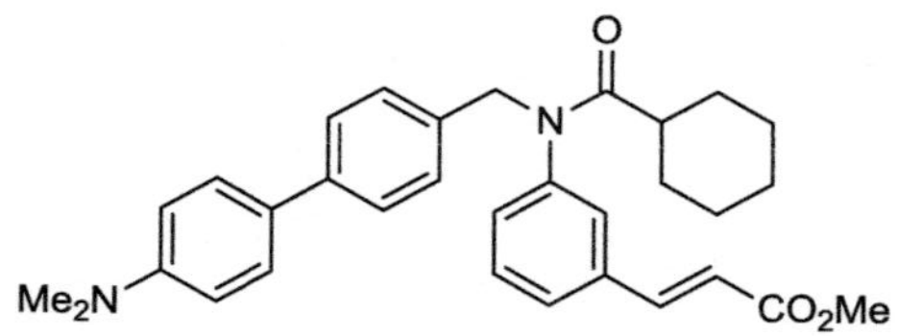	Tertiary amide fexaramine	Farnesoid X receptor (FXR) agonist; $EC_{50} =$ 25 nM	Cell-based transcription assay with an FXR-responsive luciferase reporter	33
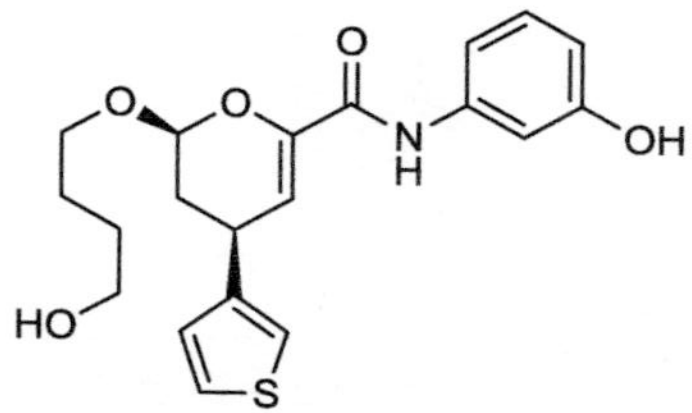	Dihydropyrancarboxamide haptamide B	Yeast transcription factor Hap3p inhibitor; $K_d = 0.33$ μM	Small-molecule microarray binding screen	34

(continued)

TABLE 18.2 (*Continued*)

Structure	Name	Activity	Assay	Ref.
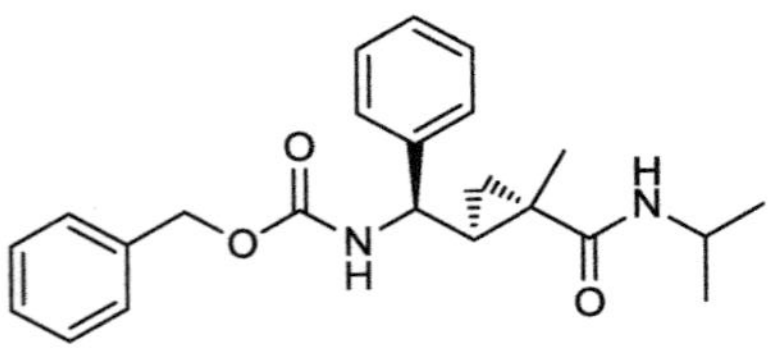	*C*-cyclopropylalkylamide analog	Human pregnane X receptor (hPXR) agonist; $EC_{50} = 400$ nM	Cell-based Gal-hPXR assay with a luciferase reporter	35
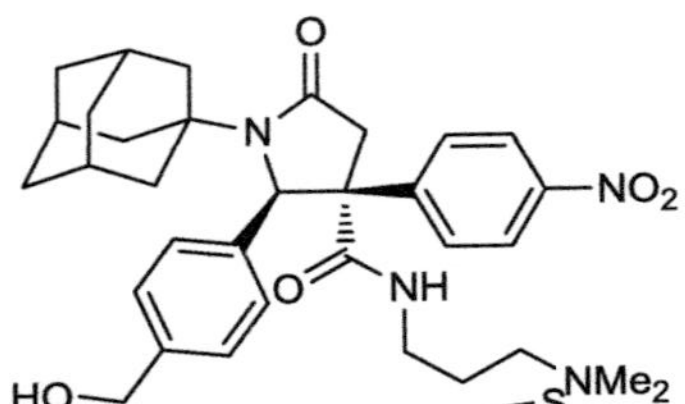	Lactam carboxamide analog	Transcription factor HOXA13 inhibitor; $IC_{50} = 6.5$ μM	Fluorescence polarization assay detecting disruption of transcription factor–DNA complex	36
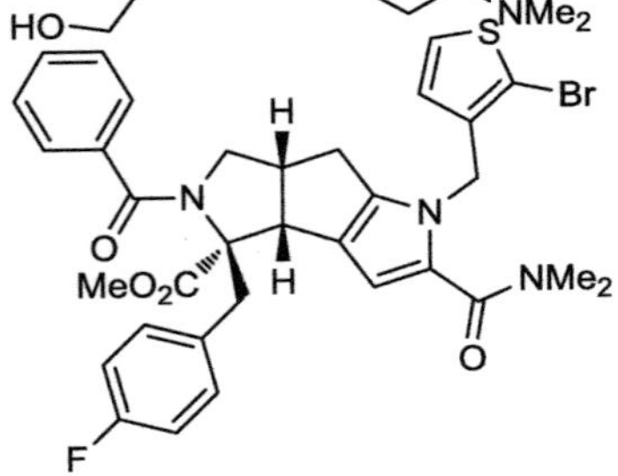	Tricyclic pyrrole carboxamide PSI2106	Mitogen-activated protein kinase phosphatase 1 (MKP-1) inhibitor; $IC_{50} = 8.0$ μM	In vitro assay using the artificial substrate 3-*O*-methylfluorescein phosphate	37

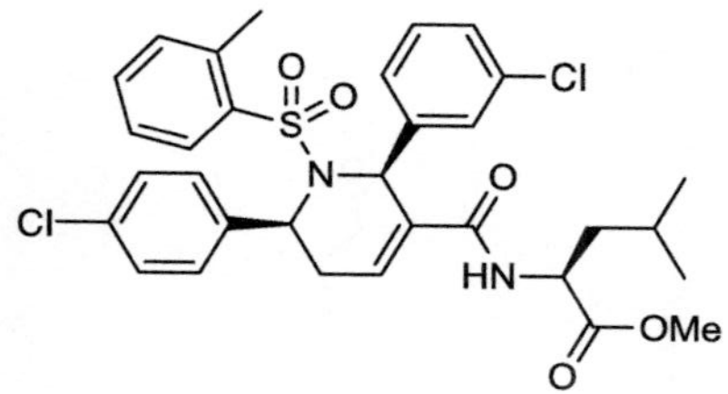	Tetrahydropyridine P61-E7	Protein geranylgeranylation inhibitor, at 5 μM; 21-fold increase in unprenylated Rap1	Detection of unprenylated Rap1 by immunoblotting analysis	38
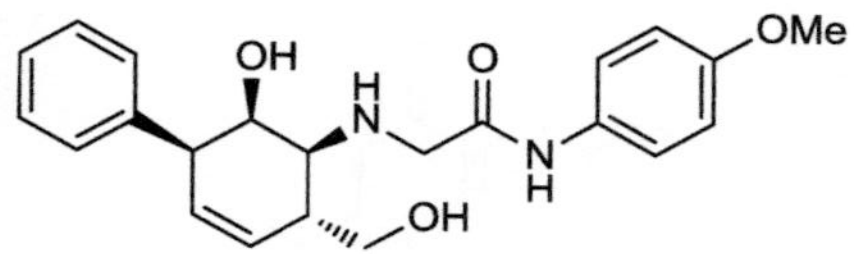	Tetrasubstituted cyclohexene analog	Bcl-x_L inhibitor	NMR-based binding assay	39
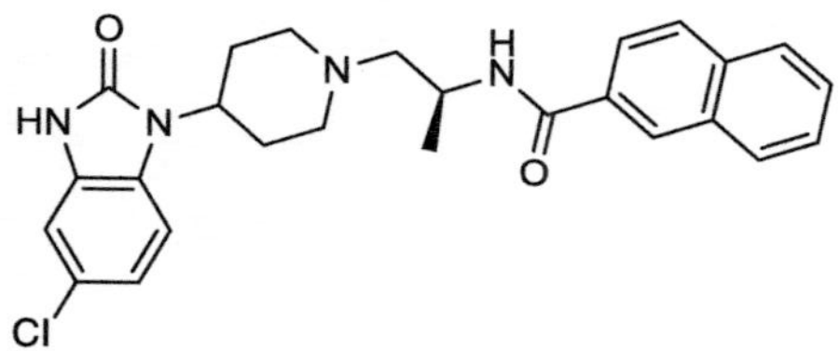	2-Hydroxybenzimidazole analog	Phospholipase D subtype 1 inhibitor; $IC_{50} = 11$ nM in vitro	Release of [methyl-^{3}H]choline from labeled DPPC	40
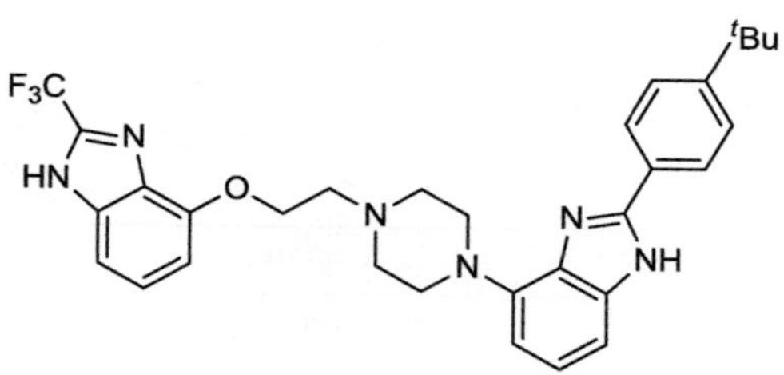	Benzimidazole analog	Gonadotropin-releasing hormone (GnRH) antagonist; $IC_{50} = 50$ nM	In vitro screen with radioactively labeled GnRH	41

(*continued*)

TABLE 18.2 *(Continued)*

Structure	Name	Activity	Assay	Ref.
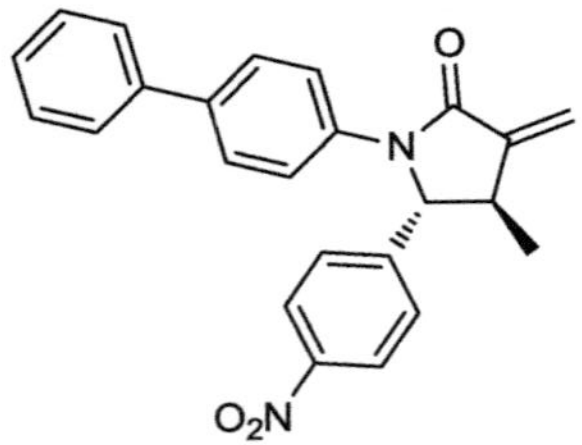	α-exo-Methylene-γ-lactam analog	Homoserine transacetylase (HTA) inhibitor; IC_{50} = 140 μM	In vitro monitoring of CoA production	42
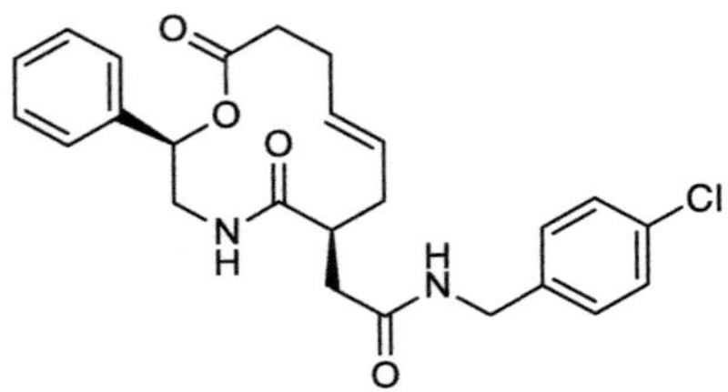	Macrocycle robotnikinin	Hedgehog pathway inhibitor; K_d = 3.1 μM binding to ShhN	Small-molecule microarray binding screen	43
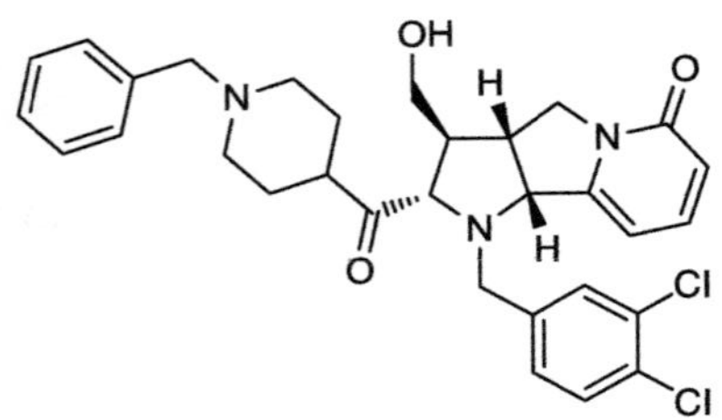	Tricyclic pyridone analog	Bcl-2 inhibitor; K_i = 1.2 μM	Competition binding assay	44

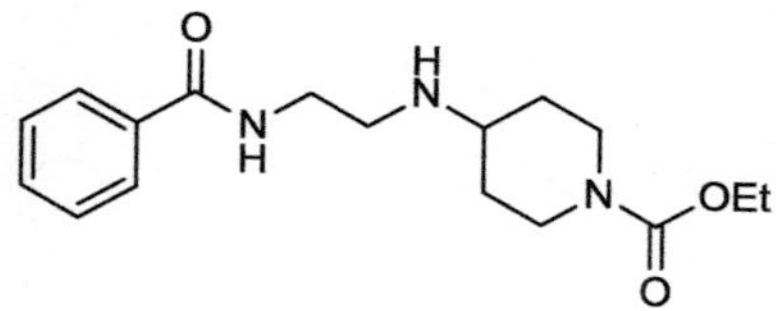	4-Aminopiperidine VU0184670	mAChR M_1 agonist; $EC_{50} = 152$ nM	Cell-based calcium flux assay using Fluo-4	45
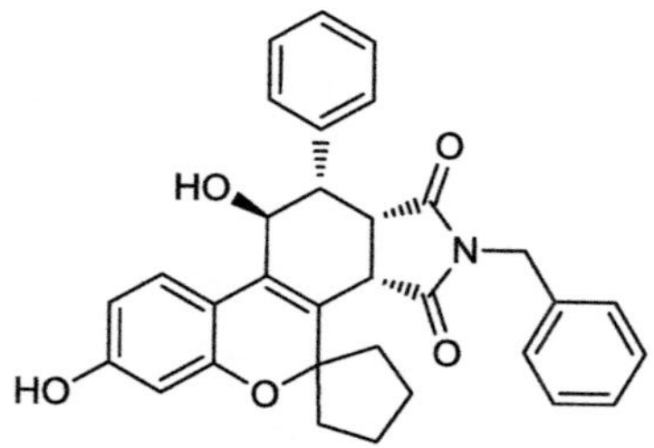	Benzopyran–succinimide analog	Androgen receptor antagonist; 15.1% of full activity at 10 μM and 10 μM DHT	Modified cell line expressing androgen receptor–binding element and a luciferase reporter	46
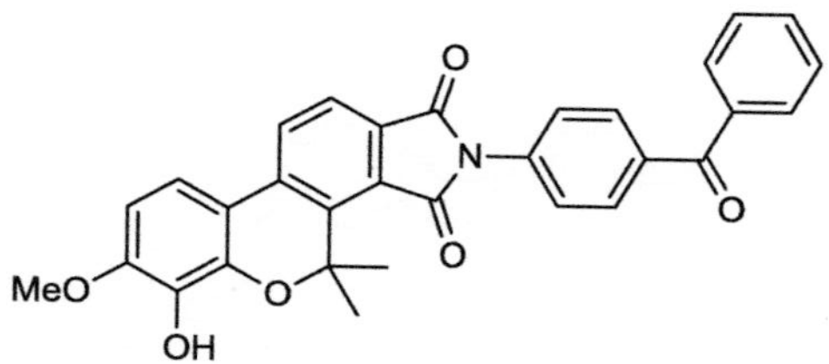	Benzopyran–maleimide ampkinone	AMPK activator 322% relative phosphorylation activity at 10 μM	Immunoblotting assay detecting phosphorylated AMPK substrate	47

(*continued*)

TABLE 18.2 (*Continued*)

Structure	Name	Activity	Assay	Ref.
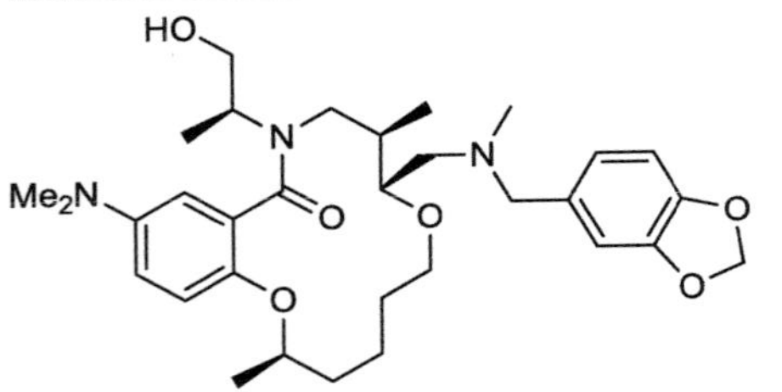	Macrolactam BRD-4805	HDAC inhibitor; IC_{50} = 2.7 μM for HDACs 1, 3; IC_{50} = 6.6 μM for HDAC 2	Histone deacetylase biochemical functional assay with fluorescent readout	48
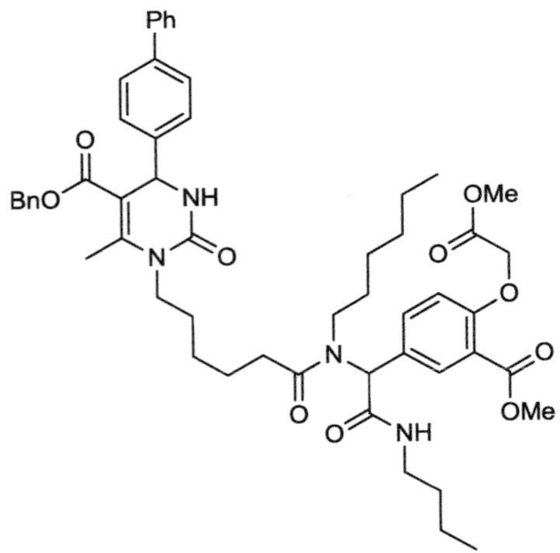	MAL3-101-dihydropyrimidinone	Fourfold decrease in TAg-stimulated Hsp70 ATPase activity at 0.3 mM	TLC separation of radiolabeled ATP and ADP	22,49

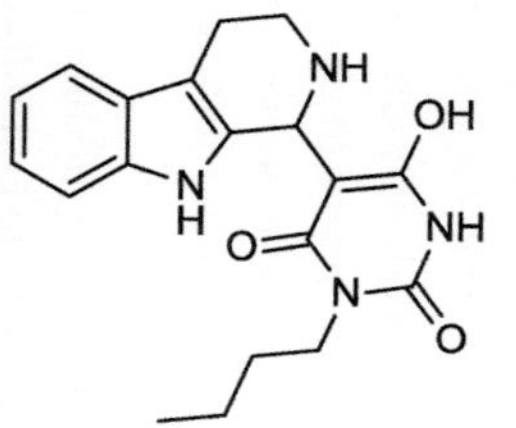	Tetrahydro-β-carboline MEL23	Mdm2 E3 ubiquitin ligase inhibitor; EC_{50} = 7.5 μM	Cell-based luciferase detection of E3 ubiquitin ligase activity	50
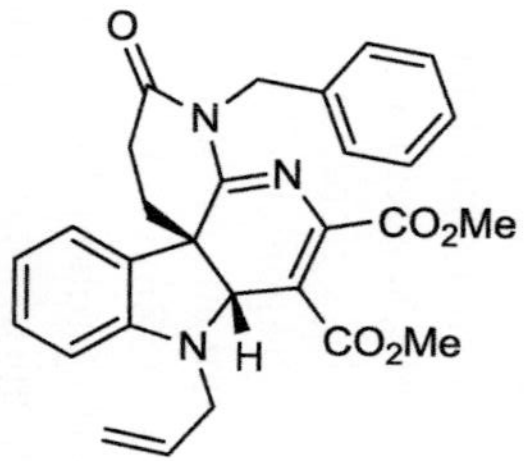	Indoline SL209	HCV core dimerization inhibitor; IC_{50} = 1.4 μM	TR-FRET assay of core106 dimerization	51

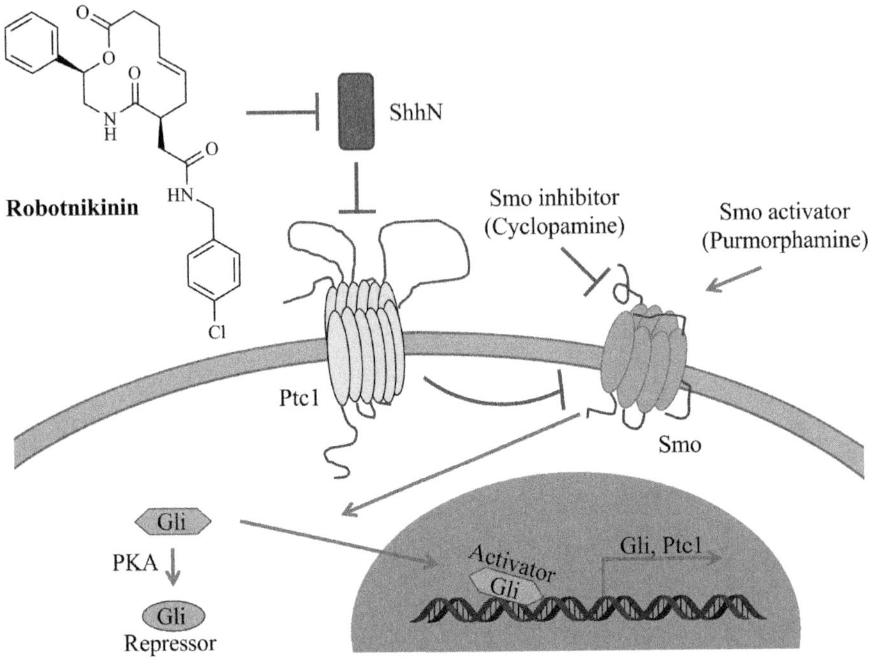

FIGURE 18.1 Robotnikinin inhibits Shh signaling by blocking Shh ligand binding to Ptc1. (*See insert for color representation of the figure.*)

tumors, such as medulloblastoma in the brain and fibromas in the heart and ovaries [52,53]. Overactivation of Shh signaling is also reported to play an important role in the development of pancreatic ductal adenocarcinoma (PDAC) [52].

Schreiber's group screened a DOS library of 2070 compounds arrayed on microarrays for binding to bacterially expressed ShhN, leading to the identification of a novel macrocyclic inhibitor of Shh signaling, robotnikinin [43,55]. An initial hit demonstrated binding to ShhN in a concentration-dependent fashion by surface plasmon resonance (SPR) assays and also showed moderate inhibition of the Shh pathway in a Gli-dependent luciferase assay with Shh-LIGHT2 cells. Optimizations led to the compound robotnikinin, which exhibited low micromolar affinity for ShhN (Figure 18.1). Robotnikinin's dose-dependent activity in the Shh signaling pathway was confirmed by its inhibition of ShhN-stimulated alkaline phosphatase induction in C3H10T1/2 cells and its repression of ShhN-stimulated Gli1 and Gli2 *m*RNA expression in human keratinocytes. These results suggested that the inhibitory effect of robotnikinin on Shh signaling is likely to be a global effect, independent of cell types and reporters. Coincubation with the Smo agonists (SAG) purmorphamine completely abolished the inhibitory effect of robotnikinin on Shh signaling, further indicating that the target of robotnikinin is upstream of Smo [43].

Prior to the discovery of robotnikinin, all of the Shh modulators reported acted on Smo or targets downstream of Smo. As the first identified inhibitor of the Shh pathway upstream of Smo, robotnikinin is a useful tool for probing earlier steps in the Shh signaling pathway that were not previously accessible, creating new opportunities for the development of anticancer drugs.

18.3.2 Inhibitor of the Secretory Pathway

The secretory pathway is a fundamental process for shuttling synthesized proteins from the rough endoplasmic reticulum (rough ER) through the Golgi apparatus and the trans-Golgi network (TGN) to the plasma membrane or a specific organelle [56]. Previous studies have established fundamental concepts relating to the biogenesis of organelles and the mechanisms of vesicle formation and fusion. However, detailed regulatory mechanisms are not fully understood in many important secretory-related pathways [57]. Small-molecule modulators of the secretory pathway are highly useful tools for studying these complicated biological processes, as the rapid movement of organelles and intracellular traffic present a challenge for traditional genetic and biochemical methods [7b,58].

The Shair and Kirchhausen groups have discovered an inhibitor, secramine, that perturbs traffic out of the Golgi apparatus (anterograde transport) in the secretory pathway through the use of a DOS library (Figure 18.2) [7]. The DOS library was generated via a biomimetic strategy inspired by the biosynthesis of the natural product galanthamine. An image-based phenotypic screening was developed to measure the inhibition of traffic out of the Golgi by monitoring the localization of the indicator vesicular stomatitis virus glycoprotein (VSVG). In this assay, the VSVG ts045 mutant strain is retained in the ER at 40°C due to reversible misfolding. When the temperature is decreased to 32°C, the protein redistributes into the Golgi complex and then into the cell membrane. This process can be observed by automated fluorescence microscopy when the strain is tagged with enhanced green fluorescent protein (EGFP) [59]. In this study, secramine was identified as a potent inhibitor of $VSVG^{ts}$-EGFP transport from the DOS library [7].

A natural product that is also known to delay $VSVG^{ts}$-EGFP transport from the Golgi apparatus, cytochalasin B, is known to prevent actin polymerization [60]. Thus, for target identification purposes, secramine was tested in an actin polymerization inhibition assay. In this assay, *Xenopus laevis* cytoplasmic egg extract was used and actin polymerization was monitored by the increase in fluorescence caused by the polymerization of actin–pyrene. The dose-dependent inhibitory effects of secramine on the rate of actin polymerization suggested that the target of secramine also exists within the actin polymerization process. Actin polymerization is activated by PIP2 liposome stimulation of Cdc42 (a member of Rho family of small GTPases) via its effector proteins, Taco-1 and N-WASP, which further recruit the actin-binding Arp2/3 complex [7b]. When actin polymerization was activated directly by the N-WASP VCA domain, secramine was found to be ineffective, which suggested that secramine interferes with upstream processes, PIP2 liposome-stimulated Cdc42 activation, and/or Cdc42 effectors of activation. Secramine was next examined to determine if it could affect the binding outcome between Cdc42 in the active, GTP-bound state and another one of its effectors, PAK1. GST-PBD, a recombinant peptide

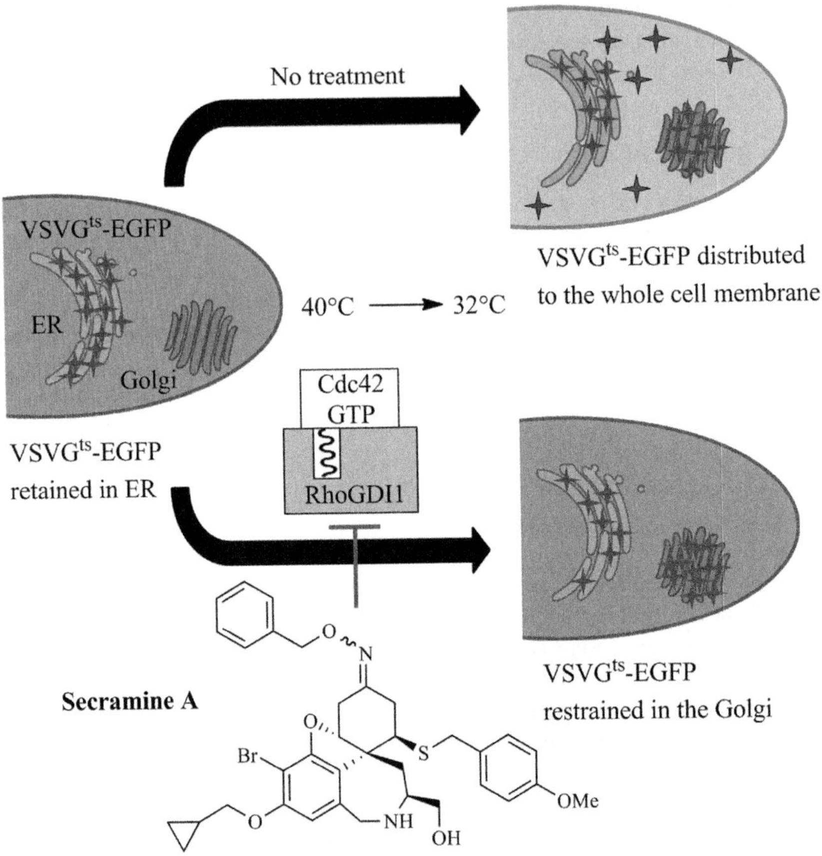

FIGURE 18.2 Secramine inhibits intracellular membrane traffic. (*See insert for color representation of the figure.*)

derived from PAK1, can be used to precipitate GTP-bound Cdc42. In the experiment, Cdc42([^{35}S]GTPγS) was made by mixing the *X. laevis* egg extract with [^{35}S]GTPγS and PIP2 liposomes in a process catalyzed by the endogenous guanine nucleotide exchange factor (GEF). Coprecipitation between Cdc42([^{35}S]GTPγS) and GST-PBD was not observed in cases when secramine was either present in the *X. laevis* egg extract or added to the pre-formed Cdc42([^{35}S]GTPγS). The fact that the coprecipitation between Cdc42([^{35}S]GTPγS) and GST-PBD was detected when secramine was added last in the assay suggested that secramine blocked the Cdc42([^{35}S]GTPγS) binding with GST-PBD indirectly in a noncompetitive manner.

To test if secramine inhibits the membrane association of Cdc42 before binding its effectors [61], the compound's effects were examined through the membrane-dependent nucleotide exchange activity of prenylated Cdc42. Secramine was found to inhibit the exchange of GDP for [^{35}S]GTPγS in a Cdc42(GDP)-RhoGDI1 complex that was prepared with isolated PIP2 liposomes and the DH-PH domain of

GEF Dbs. Nonprenylated Cdc42, which does not associated with the membrane, was unaffected by treatment with secramine, suggesting that the probe did not inhibit nucleotide exchange or interfere with GEF. Secramine also reduced the membrane association of prenylated Cdc42 with PIP2 liposomes in the form of a Cdc42(GDP)–RhoGDI1 complex, as determined by sedimentation of PIP2 liposomes after incubation with Cdc42–RhoGDI1 complex. Based on the combined results, it was concluded that secramine inhibits Cdc42's membrane association in a RhoGDI1-dependent manner.

In addition to being useful as a tool for probing the secretory pathway, secramine has been used to study a variety of other Cdc42-related biological processes, such as cAMP-induced K^+ conductance [62], cell shape and growth control in anaplastic large cell lymphoma [63], and Cdc42-dependent Golgi polarization in migrating cells [7b].

18.4 EXPANDING THE COLLECTION OF IMPORTANT BIOLOGICAL PROBES

18.4.1 Inhibitors of Heat Shock Protein 70: Probes with Higher Potency

Members of the heat shock protein 70 (Hsp70) family are operative in a wide range of protein folding and transport processes. Hsp70 is characterized by an N-terminal nucleotide-binding domain that binds ATP, a substrate-binding domain that associates with hydrophobic regions of client polypeptides, and a C-terminal helical lid that regulates access to the peptide-binding domain. Upon hydrolysis of ATP in the nucleotide-binding domain, a conformational change in the helical lid traps client proteins, allowing them to fold and preventing aggregate formation [64]. Hydrolysis of ATP is accelerated by the association of Hsp70 and a co-chaperone, heat shock protein 40 (Hsp40), through the J-domain of Hsp40 [65]. The large T antigen (TAg) of simian virus 40 (SV40) contains a similar J-domain that also can stimulate Hsp70's activity [49]. It has been reported that elevated Hsp70 expression is associated with metastasis and poor prognosis in patients after combined chemotherapy and radiation therapy treatment. The role of these heat shock proteins in oncogenesis has made them attractive targets in the development of anticancer drugs [66].

To discover new modulators of Hsp70, the Wipf and Brodsky groups assembled a library of 31 compounds based on two small molecules (deoxyspergualin and NSC-630668-R/1) known to modulate the ATPase activity of Hsp70 [49]. In assembling the library, compounds were drawn from the Developmental Therapeutics Program of the National Cancer Institute and the University of Pittsburgh Center for Chemical Methodologies and Library Development Program (UPCMLD, a DOS library). To assess the compounds' activities toward Hsp70, a thin-layer chromatography (TLC) assay was developed using a pre-formed complex of the yeast Hsp70 Ssa1p and [α-^{32}P]ATP. The complex was incubated with compound and aliquots were removed from the mixture and quenched at specified time points. Separation of [α-^{32}P]ATP from [α-^{32}P]ADP in each aliquot was accomplished by TLC, and the percentages of ATP

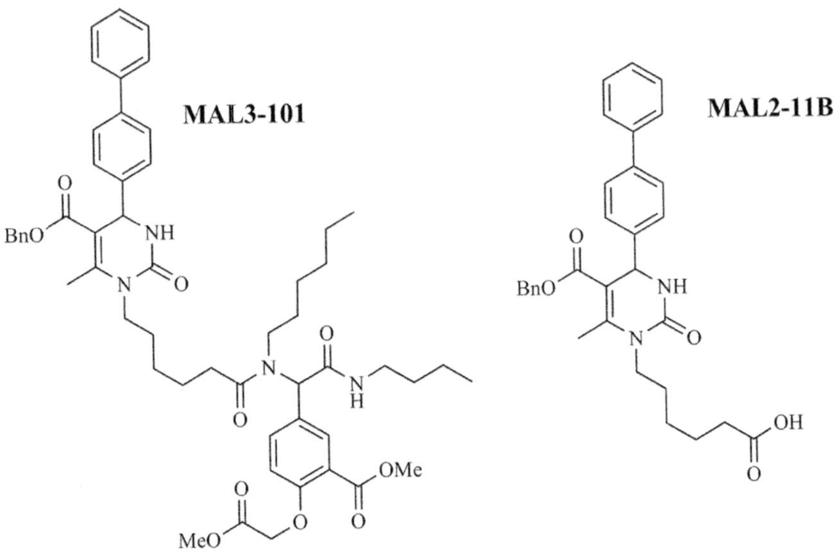

FIGURE 18.3 Inhibitors of Hsp70 ATPase activity.

hydrolyzed over the time course was used to approximate the first-order rate constant of hydrolysis. This assay was used to measure both endogenous Hsp70 ATPase activity and J-domain-stimulated Hsp70 ATPase activity. To test the latter activity, the yeast Hsp40 Ydj1p protein or the SV40 TAg were premixed with the Hsp70 Ssa1p-[α-^{32}P]ATP complex prior to addition of the compound. Several compounds in the library were identified that either enhanced or decreased endogenous or J-domain-stimulated Hsp70 ATPase activity. Most interestingly, the compound MAL3-101 was found to specifically reduce TAg-stimulated ATPase activity in a dose-dependent manner while exhibiting little impact on endogenous or Ydj1p-stimulated ATPase activity (Figure 18.3). The compound was further shown to inhibit the Hsp40/Hsp70-dependent translocation of a protein, yeast ppαF, into yeast endoplasmic reticulum (ER)–derived microsomes.

MAL3-101 proved to be useful as a probe for studying a variety of chaperone-related processes. In one study, MAL3-101 was used to examine the mode by which eight different tail-anchored (TA) proteins integrate into a mammalian ER membrane [67]. Integration of TA proteins into the ER membrane is promoted by two separate complexes: the Hsp40/Hsc70 chaperone complex (Hsc70, a member of the Hsp70 family of proteins), and the Asna-1 ATPase complex [68]. In this study, MAL3-101 was found to significantly inhibit ER integration of three TA proteins (human Bcl2, PTP1B, and Cb5), suggesting that functional Hsp40/Hsc70 is necessary for their insertion; however, the integration of five other TA proteins (rat Syb2 and Syn1A; and human Ubc6J1, Sec61b, and Ramp4) was not affected by the compound, suggesting that a different pathway, probably involving the Asna-1 ATPase complex, is sufficient for insertion of these proteins.

In another study, MAL3-101 was used as a probe to validate a screening system for identifying chemical species that are active against the causative parasite of human African trypanosomiasis, *Trypanosoma brucei* [69]. The transport of variant surface glycoproteins (VSGs) to the parasite's plasma membrane by means of the ER is critical for the parasite's survival in the host bloodstream. In contrast to mammalian systems, import into the protozoan ER is posttranslational and dependent on chaperones in the cytosol. It was hypothesized that MAL3-101 would inhibit the import of VSGs into the ER by inhibiting the protozoan Hsp70/Hsp40. A model system for the protozoan ER utilizing *T. brucei* microsomes and cytosol showed importation of 80% of the VSG protein VSG_117 posttranslationally across the membrane into the interior of microsomes. Import of the protein into the microsomes was shown by treatment of the system with proteinase K. The imported VSG_117 (i.e., 80% of the protein) was protected from proteinase K digestion by the microsomal membrane. However, in a separate experiment where MAL3-101 was included in the system, only 13% of the VSG protein was protected from digestion, indicating that the compound blocked uptake of VSG_117 into the microsomes. When assessed for activity against living *T. brucei*, MAL3-101 exhibited dose-dependent trypanocidal activity with an IC_{50} value of 1.5 μM, confirming both the activity of MAL3-101 as a trypanocidal agent and validating the assay as useful for identifying new trypanocidal compounds.

Another inhibitor of Hsp70, MAL2-11B, a precursor to MAL3-101, was also identified as a useful probe. In addition to inhibiting TAg-stimulated ATPase activity, MAL2-11B inhibited both endogenous Hsp70 ATPase activity and TAg's ATPase activity [22,70]. In subsequent studies exploring its anti-TAg properties, MAL2-11B was found to inhibit DNA replication in polyomavirus SV40 and to reduce the growth of the human BK virus in kidney cells without cytotoxic effects toward the host cells [70,71].

The identification of Hsp70 modulators from the UPCMLD library allowed for the rapid diversification of the structural space around the known Hsp70 inhibitor deoxyspergualin and facile access to second- and third-generation libraries. The structure–activity insight from these studies has provided a new class of Hsp70 probes with higher potency for the study of Hsp70-related functions [22].

18.4.2 Agonist of the Acetylcholine Muscarinic M_1 Receptor: Probes with Higher Selectivity

The five muscarinic receptors M_1 to M_5 are G-protein-coupled receptors activated by the neurotransmitter acetylcholine [72]. Among these subtypes, the M_1 receptor subtype, which is localized in the central nervous system (CNS), has attracted attention as a target for new treatments of Alzheimer's disease and schizophrenia, due to its role in cognitive processing and short-term memory [73]. However, conservation of the acetylcholine binding site, the orthosteric site, in all five of the muscarinic receptor subtypes has made the development of M_1 subtype-specific agonists difficult.

In overcoming this obstacle, Lebois et al. employed a functional high-throughput screen and a subsequent DOS optimization to prepare M_1-selective agonists that

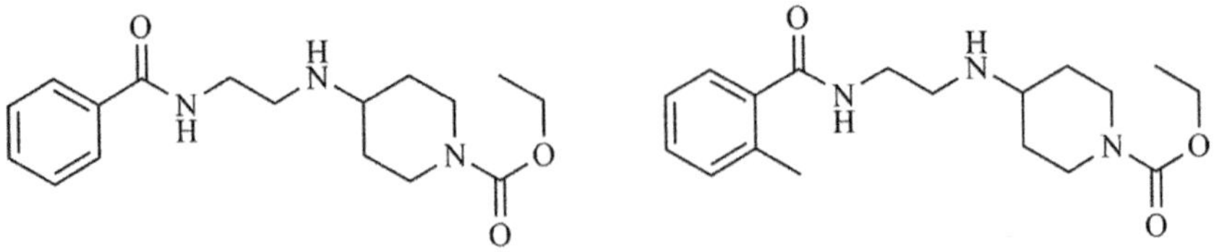

VU0184670 **VU0357017**

FIGURE 18.4 M_1 acetylcholine receptor agonists.

target an allosteric site on the receptor [45]. It is known that the M_1, M_3, and M_5 receptors function through the phosphoinositol system and the phospholipase C effector to stimulate intracellular calcium release. Based on this function, a cell-based fluorescence assay using the calcium-binding dye Fluo-4 was developed for screening 65,000 compounds in the Molecular Libraries Production Center Network (MLPCN) library to identify compounds that increase intracellular calcium levels in a CHO cell line expressing the M_1 receptor [74]. Optimization of the initial hits from the screen produced compounds VU0184670 and VU0357017 as highly subtype-selective agonists of the M_1 receptor (Figure 18.4), even at the highest compound concentrations tested (30 μM). Moreover, the compounds exhibited no significant activity in a screen of 68 GPCRs, ion channels, and transporter targets.

Site-directed mutagenesis experiments were employed to characterize the binding properties of the agonists identified. A mutation that reduced the affinity of acetylcholine and other orthosteric ligands of the receptor had little effect on VU0184670 and VU0357017 agonism [75]. The orthosteric M_1 antagonist atropine blocked the response to VU0184670 in a noncompetitive fashion, further suggesting that binding occurs at an allosteric site. Mutations in the third extracellular loop (ECL3) of the M_1 receptor reduced VU0184670-induced activation, suggesting that this loop may harbor the allosteric binding site.

The compounds were further characterized in a variety of cellular and in vivo experiments. VU0184670 was shown to potentiate the *N*-methyl-D-aspartate glutamate receptor (NMDAR) current in whole-cell patch clamp experiments with hippocampal CA1 pyramidal cells, an effect that has been postulated to mediate cholinergic function in Alzheimer's disease and schizophrenia [76]. In murine models, both compounds were found to penetrate the CNS after intraperitoneal administration, with VU035017 being cleared more readily over time.

In a rodent model, VU035017 was shown to reverse the cognitive deficits induced by the muscarinic receptor antagonist scopolamine. Rats were conditioned to exhibit freezing behavior upon exposure to a neutral conditioned stimulus by undergoing a training period in which the neutral conditioned stimulus was coupled with a mild footshock. Scopolamine interfered with the acquisition of this response. However, VU035017 administered intraperitoneally at 10 mg/kg reversed the scopolamine-induced effects, with animals receiving both compounds acquiring the contextual fear conditioning response.

The discovery of VU0184670 and VU035017 has provided highly subtype-selective agonists for the M_1 acetylcholine receptor. The compounds discovered supplement known allosteric M_1 modulators. Moreover, the compounds' clean ancillary pharmacology contrasts with the off-target activity observed for M_1 agonists described previously and for the benchmark allosteric M_1 agonist TBPB [77]. With a well-characterized binding mechanisms and demonstrated efficacy in animal models, these compounds may prove to be useful probes for expanding biological studies of the acetylcholine muscarinic M_1 receptor and for developing new treatments for cholinergic-related cognitive disorders.

18.4.3 Inhibitors of Protein Prenylation: Probes with Novel Structures

Inhibitors of protein prenylation have received much attention as potential anticancer agents due to a number of oncogenic proteins (e.g., Ras proteins) that undergo a posttranslational farnesylation or geranylgeranylation that facilitates their anchoring to the plasma membrane [78]. It is known that when FTase-catalyzed farnesylation of the Ras isoform K-RasB is blocked, GGTase-catalyzed geranylgeranylation of the protein can substitute functionally for farnesylation [79]. Consequently, inhibitors of both farnesyltransferase (FTase) and geranylgeranyltransferase (GGTase) may be necessary to fully block prenylation of oncogenic Ras proteins. Whereas extensive efforts have been made to develop FTase inhibitors, there have been fewer studied aimed at identifying inhibitors for GGTase. Two varieties of GGTase have been characterized. The first, GGTase-I, catalyzes the geranylgeranylation of Rho proteins of the Ras superfamily and heterotrimeric G-proteins at the cysteins of C-terminal CAAX sequences (A is an aliphatic amino acid, and X is leucine or phenylalanine) [80]. The second, RabGGTase, catalyzes the geranylgeranylation of two cysteins on the C-termini of the secretory and endocytosis-related Rab proteins, members of the Ras superfamily, at CC or CXC sequences [81].

Kwon and co-workers have developed a new class of geranylgeranylation inhibitors by screening a DOS library of 4288 compounds against GGTase-I-catalyzed geranylgeranylation of the two Ras superfamily proteins RhoA and K-Ras4B [38]. In the screen, the incorporation of tritiated (^{3}H) geranylgeranyl groups into the Ras proteins was measured. The compounds P5-H6 and P3-E5, the two most active hits (Figure 18.5), were used in a follow-up study with HEK-293 cells transfected with the Ras oncogene Rheb-CSVL. Both compounds were found to markedly decrease the levels of prenylated Rheb-CSVL. Furthermore, these two compounds exhibited a significant specificity for GGTase-I over both RabGGTase and FTase.

Optimizations of P5-H6 led to a more potent compound, P61-A6, which showed a 10-fold increased antiproliferative activity in K562 cells and which appreciably inhibited geranylgeranylation of Rap1, another member of the Ras superfamily that undergoes GGTase-I-catalyzed geranylgeranylation, at 2.5 μM [38b]. Follow-up studies suggested that the greater cellular activity of P61-A6 may have resulted from improved stability or absorption rather than inherent affinity for GGTase-I [82]. When P61-A6 was used as a GGTase-I inhibitor to treat xenograft pancreatic (PANC-1)

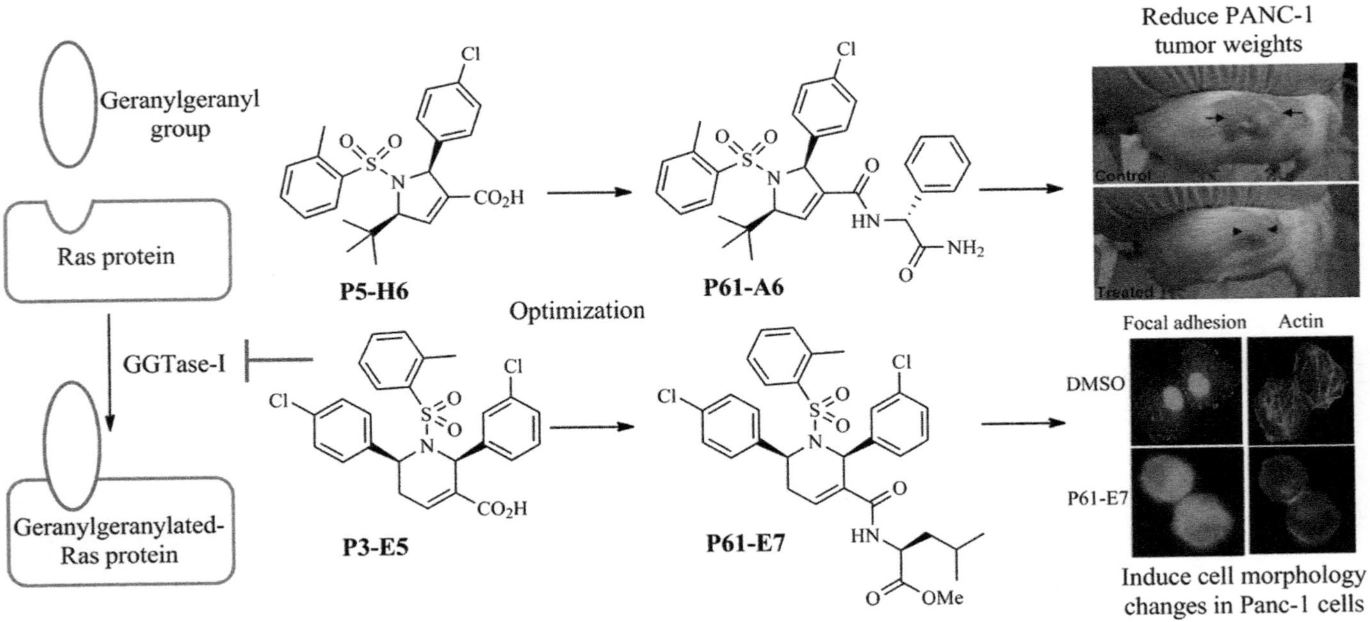

FIGURE 18.5 Small-molecule inhibitors of protein geranylgeranylation. (*See insert for color representation of the figure.*)

tumors in mice, it exhibited dose-dependent tumor-suppressing effects without severely toxic side effects [38c]. When tumor-implanted mice were treated with 1.16 mg/kg of P61-A6 six times per week for 44 days, tumor weights were reduced to approximately 32% of control. Moreover, the cytosolic fraction of RhoA and Rap1 was significantly higher in tumor samples from P61-A6-treated animals than in samples taken from untreated animals, suggesting that the inhibitor was reaching the tumors. P61-A6 was found to affect neither the vasculature nor levels of apoptosis in the tumors, and the levels of DNA synthesis measured in tumor cells suggests that the compound exerts its effects through inhibition of cellular proliferation.

Optimization of the second initial hit, P3-E5, led to the discovery of P61-E7, a more potent antiproliferative inhibitor than P61-A6 (e.g., 49% inhibition at 1.25 μM for Panc-1 cells compared with 10% for P61-A6) [38d]. Treatment with P61-E7 increased the levels of the cyclin-dependent kinase inhibitors $p21^{CIP1/WAF1}$ and $p27^{Kip1}$ and induced a variety of morphology changes in Panc-1 cells, such as disorganization of the actin cytoskeleton, cell rounding, inhibition of focal adhesion assembly, and impairment of anchorage-independent growth. When a farnesyl-accepting RhoA mutant was introduced, the effects of the P61-E7 on $p21^{CIP1/WAF1}$ and $p27^{Kip1}$ levels and cell proliferation were partially reversed and the cell morphology changes were minimized.

These new inhibitors of protein prenylation were derived from a DOS library and developed in a series of studies to expand the collection of nonpeptidomimetic inhibitors of protein prenylation. The new compounds have utility both as probes for studying geranylgeranylation of prenylation substrates and as lead compounds for the development of new anticancer therapies.

18.4.4 Inhibitors of Core Self-Association in Hepatitis C Virus: Probes with a Novel Mechanism

Hepatitis C virus (HCV) poses a significant public health burden, with a 2.2 to 3.0% global prevalence of infection [83]. The chronic liver disease caused by the virus can develop into cirrhosis or liver cancer. The current standard of treatment is suboptimal, being effective in only about half of patients, and side effects have resulted in the discontinuation of treatment in 10 to 20% of patients [84]. While the HCV protease inhibitors boceprevir and telaprevir have recently expanded the options for treatment, there remains a need to develop new therapeutics that target different stages of the HCV life cycle.

Penin, Lindenbach, Kota, and co-workers have used a DOS-based strategy to discover alkaloid-like small molecules that inhibit dimerization of the HCV nucleocapsid protein core. HCV core is one of 10 mature proteins coded by the virus, and dimerization of the protein is a crucial step in the HCV life cycle [85]. To model core dimerization, a glutathione *S*-transferase (GST)-core106/FLAG-core106 system was established using core106 (a more soluble, truncated version of the core protein) and further developed into a time-resolved fluorescence resonance energy transfer (TR-FRET) high-throughput assay (Figure 18.6) [51b–d,85c]. The TR-FRET assay was initiated by mixing GST-core106 protein and FLAG-core106 protein. Next, two

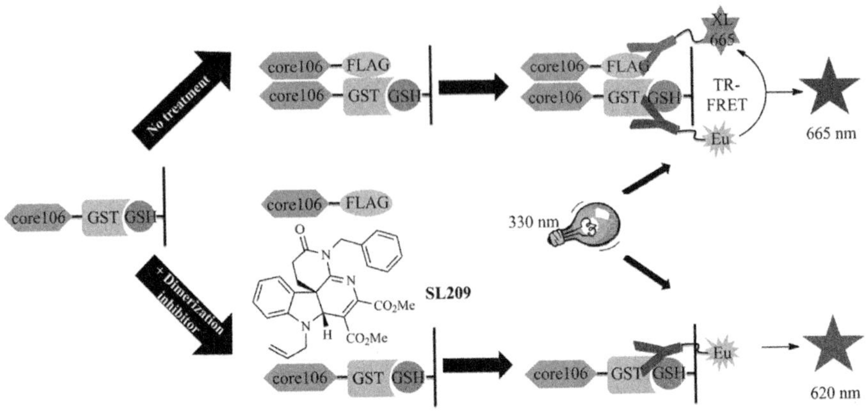

FIGURE 18.6 SL209 inhibit heterodimerization of GST-core106/FLAG-core106. (*See insert for color representation of the figure.*)

antibodies fused to fluorophores were added, an anti-GST-europium cryptate (Eu) and an anti-FLAG-allophycocyanin (XL-665). Finally, each well was irradiated with 330-nm light to excite the Eu fluorophore. The interaction between the two fluorophore-tagged proteins led to transfer of fluorescence energy and emission at 665 nm in the negative control. Small-molecule inhibitors of heterodimerization diminished FRET, which resulted in Eu-specific emission at 620 nm.

The TR-FRET assay was used to screen a DOS library of 2240 compounds developed at the Center for Chemical Methodology for Library Development at Boston University (CMLD-BU) for inhibitors of core heterodimerization [51b]. Based on an initial hit, the compound SL209 was identified as an inhibitor of core106 dimerization that exhibited low toxicity toward uninfected hepatoma cells (Figure 18.6) [51b]. The dose-dependent inhibitory activity of the compound was confirmed in an amplified luminescent proximity homogeneous assay (ALPHA).

To validate core as the target of SL209, an SL209–biotin conjugate was prepared as an affinity probe. The conjugate retained the activity of the untagged compound in the ALPHA assay. In affinity-isolation experiments with hepatoma cells, the SL209–biotin conjugate was found to associate directly with the core protein. In addition to core, several other HCV proteins were pulled down by the conjugate, including the NS3 viral helicase and the NS5A viral regulatory protein. When the HCV pSGR subgenomic replicon, which produces NS3 and NS5A but not core, was used in pull-down experiments, neither NS3 nor NS5A was retained. However, capture of NS3 and NS5A was rescued upon addition of core106 to the replicon lysate, strongly suggesting that these proteins associate with core in the cell and that the SL209–biotin interaction is specific to core. Further studies showed that SL209–biotin co-localizes with core on the lipid droplets that act as surfaces for core-directed assembly of HCV viral particles [51a].

The discovery of SL209 as a new dimerization inhibitor of the HCV nucleocapsid protein core exemplifies the utility of DOS libraries in developing small-molecule probes that target novel mechanisms relevant to human diseases. In the study, a DOS strategy was applied to construct a library and to prepare a secondary, focused sublibrary for lead optimization. Target validation confirmed SL209 as a useful probe for studying the assembly and disassembly of the HCV viral particle.

18.5 DEVELOPING PROBES FOR THERAPEUTICALLY DESIRABLE PHENOTYPES

Screening DOS libraries for therapeutically desired phenotypes has emerged as a powerful method for discovering novel lead compounds and for understanding disease biology. In addition to the representative studies presented below, several important examples of this approach are discussed in Chapter 17.

18.5.1 Inhibitors of Glucose Transport

The reliance of cancer cells on glycolysis in preference to oxidative phosphorylation (OXYPHOS) for ATP synthesis, even in the presence of oxygen, is a well-known phenomenon [86]. Consequently, molecules that interfere with glucose metabolism or transport are of interest as probes in the study of neoplastic disorders [87]. A range of molecules are known to inhibit various steps of glycolysis, such as 2-deoxyglucose, arsenate compounds, 3-bromopyruvate, and iodoacetate [88]. An alternative approach to perturbing glycolysis is to modify the activity of glucose transporters (GLUT proteins) that facilitate the passive transport of glucose across cell membranes [89]. The natural product cytochalasin B is a well-studied inhibitor of transmembrane glucose transport [90]. However, cytochalasin B also depolymerizes actin, complicating its use as a probe in studying complex cellular systems [24]. Therefore, glucose transport inhibitors that do not affect other important cellular processes are desired as probes for studying the relationship between glycolysis and neoplastic disorders.

Ulanovskaya et al. employed a pairwise chemical genetic screen to identify molecules that modulate glucose metabolism based on the two available pathways for ATP synthesis in cells: glycolysis and OXYPHOS [24]. When one pathway is shut down (e.g., by a small-molecule inhibitor), a cell will rely on the other pathway for ATP synthesis, and cellular ATP production becomes sensitized to inhibitors of the alternative pathway. This approach was validated by studying several pairs of molecules, where one molecule inhibits glycolysis or glucose transport (2-deoxyglucose, sodium iodoacetate, or cytochalasin B) and the other molecule inhibits OXYPHOS (e.g., antimycin A). An endogenous ATP level-dependent luciferase reporter assay was developed to measure the effect of a compound or a pair of compounds on ATP levels, and the assay was used to screen a library of 955 structurally diverse compounds against antimycin A–treated A549 cells to identify new inhibitors of glycolysis and glucose transport. Two compounds (Figure 18.7) that reduced ATP levels were identified. Neither compound reduced ATP levels appreciably in the absence of antimycin A, and both compounds exhibited ATP suppression

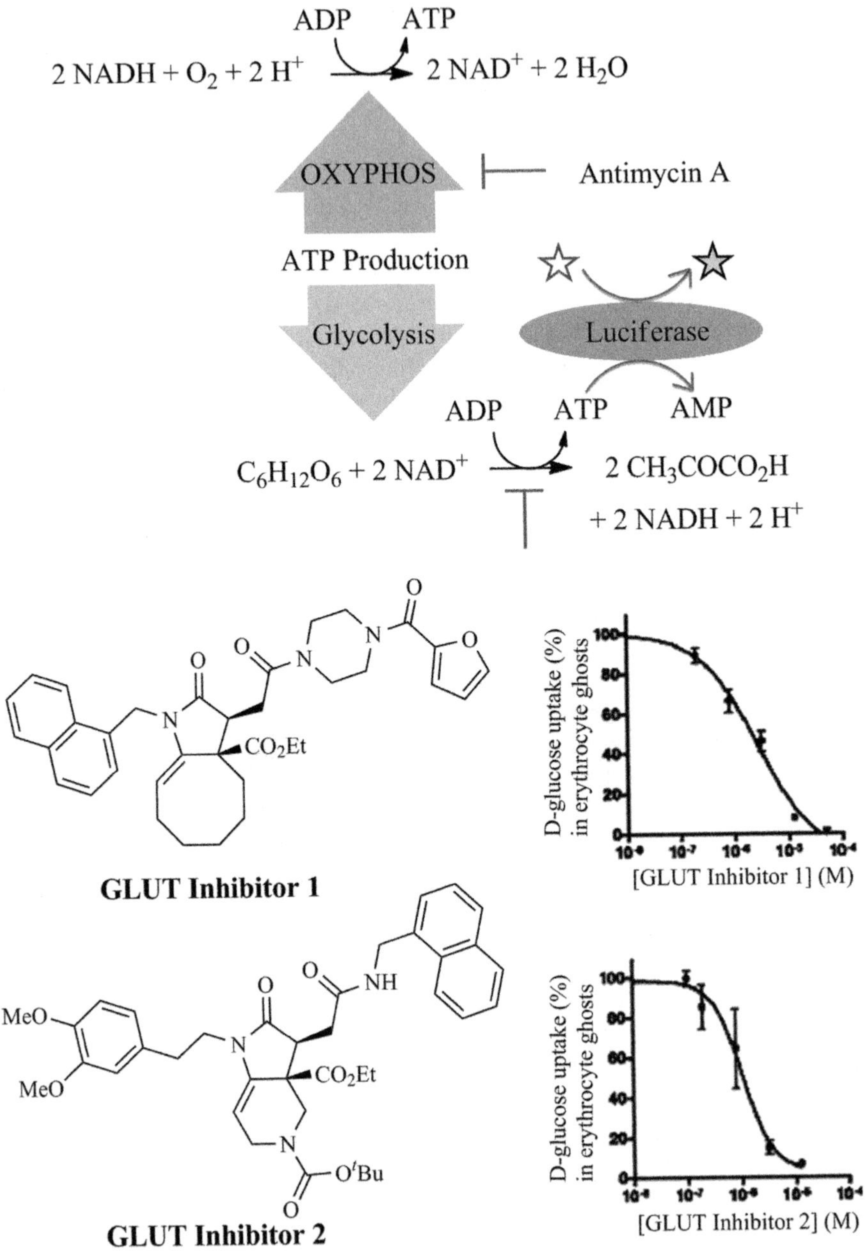

FIGURE 18.7 Pairwise chemical genetic screening identified glucose transport inhibitors.

activity similar to that of cytochalasin B, suggesting that they might function through a glucose transport inhibitory mechanism. Chinese hamster ovary cells (CHO-K1) were more sensitive to the synergistic effect than A549 cells, with GLUT inhibitors 1 and 2 exhibiting IC_{50} values of 2 and 1 μM for ATP levels in the presence of 10 nM antimycin A.

Glycolytic flux through mammalian cells can be assessed by measuring levels of lactate production, with higher levels of lactate corresponding to higher levels of glycolytic activity. GLUT inhibitors 1 and 2 exhibited dose-dependent inhibition of cellular lactate (IC_{50} values of 3 and 1.5 μM, respectively). These effects were similar to those of 2-deoxyglucose and cytochalasin B. GLUT inhibitor 1 was also found to inhibit uptake of radiolabeled glucose analogs into CHO-1K cells with low micromolar IC_{50} values. The mechanism of action of the two compounds was verified by measuring glucose uptake into purified and sealed erythrocyte membranes expressing GLUT1 exclusively. Both compounds inhibited glucose uptake with low micromolar IC_{50} values and exhibited a kinetic profile consistent with a noncompetitive mode of inhibition, similar to that of cytochalasin B. However, neither compound altered F-actin distribution, distinguishing the newly discovered inhibitors from cytochalasin-based inhibitors of glucose transport.

18.5.2 Inhibitors of Osteoclastogenesis

The integrity of bone structure relies on the interplay between osteoblast-regulated bone formation and osteoclast-regulated bone resorption. Zhu et al. have screened a DOS library of more than 2000 compounds to identify small-molecule inhibitors that target the formation and maintenance of osteoclasts, which, in disease states, contribute to the conditions of osteoporosis, rheumatoid arthritis, and metastatic cancer [13]. A cell-based, high-throughput assay was developed to detect the levels of tartrate-resistant acid phosphatase (TRAP), an enzyme expressed by osteoclasts, to monitor the receptor of NF-κB ligand (RANKL)–induced differentiation of preosteoclastic RAW264.7 cells into TRAP-forming osteoclasts [91]. Inhibitors of osteoclastogenesis are expected to reduce the levels of TRAP upon stimulation of RAW264.7 cells with RANKL. Based on the initial hits from the screening, a focused library of compounds containing a benzopyranyl skeleton was prepared. The most active compound was shown to reduce TRAP activity in a dose-dependent manner with an IC_{50} value of 5.09 μM (Figure 18.8). The one-day treatment of cells with this compound resulted in 78% viability compared with the control. When the compound was examined for its effects on genes downstream of RANK, both ERK phosphorylation and NF-κB activation were found to be affected in RAW264.7 cells. Moreover, *m*RNA levels of several biomarkers of osteoclastogenesis such as MMP-9 and s-Src were found to be reduced upon treatment of RAW264.7 cells with the compound at 10 μM.

The modulators of bone resorption identified in this work are valuable tools for identifying novel targets involved in the formation osteoclasts and for understanding the contribution of osteoclasts to the etiology of diseased states.

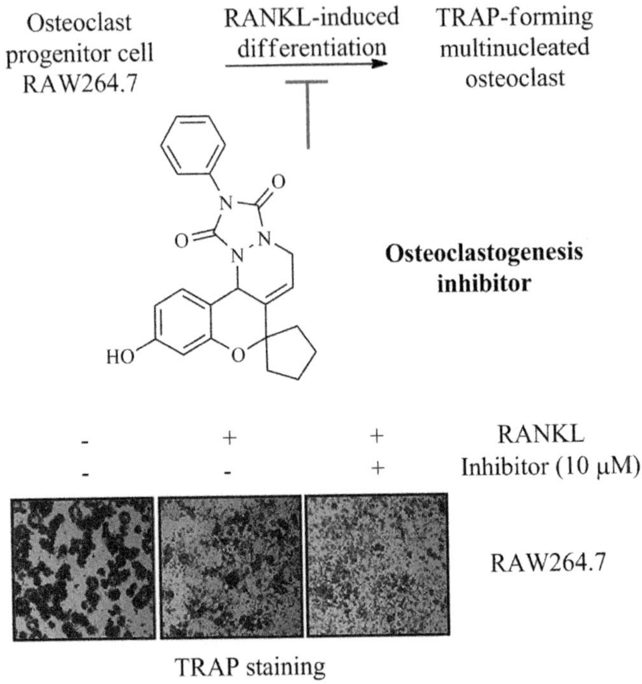

FIGURE 18.8 Benzopyranyl inhibitor of osteoclast development.

18.6 NATURAL PRODUCT–INSPIRED SMALL-MOLECULE PROBES DEVELOPED FROM DOS AND BIOLOGY-ORIENTED SYNTHESIS

Over the past decade, the DOS approach has been extended to the exploration of specific natural product chemotypes [92]. Examples of probes discovered through this approach are shown in Table 18.3. Also shown are examples of compounds developed through biology-oriented synthesis (BIOS), a method of probe discovery that has emerged in parallel with DOS for exploring the chemical space around natural product hubs.

18.6.1 Activator of the Wnt Pathway Derived from BIOS

The Wnt signaling pathways plays an important role in tissue regeneration and cancer [104]. In the canonical Wnt pathway, the cytoplasmic level of β-catenin is kept low in the absence of Wnt ligand stimulation, as a consequence of β-catenin's phosphorylation by glycogen synthase kinase-3β (GSK-3β), subsequent ubiquitination by the Axin complex, and degradation in the proteasome [105]. Activation of the canonical Wnt pathway inhibits GSK-3β and stabilizes β-catenin in the cytoplasm. After translocating to the nucleus, β-catenin binds to the transcription factors that stimulate transcription of the downstream Wnt genes. The canonical Wnt pathway

TABLE 18.3 Representative Natural Product–Inspired Probes Developed from DOS and BIOS

Structure	Name	Activity	Assay	Ref.
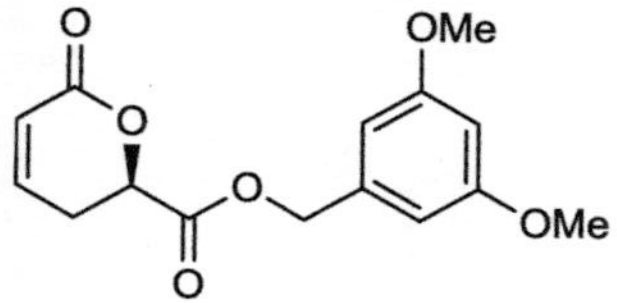	α,β-Unsaturated δ-lactone analog	Entry of vesicular stomatitis virus >3-fold reduction in infected cells	Cell-based measure of viral infection with VSV-expressing GFP	93
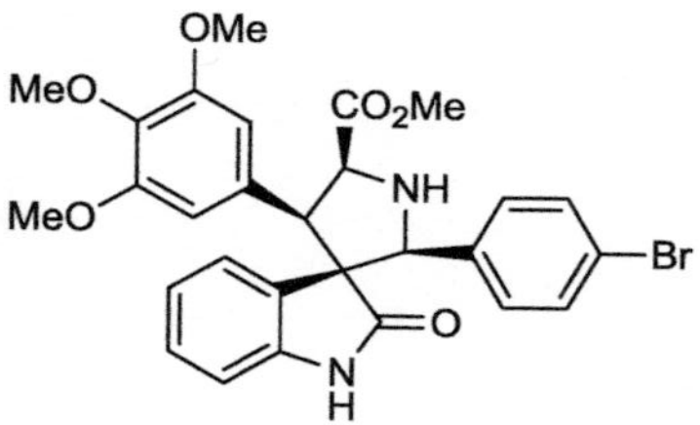	Spirooxindole analog	Inhibition of microtubule regrowth at 10 μM	Phenotypic changes associated with mitotic arrest	94
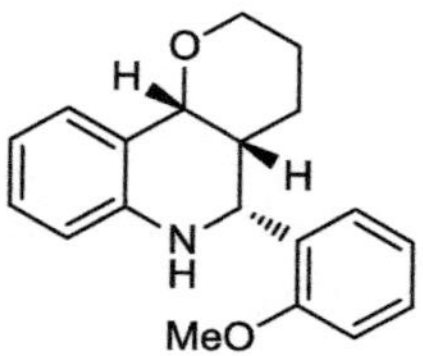	Tetrahydroquinoline analog	*Mycobacterium tuberculosis* (MTB) inhibitor; MIC = 6.25 μg/mL	MIC assay	95

(*continued*)

TABLE 18.3 ***(Continued)***

Structure	Name	Activity	Assay	Ref.
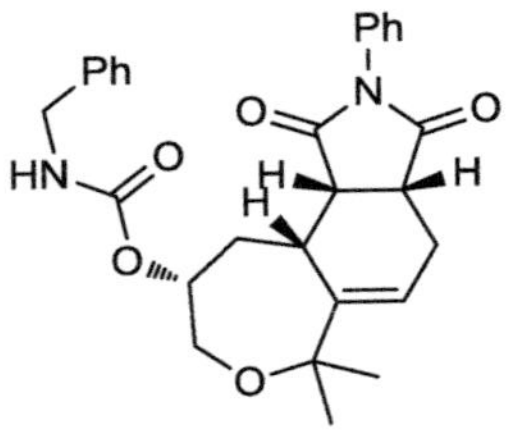	Oxepane Wntepane 1	Wnt pathway activator; $ED_{50} = 1.8\ \mu M$	Luciferase expressing reporter cell line responsive to Wnt3a	96
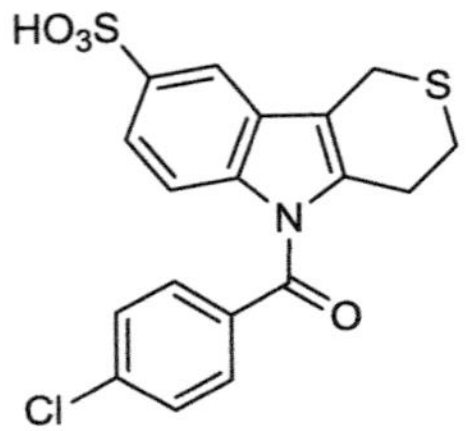	Indomethacin analog	Angiogenesis-related receptor tyrosine kinase inhibitor; $IC_{50} = 3\ \mu M$ for Tie-2	Immunosorbent assay of the amount of phosphorylated substrate	97
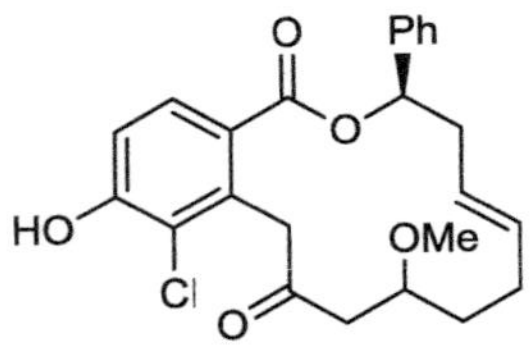	Pochonin analog	Kinase inhibitor; $IC_{50} = 8\ \mu M$ for Src	Inhibitory activity against a panel of kinases	98
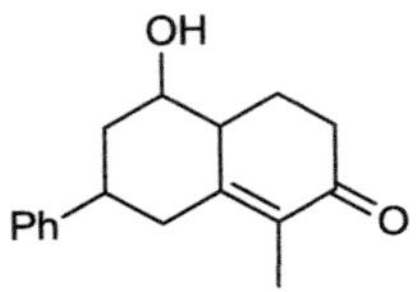	Dysidiolide analog	Acetylcholinesterase inhibitor; $IC_{50} = 3.7\ \mu M$	In vitro spectrophotometric assay [99]	100

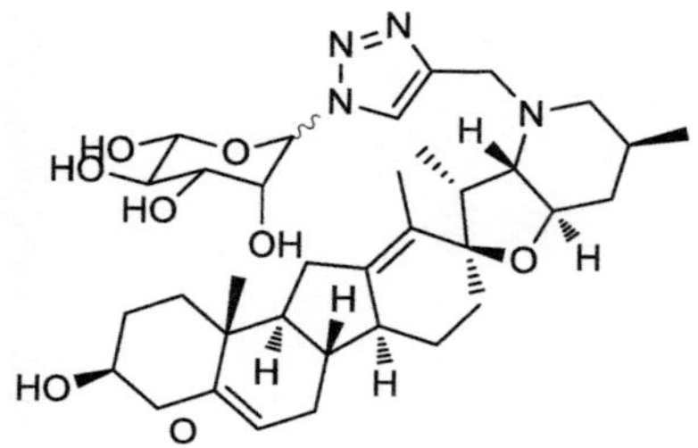	Cyclopamine analog	Hedgehog pathway signaling inhibitor; $IC_{50} = 33\ \mu M$	MTS proliferation assay of A549 cells	101
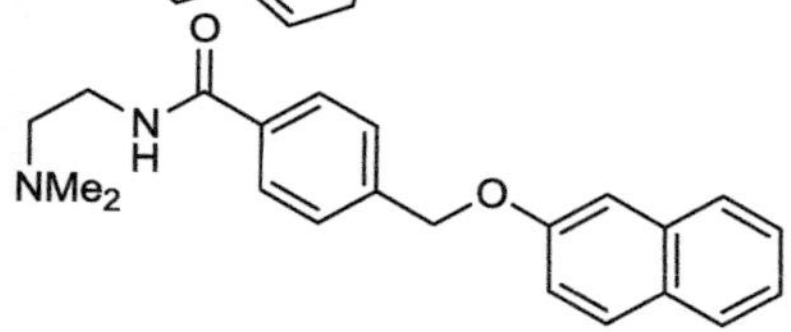	Three-ring 2-naphthol analog	5-Lipoxygenase inhibitor; $IC_{50} = 3\ \mu M$	Human blood cell assay	102
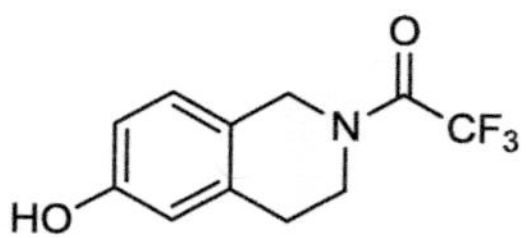	Tetrahydroisoquinoline analog	Estrogen receptor agonist; $EC_{50} = 20\ \mu M$ for ERα, $EC_{50} = 4.6\ \mu M$ for ERβ	Recruitment of a fluorescent binder of the agonist-bound receptor	102
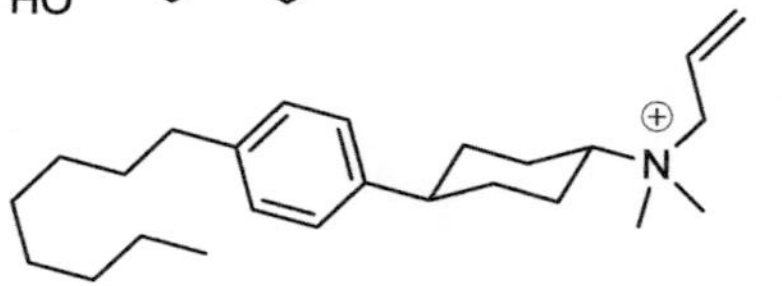	Fingolimod and myriocin analog	Sphingosine kinase 2 inhibitor; 2.1% SphK2 activity at 100 μM	Scintillation spectrometry of ^{32}P	103

ultimately regulates cellular proliferation and differentiation [105b]. The noncanonical Wnt pathways regulate downstream gene expression in a β-catenin-independent manner and are primarily involved in cell migration. These pathways include the planar cell polarity (PCP) pathway, the Wnt/calcium(Ca^{2+}) pathway, the steroid receptor binding, and the atypical receptor tyrosine kinase pathway [105a]. Abnormal up-regulation of Wnt pathways has been reported in many types of cancer [106], and small-molecule inhibitors of the Wnt pathways have been developed as potential anticancer therapeutics [104a,106b,107]. Meanwhile, diminished Wnt pathway activity has also been found to be closely associated with diseases such as osteoporosis and neurodegenerative disorders [108], suggesting that small-molecule activators of the Wnt pathway would be useful. However, only a few such activators have been reported [109].

The Waldmann group has identified a new class of Wnt pathway activators, designated as Wntepanes, using BIOS [96]. To measure the activation of the Wnt pathway, a stably transfected luciferase reporter cell line derived from HEK293 cell was made by cotransfection of a plasmid containing Tcf/Lef (Wnt pathway transcription factors) binding sites followed by the luciferase gene, and a human Frizled-1 (Fz, cell surface Wnt ligand receptor for enhancing the cells' sensitivity to the Wnt ligand. The screening of BIOS compounds in this luciferase assay led to the identification of several hits (the most potent hit is shown in Figure 18.9) that stimulated the signal synergistically with the canonical Wnt ligand Wnt3a protein. After further quantitative structure–activity relationship (QSAR) optimization, Wntepane 1 was found to be the most active Wnt activator, with an ED_{50} value of 1.8 ± 0.9 μM (Figure 18.9).

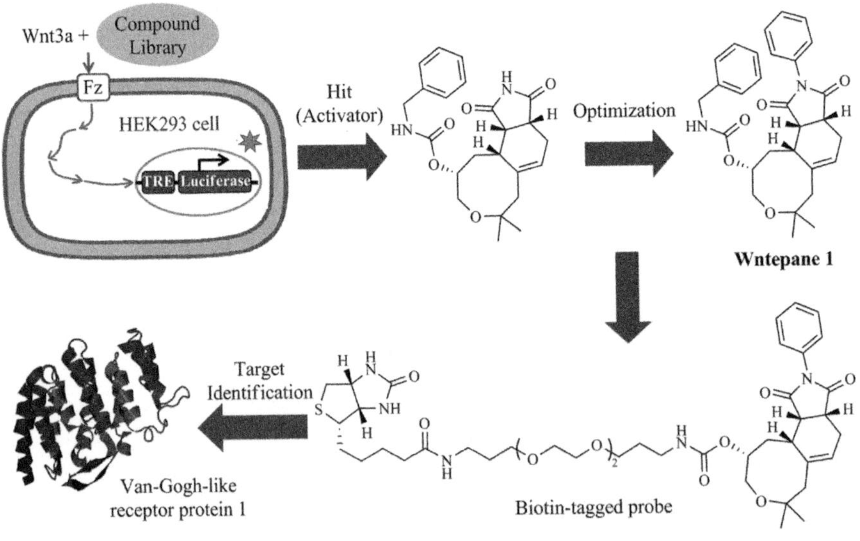

FIGURE 18.9 Novel activator of the Wnt pathway developed using BIOS [110].

A chemical proteomics approach was used to identify the target protein of Wntepane 1. A set of biotin-tagged bait compounds, including an affinity probe (Figure 18.9), was synthesized and immobilized on streptavidin-coated magnetic beads. After treatment with HEK293 cell lysate, the proteins bound to the beads were eluted with nonbiotinylated Wntepane 1 and separated by polyacrylamide gel electrophoresis (PAGE). Mass spectrometry identified the van Gogh–like receptor protein 1 (Vangl1) as the most plausible target. The identification of Vangl1 as the target of the Wntepanes is notable because Vangl1 works primarily on the noncanonical Wnt/PCP pathway [111]. Interestingly, the activators targeting Vangl1 were found work synergistically with the canonical Wnt ligand Wnt3a.

As the first activators of Wnt signaling that target Vangl1 that have been identified, the Wntepanes can serve as novel probes for studying the activation of the Wnt pathways through Vangl1, the relationship between the noncanonical Wnt/PCP pathway and the canonical Wnt pathway, and Wnt pathway protein activities in cancer and stem cell biology.

18.7 SUMMARY AND OUTLOOK

DOS strategies have greatly facilitated the discovery of new and important biological probes in the past decade. This success has yielded new insight into complex biological processes and afforded leads for developing novel therapies. We can expect that the potential of the DOS approach to probe development will continue to emerge in future studies, especially studies aimed at developing high-quality probes that affect a challenging protein target or biological process [112]. An overarching aim of the DOS approach to probe development is to provide a bridge between basic biology and human diseases. This aim will become a reality when a broad range of probes has become accessible to the research and medical communities.

REFERENCES

1. (a) S. L. Schreiber, *Chem. Eng. News* **2003**, *81*, 51–60; (b) A. B. Castoreno, U. S. Eggert, *ACS Chem. Biol.* **2011**, *6*, 86–94.
2. (a) F. E. Koehn, G. T. Carter, *Nat. Rev. Drug. Discov.* **2005**, *4*, 206–220; (b) E. E. Carlson, *ACS Chem. Biol.* **2010**, *5*, 639–653.
3. J. Clardy, C. Walsh, *Nature* **2004**, *432*, 829–837.
4. (a) B. R. Stockwell, *Nat. Rev. Genet.* **2000**, *1*, 116–125; (b) C. J. O'Connor, L. Laraia, D. R. Spring, *Chem. Soc. Rev.* **2011**, *40*, 4332–4345.
5. T. E. Nielsen, S. L. Schreiber, *Angew. Chem. Int. Ed.* **2008**, *47*, 48–56.
6. (a) D. S. Tan, *Nat. Chem. Biol.* **2005**, *1*, 74–84; (b) C. J. O'Connor, H. S. G. Beckmann, D. R. Spring, *Chem. Soc. Rev.* **2012**, *41*, 4444–4456.

7. (a) H. E. Pelish, N. J. Westwood, Y. Feng, T. Kirchhausen, M. D. Shair, *J. Am. Chem. Soc.* **2001**, *123*, 6740–6741; (b) H. E. Pelish, J. R. Peterson, S. B. Salvarezza, E. Rodriguez-Boulan, J. L. Chen, M. Stamnes, E. Macia, Y. Feng, M. D. Shair, T. Kirchhausen, *Nat. Chem. Biol.* **2006**, *2*, 39–46.
8. L. F. Peng, S. S. Kim, S. Matchacheep, X. Lei, S. Su, W. Lin, W. Runguphan, W. H. Choe, N. Sakamoto, M. Ikeda, N. Kato, A. B. Beeler, J. A. Porco, Jr., S. L. Schreiber, R. T. Chung, *Antimicrob. Agents Chemother.* **2007**, *51*, 3756–3759.
9. E. E. Wyatt, W. R. Galloway, G. L. Thomas, M. Welch, O. Loiseleur, A. T. Plowright, D. R. Spring, *Chem. Commun.* **2008**, 4962–4964.
10. G. L. Thomas, R. J. Spandl, F. G. Glansdorp, M. Welch, A. Bender, J. Cockfield, J. A. Lindsay, C. Bryant, D. F. Brown, O. Loiseleur, H. Rudyk, M. Ladlow, D. R. Spring, *Angew. Chem. Int. Ed.* **2008**, *47*, 2808–2812.
11. L. Alvey, S. Prado, B. Saint-Joanis, S. Michel, M. Koch, S. T. Cole, F. Tillequin, Y. L. Janin, *Eur. J. Med. Chem.* **2009**, *44*, 2497–2505.
12. A. Kumar, S. Sharma, R. A. Maurya, J. Sarkar, *J. Comb. Chem.* **2010**, *12*, 20–24.
13. M. Zhu, M. H. Kim, S. Lee, S. J. Bae, S. H. Kim, S. B. Park, *J. Med. Chem.* **2010**, *53*, 8760–8764.
14. A. M. Socha, N. Y. Tan, K. L. LaPlante, J. K. Sello, *Bioorg. Med. Chem.* **2010**, *18*, 7193–7202.
15. S. Kantevari, S. R. Patpi, D. Addla, S. R. Putapatri, B. Sridhar, P. Yogeeswari, D. Sriram, *ACS Comb. Sci.* **2011**, *13*, 427–435.
16. Z. Wang, S. Castellano, S. S. Kinderman, C. E. Argueta, A. B. Beshir, G. Fenteany, O. Kwon, *Chem. Eur. J.* **2011**, *17*, 649–654.
17. T. A. Wencewicz, B. Yang, J. R. Rudloff, A. G. Oliver, M. J. Miller, *J. Med. Chem.* **2011**, *54*, 6843–6858.
18. H. Oguri, T. Hiruma, Y. Yamagishi, H. Oikawa, A. Ishiyama, K. Otoguro, H. Yamada, S. Omura, *J. Am. Chem. Soc.* **2011**, *133*, 7096–7105.
19. S. Oh, S. W. Cho, J.-Y. Yang, H. J. Sun, Y. S. Chung, C. S. Shin, S. B. Park, *Med. Chem. Commun.* **2011**, *2*, 76–80.
20. B. O. Okandeji, D. M. Greenwald, J. Wroten, J. K. Sello, *Bioorg. Med. Chem.* **2011**, *19*, 7679–7689.
21. F. Shi, A. X. Dai, X. H. Zhang, B. Jiang, S. J. Tu, *ACS Comb. Sci.* **2011**, *13*, 147–153.
22. D. M. Huryn, J. L. Brodsky, K. M. Brummond, P. G. Chambers, B. Eyer, A. W. Ireland, M. Kawasumi, M. G. Laporte, K. Lloyd, B. Manteau, P. Nghiem, B. Quade, S. P. Seguin, P. Wipf, *Proc. Natl. Acad. Sci. U.S.A.* **2011**, *108*, 6757–6762.
23. L. E. Brown, K. Chih-Chien Cheng, W. G. Wei, P. Yuan, P. Dai, R. Trilles, F. Ni, J. Yuan, R. MacArthur, R. Guha, R. L. Johnson, X. Z. Su, M. M. Dominguez, J. K. Snyder, A. B. Beeler, S. E. Schaus, J. Inglese, J. A. Porco, Jr., *Proc. Natl. Acad. Sci. U.S.A.* **2011**, *108*, 6775–6780.
24. O. A. Ulanovskaya, J. Cui, S. J. Kron, S. A. Kozmin, *Chem. Biol.* **2011**, *18*, 222–230.
25. D. Cruz, Z. Wang, J. Kibbie, R. Modlin, O. Kwon, *Proc. Natl. Acad. Sci. U.S.A.* **2011**, *108*, 6769–6774.
26. D. H. C. Chou, J. R. Duvall, B. Gerard, H. B. Liu, B. A. Pandya, B. C. Suh, E. M. Forbeck, P. Faloon, B. K. Wagner, L. A. Marcaurelle, *ACS Med. Chem. Lett.* **2011**, *2*, 698–702.

27. K. Dower, C. M. Filone, E. N. Hodges, Z. B. Bjornson, K. H. Rubins, L. E. Brown, S. Schaus, L. E. Hensley, J. H. Connor, *J. Virol.* **2012**, *86*, 2632–2640.
28. R. W. Heidebrecht, Jr., C. Mulrooney, C. P. Austin, R. H. Barker, Jr., J. A. Beaudoin, K. C. Cheng, E. Comer, S. Dandapani, J. Dick, J. R. Duvall, E. H. Ekland, D. A. Fidock, M. E. Fitzgerald, M. Foley, R. Guha, P. Hinkson, M. Kramer, A. K. Lukens, D. Masi, L. A. Marcaurelle, X. Z. Su, C. J. Thomas, M. Weiwer, R. C. Wiegand, D. Wirth, M. Xia, J. Yuan, J. Zhao, M. Palmer, B. Munoz, S. Schreiber, *ACS Med. Chem. Lett.* **2012**, *3*, 112–117.
29. F. Shi, X. N. Zeng, X. D. Cao, S. Zhang, B. Jiang, W. F. Zheng, S. J. Tu, *Bioorg. Med. Chem. Lett.* **2012**, *22*, 743–746.
30. S. J. Haggarty, K. M. Koeller, J. C. Wong, R. A. Butcher, S. L. Schreiber, *Chem. Biol.* **2003**, *10*, 383–396.
31. J. C. Wong, R. Hong, S. L. Schreiber, *J. Am. Chem. Soc.* **2003**, *125*, 5586–5587.
32. F. G. Kuruvilla, A. F. Shamji, S. M. Sternson, P. J. Hergenrother, S. L. Schreiber, *Nature* **2002**, *416*, 653–657.
33. K. C. Nicolaou, R. M. Evans, A. J. Roecker, R. Hughes, M. Downes, J. A. Pfefferkorn, *Org. Biomol. Chem.* **2003**, *1*, 908–920.
34. A. N. Koehler, A. F. Shamji, S. L. Schreiber, *J. Am. Chem. Soc.* **2003**, *125*, 8420–8421.
35. Y. Mu, C. R. Stephenson, C. Kendall, S. P. Saini, D. Toma, S. Ren, H. Cai, S. C. Strom, B. W. Day, P. Wipf, W. Xie, *Mol. Pharmacol.* **2005**, *68*, 403–413.
36. P. Y. Ng, Y. Tang, W. M. Knosp, H. S. Stadler, J. T. Shaw, *Angew. Chem. Int. Ed.* **2007**, *46*, 5352–5355.
37. J. S. Lazo, J. J. Skoko, S. Werner, B. Mitasev, A. Bakan, F. Koizumi, A. Yellow-Duke, I. Bahar, K. M. Brummond, *J. Pharmacol. Exp. Ther.* **2007**, *322*, 940–947.
38. (a) S. Castellano, H. D. Fiji, S. S. Kinderman, M. Watanabe, P. Leon, F. Tamanoi, O. Kwon, *J. Am. Chem. Soc.* **2007**, *129*, 5843–5845; (b) M. Watanabe, H. D. Fiji, L. Guo, L. Chan, S. S. Kinderman, D. J. Slamon, O. Kwon, F. Tamanoi, *J. Biol. Chem.* **2008**, *283*, 9571–9579; (c) J. Lu, L. Chan, H. D. Fiji, R. Dahl, O. Kwon, F. Tamanoi, *Mol. Cancer Ther.* **2009**, *8*, 1218–1226; (d) L. N. Chan, H. D. Fiji, M. Watanabe, O. Kwon, F. Tamanoi, *PLoS ONE* **2011**, *6*, e26135.
39. S. Di Micco, R. Vitale, M. Pellecchia, M. F. Rega, R. Riva, A. Basso, G. Bifulco, *J. Med. Chem.* **2009**, *52*, 7856–7867.
40. S. A. Scott, P. E. Selvy, J. R. Buck, H. P. Cho, T. L. Criswell, A. L. Thomas, M. D. Armstrong, C. L. Arteaga, C. W. Lindsley, H. A. Brown, *Nat. Chem. Biol.* **2009**, *5*, 108–117.
41. D. M. Green, I. Goljer, D. S. Andraka, M. Chengalvala, L. Shanno, W. Hurlburt, J. C. Pelletier, *J. Comb. Chem.* **2009**, *11*, 117–125.
42. T. G. Elford, A. Ulaczyk-Lesanko, G. De Pascale, G. D. Wright, D. G. Hall, *J. Comb. Chem.* **2009**, *11*, 155–168.
43. B. Z. Stanton, L. F. Peng, N. Maloof, K. Nakai, X. Wang, J. L. Duffner, K. M. Taveras, J. M. Hyman, S. W. Lee, A. N. Koehler, J. K. Chen, J. L. Fox, A. Mandinova, S. L. Schreiber, *Nat. Chem. Biol.* **2009**, *5*, 154–156.
44. L. A. Marcaurelle, C. Johannes, D. Yohannes, B. P. Tillotson, D. Mann, *Bioorg. Med. Chem. Lett.* **2009**, *19*, 2500–2503.

45. E. P. Lebois, T. M. Bridges, L. M. Lewis, E. S. Dawson, A. S. Kane, Z. Xiang, S. B. Jadhav, H. Yin, J. P. Kennedy, J. Meiler, C. M. Niswender, C. K. Jones, P. J. Conn, C. D. Weaver, C. W. Lindsley, *ACS Chem. Neurosci.* **2010**, *1*, 104–121.
46. S. Oh, H. J. Nam, J. Park, S. H. Beak, S. B. Park, *ChemMedChem* **2010**, *5*, 529–533.
47. S. Oh, S. J. Kim, J. H. Hwang, H. Y. Lee, M. J. Ryu, J. Park, S. J. Kim, Y. S. Jo, Y. K. Kim, C. H. Lee, K. R. Kweon, M. Shong, S. B. Park, *J. Med. Chem.* **2010**, *53*, 7405–7413.
48. L. A. Marcaurelle, E. Comer, S. Dandapani, J. R. Duvall, B. Gerard, S. Kesavan, M. D. T. Lee, H. Liu, J. T. Lowe, J. C. Marie, C. A. Mulrooney, B. A. Pandya, A. Rowley, T. D. Ryba, B. C. Suh, J. Wei, D. W. Young, L. B. Akella, N. T. Ross, Y. L. Zhang, D. M. Fass, S. A. Reis, W. N. Zhao, S. J. Haggarty, M. Palmer, M. A. Foley, *J. Am. Chem. Soc.* **2010**, *132*, 16962–16976.
49. S. W. Fewell, C. M. Smith, M. A. Lyon, T. P. Dumitrescu, P. Wipf, B. W. Day, J. L. Brodsky, *J. Biol. Chem.* **2004**, *279*, 51131–51140.
50. A. G. Herman, M. Hayano, M. V. Poyurovsky, K. Shimada, R. Skouta, C. Prives, B. R. Stockwell, *Cancer Discov.* **2011**, *1*, 312–325.
51. (a) S. Kota, V. Takahashi, F. Ni, J. K. Snyder, A. D. Strosberg, *PLoS One* **2012**, *7*, e32207; (b) W. Wei, C. Cai, S. Kota, V. Takahashi, F. Ni, A. D. Strosberg, J. K. Snyder, *Bioorg. Med. Chem. Lett.* **2009**, *19*, 6926–6930; (c) S. Kota, L. Scampavia, T. Spicer, A. B. Beeler, V. Takahashi, J. K. Snyder, J. A. Porco, P. Hodder, A. D. Strosberg, *Assay Drug Dev. Technol.* **2010**, *8*, 96–105; (d) A. D. Strosberg, S. Kota, V. Takahashi, J. K. Snyder, G. Mousseau, *Viruses* **2010**, *2*, 1734–1751.
52. B. Z. Stanton, L. F. Peng, *Mol. Biosyst.* **2010**, *6*, 44–54.
53. T. L. Lin, W. Matsui, *Oncol. Targets Ther.* **2012**, *5*, 47–58.
54. N. C. Bambakidis, K. Onwuzulike, *Vitam. Horm.* **2012**, *88*, 379–394.
55. L. F. Peng, B. Z. Stanton, N. Maloof, X. Wang, S. L. Schreiber, *Bioorg. Med. Chem. Lett.* **2009**, *19*, 6319–6325.
56. R. Vazquez-Martinez, A. Diaz-Ruiz, F. Almabouada, Y. Rabanal-Ruiz, F. Gracia-Navarro, M. M. Malagon, *Gen. Comp. Endocrinol.* **2012**, *175*, 1–9.
57. H. Farhan, C. Rabouille, *J. Cell Sci.* **2011**, *124*, 171–180.
58. (a) A. Dinter, E. G. Berger, *Histochem. Cell Biol.* **1998**, *109*, 571–590; (b) Y. Feng, S. Yu, T. K. Lasell, A. P. Jadhav, E. Macia, P. Chardin, P. Melancon, M. Roth, T. Mitchison, T. Kirchhausen, *Proc. Natl. Acad. Sci. U.S.A.* **2003**, *100*, 6469–6474.
59. J. F. Presley, N. B. Cole, T. A. Schroer, K. Hirschberg, K. J. M. Zaal, J. Lippincott-Schwartz, *Nature* **1997**, *389*, 81–85.
60. K. Hirschberg, C. M. Miller, J. Ellenberg, J. F. Presley, E. D. Siggia, R. D. Phair, J. Lippincott-Schwartz, *J. Cell Biol.* **1998**, *143*, 1485–1503.
61. B. Olofsson, *Cell. Signal.* **1999**, *11*, 545–554.
62. H. E. Pelish, W. Ciesla, N. Tanaka, K. Reddy, M. D. Shair, T. Kirchhausen, W. I. Lencer, *Biochem. Pharmacol.* **2006**, *71*, 1720–1726.
63. C. Ambrogio, C. Voena, A. D. Manazza, C. Martinengo, C. Costa, T. Kirchhausen, E. Hirsch, G. Inghirami, R. Chiarle, *Cancer Res.* **2008**, *68*, 8899–8907.
64. M. P. Mayer, B. Bukau, *Cell. Mol. Life Sci.* **2005**, *62*, 670–684.
65. M. E. Cheetham, A. J. Caplan, *Cell Stress Chaperones* **1998**, *3*, 28–36.
66. C. Jolly, R. I. Morimoto, *J. Natl. Cancer Inst.* **2000**, *92*, 1564–1572.
67. C. Rabu, P. Wipf, J. L. Brodsky, S. High, *J. Biol. Chem.* **2008**, *283*, 27504–27513.

68. (a) S. Stefanovic, R. S. Hegde, *Cell* **2007**, *128*, 1147–1159; (b) B. M. Abell, C. Rabu, P. Leznicki, J. C. Young, S. High, *J. Cell Sci.* **2007**, *120*, 1743–1751; (c) V. Favaloro, M. Spasic, B. Schwappach, B. Dobberstein, *J. Cell Sci.* **2008**, *121*, 1832–1840.
69. B. Patham, J. Duffy, A. Lane, R. C. Davis, P. Wipf, S. W. Fewell, J. L. Brodsky, K. Mensa-Wilmot, *Biochem. J.* **2009**, *419*, 507–517.
70. C. M. Wright, S. P. Seguin, S. W. Fewell, H. Zhang, C. Ishwad, A. Vats, C. A. Lingwood, P. Wipf, E. Fanning, J. M. Pipas, J. L. Brodsky, *Virus Res.* **2009**, *141*, 71–80.
71. M. Jiang, J. R. Abend, S. F. Johnson, M. J. Imperiale, *Virology* **2009**, *384*, 266–273.
72. L. A. Volpicelli, A. I. Levey, *Prog. Brain Res.* **2004**, *145*, 59–66.
73. J. Wess, R. M. Eglen, D. Gautam, *Nat. Rev. Drug Discov.* **2007**, *6*, 721–733.
74. J. E. Marlo, C. M. Niswender, E. L. Days, T. M. Bridges, Y. Xiang, A. L. Rodriguez, J. K. Shirey, A. E. Brady, T. Nalywajko, Q. Luo, C. A. Austin, M. B. Williams, K. Kim, R. Williams, D. Orton, H. A. Brown, C. W. Lindsley, C. D. Weaver, P. J. Conn, *Mol. Pharmacol.* **2009**, *75*, 577–588.
75. C. Sur, P. J. Mallorga, M. Wittmann, M. A. Jacobson, D. Pascarella, J. B. Williams, P. E. Brandish, D. J. Pettibone, E. M. Scolnick, P. J. Conn, *Proc. Natl. Acad. Sci. U.S.A.* **2003**, *100*, 13674–13679.
76. M. J. Marino, S. T. Rouse, A. I. Levey, L. T. Potter, P. J. Conn, *Proc. Natl. Acad. Sci. U.S.A.* **1998**, *95*, 11465–11470.
77. (a) T. A. Spalding, C. Trotter, N. Skjaerbaek, T. L. Messier, E. A. Currier, E. S. Burstein, D. Li, U. Hacksell, M. R. Brann, *Mol. Pharmacol.* **2002**, *61*, 1297–1302; (b) C. J. Langmead, N. E. Austin, C. L. Branch, J. T. Brown, K. A. Buchanan, C. H. Davies, I. T. Forbes, V. A. Fry, J. J. Hagan, H. J. Herdon, G. A. Jones, R. Jeggo, J. N. Kew, A. Mazzali, R. Melarange, N. Patel, J. Pardoe, A. D. Randall, C. Roberts, A. Roopun, K. R. Starr, A. Teriakidis, M. D. Wood, M. Whittington, Z. Wu, J. Watson, *Br. J. Pharmacol.* **2008**, *154*, 1104–1115; (c) C. K. Jones, A. E. Brady, A. A. Davis, Z. Xiang, M. Bubser, M. N. Tantawy, A. S. Kane, T. M. Bridges, J. P. Kennedy, S. R. Bradley, T. E. Peterson, M. S. Ansari, R. M. Baldwin, R. M. Kessler, A. Y. Deutch, J. J. Lah, A. I. Levey, C. W. Lindsley, P. J. Conn, *J. Neurosci.* **2008**, *28*, 10422–10433.
78. N. M. Appels, J. H. Beijnen, J. H. Schellens, *Oncologist* **2005**, *10*, 565–578.
79. E. C. Lerner, Y. Qian, A. D. Hamilton, S. M. Sebti, *J. Biol. Chem.* **1995**, *270*, 26770–26773.
80. K. T. Lane, L. S. Beese, *J. Lipid Res.* **2006**, *47*, 681–699.
81. K. F. Leung, R. Baron, M. C. Seabra, *J. Lipid Res.* **2006**, *47*, 467–475.
82. A. Vogt, J. Sun, Y. Qian, A. D. Hamilton, S. M. Sebti, *J. Biol. Chem.* **1997**, *272*, 27224–27229.
83. D. Lavanchy, *Liver Int.* **2009**, *29*, 74–81.
84. M. G. Ghany, D. B. Strader, D. L. Thomas, L. B. Seeff, *Hepatology* **2009**, *49*, 1335–1374.
85. (a) F. Penin, J. Dubuisson, F. A. Rey, D. Moradpour, J. M. Pawlotsky, *Hepatology* **2004**, *39*, 5–19; (b) B. D. Lindenbach, C. M. Rice, *Nature* **2005**, *436*, 933–938; (c) S. Kota, C. Coito, G. Mousseau, J. P. Lavergne, A. D. Strosberg, *J. Gen. Virol.* **2009**, *90*, 1319–1328.
86. O. Warburg, *Science* **1956**, *123*, 309–314.
87. D. A. Tennant, R. V. Duran, E. Gottlieb, *Nat. Rev. Cancer* **2010**, *10*, 267–277.
88. Z. Chen, W. Lu, C. Garcia-Prieto, P. Huang, *J. Bioenerg. Biomembr.* **2007**, *39*, 267–274.

89. A. Evans, V. Bates, H. Troy, S. Hewitt, S. Holbeck, Y. L. Chung, R. Phillips, M. Stubbs, J. Griffiths, R. Airley, *Cancer Chemother. Pharmacol.* **2008**, *61*, 377–393.
90. R. D. Taverna, R. G. Langdon, *Biochim. Biophys. Acta* **1973**, *323*, 207–219.
91. H. An, S. J. Eum, M. Koh, S. K. Lee, S. B. Park, *J. Org. Chem.* **2008**, *73*, 1752–1761.
92. S. Wetzel, R. S. Bon, K. Kumar, H. Waldmann, *Angew. Chem. Int. Ed.* **2011**, *50*, 10800–10826.
93. T. Lessmann, M. G. Leuenberger, S. Menninger, M. Lopez-Canet, O. Muller, S. Hummer, J. Bormann, K. Korn, E. Fava, M. Zerial, T. U. Mayer, H. Waldmann, *Chem. Biol.* **2007**, *14*, 443–451.
94. A. P. Antonchick, C. Gerding-Reimers, M. Catarinella, M. Schurmann, H. Preut, S. Ziegler, D. Rauh, H. Waldmann, *Nat. Chem.* **2010**, *2*, 735–740.
95. A. Kumar, S. Srivastava, G. Gupta, V. Chaturvedi, S. Sinha, R. Srivastava, *ACS Comb. Sci.* **2011**, *13*, 65–71.
96. S. Basu, B. Ellinger, S. Rizzo, C. Deraeve, M. Schurmann, H. Preut, H. D. Arndt, H. Waldmann, *Proc. Natl. Acad. Sci. U.S.A.* **2011**, *108*, 6805–6810.
97. C. Rosenbaum, P. Baumhof, R. Mazitschek, O. Muller, A. Giannis, H. Waldmann, *Angew. Chem. Int. Ed.* **2004**, *43*, 224–228.
98. E. Moulin, S. Barluenga, F. Totzke, N. Winssinger, *Chem. Eur. J.* **2006**, *12*, 8819–8834.
99. G. L. Ellman, K. D. Courtney, V. Andres, Jr., R. M. Feather-Stone, *Biochem. Pharmacol.* **1961**, *7*, 88–95.
100. M. Scheck, M. A. Koch, H. Waldmann, *Tetrahedron* **2008**, *64*, 4792–2802.
101. J. Zhang, M. Garrossian, D. Gardner, A. Garrossian, Y. T. Chang, Y. K. Kim, C. W. Chang, *Bioorg. Med. Chem. Lett.* **2008**, *18*, 1359–1363.
102. S. Renner, W. A. van Otterlo, M. Dominguez Seoane, S. Mocklinghoff, B. Hofmann, S. Wetzel, A. Schuffenhauer, P. Ertl, T. I. Oprea, D. Steinhilber, L. Brunsveld, D. Rauh, H. Waldmann, *Nat. Chem. Biol.* **2009**, *5*, 585–592.
103. M. R. Raje, K. Knott, Y. Kharel, P. Bissel, K. R. Lynch, W. L. Santos, *Bioorg. Med. Chem.* **2012**, *20*, 183–194.
104. (a) B. Chen, M. E. Dodge, W. Tang, J. Lu, Z. Ma, C. W. Fan, S. Wei, W. Hao, J. Kilgore, N. S. Williams, M. G. Roth, J. F. Amatruda, C. Chen, L. Lum, *Nat. Chem. Biol.* **2009**, *5*, 100–107; (b) N. Chung, S. Marine, E. A. Smith, R. Liehr, S. T. Smith, L. Locco, E. Hudak, A. Kreamer, A. Rush, B. Roberts, M. B. Major, R. T. Moon, W. Arthur, M. Cleary, B. Strulovici, M. Ferrer, *Assay Drug Dev. Technol.* **2010**, *8*, 286–294; (c) J. L. Green, T. Inoue, P. W. Sternberg, *Cell* **2008**, *134*, 646–656; (d) D. C. Lie, S. A. Colamarino, H. J. Song, L. Desire, H. Mira, A. Consiglio, E. S. Lein, S. Jessberger, H. Lansford, A. R. Dearie, F. H. Gage, *Nature* **2005**, *437*, 1370–1375.
105. (a) R. Widelitz, *Growth Factors* **2005**, *23*, 111–116; (b) A. Kikuchi, H. Yamamoto, A. Sato, S. Matsumoto, *Int. Rev. Cell. Mol. Biol.* **2011**, *291*, 21–71.
106. (a) R. H. Giles, J. H. van Es, H. Clevers, *Biochim. Biophys. Acta* **2003**, *1653*, 1–24; (b) S. Park, J. Gwak, M. Cho, T. Song, J. Won, D. E. Kim, J. G. Shin, S. Oh, *Mol. Pharmacol.* **2006**, *70*, 960–966; (c) M. Katoh, *Oncol. Rep.* **2005**, *14*, 1583–1588.
107. (a) R. K. Gandhirajan, P. A. Staib, K. Minke, I. Gehrke, G. Plickert, A. Schlösser, E. K. Schmitt, M. Hallek, K. Kreuzer, *Neoplasia* **2010**, *12*, 326–335; (b) M. Lepourcelet, Y. N. Chen, D. S. France, H. Wang, P. Crews, F. Petersen, C. Bruseo, A. W. Wood, R. A. Shivdasani, *Cancer Cell* **2004**, *5*, 91–102; (c) K. H. Emami, C. Nguyen, H. Ma, D. H.

Kim, K. W. Jeong, M. Eguchi, R. T. Moon, J. L. Teo, H. Y. Kim, S. H. Moon, J. R. Ha, M. Kahn, *Proc. Natl. Acad. Sci. U.S.A.* **2004**, *101*, 12682–12687.

108. F. Verkaar, G. J. Zaman, *Drug Discov. Today* **2011**, *16*, 35–41.

109. (a) M. P. Coghlan, A. A. Culbert, D. A. Cross, S. L. Corcoran, J. W. Yates, N. J. Pearce, O. L. Rausch, G. J. Murphy, P. S. Carter, L. Roxbee Cox, D. Mills, M. J. Brown, D. Haigh, R. W. Ward, D. G. Smith, K. J. Murray, A. D. Reith, J. C. Holder, *Chem. Biol.* **2000**, *7*, 793–803; (b) E. J. Park, S. J. Choi, Y. C. Kim, S. H. Lee, S. W. Park, S. K. Lee, *Bioorg. Med. Chem. Lett.* **2009**, *19*, 2282–2284; (c) Q. Zhang, M. B. Major, S. Takanashi, N. D. Camp, N. Nishiya, E. C. Peters, M. H. Ginsberg, X. Jian, P. A. Randazzo, P. G. Schultz, R. T. Moon, S. Ding, *Proc. Natl. Acad. Sci. U.S.A.* **2007**, *104*, 7444–7448.

110. http://www.proteinmodelportal.org/?pid=modelDetail&pmpuid=1000015133159&range_from=1&range_to=524&ac=Q8TAA9&zid=async.

111. C. P. Heisenberg, M. Tada, *Curr. Biol.* **2002**, *12*, R126–R128.

112. (a) S. L. Schreiber, *Proc. Natl. Acad. Sci. U.S.A.* **2011**, *108*, 6699–6702; (b) S. V. Frye, *Nat. Chem. Biol.* **2010**, *6*, 159–161.

INDEX

Diversity-Oriented Synthesis: Basics and Applications in Organic Synthesis, Drug Discovery, and Chemical Biology, First Edition. Edited by Andrea Trabocchi.

Printed and bound by CPI Group (UK) Ltd, Croydon, CR0 4YY
16/09/2026
14963421-0007